Rachel A. Yoho, Ram Lalgudi (Eds.)

BioProducts

Also of Interest

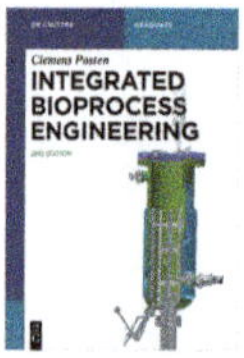

Integrated Bioprocess Engineering
Posten, 2024
ISBN 978-3-11-077334-7, e-ISBN (PDF) 978-3-11-077335-4,
e-ISBN (EPUB) 978-3-11-077394-1

Bioprocess Intensification
Dirk Holtmann (Eds.), 2024
ISBN 978-3-11-076032-3, e-ISBN (PDF) 978-3-11-076033-0,
e-ISBN (EPUB) 978-3-11-076037-8

Sustainable Products.
Life Cycle Assessment, Risk Management, Supply Chains, Ecodesign
Has, 2024
ISBN 978-3-11-131482-2, e-ISBN (PDF) 978-3-11-131546-1,
e-ISBN (EPUB) 978-3-11-131566-9

Cyclic Economy.
Policies in Major Countries and the EU
Aresta, Angela Dibenedetto, Franck Dumeignil (Eds.), 2024
ISBN 978-3-11-076707-0, e-ISBN (PDF) 978-3-11-076717-9,
e-ISBN (EPUB) 978-3-11-076719-3

Chemistry for Biomass Utilization
Alén, 2024
ISBN 978-3-11-060834-2, e-ISBN (PDF) 978-3-11-060836-6,
e-ISBN (EPUB) 978-3-11-060848-9

BioProducts

Green Materials for an Emerging Circular and Sustainable Economy

Edited by
Rachel A. Yoho, Ram Lalgudi

2nd, Revised and Extended Edition

DE GRUYTER

Editors
Rachel A. Yoho
College of Southern Nevada,
Las Vegas, Nevada 89146,
USA

Ram Lalgudi
Aries Science & Technology LLC
Columbus, OH
USA

ISBN 978-3-11-914598-5
ISBN 978-3-11-221214-1 (PDF)
ISBN 978-3-11-221249-3 (EPUB)
DOI https://doi.org/10.1515/9783112212141

Library of Congress Control Number: 2026933669

Bibliographic information published by the Deutsche Nationalbibliothek
The Deutsche Nationalbibliothek lists this publication in the Deutsche Nationalbibliografie; detailed bibliographic data are available on the Internet at http://dnb.dnb.de.

De Gruyter and Walter de Gruyter GmbH are part of De Gruyter Brill.
www.degruyterbrill.com

Questions about General Product Safety Regulation:
productsafety@degruyterbrill.com

Cover illustration: Thana Prasongsin/Getty Images Plus

Preface

In spring 2025, after the sudden passing of Dr Bhima Vijayendran, a group of authors from the first edition of this book were approached by De Gruyter to lead the second edition. We (Dr Ram Lalgudi and Dr Rachel Yoho) took up the lead of this edition to honor the lasting career impact and positive professional influence that Bhima had on our lives and careers.

In the first edition preface, Dr Vijayendran expressed his initial reactions to the scope of the content that would need to be covered to do justice to the breadth of the broadly defined field of bioproducts. We echo that concern, but build on an established platform from the incredible impact of the first edition. This edition continues the global presence of researchers working in different areas of research and time zones, coming together to share their insights.

From the first edition, we preserve much of the structure for the overall book. The first section continues to set the foundation for both the depth and breadth of the discussion on bioproducts with introductory content and new analyses. Updates for this edition include new content in the first chapter for the greatest scope of introductory content and refreshed versions of each of the chapters. Updates in this edition include two brand new chapters: one on life cycle assessment and the other on climate and policy influences on bioproducts. These two chapters provide valuable insights into spaces previously overlooked, but of increasing importance in the bioproducts space, from leading researchers and policy experts.

The second section again explores early-stage research and development projects, with recent advancements detailed from many of these areas. Certainly, some of these topics may be considered slightly less "early stage" in this edition, but still represent key areas of research and development. Notable updates in this edition include Chapters 11, 12, and 13 on ethylene glycol, soybean oil-based surfactants, and microbial electrochemical systems, respectively.

The third section on synthetic biology maintains the strength of the first edition, while providing key updates, particularly in hydrogen bacteria (Chapter 15). This edition continues to feature notable perspectives on translating research into practice, including the scale-up of technologies into industrial leaders, particularly with Chapter 14 on clean chemistry.

The fourth section builds on some of the processes introduced in the third section, highlighting commercial bioproducts. Notable updates in this edition include Chapters 17 and 18 on microbially produced green biosurfactants and polyvinyl chloride applications, respectively. The global construction industry (Chapter 19) and a deeper dive into green hydrogen (Chapter 20) are maintained from the first edition, as they provide valuable examples and case studies for applications.

The fifth section highlights bioplastics and biopolymers, with both a depth of information and wide range of content covered. Leading experts provide their insights with updates across many of the chapters, exploring green bioplastics.

 | https://doi.org/10.1515/9783112212141-202

The sixth section continues the discussion with sustainable green biocomposites. Additional depth on bioplastics are explored in Chapters 26 and 27, with insights on the emerging bioeconomy, fibers, and composites. Chapters 28 and 29 highlight innovators' own accounts of development and innovation processes and products.

The seventh section of the book provides new insights from recent developments related to artificial intelligence. This is a significant development for the second edition, as the time since the writing and publication of the first edition showed large advancements in artificial intelligence and this entered the mainstream, both across research and industries, as well as even into widespread personal use.

As we considered how to build on the first edition's impact, we asked ourselves, "What could we possibly add here to what Bhima already created?" The answer lies in recent innovations, evolving policy landscapes, and the increasing role of life cycle analyses and artificial intelligence. We hope this book continues Dr Vijayendran's tradition of bringing researchers together from around the world, strengthening the global community of bioproducts scholars. His vision lives on through these pages and through the collaboration and knowledge that drive the field forward.

Rachel Yoho
Fairfax, VA, USA

Ram Lalgudi
Columbus, OH, USA

December 2025

Contents

Section II: **Some early-stage R&D bioproducts**

Section III: **Synthetic biology**

List of contributing authors

Bhima R. Vijayendran†
Redwood Innovation Partners LLC
Carlsbad, CA 92011
USA

Ram Lalgudi
Aries Science and Technology LLC
Columbus, OH 43081
USA
lalgudir@ariesst.com

Rachel A. Yoho
College of Southern Nevada
Las Vegas
Nevada 89146
USA
rachel.yoho@csn.edu

Michael A. Schultz
Managing Director, PTI Global Solutions LLC
Indianapolis, IN
USA
mike@ptisolns.com
+1-630-464-6505

John McArdle
Redwood Innovation Partners LLC
Boston, MA
USA
johnm@redwdinnov.com

Camila González Arango
Department of Agricultural and Biological Engineering
The Pennsylvania State University
State College, PA 16803
USA

Camila Valderrama
Department of Agricultural and Biological Engineering
The Pennsylvania State University
State College, PA 16803
USA

Johannes Ali
Department of Agricultural and Biological Engineering
The Pennsylvania State University
State College, PA 16803
USA

Juliana Vasco-Correa
Department of Agricultural and Biological Engineering
The Pennsylvania State University
State College, PA 16803
USA
julianavasco@psu.edu

Forrest Stagner
Department of Agricultural and Biological Engineering
The Pennsylvania State University
University Park
State College, PA 16803
USA
fws11@psu.edu

Chris Morris
Iowa State University
Ames, IA
USA

Michael Helbing
Department of Agricultural and Biological Engineering
The Pennsylvania State University
University Park
State College, PA 16803
USA

Frank Montabon
Iowa State University
Ames, IA
USA

 | https://doi.org/10.1515/9783112212141-204

Christine Costello
Department of Agricultural and Biological Engineering
The Pennsylvania State University
University Park
State College, PA 16803
USA

Lara Fowler
Department of Agricultural and Biological Engineering
The Pennsylvania State University
University Park
State College, PA 16803
USA

Herman P. Benecke (retired)
Columbus, OH
USA

Daniel B. Garbark
formerly Lead Researcher at Battelle
Columbus, OH
USA
dangarbark@yahoo.com

Claudia Creighton
Carbon Nexus at the Institute for Frontier Materials
Deakin University
75 Pidgons Rd
Waurn Ponds, VIC 3216
Australia

Nguyen Duc Le
Carbon Nexus at the Institute for Frontier Materials
Deakin University
75 Pidgons Rd
Waurn Ponds, VIC 3216
Australia

Huma Khan
Carbon Nexus at the Institute for Frontier Materials
Deakin University
75 Pidgons Rd
Waurn Ponds, VIC 3216
Australia

Srinivas Nunna
Carbon Nexus at the Institute for Frontier Materials
Deakin University
75 Pidgons Rd
Waurn Ponds, VIC 3216
Australia

Russell J. Varley
Carbon Nexus at the Institute for Frontier Materials
Deakin University
75 Pidgons Rd
Waurn Ponds, VIC 3216
Australia
russell.varley@deakin.edu.au

Cai Li Song
Petronas Research Sdn Bhd
Lot 3288 and 3289
Off Jalan Ayer Itam
Kawasan Institusi Bangi
43000 Bangi, Selangor
Malaysia
song.caili@petronas.com

Mohamad Fakhrul Ridhwan Samsudin
PETRONAS Research Sdn Bhd
Lot 3288 and 3289
Off Jalan Ayer Hitam
Kawasan Institusi Bangi
43000 Kajang, Selangor
Malaysia

Nur Amalina Samsudin
PETRONAS Research Sdn Bhd
Lot 3288 and 3289
Off Jalan Ayer Hitam
Kawasan Institusi Bangi
43000 Kajang, Selangor
Malaysia

Wong Mee Kee
PETRONAS Research Sdn Bhd
Lot 3288 and 3289
Off Jalan Ayer Hitam
Kawasan Institusi Bangi
43000 Kajang, Selangor
Malaysia

M. Syamzari B. Rafeen
PETRONAS Research Sdn Bhd
Lot 3288 and 3289
Off Jalan Ayer Hitam
Kawasan Institusi Bangi
43000 Kajang, Selangor
Malaysia
syamzari_rafeen@petronas.com.my

Nur Liyana Ismail
PETRONAS Research Sdn. Bhd.
Lot 3288 and 3289
Off Jalan Ayer Itam
Kawasan Institusi Bangi
43000 Kajang, Selangor
Darul Ehsan
Malaysia

Siti Fatihah Salleh
PETRONAS Research Sdn. Bhd.
Lot 3288 and 3289
Off Jalan Ayer Itam
Kawasan Institusi Bangi
43000 Kajang, Selangor
Darul Ehsan
Malaysia

Jofry Othman
PETRONAS Research Sdn. Bhd.
Lot 3288 and 3289
Off Jalan Ayer Itam
Kawasan Institusi Bangi
43000 Kajang, Selangor
Darul Ehsan
Malaysia

Sara Shahruddin
PETRONAS Research Sdn. Bhd.
Lot 3288 and 3289
Off Jalan Ayer Itam
Kawasan Institusi Bangi
43000 Kajang, Selangor
Darul Ehsan
Malaysia
sarashahruddin@petronas.com

Michelle Young
Carollo Engineers, Inc.
390 Interlocken Crescent, Suite 800
Broomfield, CO 80021
USA
(did not contribute to second edition)

Aditi Hoffman
Department of Civil Engineering
Kansas State University
2118 Fiedler Hall
1701C Platt Street
Manhattan, KS 66502
USA
aditip@ksu.edu

Beth Bannerman
Formerly of Amyris, Inc.
San Francisco, CA
USA
info@amyris.com

Sunil Chandran
Amyris, Inc.
San Francisco, CA
USA
info@amyris.com

Crispinus Omumasaba
Utilization of Carbon Dioxide Institute Co. Ltd.
2-3-1 Daiba, Minato-ku
Tokyo 135-0091
Japan

Alain A. Vertés
Utilization of Carbon Dioxide Institute Co. Ltd.
2-3-1 Daiba, Minato-ku
Tokyo 135-0091
Japan

F. Blaine Metting
Pacific Northwest National Laboratories, retired
Utilization of Carbon Dioxide Institute Co. Ltd.
2-3-1 Daiba, Minato-ku
Tokyo 135-0091
Japan

Hideaki Yukawa
Utilization of Carbon Dioxide Institute Co. Ltd.
2-3-1 Daiba, Minato-ku
Tokyo 135-0091
Japan

Vikas Kumar
Department of Animal, Veterinary and Food Sciences
Aquaculture Research Institute
University of Idaho
Moscow, ID 83844-2330
USA

Anisa Mitra
Department of Zoology
Sundarban Hazi Desarat College
Pathankhali
South 24 Parganas 743611
West Bengal
India

Dan Derr
Integrity BioChem
Cresson, TX
USA

Huu Tuan Tran
Kansas State University
Manhattan, KS 66506
USA
huutuan@ksu.edu

Jacyr Quadros
Innoleics Serv. e Cons. Ltda.
São Paulo
Brazil
jquadros@innoleics.com

Gunnar Lynum
SMD Products Company, Inc. and Strategic Market Development, LLC
P.O. Box 1634
Manchester, MO 63021
USA
customerservice@smdproductscompany.com
www.smdproductscompany.com

Steven Lynum
SMD Products Company, Inc. and Strategic Market Development, LLC
P.O. Box 1634
Manchester, MO 63021
USA
customerservice@smdproductscompany.com
www.smdproductscompany.com

Jonathan Cristiani
Black and Veatch Corporation
11401 Lamar Avenue
Overland Park, KS 66211
USA
+1 919-463-3043
CristianiJM@BV.com

Grace Dearnley
Black and Veatch Corporation
11401 Lamar Avenue
Overland Park, KS 66211
USA

Justin Distler
Black and Veatch Corporation
11401 Lamar Avenue
Overland Park, KS 66211
USA

Andrew Doerflinger
Black and Veatch Corporation
11401 Lamar Avenue
Overland Park, KS 66211
USA

Vincent Mazzoni
Black and Veatch Corporation
11401 Lamar Avenue
Overland Park, KS 66211
USA

Ashok Adur
Everest International Consulting (LLC)
Jacksonville, FL 32256
USA
aadur@outlook.com

Pramod Kumbhar
Praj Matrix-R&D Center (Division of Praj Industries Ltd)
403/403/1098, Urawade
Pirangut, Mulshi
Pune 412115, Maharashtra
India
pramodkumbhar@praj.net

Anand Ghosalkar
Praj Matrix-R&D Center (Division of Praj Industries Ltd)
GAT#1098, 402, 403, Urawade
Pirangut, Mulshi
Pune 412115, Maharashtra
India

Yogesh Nevare
Praj Matrix-R&D Center (Division of Praj Industries Ltd)
403/403/1098, Urawade
Pirangut, Mulshi
Pune 412115, Maharashtra
India

Sandeep Kulkarni
KoolEarth Solutions, Inc.
Alpharetta, GA
USA
skulkarni@koolearthsolutions.com

Ray Bergstra
MTN Consulting Associates
Edmonton, Alberta
TerraVerdae Bioworks Inc
Canada
rjbergstra@mtnconsulting.ca

Camille Sobrain Saltman
INCA Renewtech
Kelowna
British Columbia V1Y 2E4
Canada
csaltman@incarenewtech.com
250-300-5254

David Saltman
INCA Renewtech
Kelowna
British Columbia V1Y 2E4
Canada
dsaltman@incarenewtech.com
250-300-5253

Robert C. Joyce
FibreTuff
Toledo, OH
USA
robert@fibertuff.us

Parthiban Pannerselvam
University of the Cumberlands
6178 College Station Drive
Williamsburg, KY 40769
ppannerselvam39788@ucumberlands.edu

Section I: **Introductory chapters**

Rachel A. Yoho, Ram Lalgudi*, Bhima R. Vijayendran

Chapter 1 Bioproducts research, development, and applications today: an introduction and overview

1.1 Introduction

Since humans settled down almost 10,000 years ago, agriculture and plant-derived products have played a significant role in all aspects of their lives and economies. This changed dramatically in the last two centuries with the increased use of fossil-derived products taking a major share of the economy due to advances in basic science, technology, engineering, and manufacturing. Many benefits have resulted in this transition based on fossil sources. One of the major impacts was the relatively inexpensive energy and products derived from them that propelled much of our way and quality of life globally.

Due to the indiscriminate, addictive, and uncontrolled reliance on nonrenewable fossil sources, the progress has come with significant cost, namely, damage to the environment with attendant pollution, greenhouse gas emissions, and global warming posing serious challenges for continued sustainability. For instance, carbon dioxide concentration in the atmosphere has risen from 280 parts per million (ppm) before the industrial era to about 420 ppm as of the first edition of this book, and it is currently rising at more than 2 ppm per year [1] and approximately 430 ppm today for the second edition [2]. In practical terms, one ppm of carbon dioxide in the atmosphere is about 7.8 billion metric tons. There are other greenhouse gases such as methane with even more potent global warming effects. It is estimated that the average methane level in 2021 was 1,896 parts per billion (ppb), around 162% greater than the preindustrial levels [1] and increased to approximately 1,934 ppb as of mid-2025 [2]. One can easily relate to the scope and magnitude of the problem. Fortunately, there is also an opportunity to make use of these polluting emissions as feedstocks for renewable and sustainable biofuels and bioproducts and reduce the current reliance on polluting fossil sources. As we continue to see, this opportunity necessitates the confluence of interest, funding, policy, and global collaboration.

***Corresponding author: Ram Lalgudi**, Aries Science & Technology, LLC, Columbus, OH 43212, USA, e-mail: lalgudir@ariesst.com
Rachel A. Yoho, (formerly) George Mason University, Fairfax, Virginia, USA, (currently) College of Southern Nevada, Las Vegas, Nevada, e-mail: rachel.yoho@csn.edu
Bhima R. Vijayendran, Redwood Innovation Partners LLC, Carlsbad, CA 92011

 | https://doi.org/10.1515/9783112212141-001

In the last quarter century or so, these environmental and sustainability issues have received attention from all segments of human society on a global scale. Many passionate, dedicated, and motivated individuals, entrepreneurs, start-ups, established businesses, governments, leading research and development (R&D) organizations, and other stakeholders have been working diligently in solving many of these issues.

One approach that reduces greenhouse gas emissions is the adoption of cost-effective bioproducts – products derived from renewable resources processed by conventional catalytic thermochemical processes and/or fossil sources processed by emerging bioprocessing techniques including fermentation and synthetic biology. The global economy with supportive public policies and consumer activism is poised for transitioning into a circular one with proactive initiatives to mitigate waste, greenhouse gas emissions, reuse and recycle of resources in a profitable and sustainable manner. The transition from a linear to a more circular economy continues to be highlighted by opportunity and fraught with complex challenges.

The book highlights some of these developments: from discovery, lab feasibility, scale-up, and eventual commercialization of bioproducts. The second edition provides an update to the global and policy context, as well as additional insights into topic-based research, life cycle analyses, and more. Given the very vast scope of the field, their global nature, topics, and contributors are selected based on the involvement in the first edition of the book, largely Dr Vijayendran's past collaborators and colleagues and his role in R&D in the field, commercialization of bioproducts, global relationships with several contributors in the field and a passion to make a difference in advancing bioproducts and contribute toward a sustainable and circular economy. The second edition adds collaborators and colleagues from the networks and extended networks of the new editors, Dr Ram Lalgudi and Dr Rachel Yoho, providing additional depth and insight into the current state of bioproducts, development, and implications.

The book covers the challenges and frustrations of introducing new bioproducts despite committed efforts and investments. Fortunately, the overall environment for sustainable bioproducts is going through a sea change as consumers are starting to pay attention to greenhouse emissions and global warming. Businesses are taking note of this change in consumer behavior and are responding to meet their expectations with sustainable and environmentally friendly offerings. Yet, the research environment and funding (notably in the United States) is changing rapidly, precipitating impacts on research projects, scientists' careers, and global researcher migration.

1.2 Some significant developments over the last quarter century

Several major developments in bioproducts related to advances in R&D, processing, funding, commercialization, consumer awareness perception and acceptance, and public policy and legislative actions have occurred. Some of these are covered in this chapter.

1.2.1 An evolutionary process

The progress in the field has been an evolutionary one with cycles and phases [3]. The initial focus was based on the expectation that green bioproducts would have a premium and the consumers would be willing to pay a higher price. This was a misguided one, a lesson learned, somewhat slowly, with serious consequences in the commercialization of early bioproducts. The second cycle or wave was focused on fungible or direct petroleum substitute products. Several fungible bioproducts were developed, and their impact on reducing carbon footprint reduction has been reported [4]. One example is bio-monoethylene glycol (MEG) by India Glycols Ltd. Bio-MEG is projected to avoid 407 kg of CO_2/ton of the bioproduct used in polyester soft drink bottles (see Chapter 11 for details on bio-MEG). Another focus has been on drop-in replacements such as bioethylene and bioglycols. One more approach has been to develop functionally equivalent bioproducts such as polyesters based on 2,5-furandicarboxylic acid. A few of these are covered in Chapter 2. However, commercial impact of such fungible, direct replacement, and functional equivalent bioproducts has been somewhat muted to date.

A more recent trend has been to focus on higher value, but smaller volume functional molecules of interest in nutraceuticals, cosmetics, pharma, and others (see Chapter 14). Another successful approach has been to develop molecules with desirable target functionality, not readily achievable by fossil-based products. A commercially successful example of this is 1,3-propanediol (PDO) from sugar-based feedstock and fermentation using engineered *E. coli* organism (see Chapter 3 for details).

A promising emerging trend is to use greenhouse gases and industrial wastes as feedstocks to produce higher value bioproducts (see Chapter 2).

1.2.2 Product and process: R&D

Significant advances have been made in basic sciences – synthetic and systems biology, analytical tools including data analytics/machine learning, generative artificial intelligence (AI) , and instrumentation – thermochemical catalyst technologies and processing, enzyme design and bioprocessing, and manufacturing at scale. Such advances have occurred due to significant and persistent investments from various stakeholders. In the United States, major investments by the US Department of Energy (DOE), the US Department of Agriculture (USDA), and the National Science Foundation (NSF) are worth noting. It is also worth mentioning that similar sustained efforts have been happening in Europe, Asia, and elsewhere. A few of the US-based initiatives and programs are covered in the following two sections.

1.2.2.1 United States Department of Energy (DOE) initiatives

The US DOE has been a major driver in the renewable and sustainable fields for decades. A few specific ones are listed further:

- The DOE National Laboratories such as Pacific Northwest National Labs, (the US national lab formerly known as) National Renewable Energy Labs, Oak Ridge National Laboratory, and Idaho National Labs have been active in advancing various aspects of bioproduct development. Their websites have full details of their research activities in the field. It is worth noting that Battelle Memorial Institute (BMI) based in Columbus, OH, an independent nonprofit R&D organization that manages or comanages many of these DOE labs, has been involved in bioproducts since the mid-1980s. See more on this later.
- DOE has been actively supporting innovative technology and product development in the field through periodic solicitations related to improved catalysts, scale-up and manufacturing at scale, use of industrial wastes such as CO_2 as feedstock to make bioproducts, enzyme design, and bioprocessing (see https://eere-Exchange,enrgy.gov/).
- DOE has also been funding regional lab facilities and trained staff to facilitate start-up companies in the field with their product development and scale-up needs (see http://abpdu.lbl.gov/).
- A good example of strategic leadership is the roadmap effort of DOE in identifying top 12 target molecules for the emerging bioproducts industry in the mid-2000s [5]. See Chapter 2 for further details.

DOE has prioritized the development of performance-advantaged bio-based chemicals (PABCs) [6] to accelerate the transition from petroleum-derived feedstocks to sustainable, high-value bioproducts. Unlike conventional drop-in replacements, PABCs are designed to leverage the inherent chemical diversity of biomass – such as oxygenated and heteroatom-rich structures – to deliver superior functionality, environmental benefits, and cost competitiveness. DOE's strategy integrates computational screening, predictive modeling, and experimental synthesis to identify and validate candidate molecules derived from lignin and sugar intermediates. Efforts, including the NREL-led project (FY2021–2023) [7], have demonstrated polymers and small molecules with enhanced thermal stability, biodegradability, and reduced toxicity, alongside technoeconomic analysis and life cycle assessments to ensure scalability and sustainability. By fostering public-private partnerships and advancing platform technologies, DOE aims to establish a robust bio-based chemical supply chain that supports rural economies, reduces greenhouse gas emissions, and enables novel material applications beyond the capabilities of petrochemical analogs.

DOE through its Bioenergy Technologies Office (BETO), has established a comprehensive strategy to advance bioproduct development as part of the US bioeconomy and decarbonization goals [8]. By 2030, DOE aims to enable commercial scale production of renewable chemicals and materials with at least 70% greenhouse gas (GHG) reductions, strategic initiatives include the Clean Fuels & Products Shot™, the SAF Grand Challenge, and alignment with the National Biotechnology & Biomanufacturing Initiative. These efforts emphasize technology de-risking, feedstock innovation, and cross-sector integration to decarbonize aviation, marine, and heavy-duty transport. DOE's roadmap combines R&D funding, public-private partnerships, and scale-up investments to accelerate commercialization of bioproducts, contributing to a resilient, low-carbon economy and net-zero emissions by 2050.

1.2.2.2 United States Department of Agriculture (USDA) initiatives

USDA has been an active funding source for much of the impactful work in the field. A few efforts are highlighted here:

- USDA has several regional lab facilities such as in Illinois, Pennsylvania, and California where dedicated scientists are involved in all aspects of bioproduct R&D. A couple of specific examples are the work on xanthan gum, an extracellular microbial polysaccharide polymer, and novel estolide compound-based fatty acid chemistry. Further details of these products are given in Chapter 2.
- USDA has also pioneered in facilitating commercialization of bioproducts through their USDA-Certified Bio-based Product Label Program started with 2002 Farm Bill and Reauthorized in 2018 [9, 10]. Several hundred products from numerous companies from all of the 50 states covering diverse products from paints, industrial products, cosmetics, cleaners, and so on are listed in the program and the list is growing rapidly. The construction products described in Chapter 19 and plasticizer additives covered in Chapter 18 are some examples.

1.2.2.3 Farmer-based organizations

Several farmer-based organizations at the national and state levels are active in funding research and product development derived from crops. An example is the National Soybean Checkoff program that supports funding for bioproducts development at the national level (United Soybean Board, USB) and soybean-growing state associations such as the Ohio Soybean Council (OSC). They support bioproducts R&D, and OSC opened a research laboratory dedicated to developing soy-based industrial bioproducts (see https://www.airableresearchlab.com/).

1.2.2.4 A few other organizations

There are several active organizations in Europe, Canada, and Australia that are also supporting R&D activities and commercialization of bioproducts. A few of them are listed:
- Nova Institute in Germany – www.bio-based.eu
- EuropaBio – Industrial Biotechnology Enabling a Circular BioEconomy in Europe – www.eurobio.org
- Cosun Biobased Products – www.cosunbiobased.com
- BIOTECCanada – www.Biotech.ca
- Ontario East Economic Development Commission – www.ontarioeast.ca
- Queensland University of Technology – www.qui-edu.au/research/ctcb

1.3 Role of bioproducts

The role and importance of bioproducts that are not fuel- and/or energy-related – the focus of this book – in the evolving sustainable green circular economy are worth reviewing. As shown in Chapter 2, only about 10% of the nonrenewable carbon from fossil sources is used for chemical production. Thus, the bulk of polluting greenhouse gas emissions is from the energy sector for use in transportation and power generation industries. Yet, in a typical fossil industry composed of energy and chemical products, the output from this 10% of polluting carbon provides the needed profitability as measured by return on investment for the entire petrochemical enterprise due to the higher value of chemical products in the marketplace. In other words, the petrorefinery complex becomes a profitable venture as every carbon atom from fossil resources is converted to higher value and profitable outcomes. It is expected that in the emerging bioeconomy, a similar biorefinery model would evolve and thus make bioindustry an economically sustainable business. Furthermore, eliminating even part of the 10% of polluting fossil carbon-based products by bioproducts should have a significant impact on reducing greenhouse emissions. Use of fossil fuel combustion products as bioproduct feedstocks is especially attractive as higher value products are created while capturing greenhouse gases.

There are other benefits as well. Bioproducts such as bioplastics have the potential to be biodegradable and offer additional benefit in improved waste management. The option for recycling is also an attractive feature. Issues, challenges, and opportunities in this topic are covered notably in Chapters 21 and 23.

1.4 A few takeaways from the quarter century of development in bioproducts

One of the questions that practitioners in the field are asked is what has been accomplished. It is a fair challenge from two perspectives. One is to get a realistic gut check and the other is to build from lessons learned to guide and redirect future efforts, as needed. However, answers to this question are not simple and straightforward. As mentioned earlier, the activities over the last quarter century have been quite diverse, covering a lot of ground. Perhaps, it would be helpful to break this analysis into a few tangible and measurable parts. An attempt is made in the following section.

1.4.1 Advances in basic science relevant to bioproducts

Even a casual review of the literature would clearly show the tremendous advances made in this area, especially in areas of synthetic biology and systems biology and their impact on industrial biotechnology-derived bioproducts.

It is often said that the twenty-first century is the century of synthetic biology, as late nineteenth- and much of early twentieth centuries were for synthetic chemistry that propelled the growth of fossil-based energy and petrochemical industries. Much is expected from the disruptive synthetic biology: technology-innovative approaches for engineering new biological processes or redesigning existing ones for purpose-driven outcomes. Synthetic biology has been described as the heart of the emerging bioeconomy, capable of delivering novel solutions to global healthcare, agriculture, manufacturing, and environmental challenges [10, 11]. Please visit the Synbiobeta website for a comprehensive overview of the activities in the field (www. Synbiobeta.com).

An earlier report [12] highlights some of the advances and their potential impact on the bioeconomy:

- Breakthrough advances in gene sequencing, tools, and techniques to manipulate metabolic pathways of interest in myriad fields including developing industrial and consumer products.
- Database powered by AlphaFold algorithm's ability to predict over 200 million protein structures from over 10 million species. It is expected that the expanded database will aid countless researchers in their work and open new avenues of scientific discovery [13]. One potential area of interest in bioproducts is the design of enzymes that can decompose plastic pollution into harmless substances.
- Lower capex biomanufacturing facilities to produce high-value bioproducts with more benign environmental footprint.
- Improved next-generation pesticides and crop control agents with improved safety, efficacy, and handling.

- Enzymes that improve efficiency and reduce energy use in traditional industries such as pulp and paper, textiles, and food processing.
- Cosmetics and personal care items made from sustainable bio-produced components that do not use ingredients from threatened and endangered species (see Chapter 14).

1.4.2 Deployment of improved and cost-effective manufacturing at scale

In this area, there are several success stories, and a few are listed further for illustrative purposes:
- Commercial-scale production of several diols including propylene glycol and 1,3-butylene glycols. See Chapter 3 for details.
- Commercial-scale production of 1,4-butanediol for nylon intermediates and other polymers. See Chapters 23 and 24 for details.
- Full-scale commercialization of PDO for improved polyester polymer. See Chapter 3 for details.
- Continued global expansion of polylactic acid (PLA) manufacturing, including the recent announcement by ADM and LG Chemicals [13].
- Announcement by Avantium and Covation, a Du Pont spin out, to produce furanic-based polymers. See Chapter 7 for details.

This list goes on. However, still the market penetrations of bioplastics and biopolymers are small.

1.4.3 Commercialization challenges

Commercialization has certainly been an uphill battle and continues to be one, for many reasons. A few of them are explored later. The obvious ones are higher cost, undeveloped/nonexisting supply chains, unproven performance and associated risks, poor compatibility with existing industry infrastructure, and sunken investments. All these issues are formidable, leading often to the premature death of a new bioproduct, especially if the new product is a replacement or substitute product for a well-established fossil product in a relatively mature market. Lack of consumer and market pull for many of these failed products have been a common theme, as bioproducts do not have any subsidy unlike the biofuels.

It is probably worth looking at a few bioproducts that have been highly successful in the marketplace. Some like the first two – guar and xanthan gum – have been commercial for decades.

1.4.3.1 Guar

A nonionic high-molecular-weight galactomannan water-soluble polymer derived from the plant source. See Chapter 22 for details. Guar has myriad uses in diverse industrial and food applications. The major application that has contributed to its growth is in oil and gas well fracturing. Due to its structure and functionality such as high viscosity, easy cross-link ability, and breakdown with enzymes in the downhole environment, guar is uniquely suited in the fracking application. Low cost and reliable supply from India and Pakistan are also contributing factors. The first plant to make derivatized propylene-oxide-modified guar in the mid-1970s after the first oil shock had a capacity of 10 million pounds/year. This quickly grew to 100 million pounds by the end of that oil shock period. Today, guar consumption is approaching a billion dollar in sales and growing at 7.9% per year.

1.4.3.2 Xanthan gum

Xanthan gum is an extracellular anionic water-soluble polymer produced by fermentation (see Chapter 22 for details). It has excellent viscosifying and thickening properties of value in myriad food and industrial applications. No other polymer has commensurate properties and is hence used in large quantities even though the cost is relatively high. The market size is estimated at 960 million dollars/year growing at 6% per year.

1.4.3.3 Wood plastic composite (WPC)

A very successful wood biocomposite product introduced about 25 years ago for the building and construction industry. Its main use is in decking, as replacement for pressure-treated lumber. This product is cost-competitive and compatible with the existing plastic manufacturing process, infrastructure, and supply chain. Further, the product is more durable, aesthetically pleasing, and the manufacturers could provide 25 years or longer lifetime warranty. Also, it does not have the issue of treated lumber with toxic chemicals and their leaching. Current sales of WPC are in several billion dollars/year and growing at a high rate. See Chapter 26 for details.

1.4.3.4 Polylactic acid (PLA)

PLA has been studied and been under development for a long time due to its inherent biodegradability. However, it has some serious deficiencies in its properties and processing in the existing equipment. It has a long and circuitous history. Early work started in the late 1980s at BMI funded by Du Pont and Dainippon Inc. The work led to over 30 patents, where the main development was the stereoselective vapor-phase cyclization of L-lactic acid mixtures directly to L-lactide that effectively underwent ring-opening polymerization to L-polylactic acid. Unfortunately, the sponsors lost interest and did not move forward with the technology developed at BMI. The patents were acquired by Chronopol, a venture arm of Coors Beer company. They tried for a few years and the patents were acquired by Cargill, which later formed a joint venture with Dow to further develop and commercialize PLA. The venture exited PLA development and Cargill took on the task and challenge of further developing and commercializing PLA. Today, PLA is a multimillion pound/year commercial product with several players including Total Energies-Corbion (www.totalenergies-corbion.com) and the recent ADM-LG Chemicals announcement [13] with manufacturing plants spread across the globe (see Chapters 21, 22, and 23 for details).

The story of PLA shows challenges in introducing a new product, be it bio or otherwise, that has marginal properties and a tenuous value proposition, namely, biodegradability of unproven status in practical real-world environments. Further, there is a need for patience and the ability to weather the inevitable valley of death with financial and strategic commitments.

1.4.3.5 1,3-Propanediol (PDO)

This has been mentioned earlier and in Chapter 3. In this case, the fermentation approach using an engineered organism and low-cost sugars as feedstock led to cost-effective manufacturing of the monomer of value in the polyester and other industrial markets. Having commitment and support from a major company, namely, DuPont with extensive knowledge of the end use markets, customers, and supply chain were factors in its successful commercialization. The 2024 estimated global sales of bio-PDO were 250 million dollars/year.

1.4.3.6 Some key takeaways

A few lessons learned from the commercialization initiatives are worth noting, even though they may be obvious at least in hindsight. Some of them are highlighted:

- Commercialization of any new product including bioproducts is not for the faint of heart – it requires sustained financial resources and commitment.
- Having a clear and easy to translate value proposition to the end user in the marketplace is critical. Having a lower price as the main value proposition is a bit problematic, especially in mature and well-established replacement applications and markets. Relying on somewhat hard-to-quantify environmental benefits will not be sufficient but could be beneficial if price is equivalent.

- Bioproducts with enhanced functionality providing high-value performance attributes in final products/applications have a better shot.
- Compatibility with the existing infrastructure and supply chains and/or the ability to develop one facilitates commercialization and reduces external hurdles.
- As noted in a 2022 Barron's article (July 12, 2022), the transition to bio needs to hit a Trifecta:
  - The sources that should replace fossil need to be simultaneously clean, affordable, and reliable.
  - The transition, especially for energy, will require vast amounts of resources (new technology and infrastructure developments, manufacturing, supply chain, etc.) that may impact sustainability of renewables.
  - The transition will require new partnerships, alliances, and business models.

Many bioproducts currently under development including those described in this book have not achieved this trifecta, but many are getting closer. It will neither be easy nor a linear path but overall, many are on an encouraging trajectory.

1.5 Public awareness and policy advances related to bioproducts

One of the significant and major developments over the last decade or so is the increased awareness of the environmental movement and sensitivity of the harmful effects of global warming that threatens our future. Consumers are becoming more proactive in their purchasing behaviors, and the leading manufacturers of products are starting to pay increased attention in their product development and marketing strategies. Several major brand name companies in consumer products and others are partnering with raw material and chemical intermediate suppliers to drive the supply chains toward bioproducts.

At the public policy and legislative levels, there is a global trend to encourage products including bioproducts that address and mitigate greenhouse gas emissions and global warming. In some cases, tax and other financial incentives such as carbon credits are being tried to accelerate the transition to a green and sustainable economy. A very positive and encouraging recent development is the enactment of Inflation Reduction Act (IRA) passed by the US Congress in August 2022. Over $390 billion is allocated for renewables, including bioproducts and manufacturing. As a key addition to the second edition of this Bioproducts book, Chapter 6 provides a great depth of analysis on public policy and the policy areas relevant to Bioproducts.

The investment and financial community closely watched these developments and started to view this area of green and sustainable products and markets favorably as a long-term attractive option.

Overall, the public perception and support for bioeconomy are very bullish, yet likely partisan in the United States.

1.6 The role of artificial intelligence (AI) in advancing bioproducts innovation

The second edition includes chapters that discuss the transformative role of AI in advancing bioproduct innovation. Chapter 29 specifically examines the impact of AI on synthetic biology, bioprocess optimization, and bioinformatics. This discussion, as well as other sections among the topics covered in other chapters, provides examples of how AI has aided in the discovery and commercialization of sustainable bioproducts, such as bio-based coatings and environmentally friendly microencapsulation materials. This highlights AI's essential contribution to reducing environmental impacts and promoting a circular economy.

1.7 Some trends for bioproducts

A few major trends emerging in the field are listed:
- As synthetic biology with its rich and vast potential matures, more and more industrial chemical products/intermediates will be produced through biological routes, perhaps, as a hybrid approach with a traditional thermochemical catalytic process.
- Increased use of waste feedstock as carbon sources by taking advantage of potential negative raw material cost in the overall production cost of bioproducts. This is important as over 50%, in some cases even more, of manufacturing cost is from raw materials.
- A stepwise progressive transition to green products seems to be evolving. This involves moving from current gray fossil polluting sources to blue (where the CO_2 is captured from the conventional fossil industry processes) and then to green, where no fossil carbon is used. A good example is current H_2 production (gray hydrogen) from steam reforming of methane that releases massive amount of CO_2 and moving to blue hydrogen, which captures the polluting CO_2 derived from fossil sources, and finally making green hydrogen from water electrolyzers using solar and wind energy. The blue and green hydrogen can be used to make products such as blue and green ammonia (see Chapter 20 for details). It is estimated

that global production of ammonia accounts for about 1% of greenhouse emissions. As hydrogen is used in numerous chemical processes, one can easily envision similar transitioning with other chemical products.
- As discussed earlier under commercialization, there is increased attention to revisit business models in developing and commercializing bioproducts:
  - The first one is to build capabilities and tools that can be deployed across promising and attractive products through strategic partnerships. An example is the acquisition of Zymergen, a strong biotech company, with product focus by Ginkgo Bioworks, a powerhouse in the field.
  - The second is to develop product and process technologies with a strong patent portfolio and then license them for large-scale manufacture and strategic partnerships with potential end users.
  - The third is to focus on low-volume specialty/fine chemical markets and high-value luxury niche applications.
- Now there is recognition that it is important to incorporate how science and engineering affects citizens' safety and health. This has given rise to "science and technology studies" or variations thereof, and European Innovation and Patent regimes, considering these aspects in the development phase. NAS and NSF have been active as well, as are funding organizations, for example, Technology in the Public Interest Program Strategy – MacArthur Foundation (see: macfound.org). Current changes in the federal research and development budgets, grants provided by the NSF and other US federal agencies, continue to be an evolving area of focus, with impacts on basic science and research innovation, higher education, and global movement of highly skills researchers and scientists.
- Generative AI is one of the next frontiers in research in many areas, with profound impacts already demonstrated in many aspects of society. Bioproducts research and innovation is no exception. Several chapters begin to explore this, including a notable section in Chapter 13, as well the new Chapter 29.
- Transitioning to bioproducts, including fuels and energy, is not going to be a linear path but a bumpy and a painful one, costing a lot more than we are all used to.

In summary, prospects portend well for bioproducts, yet new challenges continue to arise.

References

[1] Spinard R Increase in Atmospheric Methane Set Another Record During 2021, National Oceanic and Atmospheric Administration, U.S. Department of Commerce; April 7, 2022.

[2] NOAA Global Monitoring Laboratory. In Trends in CO_2, CH_4, N_2O, SF_6, National Oceanic and Atmospheric Administration, U.S. Department of Commerce; June 10, 2025.

[2] Vijayendran BR A commentary bio products from bio refineries-trends, challenges and opportunities. J Bus Chem 2010;7:3.
[3] www.icca.chem.org
[4] Top Value-Added Chemicals from Biomass Volume 1- Results of Screening for Potential Candidates from Sugars and Synthesis Gas. Northwest National Laboratory (PNNL), National; Energy Laboratory (NREL). Office of Biomass Program (EERE); Werpy, T and Petersen, G., Series editors, 2002.
[5] help@usdabioproducts.net *or see 2024 report*: U.S. Department of Agriculture. 2024. Building a Resilient Biomass Supply: A Plan to Enable the Bioeconomy in America. https://www.usda.gov/sites/default/files/documents/biomass-supply-chain-report.pdf
[6] U.S. Department of Energy, Office of Energy Efficiency & Renewable Energy. June 2018. Moving Beyond Drop-In Replacements: Performance-Advantaged Biobased Chemicals. Workshop Summary Report. https://www.energy.gov/sites/prod/files/2018/06/f53/Performance-Advantaged%20Biobased%20Chemicals%20Workshop%20Report.pdf
[7] NREL. Synthesis and Analysis of Performance-Advantaged Bioproducts. Technology session review area: Performance-advantaged bioproducts, bioprocessing, separations, and plastics. PI: Gregg T. Beckham, National Renewable Energy Laboratory. https://www.nrel.gov/docs/fy23osti/79481.pdf
[8] https://www.energy.gov/eere/bioenergy/bioenergy-technologies-office
[9] www.Biopreferredgov/BioPreferred/.
[10] El-Karoui M, Hoyos-Flight M, Fletcher L Future trends in synthetic biology – A report. Front Bioeng Biotechnol 2019.
[11] The US Bioeconomy: Charting a course for a Resilient and Competitive Future. http://www.HTTPS://DOI.ORG/10.55879/D2HRS72WC.
[12] C&ENews. DeepMind releases Structure predictions for nearly every known protein, Aug 4, 2022
[13] C&ENews. LG Chem and ADM detail lactic acid plants, August 22, 2022.

Bhima R. Vijayendran, Ram Lalgudi

Chapter 2 Feedstock for bioproducts

2.1 Introduction

Feedstock in the development and commercialization of bioproducts is derived from various sources. Useful and economically viable ones have several critical features, such as low cost, reliable supply chain, amenable to industrial-scale conversion processes and functionality, which can impart properties and performance in the finished bioproducts of value in the marketplace. Historically, renewable biomass from agriculture and forest products has been the main stay of feedstock to produce chemical intermediates and products. Over the last 100 years or so, the feedstock picture has changed and shifted to nonrenewable fossil-based sources. The main drivers are abundant supply, low cost, and advances in process engineering, including catalyst technology, which propelled its dramatic growth. Today, only about less than 10% of the global industrial chemical market is based on renewable feedstock sources. As noted in Chapter 1, the picture is changing again due to numerous factors such as increased awareness and concerns from greenhouse gas emissions from fossil sources, progress in science and technology related to renewable-based feedstocks, process advances and catalyst innovations, and regulatory and public policy pressures.

In this chapter, an overview of feedstock for bioproducts is covered. Specifically, the following sources of feedstock are reviewed:

- Plant seed oils
- Sugars
- Carbohydrates
- Lignin
- Waste and coproducts
- Miscellaneous

2.1.1 Feedstock overview

An overview of fossil-derived and bio-derived feedstock is shown in Figure 2.1 and Table 2.1. Table 2.1 highlights a few characteristics of fossil-based and bio-based feedstocks. Less than 10% of fossil oil and gas production is used in the manufacture of

Bhima R. Vijayendran, Redwood Innovation Partners LLC, Carlsbad, CA 92011
Ram Lalgudi, Aries Science and Technology, LLC, Columbus, OH 43212,
e-mail: lalgudir@ariesst.com

 | https://doi.org/10.1515/9783112212141-002

petrochemical products, while Figure 2.1 presents the relative scale of global fossil and plant oil production and utilization in 2022 [1].

Table 2.1: Some features of fossil and biofeedstock.

Features	Fossil	Bio
Resource distribution	A few regions	More distributed
Reserves	Finite	Potentially infinite
Sustainability	Increasingly challenging	Manageable
Current global use	Over 35 billion barrels/year	Less than 6–10 billion boe*/year
Production technology	Well established	Emerging
Infrastructure	Well established	Emerging
Price	About $100/barrel	About $20/boe**
Chemistry	Simple hydrocarbon	Complex oxygenated hydrocarbons
Conversion processes	Catalytic thermochemical – well established	Emerging
Environmental footprint	High	Low

*boe, barrel of oil equivalent.
**$71/ton of dry biomass.

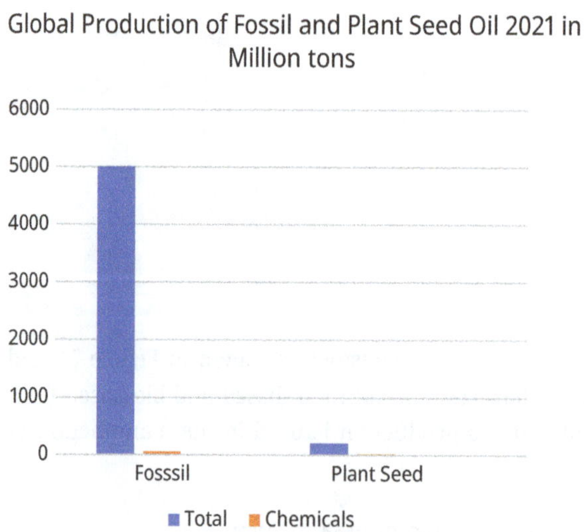

Figure 2.1: Indicative relative scale of global fossil and plant oil production and utilization in 2022 [1].

2.1.2 Plant seed oils

Global production of vegetable and plant oils is over 200 million tons/year [1]. There are many sources of plant oils from various parts of the world. The main oils of commercial interest are soybean, palm, sunflower, canola, and castor oil. There are some specialty oils such as *Vernonia*, cashew nut, and the emerging *Camelina* oil.

A few of these are shown in Figures 2.2–2.7.

Figure 2.2: Soybean (courtesy of Merideth at https://unsplash.com/).

Figure 2.3: Palm (courtesy of Szewczyk at https://unsplash@com/).

Figure 2.4: Sunflower (courtesy of Bonnie Kittel at https://unsplash.com).

Figure 2.5: Castor seed (courtesy of CDC at https://unsplash.com/).

Figure 2.6: *Camelina* (courtesy of Christina Eynck at Canada Government).

Figure 2.7: Canola (courtesy of Michael Ohey Sekara at http://unsplash.com/).

2.2 Composition of plant oils

Chemical composition of the three significant seed oils, namely, soybean, palm, and sunflower, is shown in Table 2.2. Triglyceride is shown in Figure 2.8.

Table 2.2: Triglyceride compositions and functionality [2].

Triglyceride	Soybean oil	Palm oil	Sunflower
Stearic	6.5%	5.4	4.2
Palmitic	12.2	39.6	5.6
Oleic	29.1	42.9	26.2
Linoleic	44.1	8.9	64.1
Linolenic	2.7		
Higher chain length	<1%	<0.5%	<1%

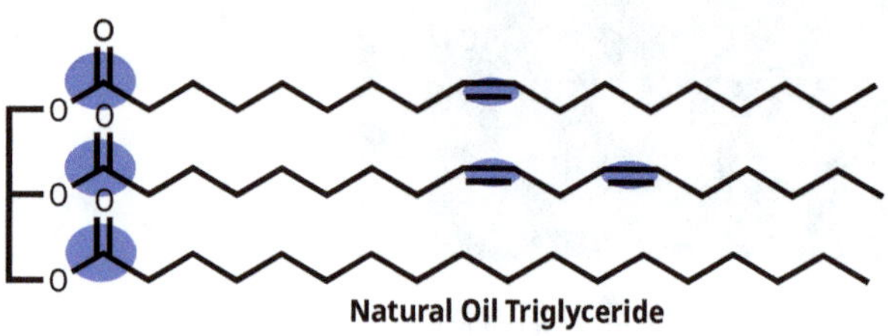

Figure 2.8: Functionality.

2.2.1 Functionality modifications

Plant oils are quite versatile in their functionality, being amenable to a myriad of chemical modifications and functionalization using cost-effective industrial processes and existing infrastructure (Figure 2.8).

Various chemical modifications are extensively covered in the literature [3–7]. A few of them are highlighted further. A schematic of this is shown in Figure 2.9.

An area worth noting is the use of plant breeding and recently introduced genetically modified soybean oil with high oleic acid content. This oil, having a high-level oleic acid and concomitant reduction in saturated fatty acids, imparts useful functionality for chemical conversion to several industrial bioproducts such as bioplasticizers, biolubricants, surfactant, and coatings [8].

Myriad of useful and commercial industrial products such as plasticizers, surfactants, lubricants, coatings and adhesives, polymers, and plastics have been successfully developed from plant oils. Chapters 8, 10, 12, 17, and 18 cover some of these bioproducts.

2.2.2 Specialty oils

A few specialty oils are also of interest and value commercially. They are covered further.

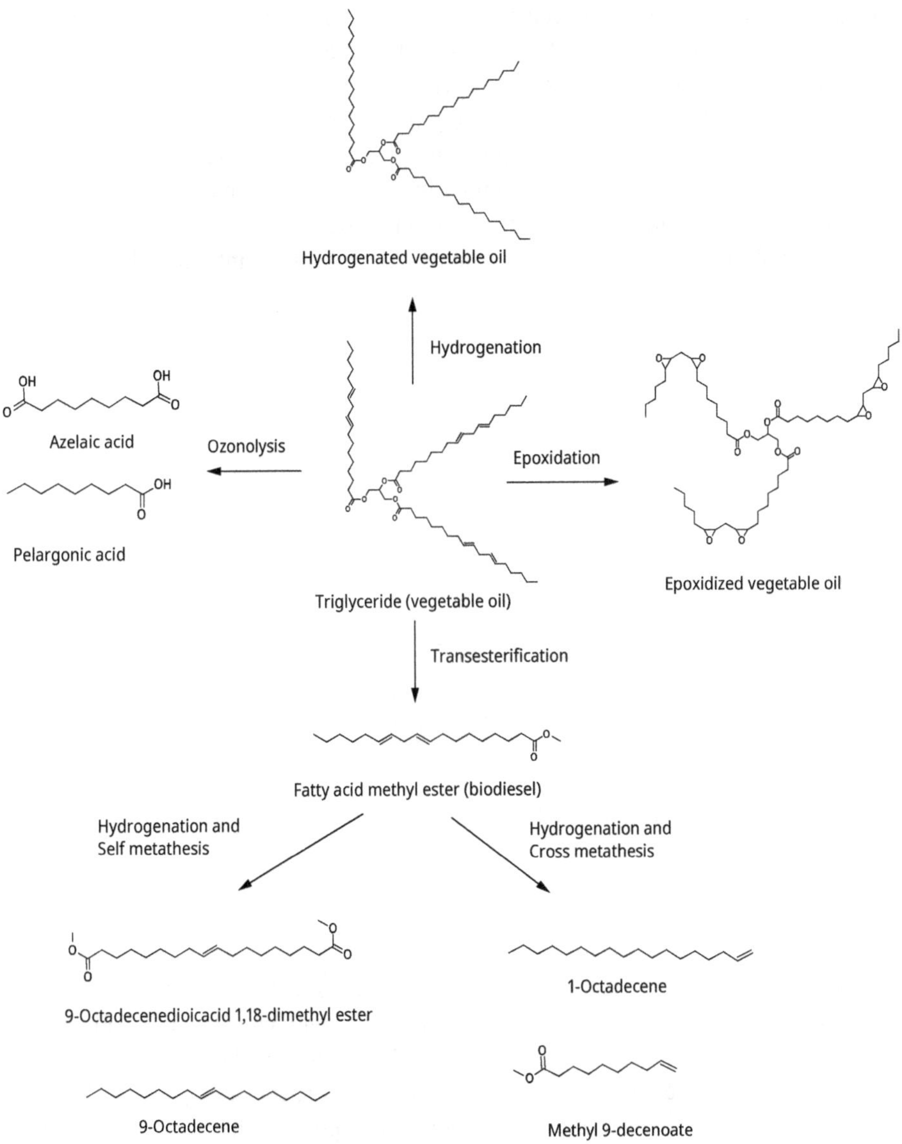

Figure 2.9: Overview of triglyceride functionalization and chemistry [3–7].

2.2.2.1 Castor oil

Castor is a nonfood industrial crop mainly cultivated in Africa, South America, and India. It grows well with high oil content in tropical regions with warm climate conditions, typically below 38 °C. Frosting weather is known to adversely affect the plant growth. Attempts to grow and process castor in the United States have not been successful due to unfavorable climate for castor plantation, high labor cost, undesirable oil yield and the process safety associated with the ricin toxicity, among others [9].

It is a vegetable oil obtained by pressing the seeds of the castor oil plant (*Ricinus communis* L.); castor oil is a nondrying oil with a unique triglyceride derived from 12-hydroxy-9-*cis*-octadecenoic acid or commonly referred to as ricinoleic acid (RA; Figure 2.10). The high content of RA triglyceride in castor oil allows them to be directly used as polyol for making coatings [10], foams [11], and surfactants [12–14].

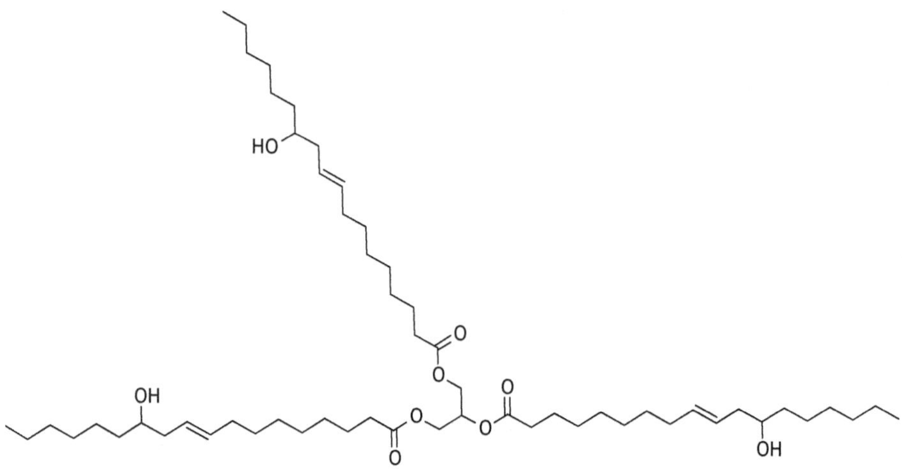

Figure 2.10: Chemical structure of castor oil.

2.2.2.2 *Vernonia* oil

Vernonia is an industrial crop that can grow as weeds in fields or in woodlands of Africa under wide agroecological conditions [15]. *Vernonia* crop has naturally occurring epoxy fatty acids in the triglyceride (Figure 2.11), which makes it an economically attractive candidate for plasticizer [16]. Typically, plasticizers are generally derived from epoxidation of soybean oil. Furthermore, the epoxide ring in the *Vernonia* oil can be hydroxylated, and the resulting polyol can be used to make bio-based polyurethane coating [17, 18]. The wide industrial application of neat *Vernonia* oil has created significant interest in domestication of *Vernonia* as an industrial oil seed crop in the United States [19]. *Vernonia* oil is still not available at scale for industrial use, and significant improvement in agronomics is needed for growing *Vernonia* crop in the United States.

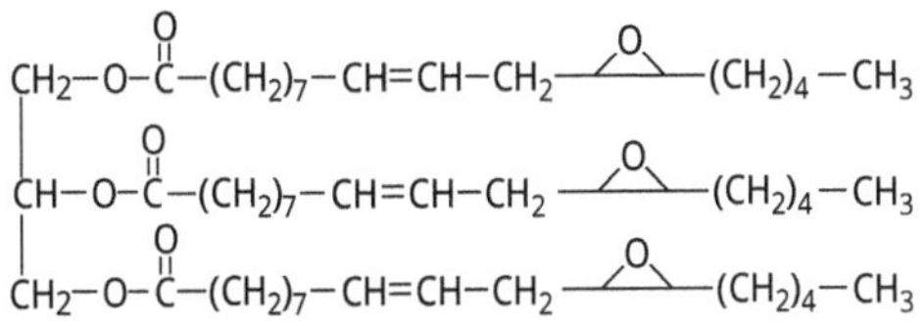

Figure 2.11: Idealized representations of *Vernonia* oil (structure adopted from Trumbo et al. [6]).

2.2.2.3 Cashew oil

Cashew nutshell liquid (CNSL) is a dark brown viscous liquid, which is a very important agricultural byproduct of cashew nut production. Natural CNSL is a mixture of phenolic compounds [20] (Figure 2.12) with aliphatic side chains, and typically comprises 70% anacardic acid, 5% cardanol, and 18% cardol. Solvent treatment, followed by vacuum distillation, is the most used extraction method to produce CNSL [21].

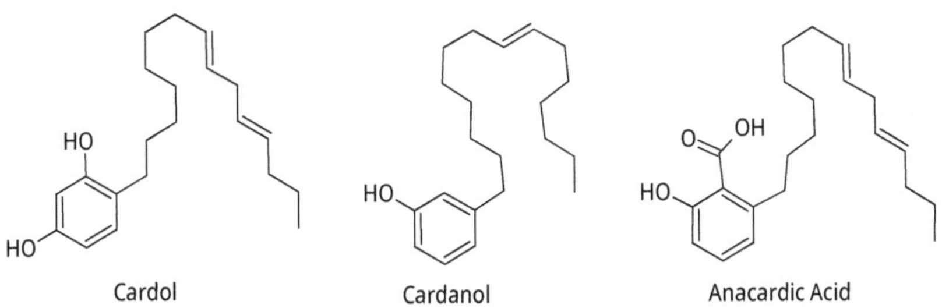

Figure 2.12: Typical chemical composition of oils extracted from cashew nuts.

CNSL finds several industrial uses, owing to its attractive chemical structure. The phenolic moiety has been utilized to prevent oxidation [22] and corrosion [23] and to produce thermosetting resins by reacting with formaldehyde [24]. The hydrophobic structure of cardanol resembles petroleum-derived nonylphenol. Ethoxylation of cardanol produces a bio-based nonionic surfactant that can replace toxic petroleum-derived nonylphenol-based nonionic surfactants [25].

2.2.2.4 Cyclopropane and cyclopropene fatty acids

Cyclopropane-containing fatty acids (CFAs) and their unsaturated analogs, cyclopropene fatty acids (CPFAs), constitute a structurally distinct class of plant- and microbially derived lipids, characterized by a highly strained three-membered carbocyclic ring embedded within the aliphatic chain. This ring system, typically formed by enzymatic cyclopropanation of a *cis*-double bond, imparts unique physicochemical and biological properties, relative to conventional long-chain fatty acids. In microbial systems, CFAs are synthesized by cyclopropane-fatty-acyl-phospholipid synthase using *S*-adenosylmethionine as the methyl donor, and they play a critical role in modulating membrane rigidity and stress tolerance under acidic or thermal stress conditions.

In higher plants, CPFAs such as malvalic acid and sterculic acid (see Figures 2.13 and 2.14) occur predominantly in members of the Malvaceae family, including *Sterculia foetida*, *Gossypium* species, and *Adansonia digitata* (baobab). These CPFAs are typically present at low to moderate levels in crude seed oils and are largely removed or degraded during high-temperature refining. Their strained ring system makes them chemically reactive, and they are known inhibitors of stearoyl-CoA desaturase, a key enzyme in lipid metabolism. As a result, CPFAs are considered antinutritional in edible oils, but they are of mechanistic interest in lipid chemistry and polymer science due to their unusual reactivity and potential for ring-opening functionalization.

From a bioproducts perspective, the presence of cyclopropane or cyclopropene rings creates opportunities for selective chemical transformations that are not accessible with conventional unsaturated fatty acids. The ring strain can be exploited in controlled ring-opening reactions to generate functionalized intermediates, whereas the inherent rigidity of the cyclopropane motif may confer unique properties in specialty polymers or surfactants. Although their industrial use is currently limited, CFAs and CPFAs represent a niche but chemically valuable subset of plant-derived lipids, whose structural features expand the functional landscape of renewable feedstocks.

Figure 2.13: Chemical structure of malvalic acid ($C_{18}H_{32}O_2$).

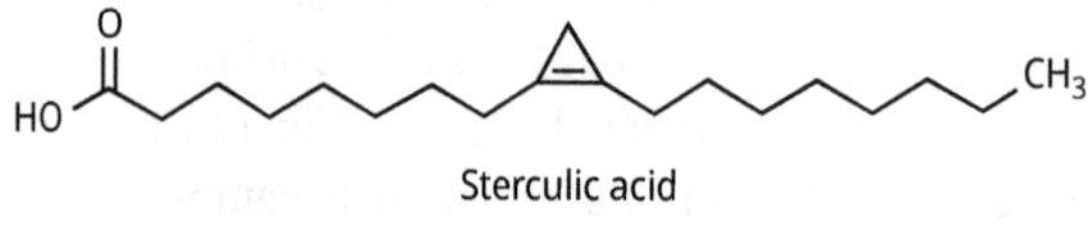

Figure 2.14: Chemical structure of sterculic acid ($C_{19}H_{34}O_2$).

2.2.3 Biotechnology advances

Advances in synthetic biology and industrial biotechnology have started to play a bigger role in bio-derived feedstocks.

Plant biotechnology has played and continues to play a bigger role in crop and plant-based feedstocks. The impacts have been in several areas such as increased yield, lower input costs, and improved drought and insect resistance. One area of particular interest from feedstock perspective is trait modifications and increased functionality of value in myriad bioproduct applications.

2.3 Sugars and carbohydrates

Simple C_5 and C_6 sugars (Figure 2.15) and complex ones provide a reliable source for renewable feedstock for bioproducts.

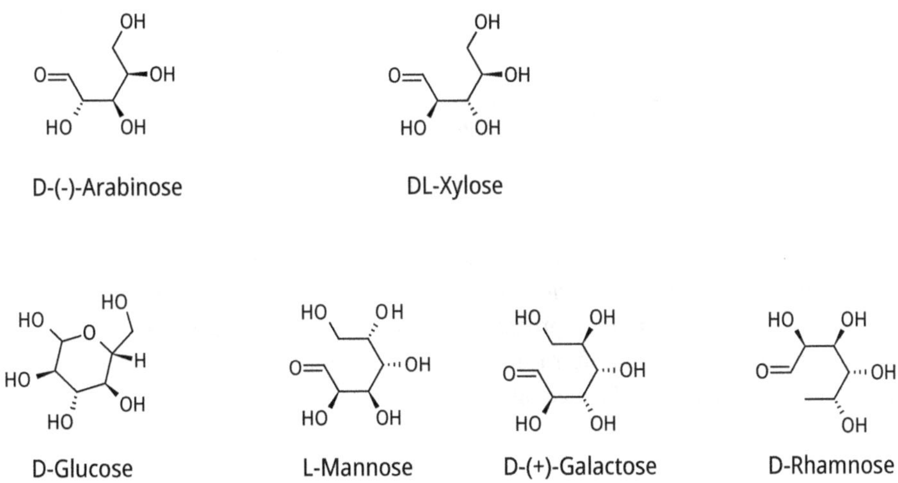

Figure 2.15: Structures of simple sugars.

A study by the Department of Energy (DOE) [26] highlighted top 11 chemicals that could be derived from sugars through conventional catalytic thermochemical and emerging bioprocessing schemes.

Some of these target molecules such as 1,3-propanediol, propylene glycol, and 1,3-butanediol have moved on to full commercial scale. Others such as acrylic acid and glucaric acid are still in the development stage. A few are covered in Chapters 9 and 19.

> Sugar- and carbohydrate-based feedstocks have been converted to functional C_4 intermediates such as succinic acid for use as platform intermediates for further modifications to useful chemical products.

A couple based on succinic acid and furans are shown later from reference [26] (Figures 2.16 and 2.17).

Chapter 7 on furanic-based fiberglass binders has some recent updates on the furanic platform.

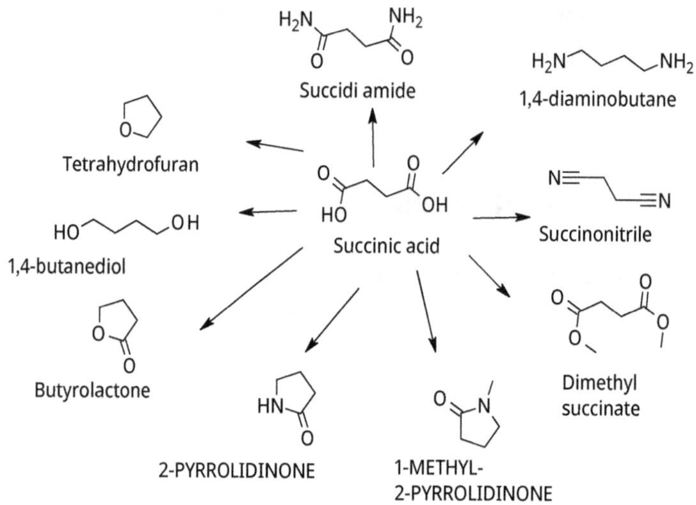

Figure 2.16: Succinic acid platform [26].

Some recent works have successfully converted simple sugars to high-value chemical intermediates such as fatty acid and hydroxy long-chain molecules using engineered *E. coli* in a fermentation process [27]. Start-up companies are developing high-value plant oils for food and nutraceutical applications with engineered organisms in bioreactors, touted as zero-farmed oils.

2.4 Lignin

A ubiquitous biomass source is available in large quantity, having a complex condensed polyphenolic structure. Lignin is a natural phenolic-based polymer that is abundant in the cell walls of plants, especially in land-based plants. It provides mechanical support for plant growth as well as in assisting in the long-distance transport of water in xylem, resisting a variety of stresses, and providing protection against pathogens. The chemical structure of native lignin comprises three *p*-hydroxycinnamyl alcohol monomers, referred to as monolignols. These are *p*-coumaryl alcohol (*p*-hydroxyphenyl, H unit), coniferyl alcohol (guaicyl, G unit) and sinapyl alcohol (syringyl, S unit), which are characterized by their degree of methoxyl substitution at the third and fifth carbon (C_3 and C_5) positions of the aromatic ring, that is, *p*-coumaryl alcohol has no methoxyl group, whereas coniferyl alcohol has a methoxyl group at the C_3 position, and sinapyl alcohol has a methoxyl group at both the C_3 and C_5 positions. The most predominant intermonomer linkage within lignin gives rise to its complex three-dimensional amorphous network structure called phenylpropane β-aryl ether (β-*O*-4). Other interunit linkages that have been identified to date are biphenyl and dibenzodioxocin (5–5), phyenylcoumarin

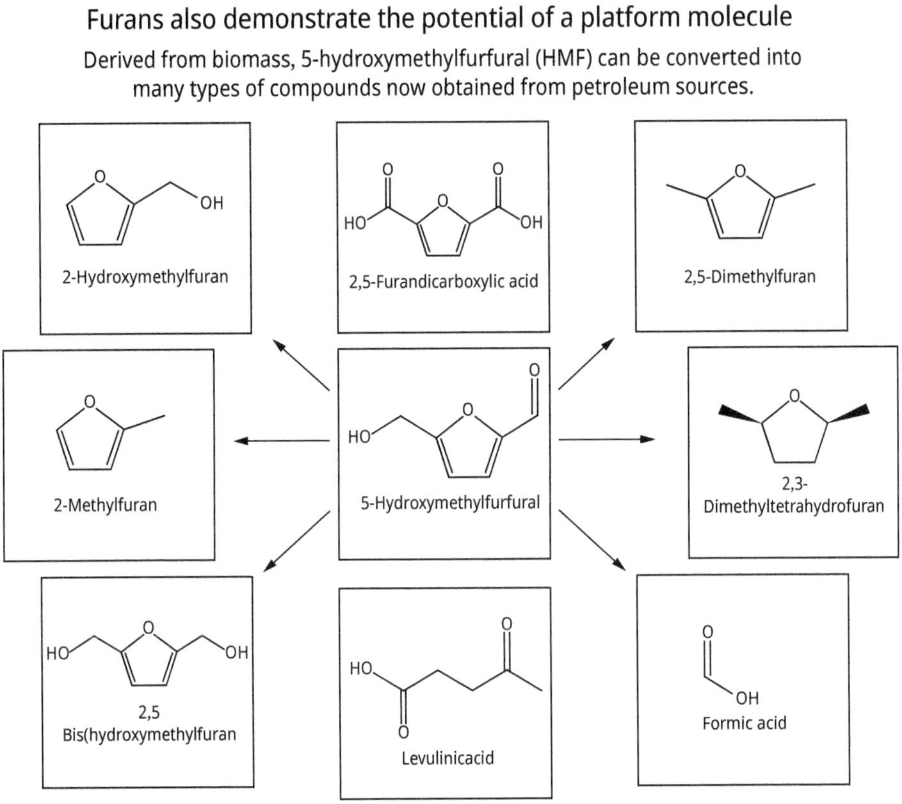

Figure 2.17: Furanic platform [26].

(β-5), 1,2-diaryl propane (β-1), phenylpropane α-aryl ether (α-*O*-4), diaryl ether (4-*O*-5), and β–β-linked structures (β–β). See Chapter 9 for additional details on lignin structure and chemistry.

Despite lignin's low cost and huge supply, its use in bioproducts has been extremely limited despite numerous attempts and substantial R&D investments by large companies, academic institutions, and others. This is mainly due to the rather complicated and involved process in deconstructing lignin structure to useful building blocks at scale in an economic way.

A few examples of lignin-based bioproducts are in lignin-derived carbon fiber, described in Chapter 9.

2.5 Miscellaneous

Glycerin, a coproduct in the manufacture of soaps and biodiesel manufacture, shows promise as a cost-effective building block for several industrial products such as polyols [28], propylene glycol, and epichlorohydrin (Figure 2.18).

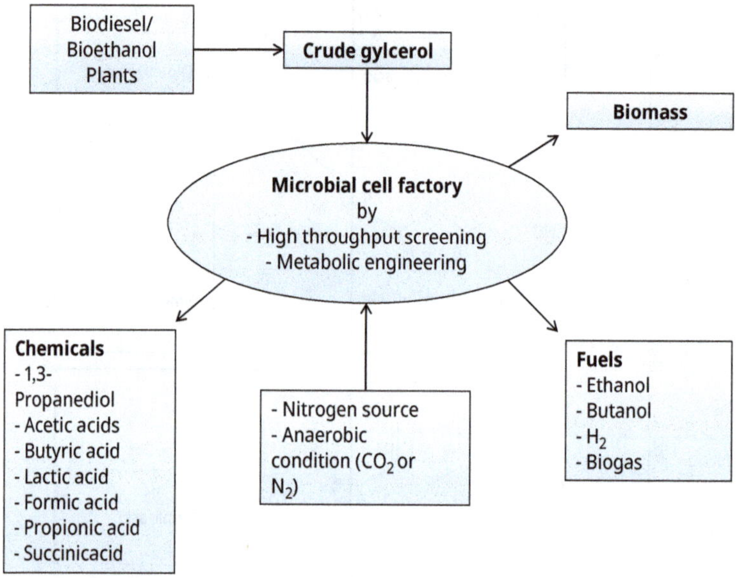

Figure 2.18: Glycerin platform.

2.6 Waste-derived feedstock

Waste-derived feedstocks represent an increasingly important class of renewable raw materials that align with circular economy principles by converting low-value residues into functional intermediates for energy, materials, and chemical production. These feedstocks originate from diverse waste streams – including agricultural residues, food waste, municipal biosolids, yard trimmings, and industrial organic by-products – and offer a dual benefit: reducing environmental burdens associated with waste disposal, while supplying cost-effective inputs for biobased manufacturing. Their heterogeneous composition, variable moisture content, and varying degrees of biodegradability require tailored processing strategies, yet they collectively provide a robust, underutilized resource base for sustainable materials development.

Thermochemical, biochemical, and physicochemical conversion pathways can transform these wastes into a spectrum of value-added products, including bio-oils, syngas, compost, biogas, and carbonaceous solids. Among these, biochar has emerged as one of the most versatile and technically significant waste-derived materials [29]. Produced through pyrolysis under oxygen-limited conditions, biochar serves as both a functional product and a platform material, whose properties can be tuned through feedstock selection and processing conditions. Its high carbon content, structural stability, and surface reactivity make it suitable for soil amendment, contaminant adsorption, carbon sequestration, and as a precursor for engineered carbon materials. As such, biochar provides a compelling example of how waste-derived feedstocks can be leveraged to create high-performance, environmentally beneficial materials.

Biochar is a carbon-rich, fine-grained solid produced through the thermochemical conversion of biomass under oxygen-limited conditions. Waste-derived feedstocks – including agricultural residues, food waste, yard waste, sewage sludge, and industrial organic by-products – are increasingly used for biochar production due to their abundance, low cost, and the opportunity to valorize materials that would otherwise require disposal. Slow pyrolysis at moderate temperatures (typically 250–700 °C) is the most common production route, yielding biochar with high fixed-carbon content, substantial surface area, and a porous microstructure that supports a wide range of environmental and industrial applications.

The physicochemical properties of biochar depend strongly on feedstock composition and pyrolysis conditions. Lignocellulosic wastes generally produce high-carbon Class I biochars, with carbon contents exceeding 60%, whereas sewage-sludge-derived biochars often fall into Class III due to lower organic carbon content and higher mineral fractions. Increasing pyrolysis temperature typically decreases yield but enhances aromaticity, structural stability, and surface area. Waste-based feedstocks such as tires, corn stalks, and municipal residues can produce biochar yields ranging from 20 to 52 wt.% under slow pyrolysis and 29 to 48 wt.% under rapid pyrolysis, depending on the temperature and residence time.

Biochar's functional attributes – cation-exchange capacity, surface reactivity, and structural robustness – enable its use in soil amendment, carbon sequestration, contaminant adsorption, wastewater treatment, and as a precursor for engineered carbon materials [30]. However, waste-derived biochars may contain volatile organic compounds or trace contaminants that can affect soil health and necessitate careful characterization and quality control. Advanced analytical techniques such as XRD, XRF, FTIR, TG, and SEM are routinely used to assess the mineralogy, functional groups, thermal stability, and morphology of biochars from diverse waste streams.

As a waste-derived feedstock, biochar exemplifies circular economy principles by transforming low-value residues into functional carbon materials with environmental and industrial utility. Its tunable properties, broad feedstock compatibility, and potential for carbon-negative production make it an increasingly important component of sustainable materials development. Table 2.3 highlights the key characteristics of major waste-derived feedstocks, showcasing their potential for sustainable applications.

Table 2.3: Major waste-derived feedstocks and their characteristics.

Feedstock type	Typical composition	Conversion pathways	Key products	Advantages	Constraints
Agricultural residues	Lignocellulosic biomass (cellulose, hemicellulose, and lignin)	Pyrolysis, gasification, and fermentation	Biochar, bio-oil, syngas, and bioethanol	Abundant, renewable, and low cost	Seasonal availability and variable moisture
Food waste	Carbohydrates, proteins, lipids, and high moisture	Anaerobic digestion, composting, and pyrolysis	Biogas, compost, and biochar	High biodegradability and methane reduction	Rapid spoilage and logistics challenges
Municipal biosolids	Organic matter, nutrients, and minerals	Pyrolysis and composting	Biochar and soil amendments	Nutrient-rich, and waste diversion	Heavy metals and regulatory limits
Yard waste	Lignocellulosic biomass	Compostingand pyrolysis	Compost and biochar	Widely available and low cost	Heterogeneous composition
Industrial organic waste	Process-specific organics	Pyrolysis and anaerobic digestion	Biochar, biogas, and specialty chemicals	Valorizes difficult wastes	Requires pretreatment

References

[1] Statista Research Department, Feb 11, 2022, and https://Www.Ees/Org.

[2] Alves AQ, da Silva VAJ, Góes AJS, Silva MS, De Oliveira GG, Bastos IVGA, De Castro Neto AG, Alves AJ. The fatty acid composition of vegetable oils and their potential use in wound care. Adv Skin Wound Care 2019;32(8):1–8.

[3] Armylisas AHN, Hazirah MFS, Yeong SK, Hazimah AH. Modification of olefinic double bonds of unsaturated fatty acids and other vegetable oil derivatives via epoxidation: A review. Grasas Y Aceites 2017;68:174.

[4] Tang Q. Bio-based epoxy resin from epoxidized soybean oil, Chen Y, Soybean, Li Q, Translator, Huizhi Gao, Series Ed., Rijeka: IntechOpen; 2019, p. Ch 8. https://doi.org/10.5772/intechopen.81544.

[5] Graiver D, Patil M, Narayan R. Recent advances in ozonation of vegetable oils. Recent Pat Mater Sci 2010;3(3):203–218. https://doi.org/10.2174/1874464811003030203.

[6] Marvey BB. Sunflower-based feedstocks in nonfood applications: Perspectives from olefin metathesis. Int J Mol Sci 2008;9(8): https://doi.org/10.3390/ijms9081393.

[7] Petrović ZS, Javni I, Ionescu M. Biological oils as precursors to novel polymeric materials. J Renewable Mater 2013;1:167–186.

[8] Heggs R. Chapter 10 – Industrial uses of High-Oleic oils, Flider FJ. editor. High Oleic Oils, AOCS Press; 2022, pp. 245–259. https://doi.org/10.1016/B978-0-12-822912-5.00009-5.

[9] Patel VR, Dumancas GG, Kasi Viswanath LC, Maples R, Subong BJJ. Castor oil: Properties, uses, and optimization of processing parameters in commercial production. Lipid Insights 2016;9:1–12. https://doi.org/10.4137/LPI.S40233.

[10] Thakur S, Karak N. Castor oil-based hyperbranched polyurethanes as advanced surface coating materials. Prog Org Coat 2013;76(1):157–164. https://doi.org/10.1016/j.porgcoat.2012.09.001.

[11] Wang HJ, Rong MZ, Zhang MQ, Hu J, Chen HW, Czigány T. Biodegradable foam plastics based on castor oil. Biomacromolecules 2008;9(2):615–623. https://doi.org/10.1021/bm7009152.

[12] Zhou J, Sun Y, Zhu K, Di Serio M, Zhang Y, Sun J, Wu H, Ding L, Liang H. Influence of sulfonic acid group on the performance of castor oil acid based methyl ester ethoxylate sulfonate. Null 2018;39 (12):1693–1698. https://doi.org/10.1080/01932691.2018.1461641.

[13] Zhang Q, Sun Y, Zhi L, Zhang Y, Di Serio M. Properties of ethoxylated castor oil acid methyl esters prepared by ethoxylation over an alkaline catalyst. J Surfactants Deterg 2015;18(2):365–370. https://doi.org/10.1007/s11743-014-1657-3.

[14] Zhang -Q-Q, Cai B-X, Xu W-J, Gang H-Z, Liu J-F, Yang S-Z, Mu B-Z. Novel zwitterionic surfactant derived from castor oil and its performance evaluation for oil recovery. Colloids Surf A Physicochem Eng Asp 2015;483:87–95. https://doi.org/10.1016/j.colsurfa.2015.05.060.

[15] Baye T, Becker HC. Exploration of Vernonia galamensis in Ethiopia, and variation in fatty acid composition of seed oil. Genet Resour Crop Evol 2005;52(7):805–811. https://doi.org/10.1007/s10722-003-6086-5.

[16] Krewson CF, Riser GR, Scott WE. Euphorbia and vernonia seed oil products as plasticizer-stabilizers for polyvinyl chloride. J Am Oil Chem Soc 1966;43(6):377–379. https://doi.org/10.1007/BF02646792.

[17] Trumbo DL, Rudelich JC, Mote BE. Applications of Vernonia oil in coatings. In Janick J, editor. Perspectives on New Crops and New Uses, Alexandria, VA: ASHS Press; 1999. pp. 267–271.

[18] Dirlikov S, Frischinger I, Islam MS, Lepkowski TJ. Vernonia oil: A new reactive monomer. In Gebelein CG, editor. Biotechnology and Polymers, Boston, MA: Springer; 1991. pp. 79–93.

[19] Thompson AE, Dierig DA, Johnson ER, Dahlquist GH, Kleiman R. Germplasm development of Vernonia galamensis as a new industrial oilseed crop. Ind Crops Prod 1994;3(3):185–200. https://doi.org/10.1016/0926-6690(94)90066-3.

[20] Morais SM, Silva KA, Araujo H, Vieira IGP, Alves DR, Fontenelle ROS, Silva AMS. Anacardic acid constituents from cashew nut shell liquid: NMR characterization and the effect of unsaturation on its biological activities. Pharmaceuticals 2017;10(1): https://doi.org/10.3390/ph10010031.

[21] Nyirenda J, Zombe K, Kalaba G, Siabbamba C, Mukela I. Exhaustive valorization of cashew nut shell waste as a potential bioresource material. Sci Rep 2021;11(1):11986. https://doi.org/10.1038/s41598-021-91571-y.

[22] Andradedos de T. J. A. S.Araújo BQ, Amdgl C, Da Silva J, Saffi J, Richter MF, Adbf F. Antioxidant properties and chemical composition of technical cashew nut shell liquid (TCNSL). Food Chem 2011;126(3):1044–1048. https://doi.org/10.1016/j.foodchem.2010.11.122.

[23] Ghosh T, Karak N. Cashew nut shell liquid terminated self-healable polyurethane as an effective anticorrosive coating with biodegradable attribute. Prog Org Coat 2020;139:105472. https://doi.org/10.1016/j.porgcoat.2019.105472.

[24] Chuayjuljit S, Rattanametangkool P, Potiyaraj P. Preparation of cardanol–formaldehyde resins from cashew nut shell liquid for the reinforcement of natural rubber. J Appl Polym Sci 2007;104(3):1997–2002. https://doi.org/10.1002/app.25915.

[25] Roy A, Fajardie P, Lepoittevin B, Baudoux J, Lapinte V, Caillol S, Briou B. CNSL, a promising building block for sustainable molecular design of surfactants: A critical review. Molecules 2022;27(4): https://doi.org/10.3390/molecules27041443.

[26] Top Value Added Chemicals from Biomass Volume I – Results of Screening for Potential Candidates from Sugars and Synthesis Gas Produced by the Staff at Pacific Northwest National Laboratory (PNNL), National Renewable Energy Laboratory (NREL), Office of Biomass Program (EERE); Werpy T, Petersen G, Series Eds., 2004.

[27] Del Cardyre SB, Brubaker S, Keasling JD Modified Organisms and Uses Therefor, U.S. Pat 8,535,916, Feb 13, 2007.

[28] Benecke HP, Vijayendran BR, Garbark, Mitchell K. Low cost and highly reactive polyols: A co-product of the emerging biorefinery economy. Clean Air Water 2008;36(8):694–699.

[29] Ahmed SF, Mehejabin F, Chowdhury AA, Almomani F, Khan NA, Badruddin IA, Kamangar S. Biochar produced from waste-based feedstocks: Mechanisms, affecting factors, economy, utilization, challenges, and prospects. GCB Bioenergy 2024;16(8):e13175. https://doi.org/10.1111/gcbb.13175.

[30] Liu W-J, Jiang H, Yu H-Q. Development of biochar-based functional materials: Toward a sustainable platform carbon material. Chem Rev 2015;115(22):12251–12285. https://doi.org/10.1021/acs.chemrev.5b00195.

Michael A. Schultz*

Chapter 3 Process scale-up for bioproducts: enabling the emerging circular economy

Abstract: This chapter will provide guidelines and best practices for scale-up of bioproducts and bioprocessing technology, with the aim of reducing the time, cost, and risk of scale-up. Experimental planning, modeling, and techno economics along with early-stage engineering are critical to this effort. Similarities and differences in the scale-up of bioproducts compared to the scale-up of more conventional thermochemical processes using petroleum-based feedstocks will be presented, along with a case study and commercial success stories for the scale-up of bioproducts.

3.1 Introduction

We have seen a growth in bioproducts over the last number of years [1]. The drivers for this growth include an increasing focus on sustainability, with the goal to reduce emissions of CO_2 and other air and water pollutants, and to enable circularity through reuse of carbon and carbon-based products. Some cases offer the potential for a lower cost of carbon, through use of waste feedstocks from industrial gaseous emissions, end-of-life plastic, and waste biomass. In addition, in some cases, the bioproduct is actually a better product than the petroleum-based version, coming with a cheaper, safer processing route and performance advantages over traditional materials.

While fossil fuels still dominate, bioprocessing represents an opportunity to be a key piece of the puzzle to reducing CO_2 emissions from fossil-based products. Market transformation is only possible by continuing to push boundaries, including:

- Growth in installed capacity: Production of biodiesel, renewable diesel, and ethanol [2] is increasing dramatically [3]. Sustainable Aviation Fuel (SAF) is an emerging alternative fuel [4]. Markets, supply chains, equipment manufacture, and technology know-how in this space are following suit.
- Emergence of tools to support industrial bioprocessing: Advanced tools are emerging, mirroring the past advances in thermochemistry: advanced analytics, high-throughput screening, and genetic tools for custom microbes.
- Scale-up methodologies: Scale-up methodologies are critical to advancing existing technologies and developing the next generation.

*Corresponding author: **Michael A. Schultz**, Managing Director, PTI Global Solutions LLC Indianapolis, IN, USA, e-mail: mike@ptisolns.com

 | https://doi.org/10.1515/9783112212141-003

3.2 Challenges with scale-up of emerging bioprocessing technology

Scale-up of emerging technology for production of bioproducts faces many hurdles along the way. One of these is the need to reduce technology risk through measured approaches of increasing scale, starting at the lab or bench scale (perhaps with several lab scales from milliliter quantities on plates or in shake flasks, to 1 L CSTRS and then 3–10 L CSTRS), moving to a pilot, then demonstration, and finally commercial. Of course, not only are the volumes increasing, but the time to build each scale, and the capital cost associated with each scale increases dramatically as well (see Figure 3.1).

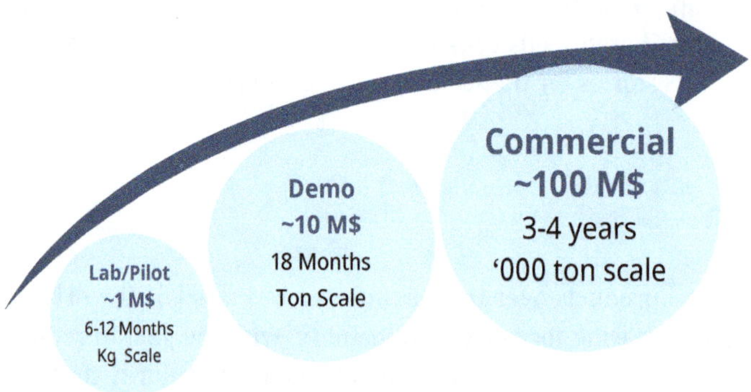

Figure 3.1: Typical scale-up stages.

Often a key challenge is to significantly *reduce* cost even while scaling up. Finally, because the first commercial scale is typically small relative to the conventional technology, it is difficult to compete on a unit cost basis with incumbent products because of the higher capital cost factor.

Bioprocessing also brings scale-up challenges that are unique to this area relative to conventional fuel and chemical technology:

- **Lack of Established Data and Models:** e.g., bacterial growth kinetics and flux models
- **New Separation Challenges:** CO_2 from air, intercellular bioproducts
- **New Optimization Criteria:** Carbon footprint and other environmental, social, and governance (ESG) and life cycle assessment (LCA) metrics
- **Bio-based Catalysis:** Biological catalysts, whether using enzymes or whole cells require different handling than conventional heterogeneous catalysis using zeolites, other inorganics, etc.

3.3 Commercial-scale project development: bioprocessing technology

The project development cycle for an established process technology is well-known, with an initial conceptual design phase to define the project, develop a block flow diagram, and generate a cost estimate that is typically ±50%. Basic engineering, detailed engineering, procurement and construction, and startup then follow.

The conceptual design phase for an established technology can generally be completed in 2–4 months. However, for a new bioprocessing technology much more time is needed to get this right! To do this creative process engineering should be brought into the picture as early as possible, even during discovery R&D. In fact, by starting with the conceptual design, or process concept, this framework can be used to drive new technology development, scale-up and commercialization. This process concept is not set in stone, and, in fact, should be reviewed and updated throughout the scale-up effort.

Ultimately, this approach must result in a design to move into the basic engineering phase while reducing technical risk and optimizing economics. The process concept is established to drive the scale-up effort, not just inform it. This then enables the innovation to be directed to areas that create the greatest value from breakthrough and disruptive ideas, by identifying challenges early, failing fast when it is cheaper and quicker, and making sure efforts are focused on solving commercially relevant problems.

The process concept is developed, and frequently revised, through a combination of creative process engineering, multi-scale experimental data, and modeling and analysis.

Creative Process Engineering: The flow scheme is developed, the material balance is estimated, and key process design decisions are identified to establish the best process flowsheet for the technology.

Modeling and Analysis: A good model can save time and resources in the lab. Coupled with the right analysis, the scale-up team can prioritize objectives in the lab, pilot, and demo units. As a cautionary note, useful models are more important than perfect models!

Experimental Data: The right data is needed to prove out breakthrough ideas, secure partners and investors, and develop engineering data for equipment design. Multi-scale data is critical to this effort, and with good planning, multiple assets and external resources can be leveraged.

The key benefits to this approach are:
- Prioritization of R&D to de-risk and optimize new technology.
- Identification of cost reduction opportunities throughout the scale-up effort.
- Anticipation of engineering needs as early as possible.

In this way the team can reduce risk and optimize economics of the new design, while efficiently managing the time, cost, and deployment of limited technical and financial resources of its efforts.

3.4 Key element #1: creative process engineering

Creative process engineering (see Figure 3.2) is critical for success for first of its kind technology. Creative process engineers understand commercial plant design and can also accommodate the ambiguity that is common with any new technology.

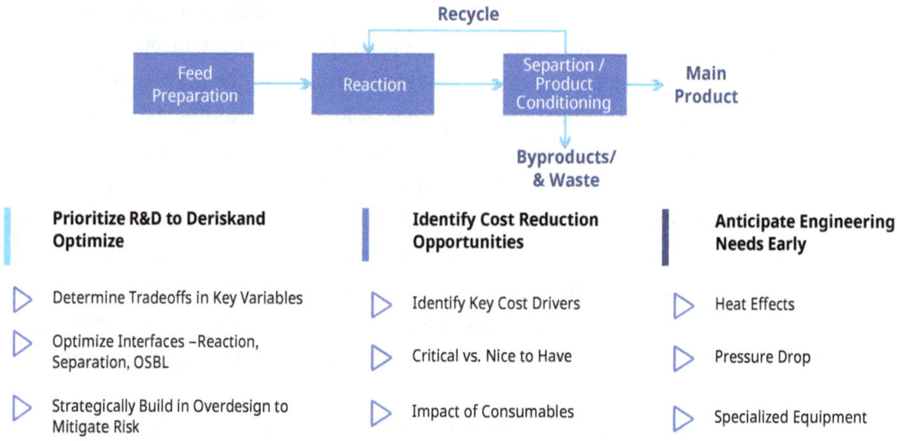

Figure 3.2: Creative process engineering.

This creativity enables the engineers working closely with the science experts to develop the process concept, establish the material balance, and make key process design decisions to set the framework for the evolving novel technology. These decisions fall into categories such as:

- Product requirements, product quality, waste versus by-product, batch versus continuous
- Catalyst: Composition, biological versus thermochemical
- Size/shape, heterogeneous versus homogeneous
- Major unit operations, reactor concept, feedstock processing, separation processes.
- Major equipment, standard or custom, pump/exchanger/compressor type
- Design conditions temperature, pressure, product specifications

The challenge of translating discoveries from the lab into viable process flowsheets has been described by Douglas [5] to require assumptions: (1) that fix parts of the process flowsheet, (2) that fix some of the design variables, and (3) that fix the connections to the environment. Douglas estimates that more than one million process flowsheets can be generated just from the varied assumptions associated with the first process flowsheet. Clearly it is not feasible to evaluate all of these alternatives. The good news is that the number of alternatives can be quickly reduced to a more manageable number, but good engineering judgment is needed to make decisions with relatively little information. This is where the *creative* aspect of process engineering is critical.

In practicality, it is best to identify the reactor concept and separation scheme that are the best options and then build the flow scheme around these. Often, this is a case of screening out the "bad options" resulting in several process concepts that make sense for further evaluation.

Separations for bioprocesses (often called downstream processing, or DSP) bring particular challenges as compared to conventional chemical and petrochemical processes, especially for bioproducts with high value and/or strict purification requirements (nutraceuticals, food, monomers, etc.). Typically, product isolation and purification require several processing steps, which add cost to the flowsheet, and the potential for product loss in each step. This is often underestimated at the planning stage, leading to added cost and development time. Scoping these needs out early and evaluating optimization with the reaction section can increase the chances of scale-up success.

The following represents a process design algorithm that has proven to be successful in the scale-up of new bioprocessing technology:
- Determine system requirements
- Size reactor system
- Size supporting equipment (pumps, heat exchangers, etc.)
- Integrate separation system
- Develop cost model
- Evaluate trade-offs and set targets for further data generation

3.5 Key element #2: efficient experimentation

Experimental data is clearly the lifeblood of any new technology. Getting data to prove out an invention can be the key to obtaining an important patent, generating early-stage investment, and securing key partnerships. However, generating data is expensive and time consuming, particularly as scale increases, making it critical to ensure that the right data is generated to make the best use of available resources.

The entire scale-up effort can be viewed as one integrated data gathering exercise, with the overall goal of generating the necessary data to define the commercial process design. Along the way data is also needed to demonstrate a reduction in technical risk and allow optimization of the process economics. This is a bit of a different mindset from trying to prove out a "result" at each scale (e.g., proving conversion of raw materials A and B into product C with desired efficiency X in the lab, then the lab-pilot, then the pilot, and finally the demo).

Upfront planning can ensure that the right data is gathered. This planning effort will yield a scale-up plan with experiments designed to generate the necessary design data and identify the parameters that have the greatest impact on economics and technical risk. In fact, the product of this effort is data, rather than the bioproduct itself.

A key aspect of this early-stage planning is decoupling these parameters, understanding that "science parameters" such as reaction kinetics and separation factors can, and should, be explored at the lab stage. Conversely, a lab-scale effort to evaluate issues related to heat and mass transfer or pressure drop will be a futile effort at best leading to inconclusive or even incorrect results and is best done at a larger scale. This decoupling is illustrated in Table 3.1.

Table 3.1: Decoupling scale-up parameters.

	Chemistry/biology/electrochemistry	**Engineering**
	Process inputs (temperature and pressure), kinetics, contaminants, etc.	Hydrodynamics, fouling, mass transfer, heat transfer, etc.
Lab scale	X	
Lab-pilot prototype	X	X
Pilot scale		X
Large demo/ commercial		X

Multi-scale data is beneficial for many additional reasons:

- Model development: Data at multiple scales enables the generation of robust models for process development and equipment design.
- Troubleshooting: The smaller lab and pilot rigs can be instrumental to troubleshooting challenges in the larger units. If possible, it is worth the investment to keep these smaller units operating in support of the larger scale operations.
- Continuous improvement: Continuous improvement is often needed while scaling new technology to meet aggressive timelines and cost targets. These improvements can be identified and scaled in parallel to ensure that the first commercial unit has the benefit of the learnings from several generations of technology improvements that are identified and de-risked in multiscale operations.

3.6 Key element #3: modeling and analysis

Modeling and analysis is used to set targets for economic and sustainability performance, encapsulate experimental data into engineering models, and design process equipment. However, it is critical to recognize the limitations of models. British statistician George Box liked to say that all models are *wrong*, but some are *useful* [6]. For scale-up purposes, models should be useful tools to support process development, scale-up, and design, rather than exact replications of the system in question.

An effective ("useful") modeling effort typically starts off with something simple, adding more detail as scale-up progresses. A simple overall mass balance using a spreadsheet is a great place to start! Additional detail is later added to this simple model, with additional types of models developed depending on the requirements. Examples of additional types of useful models include:
- Kinetic models for chemical and biological reaction systems
- Reactor design models for common reactor types, such as packed bed, trickle flow, fluidized bed, and external loop
- Phase equilibrium models to support design of separation systems
- Life cycle assessment models for sustainability analysis
- Technoeconomic models for economic analysis
- Process simulation models for flowsheet and equipment design

The level of detail needed is driven by the requirements of the task at hand (see Figure 3.3).

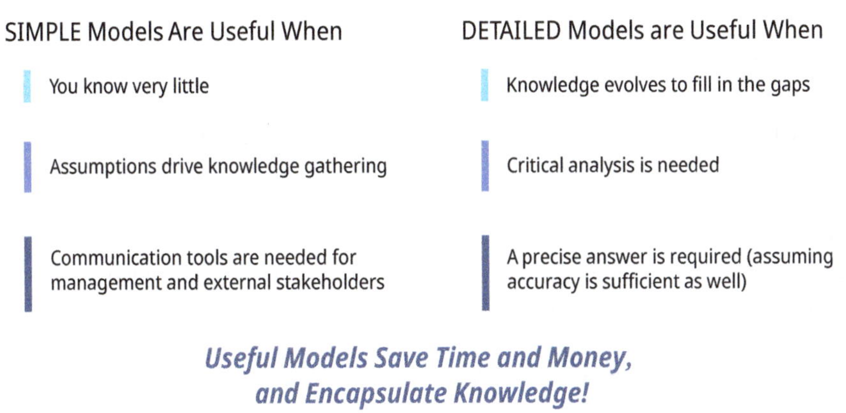

Figure 3.3: Useful models for scale-up.

Where data does not exist, or is inconclusive, assumptions can be used to establish a working model. The criticality of these assumptions to the system in question is evaluated by exploring sensitivities. If the answer is "very critical," this result can be used to inform upcoming experimental activities. This interplay between engineering design, modeling, and experimentation is quite important. When modeling is done in a vacuum, with little or no interaction with experimentalists, the result is often a very beautiful model with limited value. Similarly, some experimentalists insist it is impossible to model their system and find no value in the results from a model. The reality is that a useful model can, and should, complement experimentation to reduce the time and cost of scale-up, providing insight as to when additional data is needed to enhance understanding. A great model can also produce results and understanding that may be too time-consuming, costly, or just not possible through additional experimentation. The models can also direct future opportunities for experimental programs.

The models should then be refined as more data is collected. This data should be generated at multiple scales to enhance the robustness and utility of the model.

3.7 Process safety and operation: design issues to consider during bioprocess scale-up

The field of safety in R&D facilities and manufacturing plants is necessarily quite broad. Much of this is driven by regulations at the local and national levels, but best practices have developed in the process industries with the aim of designing and operating lab-, pilot-, and commercial-scale facilities in a manner that minimizes risk of to the workers and the surrounding environment from both expected and unexpected process conditions. The best-run facilities in regard to worker safety have management commitment to providing a safe workspace, and necessary work practices, procedures, and equipment to follow these best practices. In turn, the workers are obligated to follow these practices, wear appropriate safety equipment, ensure their work area is free of hazards, and ensure that their fellow workers are safe as well.

First some definitions (many taken from the AIChE Center for Chemical Process Safety (CCPS) Glossary [7], noted in quotes with (CCPS)):

- Fault Tree Analysis (FTA): *(CCPS)* "A method used to analyze graphically the failure logic of a given event, to identify various failure scenarios (called cut-sets), and to support the probabilistic estimation of the frequency of the event."
- Failure Mode and Effect Analysis (FMEA): *(CCPS)* "A hazard identification technique in which all known failure modes of components or features of a system are considered in turn and undesired outcomes are noted."
- Hazard Identification (HAZID): (*CCPS)* "Part of the Hazards Identification and Risk Analysis (HIRA) method in which the material and energy hazards of the process, along with the siting and layout of the facility, are identified so that a risk analysis can be performed on potential incident scenarios."
- Hazard and Operability Study (HAZOP): (*CCPS*) "A systematic qualitative technique to identify process hazards and potential operating problems using a series

of guide words to study process deviations. A HAZOP is used to question every part of a process to discover what deviations from the intention of the design can occur and what their causes and consequences may be. This is done systematically by applying suitable guidewords. This is a systematic detailed review technique, for both batch and continuous plants, which can be applied to new or existing processes to identify hazards."

- Layers of Protection Analysis (LOPA): *(CCPS)* "The extension of a HAZOP Study to include aspects of a LOPA, including selecting identified scenarios for further analysis; evaluating the initiating event frequency, consequence severity and effectiveness of Independent Protection Layers (IPLs) on an order-of-magnitude basis; considering enabling conditions and/or conditional modifiers as appropriate when evaluating scenario risk; and comparing the calculated scenario risk to a risk goal to determine the adequacy of existing risk control measures."
- Management of Change (MOC): *(CCPS)* "A management system to identify, review, and approve all modifications to equipment, procedures, raw materials, and processing conditions, other than replacement in kind, prior to implementation to help ensure that changes to processes are properly analyzed (for example, for potential adverse impacts), documented, and communicated to employees affected."
- Pre-start-up Safety Review (PSSR): *(CCPS)* "A systematic and thorough check of a process prior to the introduction of a highly hazardous chemical to a process. The PSSR must confirm the following: Construction and equipment are in accordance with design specifications; Safety, operating, maintenance, and emergency procedures are in place and are adequate; A process hazard analysis has been performed for new facilities and recommendations have been resolved or implemented before startup, and modified facilities meet the management of change requirements; and training of each employee involved in operating a process has been completed."
- Process Hazard Analysis (PHA): A generic term for any number of hazard assessments that can be carried out for a process unit. In some cases, the term "PHA" is used interchangeably with "HAZOP." It is important to make sure that if the term PHA is used in an organization then there is alignment on what type of assessment is to be carried out.
- Root Cause Analysis (RCA): *(CCPS)* "A formal investigation method that attempts to identify and address the management system failures that led to an incident. These root causes often are the causes, or potential causes, of other seemingly unrelated incidents. Identifies the underlying reasons the event was allowed to occur so that workable corrective actions can be implemented to help prevent recurrence of the event (or occurrence of similar events)."
- What-If Analysis: *(CCPS)* "A scenario-based hazard evaluation procedure using a brainstorming approach in which typically a team that includes one or more persons familiar with the subject process asks questions or voices concerns about what could go wrong, what consequences could ensue, and whether the existing safeguards are adequate."

The number of assessments defined above can lead to confusion and an "alphabet soup" of acronyms that cloud the overall goal of an inherently safe process unit. In this section we will review the types of assessments that can be carried out at various stages of new engineering projects. Please refer to the definitions above for the various hazard assessment techniques identified below. In addition, references at the end of this document contain a more comprehensive description of which type of assessment tool to use for a particular scenario:

- Research/bench scale
  - Typically, at this scale we are screening different reaction chemistries. We can identify potential raw materials, catalysts, solvents, intermediates, products and by-products, and use Material Safety Data sheets (MSDS) for each to identify major hazards. We can also assess process conditions to identify high pressure or temperature hazards. We should avoid, however, the temptation to start going too far into process design at this stage. As an example, the hazard review (such as an FTA) could identify a particularly hazardous solvent. The technical personnel assigned to the project may then wish to perform additional experiments to evaluate alternative solvents. Those solvents may be less effective chemically, but with a lower hazard rating may lead to a more inherently safe process design. The goal is to generate sufficient data that the design engineer can use to carry out trade-offs and propose the best design.
- Conceptual design (FEL-1)
  - At this point we are able to develop a process flow diagram (PFD), preliminary heat and material balance and preliminary equipment list. It is useful at this time to carry out a HAZID. The goal is to understand potential safety hazards, and adjust the PFD if needed, or perhaps to generate process alternatives for further evaluation. We also would like to identify flammable raw materials, intermediates, and solvents, which may require design of hazardous classification areas and special considerations for explosion proof equipment, all of which can add cost to the design.
- Basic engineering (FEL-2)
  - In this stage, the piping and instrumentation diagram (P&ID) is developed, along with the detailed heat and material balance and control scheme. At this point, we can carry out the HAZOP. For the HAZOP to be effective, the P&ID should be complete from a process perspective, with control schemes defined and operating instructions developed. The HAZOP is best executed with an independent facilitator, and key disciplines represented from process chemistry/biology, design engineering, plant engineering, operations, and analytical. Recommendations from the HAZOP may result in additional design changes to the P&ID. If this is the first time in the design process that safety hazards have been reviewed, it is possible, in fact likely, that somewhat major design changes are required that can impact both the cost and project schedule. To prevent this, it is worth taking the time earlier in the design process to review

hazards as discussed above. Note that the overall goal is not to simply complete a HAZOP document – the assessment should inform design decisions and future safety reviews and instructions. If HAZOP is completed and put on the shelf never to be reviewed again it has not been useful.
- Start-up/commissioning
  - The pre-start-up safety review (PSSR) is a critical piece of any pre-commissioning activities. During the PSSR, the team (ideally the same team that was part of the HAZOP) reviews that the plant was built as designed, safety/operating/ maintenance/emergency procedures are in place, and that all recommendations from the HAZOP have been implemented and resolved. The plant "owner," meaning the engineer or other individual responsible for plant start-up should take ownership of this review to ensure that the review is complete, and the plant is ready for start-up.
- Post start-up, existing facilities
  - Process safety considerations do not stop once the plant has started up. A working "Management of Change" process is important to determine how the plant will respond to changes introduced that are outside the scope of the original HAZOP. These changes may include introduction of new raw materials, changing operating conditions, or mechanical changes to the process itself. A robust MOC process is used to determine what type of safety review is warranted for a particular change, ranging from a "What If" analysis to a revised HAZOP. Finally, root cause and fault tree analyses are used to determine the causes of a process failure that occurs during operation.

This section touches on key elements of process safety but is by no means exhaustive. The reader is referred to additional references, which are much more comprehensive [8, 9].

3.8 A case study

This chapter has presented some guidelines for practical scale-up of bioprocessing technology and the bioproducts produced by these technologies. These concepts will be illustrated with a case study. Consider the conversion of lignin to a surfactant product using aqueous caustic reaction chemistry.

Section 3.4 presented an algorithm to develop and optimize the process flow scheme:
- Determine system requirements
- Size reactor system
- Size-supporting equipment (pumps, heat exchangers)
- Integrate separation system
- Develop cost model
- Evaluate trade-offs and set targets for further data generation

The block flow for this process is initially developed, as shown in Figure 3.4. By identifying the key processing steps in a block flow first, before advancing to more detail in a process flow diagram, the key reaction and separation steps are clearly identified, and an overall mass balance can be prepared.

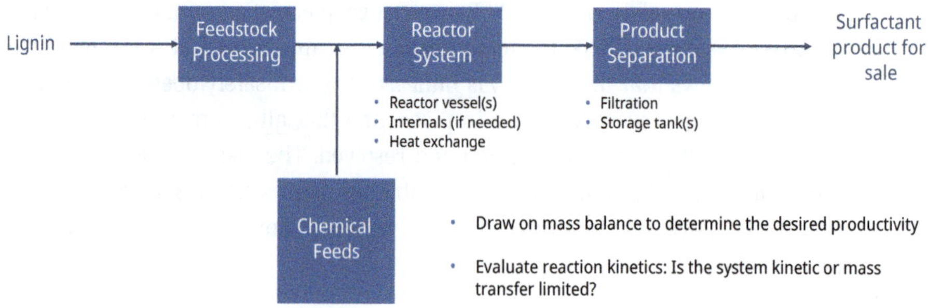

Figure 3.4: Block flow for new process.

The next step is to size the reactor system. Figure 3.5 shows a graph of conversion versus residence time. Other parameters such as temperature and feedstock properties may be explored using design of experiments.

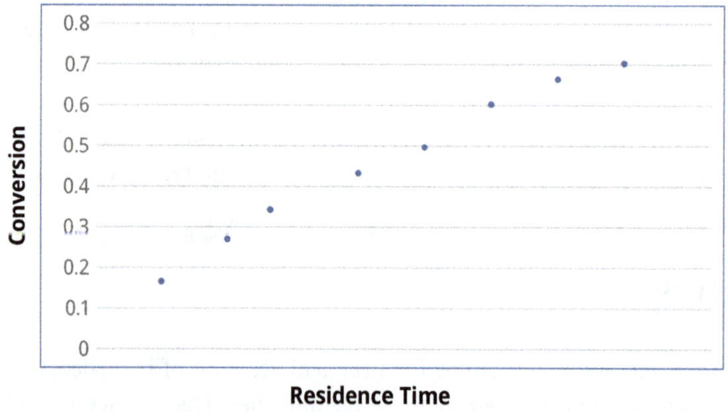

Figure 3.5: Conversion versus residence time.

The separation system can then be integrated into the flow scheme, so that reaction/ separation trade-offs can be explored and optimized. For instance, operating at a very high conversion may lead to a greater yield of nonselective reaction by-products, increasing the cost of the separation system. It may be better to operate at a lower conversion, reducing the load on the separation system. By carrying out the reaction experiments described above, these trade-offs can be explored along with separation scheme.

A techno-economic model is developed to enable these trade-offs to be explored. For instance, Figure 3.6 shows the cost performance of two different cases relative to the overall cost targets.

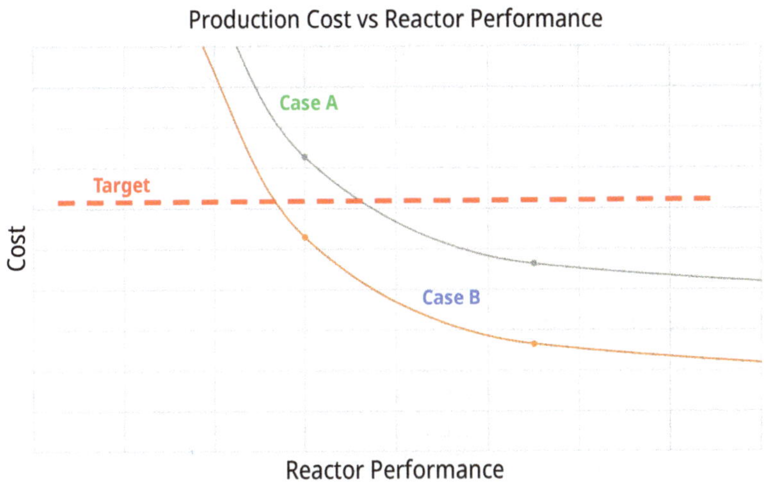

Figure 3.6: Techno-economic trade-offs for new bioprocesses.

3.9 Gas fermentation: an emerging technology in the bioproducts space

Gas fermentation represents a transformative industrial biotechnology that addresses the growing need for sustainable chemical and fuel production. This emerging technology leverages readily available, low-cost carbon sources including methane, carbon monoxide, and carbon dioxide to produce diverse products ranging from ethanol and ethylene to triglycerides, proteins, and polyesters. As the bioeconomy continues to grow, gas fermentation offers a compelling alternative to traditional petrochemical processes, particularly in applications requiring moderate operating conditions and feedstock flexibility.

3.9.1 Technology overview and pathways

Gas fermentation operates through several conversion pathways. For the purposes of CO_2 reduction, these can be classified into two broad categories. The first involves direct CO_2 conversion, typically utilizing hydrogen and sometimes oxygen as co-feeds to produce triglycerides for food, materials, and fuel applications, along with chemicals such as acetic acid and ethylene. Single-cell proteins for animal feed and alternative protein applications represent another significant product category through this route.

The second pathway focuses on converting CO_2 precursors, specifically carbon monoxide or methane, into valuable products. This approach enables the production of alcohols including ethanol, methanol, and isopropanol, as well as biopolymers such as polyhydroxyalkanoate (PHA). The technology's versatility in handling different feedstock compositions and ratios provides significant operational advantages over more rigid thermochemical processes.

Pathways that produce gases such as methane or hydrogen are also sometimes classified as gas fermentation but will be considered out of scope for the purposes of this chapter.

3.9.2 Competitive advantages

Gas fermentation offers several compelling advantages over conventional processing routes. The technology operates under moderate conditions at ambient temperature and low pressure (less than 10 bar), significantly reducing energy requirements compared to high-temperature thermochemical alternatives. Its inherent robustness allows for fluctuations in feed rate and composition while maintaining tolerance to contaminants in gaseous feeds – characteristics that prove particularly valuable when processing waste streams or variable feedstock sources.

3.9.3 Technical challenges and solutions

The primary technical challenge in gas fermentation centers on achieving cost-effective reactor systems with high mass transfer coefficients for gaseous feedstocks into aqueous media. The mass transfer rate must exceed the biological gas uptake rate to avoid mass transfer limitations that could constrain overall process performance.

Multiple reactor configurations have been developed to address this challenge, including bubble columns, stirred tank reactors, and airlift systems. Each design presents trade-offs between mass transfer efficiency, capital cost, and operational complexity. The selection of appropriate reactor technology requires careful consideration of the specific microbial system and target products, with mass transfer coefficient ($k_L a$) serving as a critical design parameter for ensuring economic viability.

3.9.4 Commercialization barriers

Several significant challenges must be addressed for successful commercial deployment of gas fermentation technology. The lack of established data and models, particularly regarding bacterial growth kinetics and flux models, contrasts sharply with the extensive knowledge base available for petrochemical processes. This knowledge gap drives substantial investment in fundamental research and development.

Additional challenges include:

- New equipment: To address mass transfer challenges, novel gas fermentation reactors are an active area of development.
- New separation challenges: Recovery of extracellular products (such as ethanol or acetic acid) and intracellular products (including triglycerides and proteins) from fermentation broths.
- New supply chain challenges: Manufacture and distribution of specialized microbial catalysts at commercial scale.
- Additional optimization: Carbon footprint and ESG/LCA metrics now complement traditional economic considerations of operating and capital costs in process design decisions.

3.9.5 Future development opportunities

The future of gas fermentation technology lies in advancing several key areas. Enhanced microbial modeling, including bacterial growth kinetics and flux models, will enable more sophisticated understanding of biological reactor systems and support custom reactor designs optimized for specific microbial platforms.

Strain development efforts focus on reducing by-product formation, increasing contaminant tolerance, and enabling operation under more extreme conditions such as higher temperatures to increase reaction rates. Parallel advances in reactor design and scale-up continue pushing the boundaries of achievable reactor sizes while optimizing mass transfer performance.

Product recovery innovations, including in situ product recovery systems and novel separation technologies, promise to reduce the cost and complexity of downstream processing. Emerging concepts such as biofilm-based and cell-free reactor systems may revolutionize the technology's implementation in the coming years.

3.9.6 Gas fermentation conclusion

Gas fermentation stands at the forefront of sustainable technology innovation, offering a pathway to reduce industrial carbon emissions while producing essential chemicals and fuels. Despite current commercialization challenges, the technology's fundamental advantages in feedstock flexibility, moderate operating conditions, and environmental performance position it as a key contributor to the growing bioeconomy. Continued investment in research and development, particularly in microbial modeling, reactor optimization, and product recovery systems, will be essential for realizing the full commercial potential of this promising technology.

3.10 Commercial success stories

In recent history, a number of bioproducts, whether using bio-based feedstocks, a biological catalyst, or both, have achieved commercial success. A few examples are given below, meant to be illustrative rather than exhaustive, as of Q2 2025:

- 1,3-Propane diol (PDO): PDO has been made conventionally by hydration of acrolein or hydroformylation of ethylene oxide. More recently, a bioprocessing-based route has been developed by modifying *E. coli* to enable production by fermentation of feedstocks such as glucose and glycerol. Commercial producers now include Covation Bio (formally DuPont Biomaterials) with a capacity of 77,000 ton/year. The PDO is then converted by Covation Bio to produce a range of bioproducts under the trade names Susterra® and Zemea® [10]. Additional producers of 13PDO through the fermentation route include companies such as Shenghong Group Holdings.
- Polylactic acid (PLA): PLA was produced by Natureworks, a joint venture established by Cargill and Dow Chemical, with the first commercial production coming onstream in 2001. Natureworks now is a joint venture between PTT Global Chemical and Cargill. Lactic acid is produced by fermentation, and the PLA is then produced through a thermochemical condensation reaction. Currently, Natureworks has a production capacity of 150,000 tons/year at their Nebraska facility, with additional capacity under construction in Thailand [11]. Other manufacturers include Total Corbion (75,000 ton/year), and several manufacturers in China, Japan, and Europe with capacities of 1,000–15,000 ton/year [12]. PLA is used to make biodegradable plates, cups, cutlery and containers, along with a variety of films, fibers, and 3D printing applications.
- Geno has commercialized a fermentation-based route to produce14 butanediol (14 BDO). Like PDO, 14 BDO has been produced historically through thermochemical routes using fossil-based feedstocks. As a technology licensor, Genomatica now has licensed technology for two commercial plants, for a total global capacity of more than 100,000 ton/year [13]. 14 BDO is used in downstream applications for conventional products such as polyester and polyurethanes. In 2025 the production facility operated by Qore, using licensed Geno technology commenced operation.
- Global ethanol production in 2023 was 29.9 billion gallons (~88 million tons) [14], primarily used for blending into fossil-based gasoline. Virtually all of this uses sugarcane or corn as a starting material and is known as first generation ethanol. To further reduce the carbon intensity of this alternative fuel and to avoid competition with food sources, technology for producing ethanol from alternate feedstocks (second generation, 2G ethanol) is starting to come on stream. Examples include:

- Raizen (Brazil, 21 million gallons/year, 63,000 ton/year) [15]
- Praj (India, 250,000 gal/year, 740 ton/year) [16]
- In addition, LanzaTech has six facilities converting a variety of industrial waste gases into ethanol, for a total capacity of 300,000 ton/year [17].

3.11 Conclusions

It is critical to "start with the end in mind" using a technology concept (or process concept) that is used as a framework to drive new technology development, scale-up, and commercialization (see Figure 3.7). This technology concept is not set in stone, and, in fact, should be reviewed and updated as we progress throughout the scale-up effort. We establish the technology concept to drive the scale-up effort, not just inform it. This then enables us to direct the innovation to create the ***greatest value from breakthrough and disruptive ideas***, as we identify challenges early, fail fast when it is cheaper and quicker, and make sure our efforts are focused on solving commercially relevant problems.

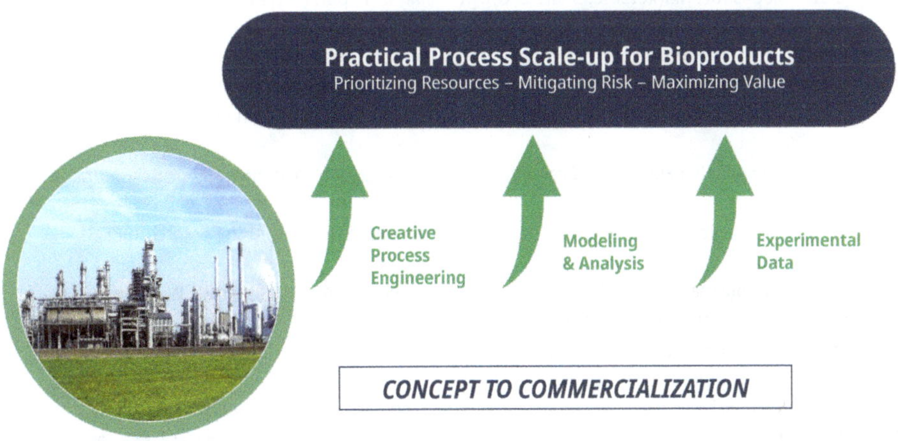

Figure 3.7: Practical process scale-up.

The technology concept is developed, and iteratively revised, through a combination of ***creative process engineering, multi-scale experimental data, and modeling and analysis.***

Creative Process Engineering: The flow scheme is developed, the material balance is estimated, and key process design decisions are identified so that we can establish the best process flowsheet for the technology.

Modeling and Analysis: A good model can save time and resources in the lab. Coupled with the right analysis, this can be used to prioritize objectives in the lab, pilot, and demo units. Cautionary note–useful models are more important than perfect models!

Experimental Data: We need the right data to prove out breakthrough ideas, secure partners and investors, and develop engineering data for equipment design. Multi-scale data is critical to this effort, and with good planning, multiple assets and external resources can be leveraged.

The key benefits to this approach are:
- Prioritization of R&D to de-risk and optimize a new technology.
- Identification of cost reduction opportunities throughout the scale-up effort.
- Anticipation of engineering needs as early as possible.

In this way we can ***reduce risk and optimize economics*** of our new design, while efficiently managing the time and cost of our efforts.

References

[1] Global Bioproduct Market Size, Share & Growth Analysis Report (bccresearch.com), https://www.bccresearch.com/market-research/energy-and-resources/biorefinery-products-markets-report.html.
[2] https://ethanolrfa.org/resources/ethanol-biorefinery-locations
[3] Global biofuel production in 2019 and forecast to 2025 – Charts – Data & Statistics – IEA, https://www.iea.org/data-and-statistics/charts/global-biofuel-production-in-2019-and-forecast-to-2025
[4] Sustainable Aviation Fuels (SAF) (icao.int), https://www.icao.int/environmental-protection/pages/SAF.aspx
[5] Douglas JM. A Hierarchical Decision Procedure for Process Synthesis. AIChE J March 1985;31 (3):353–362.
[6] Box GEP. Robustness in the strategy of scientific model building. In Launer RL, Wilkinson GN. Robustness in Statistics, Academic Press; 1979. pp. 201–236.
[7] https://www.aiche.org/ccps/resources/glossary
[8] Guidelines for Risk Based Process Safety, by the AIChE Center for Chemical Process Safety, https://www.aiche.org/resources/publications/books/guidelines-risk-based-process-safety
[9] https://www.process-improvementinstitute.com/_downloads/Selection_of_Hazard_Evaluation_Techniques.pdf
[10] https://covationbiopdo.com/
[11] https://www.natureworksllc.com/About-NatureWorks
[12] https://www.gianeco.com/en/faq-detail/1/37/where-to-buy-pla
[13] https://www.genomatica.com/news-content/second-commercial-plant-will-expand-capacity-to-over-100000-tons
[14] https://ethanolrfa.org/markets-and-statistics/annual-ethanol-production
[15] https://www.reuters.com/business/energy/brazils-razen-build-second-cellulosic-ethanol-plant-filing-2021-06-25/
[16] https://www.praj.net/businesslines/2g-ethanol
[17] Biorefining – LanzaTech, https://lanzatech.com/biorefining/

John McArdle*

Chapter 4 Bioproduct projects: financial and investment analysis tools

4.1 Overview

This chapter provides a brief introduction to financial and investment analytical methods that are used by early-stage companies to attract investment funding. These analytical methods can be used by entrepreneurs, researchers, and students involved in the development of novel green material bioproducts.

4.2 Background

Differentiated technologies and products with high value are very important in any early-stage company. Equally, if not more critical, is the ability to secure needed funding to realize the potential of the company with a compelling economic return financial story.

It is not always possible for early-stage companies to fully achieve their potential for growth by self-financing or operating cash flow. Accordingly, outside funding is often required. However, early-stage companies do not often have access to the funding sources that are available to more established companies.

For example, traditional banks are not willing to lend money to companies that do not have an established line of credit or sufficient collateral, both of which can be in short supply with early-stage companies. Accordingly, these companies often rely on private "angel" investors and venture capitalists (VCs) for the funding needed to support growth.

Investors do not fund early-stage companies because they intend to operate such companies. Instead, they provide funding because of their expectation of high financial return with the sale of the company.

Investor returns are primarily, but not exclusively, based on the valuation and ownership share of the early-stage company that are negotiated by the investor and principals of the company at the time of the funding.

*Corresponding author: John McArdle, Redwood Innovation Partners, LLC, Boston, MA, USA, e-mail: johnm@redwoodinnov.com

 | https://doi.org/10.1515/9783112212141-004

4.3 Investment analysis

Investment analysis is used to identify attractive funding opportunities that are believed to provide a reasonable chance of high financial reward relative to their financial risk. Investment analysis can also be used to rank the attractiveness of individual investment opportunities under capital constraints.

Payback, internal rate of return (IRR), and net present value (NPV) are methods of investment analysis. Payback and IRR methods have shortcomings that can lead to less-than-optimal investment decisions. When properly applied, NPV will reliably identify attractive investment opportunities.

Attractive investment opportunities in early-stage companies are not always obvious. First, there can be significant uncertainty in projecting sales revenues and operating costs. Second, different methods of investment analysis, in some scenarios, provide unhelpful or contradictory results.

Table 4.1 shows projected cash flows for two hypothetical investments. Funding is $1.0 million for each investment, with different cash flows projected over 4 years. Investment A is more attractive based on a shorter payback period and a higher IRR. However, investment B is more attractive based on a higher NPV at 10% discount factor.

Table 4.1: Investment analysis methods.

Year	**Investment $, out**	**Cash flow $, in**			
	0	**1**	**2**	**3**	**4**
Investment A	1,000,000	1,000,000	500,000	100,000	50,000
Investment B	1,000,000	500,000	500,000	500,000	500,000
Investment analysis method	**Payback year**	**IRR %**	**NPV @ 10% $**		
Investment A	1	42	432,000		
Investment B	2	35	584,000		

It is apparent from these contradictory results that the output from different methods of investment analyses should be carefully interpreted to properly identify optimal investment opportunities.

4.3.1 Net present value

Net present value (NPV) is the sum of an initial investment plus the discounted future cash flows that are projected to arise from such investment over a defined period of time. Investments that have positive NPVs are attractive. Conversely, investments that have negative NPVs are not attractive.

Projected future cash flows are adjusted to present values by the discount rate. Discount rate represents the financial return that can be expected from an alternative investment that has risk equivalent to that of the investment being considered. For this reason, the discount rate is also referred to as the opportunity cost of capital.

Higher and lower discount rates are used for NPV calculations with more risky and less risky investment opportunities, respectively. The long-term, pre-tax financial return of a well-diversified portfolio of common stock has been about 10% per year, including capital gains and dividends. This historical stock market return provides a useful benchmark that can be used to select the discount rates for other investment opportunities, taking into account the risk of such investments.

Investments in early-stage companies often have higher financial risks than an investment in a diversified portfolio of common stocks. Accordingly, investors target higher financial returns for these companies than can be expected based on the historical performance of a diversified portfolio of common stocks.

4.4 Investors

Investors recognize the risk of investing in early-stage companies. In fact, they expect that the majority of their investments made in these companies will have minimal or negative financial return.

Investors often invest in a portfolio of early-stage companies, with each company in the portfolio believed to have the potential for a high financial return. In this manner, the high financial returns from a few successful investments will balance out the majority of those that have minimal or no return. In theory, this will provide an overall average financial return for the investment portfolio that is higher than can be expected from other less risky investments, including, for example, a well-diversified portfolio of common stocks.

Financial return for a private investor in an early-stage company is realized with the sale of the company to a third party or an initial public stock offering. Typically, the investment period for private investors is 5–8 years.

4.5 Company valuation

Company valuation is the amount of money that transfers from the buyer to the seller with the sale of a company. Until the actual sale of a company, the calculation of company valuation is an estimate.

Investors funding early-stage companies negotiate the company valuation and percentage of company ownership with the principals of the company at the time of a funding round. These parameters establish the basis for the investor's financial return with the sale of the company as well as the compensation for the ideas and sweat equity of the founders.

Valuation of an early-stage company is primarily based on the expectation of company growth and profitability. Company valuation must increase during the funding period if the investor is to achieve a positive return on their investment.

Investors take into account many factors when evaluating an early-stage company. These include the current financial performance of the company as well as the expertise and track record of the management team, size of the market opportunity, strength of the competition, commercial readiness of the product, and maturity of the sales channels.

Early-stage, pre-revenue companies have minimum or no-sales revenue and partly developed products and sales channels. For these companies, valuation is based on a more subjective analysis of the company's potential to grow into a profitable business.

Early-stage, post-revenue companies have some sales revenue and more developed products and sales channels. For these companies, valuation is based on a more quantitative assessment of the company's projected growth and profitability.

4.5.1 Pre-revenue company valuation calculation method

Valuations of pre-revenue companies can be based on the reported average valuation of other previously funded pre-revenue companies in a similar business sector, including, for example, industry, application, and region. These reported valuations are adjusted for the "target" pre-revenue company using weighted scores of valuation parameters that are considered by the investors to be most critical to achieve significant company growth and profitability.

Table 4.2 illustrates a valuation method that uses team qualifications, potential market size, competitive advantage, and prototype performance as the valuation parameters of a "target" company. Depending on investor preferences, other valuation parameters as well as different parameter weightings can be applied.

In this example, the calculated valuation of the target company is $5.5 million compared to an average valuation of $5.0 million for other previously funded pre-revenue companies in a similar business sector. The higher valuation of the target company reflects a higher total weighted score of the individual valuation parameters of this company compared to that of the previously funded companies.

Company valuations, with the valuation parameters and weightings used in this example, range from a maximum of $7.5 million with all strong 1.5 scores, to a minimum of $2.5 million with all weak 0.5 scores. Investors can rate any of these parameters as zero. However, a zero rating in any of these critical valuation parameters will probably eliminate the company from funding consideration.

4.5.2 Post-revenue company valuation calculation method

Valuation of an early-stage, post-revenue company can be based on the projected financial performance of the company taking into account sales revenue, factory cost, indirect costs, and taxes.

Table 4.3 shows the projected financial performance of a hypothetical post-revenue company. Investor funding of \$5.0 million is provided to purchase new factory equipment. This funding enables the company to increase manufacturing and associated sales revenue while at the same time maintaining positive cash flows.

Table 4.2: Pre-revenue company valuation [1].

Average company valuation	**5,000,000**			
Target company valuation	**5,500,000**			
Score:	**0.5**	**Weak**		
	1.0	**Average**		
	1.5	**Strong**		
Company score parameter	**Weight**	**Score**	**Weighted score**	**Value contribution**
Team qualifications	0.40	1.5	0.60	3,000,000
Potential market size	0.30	1.0	0.30	1,500,000
Competitive advantage	0.20	0.5	0.10	500,000
Prototype performance	0.10	1.0	0.10	500,000
Total	**1.00**		**1.10**	**5,500,000**

Net sales revenues are based on projected unit sales, price, and factory cost. Economies of scale and increased competition are assumed over time to reduce unit factory cost and reduce sales price, respectively.

Sales, general, and administrative (SG&A), research and development (R&D), engineering, and net working capital are calculated using multipliers based on total sales revenue. Depreciation of factory equipment is based on 7-year straight-line schedule. Land is not depreciated.

NPV is about \$1.7 million at 25% discount factor over a 5 year period. Sales revenue, net income, and cash flow are increasing after the initial 3-year period, indicating a successful ongoing business operation.

Company valuation is estimated at \$120 million based on a 15× EBITDA (\$8.0 MM × 15) at 5 years [2]. The financial return for the investor is based on this valuation along with ownership share.

NVP does not necessarily correlate to an attractive investment opportunity because investors are less focused on the financial performance and more focused on the valuation of a company, although these two items can be closely correlated. Investors only realize attractive financial returns with companies that over time achieve increased earnings, because this translates into higher company valuations

Table 4.3: Post-revenue company valuation.[1,2]

Investment analysis					
Net present value	$	1,667,000			
Discount rate	%	25.0			
SG&A, % total revenue	%	20.0	Minimum	$	200,000
R&D, % sales revenue	%	5.0	Minimum	$	200,000
Engineering, % sales revenue	%	5.0	Minimum	$	200,000
Working capital, % incremental revenue	%	20.0			
Tax	%	20.0			
Total investment	$	5,000,000			
Company value, multiple of EBITDA	#	15.0			
Company market value @ Year 5	$	120,000,000			

Year		**1**	**2**	**3**	**4**	**5**
Investment	$	5,000,000	–	–	–	–
Cost of new factory equipment	$	3,500,000	–	–	–	–
Cost of new land	$	250,000	–	–	–	–
Unit sales	#	500	1,000	2,500	5,000	10,000
Unit sales price	$/#	5,000	5,000	4,500	4,250	4,000
Unit factory cost	$/#	4,000	3,000	2,500	2,250	2,000
Unit sales price less Unit factory cost	$	1,000	2,000	2,000	2,000	2,000
Gross margin	%	20	40	44	47	50
Total sales revenue	$	2,500,000	5,000,000	11,250,000	21,250,000	40,000,000
Total factory cost	$	2,000,000	3,000,000	6,250,000	11,250,000	20,000,000
Net sales revenue	$	500,000	2,000,000	5,000,000	10,000,000	20,000,000
SG&A	$	500,000	1,000,000	2,250,000	4,250,000	8,000,000
R&D	$	200,000	250,000	563,000	1,063,000	2,000,000
Engineering	$	200,000	250,000	563,000	1,063,000	2,000,000
EBITDA	$	(400,000)	500,000	1,624,000	3,624,000	8,000,000
Depreciation	$	-	500,000	500,000	500,000	500,000
Taxable income	$	(400,000)	–	1,124,000	3,124,000	7,500,000
Tax	$	-	–	225,000	625,000	1,500,000
Net income after tax	$	(400,000)	–	899,000	2,499,000	6,000,000
Change in net working capital	$	500,000	500,000	1,250,000	2,000,000	3,750,000
Cash flow	$	350,000	–	149,000	999,000	2,750,000
Present value	$	280,000	–	76,000	409,000	901,000
Incremental present value	$	280,000	280,000	356,000	765,000	1,666,000

1 Appendix 1 for accounting terms.
2 Appendix 2 for accounting factors.

4.5.3 Investor target return

Investors target 6× to 10× return multiples on their funding for early-stage companies over an investment period of 5–8 years. Investors in earlier funding rounds often target higher return multiples of up to 30× to account for any possible dilution of their ownership share resulting from subsequent funding rounds.

Table 4.4 shows that IRR is 25.0–58.4% for this range of return multiples and investment periods. IRR increases with higher return multiples and shorter investment periods. NPV is based on a $1.0 million investment at 25% discount rate. NPV is zero at the lowest return multiple (6×) and longest investment period (8 years). NPV under this analysis, is, by definition, zero at 25% IRR.

High target returns are required for each individual company in an investment portfolio of early-stage companies so that in the event that the majority of these companies return minimal or no return, as expected, the overall "risk-adjusted" return of the portfolio remains attractive compared to other investment options.

Table 4.4: Investor return matrix [3].

Investment		**$1,000,000**		
Minimum target return		**25.0%**		
Years to exit	**Analysis method**	**Investment return multiple**		
		6×	**8×**	**10×**
5	IRR	43.1%	51.5%	58.4%
	NPV	$0	$1,621,000	$2,277,000
6	IRR	34.8%	41.4%	46.8%
	NPV	$573,000	$1,097,000	$1,621,000
7	IRR	29.1%	34.6%	38.9%
	NPV	$258,000	$678,000	$1,097,000
8	IRR	25.0%	29.7%	33.3%
	NPV	$0	$342,000	$678,000

4.5.4 Investor ownership

Table 4.3 shows an investment of $5 million in a hypothetical early-stage company. Table 4.4 shows that investors in early-stage companies target 6× to 10× multiple on their investment in 5–8 years. Accordingly, investors will expect to receive $30 million (6× $5 million) to $50 million (10× $5 million) from the sale of the company.

Projected valuation of this company in 5 years is $120 million based on 15 × EBITDA multiple. Investors will likely attempt to negotiate not less than 25% ownership share in this company ($30 million/$120 million). This ownership share and company valuation will provide investors with 43.1% IRR in 5 years.

If this company has the same $120 million valuation, but is instead sold in 8 years, this 25% ownership share will provide the investor with a (minimum) acceptable 25% IRR. This lower return is directly correlated to the time value of money at 25% discount rate for the longer investment period.

4.5.5 Dilution

Company valuation and ownership share negotiated at the time of a funding round establish the basis for investor return, upon sale of the company. However, as shown in this section, the impact of future funding rounds can also play a significant role.

Table 4.3 shows that the $5.0 million funding for the hypothetical early-stage company provides sufficient cash to support company growth during the initial three years of operation. In later years, cash from operations is sufficient to support the company growth. Under this scenario, no further funding round is required.

Series A Round valuation and ownership shares are shown later for one conceptual scenario. Many other scenarios are possible. Share price increases from $1.00/share to $12.00/share with $120 million company valuation (year 5). Series A investor payout is $36 million. Series A investor return is 7.2× .

Series A round

Pre-money valuation:	$10 million (negotiated)
Series A funding:	$5 million
Post-money valuation:	$15 million
Issued shares:	10 million
Pre-money share price:	$1.00/share ($10 million/10 million shares)
Series A ownership:	30% (negotiated)
Series A investor shares:	3.0 million (0.30 × 10 million shares)
Company valuation:	$120 million @ Year 5
Share price:	$12.00/share ($120 million/10 million shares)
Series A investor payout:	$36 million (3.0 million shares × $12.00/share)
Series A investor return:	7.2× ($36 million/$5 million)

Multiple funding rounds can occur for rapidly growing companies. This additional funding can be used to purchase additional manufacturing equipment and cover increased working capital requirements.

However, later funding rounds reduce the ownership share of the investors of earlier funding rounds. Ownership "dilution" can result in lower returns for the earlier investors than would otherwise be expected.

Series B round valuation and ownership shares are shown later for one conceptual scenario. Many other scenarios are possible. Series B round funding dilutes ownership share of Series A investors from 30% to 24%. Share price increases from $1.00/share to $9.60/share, instead of $12.00/share, over 5 years. Series A investor payout has been reduced from $36 million to $29 million and return reduced from 7.2× to 5.8×. Series B investor payout is $24 million with a return of 4.8×.

Series B round

Series B funding:	$ 5 million
Series B ownership:	20% (negotiated)
Series B investor shares:	2.5 million (2.5 million/(10 million + 2.5 million) = 0.20)
Post-money issued shares:	12.5 million (10 million + 2.5 million)
Series A ownership:	24% (3 million shares/12.5 million shares)
Company valuation:	$120 million @ Year 5
Share price:	$9.60/share ($120 million/12.5 million shares)
Series A investor payout:	$29 million (3.0 million shares × $9.60/share)
Series A investor return:	5.8× ($29 million/5.0 million)
Series B investor payout:	$24 million (2.5 million shares × $9.60/share)
Series B investor return:	4.8× ($24 million/$5 million)

Series B investors may be willing, at the time of Series B round, to accept a lower return multiple than the earlier Series A investors if the perceived risk of the company is lower and the investment period is shorter than for the Series A round.

4.5.6 Divergence

For the Series A round, with no later funding rounds, the 12× ratio of share price increase from $1.00/share to $12.00/share is exactly the same as the 12× ratio of company valuation increase from $10 million to $120 million. There is no "divergence" between the ratio of the increase of company share price and the ratio of the increase of company valuation [4].

For the Series B round, the 9.6× ratio of share price increase from $1.00/share to $9.60/share is less than (and diverges from) the 12× ratio of company valuation increase from $10 million to $120 million. In this scenario, the divergence ratio is 1.25 (12×/9.6×). Industry surveys indicate that divergence ratios can be 3× to 5× for early funding rounds.

Private investors use various strategies to attempt to remediate the negative impact on their investment returns arising from later funding rounds. For example, investors can increase the target return on their investment by negotiating combinations of higher ownership share and lower company valuation. Other contract terms can be negotiated at the time of the funding, including warrants to purchase additional shares, dividends, management structure, and liquidation preferences.

4.5.7 Market analysis

Estimating future valuation of an early-stage company for investment purposes involves risk, in part because of the uncertainty in projecting sales revenue for a new product; i.e., units sold × unit price. Market analysis is used to address this by projecting the amount and timing of sales revenues for different market segments.

Market analysis starts with an estimate of total addressable market (TAM); i.e., the maximum sales revenue that can potentially be captured with the new product (Figure 4.1). This analysis then proceeds with an estimate of serviceable addressable market (SAM); i.e., the percentage of TAM that can reasonably be achieved with the new product in an identified market segment. Finally, an estimate of SOM (serviceable obtainable market) is completed; i.e., the percentage of SAM that represents the initial target market for the new product. SAM is usually the market segment that has the lowest competition, least barriers to entry, and potential for the most rapid and profitable growth for the new product.

Various free and fee-based industry market studies as well as government and industry databases are available for estimating TAM, SAM, and SOM. Internal and contract market studies can also be used for market analysis.

Market analysis focuses on critical factors that should be considered to identify and rapidly sell a new product and increase early-stage company valuation. These fac-

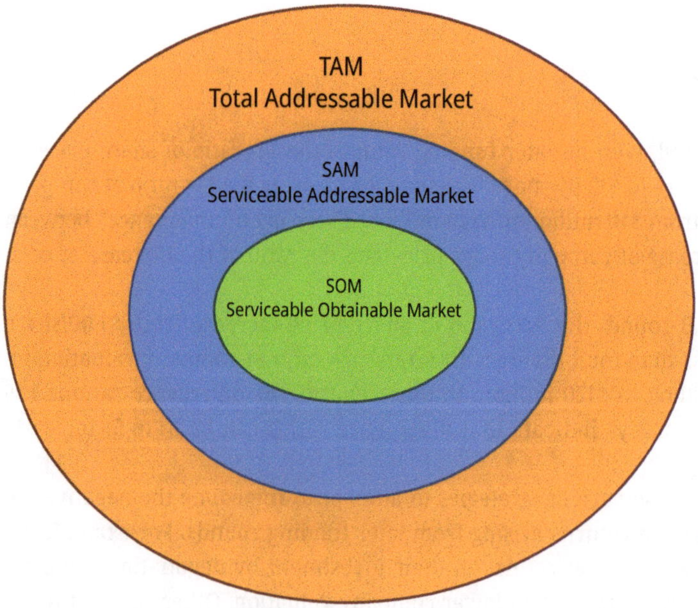

Figure 4.1: TAM (total addressable market), SAM (serviceable addressable market), and SOM (serviceable obtainable market).

tors include; for example, current and anticipated future competition, geographic limitations, logistical and political barriers, growth constraints related to lack of capital, and ability to scale production and establish sales channels.

TAM, SAM, and SOM are essential components of a business plan that are important to attract investor interest in an early-stage business. These projections must be large enough to attract investors, but also reasonable, taking into account the specific barriers of the market for a new product.

Sales revenue is made up of two factors: unit sales (# units) and unit price ($/unit). Unit sales are projected using the market analysis described above. Unit price is determined by assessing the value that buyers place on the product. This value is based on the satisfaction that buyers expect to receive from purchasing the product compared to other available purchase options.

The price that a customer is willing to pay for a product is subjective. Each individual buyer will have a price that is different from other buyers for the same product. Price also depends on the circumstances of each buyer at a specific time. Subjective value identifies buyers willing to pay the highest price for a product and helps to uncover ways to create maximum value.

In a free market, transactions will not occur unless both parties benefit. If the value to the buyer is more than the cost to a seller, a price can be agreed upon that

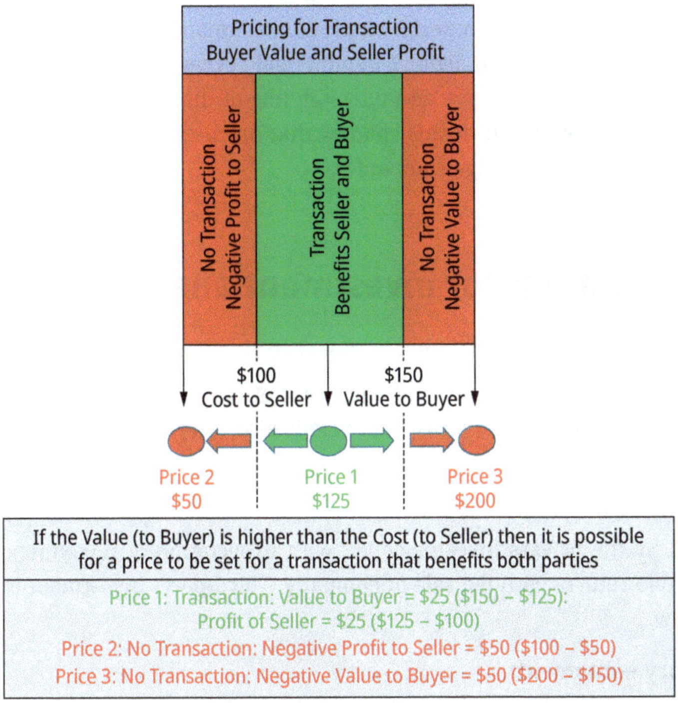

Figure 4.2: Pricing for Transaction.

allows for a transaction. In this scenario, both the buyer and the seller benefit because the buyer values the product more than the price, while at the same time, the seller can sell the product at a profit.

Figure 4.2 shows that a buyer is willing to buy at a price as high as $150 per unit and that the cost for a seller is $100 per unit. Therefore, there is an opportunity for a transaction to occur between $100 and $150 per unit. The seller wants to maximize profit at a price as close to $150/unit as possible. The buyer wants to minimize cost at a lower price but the seller cannot generate a profit at a price lower than $100/unit. Seller and buyer do not have perfect price and cost information; however, both will benefit from a transaction price between $100 and $150/unit.

Not all potential buyers are willing (or have the ability) to pay the same price for a product so the seller will attempt to price the product to maximize sales revenue. Prices are not static because of changes in market conditions and competition. Costs are not static because of changes in factory and labor costs, economies of scale, and tax policies. In the market analysis, it is important to consider possible changes in unit price and cost as these will impact sales revenue.

4.6 Summary

Early-stage companies, including green material bioproduct companies, often require funding from "angel" or VC investors to achieve commercial success. It is important that the principals of these companies have an understanding of the financial and investment analytical methods described in this chapter that are used by investors to project their financial return in early-stage companies.

Appendix 1: definitions for investment analysis

1. **Discount rate**
   Percentage that is used to discount cash for each year of the investment period; accounts for the time value of money. Discount rate is also referred to as the opportunity cost of capital.

2. **Investment**
   Funding provided by the private investor or VC, with expectation of repayment and with a profitable return, with the sale of company. Third-party investment is a positive cash flow.

3. **Cost of new factory equipment**
   Cost of new factory equipment is a negative cash flow. Factory equipment is depreciated.

4. **Cost of new land**
   Cost of new land is a negative cash flow. New land cost is not depreciated.
5. **Unit sales**
   Number of units that are sold each year.
6. **Unit sales price**
   Price of units that are sold in each year.
7. **Unit factory cost**
   Materials, factory labor, and other factory operating costs.
8. **Unit sales price less unit factory cost**
   Difference between unit sales price and unit factory cost.
9. **Gross margin**
   Percentage ratio of unit sales price less unit factory cost and unit sales price.
10. **Total sales revenue**
    Number of units sold in each year multiplied by the unit sales price.
11. **Total factory cost**
    Number of units sold in each year multiplied by the unit factory cost.
12. **Net sales revenue**
    Difference between total sales revenue and total factory cost.
13. **SG&A**
    Expense for SG&A. Industry benchmark, 20% of total revenue.
14. **R&D**
    Expense for R&D. Industry benchmark is 5% of total revenue.
15. **Engineering**
    Expense for engineering. Industry benchmark is 5% of total revenue.
16. **EBITDA**
    Earnings before interest, tax, depreciation, and amortization. Net sales revenue, less SG&A, less R&D, and less engineering.
17. **Depreciation**
    Tax shield based on 7-year straight-line depreciation of cost of new factory equipment.
    Depreciation is a positive cash flow.
18. **Taxable income**
    EBITDA less depreciation.

19. **Tax**
Corporate federal, state, and local tax as percentage of taxable income.

20. **Net income after tax**
Taxable income, less tax. Net income after tax can be a negative or positive cash flow.

21. **Change in net working capital**
Net working capital is current assets less current liabilities. Industry benchmark is 20% of change in year-to-year total sales revenue. Change in net working capital is a negative cash flow.

22. **Cash flow**
Third-party investment plus depreciation plus net income after tax, less cost of new factory equipment, less cost of new land, less change in net working capital.

23. **Present value**
Discounted cash flow for each year during the five-investment horizon.

24. **Incremental present value**
Accumulated present value for each year during the five-year investment horizon.

25. **Net present value (NPV)**
Accumulated present value over the 5-year investment horizon.

Appendix 2: accounting factors

Net working capital

Net working capital is current assets less current liabilities. Current assets are cash, accounts receivable, and inventory. Current liabilities are short-term notes and payable accounts, taxes, and interest. Investment analysis considers an increase in net working capital as a negative cash flow and a decrease in net working capital as a positive cash flow. Companies will often have an increase in net working capital with growing sales revenue. This can impede the ability of a company to grow because of cash flow constraints. Based on industry benchmarks, a change in net working capital is estimated as 20% of the year-to-year difference in total sales revenue. Change in the net working capital is assumed to be only a function of the annual increase or decrease in total sales revenue and does not take into account any change in company efficiency of managing net working capital.

Sales, general, and administrative (SG&A)

SG&A includes all commercial expenses of operations incurred during the regular course of business. SG&A includes advertising, commissions, marketing, selling, and other administrative expenses. Based on industry benchmarks, SG&A is estimated at 20% of total sales revenue. However, higher percentages of SG&A are assumed during the first few years of rapid company growth when there is a relatively higher amount of SG&A required, relative to total sales revenue.

Research and development (R&D) and engineering

R&D and engineering include all expenses that are incurred to develop, design, and improve a product. Based on industry benchmarks, R&D and engineering are each estimated at 5% of total sales revenue. However, higher percentage R&D and engineering are assumed during the first few years of rapid company growth when there is a relatively higher amount of R&D and engineering required, relative to the total sales revenue.

References

[1] Adapted from Scorecard Valuation Methodology (Rev 2019), Frontier Angels, Bill Payne. www.angelcapitalassociation.org/blog/scorecard-valuation-methodology-rev-2019-establishing-the-valuation-of-pre-revenue-start-up-companies/

[2] www.eval.tech/valuation-multiples-by-industry.

[3] Adapted from Investment Valuations of Seed and Early-Stage Ventures, Luis Villalobos; 2007. www.entrepreneurship.org/articles/2007/07/investment-valuations-of-seed-and-earlystage-ventures.

[4] Adapted from Valuation Divergence, Luis Villalobos; 2007.

Camila González Arango, Camila Valderrama, Johannes Ali,
Juliana Vasco-Correa*

Chapter 5
Carbon-negative bioproducts

5.1 Introduction

Bioproducts from biomass offer renewable alternatives to materials, chemicals, and fuels that are traditionally derived from fossil resources. By drawing on carbon fixed through photosynthesis, bioproducts provide a foundation for more sustainable production systems while also supporting rural economies. Beyond their role as renewable substitutes, an emerging concept is that certain bioproducts can be intentionally designed to contribute to carbon management by storing atmospheric carbon over extended timescales, effectively becoming carbon-negative [1]. This dual role of providing economic value while also addressing climate change positions bioproducts as a key tool for value-added solutions in climate change mitigation.

Plants capture atmospheric carbon dioxide (CO_2) via the process of photosynthesis and subsequently transform it primarily into lignocellulosic biomass, thereby generating a renewable reservoir of biogenic carbon [2]. This CO_2 capture step is the essential basis of all carbon-negative bioproducts pathways, since it ensures that the carbon incorporated into products originates from the atmosphere rather than fossil reserves [1]. Once the CO_2 is fixed in biomass, this carbon can be harvested and transformed into durable goods, chemicals, or energy carriers, where it may be retained over extended timescales [3]. In this way, biogenic carbon capture provides the starting point for strategies that position biomass not only as a renewable substitute for fossil carbon but also as a promising mechanism for long-term sequestration. Accordingly, the climate value of biomass has been argued to extend beyond its function as an energy source [4]. Through the integration of natural carbon sequestration with technology development, bioproduct systems not only mitigate greenhouse gas (GHG) emissions and reliance on fossil fuels but also generate alternatives for carbon sequestration, making them pivotal tools for achieving negative emissions [5].

The rapid increase in atmospheric CO_2 concentrations represents one of the most pressing challenges facing our society, with significant repercussions for global health, economic stability, and ecological systems [6]. There is a broad consensus within the scientific community and climate experts that mitigation and ultimately

***Corresponding author: Juliana Vasco-Correa**, Department of Agricultural and Biological Engineering, The Pennsylvania State University, University Park, PA, USA, email: julianavasco@psu.edu
Camila González Arango, Camila Valderrama, Johannes Ali, Department of Agricultural and Biological Engineering, The Pennsylvania State University

 | https://doi.org/10.1515/9783112212141-005

stopping anthropogenic climate change requires the achievement of net-zero GHG emissions across most sectors [7]. To achieve net-zero carbon emissions, we rely on a wide portfolio of strategies, including large-scale deployment of renewable electricity, investment in storage and transmission infrastructure, expansion of hydrogen production and use, and the development of renewable liquid fuels. Collectively, these efforts are referred to as the process of decarbonization [8]. Still, decarbonization does not imply that carbon is fully eliminated from energy and material systems, but rather that fossil carbon is progressively displaced by low-carbon and renewable alternatives [9].

Despite extensive efforts toward decarbonization, residual emissions are expected to persist within sectors that are challenging to decarbonize, and advancements are still required to address those emissions and the excess carbon that has already accumulated in the atmosphere. Within this broader effort, carbon dioxide removal (CDR), defined as human-driven processes that capture CO_2 directly from the atmosphere and store it durably in biological, geological, or product-based reservoirs, has emerged as indispensable for meeting the Paris Agreement targets of limiting warming to 1.5–2 °C [10]. In addition, the IPCC indicates that most viable mitigation pathways rely on removing 1–20 billion tons of CO_2 annually by 2050 [7]. Therefore, given the current limitation that clean energy technologies are not expanding at the pace required to stabilize global warming, additional emphasis is placed on the development of scalable CDR strategies.

The bioeconomy offers a pathway to achieve the dual objectives of carbon management and value generation simultaneously. By converting renewable, biogenic carbon from agriculture, forestry, and aquaculture into food, feed, fibers, polymers, bioenergy, and other value-added products, it establishes a foundation for products that can both substitute fossil resources and, when designed appropriately, act as carbon sinks [4]. From a systems perspective, several specific strategies have emerged for embedding carbon storage into bioproduct pathways. These range from natural sink enhancement in agriculture, forestry, and peatlands to carbon storage in durable goods, such as engineered timber, insulation panels, and biochar, as well as integrated energy systems, including bioenergy with carbon capture and storage (BECCS) [11]. Other frameworks, such as BiCRS (biomass carbon removal and storage) [12], highlight the diversity of approaches through which biomass can deliver durable sequestration. While these strategies are instrumental in contextualizing bioproducts within extensive carbon management frameworks, the primary emphasis of this chapter is on the conceptualization and implementation of bioproducts as instruments for achieving carbon negativity.

It is also essential to recognize that not every bioproduct contributes to a uniform impact on climate mitigation efforts. Many are defined as low-carbon because they displace fossil resources and thereby lower emissions relative to conventional products. For instance, bioplastics and biofuels generated from renewable biomass can serve as replacements for fossil fuel-based counterparts. Nevertheless, these products do not inherently eliminate carbon from the environment or guarantee its prolonged sequestration, as a significant portion of their carbon is ultimately re-released during utilization or disposal [13]. In contrast, bioproducts designed for carbon sequestration effectively capture and store atmospheric carbon, thereby achieving a net-negative carbon balance. Depending on their design, the captured carbon can be stabilized biologically (in soils, wetlands, and forests), stored in products (timber, insulation, biochar), or injected into geological formations after biomass processing (BECCS) [11]. Distinguishing between low-carbon and carbon-negative systems is therefore critical, as only the latter deliver durable carbon sinks.

The deployment of bioproducts as solutions for carbon removal demands a robust systems-level understanding of the trade-offs, synergies, and constraints embedded within their value chains. Systems analyses (e.g., technoeconomic analysis and life cycle assessment, LCA) play a critical role in quantifying the net climate benefits of bioproducts, identifying hotspots of emissions or resource use, and guiding design choices that maximize economic benefits as well as carbon storage permanence. In addition, these analyses enable the comparison of bioproduct pathways across regional contexts, feedstock types, and end-use applications, thereby informing policy frameworks and investment strategies. Without rigorous systems analysis, there is a high risk that well-intentioned bioproducts will fall short of delivering true carbon negativity or inadvertently shift environmental burdens elsewhere in the system [14]. Therefore, embedding systems thinking into the development and evaluation of bioproducts is essential for aligning bioproduct development with climate goals and ensuring that these technologies contribute meaningfully to carbon removal while bringing economic value.

This chapter explores the potential of bioproducts beyond renewable alternatives to fossil resources, showing that, when intentionally engineered, they possess the potential to function as carbon sequestration systems. When manufactured with considerations of permanence and additionality, bioproducts can transform into both economic commodities and engineered repositories of atmospheric CO_2. The discussion that follows examines the capacity of bioproducts to foster carbon negativity by evaluating their contribution to decarbonization trajectories, assessing their efficacy through LCA frameworks, and delineating methodologies to enhance durability, scalability, and sustainability. Ultimately, the exigencies of climate change require not only renewable alternatives but also bioproducts that proactively sequester and stabilize carbon, thus reinforcing ecological resilience and facilitating sustainable economic development.

5.2 Carbon sequestration pathways via bioproducts

Carbon sequestration pathways through bioproducts present different approaches for facilitating climate change mitigation. The concept of carbon-negative bioproducts derives from the intrinsic capacity of plants to act as both a source of renewable carbon and a viable long-term carbon storage mechanism [3]. Their potential rests on multiple pathways through which atmospheric CO_2 can be fixed, transformed, and retained in forms that delay or prevent re-emission. This section outlines the major categories of sequestration pathways, from biogenic carbon capture in plant and algal feedstocks to the incorporation of captured carbon into durable materials, substitution effects, and integration with broader carbon removal strategies.

5.2.1 Biogenic carbon capture: plant- and algae-based feedstocks

Terrestrial plants and algae are biological systems that effectively absorb atmospheric CO_2 through photosynthesis and convert it largely into lignocellulosic biomass (cellulose, hemicellulose, and lignin) with smaller fractions of starch, soluble sugars, proteins, and lipids [2]. Photosynthesis is the foundation of all carbon-negative bioproducts, providing a renewable reservoir of biogenic carbon available as feedstock for diverse applications, instead of relying on carbon coming from fossil carbon reservoirs [1].

Different varieties of feedstock serve as reservoirs of biogenic carbon. Terrestrial biomasses include annual crops, perennial grasses, and forestry resources. In general, annual crops capture carbon efficiently during the growing season, with the drawback of usually being used for short-lived products or food, providing short-term carbon storage. To achieve meaningful long-term storage, carbon needs to be intentionally stabilized into durable products. In contrast, perennial grasses are valued for their higher productivity and longer storage capacity, similar to forestry resources, which retain carbon in above- and below-ground biomass for decades [15].

Agricultural and forestry residues (e.g., crop husk, stalks, sawdust, pruning waste, and bark) offer additional carbon pools that can be used without new land-use conversion since they come from existing production systems [16]. However, they are easily degradable, generating a quick carbon release. Hence, they need to be stabilized through appropriate conversion processes to contribute to long-term carbon retention.

Algal feedstocks are emerging as a promising and increasingly studied carbon reservoir. Microalgae exhibit high growth rates, photosynthetic activity, and ability to accumulate large amounts of lipids, carbohydrates, and proteins [17]. They are recognized for having carbon fixation efficiencies up to 10 times greater than terrestrial plants, particularly under optimized photobioreactor settings or continuous harvesting regimes [18]. Marine-cultivated macroalgae, or seaweed, for their part, do not require freshwater or arable land, eliminating land-use pressure and allowing large-scale biomass production. In addition, algae-bacterial consortia can enhance nutrient recycling and CO_2 capture in engineered systems such as wastewater treatment [19].

The choice of feedstock is critical, as it influences the rate of carbon capture, the permanence of storage, and the overall sustainability of bioproduct systems. This outcome depends on biomass productivity, carbon content, and where the carbon is stored within the feedstock structure (e.g., roots, stems, or storage compounds) [20]. For instance, lignocellulosic biomass, due to its recalcitrant nature, is better suited for long-lived materials, while starches, lipids, and proteins are more easily transformed into fuels or chemicals, offering shorter carbon storage timescales. *C4* crops (e.g., switchgrass and corn) generally fix carbon more efficiently than *C3* species (e.g., wheat and rice), and algae usually outperform in both terms of photosynthetic efficiency and yield potential [3].

Finally, it is essential to consider environmental constraints and trade-offs associated with feedstock use. Resource demands such as land, water, and nutrients must be carefully managed to avoid trade-offs with food production, indirect land-use change, or biodiversity loss from monocultures or large-scale plantations [21]. At the same time, there are significant opportunities to mitigate these challenges. For example, valorizing residues and waste streams helps avoid new land conversion offering economic advantages while integrating algae consortia into engineered systems and optimized reactors allows for simultaneous carbon capture, nutrient recovery, and biomass production [19].

5.2.2 Carbon retention in bioproducts

Carbon retention in bioproducts through sequestration involves capturing atmospheric CO_2 through plant- or algae-based feedstocks and stabilizing it in biological materials, which can then be used to produce various products with varying lifetimes, thereby delaying the return of the carbon to the atmosphere. All such bioproducts are derived from biogenic carbon, but not all of them can be considered carbon-negative, as their climate benefit depends largely on the product's lifetime and end-of-life (EoL) disposal [22]. Furthermore, the effectiveness of carbon sequestration in bioproducts depends on the type of biomass used, the method of sequestration, and the permanence of the products created. Accordingly, carbon storage in bioproducts can be classified into three categories: shorter-lived, semi-durable, and durable products (Figure 5.1).

Shorter-lived products, also known as low-carbon or substitutional products retain carbon for only months to a few years. They play an important role by reducing reliance on fossil-based resources. Some examples of these products are biofuels, algal-sourced nutraceuticals such as algal omega-3 fatty acids, carotenoids, proteins, pigments, and specialty chemicals [23, 24]. For biofuels, combustion typically releases CO_2 back into the atmosphere, and even when produced sustainably, they can at best achieve carbon neutrality, though in practice this is rarely realized. In these cases, the carbon benefit lies in the substitution of fossil-derived products, but they do not guarantee long-term sequestration [25].

Semi-durable products provide carbon storage on the order of years to decades. Bioplastics and composites derived from cellulose, lignin, or starch can extend carbon residence time, although their environmental performance depends on durability versus biodegradability trade-offs [26]. Fibers, textiles, and coated paper also contribute to medium-term retention, especially when recycling loops are established. Packaging and insulation products fall within this category, offering moderate lifetimes while displacing fossil-derived materials [27]. Industrial and consumer products such as insulation panels, furniture, textiles, and biocomposites also contribute, though their lifetimes vary. Products used in housing or vehicles may store carbon for decades, whereas textiles and packaging typically release it within a few years, offering limited sequestration value [28]. This highlights the importance of product design and market application in determining climate benefits. Bioplastics and biopolymers illustrate the pressure between durability and biodegradability. While marketed as sustainable substitutes for fossil plastics, many bioplastics are designed for short-lived use and rapid degradation, offering little sequestration. Only when applied for durable goods, such as automotive components, electronics casings, or construction panels, are they able to provide meaningful carbon storage [13].

Durable storage products can retain carbon for decades to centuries. One of the most direct pathways to carbon negativity is the incorporation of biogenic carbon into long-lived products. Unlike fuels or short-lived packaging that re-emit carbon within months or years, durable bioproducts can delay atmospheric release for decades to centuries, depending on their composition and use [29]. Wood-based construction materials, including engineered timber, insulation panels, and building components, store carbon within the built environment while replacing emission-intensive materials such as steel and concrete [30]. Biochar represents another strong carbon-negative pathway; once applied to soils or incorporated into cement and asphalt, it can persist for centuries and deliver agronomic or structural co-benefits. More recently, mineralized biomass products such as silicon carbide, bio-cement, and carbon-infused concrete have emerged as highly stable forms of storage with significant industrial potential [31].

Several considerations shape the climate value of carbon retention of bioproducts. The timescale of storage and the EoL disposal, i.e., whether recycling, reuse, landfilling, or incineration is used with or without carbon capture, determine whether carbon remains sequestered or reenters the atmosphere. The overall effectiveness of the different pathways depends on permanence. Short-lived products provide limited sequestration, whereas materials embedded in construction or soil can serve as substantial carbon sinks.

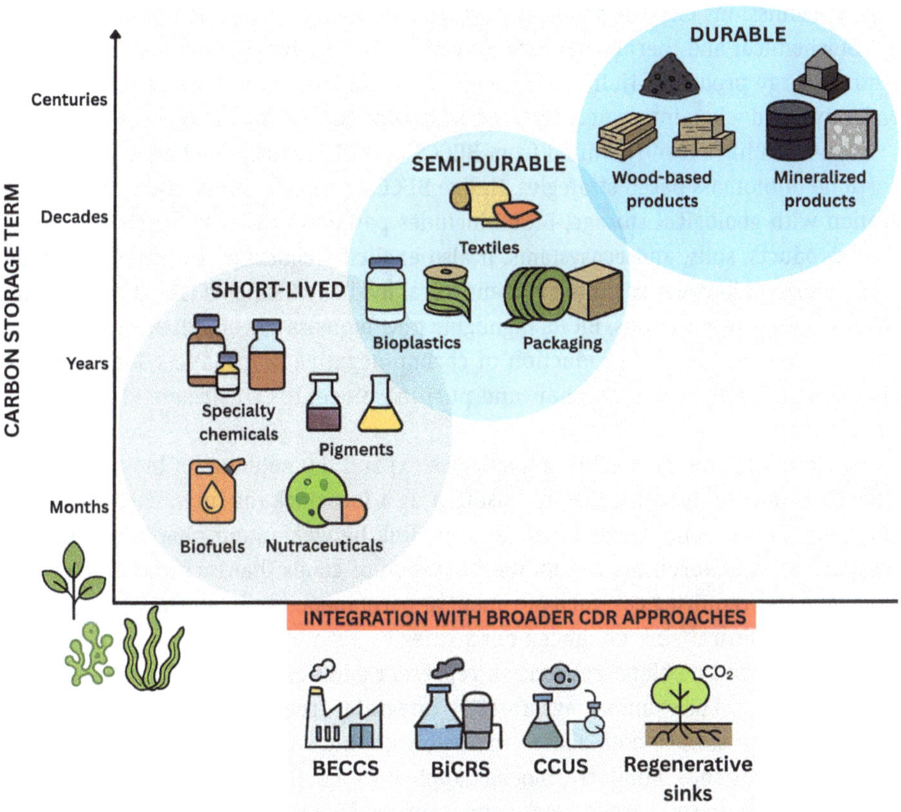

Figure 5.1: Carbon storage potential of plant- and algae-based bioproducts across product lifespans and integration pathways.

5.2.3 Integration with broader carbon removal strategies

Bioproducts can intersect with broader engineered CDR approaches by linking the biological capture of atmospheric CO_2 with technological systems for its storage or utilization. In this way, they act not only as standalone sequestration pathways but also as enablers of strategies such as bioenergy with carbon capture and storage (BECCS), biomass carbon removal and storage (BiCRS), and carbon capture, utilization, and storage (CCUS).

BECCS is designed to use biomass, often wood or agricultural energy crops, as fuel for power, heat, steam, hydrogen, or other gaseous or liquid fuels. In BECCS systems, biomass is combusted or converted into biofuels in facilities equipped with CO_2 capture technology. Contrary to bioproducts, where the carbon is stabilized in products, the captured CO_2 in BECCS is injected into deep geological formations for long-term storage [32]. If the emissions associated with biomass cultivation, supply, and processing do not outweigh the CO_2 captured during combustion, BECCS achieves net-negative emissions [32]. Importantly, BECCS technologies can be applied across different biochemical and thermochemical conversion routes, broadening the scope of potential energy products [4]. In that sense, BECCS and bioproducts are complementary pathways, and when integrated, they can maximize climate impact.

Attention has recently shifted from BECCS to BiCRS, which emphasizes a broader portfolio of biomass-based strategies. Unlike BECCS, which is centered on energy generation with geological storage, BiCRS includes pathways that store carbon in long-lived products, soils, and ecosystems. It also explicitly considers co-benefits such as food security, biodiversity protection, and rural livelihood support [12]. By expanding beyond energy applications, BiCRS highlights how biomass can simultaneously enable carbon removal and the production of carbon-negative bioproducts such as engineered wood, bioplastics, biochar, and purified biogenic CO_2 streams destined for storage [32].

In addition, emerging CCUS technologies expand the potential of bioproduct systems by capturing biogenic CO_2 and using it as a feedstock for fuels, chemicals, and polymers. These technologies create a direct link between atmospheric carbon capture and product formation, enabling the design of goods that incorporate recycled CO_2 while still requiring careful assessment of permanence and EoL emissions [33].

Finally, natural sink enhancement strategies such as agroforestry, soil carbon management, and peatland restoration represent additional bio-based pathways for carbon removal. For example, agroforestry integrates trees into farming systems, increasing carbon storage both above- and below-ground while supporting biodiversity and rural livelihoods. Similarly, biochar applied as a soil additive has been demonstrated to increase soil carbon stocks and improve fertility [11]. These practices illustrate how land-based approaches can align ecological and carbon removal objectives.

5.3 Systems-level analysis for carbon-negative bioproducts

Systems-level assessments such as LCA are essential to ensure that the environmental benefits of carbon-negative bioproducts are real, measurable, and not overstated. These tools allow for the tracking of biogenic carbon throughout the full life cycle, from feedstock production to EoL product disposal, and are the only reliable way to determine whether a bioproduct truly results in net carbon removal rather than simply delaying emissions [34].

Although all bioproducts are derived from biomass and thus incorporate biogenic carbon, their climate impact varies widely. Some products offer long-term carbon storage, while others release carbon quickly upon use or disposal [14]. In some cases, fossil emissions associated with feedstock cultivation, energy use, processing, and transportation may outweigh the carbon removal benefits. Without a systems-level analysis, such trade-offs are easily overlooked.

Systems-level assessments are usually performed against incumbent technologies and other baselines that allow for comparison and show the magnitudes of the benefits and risks of these emerging technologies. In addition, systems-level assessment allows for a comprehensive environmental evaluation of the bioproducts beyond carbon, assessing effects on land use, water, nutrients, and biodiversity. This is critical to avoid unintended consequences of emerging technologies. Thus, systems-level analyses should be the basis for process and product design, technical and economic decision-making, and policy implementation.

5.3.1 Life cycle assessment (LCA) and carbon removal

LCA provides the standard methodology for evaluating the environmental implications of products and processes across their life cycles. According to ISO 14040 and 14044, an LCA involves four iterative steps: (i) goal and scope definition, (ii) life cycle inventory (LCI) analysis, (iii) life cycle impact assessment (LCIA), and (iv) interpretation [35]. Most frameworks include indicators beyond GHG emissions, such as land, water, and nutrient use, which are essential for determining broader sustainability outcomes. For assessing carbon sequestration, however, life cycle GHG accounting is central. The most common metric is Global Warming Potential (GWP), expressed in kilograms of CO_2 equivalent (CO_2e) as a measure of the warming potential of a kilogram of CO_2 in the atmosphere for 100 years. While GWP is widely applied, debates persist regarding its adequacy for capturing long-term sequestration dynamics and the appropriate time horizons for carbon removal [36, 37]. LCA guidelines have been developed for engineered pathways such as direct air capture and storage [38] and biomass-based technologies such as BiCRS [39], but further standardization is still needed.

Importantly, LCA should not be treated as a one-time evaluation but as an evolving process that matures alongside the technology. As bioproduct systems advance from conceptual design to pilot and commercial scales, key parameters, such as feedstock sourcing, energy requirements, process yields, and EoL management become better defined and should be continuously updated in the LCA model. Embedding LCA iteratively within product development ensures that environmental assessments remain accurate, relevant, and reflective of real-world performance. Moreover, when LCA is used as a design tool rather than a retrospective validation exercise, sustainability considerations can actively inform process optimization, design choices, and trade-off evaluations [40]. For carbon-negative bioproducts, this iterative integration is particularly critical, since small changes in system configuration or emissions assumptions can substantially alter the net carbon balance. Ultimately, continuous use of LCA helps ensure that emerging bioproducts achieve their intended climate and environmental goals.

5.3.2 Evaluating sequestration performance

Evaluating the carbon sequestration performance of bioproducts requires more than tracking biogenic carbon inputs and outputs. Three criteria are particularly important: permanence, additionality, and system boundaries, each of which shapes whether a product can credibly be considered carbon-negative.

Permanence refers to the duration over which carbon remains stored outside the atmosphere. Bioproducts vary widely in this respect. Short-lived products, such as biofuels or compostable packaging, release CO_2 back into the atmosphere within weeks to years. By contrast, durable goods such as engineered timber for construction or bioplastics used in long-lived applications can retain carbon for decades or longer. Biochar represents another pathway, with the potential to stabilize carbon in soil for centuries, though its persistence depends strongly on soil conditions and application methods. Assessing permanence, therefore, requires a careful understanding of product fate, degradation rates, and EoL scenarios. Without this, claims of carbon sequestration may reflect temporary carbon delays rather than durable storage [4].

Additionality ensures that carbon removal associated with a bioproduct represents a genuine climate benefit beyond what would have occurred under a baseline or business-as-usual scenario. For example, if a crop is already cultivated for another purpose, diverting part of its biomass into a bioproduct may not represent additional sequestration unless it increases overall storage or displaces more carbon-intensive alternatives. Using waste biomass that would otherwise decompose and release CO_2 or CH_4 can qualify as additional if the bioproduct pathway leads to longer retention or reduced emissions compared to the counterfactual fate. Demonstrating additionality requires defining a credible counterfactual scenario and quantifying what emissions are truly avoided or removed by the new product system. In the context of LCA, negative emissions can result from two main sources: (i) direct removal of CO_2 from the environment through photosynthesis, which does not guarantee permanent storage and (ii) avoided emissions through substitution of more carbon-intensive products or processes, typically modeled using system expansion [38]. Although these two contributions are often aggregated, reporting them separately is critical for transparency and credibility.

System boundaries represent a third pillar of robust evaluation. While cradle-to-gate boundaries are commonly applied in bioproduct LCAs due to uncertainty in product use and disposal, they can significantly overstate sequestration benefits [14]. Only cradle-to-grave assessments adequately capture emissions from cultivation and harvest, transportation and processing, product use, and EoL management, alongside any long-term carbon storage or release [39]. In some cases, cradle-to-cradle perspectives are necessary to account for recycling or reuse (Figure 5.2). The choice of boundaries can drastically affect conclusions; Scown's recent analysis of biomanufacturing shows that cradle-to-gate assessments may suggest carbon negativity, while cradle-to-grave perspectives reveal significant re-emissions [14]. In addition, leakage effects, in-

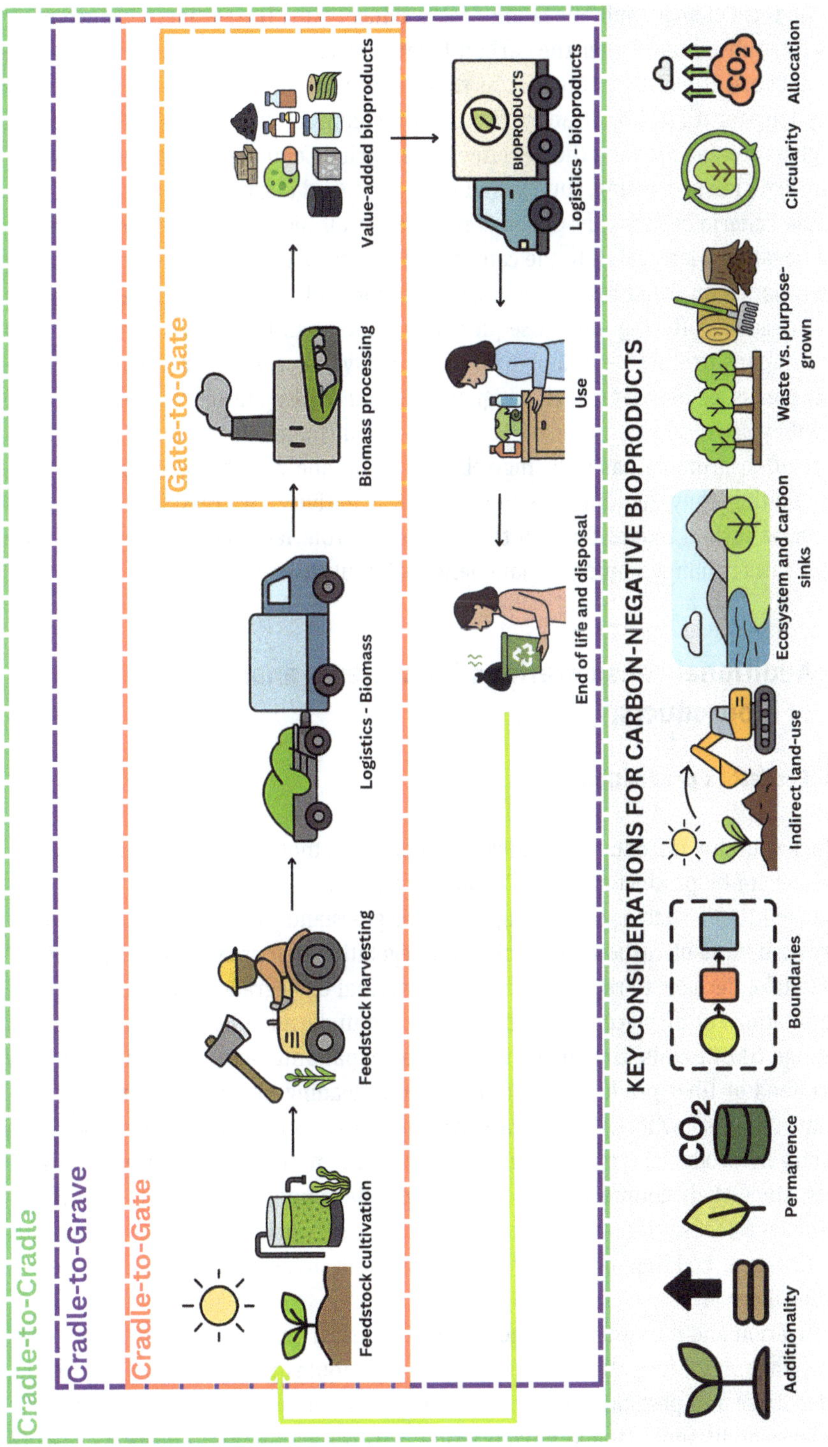

Figure 5.2: Life cycle stages and key considerations for carbon-negative bioproducts.

direct land-use change, and allocation of coproducts must be considered, since these factors can substantially alter the carbon balance. The expected use and longevity of the product are especially influential, as they determine whether storage is temporary or truly long-term [4]. Importantly, the carbon contained in the product represents only a gross storage amount; the net effect must include emissions from feedstock cultivation, logistics, processing, supply chain activities, and disposal.

These criteria collectively underscore that not all biomass-derived products are carbon-negative. Many fall into the category of low-carbon products, which primarily reduce reliance on fossil resources but re-emit most of their carbon content within years to decades, offering limited sequestration potential. In contrast, products with longer lifespans or durable immobilization pathways, such as engineered timber, insulation panels, or biochar, provide higher potential for long-term storage, depending on how they are designed and used. While bio-based commodity chemicals and materials are often promoted as reducing reliance on oil and gas, the more critical question is whether they genuinely deliver carbon negativity. Addressing this requires tracing how carbon moves through these systems, from feedstock production to use and EoL, rather than assuming climate benefits by substitution alone [14].

5.3.3 Additional considerations for systems analysis of bioproducts

5.3.3.1 Feedstock production

Feedstock sourcing introduces significant complexities that strongly influence the carbon balance of bioproduct systems. Purpose-grown feedstocks can trigger direct land-use change (dLUC), such as converting forests or grasslands to croplands, often releasing large amounts of carbon from soils and vegetation and creating a "carbon debt" that may take decades to repay. Even when marginal or degraded lands are targeted, yield limitations or displacement of other uses can undermine benefits. Indirect land-use change (iLUC) adds further uncertainty, as expanding biomass cultivation may displace food or fiber production to other regions, leading to deforestation or ecosystem conversion elsewhere. By contrast, waste- and residue-derived feedstocks, such as forestry by-products, crop residues, or manure, often provide clearer additionality benefits, since their counterfactual fate is rapid decomposition and CO_2 or CH_4 release. However, these streams also have competing uses, including soil carbon maintenance, livestock bedding, or energy production, and diverting them can shift emissions elsewhere [41].

Cultivation and harvesting also generate substantial emissions. Fossil energy used in farm machinery, irrigation, and transport contributes to the life cycle footprint, while fertilizer and pesticide production and application, particularly nitrogen fertilizer, release additional CO_2 and nitrous oxide [42]. These upstream emissions can offset much of the carbon captured in biomass, especially in low-yield systems with high input requirements.

Further complexity arises from soil carbon dynamics, nutrient cycling, and spatial variability. Removing residues can deplete soil organic carbon, reduce fertility, and increase erosion risk, while perennial grasses, cover crops, or agroforestry systems may build soil carbon through root inputs. Residues and manures also play a key role in nutrient recycling; diverting them to bioproducts may require replacement fertilizers, while some pathways return coproducts such as digestate, ash, or biochar that offset these burdens. Outcomes vary substantially across regions, soils, and management practices, and seasonal factors such as storage or drying can introduce additional emissions that are often overlooked in simplified assessments.

Finally, transport and allocation issues further shape the life cycle balance. Biomass is bulky and of low energy density, making transport a potentially significant source of emissions; distance, mode, and densification options all matter. Allocation of impacts in multipurpose systems is equally consequential: crops such as corn, sugarcane, or soy generate multiple products, and LCA results differ widely depending on whether impacts are assigned by mass, energy, or economics, or whether substitution credits are applied.

5.3.3.2 Bioproduct end-of-life (EoL)

EoL dynamics represent another critical determinant of whether bioproducts achieve durable carbon removal. Even products that store carbon during use may ultimately release it back into the atmosphere depending on how they are managed at disposal. Bioplastics, for example, can function as temporary carbon storage during their service life, but if they are composted or incinerated without carbon capture, most of the stored carbon is rapidly re-emitted [43]. Similarly, engineered timber used in construction can retain carbon for decades, but demolition practices and waste management strongly influence whether this storage is extended through recycling into particleboard or bioenergy, or lost quickly through uncontrolled combustion and decay [44, 45].

Different EoL pathways carry very different implications. Recycling and reuse can extend the duration of storage and reduce demand for virgin feedstock, while landfilling may slow degradation and prolong carbon residence, though it can also generate methane emissions if not properly captured. By contrast, incineration with energy recovery provides substitution benefits for fossil fuels but typically results in immediate re-release of biogenic carbon. For biochar, EoL considerations are somewhat different: once applied to soils, its stability is determined by interactions with soil minerals, microbial activity, and environmental conditions. Here, the critical uncertainty is not disposal practices but the timescale of decomposition, which remains one of the most debated aspects of biochar permanence [46].

Accounting for EoL is particularly important because temporary storage is not equivalent to permanent removal. For example, LCAs of compostable packaging that include disposal pathways often find that their net climate benefit is limited, since the carbon is released within months of use [47]. In contrast, assessments of bio-based insulation materials or structural wood demonstrate substantial long-term storage potential when recycling and cascading use are assumed [29]. Such cases illustrate that the same product may appear carbon-negative or carbon-positive depending on the assumed EoL fate.

Moreover, regional waste management infrastructure adds another layer of variability. Landfills in high-income countries may capture a large fraction of methane emissions, reducing climate impacts of bioproduct disposal, whereas in regions with poor methane capture, the same system could result in net emissions. Similarly, substitution benefits from energy recovery depend on the regional electricity mix; incineration may displace coal in one context but only natural gas or renewables in another, altering the overall balance.

Ultimately, robust evaluation of carbon-negative bioproducts must include plausible EoL scenarios, region-specific waste management practices, and sensitivity analyses to reflect uncertainty in product fate. Without these considerations, claims of durable carbon removal risk overstating the true climate benefit of emerging bioproduct pathways.

A further consideration in assessing carbon-negative bioproducts is their role within a *circular bioeconomy* framework. Circularity emphasizes maximizing the value of biomass resources by extending product lifetimes, cascading material use, and returning nutrients and carbon back to productive systems. From an LCA perspective, circularity can improve carbon balances by displacing virgin feedstock production, reducing waste, and prolonging storage through reuse and recycling. For example, cascading use of wood from primary construction into secondary panels or bioenergy extends carbon storage while offsetting fossil-based alternatives [48]. Similarly, residues returned as digestate, ash, or biochar can recycle nutrients and maintain soil fertility, closing material loops that might otherwise require energy-intensive inputs [49]. Evaluating carbon-negative bioproducts through a circularity lens expands the system boundary to consider not only direct sequestration but also how effectively biomass is cycled within the economy, reducing leakage and enhancing long-term climate benefits [50].

Finally, any framework for evaluating sequestration must explicitly account for sensitivity and uncertainty. Assumptions about degradation rates, counterfactual scenarios, EoL management, and iLUC can drastically alter conclusions about permanence, additionality, and net storage [4]. Sensitivity analyses are therefore essential to identify parameters with the greatest influence on results and to test the robustness of carbon-negative claims under different scenarios. Without explicit treatment of uncertainty, assessments risk overstating the climate benefits of emerging technologies and undermining their credibility.

5.3.4 Technoeconomic analysis and integration with LCA

While carbon capture and storage (CCS) solutions often depend on carbon markets or policy incentives to be economically viable, bioproduct-based approaches offer an additional opportunity: they can generate intrinsic economic value through the production of marketable goods derived from carbon-sequestering feedstocks. This opens the door to co-benefits, where carbon removal and revenue generation are achieved within the same system.

However, realizing this dual value proposition requires assessment of the economic feasibility of bioproduct pathways, independent of their sequestration potential. A process that stores carbon but is prohibitively expensive or unscalable will not be adopted. Conversely, a bioproduct with strong market potential may not be environmentally beneficial without evidence of durable carbon storage and net emission reductions.

Technoeconomic analysis (TEA) is the most common tool for evaluating feasibility. TEA is generally based on process design and synthesis, and engineering economics, to determine the bioproducts minimum selling price as well as the proposed project's financial metrics such as the net present value (NPV), internal rate of return (IRR), and payback period [51]. TEA can also be used to optimize conversion pathways [52], determine the minimum unit carbon sequestration cost [53], maps cost drivers via sensitivity/uncertainty analysis [54, 55]; tests scale, learning, and logistics effects [56, 57]; compares product slates and coproduct strategies [58, 59]; and quantifies policy dependence (e.g., credit stacking and tax incentives) [60, 61].

Although TEA and LCA are often developed separately, their integration is essential to ensure the dual purpose of carbon-negative bioproducts. TEA typically uses narrower system boundaries and focuses on process-level detail and cost structures. In contrast, LCA accounts for broader environmental impacts, including product use, disposal, and externalities such as land use change. To ensure consistency, outputs from TEA models (e.g., material and energy flows, utility demands, and coproduct yields) should be used as inputs for LCA models, and scenario assumptions (e.g., technology maturity and geographic context) and functional units must be aligned between TEA and LCA to enable meaningful cross-analysis [62, 63].

Most carbon-negative bioproduct pathways face real trade-offs that shape feasibility and climate value [22, 53, 64–66]. Pushing yields through harsher pretreatment or high solvent recycle can raise energy demand, solvent loss, and toxicity; centralizing plants lowers unit costs but expands biomass hauling, fugitive emissions, and local impacts; replacing fossil heat with electrified or biomass heat cuts GHG emissions yet increases CAPEX/OPEX; monetizing coproducts boosts return on investment but makes results highly sensitive to allocation and displacement assumptions; designing for multi-feedstock/product flexibility hedges market risk but adds idle capital; routing biogenic carbon into short-lived products yields near-term revenue but weaker long-term storage than durable goods, biochar, or geologic options; intensifying feedstock supply can stress water and biodiversity; and siting for local jobs or equity may raise unit costs while improving legitimacy. Integrated TEA-LCA should surface these tensions with Pareto curves (cost vs. CO_2e) [64–66], marginal abatement cost metrics [67, 68], and uncertainty ranges [69], clarifying where "win–wins" exist and where informed compromises are required. Table 5.1 presents some of the methods that can be used to visualize the potential trade-offs. Integrated modeling can reveal a frontier of cost–impact options rather than a single optimum; choices that improve IRR can raise GHG emissions, water use, or biodiversity risks, and vice-versa. Harmonized boundaries, functional units, and time horizons, plus uncertainty analysis, make those trade-offs explicit. Reporting IRR alongside cost per ton CO_2e avoided/stored and testing coproduct allocation and displacement assumptions, allows decision-makers to choose designs that meet climate targets at acceptable cost under specific regional and policy contexts.

Table 5.1: Methods to make trade-offs visible in integrated TEA-LCA.

Method	*What it reveals*	*How to present*	*Notes/pitfalls*
Pareto analysis	Feasible trade-offs between cost and environmental impact (no single optimum)	Scatter/curve of levelized cost versus impact (e.g., kg CO_2e avoided per functional unit); highlight frontier	Ensure consistent boundaries/assumptions across points; label key design levers
Marginal abatement cost curve	Cost-effectiveness and total abatement potential	Bars sorted by $/t$CO_2e$ avoided/stored with cumulative potential on the *x*-axis	Beware of double-counting; clarify baseline and policy credit treatment
Multi-criteria decision analysis	Choices under multiple impacts, including financial and environmental impact indicators	Weighted score table or radar chart; show sensitivity to weights	Make weights transparent; report rank reversals under alternative weight sets
Consequential versus attributional LCA	Market-mediated effects versus inventory averages	Side-by-side results with clear system diagrams and displacement assumptions	Specify marginal products displaced and regions; avoid hidden substitution credits
Dynamic LCA	Timing of emissions/ storage vs. financial flows	Plots for 20y versus 100y GWPs; storage duration curves aligned to TEA cash-flow horizon	State discounting choices; show how near-term warming changes rankings
Uncertainty propagation	Robustness and key drivers of outcomes	Monte Carlo ranges for financial indicator and CO_2e; tornado/Sobol/spider charts for sensitivities	Report input distributions and data quality; separate aleatory versus epistemic uncertainty

Ultimately, combining TEA and LCA provides a holistic framework that links environmental integrity with economic viability, offering the evidence base required to guide policy support, attract investment, and scale carbon-negative bioproducts beyond demonstration projects.

5.4 Examples of bioproducts as long-term carbon storage

5.4.1 Biochar and pyrolysis products

5.4.1.1 Production systems and properties

Biochar is produced through pyrolysis, the thermal decomposition of biomass under oxygen-limited conditions at temperatures typically between 300 and 700 °C. Its yield and properties are strongly influenced by feedstock type, reactor design, and operating temperature. High-temperature pyrolysis (>500 °C) typically produces biochar with greater aromaticity, porosity, and fixed carbon content, which enhances long-term stability in soils but reduces overall yield and nutrient retention. Lower temperatures (300–400 °C) preserve more volatile compounds and minerals, resulting in biochar that is more beneficial for soil fertility and microbial enhancement [70].

Feedstock selection also influences biochar properties. Woody biomass tends to produce more carbon-dense, stable biochars, while herbaceous residues and manures generate materials richer in ash and nutrients. Post-production treatments, such as activation or ball milling, can further enhance surface area and reactivity, but often at the cost of reduced fixed carbon. These trade-offs underscore the necessity of tailoring pyrolysis conditions and feedstock choices to specific end uses, thereby balancing carbon sequestration, agronomic performance, and energy coproducts (Table 5.2) [70, 71].

Table 5.2: Summary of how pyrolysis conditions influence biochar properties and functions.

Pyrolysis condition	Carbon yield	Stability in soil	Nutrient content	Agronomic benefits
Low temperature (300–400 °C)	Higher mass yield (45–75% C retained); more volatile compounds retained	Moderate; less aromatic, more labile fractions	High; retains more minerals and volatile organics	Improves fertility, microbial activity, crop yields
Medium temperature (400–500 °C)	Intermediate yield (35–75% C retained)	Intermediate stability	Moderate	Balanced effects on fertility and stability

Table 5.2 (continued)

Pyrolysis condition	Carbon yield	Stability in soil	Nutrient content	Agronomic benefits
High temperature (>500 °C)	Lower mass yield (25–55% C retained); greater fixed carbon content	High; highly aromatic, recalcitrant carbon structures	Lower; fewer nutrients due to volatilization	Enhances soil structure, water retention, reduces GHG emissions

Lower temperatures favor nutrient-rich biochars with fertility benefits, while higher temperatures favor carbon-dense biochars with long-term stability.

Decentralized biochar systems, particularly those linked to agricultural or agro-industrial operations, can create closed-loop cycles where biomass waste is transformed into inputs for improved soil health and climate resilience. However, high production costs and variability in biochar quality remain key challenges. Overcoming these barriers will require improvements in reactor design, better feedstock logistics, and robust standards to ensure product consistency and efficacy [70]. In their comprehensive review, Yaashikaa et al. [70] critically examined the state of biochar production technologies, stability parameters, and diverse applications, offering valuable insights into its potential and limitations in carbon sequestration strategies.

5.4.1.2 Soil applications and co-benefits

When applied to soils, biochar provides both carbon storage and agronomic benefits. These include improved soil structure, enhanced water retention, reduced GHG emissions, and more efficient nutrient cycling. A meta-analysis found that biochar application increased crop yields by an average of ~10%, with the greatest benefits in acidic and sandy soils where its liming and water-holding effects are most impactful [72]. Biochars derived from nutrient-rich feedstocks, such as poultry litter, generally provide more pronounced yield gains than those produced from woody materials. Low-temperature biochar, rich in organic matter and minerals, can boost microbial activity and crop yields. In contrast, high-temperature biochar is more effective at improving soil physical properties and reducing GHG emissions such as nitrous oxide and methane, particularly in waterlogged or degraded soils [70].

These benefits, however, are highly context-dependent, varying with soil type, crop system, and biochar properties [70]. Positive outcomes are most consistent in degraded or low-input agricultural systems, particularly in tropical and subtropical regions. This variability underscores the importance of site-specific applications and decision-support tools to guide farmers and policymakers [73].

5.4.1.3 Carbon stability and permanence

The climate mitigation value of biochar depends on its ability to resist decomposition and retain carbon in soils over extended timescales. Biochar stability is primarily governed by its chemical composition and structural properties, which are strongly influenced by feedstock and pyrolysis conditions. High-temperature pyrolysis (>500 °C) produces biochars rich in condensed aromatic structures, reflected in low hydrogen-to-carbon (H:C) and oxygen-to-carbon (O:C) molar ratios. These compounds exhibit high resistance to microbial and oxidative degradation, often persisting for centuries to millennia. In contrast, low-temperature biochars contain more labile carbon fractions that are more readily mineralized [74].

In addition to chemical recalcitrance, physical mechanisms contribute to carbon stability. Biochar's high porosity and surface complexity can physically protect carbon from microbial access, while surface oxidation during soil aging enhances sorption of organic molecules and nutrients, effectively embedding biochar in the soil matrix. These processes suggest that biochar stability is not static but may increase over time as it becomes integrated with soil minerals and aggregates.

Evidence from both laboratory and field studies support the long-term persistence of biochar. A meta-analysis of 24 studies found that, on average, only ~3% of biochar carbon is mineralized within the first year of soil application, while the rest remains highly recalcitrant, supporting its function as a long-term carbon [75]. Ancient examples such as Amazonian terra preta soils, which contain charcoal-derived carbon that has remained stable for millennia, provide compelling analogs for the potential longevity of modern biochar [71]. However, the degree to which these natural precedents reflect engineered biochars under diverse environments remains uncertain.

Biochar also interacts with native soil organic matter (SOM) in ways that affect net carbon storage. In many cases, biochar additions have shown neutral or even negative priming effects, meaning they suppress SOM decomposition and thereby enhance total soil carbon stocks. Yet, in nutrient-poor or sandy soils, biochar can stimulate microbial activity through the labile carbon it contains, leading to positive priming effects that offset some sequestration benefits. This highlights the context dependence of biochar-SOM interactions and the need for models that account for both direct carbon stabilization and indirect priming effects.

Uncertainty remains regarding the true timescale of permanence. While short-term studies consistently indicate high stability, extrapolating decay dynamics beyond decades is inherently challenging. Model projections range from several centuries to over a thousand years, depending on assumptions about environmental conditions and the fraction of highly recalcitrant carbon. For climate policy and carbon markets, this uncertainty is consequential, since durable carbon removal requires sequestration on timescales that are functionally equivalent to fossil carbon emissions. Until more multi-decadal field data are available, permanence estimates must balance optimism from chemical and empirical evidence with conservative assumptions to ensure scientific credibility.

5.4.1.4 Measuring and crediting biochar carbon

Translating biochar's carbon stability into climate accounting requires reliable methods for measurement, reporting, and verification. Because direct observation of biochar persistence over centuries is impossible, permanence estimates rely on a combination of chemical proxies, laboratory incubations, field studies, and modeling approaches.

Laboratory decay models typically separate biochar carbon into multiple pools: a small labile fraction that decomposes rapidly, and a dominant aromatic fraction that is highly recalcitrant. These models are useful for estimating near-term dynamics but often fail to capture the complexity of soil environments, where moisture, pH, microbial community composition, and mineral interactions influence decomposition rates. Short timescales (months to years) in lab studies also limit their ability to predict long-term behavior [74].

Despite these limitations, chemical and structural indicators have become widely used in permanence assessments. Ratios such as H:C and O:C are correlated with aromaticity and resistance to degradation, providing a standardized metric for classifying biochars. Many carbon crediting methodologies, including those adopted by voluntary markets, use threshold values of H:C (<0.4) to define fractions expected to persist for centuries. Complementary approaches, such as solid-state NMR and Fourier-transform infrared (FTIR) spectroscopy, provide more detailed insights into molecular structure and oxidation state, though they are less practical for routine project verification [74].

The use of random reflectance methods adapted from coal geology has been proposed to estimate biochar permanence by comparing its optical properties to highly recalcitrant inertinite. While reflectance offers promise as a standardized, relatively low-cost metric, its analogy to coal is imperfect. Unlike buried coal, soil-applied biochar is continuously exposed to oxygen, water, ultraviolet light, and microbial activity. Without extensive field validation, reliance on reflectance risks systematic overestimation of permanence [74].

Divergent methodologies have led to substantial variation in credited carbon values for the same biochar. Some protocols adopt conservative 100–200-year persistence assumptions, while others allow crediting for >1,000 years if structural proxies or reflectance values are high. This lack of harmonization introduces risks of overcrediting, which could inflate the perceived climate benefit, or undercrediting, which may undermine project viability, especially for small-scale producers who depend on carbon revenues [74].

A balanced approach is therefore needed. Best practice emphasizes:

- Grounding permanence claims in measurable properties, such as H:C ratios and pyrolysis temperature
- Applying conservative crediting rules until long-term field validation is available

- Prioritizing the most stable carbon fractions (aromatic, mineral-associated) in crediting calculations
- Integrating multi-decadal field studies to refine model predictions and reduce uncertainty

Establishing robust, science-driven reporting and verification systems will be essential for scaling biochar within carbon markets. Only with transparent and conservative permanence assessments can biochar-based carbon removal be credibly positioned alongside other negative emissions technologies.

5.4.1.5 Broader context and emerging applications

Biochar fits naturally within circular bioeconomy frameworks, transforming organic residues into a stable carbon product while co-generating renewable energy. Estimates suggest that sustainable biochar production and application could mitigate up to ~12% of current anthropogenic GHG emissions [76]. Ensuring sustainability, however, requires avoiding competition with food production, safeguarding biodiversity, and adhering to protocols that maintain carbon stability.

In addition to soil applications, emerging uses of biochar, including incorporation into building materials, asphalt, and water treatment systems, offer durable, non-soil pathways for carbon storage. These applications bypass uncertainties tied to soil decomposition and may strengthen the role of biochar in carbon removal portfolios. As biochar scales within carbon markets, robust monitoring and verification frameworks will be essential to ensure that credited climate benefits are scientifically credible and durable.

5.4.2 Bioplastics and biopolymers

Biopolymers are produced by or derived from living organisms such as plants, animals, bacteria, fungi, and yeast. Compared to synthetic polymers, their main advantage lies in their sustainability potential. There is a wide range of bio-based polymers, including proteins, cellulose, and starch [77]. Some of these polymers can be used to produce bioplastics; however, not every bioplastic is produced from biopolymers.

For a polymeric material to be considered a bioplastic, it must either be bio-based or biodegradable. Products are considered bio-based when they are composed, in whole or in significant part, of renewable agricultural or forestry materials, while biodegradability depends on chemical structure rather than feedstock source. However, due to concerns about food competition, some proposed redefining bio-based products as commercial or industrial goods different from feed and food [78, 79]. Bioplastics arise as an alternative to decarbonize the plastic industry, as they can mitigate CO_2 emissions and potentially be economically viable replacements for fossil-based plastics [80].

5.4.2.1 Key examples of biopolymers

Bioplastics can be separated into three different groups: (i) bio-based and biodegradable, (ii) bio-based and durable, and (iii) fossil-based and biodegradable (Figure 5.3).

Bio-based and Biodegradable Plastics: Polylactic acid (PLA) and polyhydroxyalkanoate (PHA) have gained attention as they are part of the bio-based and biodegradable group. They are currently commercially available, despite their limitations for large-scale production and controversial food-versus-fuel competition with first-generation PLA production [81]. PLA can be produced via fermentation of sugars such as starch (first- and second-generation bioplastics) or by engineered microorganisms such as cyanobacteria (third-generation bioplastics). PHA, by contrast, is produced by microorganisms as part of their survival mechanism (third-generation bioplastics) [81, 82]. Both are considered carbon-capture materials, as they store biogenic carbon either from photosynthetic pathways of plants used as feedstocks or through microbial carbon uptake for energy production [78, 83]. However, when they biodegrade, this biogenic carbon is released back into the atmosphere, limiting their potential to store carbon long-term. Nevertheless, biodegradability is a desirable property for other sustainability reasons, including the increasing plastic and microplastic pollution in the environment. Other commonly used bio-based and biodegradable plastics include poly-3-hydroxybutyrate (PHB), polybutylene succinate (PBS), cellulose-based, starch-based, and protein-based [84].

Bio-based and Durable Plastics: These plastics are nonbiodegradable but share the same chemical structure and technical properties as their petrochemical equivalents. Bio-polyethylene (Bio-PE), for example, is polyethylene produced from ethylene derived from biomass-based alcohol. Other examples include bio-polyethylene terephthalate (Bio-PET), polysaccharide derivatives, and bio-polyamide (Bio-PA) [85].

Fossil-Based and Biodegradable Plastics: Examples include polybutylene adipate terephthalate (PBAT) and PBS, which can be produced either from petrochemical or renewable sources.

For this section, focus remains on bio-based counterparts and their biodegradability potential and EoL treatment, as these are the factors contributing to carbon sequestration [86, 87].

5.4.2.2 Carbon dynamics and climate benefits

Bio-based plastics play a crucial role in carbon capture, as they store biogenic carbon either from photosynthetic pathways of plants used as feedstocks or through microbial carbon uptake for energy production [78, 83]. They can therefore be seen as artificial reservoirs of carbon, maintaining denser and more stable forms than raw biomass [88].

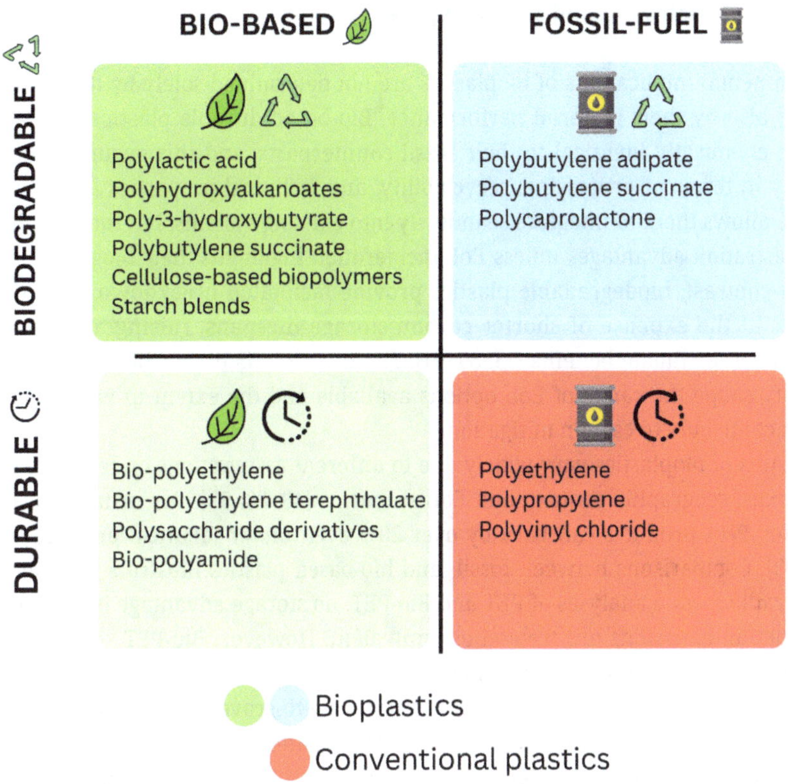

Figure 5.3: Classification of plastics by feedstock source (bio-based or fossil-based) and biodegradability (biodegradable or durable).

A central debate in LCA concerns whether and how to account for this biogenic carbon. Two main approaches exist: (i) carbon neutrality, in which biogenic CO_2 released at EoL is assumed to equal photosynthetic CO_2 fixation, resulting in a "net-zero" footprint and (ii) temporary carbon storage, in which biogenic carbon is retained for a period before release, delaying radiative forcing [13, 89].

Temporary storage benefits depend strongly on product lifespan. If a product expected to store carbon for one year fails prematurely, the sequestration target is not achieved. Conversely, longer-lived products can contribute to mitigation [90]. Nevertheless, reviews suggest that the overall effect on GWP is modest when biogenic carbon is rigorously accounted for [91]. Wide variability in reported results further illustrates that bioplastics can range from carbon-negative to net emitters depending on feedstock and production conditions [43].

5.4.2.3 End-of-life pathways and carbon implications

The environmental implications of bioplastics are not determined solely by their production but also by their material performance. Bio-based durable plastics such as Bio-PET are chemically identical to their fossil counterparts, and this ensures functional parity in terms of processing, recyclability, and technical properties [85]. This equivalence allows them to integrate seamlessly into existing systems, but limits additional sequestration advantages unless EoL incineration highlights their biogenic origin [92]. By contrast, biodegradable plastics provide facilitated breakdown, yet this often comes at the expense of shorter carbon storage lifespans, raising questions about their overall climate benefits [13, 89–91]. These trade-offs in durability and biodegradability shape the range of EoL options available and the extent to which bioplastics can contribute to carbon mitigation.

LCA results of bioplastics vary widely due to different assumptions on feedstocks, energy sources, geographic context, and EoL scenarios [43, 83, 93]. For instance, CO_2 emissions for PHA products can vary by over 20% even under identical production methods [93]. Comparisons between fossil- and bio-based plastics illustrate this variability. In cradle-to-gate analyses of PET and Bio-PET, no storage advantage is evident, since production emissions are treated as equivalent. However, Bio-PET sequesters approximately 50 kg of CO_2 per functional unit through biogenic uptake, a benefit hidden in cradle-to-gate assessments but revealed in cradle-to-grave scenarios where incineration highlights its biogenic origin [92]. A similar pattern appears in the case of polyethylene furanoate (PEF); bottles produced from Bio-PEF achieve up to 50% CO_2 savings compared to PET when incinerated at EoL, although this benefit declines to 26% if biogenic carbon uptake is excluded [94].

Beyond individual case studies, waste management strategies strongly influence carbon outcomes. The accepted hierarchy places reuse at the top, followed by mechanical recycling, chemical recycling, and biodegradation or incineration, with landfilling as the least favorable option [43]. Mechanical recycling generally shows the lowest global warming potential and the largest avoided emissions, with benefits compounding across multiple cycles [84, 94, 95]. Recycling also extends polymer lifetimes, effectively acting as temporary carbon storage. In the case of PLA, mechanical recycling can avoid approximately 0.5 million tons of CO_2-equivalent emissions, while chemical recycling and solvent-based recycling can achieve reductions of about 11.4 million tons and 29.1 million tons, respectively. These pathways outperform other EoL options, whereas industrial composting or anaerobic digestion may generate methane under uncontrolled conditions and thus negate sequestration benefits [94].

Biodegradability, while often presented as a key advantage, also carries important limitations. Biodegradable plastics fail to store carbon over long periods, and are thus limited in their potential for being carbon-negative. In addition, complete degradation typically requires controlled conditions such as those found in industrial composting facilities. Under ambient environments, bioplastics may persist for long periods and contribute to waste accumulation, much like conventional plastics [85, 86]. These dynamics demonstrate that the climate benefits of bioplastics depend not only on their chemistry but also, critically, on the systems available to manage them at the end of their life cycle.

5.4.2.4 Broader context and emerging applications

Bioplastics and biopolymers demonstrate carbon-negative potential by storing biogenic carbon and reducing reliance on fossil feedstocks. However, their climate benefit is highly dependent on feedstock choice, production energy mix, and especially EoL management. Current research and LCA studies remain concentrated on PLA, limiting insights into the broader spectrum of bio-based materials. Future work must focus on standardizing carbon accounting methods in LCAs to enable comparability across studies [13, 89–91]. Expanding analyses to underrepresented polymers is equally important to avoid PLA-centric conclusions [43]. Moreover, greater emphasis on testing real-world degradation and recycling pathways is needed to validate laboratory claims [94]. Finally, advances in microbial engineering and integration with carbon capture strategies hold promise for broadening the portfolio of carbon-sequestering polymers and scaling their production [81, 82].

5.4.3 Wood products, biocomposites, and construction materials

The construction sector has developed new materials to lower embodied emissions and serve as CCS strategies. Embodied emissions refer to the lifetime of the material, from production, transportation, construction, and EoL treatment [96]. CO_2 can be stored in construction materials either through in situ carbonation processes in cementitious materials (e.g., concrete and mortar) or through photosynthetic materials (e.g., wood, bamboo, and agricultural products) [97]. Since buildings can have a lifespan of up to 70 years, they also represent effective tools for long-term carbon storage. Substituting carbon-intensive materials such as steel and cement with bio-based alternatives can therefore play a central role in reducing embodied emissions.

5.4.3.1 Key bio-based construction materials

Carbonation-Based Cementitious Composites: The carbonation process is a natural process where concrete sequesters CO_2 from the environment as it interacts with calcium hydroxide and other hydrates in the mixture. This process lowers the pH of the concrete, negatively affecting the material and initiating corrosion in reinforced concrete [98]. As this process is inevitable, carbon credits can be accounted for in these materials; however, the uptake is very slow and sometimes incomplete. Therefore, accelerated carbonation during different cementitious phases is a more viable method to store CO_2. In this method, CO_2 is added during the mixing to form $CaCO_3$, accelerating the reaction and strengthening the material, as well as lowering cement content [99].

Solid Wood Products and Engineered Timber: Engineered timber has emerged as a cornerstone of carbon-storing construction. Products such as sawn timber, plywood, oriented strand board (OSB), and laminated veneer lumber (LVL) have long been used in buildings, but the development of cross-laminated timber (CLT) has expanded possibilities for large-scale applications. CLT panels, composed of layers of lumber boards glued at right angles, provide high strength-to-weight ratios, dimensional stability, and fire resistance suitable for mid- and high-rise buildings [100]. The substitution of concrete and steel with CLT in structural applications can reduce embodied emissions by 20–60%, depending on system boundaries and regional energy mixes [101]. Because CLT elements often remain in service for 50–100 years, they act as long-term carbon reservoirs, while also displacing energy-intensive materials. Challenges remain in adhesive sustainability, durability under humid conditions, and building code acceptance, but CLT has already been adopted in landmark projects across Europe, North America, and Australia, illustrating its scalability in carbon-negative architecture [100].

Bamboo-Based Products: Bamboo offers another promising bio-based alternative due to its fast growth, high tensile strength, and rapid carbon fixation. Laminated bamboo products, flooring, and structural beams have been developed to replace wood or steel in both residential and infrastructure projects. Bamboo culms can reach maturity within three to five years, providing shorter rotation cycles compared to traditional timber species and enabling continuous carbon uptake when cultivated sustainably [102]. Laminated bamboo lumber (LBL) has been shown to perform comparably to engineered wood products in terms of strength and stiffness, expanding its potential for structural use. In addition to carbon storage during its service life, bamboo cultivation contributes to rural economies in Asia, Latin America, and Africa, where it is already used for scaffolding, bridges, and housing [103]. However, the use of adhesives in engineered bamboo, durability under humid conditions, and limited global certification systems present challenges for widespread adoption.

Hempcrete: Hempcrete is an example of a biocomposite material that can store carbon through both carbonation and photosynthesis. It is a mixture of hemp shiv, a by-product of hemp fiber production, with a lime-based binder such as hydrated lime and natural hydraulic lime or ordinary Portland cement. These hydraulic binders accelerate the set time and are responsible for the carbonation process of hempcrete. When accounting for carbonation and biogenic carbon stored, hempcrete often shines as a carbon-negative material; however, depending on the binder chosen for the mix design, this is not always the case. It is important to note that this mixture has lower strength compared to other construction materials. Therefore, hempcrete is often utilized as an insulation material rather than as a structural material due to its low thermal conductivity [97].

Mycelium-Based Composites: Mycelium biocomposites are also used for insulation purposes, emerging as an alternative to synthetic insulation materials. This biodegradable material is produced by allowing fungi to bind together organic matter, such as agricultural and forestry by-products, with the mycelium. Studies have revealed that mycelium-wood biocomposites sequester more CO_2 than the emissions caused by the growth of the mycelium and material operation. Some LCA analyses do not count the metabolic CO_2 emitted by the fungi because it is of biogenic source; but when accounted, it represents 21% of the total embodied CO_2. Nevertheless, the material is still recognized as a carbon-negative product as it results in –39.5 kg/m^3 [104]. Since the material has a higher conductivity than most insulation products, more material quantity is needed to thicken the insulation layer and therefore, more CO_2 is stored. However, this material still has some limitations regarding the space needed to be functional [105].

Other Biocomposites: A biocomposite refers to a material composed of a matrix, derived from sustainable and/or nonrenewable resources, and a bio-based reinforcing component. This mixture is formed from two or more distinct constituent materials, improving performance over a single constituent material. In this structure, the matrix (e.g., bioplastic matrix) binds the reinforcing component together (e.g., natural fiber) to provide mechanical support and strength, allowing the material to be utilized in more applications. Determining the biomass used in biocomposites also has an effect on carbon storage. When using fast-growing plants like hemp or straw, shorter rotation periods are needed to grow more biomass, allowing them to store more CO_2 and reach carbon-negative results briefly after installation. Alternatively, wood-derived products provide advantages in mechanical properties and also provide high-quality nutrients for mycelium composites [106].

5.4.3.2 Carbon storage dynamics

Biogenic carbon fixation and its transfer into construction products reinforce the climate benefits of bio-based materials. Hempcrete and mycelium composites demonstrate how both carbonation and photosynthetic uptake can be combined, while wood and bamboo offer long-term storage in structural applications. Short-lived products such as insulation panels store carbon temporarily, whereas long-lived materials such as CLT or laminated bamboo beams can act as durable reservoirs for several decades. Carbon accounting differs depending on perspective: cradle-to-gate assessments may overlook biogenic uptake, while cradle-to-grave boundaries capture substitution effects, reuse potential, and delayed release.

5.4.3.3 Performance and durability considerations

The performance of bio-based construction materials depends on their structural properties and durability. Hempcrete, while effective for thermal insulation, has lower mechanical strength than conventional materials. Mycelium composites face limitations related to conductivity and thickness requirements. Wood-derived products and bamboo offer superior structural performance but require protective treatments to address risks such as fire, moisture, and biodegradation. EoL pathways, including reuse, recycling, energy recovery, and biodegradation, further influence the climate balance of these materials, with reuse and recycling generally offering the most favorable outcomes.

5.4.3.4 Case studies in carbon-storing architecture

Overall, integrating biocomposites into construction offers an opportunity to transform buildings and infrastructure into long-term carbon reservoirs. Materials such as hempcrete and mycelium composites demonstrate how carbon can be stored through both biogenic uptake and material carbonation, while also expanding the range of sustainable insulation options. A case study on bamboo bio-concrete illustrates that when used as reinforcement in conventional concrete, bamboo fibers extend service life and enable mineralization of embedded biomass, resulting in a 36% reduction in carbon footprint compared to traditional mixes. It is concluded that the use of bio-based components improves carbon emissions and should be extended to other construction components [107]. Such findings highlight the potential of bio-based construction materials not only to reduce embodied emissions but also to act as carbon sinks, making them a strategic component in the global effort to achieve climate goals.

5.4.3.5 Broader context and emerging applications

Despite their potential, several challenges remain for bio-based construction materials. Biomass sourcing raises land-use concerns, such as sustainable forestry, bamboo cultivation, and hemp production must avoid competition with food systems and biodiversity loss. Material certification and building code acceptance are also hurdles, since many bio-based materials fall outside established regulatory frameworks. In addition, transport and processing emissions may reduce net climate benefits, while durability issues such as moisture sensitivity or limited mechanical strength demand careful engineering solutions.

At the same time, wood products, biocomposites, and other bio-based materials can significantly contribute to carbon-negative strategies by storing biogenic carbon and substituting carbon-intensive materials. However, consistent and standardized carbon accounting remains a challenge, particularly in construction LCAs, where methodological inconsistencies limit comparability. Expanding research to cover underrepresented bio-based products and to test real-world performance under diverse climatic and infrastructural conditions is therefore essential. Looking ahead, promising directions include integrating bio-based materials with CCS, applying circular design principles to extend material lifespans, and developing multifunctional composites that combine structural, insulation, and carbon-storage functions.

5.5 Outlook and future directions

5.5.1 Summary of implementation challenges

Throughout this chapter, we have discussed the potential implementation of bioproducts that not only provide added value but also allow for carbon removal. The challenges for the realization of these benefits have also been covered. In summary, the implementation of carbon-negative bioproducts is limited, among others, by technical hurdles, feedstock availability and sustainability, and transparency and validation of monitoring and assessment tools.

Many pathways face technical constraints related to yield, process efficiency, and the stability of stored carbon. Biochar production, for example, varies widely with feedstock and reactor conditions, making long-term permanence uncertain. Bioplastics and biocomposites are often restricted by high production costs, modest performance compared to fossil counterparts, and limited data on their carbon storage potential at scale. Scalability remains a major hurdle, as decentralized systems must balance economies of scale with transportation emissions and logistics.

The availability of biomass that is produced sustainably is also a challenge. Residues and wastes are widely cited as "ideal" feedstocks but are geographically uneven and often serve competing uses, including soil fertility, livestock bedding, and bioenergy. Purpose-grown crops can increase supply but raise risks of land-use change, biodiversity loss, and water demand.

LCA is central to evaluating climate impacts, yet inconsistent system boundaries, allocation rules, and functional units create wide variation in reported results. A particularly contentious issue is defining the starting point for carbon accounting, whether to count removals beginning at photosynthesis, harvest, or post-processing, since different assumptions can substantially alter sequestration claims. Limited access to open data and models further reduces reproducibility and credibility.

5.5.2 Policy and markets

Scaling carbon-negative bioproducts requires more than technological progress; it depends on enabling standards, incentives, and procurement strategies. These are critical to not only ensure the realization of the technologies at scale, but also to combat public skepticism of carbon markets and uncertainty around permanence, which undermines confidence in carbon crediting.

Credible measurement, reporting, and verification protocols are essential to ensure that product-based carbon removal claims are robust. Protocols for biochar, engineered timber, and biogenic CO_2 streams must address permanence, additionality, and EoL fate. Harmonized rules on system boundaries and time horizons are needed to align LCA with carbon crediting frameworks and avoid double-counting.

Bioproducts often rely on stacked value streams, which include product revenues, low-carbon premiums, and carbon credits. Effective policy tools include production tax credits, contracts-for-difference, and public procurement programs that reward durability and verified storage. Carbon markets are beginning to recognize biochar and biogenic CO_2 storage, but consistent rules across voluntary and compliance markets are still evolving. Building codes, material standards, and green public procurement can rapidly expand markets for bio-based construction materials. Embodied carbon reporting and carbon storage "adders" for public projects can accelerate adoption of engineered timber, hempcrete, and biocomposites.

Expanding supply requires investment in feedstock logistics, modular conversion technologies, and material quality standards. At the same time, safeguards must prevent negative impacts from land-use change, biodiversity loss, and water use. Prioritizing residues and marginal-land crops, and embedding circular design principles (reuse, cascading use, nutrient recycling), can mitigate risks while improving local acceptance.

5.6 Future research directions

Future progress on carbon-negative bioproducts requires strengthening the technical and empirical evidence base. Long-term field data are needed to verify the stability of biochar in soils, the service lifetimes of engineered timber and bamboo in construction, and the durability of bio-based composites under real-world conditions. At the same time, more empirical data on EoL outcomes, including recycling rates, landfill gas capture, and incineration with CCS, are critical to replace generic assumptions in LCA models.

Embedding bioproducts into carbon-smart infrastructure and durable consumer goods represents another key opportunity. The built environment is especially promising, as engineered timber, bamboo, hempcrete, and biochar-concrete can serve as long-term carbon reservoirs while displacing emission-intensive materials such as steel and cement. Consumer goods designed for long service lives and recyclability can also extend storage and reduce dependence on virgin feedstocks.

Integrated TEA-LCA frameworks should become standard practice, aligning assumptions on boundaries, allocation, and functional units while explicitly addressing uncertainty. More open-source models and transparent datasets are needed to improve comparability and build confidence in the climate benefits of bioproducts. Improved accounting methods that reflect the timing of carbon storage and release, such as dynamic LCA, will make sequestration claims more accurate and comparable. Importantly, future work must also consider the socioeconomic and equity dimensions of scaling these systems, ensuring that rural communities benefit from new markets, land-use competition does not exacerbate inequities, and policies support fair value distribution across supply chains.

Emerging fields such as synthetic biology and artificial intelligence can further expand the potential of carbon-negative bioproducts. Synthetic biology offers new routes to engineer microbes for enhanced production of biopolymers, fuels, and specialty chemicals, including the development and improvement of photosynthetic microorganisms that can produce high-value products. AI can accelerate process discovery, optimize conversion systems, and improve predictive modeling for process development or systems analysis. Combined with innovations in process engineering, these tools can drive efficiency gains and cost reductions while supporting rigorous environmental performance.

Looking forward, the overarching goal is to position bioproducts as cornerstones of net-negative supply chains. Linking carbon sequestration with circular economy models, through cascading material use, recycling loops, and nutrient return, can amplify climate benefits while generating valuable commodities. Realizing this vision will also require advances in long-term monitoring and verification, including cost-effective sensing, tracking, and reporting systems that ensure storage claims remain credible over decades. By embedding carbon storage directly into industrial systems, bioproducts can evolve from niche applications into durable pillars of climate mitigation.

References

[1] Venkata Mohan S, Modestra JA, Amulya K, Butti SK, Velvizhi G. A circular bioeconomy with biobased products from CO_2 sequestration. Trends Biotechnol 2016.

[2] Harris ZM, Milner S, Taylor G. Biogenic carbon – capture and sequestration. Greenh Gas Bal Bioenergy Syst 2018;55–76.

[3] Yang X, Liu D, Lu H, Weston DJ, Chen JG, Muchero W, Martin S, Liu Y, Hassan MM, Yuan G, Kalluri UC, Tschaplinski TJ, Mitchell JC, Wullschleger SD, Tuskan GA. Biological parts for plant biodesign to enhance land-based carbon dioxide removal. BioDesign Res 2021.

[4] Dees JP, Sagues WJ, Woods E, Goldstein HM, Simon AJ, Sanchez DL. Leveraging the bioeconomy for carbon drawdown. Green Chem 2023;25(8):2930–57.
[5] Jiang P, Zhao G, Liu L, Zhang H, Mu L, Lu X, Zhu J. A negative-carbon footprint process with mixed biomass feedstock maximizes conversion efficiency, product value and CO_2 mitigation. Bioresour Technol 2022;351:127004.
[6] Schweitzer H, Aalto NJ, Busch W, Chat Chan DT, Chiesa M, Elvevoll EO, Gerlach R, Krause K, Mocaer K, Moran JJ, Noel JP, Patil SK, Schwab Y, Wijffels RH, Wulff A, Øvreås L, Bernstein HC. Innovating carbon-capture biotechnologies through ecosystem-inspired solutions. One Earth 2021;4(1):49–59.
[7] Calvin K, Dasgupta D, Krinner G, Mukherji A, Thorne PW, Trisos C, Romero J, Aldunce P, Barrett K, Blanco G, Cheung WWL, Connors S, Denton F, Diongue-Niang A, Dodman D, Garschagen M, Geden O, Hayward B, Jones C, Jotzo F, Krug T, Lasco R, Lee -Y-Y, Masson-Delmotte V, Meinshausen M, Mintenbeck K, Mokssit A, Otto FEL, Pathak M, Pirani A, Poloczanska E, Pörtner H-O, Revi A, Roberts DC, Roy J, Ruane AC, Skea J, Shukla PR, Slade R, Slangen A, Sokona Y, Sörensson AA, Tignor M, van Vuuren D, Wei Y-M, Winkler H, Zhai P, Zommers Z, Hourcade J-C, Johnson FX, Pachauri S, Simpson NP, Singh C, Thomas A, Totin E, Alegría A, Armour K, Bednar-Friedl B, Blok K, Cissé G, Dentener F, Eriksen S, Fischer E, Garner G, Guivarch C, Haasnoot M, Hansen G, Hauser M, Hawkins E, Hermans T, Kopp R, Leprince-Ringuet N, Lewis J, Ley D, Ludden C, Niamir L, Nicholls Z, Some S, Szopa S, Trewin B, van der Wijst K-I, Winter G, Witting M, Birt A, Ha M. IPCC, 2023: Climate Change 2023: Synthesis Report. Contribution of Working Groups I, II and III to the Sixth Assessment Report of the Intergovernmental Panel on Climate Change [Core Writing Team, H. Lee and J. Romero (Eds.)], *Geneva, Switzerland*: *IPCC*; 2023.
[8] Markard J, Rosenbloom D. Phases of the net-zero energy transition and strategies to achieve it. In: Araújo K, editor. Routledge Handbook of Energy Transitions, London: Routledge; 2022. pp. 102–23.
[9] Rockström J, Gaffney O, Rogelj J, Meinshausen M, Nakicenovic N, Schellnhuber HJ. A roadmap for rapid decarbonization. Science 2017;355(6331):1269–71.
[10] Rueda O, Mogollón JM, Tukker A, Scherer L. Negative-emissions technology portfolios to meet the 1.5 °C target. Glob Environ Change 2021;67:102238.
[11] Wollnik R, Borchers M, Seibert R, Abel S, Herrmann P, Elsasser P, Hildebrandt J, Meisel K, Hofmann P, Radtke K, Selig M, Kazmin S, Szarka N, Thrän D. Dynamics of bio-based carbon dioxide removal in Germany. Sci Rep 2024;14(1).
[12] Sandalow D, Aines R, Friedmann J, McCormick C, Sanchez D. Biomass Carbon Removal and Storage (BiRCS) Roadmap, Livermore, CA (United States); 2021.
[13] Narayan R. Carbon footprint of bioplastics using biocarbon content analysis and life-cycle assessment. MRS Bull 2011;36(9):716–21.
[14] Scown CD. Prospects for carbon-negative biomanufacturing. Trends Biotechnol 2022;40 (12):1415–24.
[15] Ferchaud F, Vitte G, Mary B. Changes in soil carbon stocks under perennial and annual bioenergy crops. GCB Bioenergy 2016;8(2):290–306.
[16] Tripathi N, Hills CD, Singh RS, Atkinson CJ. Biomass waste utilisation in low-carbon products: Harnessing a major potential resource. Npj Clim Atmos Sci 2019;2(1):1–10.
[17] Tomar S, Agarwal S, Singh H, Kumar R, Qureshi KA, Jaremko M, Emwas AH, Rai PK. Microalgae: A promising source for biofuel production. Biocatal Agric Biotechnol 2023;53:102877.
[18] Katiyar R, Gurjar BR, Biswas S, Pruthi V, Kumar N, Kumar P. Microalgae: An emerging source of energy based bio-products and a solution for environmental issues. Renew Sustain Energy Rev 2017;72:1083–93.
[19] Fallahi A, Rezvani F, Asgharnejad H, Khorshidi E, Hajinajaf N, Higgins B. Interactions of microalgae-bacteria consortia for nutrient removal from wastewater: A review. Chemosphere 2021;272:129878.
[20] Kalluri UC, Yang X, Wullschleger SD. Plant biosystems design for a carbon-neutral bioeconomy. BioDesign Res 2020;2020:7914051.

[21] Bonin C, Lal R. Agronomic and ecological implications of biofuels. Adv Agron 2012;117:1–50.

[22] Zuiderveen EAR, Kuipers KJJ, Caldeira C, Hanssen SV, van der Hulst MK, de Jonge MMJ, Vlysidis A, van Zelm R, Sala S, Huijbregts MAJ. The potential of emerging bio-based products to reduce environmental impacts. Nat Commun 2023;14(1):1–7.

[23] Bazyar P, Sheidaee E. Algae-based carbon sequestration optimizing renewable energy and climate strategy. Adv AI Simul Optim Energy Syst 2025;83–108.

[24] Sarwer A, Hamed SM, Osman AI, Jamil F, Al-Muhtaseb AH, Alhajeri NS, Rooney DW. Algal biomass valorization for biofuel production and carbon sequestration: A review. Environ Chem Lett 2022;20 (5):2797–851.

[25] Prasad S, Ingle AP. Impacts of sustainable biofuels production from biomass. Sustain Bioenergy Adv Impacts 2019;327–46.

[26] Negrete-Bolagay D, Guerrero VH. Opportunities and challenges in the application of bioplastics: Perspectives from formulation, processing, and performance. Polymers 2024;16(18):2561.

[27] Sudheshwar A, Vogel K, Nyström G, Malinverno N, Arnaudo M, Gómez Camacho CE, Beloin-Saint-Pierre D, Hischier R, Som C. Unraveling the climate neutrality of wood derivatives and biopolymers. RSC Sustain 2024;2(5):1487–97.

[28] Zhao J, Wei X, Li L. The potential for storing carbon by harvested wood products. Front For Glob Change 2022;5:1055410.

[29] Pomponi F, Hart J, Arehart JH, D'Amico B. Buildings as a global carbon sink? A reality check on feasibility limits. One Earth 2020;3(2):157–61.

[30] Gu H, Nepal P, Arvanitis M, Alderman D, Gu H, Nepal P, Arvanitis M, Alderman D. Carbon impacts of engineered wood products in construction. Eng Wood Prod Constr 2021.

[31] Fawzy S, Osman AI, Yang H, Doran J, Rooney DW. Industrial biochar systems for atmospheric carbon removal: A review. Environ Chem Lett 2021;19(4):3023–55.

[32] Makepa DC, Chihobo CH. Sustainable pathways for biomass production and utilization in carbon capture and storage – A review. Biomass Convers Biorefin 2024;15(8):11397–419.

[33] Osman AI, Hefny M, Abdel Maksoud MIA, Elgarahy AM, Rooney DW. Recent advances in carbon capture storage and utilisation technologies: A review. Environ Chem Lett 2021.

[34] Aghbashlo M, Hosseinzadeh-Bandbafha H, Shahbeik H, Tabatabaei M. The role of sustainability assessment tools in realizing bioenergy and bioproduct systems. Biofuel Res J 2022;9(3):1697–706.

[35] International Organization for Standardization. ISO 14040:2006 – Environmental Management – Life Cycle Assessment – Principles and Framework, 2006.

[36] Van Den Berg M, Hof AF, Van Vliet J, Van Vuuren DP. Impact of the choice of emission metric on greenhouse gas abatement and costs. Environ Res Lett 2015;10(2):024001.

[37] Miller JS, Dreyfus GB, Daniel JS, Xu Y. Assessing alternative methane emission metrics conducive to quantifying global warming and setting near-term climate goals. Environ Res Lett 2025;20 (5):054052.

[38] U.S. DOE. Best Practices for Life Cycle Assessment (LCA) of Direct Air Capture with Storage (DACS), Washington, DC; 2022.

[39] U.S. DOE. Best Practices for Life Cycle Assessment (LCA) of Biomass Carbon Removal and Storage (Bicrs) Technologies, Washington, DC: SW; 2025.

[40] Sander-Titgemeyer A, Risse M, Weber-Blaschke G. Applying an iterative prospective LCA approach to emerging wood-based technologies: Three German case studies. Int J Life Cycle Assess 2023;28 (5):495–515.

[41] O'Malley J, Searle S, Pavlenko N. Indirect Emissions from Waste and Residue Feedstocks: 10 Case Studies from the United States, Washington, DC; 2021.

[42] Crutzen PJ, Mosier AR, Smith KA, Winiwarter W. N_2O release from agro-biofuel production negates global warming reduction by replacing fossil fuels. In: Paul J. Crutzen: A Pioneer on Atmospheric Chemistry and Climate Change in the Anthropocene, Springer International Publishing Cham; 2016. pp. 227–38.

[43] Spierling S, Venkatachalam V, Mudersbach M, Becker N, Herrmann C, Endres H-J. End-of-life options for bio-based plastics in a circular economy – status quo and potential from a life cycle assessment perspective. Resources 2020;9(7):90.

[44] Qian Y, Gunawardena T, Mendis P, Aye L. Carbon footprint variability in engineered wood products for timber buildings: A systematic review of carbon accounting methodologies. Sustainability 2025;17(11):4804.

[45] Farjana SH, Tokede O, Tao Z, Ashraf M. Life cycle assessment of end-of-life engineered wood. Sci Total Environ 2023;887:164018.

[46] Han L, Sun K, Yang Y, Xia X, Li F, Yang Z, Xing B. Biochar's stability and effect on the content, composition and turnover of soil organic carbon. Geoderma 2020;364:114184.

[47] Walker S, Rothman R. Life cycle assessment of bio-based and fossil-based plastic: A review. J Clean Prod 2020;261:121158.

[48] Navare K, Arts W, Faraca G, Bossche GVD, Sels B, Acker KV. Environmental impact assessment of cascading use of wood in bio-fuels and bio-chemicals. Resour Conserv Recycl 2022;186:106588.

[49] Knapp BA, Insam H. Recycling of biomass ashes: Current technologies and future research needs. Recycl Biomass Ashes 2011;1–16.

[50] Lower L, Cunniffe J, Cheng JJ, Sagues WJ. Coupling circularity with carbon negativity in food and agriculture systems. J ASABE 2022;65(4):849–64.

[51] Scown CD, Baral NR, Yang M, Vora N, Huntington T. Technoeconomic analysis for biofuels and bioproducts. Curr Opin Biotechnol 2021;67:58–64.

[52] Wu Y, Restrepo-Flórez JM, Vasco-Correa J. A superstructure-based lignin valorization process optimization model for lignocellulosic biorefineries through biological upgrading. ACS Sustain Chem Eng 2024;12(18):6913–26.

[53] Gong J, You F. Optimal design and synthesis of algal biorefinery processes for biological carbon sequestration and utilization with zero direct greenhouse gas emissions: MINLP model and global optimization algorithm. Ind Eng Chem Res 2014;53(4):1563–79.

[54] Strunge T, Renforth P, Van der Spek M. Uncertainty quantification in the techno-economic analysis of emission reduction technologies: A tutorial case study on CO_2 mineralization. Front Energy Res 2023;11:1182969.

[55] Roussanaly S, Ouassou JA, Anantharaman R, Haaf M. Impact of uncertainties on the design and cost of CCS from a waste-to-energy plant. Front Energy Res 2020;8:504262.

[56] Li Q, Hu G. Techno-economic analysis of biofuel production considering logistic configurations. Bioresour Technol 2016;206:195–203.

[57] Faber G, Ruttinger A, Strunge T, Langhorst T, Zimmermann A, van der Hulst M, Bensebaa F, Moni S, Tao L. Adapting technology learning curves for prospective techno-economic and life cycle assessments of emerging carbon capture and utilization pathways. Front Clim 2022;4:820261.

[58] Biddy MJ, Davis R, Humbird D, Tao L, Dowe N, Guarnieri MT, Linger JG, Karp EM, Salvachúa D, Vardon DR, Beckham GT. The techno-economic basis for coproduct manufacturing to enable hydrocarbon fuel production from lignocellulosic biomass. ACS Sustain Chem Eng 2016;4 (6):3196–211.

[59] Kosamia NM, Sanchez A, Rakshit SK. Scenario-based techno-economics and heat integration feasibility assessment of integrated multiproduct biorefineries with biosuccinic acid as the main product and various byproduct options. Biomass Convers Biorefin 2024;14(7):8729–43.

[60] Yang M, Baral NR, Anastasopoulou A, Breunig HM, Scown CD. Cost and life-cycle greenhouse gas implications of integrating biogas upgrading and carbon capture technologies in cellulosic biorefineries. Environ Sci Technol 2020;54(20):12810–19.
[61] Stewart DW, Cortés-Peña YR, Li Y, Stillwell AS, Khanna M, Guest JS. Implications of biorefinery policy incentives and location-specific economic parameters for the financial viability of biofuels. Environ Sci Technol 2023;57(6):2262–71.
[62] Mahmud R, Moni SM, High K, Carbajales-Dale M. Integration of techno-economic analysis and life cycle assessment for sustainable process design – A review. J Clean Prod 2021;317:128247.
[63] Meramo S, Fantke P, Sukumara S. Advances and opportunities in integrating economic and environmental performance of renewable products. Biotechnol Biofuels Bioprod 2022;15(1):1–18.
[64] Vasilakou K, Billen P, Van Passel S, Nimmegeers P. A Pareto aggregation approach for environmental-economic multi-objective optimization applied on a second-generation bioethanol production model. Energy Convers Manag 2024;303:118184.
[65] You F, Tao L, Graziano DJ, Snyder SW. Optimal design of sustainable cellulosic biofuel supply chains: Multiobjective optimization coupled with life cycle assessment and input–output analysis. AIChE J 2012;58(4):1157–80.
[66] Liu Z, Qiu T, Chen B. A study of the LCA based biofuel supply chain multi-objective optimization model with multi-conversion paths in China. Appl Energy 2014;126:221–34.
[67] Bose A, O'Shea R, Lin R, Long A, Rajendran K, Wall D, De S, Murphy JD. The marginal abatement cost of co-producing biomethane, food and biofertiliser in a circular economy system. Renew Sustain Energy Rev 2022;169:112946.
[68] Shackley S, Hammond J, Gaunt J, Ibarrola R. The feasibility and costs of biochar deployment in the UK. Carbon Manag 2011;2(3):335–56.
[69] Beattie A, Vermaas W, Darzins A, Holland SC, Li S, McGowen J, Nielsen D, Quinn JC. A probabilistic economic and environmental impact assessment of a cyanobacteria-based biorefinery. Algal Res 2021;59:102454.
[70] Yaashikaa PR, Kumar PS, Varjani S, Saravanan A. A critical review on the biochar production techniques, characterization, stability and applications for circular bioeconomy. Biotechnol Rep 2020.
[71] Lehmann J. Bio-energy in the black. Front Ecol Environ 2007;5(7):381–87.
[72] Jeffery S, Verheijen FGA, van der Velde M, Bastos AC. A quantitative review of the effects of biochar application to soils on crop productivity using meta-analysis. Agric Ecosyst Environ 2011.
[73] Schmidt HP, Kammann C, Hagemann N, Leifeld J, Bucheli TD, Sánchez Monedero MA, Cayuela ML. Biochar in agriculture – A systematic review of 26 global meta-analyses. GCB Bioenergy 2021.
[74] Belmont EL, Sanchez DL. Unraveling the Science of Carbon Permanence, 2025.
[75] Wang J, Xiong Z, Kuzyakov Y. Biochar stability in soil: Meta-analysis of decomposition and priming effects. GCB Bioenergy 2016.
[76] Woolf D, Amonette JE, Street-Perrott FA, Lehmann J, Joseph S. Sustainable biochar to mitigate global climate change. Nat Commun 2010;1(5).
[77] Liu L, Liang W, Zhang Y, Fu Q. Nanoencapsulation in polymeric materials: Weaving magical coats for microorganisms. Nano Today 2023.
[78] Mekonnen T, Mussone P, Khalil H, Bressler D. Progress in bio-based plastics and plasticizing modifications. J Mater Chem A 2013;1(43):13379–98.
[79] Greene JP. Degradation and biodegradation standards for biodegradable food packaging materials. In: Reference Module in Food Science, Elsevier; 2019.
[80] Nanda S, Patra BR, Patel R, Bakos J, Dalai AK. Innovations in applications and prospects of bioplastics and biopolymers: A review. Environ Chem Lett 2022.
[81] Koch M, Spierling S, Venkatachalam V, Endres HJ, Owsianiak M, Vea EB, Daffert C, Neureiter M, Fritz I. Comparative assessment of environmental impacts of 1st generation (corn feedstock) and 3rd

generation (carbon dioxide feedstock) PHA production pathways using life cycle assessment. Sci Total Environ 2023;863.

[82] Tan C, Tao F, Xu P. Direct carbon capture for the production of high-performance biodegradable plastics by cyanobacterial cell factories. Green Chem 2022;24(11):4470–83.

[83] Zong Z, Rao C, Du C, Lu R, Upham DC. Polyhydroxyalkanoates (PHA) production in a circular CO_2 economy: It's role in mitigating global CO_2 emissions. Resour Conserv Recycl 2025;219.

[84] Changwichan K, Silalertruksa T, Gheewala SH. Eco-efficiency assessment of bioplastics production systems and end-of-life options. Sustainability 2018;10(4).

[85] Nizamuddin S, Baloch AJ, Chen C, Arif M, Mubarak NM. Bio-based plastics, biodegradable plastics, and compostable plastics: Biodegradation mechanism, biodegradability standards and environmental stratagem. Int Biodeterior Biodegrad 2024.

[86] Fredi G, Dorigato A. Recycling of bioplastic waste: A review. Adv Ind Eng Polym Res 2021.

[87] Luo C, Zhou Y, Chen Z, Bian X, Chen N, Li J, Wu Y, Yang Z. Comparative life cycle assessment of PBAT from fossil-based and second-generation generation bio-based feedstocks. Sci Total Environ 2024;954.

[88] Sun X, Xie M, Mai L, Zeng EY. Biobased plastic: A plausible solution toward carbon neutrality in plastic industry? J Hazard Mater 2022;435.

[89] Pawelzik P, Carus M, Hotchkiss J, Narayan R, Selke S, Wellisch M, Weiss M, Wicke B, Patel MK. Critical aspects in the life cycle assessment (LCA) of bio-based materials – Reviewing methodologies and deriving recommendations. Resour Conserv Recycl 2013;73:211–28.

[90] Jørgensen SV, Hauschild MZ, Nielsen PH. The potential contribution to climate change mitigation from temporary carbon storage in biomaterials. Int J Life Cycle Assess 2015;20(4):451–62.

[91] Bishop G, Styles D, Lens PNL. Environmental performance comparison of bioplastics and petrochemical plastics: A review of life cycle assessment (LCA) methodological decisions. Resour Conserv Recycl 2021.

[92] Keyes A, Saffron CM, Manjure S, Narayan R. Biobased compostable plastics end-of-life: Environmental assessment including carbon footprint and microplastic impacts. Polymers 2024;16(21).

[93] Lee S, Lee I, Seo D, Kim H, Joo G, Lee S, Park K. Life cycle assessment of aPHA production. ACS Sustain Chem Eng 2024;12(1):72–84.

[94] Stegmann P, Gerritse T, Shen L, Londo M, Puente Á, Junginger M. The global warming potential and the material utility of PET and bio-based PEF bottles over multiple recycling trips. J Clean Prod 2023;395.

[95] Merchan AL, Fischöder T, Hee J, Lehnertz MS, Osterthun O, Pielsticker S, Schleier J, Tiso T, Blank LM, Klankermayer J, Kneer R, Quicker P, Walther G, Palkovits R. Chemical recycling of bioplastics: Technical opportunities to preserve chemical functionality as path towards a circular economy. Green Chem 2022.

[96] Amiri A, Emami N, Ottelin J, Sorvari J, Marteinsson B, Heinonen J, Junnila S. Embodied emissions of buildings – A forgotten factor in green building certificates. Energy Build 2021;241:110962.

[97] Arehart JH, Nelson WS, Srubar WV. On the theoretical carbon storage and carbon sequestration potential of hempcrete. J Clean Prod 2020;266.

[98] Bajoria A, Kanpariya J, Bera A. Introduction to carbonation mechanisms and applications. In: Advances and Technology Development in Greenhouse Gases: Emission, Capture and Conversion, Elsevier; 2024. pp. 359–76.

[99] Winnefeld F, Leemann A, German A, Lothenbach B. CO_2 storage in cement and concrete by mineral carbonation. Curr Opin Green Sustain Chem 2022.

[100] Pei S, Lindt JWVD, Popovski M, Berman JW, Dolan JD, Ricles J, Sause R, Blomgren H, Rammer DR. Cross-laminated timber for seismic regions: Progress and challenges for research and implementation. J Struct Eng 2014;142(4):E2514001.

[101] Chen Z, Gu H, Bergman RD, Liang S. Comparative life-cycle assessment of a high-rise mass timber building with an equivalent reinforced concrete alternative using the athena impact estimator for buildings. Sustainability 2020;12(11):4708.

[102] Pan C, Zhou G, Shrestha AK, Chen J, Kozak R, Li N, Li J, He Y, Sheng C, Wang G. Bamboo as a nature-based solution (NbS) for climate change mitigation: Biomass, products, and carbon credits. Climate 2023;11(9):175.

[103] Terefe R, Jian L, Kunyong Y. Role of bamboo forest for mitigation and adaptation to climate change challenges in China. J Sci Res Rep 2019;1–7.

[104] Livne A, Wösten HAB, Pearlmutter D, Gal E. Fungal mycelium bio-composite acts as a CO_2-sink building material with low embodied energy. ACS Sustain Chem Eng 2022;10(37):12099–106.

[105] Candido A, Amiri A, Junnila S, Pittau F. Mycelium-wood composites as a circular material for building insulation. Front Sustain Cities 2024;6.

[106] Habert G, Iannaccone G, Pittau F. The renovation of the building stock in Europe: An essential opportunity to store carbon in buildings. Rivista Tema 2021;05(02).

[107] Correa de Melo P, Caldas LR, Masera G, Pittau F. The potential of carbon storage in bio-based solutions to mitigate the climate impact of social housing development in Brazil. J Clean Prod 2023;433.

Forrest Stagner*, Chris Morris, Michael Helbing, Frank Montabon, Christine Costello, and Lara Fowler

Chapter 6 Bioproduct policy: an overview and case study of farm-based biogas production in the United States

6.1 Introduction

6.1.1 Defining policy

The acceptance and integration of a new technology into a market or population cannot be fully understood without also understanding the interconnected political, economic, and social contexts related to the technology. Among a multitude of contextual factors, one of the most influential can be that of the role of policy. A challenge exists in defining the term "policy" due to the many various contexts in which the word is used. Policy can be formal or informal, legal or nonlegal, private or public, and policy can be created at a variety of scales (e.g., local, state, national, international) [1]. For the purposes of this chapter, policy is defined as "[a] standard course of action that has been officially established by an organization, business, political party, etc." [2]. A related but distinct term is law, which refers to "[t]he set of rules or principles dealing with a specific area of a legal system" or a "statute" [2]. Policies are often codified as law, but this is not always the case (such as with private sector policy established within a corporation). Public policy at the state and national levels is most often created by elected officials (e.g., the U.S. Congress) or an elected or appointed governing body (e.g., the California Air Resources Board). Policy created by these bodies is then usually interpreted and implemented by state or federal agencies or other political organizations.

6.1.2 The role of policy in technology development

Policy can be critically important to the development and use of a particular technology because it often defines and regulates the way in which the technology can be produced, distributed, sold, and used [3]. Policy can also influence the development of

***Corresponding author: Forrest Stagner**, The Pennsylvania State University, University Park, PA, USA, e-mail: fws11@psu.edu
Chris Morris, Frank Montabon, Kansas State University, Manhattan, KS
Michael Helbing, Christine Costello, Lara Fowler, The Pennsylvania State University, University Park, PA

 | https://doi.org/10.1515/9783112212141-006

particular technologies through prioritizing research funding for certain types of projects [3]. Additionally, policy can influence the economic conditions surrounding a technology, for instance, by offering financial incentives for individuals or organizations who are involved in the creation of a technology or product [3]. Policy can also offer financial incentives for consumers, such as federal rebates and tax breaks for purchasing electric vehicles. For example, a federal tax credit had been available for purchasers of electric vehicles through September 30, 2025 (see 26 U.S.C. § 30D). That credit was discontinued by the One Big Beautiful Bill Act for purchases after September 30, 2025 (see One Big Beautiful Bill Act, ch. 5, subch. A, § 70,502, 139 STAT. 72, 250 (July 4, 2025) (codified at 26 U.S.C. § 30D)). Policies can even institute mandates on how much of a particular technological product must be used in a given market, such as the state of Iowa requiring that at least one dispenser at gas stations sell gasoline blended with 15% ethanol [4]. Importantly, these types of policy functions mean that, in a potential market of competing technologies, policy can, either intentionally or inadvertently, favor one type of technology over another. This means that policy can have a substantial impact on influencing which technologies rise to prominence, ultimately helping to shape the everyday lives of citizens.

Given the considerable impact that policy can have on the development of technologies, their prominence in markets, and their integration into the lives of citizens, it is crucially important that policies are developed using science-based recommendations. For instance, there has been consternation regarding the validity of "carbon offset" programs, which critics accuse of overpromising on climate benefits [5]. On the other hand, [6] found that air pollution in the United States has decreased over the past several decades as a result of science-informed policy even as the economy has grown. There is a role, therefore, for researchers to inform both the creation of new policy and the evaluation of the effectiveness of existing policy. Such research can use stakeholder perspectives, economic analyses, life cycle analysis, and other methodologies to investigate the potential or actual social, economic, and environmental effects of a specific policy and the likelihood of the policy to achieve its stated goals. Policy-oriented research can be interdisciplinary, in which researchers from different academic disciplines (such as sociology, economics, and engineering) work together on a project to investigate the multiple factors related to a policy. This research can also be transdisciplinary, in which academic researchers work alongside partners from the public (such as government agencies) or private (such as companies or consumer advocacy groups) sectors. Such interdisciplinary and transdisciplinary partnerships in research incorporate a diversity of perspectives and methodologies, which can increase the breadth, depth, and validity of the resulting research findings. Policy resulting from such research is known as evidence-based policy [7]. Evidence-based policy has positively contributed to many areas of social life, such as vaccination policies based on clinical trials, education policies based on learning and development research, and poverty reduction policies based on social research [8]. When policymakers incorporate policy-oriented research findings to inform the creation and long-term development of policy, the resulting evidence-informed policy has an increased potential to be more effective and equitable [9].

Having discussed the important role of policy in the acceptance and integration of new technologies in general, we will now focus more specifically on public policy related to bioproducts. First, we will present a brief overview of bioproduct-related policies established in different economic blocs across the world. Then, in order to examine the implications of policy on bioproducts in greater detail, we will focus on the case study of on-farm anaerobic digestion (OFAD), a technology that has recently gained increased attention for its ability to produce biogas and other bioproducts from agricultural sources.

6.2 Bioproduct policies

As a category, bioproducts can be broken down into subcategories of food and feed ingredients, chemicals, materials, and energy. Bioproducts are derived from biomass and characterized by product-driven or energy-driven biorefining processes [10]. Bioproducts can replace fossil fuel-based products (e.g., plastics) and energy sources (e.g., natural gas) and are therefore seen as a potential pathway to reducing carbon emissions [11]. Aside from displacing fossil fuel-derived energy and products, the cultivation practices employed to generate biomass for bioproducts can provide additional ecosystem services without impacting food crop production [12]. For example, double cropping (i.e., planting a second crop following the primary crop) can provide soil cover and nutrient retention as well as biomass for use as a product or energy [13, 14]. Likewise, perennial grasses planted standalone or within agricultural landscapes provide habitat for wildlife, sequester carbon, and can mitigate runoff of pesticides and nutrients to waterways [15].

6.2.1 Global approaches to bioproduct policy

From a policy standpoint, bioproducts tend to be subsumed within a larger bioeconomy framework. Though there is not a consensus definition of the "bioeconomy," the term usually includes the concepts of the usage of biological resources for generating scientific knowledge, conversion into power or products, or ecosystem services [16]. At the international level, there has been a recognition of the role of the bioeconomy, and for bioproducts specifically, in mitigating climate change (the Paris Agreement, 2015) [17], contributing to the equitable distribution of clean energy (the UN Sustainable Development Goals, 2015) [18], and promoting biodiversity (Convention on Biological Diversity) [19].

More targeted initiatives include the Biofuture Platform [20], a coalition of countries, including the United States, EU member states, China, and Brazil, working to scale up bioenergy and advanced bioproducts. Meanwhile, the International Civil Aviation Organization has implemented a cap on aviation emissions, creating demand for sustainable aviation fuels, which are often biofuels [21].

Though international agreements can provide targets and structure for action, policies will largely be enacted and carried out at the country level. We therefore consider the approaches of the three largest global economies, the United States, EU, and China, to bioproduct policy.

6.2.1.1 United States

In the United States, bioproducts are defined through the Department of Agriculture (USDA) as products that "are produced from renewable matter, usually found in the form of living plants. Commonly, biomass can be transformed into products, packaging, or alternative power, but must meet approval of the USDA [22]." The US incentivizes several bioproduct categories. Perhaps most prominently, biofuel production is subsidized through the Renewable Fuel Standard (RFS) program, which both mandates that a certain amount of transportation fuel is blended with biofuels and provides a market incentive for production of biofuels [23]. The Inflation Reduction Act (IRA) of 2022 also provides tax credits and other incentives for producing low-carbon fuels such as sustainable aviation fuel and biogas [24]. The USDA BioPreferred Program requires government agencies to purchase bio-based products across multiple categories [25]. The future of these programs under the current Trump administration is unclear, as the administration has expressed hostility to the IRA and climate change research, adaption, and mitigation [26, 27]. Immediately upon taking office, the president halted IRA spending through executive action, leading to court challenges [28, 29]. More damaging to the IRA was the One Big Beautiful Bill signed into law in July 2025, which will phase out several clean energy tax credits by the end of 2025 (see One Big Beautiful Bill Act, Pub. L. No. 119–21, 139 Stat. 72.). Biofuels, however, are less politically polarizing than other clean energy sources – perhaps because of their ties to the agriculture and fossil fuel industries [30] – and so several provisions within the IRA to promote biofuels remain in place.

6.2.1.2 European Union

By comparison, EU policies are more developed, though with some notable weaknesses. The EU has a comprehensive Bioeconomy Strategy meant to advance a sustainable, circular bioeconomy in Europe [31]. The strategy was launched in 2012 and updated in 2018 and links to the European Green Deal's goal of carbon neutrality by

2050 [31]. The EU has set specific targets for integrating biofuels into transport by 2030 through the Renewable Energy Directive (RED II/III) [32]. RED also requires that biofuels meet certain criteria such as contributing to greenhouse gas (GHG) reductions through their production [33]. The EU has also promoted bioplastics to reduce plastic waste via the Circular Economy Action Plan & Plastics strategy. Other avenues for bioproduct development include the Common Agricultural Policy, which provides funding for energy crops [34], and the Green Public Procurement mechanism, through which the EU also promotes public purchasing of bioproducts [35].

[36] notes that sometimes EU-level classification of relevant materials clashes with local-level classification, thus inhibiting the achievement of a true circular economy. They also point out differences in how developed waste collection and separation systems are. [10] could be described as being more optimistic about EU bioproduct policy, though they note it is limited "to R&D grants, standardisation and education, and outreach tools, compared with the bioenergy, biofuels and petrochemical sectors [p. 11]." More direct policy tools such as incentives, taxation instruments, and public procurement policies or price guarantees for bioproducts are "limited", though the authors argue this is in line with economic theory "that economic policy instruments aren't effective during the early stages of development until stability and confidence in the market has been established and commercial viability proven [p. 11]."

6.2.1.3 China

China's bioproduct strategy, meanwhile, falls under its broader bioeconomy approach, outlined in the country's 14th 5-year plan (2021–2025). In terms of biomanufacturing, China has long prioritized biogas production, and this effort continues under the current 5-year plan [37, 38]. As of 2022, more than 7,700 biogas production facilities had been built [38]. There have been moves made within China to mandate E10 ethanol, however, the mandate was suspended following concerns about feedstock shortages [39]. China has also banned single-use plastics but made an exception for bioplastics, which has been a boon to the bioplastics industry there [40].

6.2.2 Differing policy approaches

As evidenced by the examples discussed here, different jurisdictions take varying approaches to bioproducts and biofuels. While some countries try to promote bioproduct development using incentives ("carrots") like many of the United States' IRA policies, others use mandates ("sticks") like the EU's Renewable Energy Directive. Other countries may take still different approaches by mandating information disclosure and/or developing regulatory oversight procedures to more subtly influence private decision-making. And some countries will take a laissez-faire approach to allow private actors to determine whether and how bioproducts are incorporated into the economy.

In our global economy, this mixture of policies will come together to impact whether, how, and to what extent bioproducts will be developed and incorporated into supply chains. As political leadership within each country changes and as the global economy reacts to stressors, these policies can be expected to change and adapt, creating a dynamic environment for bioproduct development. The ever-shifting political and economic environment poses a challenge for multinational corporations who have to negotiate a dynamic maze of policies when choosing which projects to fund.

6.3 Case study: on-farm anaerobic digestion in the United States

6.3.1 Anaerobic digestion overview

Anaerobic digestion (AD) is a natural biological process in which microorganisms break down organic material in a low-to-no oxygen environment, producing biogas (largely composed of carbon dioxide and methane, alongside other trace molecules) and digestate [41]. The raw biogas can be burned in a generator to produce heat and electricity, or it can be cleaned of impurities to create renewable natural gas (RNG), a bioproduct essentially identical to conventional natural gas [41]. The digestate (i.e., the leftover organic materials) has both solids and liquids which can be separated and used for different purposes [41]. The nutrient-rich liquid digestate can be applied on or injected into crop fields to provide plants with nutrients like nitrogen and phosphorous, while the solid digestate works well as livestock bedding [41].

AD facilities are typically deployed at landfills, water treatment facilities, and on livestock farms – areas with large amounts of organic wastes in need of management [42]. Anaerobic conditions encourage methanogenic bacteria to utilize carbon in organic materials, breaking down larger molecules (e.g., sugars and proteins) and producing carbon dioxide (CO_2) and methane (CH_4). Landfill sites may opt to have AD onsite for food and yard waste, and many modern landfills are designed to facilitate biomethane formation and collection systems, which can be burned onsite to produce electricity [43]. In wastewater treatment facilities (WWTF), the primary function of AD is to reduce the volume of solid material and remove pathogens to minimize the need for sludge disposal in landfills. As of 2022, an estimated 10% of WWTF use AD [44].

Livestock operations typically do not have AD onsite, but the numbers are increasing [45]. Instead, manure is often stored in open lagoons, where methane, ammonia (NH_3), and nitrous oxide (N_2O) emissions are substantial. Regardless of the waste stream, by capturing the biogas at these locations, methane is prevented from reaching the atmosphere, where, over a 100-year period, it is 27–30 times more potent than carbon dioxide in terms of global warming potential [46]. When burned, methane produces energy and is converted into carbon dioxide. In theory, then, capturing methane at landfills, WWTF, and livestock farms can both reduce the global warming potential (GWP) of these critical sites and reduce GWP from the energy sector by replacing fossil fuels.

In addition to methane, sulfurous compounds like hydrogen sulfide (H_2S), ammonia volatilization, and nitrous oxide emissions can also be avoided with biodigesters [47]. Ammonia volatilization leads to multiple deleterious outcomes as it redeposits to land, contributing to indirect nitrous oxide emissions and nitrate formation (NO_3) [48]. While nitrate is critical for crop growth, in excess, it leaches from soils or runs off of cropland, leading to eutrophication, dead zones, and other water quality issues [49]. In addressing hydrogen sulfide and ammonia volatilization, AD can also reduce odor from livestock farms [50, 51].

Lately, there has been more interest in energy and/or value-added molecule production via AD. Biomethane, just like fossil-derived methane, is a basic hydrocarbon that can be used as a feedstock for a variety of thermochemical and/or catalytically induced conversions to produce liquid fuels [52] or a variety of other molecules via catalysis [53–56].

6.3.2 Previous research

Since 2020, a team of researchers based at Iowa State University and Penn State University have been leading the Consortium for Cultivating Human And Naturally reGenerative Enterprises, (C-CHANGE), funded by a USDA-National Institute for Food and Agriculture (NIFA) grant. C-CHANGE is tasked with studying a proposed herbaceous feedstock-based anaerobic digestion value chain known as Grass2Gas (G2G). This value chain would incorporate annual crops grown in the off-season (similar to cover crops) and perennial grasses grown on marginal agricultural land into manure biodigesters. Increasing herbaceous groundcover and maintaining perennials on agricultural landscapes can reduce soil erosion and absorb excess nutrients before they volatilize, run off, or leach from fields, polluting the air and water [13, 15]. Digesting these crops could potentially provide an economic incentive to grow them and generate energy capable of displacing fossil fuels.

Through the G2G project, this team has developed a robust understanding of the policy framework supporting on-farm AD (OFAD) in the United States, how this framework entrenches large-scale, manure-based OFAD, and how alternative forms of OFAD, such as G2G, struggle to gain traction among farmers and biogas producers. Below, we describe this US policy framework. We then contrast it with the OFAD framework of the European Union to demonstrate how different policy choices are both possible and lead to different outcomes.

6.3.3 US policy framework

There are numerous policies that operate in different ways to support the development of OFAD in the United States. A recent C-CHANGE publication in *Energy Research and Social Science* identified multiple policy drivers of manure-based OFAD on large, concentrated livestock feeding operations [57]. Some of the most significant are programs that shape market forces to create incentives for investment in OFAD technology.

Most notably, California's Low Carbon Fuel Standard (CA-LCFS) drives much RNG production in the United States. Under the CA-LCFS, the California Air Resources Board establishes an average carbon intensity baseline for California transportation fuels that decreases over time [58]. Transportation fuels that have a lifecycle carbon intensity higher than CARB's carbon intensity baseline, generate deficits, while fuels with carbon intensity lower than the baseline, generate credits [58]. Obligated parties, with an overall deficit for a compliance year, may purchase credits to offset emissions from their operations to come into compliance with their obligations under the CA-LCFS [58]. Thus, through the program, if RNG producers can interconnect with a natural gas pipeline that is itself connected to pipelines in California (regardless of whether the RNG is actually burned there), then producers can receive a credit based on the carbon intensity of the RNG. Under the program, the lower the carbon intensity of the RNG, the higher is the value of the credit. Without policy support, RNG is generally unprofitable to produce, given the high costs associated with separating out impurities in raw biogas. The CA-LCFS credit helps producers close the gap on profitability in order to get RNG onto the market.

On the federal level, the EPA's Renewable Fuel Standard (RFS), also provides incentives for biofuel production. EPA establishes an annual volume of renewable fuels that obligated parties – generally refiners or importers of gasoline or diesel fuel – must incorporate into the fuel supply [23]. The agency establishes different volume obligations for four different categories of renewable fuels – biomass-based diesel, cellulosic biofuel, advanced biofuel, and renewable fuel [23]. Producers of fuels that satisfy regulatory criteria generate renewable identification numbers (RINs) appropriate for the category of fuel produced [23]. Historically, the most valuable of these has been the D3 RIN, and fuel produced from animal manure automatically qualifies for this RIN [59]. Fuel producers can retire RINs to satisfy their compliance obligations or sell them in the market [23].

Other policies operate to reduce the initial up-front cost of OFAD development. The initial capital required to establish a digester tends to be high, thus creating a barrier to entry, especially for smaller farms. Policies that help to bring down the effective cost of installing an OFAD system can increase uptake. For example, the federal government provides loans and grants to farmers and small businesses for renewable energy systems, including OFAD through the Rural Energy for America Program (REAP) [60]. Such loans can be obtained for up to 75% of eligible costs (up to $25 million), while grants can be obtained for 25% of eligible costs (up to a maximum of $1 million) [60]. Another federal program that can help offset the initial cost of developing bioproducts is the Environmental Quality Incentives Program (EQIP), which provides incentive payments to farmers for installing conservation practices (including some practices related to OFAD, such as digesters). Payments are fixed, calculated regionally, and are meant to cover part of the cost of a practice (and total contract payments for all practices cannot exceed $450,000 per individual or legal entity per year) [61].

Another policy approach that the federal government has taken is to invest in research and development for OFAD generation. By funding research, the government can facilitate technological improvements that can make bioproduct projects, including anaerobic digestion more feasible for farmers and businesses. For example, as part of its $79 million Bioenergy Research and Development program, the U.S. Department of Energy performed research designed to rethink anaerobic digestion in 2019 [62]. That research was intended to develop "anaerobic processes or alternative strategies to enhance carbon conversion efficiency and lower costs of smaller scale wet waste systems [62]."

Finally, policies that promote information disclosure can indirectly incentivize the uptake of OFAD. By requiring businesses to publicly release information about environmental impacts, such as GHG emissions, policy can create public awareness that can put pressure on businesses to take action designed to improve its public image. When attention is drawn to environmental impacts, the pressure can result in businesses turning to greener products such as bioproducts. Examples of such policies are California's Climate Corporate Data Accountability Act and Climate-Related Financial Risk Act [63, 64]. Together, these laws require certain corporations to make public disclosures about their GHG emissions and climate-related financial risks [63, 64].

Cumulatively, these programs create an incentive for large operations to capture methane emissions from livestock manure and produce RNG. The larger the operation, the easier and more profitable this investment becomes. Large operations not only have access to higher levels of feedstock (i.e., manure), but hiring employees to do the paperwork associated with governmental incentive programs and to run the digesters takes a smaller bite out of the operations' budgets than it does on small farms. Investing in digesters and the equipment to upgrade biogas to RNG is also more feasible for a large farm than for a small one. Finally, these programs tend to focus on RNG production as a way to offset GHG emissions, but generally fail to account for other sustainability concerns such as those caused by concentration of nitrogen and phosphorous on farms, application of pesticides, biodiversity loss, and soil erosion. As noted previously, many of these sustainability challenges could be improved by maintaining more herbaceous living cover on agricultural land, but the current policy orientation does not value these aspects of sustainability as highly as GHG reduction.

6.3.4 European Union policies

The European Union and individual countries within the EU have implemented policies that may directly or indirectly incentivize the development of anaerobic digestion. The first, and potentially most significant, is the EU's Renewable Energy Directive [65]. Originally enacted in 2009, the directive mandates that the EU obtain a minimum percentage of its energy consumption from renewable sources over a certain time period [65]. The directive has been modified over time to strengthen renewable energy targets [65]. In its current form, the Renewable Energy Directive requires that the EU obtain 42.5% of its energy from renewable sources by 2030 [65]. By creating a binding mandate to increase reliance on renewable energy sources, the Renewable Energy Directive spurs demand for bioproducts that provide energy. To date, the share of EU energy generated from renewable sources has increased from 12.5% in 2010 to 24.5% in 2023 [65].

Individual countries within the EU have also adopted policies that can lead to increased interest in bioproducts, such as those generated from anaerobic digestion. For example, a number of countries have established feed-in tariffs, which require utilities to buy energy generated from renewable sources (sometimes at increased rates) [66]. These policies help to increase the financial viability of bioproducts by enhancing returns on investment and improving investor confidence [66].

Poland enacted the Act of July 13, 2023, on Facilitating the Preparation and Implementation of Investments in Agricultural Biogas Plants and Their Operation (Dz.U. 2023, item 1597), which is designed to make biogas development easier by simplifying the steps needed to build and operate a biogas facility [67, 68]. It streamlines permitting and other administrative functions, reduces barriers to grid interconnection, and facilitates siting [67, 68]. The Polish government predicts that these changes will lead to 2,000 new agricultural biogas plants being built within the country [69].

Even policies that do not directly relate to renewable energy can have an impact on the uptake of bioproducts. The Netherlands requires that manure basins be covered [70]. By requiring covered manure basins, the Netherlands reduces the cost differential between farms that adopt anaerobic digestion and those that do not, thus minimizing the financial disincentive to develop bioproducts.

Viewed together, the policies enacted in the European Union provide a menu of options available to policymakers seeking to incentivize biofuel production. Establishing mandates for renewable fuel production, creating preferential pricing for fuels from certain feedstocks, reducing administrative barriers for new technologies, and reducing the cost differential between new and legacy technologies all present mechanisms for incentivizing the uptake of biofuels.

6.3.5 Comparison of US and EU approaches

While programs such as the CA-LCFS and RFS have expanded biogas and RNG production on farms in the United States in recent years, the United States still lags far behind Europe in renewable energy production (nine percent in the United States versus 24.5% in the EU in 2023) [65, 71]. Europe also has a massive lead in biogas production where there are thousands of biodigesters on farms compared to only hundreds in the United States [57]. It appears that the discrepancy is largely due to prioritizing biogas production earlier in the EU than in the United States. The United States has also been able to take advantage of large natural gas deposits due to the growth in hydraulic fracturing technology whereas the EU is largely reliant on Russia for natural gas [72–74]. Given Russia's aggression in Europe – invading Georgia in 2008, Crimea in 2014, and now pursuing a full-scale invasion of Ukraine – Europeans have sought to reduce their reliance on Russian natural gas, though as recently as 2022, 40% of Europe's natural gas still came from Russia [72]. Whether the United States follows Europe's lead on biogas production or whether the current biogas boom in the United States fizzles out will depend on the continuation and expansion of the biogas policy environment in the United States. In the final section below, we outline the challenges facing biogas production in the United States and provide a series of recommendations for overcoming these challenges.

6.4 Current policy barriers to US OFAD production and recommendations

Increasing the use of bioproducts can have positive environmental effects, especially when such products can replace fossil-derived and/or synthetic alternatives. In the case of AD, the bioproducts – biogas and digestate – can replace fossil fuels and synthetic fertilizers. Life cycle assessments on anaerobic digestion of manure are typically favorable when the energy produced is compared to a fossil-based alternative [75]. However, criticisms have been raised about the importance of the relationship with the livestock sector, that is, does integration of an energy technology incentivize expansion of the livestock industry such that benefits could be overshadowed at the larger systems level [76]. For example, [77] examined how the production of corn responded to the Low Carbon Fuel Standard and report that corn-derived ethanol ultimately expanded corn production nationally, resulting in a net increase in carbon intensity over fossil-based gasoline.

While climate concerns tend to dominate environmental sustainability discourse, a singular focus on GHG emissions may negatively impact other aspects of ecological sustainability such as air and water quality, biodiversity, soil health, acidification, and so forth [78, 79]. A 2024 review of anaerobic digestion to reach zero-carbon emissions targets at municipal WWTF found that narrow efforts to reduce GHG emissions at these facilities can have environmental co-benefits, but some level of tradeoff with other environmental goals seems inevitable [80]. A systems approach can reduce some of these tradeoffs – for example, via resource recovery processes [80]. This reduction in tradeoffs, in pursuit of increased GHG mitigation, would not necessarily be apparent if mitigation were the only target considered by policymakers.

For AD, there is an urgent need for attention to nutrients. Careful analysis of the feedstock-to-energy conversion technology, and how it could impact the broader agricultural and waste management sectors in which it is situated must be done with relevant stakeholders involved. Analysis should include more impacts than global warming potential, particularly nutrient cycling and potential impact [81].

Finding policy solutions for bioproducts is particularly challenging, as it requires collaboration across multiple agencies and offices, with fundamentally different mandates. These legal bodies operate at multiple levels (federal, state, and municipal) and interact with other interested stakeholders, including nongovernmental organizations, commodity groups, businesses, and farmers, to name a few.

6.4.1 Policy challenges

Part of the policy struggle over AD, in light of climate concerns, is the fact that the technology is a potential means for reducing GWP that is being treated as an end in and of itself. If policymakers wanted to rapidly reduce carbon and other GHG emissions and transition to cleaner energy sources, they would presumably introduce legislation and regulation to do so directly through mechanisms like carbon tax or creating and enforcing caps on GHG production. The current policy landscape in the United States, however, consists of a mix of state- and federal-level incentives and regulations. This framework results in inefficient movement toward climate and other environmental goals and generates perverse incentives such as keeping more livestock on smaller parcels of land.

We note that the Organisation for Economic Co-operation and Development stated that a timeframe of 15 to 25 years is needed for bioproducts to develop and establish a competitive advantage over fossil-fuel-based products [36]. This requires stability in government policies, which can be challenging.

In the United States, there remain a number of policy barriers that limit the growth of the bioproducts industry. These obstacles, many of which simply reflect the fact that law and policy have not evolved at the same rate as technology, may present opportunities for policymakers to accelerate the development of bioproducts with only modest government investments. Policy barriers include inadequate market signals, complex (and sometimes redundant) permitting systems, funding uncertainty, and interconnection difficulties, among other things.

6.4.1.1 Inadequate market signals

Many market signals in the United States are inconsistent or insufficient, creating disincentives to adoption of technology, even when such adoption would be socially beneficial. Inadequate market signals reduce the incentive for farmers and industry to adopt practices (such as AD) that would reduce the impact of externalities from their operations.

Most notably, although there are some programs that incentivize GHG reductions (e.g., Renewable Fuel Standard and California's Low Carbon Fuel Standard, discussed above), there is no general mechanism (such as a carbon tax) that puts a price on GHG emissions – either for the agricultural industry or the economy more generally. By failing to ensure that GHG generators internalize the cost of those emissions, the United States' system significantly decreases the incentive for private parties to invest in technologies that would reduce GHG.

Further, many environmental laws in the United States (e.g., Clean Water Act, Clean Air Act) regulate emissions from agricultural practices less strictly than comparable emissions from other industries. For example, there are minimal controls on nutrient management from manure spreading, especially for agricultural operations, which are not considered "concentrated animal feeding operations" and do not require a National Pollutant Discharge Elimination System (NPDES) permit under the Clean Water Act. Increasing controls on emissions from agricultural operations would increase the incentive for parties to adopting technologies like anaerobic digestion that can help to control air and water pollutants.

6.4.1.2 Complexity of permitting

Even for stakeholders that are interested in developing bioproducts, the practical effort required to establish one can sometimes be too much to overcome. In many American states, a significant number of permits is required to initiate an anaerobic digestion system. These permits often come from different offices throughout federal, state, and local government that do not coordinate with each other. This can result in confusion over which permits are required and the need to comply with redundant conditions. Some of these permits (e.g., NPDES permits under the Clean Water Act) require a lot of effort to apply for and can take a significant amount of time for review and approval. The costs associated with these permits can be significant.

6.4.1.3 Funding inadequacy

The capital expenditure required to establish an AD system is significant, especially for a small farming operation. Obtaining the necessary financing for these projects can be difficult and time-consuming. As discussed above, there are a number of federal and state programs in the United States (e.g., EQIP and REAP) that provide grants, loans, and/or cost-share for AD projects and related practices. Few of these funding opportunities fully finance a project, requiring developers to piece together multiple funding sources, which can be difficult and time-consuming. Further, government incentives (especially tax credits) are rarely guaranteed long-term, which leads to uncertainty in financial planning.

Requirements in specific funding sources limit their usefulness. Some funding sources have timing requirements that can be difficult to satisfy, given the realities of obtaining permission to operate. And in some programs, applicants may be limited by caps on total Farm Bill funding or adjusted gross income restrictions. One specific example of a funding requirement that creates a barrier to adoption is in the Conservation Reserve Program, which imposes a 25% penalty on rental income if plant biomass is hayed or harvested, instead of mowing, for vegetation management [82].

6.4.1.4 Interconnection difficulties

For projects that provide energy off-site, project operators often must connect either to the electric grid (for electricity projects) or to a natural gas pipeline (for RNG projects). These connections can be difficult and lengthy. As an initial matter, proximity to an eligible connection point is a significant factor that often cannot be controlled by the AD operator. Although policymakers are unlikely to be able to readily change existing infrastructure, they can develop processes to make it easier for developers to locate information about possible interconnection locations as well as the process for making the interconnection.

Further, each type of project can come with its own challenges. For RNG pipelines, different pipeline operators can have different standards for natural gas distributed through the pipeline, leading to confusion and difficulty during the planning process. For electricity projects, permission to connect to the electric grid can be a long and expensive process. Current limitations in the regional transmission grid queue for new renewable energy projects can exacerbate these challenges. In some situations, grid-connected electric systems may suffer from the additional challenge of being required to shut down their electricity generation in the case of a grid outage. In those cases, this prevents the project operator from using the system as a back-up power source.

6.4.1.5 Structure for improved framework

To encourage the growth of bioproducts from OFAD, a mix of sticks (taxes, mandates, fines, etc.) and carrots (incentives, subsidies, etc.) should be used, as no single approach will be as effective without the other. We note that in their literature review of policies to promote food loss and waste conversion into bioproducts, [10] discussed carrots more than sticks. Based on our research, we have developed several recommendations for both policymakers and practitioners. These recommendations are based on our research in the United States, but much of what is provided here will be applicable in other contexts.

First, livestock farms should be required at the very least to cover manure storage facilities and burn off the methane in order to reduce the GHG intensity of livestock agriculture. Second, livestock farmers should be encouraged to produce biogas, which would enable them to meet most or all of the energy they need on-farm, while reducing fertilizer and livestock bedding costs, in addition to GHG. Government policies to encourage biogas production should include subsidies for digester construction and generator purchases as well as guaranteed pricing or purchasing of the biogas or electricity produced from it. Third, farmers should be encouraged to maintain living cover on agricultural land, either in the form of cover crops, winter crops, or perennials. To incentivize more growth of perennials, the RFS' D3 RIN could be expanded to include locally appropriate perennials such as prairie or switchgrass as eligible feedstocks.

Fourth, given that there is not nearly enough biogas potential to replace natural gas at its current level, biogas should largely be thought of as a supplement to other forms of renewable energy such as solar and wind. To make this shift toward renewables, energy infrastructure will need to be overhauled to allow for more diffuse energy production. New infrastructure will also require greater versatility in energy transmission due to intermittency of wind and solar. Areas that rely on solar energy production during the day, for example, may need to pull energy from biogas facilities at night.

Fifth, policymakers should also take nutrient management into account when considering OFAD incentives and regulations. Conceiving of OFAD as primarily a climate solution – with perhaps some economic benefits mixed in – will not address, and may exacerbate, nutrient runoff, leaching, and volatilization from farms. We have seen that without maintaining herbaceous living cover on agricultural fields, nutrients applied to the land will be lost to the surrounding air and water. Moreover, if food waste from surrounding areas is added to digesters, more nutrients will be concentrated on farms, adding to nutrient losses to the environment, leading to challenges such as eutrophication and dead zones. Therefore, regulatory systems must be put in place if we are to achieve sustainable nutrient management. Just as importantly, farmers must be provided with the knowledge and resources to manage their nutrients. Agricultural operations should be financially remunerated for maintaining living ground cover and penalized for avoidable nutrient losses. Moreover, government and businesses should work toward improved biorefining of liquid digestate to meet crop demands and develop distribution pathways to return nutrients to back to the fields of origin. This approach would improve circularity and drastically reduce the need for synthetic fertilizers.

6.5 Conclusion

Emergent technologies do not exist in a vacuum and therefore cannot be separated from the social, economic, and political contexts in which they are situated. Consequently, to understand the development and social acceptance of a technology, it is imperative to have an understanding of the associated policies that act upon it. Policies created at local, national, and international levels can have a significant impact on both a technology's development and its acceptance and dissemination throughout a population over time.

With regard to bioproducts, a variety of policies have been designed and implemented across the globe with the overall goal of increasing the development of bioproducts in the energy sector as at least a partial replacement for fossil fuels. Several different approaches have been implemented, from more incentive-based polices like those in the United States, to more regulatory policies like those in the EU, or a combination of these, in addition to other approaches. In a global economy, this has resulted in a dynamic landscape for bioproduct policy, with significant effects on the growth in use of these products.

OFAD serves as an illustrative case study for discussing the effect of policy on the development of a bioproduct technology. Located at the nexus between the industries of energy and agriculture, much attention has been devoted to OFAD as a means of increasing the sustainability of both of these industries. Recent research has investigated how recent growth in OFAD across the United States has been heavily influenced by incentive-based federal energy policy as well as state-level renewable energy policy from California. Despite this recent increase in development, the growth in use of this particular technology significantly lags behind that of the EU, mainly due to the EU's reliance on more regulatory-based policies and its much earlier prioritization of biogas production.

Several barriers exist to limit the growth of this technology in the United States, including inadequate market signals for incentivizing OFAD, complexity in permitting for building OFAD systems, lack of funding for developing these systems, and difficulties connecting systems to the existing energy grid. The development of policies to overcome these barriers would potentially involve a combination of regulations and financial incentives to encourage livestock farms to trap methane emissions, use them to produce biofuel, and incorporate perennial feedstocks, while adequately managing the associated nutrient applied to cropland. Policy should also encourage the growth in renewable energy infrastructure in general. Finally, as bioproduct technology continues to evolve among an array of ever-changing environmental, economic, and social factors, it will be critical to rely on scientific research to help inform the development of evidence-based policy for successfully addressing society's needs.

References

[1] Colebatch H. Policy, Maidenhead, UK New York: Open University Press; 2009.

[2] Garner BA, editor. Black's Law Dictionary, 12th Edition, Thomson Reuters 2024.

[3] Steinmueller WE. Chapter 28 – Economics of technology policy. In Hall BH, Rosenberg N, editor. Handbook of the Economics of Innovation, North-Holland; 2010. Vol. 2, pp. 1181–1218. doi: 10.1016/S0169-7218(10)02012-5.

[4] "Alternative Fuels Data Center: RetailE15Access Requirements." Accessed: Oct. 07, 2025. [Online]. Available: https://afdc.energy.gov/laws/12998

[5] Mia BMD, "Carbon offsets are climate scams. Wildfires will make them worse.," Food & Water Watch. Accessed: Oct. 07, 2025. [Online]. Available: https://www.foodandwaterwatch.org/2024/05/14/carbon-offsets-wildfires/

[6] Sullivan TJ, et al. Air pollution success stories in the United States: The value of long- term observations. Environ Sci Policy June 2018;84:69–73. doi: 10.1016/j.envsci.2018.02.016.

[7] Young K, Ashby D, Boaz A, Grayson L. Social Science and the Evidence-based Policy Movement. Soc Policy Soc July 2002;1(3):215–224. doi: 10.1017/S1474746402003068.

[8] Baron J. A brief history of evidence-based policy. Ann Am Acad Pol Soc Sci July 2018;678(1):40–50. doi: 10.1177/0002716218763128.

[9] Bowers J, Testa PF. Better government, better science: The promise of and challenges facing the evidence-informed policy movement. Annu Rev Polit Sci May 2019;22(1):521–542. doi: 10.1146/annurev-polisci-050517-124041.

[10] Churton H, McCabe BK. Advancing a food loss and waste bioproduct industry: A critical review of policy approaches for application in an Australian context. Heliyon June 2024;10(12):e32735. doi: 10.1016/j.heliyon.2024.e32735.

[11] Usman IMT, Ho Y-C, Baloo L, Lam M-K, Sujarwo W. A comprehensive review on the advances of bioproducts from biomass towards meeting net zero carbon emissions (NZCE). Bioresour Technol Dec. 2022;366:128167. doi: 10.1016/j.biortech.2022.128167.

[12] Langeveld JWA, Dixon J, Van Keulen H, Quist-Wessel PMF. Analyzing the effect of biofuel expansion on land use in major producing countries: Evidence of increased multiple cropping. Biofuels Bioprod Biorefining 2014;8(1):49–58. doi: 10.1002/bbb.1432.

[13] Schulte LA, et al. Meeting global challenges with regenerative agriculture producing food and energy. Nat Sustain May 2022;5(5):384–388, doi: 10.1038/s41893-021-00827-y.

[14] Moore KJ, Karlen DL. Double cropping opportunities for biomass crops in the north central USA. Biofuels Nov 2013;4(6):605–615. doi: 10.4155/bfs.13.50.

[15] Schulte LA, et al. Prairie strips improve biodiversity and the delivery of multiple ecosystem services from corn–soybean croplands. Proc Natl Acad Sci Oct. 2017;114(42):11247–11252, doi: 10.1073/pnas.1620229114.

[16] E. National Academies of Sciences, et al. Defining the U.S. bioeconomy. In Safeguarding the Bioeconomy, National Academies Press (US); 2020. Accessed: Aug. 13, 2025. [Online]. Available https://www.ncbi.nlm.nih.gov/books/NBK556424/.

[17] "The Paris Agreement | UNFCCC." Accessed: Aug. 15, 2025. [Online]. Available: https://unfccc.int/process-and-meetings/the-paris-agreement

[18] "THE 17 GOALS | Sustainable Development." Accessed: Aug. 15, 2025. [Online]. Available: https://sdgs.un.org/goals

[19] "Home," Convention on Biological Diversity. Accessed: Aug. 15, 2025. [Online]. Available: https://www.cbd.int/

[20] "BioFuturePlatform – Accelerate the transition to a global bioeconomy." Accessed: Aug. 15, 2025. [Online]. Available: https://biofutureplatform.org/

[21] "Alternative Fuels Data Center: Sustainable Aviation Fuel." Accessed: Aug. 15, 2025. [Online]. Available: https://afdc.energy.gov/fuels/sustainable-aviation-fuel

[22] Thangari P, Gordon B, Zilberman D, Wang D, Hochman G, "Biomass to Bioproducts: Economic and Environmental Viability," USDA. [Online]. Available: https://www.usda.gov/sites/default/files/documents/Bioproduct_Factsheet.pdf

[23] US EPA, "Overview of the Renewable Fuel Standard Program." Accessed: Oct. 07, 2025. [Online]. Available: https://www.epa.gov/renewable-fuel-standard/overview-renewable-fuel-standard-program

[24] "Clean Fuel Production Credit | Internal Revenue Service." Accessed: Oct. 07, 2025. [Online]. Available: https://www.irs.gov/credits-deductions/clean-fuel-production-credit

[25] "BioPreferred | About." Accessed: Oct. 07, 2025. [Online]. Available: https://www.biopreferred.gov/BioPreferred/faces/pages/AboutBioPreferred.xhtml

[26] "Trump Administration Halts Climate Research." Accessed: Oct. 10, 2025. [Online]. Available: https://www.nrdc.org/press-releases/trump-administration-halts-climate-research

[27] "Federal Agencies Ordered to Pause Spending of Inflation Reduction Act, Infrastructure Investment and Jobs Act Funds." Accessed: Oct. 10, 2025. [Online]. Available: https://www.morganlewis.com/pubs/2025/01/federal-agencies-ordered-to-pause-spending-of-inflation-reduction-act-infrastructure-investment-and-jobs-act-funds

[28] "New executive order pauses certain funding under IRA and IIJA | Baker Tilly." Accessed: Oct. 10, 2025. [Online]. Available: https://www.bakertilly.com/insights/new-executive-order-pauses-funding-under-ira-iija

[29] "Trump Administration Freezes Billions of Dollars in Federal Grants and Loans | Sabin Center for Climate Change Law." Accessed: Oct. 10, 2025. [Online]. Available: https://climate.law.columbia.edu/content/trump-administration-freezes-billions-dollars-federal-grants-and-loans

[30] De L. T. Oliveira G, McKay B, Plank C. How biofuel policies backfire: Misguided goals, inefficient mechanisms, and political-ecological blind spots. Energy Policy Sept. 2017;108:765–775. doi: 10.1016/j.enpol.2017.03.036.

[31] "Bioeconomy Strategy – Environment – European Commission." Accessed: Oct. 07, 2025. [Online]. Available: https://environment.ec.europa.eu/strategy/bioeconomy-strategy_en

[32] "Biofuels." Accessed: Oct. 07, 2025. [Online]. Available: https://energy.ec.europa.eu/topics/renewable-energy/bioenergy/biofuels_en

[33] Von Witzke K, "Biofuel Mandates in the EU by Member State – 2025." Accessed: Oct. 07, 2025. Accessed: Oct. 07, https://apps.fas.usda.gov/newgainapi/api/Report/DownloadReportByFileName?fileName=Biofuel%20Mandates%20in%20the%20EU%20by%20Member%20State%20-%202025_Berlin_European%20Union_E42025-0004

[34] "CAP and bioenergy – European Commission." Accessed: Oct. 09, 2025. [Online]. Available: https://agriculture.ec.europa.eu/cap-my-country/sustainability/economic-sustainability/bioeconomy/cap-and-bioenergy_en

[35] "Green procurement – European Commission." Accessed: Oct. 09, 2025. [Online]. Available: https://commission.europa.eu/funding-tenders/tools-public-buyers/green-procurement_en

[36] "Realising the circular bioeconomy," OECD Science, Technology and Industry Policy Papers 60, Nov. 2018. doi: 10.1787/31bb2345-en.

[37] He PJ. Anaerobic digestion: An intriguing long history in China. Waste Manag Apr. 2010;30(4):549–550. doi: 10.1016/j.wasman.2010.01.002.

[38] Zhang X, Zhao C, Shao M, Chen Y, Liu P, Chen G. The roadmap of bioeconomy in China. Eng Biol Dec. 2022;6(4):71–81. doi: 10.1049/enb2.12026.

[39] Muyu HGX, Singh S, "Exclusive: China suspends national rollout of ethanol mandate – sources," Reuters, Jan. 08, 2020. Accessed: Oct. 07, 2025. [Online]. Available: https://www.reuters.com/article/business/exclusive-china-suspends-national-rollout-of-ethanol-mandate-sources-idUSKBN1Z71R3/

[40] "China Leads as New Regulations Redefine the Bioplastics Market," IDTechEx. Accessed: Oct. 07, 2025. [Online]. Available: https://www.idtechex.com/en/research-article/china-leads-as-new-regulations-redefine-the-bioplastics-market/31689
[41] US EPA, "How Does Anaerobic Digestion Work?" Accessed: Oct. 07, 2025. [Online]. Available: https://www.epa.gov/agstar/how-does-anaerobic-digestion-work
[42] "Biogas Industry Market Snapshot | American Biogas Council." Accessed: Oct. 10, 2025. [Online]. Available: https://americanbiogascouncil.org/biogas-market-snapshot/
[43] US EPA, "Basic Information about Landfill Gas." Accessed: Oct. 10, 2025. [Online]. Available: https://www.epa.gov/lmop/basic-information-about-landfill-gas
[44] Ha M, Gutenberger G, Ou L, Cai H, Hawkins T, "Opportunities for recovering resources from municipal wastewater," ANL/ESD-21/11, 1876441, 176194, July 2022. doi: 10.2172/1876441.
[45] US EPA, "AgSTAR Data and Trends." Accessed: Aug. 15, 2025. [Online]. Available: https://www.epa.gov/agstar/agstar-data-and-trends
[46] US EPA, "Understanding Global Warming Potentials." Accessed: July 17, 2025. [Online]. Available: https://www.epa.gov/ghgemissions/understanding-global-warming-potentials
[47] Sihlangu E, Luseba D, Regnier T, Magama P, Chiyanzu I, Nephawe KA. Investigating Methane, Carbon Dioxide, Ammonia, and Hydrogen Sulphide content in agricultural waste during biogas production. Sustainability Jan. 2024;16(12):5145. doi: 10.3390/su16125145.
[48] "Publications – IPCC-TFI." Accessed: Oct. 30, 2024. [Online]. Available: https://www.ipcc-nggip.iges.or.jp/public/2006gl/vol4.html
[49] Bricker SB, et al. Effects of nutrient enrichment in the nation's estuaries: A decade of change. Harmful Algae Dec 2008;8(1):21–32. doi: 10.1016/j.hal.2008.08.028.
[50] Ostlie J, "Livestock Odor Facts", [Online]. Available: https://www.pca.state.mn.us/sites/default/files/wq-f2-04.pdf
[51] "Odors From Livestock Operations: Causes and Possible Cures | MU Extension." Accessed: Aug. 15, 2025. [Online]. Available: https://extension.missouri.edu/publications/g1884
[52] Chiaramonti D, Testa L. Deploying EU biomethane potential for transports: Centralized/decentralized biogas refinery schemes to SAF and maritime fuels. Appl Energy July 2024;366:123306. doi: 10.1016/j.apenergy.2024.123306.
[53] Mishra S, et al. Green synthesis of biomethanol – Managing food waste for carbon footprint and bioeconomy. Biomass Convers Biorefin May 2022;12(5):1889–1909, doi: 10.1007/s13399-021-02188-0.
[54] Chiaramonti D, Testa L. Deploying EU biomethane potential for transports: Centralized/decentralized biogasrefinery schemes to SAF and maritime fuels. Appl Energy July 2024;366:123306. doi: 10.1016/j.apenergy.2024.123306.
[55] Pieja AJ, Morse MC, Cal AJ. Methane to bioproducts: The future of the bioeconomy?. Curr Opin Chem Biol Dec. 2017;41:123–131. doi: 10.1016/j.cbpa.2017.10.024.
[56] Van Dyk S, Saddler J. Progress in Commercialization of Biojet /Sustainable Aviation Fuels (SAF): Technologies and policies. International Energy Agency 2024. Available: https://www.ieabioenergy.com/wp-content/uploads/2024/06/IEA-Bioenergy-Task-39-SAF-report.pdf
[57] Morris C, et al. Growing gas on the farm? Stakeholder perspectives on divergent models of biogas production in the United States. Energy Res Soc Sci Sept. 2025;127:104255. doi: 10.1016/j.erss.2025.104255.
[58] "Low Carbon Fuel Standard | California Air Resources Board." Accessed: Oct. 07, 2025. [Online]. Available: https://ww2.arb.ca.gov/our-work/programs/low-carbon-fuel-standard/about
[59] EPA US, "Renewable Natural Gas from Agricultural-Based AD/Biogas Systems." Accessed: Oct. 07, 2025. [Online]. Available: https://www.epa.gov/agstar/renewable-natural-gas-agricultural-based-adbiogas-systems
[60] Korkut E, Helbing M, Moore LS, Arbuckle JG, Stagner F, Montabon F. The Regulatory Landscape of Livestock Anaerobic Digesters and Renewable Natural Gas (RNG). J Anim Environ Law 2023;15(1):

Accessed: Aug. 06, 2025. [Online]. Available:, https://drive.google.com/file/d/1J8-5a0GM5mxNc5Ad4BT5MiuDjlYnyFU3/view?usp=embed_facebook.
[61] "Environmental Quality Incentives Program (EQIP) | Natural Resources Conservation Service." Accessed: Oct. 10, 2025. [Online]. Available: https://www.nrcs.usda.gov/programs-initiatives/environmental-quality-incentives-program
[62] "DOE Announces $79 Million for Bioenergy Research and Development," Energy.gov. Accessed: Aug. 06, 2025. [Online]. Available: https://www.energy.gov/articles/doe-announces-79-million-bioenergy-research-and-development
[63] "2024 California Code: Health and Safety Code – HSC: DIVISION 25.5 – CALIFORNIA GLOBAL WARMING SOLUTIONS ACT OF 2006: PART 2 – MANDATORY GREENHOUSE GAS EMISSIONS REPORTING: Section 38532.," Justia Law. Accessed: Aug. 06, 2025. [Online]. Available: https://law.justia.com/codes/california/code-hsc/division-25-5/part-2/section-38532/
[64] "2024 California Code: Health and Safety Code – HSC: DIVISION 25.5 – CALIFORNIA GLOBAL WARMING SOLUTIONS ACT OF 2006: PART 2 – MANDATORY GREENHOUSE GAS EMISSIONS REPORTING: Section 38533.," Justia Law. Accessed: Aug. 06, 2025. [Online]. Available: https://law.justia.com/codes/california/code-hsc/division-25-5/part-2/section-38533/
[65] "Renewable Energy Directive." Accessed: Aug. 06, 2025. [Online]. Available: https://energy.ec.europa.eu/topics/renewable-energy/renewable-energy-directive-targets-and-rules/renewable-energy-directive_en
[66] "Anaerobic Digestion Screening Tool User Manual | Global Methane Initiative," https://www.globalmethane.org/. Accessed: July 09, 2024. [Online]. Available: https://www.globalmethane.org/resources/details.aspx?resourceid=5174
[67] Łukomska A, Pulka J, Broński M, Dach J. Demand-driven biogas plants in poland – potential and growth perspectives. J Ecol Eng Nov. 2024;25(11):236–248. doi: 10.12911/22998993/193262.
[68] "Biogas & Biomethane in Poland," Dudkowiak & Putyra. Accessed: Aug. 11, 2025. [Online]. Available: https://www.dudkowiak.com/environmental-law-in-poland/biogas-and-biomethane/
[69] Łukomska A, Pulka J, Broński M, Dach J. Demand-driven biogas plants in poland – potential and growth perspectives. J Ecol Eng Nov. 2024;25(11):236–248. doi: 10.12911/22998993/193262.
[70] Rvo NEA, "Building regulations for manure basins," business.gov.nl. Accessed: Aug. 11, 2025. [Online]. Available: https://business.gov.nl/regulation/structural-regulations-manure-basins/
[71] "U.S. energy facts explained – consumption and production – U.S. Energy Information Administration (EIA)." Accessed: Oct. 08, 2025. [Online]. Available: https://www.eia.gov/energyexplained/us-energy-facts/
[72] "Europe's messy Russian gas divorce," Brookings. Accessed: Oct. 08, 2025. [Online]. Available: https://www.brookings.edu/articles/europes-messy-russian-gas-divorce/
[73] "Where our natural gas comes from – U.S. Energy Information Administration (EIA)." Accessed: Oct. 08, 2025. [Online]. Available: https://www.eia.gov/energyexplained/natural-gas/where-our-natural-gas-comes-from.php
[74] S. P. A. B. Yucel Mine K., "The Shale Gas and Tight Oil Boom | Council on Foreign Relations." Accessed: Oct. 08, 2025. Accessed: Oct. 08, https://www.cfr.org/report/shale-gas-and-tight-oil-boom
[75] Meena PK, Kumar D, Singh L, Sharma S, Didwania M. Evaluating the sustainability of biohydrogen, biogas, and biohythane generation from agricultural, industrial, and municipal waste sources. Biofuels Bioprod Biorefining 2025;19(2):485–507. doi: 10.1002/bbb.2713.
[76] Magnolo F, Candel J, Speelman S. Biomethane from manure in the RePowerEU: A critical perspective on the scale-up of renewable energy production from the livestock sector. Energy Res Soc Sci Dec. 2024;118:103793. doi: 10.1016/j.erss.2024.103793.
[77] Lark TJ, et al. Environmental outcomes of the US Renewable Fuel Standard. Proc Natl Acad Sci Mar 2022;119(9):e2101084119. doi: 10.1073/pnas.2101084119.

[78] Konietzko J, “Moving beyond carbon tunnel vision with a sustainability data strategy,” www.cognizant.com. Accessed: July 22, 2025. [Online]. Available: https://www.cognizant.com/us/en/insights/insights-blog/moving-beyond-carbon-tunnel-vision-with-a-sustainability-data-strategy-codex7121
[79] Laurent A, Olsen SI, Hauschild MZ. Limitations of Carbon footprint as indicator of environmental sustainability. Environ Sci Technol Apr 2012;46(7):4100–4108. doi: 10.1021/es204163f.
[80] Zheng X, Lam KL. An overview of environmental co-benefits and trade-offs to reduce greenhouse gas emissions in municipal wastewater management. Sustain Prod Consum May 2024;46:1–10. doi: 10.1016/j.spc.2024.02.002.
[81] Khoshnevisan B, et al. A critical review on livestock manure biorefinery technologies: Sustainability, challenges, and future perspectives. Renew Sustain Energy Rev Jan. 2021;135:110033. doi: 10.1016/j.rser.2020.110033.
[82] “Conservation Reserve Program,” Federal Register. Accessed: Oct. 10, 2025. [Online]. Available: https://www.federalregister.gov/documents/2015/07/16/2015-17317/conservation-reserve-program

Section II: **Some early-stage R&D bioproducts**

Herman P. Benecke

Chapter 7 Formaldehyde-free binders

7.1 Background

Typical US homes may contain insulation in their attic and other locations consisting of glass fiber mats that are bound together by a coating consisting of mixed phenol/formaldehyde (PF) and urea/formaldehyde (UF) resins. Glass fibers are bound to each other on a porous continuous conveyer belt by evaporating aqueous solutions of mixed PF/UF binders at temperatures up to 400 °F within air-blown ovens where evaporation of water occurs initially and the resin is at this temperature for about 20 s. A vacuum is applied to the bottom of the belt so a continuous glass fiber mat is collected on the conveyor belt. During this drying process, the binder collects where glass fibers contact each other, thereby generating a cross-linked and semirigid matrix. Furniture as well as molded media and structural board found in homes may be prepared from wood fibers that use PF/UF binders to cement these fibers together.

However, these binders pose significant health risks since the UF component is known to slowly release formaldehyde [1], which has been classified as a known carcinogen by the State of California [2] and the World Health Organization [3]. Allergic contact dermatitis is also caused by formaldehyde exposure [1].

7.2 Early formaldehyde-free binders

Companies producing PF/UF binders responded to these health hazards by developing alternate formaldehyde-free binders. Early work involved the esterification reactions at about 400 °F of aqueous solutions containing glass fibers, the polymeric carboxylic acid polyacrylic acid (PAA), polyols (compounds containing two or more hydroxyl groups) such as glycerin [4] or triethanolamine [5] when using sodium hypophosphite as the esterification catalyst. These processes generate polyester networks that also effectively encapsulate glass fibers as the water evaporates. These binder systems do not evolve formaldehyde but major deficiencies still exist. One is that their curing times are relatively long, leading to decreased production of cured glass fiber mats requiring increased production lines and increased capital and plant operating costs. These binders also have relatively high acidities due to the required high concentration of carboxylic acid functionality that leads to erosion of binder curing equipment or the need for expensive corrosion-resistant steel.

Herman P. Benecke, Columbus, OH, e-mail: beneckefl@gmail.com

 | https://doi.org/10.1515/9783112212141-007

7.3 Improved formaldehyde-free binders

An improved approach has been developed that involves the synthesis of polyols that serve as one component of an alternate and improved binder system [6–8]. These polyols are partially derived from 2,5-furandicarboyxlic acid (FDCA), a bio-based and renewable substance that can be derived from fructose, which is in over-supply partly due to the popularity of noncaloric sweeteners. As shown in Figure 7.1, one approach to prepare FDCA involves the initial acid-catalyzed cyclization of fructose to 5-hydroxymethylfurfural (HMF) followed by its oxidation to FDCA.

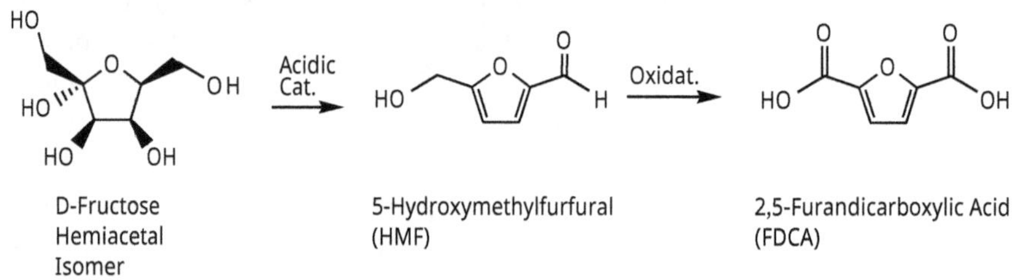

Figure 7.1: Conversion of D-fructose to 5-hydroxymethylfurfural to 2,5-furandicarboyxlic acid.

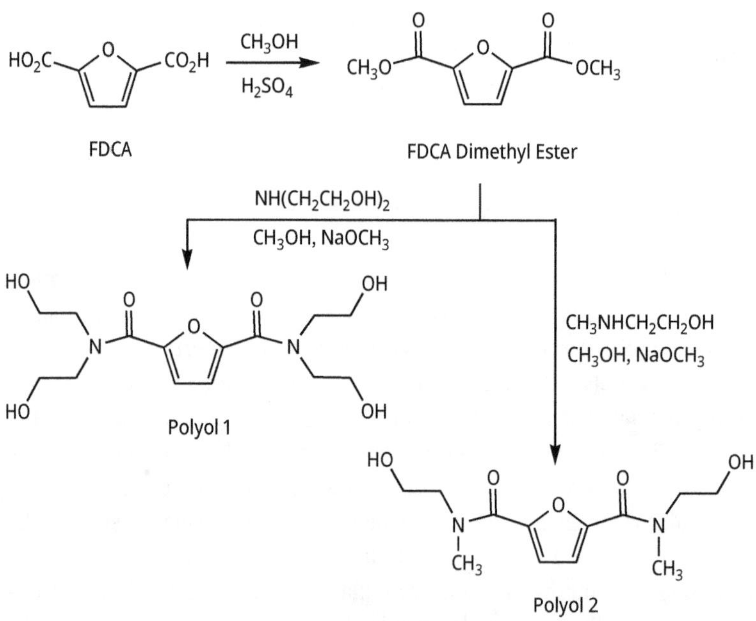

Figure 7.2: Conversion of FDCA to polyols 1 and 2.

The preparation of two representative next-generation polyols is shown in Figure 7.2, wherein FDCA is first esterified with methanol to produce FDCA dimethyl ester. This ester is amidified with diethanolamine in methanol using sodium methoxide as catalyst to produce polyol 1 or with *N*-methyl-*N*-ethanolamine with the same reactive system to produce polyol 2. The main reason for producing and testing polyols 1 and 2 is that they incorporate the tertiary beta-hydroxyethyl amide structure whose primary hydroxyl groups undergo esterification reactions faster than nonfunctionalized primary alcohols [9]. As described later, the enhanced esterification rates of these hydroxyl groups incorporated in tertiary beta-hydroxyethyl amide functionality provided desired faster cure rates.

7.4 Cure times for reaction of PAA with glycerin versus polyol 1

7.4.1 Testing procedure and stroke/cure tests

To obtain comparative standards, aqueous binder test solutions containing glycerin and PAA with the esterification catalyst sodium hypophosphite were prepared. Aqueous test solutions containing polyol 1 and PAA with and without this catalyst were also prepared. All solutions had the same carboxylic acid to hydroxyl mole ratios (1.65:1), the same final water content (69.1%), and the same weight ratios of catalyst to reactants. The pH of fully formulated binder solutions was determined before and after curing. One purpose of these tests was to determine the comparative times needed for curing of binders based on glycerin/PAA versus polyol 1/PAA.

7.4.2 Stroke/cure tests

Binder compositions were evaluated in a preliminary manner by placing equivalent quantities at time zero on a hot plate maintained at 180 °C. These samples were stroked back and forth in a regular fashion with a spatula in “stroke/cure” tests, and the times needed to first obtain a viable and self-supporting string were noted as well as the slightly longer times needed for the entire mass to solidify into a tight mass. These later times were recorded as the actual cure times.

7.4.3 Catalyzed cure times of glycerin with PAA

The average cure times obtained in seven experiments for the curing of glycerin with PAA in the presence of the esterification catalyst sodium hypophosphite was 102s with a standard deviation of 3.3 s. The initial percentage of PAA in these mixtures was 22.3%, and the initial pH values of each solution were in the range of 1–2.

These experiments were performed to determine the variability involved in the stroke/cure tests and also to determine the cure times for the reaction of glycerin with PAA as a standard to surpass in the development of improved processes.

7.4.4 Catalyzed and uncatalyzed reactions of polyol 1 with PAA

The average cure time of two experiments involving curing polyol 1 with PAA in the presence of the catalyst sodium hypophosphite was 84 s. To maintain the same carboxylic acid-to-hydroxyl mole ratios in the reaction of polyol 1 with PAA compared to the reaction of glycerin with PAA, the initial weight percentage of PAA in the polyol 1 reaction had been decreased to 17.1% while the weight percent of polyol 1 was increased. This compositional shift is due to the higher molecular weight of the tetraol polyol 1 (330.3 g/mol) where 82.6 g is needed to provide 1 mole of hydroxyl groups, while only 30.7 g of the lower molecular weight triol glycerin (92.1 g/mol) is needed to provide 1 mole of hydroxyl groups. Thus, PAA is more dilute in its reaction with polyol 1 reaction, which causes the initial pH values of polyol 1-based solutions to be in the 2–3 pH range (less acidic than glycerin).

When polyol 1 was esterified twice with PAA, but without the esterification catalyst, the average cure time was 87 s and the pH was the same as when performed with the esterification catalyst. Determination of these very similar cure times of polyol 1 with PAA, with and without the catalyst sodium hypophosphite, and given the standard deviation of 3.3 s in determining the catalyzed cure times of glycerin with PAA with the same stroke/cure test, indicates that these two cure times are essentially the same. Thus, use of this catalyst was redundant and did not provide meaningful catalytic acceleration.

7.4.5 Implication of cure time differences between glycerin and polyol 1

Comparison of the cure time of glycerin with PAA in the presence of esterification catalyst sodium hypophosphite (102 s) versus the cure time of polyol 1 with PAA in the absence of catalyst (87 s) implies that the hydroxyl groups of polyol 1 are activated and had access to an efficient esterification mechanism that enhanced their rate of esterification (see later) while glycerin did not.

7.4.6 Mechanism involved in esterification of polyol 1 with PAA

As shown in Figure 7.3, the presumed cure mechanism for the enhanced esterification rate of activated polyol 1 (and presumably activated polyol 2) with PAA is that in the presence of a carboxylic acid, the tertiary beta-hydroxyethyl amide functionality undergoes rapid cyclization to a hydroxy-substituted oxazolidine [9]. This intermediate undergoes the shown reaction to convert the carboxylic acid to a negatively charged carboxylate anion that is paired with a positively charged resonance-stabilized oxazolinium cation, which efficiently react with each other to generate polyol 1 or 2 esterified with any carboxylic acid such as PAA.

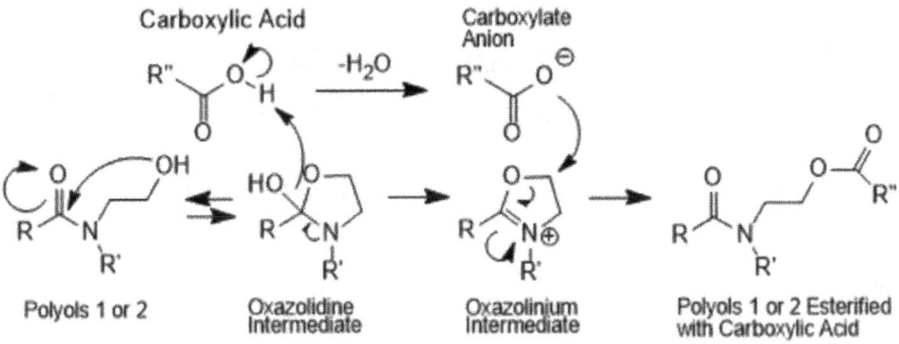

Figure 7.3: Potential mechanism providing accelerated esterification of tertiary beta-hydroxyethyl amide primary hydroxyl groups.

7.4.7 Implication of testing results of activated polyol 1/PAA-based binder and expected properties

The reduced cure times of polyol 1/PAA-based binder compared to the cure times of the glycerin/PAA-based binder indicate that the polyol 1-based fiberglass insulation production time rate for the polyol 1-based binder would provide a cure time that is 85% of the glycerin/PAA cure time without use of an esterification catalyst in a binder production plant. Not needing an esterification catalyst should lead to very reproducible binder properties with no need for exact replication of esterification catalyst concentrations as needed for ester-based binders incorporating glycerin or other similar nonactivated polyols.

That the polyol 1-based binder required appreciably less PAA than the glycerin/PAA-based binder should provide an economic advantage since PAA is a petroleum-based product whose cost will mirror the potential price increases of crude oil. Another advantage of the polyol 1/PAA binder system is that pH values before and during curing were higher (less acidic) than the glycerin/PAA system, which should cause decreased production line corrosion and repair and maintenance expenses.

Blending polyol 1 and polyol 2 (that has half the hydroxyl groups compared to polyol 1) would allow adjustment of binder cross-linking that will appreciably influence the viscosity of binder reaction solutions and the rigidity of cured binder mats.

The presence of furanic rings within the cured binder should impart reduced flammability since it is known that furanic compounds are excellent char-formers [10]. Char formation is thought to prevent the spreading of fire and/or cellular degradation because the char has enough integrity to stay in place, thus reducing the heat transfer from the fire to nonburned material.

7.5 Preparation and testing of FDCA/polyol 1 oligomers as binder components

7.5.1 Rationale for preparing and evaluating polyol 1/FDCA oligomers

Another related-type binder system was obtained by pre-esterifying polyol 1 with the diacid FDCA to obtain derived oligomers. Polyol 1 was found to undergo efficient esterification with FDCA to form an oligomer of the type shown in Figure 7.4 [6–8]:

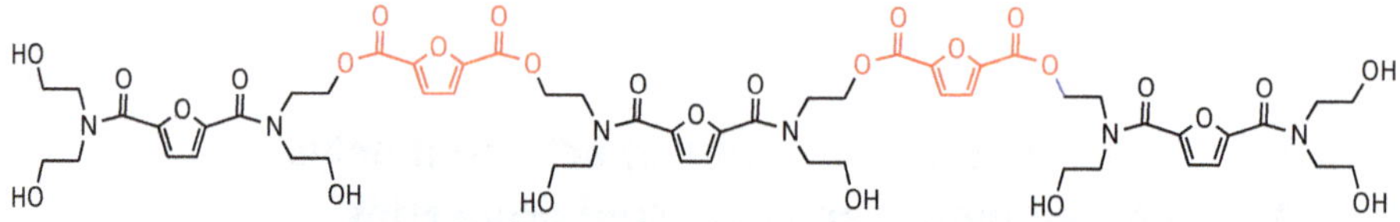

Figure 7.4: Potential oligomer from esterification of polyol 1 with FDCA when using a 3:2 molar ratio.

As an aid to understand the oligomer structure shown above, polyol 1-derived components are colored black while the FDCA-derived components are colored red. This 3:2 oligomer would be prepared by heating this molar ratio of polyol 1 and FDCA, small amounts of sodium hypophosphite catalyst, and water in a flask and slowly increasing the reaction temperature over time while passing an inert gas such as nitrogen or argon through the flask to sweep out water to drive the esterification reactions to completion [8]. The structures of isolated oligomers were ascertained primarily by nuclear magnetic resonance (NMR) spectroscopy and infrared (IR) spectroscopy. Stroke/cure testing was performed to obtain cure times.

The oligomeric structure shown above was arbitrarily drawn in a linear fashion with a limited molecular weight due to drawing space restrictions, while branching, cross-linking of molecular chains, as well as variations in chain lengths would also have been expected. The molecular weights of these oligomers would be influenced by the molar ratio of FDCA to polyol 1, and the closer this ratio is to 1:1, the higher the expected oligomer molecular weights.

Use of these types of previously prepared oligomers represents a type of precuring that was expected would decrease the binder curing times when esterified with PAA compared to the curing times involved in direct esterification of polyol 1 with PAA as previously described. Efficient esterification of the activated oligomeric primary alcohol sites with PAA during curing should rapidly induce cross-linking and desired rigidity increases.

The presence of multiple hydroxyl groups in these oligomers leads to their having high water solubility caused by the water-solubilizing effect of hydroxyl groups due to hydrogen bonding effects. Binder candidates are required to have high water solubilities since they would be applied to fiberglass mats from relatively concentrated aqueous solutions.

7.5.2 Determining polyol 1/FDCA oligomer cure times with PAA

Oligomer cure times for reacting with PAA were determined by stroke/cure tests on a hot plate maintained at 180 °C. As previously described, the moles of carboxylic acid provided by PAA relative to the moles of hydroxyl groups provided by different oligomers were maintained very close to 1.65:1, and these compositions had the same water percentage of 69.1% and also the standard weight ratio of catalyst to reactants.

7.5.3 Cure times for reaction of polyol 1: FDCA oligomers with PAA

Stroke/cure tests were performed by reacting aqueous solutions of oligomers of polyol 1: FDCA having molar ratios ranging from 1:1 to 1:0.6 with PAA. The initial PAA percentages were in the 12–15% range and thus would provide lower corrosion potential than the polyol 1/PAA as well as the glycerin/PAA curing systems.

Some test mixtures had insoluble components and only those test results obtained from solutions that were completely homogeneous are listed in Table 7.1. As mentioned in Section 7.4.4 concerning the curing of polyol 1 with PAA, the fact that cure times were almost the same with and without the catalyst sodium hypophosphite indicates that the use of this catalyst was redundant and had minimal catalytic effect. Thus, if sodium hypophosphite had been omitted, the same results should have been obtained.

It can be seen that oligomers 2 and 3 with polyol 1: FDCA compositions of 1:0.7 and 1:0.6 had the shortest cure times of 57 s. Continued development and testing of the polyol 1: FDCA oligomer approach would focus on these two oligomer compositions.

Table 7.1: Stroke/cure test results for reaction of polyol 1: FDCA oligomers with PAA.

Oligomer number	Oligomer molar composition		Cure time (s)
	Polyol 1	FDCA	
1	1	0.8	83
2	1	0.7	57
3	1	0.6	57

7.6 Overall comparison of cure times for reactions of glycerin, polyol 1, and polyol 1/oligomers with PAA

Table 7.2 provides a comparison of the cure times of glycerin, polyol 1, and polyol 1/FDCA oligomers when cured with PAA. It is apparent that polyol 1: FDCA-based oligomers 2 and 3 have by far the shortest cure times, and these results validate the advantage of this type of pre-curing. Further advantages of this oligomer approach are the reduced use of petroleum-derived PAA in terms of decreased PAA expense and equipment corrosion and consequential equipment repair or replacement.

Table 7.2: Stroke/cure test results of binder compositions with PAA at 180 °C.

Samples	Cure times (s)	Percent of glycerin cure time
Glycerin	102	–
Polyol 1	87	85
Polyol 1: FDCA	57	56
Oligomers 2 and 3	57	56

Additionally, an oligomer of polyol 1 and malic acid (2-hydroxybutanedioic acid) had a 46-s cure time when cured with PAA but malic acid is not being promoted since it is not commercially available at scale at this time [8].

7.7 FDCA availability

The relatively low availability and high price of FDCA have provided a severe economic obstacle to the commercialization of products incorporating FDCA and the polyol 1-based compounds and oligomers described in this chapter. However, Avantium N.V., a Dutch producer of numerous bio-based products, announced in a press release on April 7, 2021, that it had plans to open the world's first FDCA plant in 2023. The planned FDCA flagship plant is set to economically produce 5 kton per annum of FDCA, the key building block for the 100% fossil-free, recyclable polymer polyethylene furanoate that is expected to replace high-volume petroleum-based polyethylene terephthalate (PET). Avantium's plans to produce a wide variety of bio-based and renewable products have been published [11].

DuPont recently sold its biomaterial business to China's Huafon Group, and the resulting Covation Biomaterials plans to work with Huafon to produce furan dicarboxylic acid methyl ester that is an intermediate shown in Figure 7.2 for producing polyol 1 and polyol 2 [12].

7.8 Potential development and marketing activities for formaldehyde-free binders

Continued development of formaldehyde-free binders will primarily depend on the economic production of FDCA, diethanolamine and N-methyl-N-ethanolamine, which are the central components of polyol 1, polyol 2, and polyol 1/FDCA oligomers, respectively. As indicated earlier, the availability of FDCA depends on Avantium, Covation/Huafon, or another company successfully opening an FDCA plant.

A more accurate method to determine cure times could be developed by using solid-state NMR spectroscopy or IR spectroscopy to determine cure times to replace useful but visually-judged stroke/cure tests. Tensile strength testing methods need to be developed to determine bonding strength between glass fibers or wood particles bonded together by the described formaldehyde-free binders.

US patents 7638592, 8309676, and 9290698 will remain in effect until 2026, 2029, and 2033, respectively, to allow interested companies time to recoup continued research cost before patent expiration. Expiration dates were calculated by adding 17 years to the issuance date.

7.9 Summary

The properties and main features of proposed binders that completely avoid the release of formaldehyde and have curing times with PAA that are appreciably shorter than that of glycerin and other nonactivated polyols are as follows:

- Polyol 1/FDCA oligomers have been developed with cure times with PAA that are appreciably less than polyol 1/PAA-based binders and glycerin/PAA-based binders.
- Polyol 1/FDCA oligomers require reduced use of PAA compared to polyol 1/PAA and glycerin/PAA systems to help moderate future increased prices of petroleum.
- Polyol 1-FDCA oligomers have higher pH values (less acidic) when cured with PAA than polyol 1/PAA and glycerin/PAA-based binders that would lead to decreased production-line corrosion and repair expenses.
- The fact that polyol 1/FDCA oligomers and polyol 1-based binders do not require esterification catalysts for reaction with PAA should lead to very reproducible binder properties with no need for exact replication of catalyst concentrations as needed for binders based on glycerin or similar nonactivated polyols.
- Use of polyol 2 that has only two hydroxyl groups per molecule in combination with polyol 1-based systems should lead to adjustable reaction viscosities and mat rigidities.
- Use of furan-containing polyols in various binder approaches should lead to reduced binder flammability.
- Biobased and renewable fructose is an essential and available starting material needed to synthesize FDCA and polyols 1 and 2.

References

[1] Latorre N, Silvestre JF, Monteagudo AF. Allergic contact dermatitis caused by formaldehyde and formaldehyde releasers. Actas Dermosifiliogr 2011;102(2):86–97.

[2] State of California Environmental Protection Agency, Office of Environmental Health Hazard Assessment. Safe drinking water and toxic enforcement act of 1986-chemicals known to the state to cause cancer or reproductive toxicity, September 29, 2006.

[3] World Health Organization (WHO). International agency for research on cancer monographs on evaluation of carcinogenic risks for humans by formaldehyde.

[4] Strauss CR. Fibrous glass binders, US Patent 5318990.

[5] Arkens CT. Curable aqueous composition and use as fiberglass non-woven binder, US Patent 5763524.

[6] Benecke HP, Garbark DB, Kawczak AW, Clingerman MC. Formaldehyde free binders, US Patent 7638592.

[7] Benecke HP, Garbark DB, Kawczak AW, Clingerman MC. Formaldehyde free binders, US Patent 8309676.

[8] Benecke HP, Garbark DB, Kawczak AW, Clingerman MC. Formaldehyde free binders, US Patent 8378056 (abandoned).

[9] Wicks Zw Jr., Appelt MR, Soleim JC. Reaction of N-(2-hydroxyethyl) amido compounds. J Coat Technol 1985;57(726):51–61.

[10] Benecke HP, Garbark DB. Biobased polyols for potential use as flame retardants in polyurethane and polyester applications, US Patent 9,290,698.

[11] Reisch MS. Avantium tries for an encore. Chem Eng News 2018;11:24–25.

[12] McCoy M. DuPont biomaterials emerges as Covation. Chem Eng News 2022;100(20):10. https://cen.acs.org/business/biobased-chemicals/DuPont-biomaterial-business-emerges-new/100/i20?ref=search_results

Herman P. Benecke, Daniel B. Garbark

Chapter 8
Bio-based lubricants and hydraulic fluids

8.1 Background and advantages of soybean oil

Vegetable oils such as soybean oil (SBO) are composed of the triol glycerin that is esterified with three fatty acids to generate triglycerides that have very favorable lubricant properties. This lubricity is based on their excellent film-forming properties due to the strong association of their ester carbonyl groups with metal surfaces. Other advantages of SBO include its high fume point and flash point of approximately 200 °C and 300 °C, respectively [1]. These performance properties and the fact that SBO is bio-based, sustainable, and biodegradable has led to its increased use as lubricants and hydraulic fluids over the last 15–20 years [2].

SBO is the major vegetable oil harvested in the USA with a 2021 production rate of about 27 million tons. Many varieties of SBO are grown in the USA, which are characterized by the relative proportion of the unsaturated fatty acids (oleic acid), the polyunsaturated fatty acids (linoleic and linolenic acids), and the saturated fatty acids (stearic and palmitic acids). Based mainly on genetic modification and minor amount of crop breeding, SBO production has been shifting toward higher oleic SBO due to the enhancement of its lubricant properties with increased oleic acid and other properties, as described in this chapter. The fatty acid compositions of these SBO grades are shown in Table 8.1, where we refer to the original starting composition as "commodity"-grade SBO.

Table 8.1: Percentages of oleic, polyunsaturated fatty acids, and saturated fatty acids in US soybean oils.

Description	% Oleic	% Polyunsaturated	% Saturated
Commodity	22	61	16
Low saturates	22	69	9
Mid oleic	53	32	15
High oleic	77	11	12
90% oleic	90.5	1.5	8.0

Herman P. Benecke, Columbus, OH, e-mail: beneckefl@gmail.com
Daniel B. Garbark, Columbus, OH, e-mail: dangarbark@yahoo.com

 | https://doi.org/10.1515/9783112212141-008

8.2 Performance limitations of SBO

A serious liability of commodity-grade SBO is the oxidative instability of its allylic methylene groups that are adjacent to the lone double bond in oleic acid and especially the double allylic methylene groups, flanked by two double bonds, as found in linoleic and linolenic acids. Hydrogen atoms on these methylene groups are readily abstracted by molecular oxygen (a diradical) at normal, and especially at elevated, temperatures. Hydrogen abstraction is enabled, as the resulting radicals are stabilized by resonance with one double bond, as found in oleic acid, and by two double bonds, as found in linoleic and linolenic acids. These carbon radicals initiate fatty acid bridging processes as well as scission, leading to undesired deposits and volatilization [3].

Other SBO liabilities, when functioning as lubricants, include the hydrolysis of ester linkages, relatively low viscosities, and relatively high pour points (temperature below which the oil does not flow). These liabilities and solutions are addressed by different chemical modifications described in this chapter.

One tactic used to address the oxidative instability of SBO had been to employ heavy metal-based antioxidants but these materials can be harmful since about 50% of used motor oil worldwide is dumped in the environment [3]. As mentioned, mainly genetic modification and some crop breeding have been used to reduce the relative proportion of linoleic and linolenic acids in SBO that are most reactive toward oxygen and also to reduce saturated fatty acids to help reduce oil pour points. However, the oxidative and hydrolytic stabilities of relatively high oleic SBOs, achieved by these methods, still do not measure up to the properties provided by petroleum-based lubricants and hydraulic fluids that have minimal oxidative and hydrolysis issues.

8.3 Chemical modifications to enhance the oxidative stability of SBO

Two related SBO molecular modifications were developed to address the oxidative instability that involves adding specific substituents to the original double-bonded carbon atoms in SBO's unsaturated fatty acids, thus removing double bonds that are the cause of their oxidative vulnerability [1, 3]. In one approach, two ester groups are added across each double bond, and in the second approach, one ester group and one hydrogen atom are added across each double bond. Both approaches are initiated by fully epoxidizing all SBO double bonds to their epoxidized derivatives by reaction with formic acid and hydrogen peroxide.

8.3.1 Preparation of triglyceride vicinal diesters

Figure 8.1 illustrates the reaction of epoxidized linoleic acid with carboxylic acid anhydrides in the presence of the base potassium carbonate. This process leads to positioning two vicinal (adjacent) ester groups along the fatty acid, at each original olefinic carbon atom, to form "diesters." These modifications shown with epoxidized linoleic acid will also be taking place with oleic acid and epoxidized linolenic acid so that, essentially, no double bonds remain to activate methylene group oxidation.

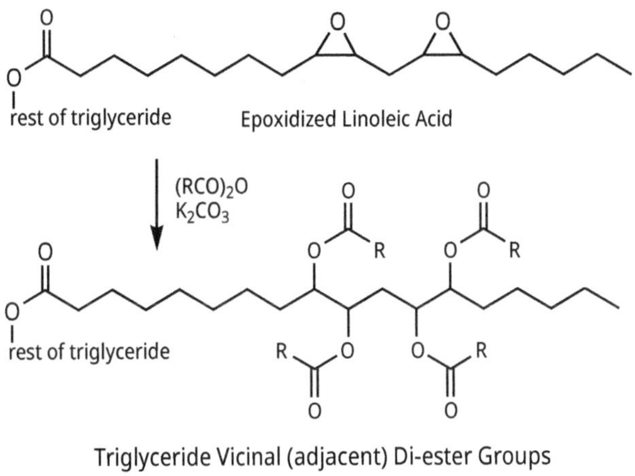

Figure 8.1: Preparation of triglyceride vicinal diesters.

When the term "diester" or a more specific term such as "dihexanoate" is used, we are referring to the vicinal diesters positioned at each end of all original double bonds, as implied in Figure 8.1. The size of "R" groups may be varied to achieve variations in viscosities and physical properties to influence their lubricant and hydraulic fluid applicability.

8.3.2 Preparation of triglyceride monoesters

Figure 8.2 shows a two-step reaction sequence, starting with epoxidized linoleic acid that involves the hydrogenolysis of each epoxide ring, to deposit a hydroxyl group and a hydrogen atom at either end of the original double bond. One of four possible regioisomers is shown in Figure 8.2 but other regioisomers (where hydroxyl and hydrogen groups are switched) would also be generated. The hydroxyl groups are typically acylated with acid chlorides to generate "monoester" functionality, positioned along the fatty acids at their original double-bonded carbon atom sites.

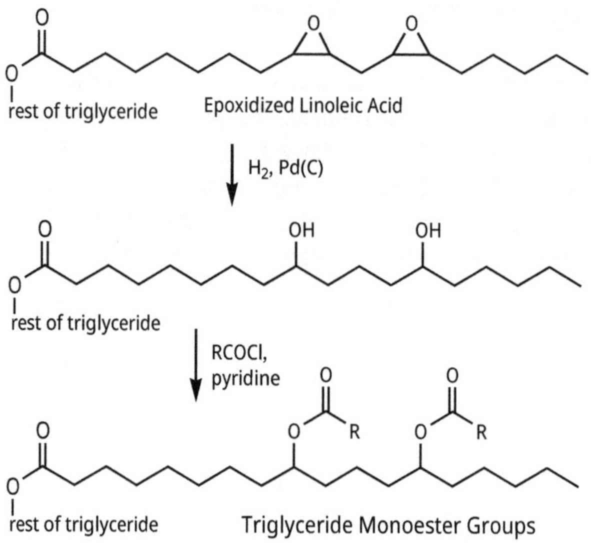

Figure 8.2: Preparation of triglyceride monoesters.

The size of "R" groups may also be varied to influence their lubricant and hydraulic fluid applicability.

8.4 Determining oxidative resistance

Two tests were used to determine the oxidative resistance of SBO and the ester derivatives described earlier, when exposed to specified high temperatures and heated air flows for specified time periods, to determine their suitability for use as lubricants and hydraulic fluids.

8.4.1 Noack test

The Noack test (described in ASTM D5800) determines the loss by volatilization of samples heated at 250 °C for 60 min to initiate oxidative processes, with a constant flow of heated air over the samples that also removes volatile components. This test data is relevant for combustion engine oil candidates since it measures the weight loss of lighter components formed by oxidative degradation that would increase the viscosity of remaining oil, thus contributing to poor engine circulation and reduced lubricity. In general, the Noack test for engine oils requires volatility losses of less than 15%, for certification for its use as engine oil. Multiple Noack tests were run with a variety of derivatized SBO diester and monoester samples and all received passing scores, thus helping qualify their potential use as motor oils. Hydraulic fluid candidates may also be characterized by this test.

8.4.2 Penn. state microoxidation test

In the Penn. state microoxidation test, specified amounts of samples are placed on stainless steel coupons and heated at 180 °C, and specified volumes of air are passed over the sample for specified times. Samples are weighed afterward to determine the amount of volatilized sample, and washed with the nonpolar solvent tetrahydrofuran, to dissolve the oil, leaving behind potential deposits that are weighed to determine their quantity [3].

8.5 Performance evaluation of SBO diesters and monoesters

In a type of test described later, SBO lubricities and SBO triglyceride diesters and monoesters were determined by measuring the scars incurred in four-ball wear tests, as described in ASTM D4172. In these tests, stainless steel balls are rotated past each other under high load while in contact with lubricant candidates.

Seal swelling is not desired in lubricant and hydraulic systems. A testing approach to determine the extent of seal swelling of lubricant and hydraulic oil candidates was developed, as described further [3].

8.5.1 SBO ester derivative lubricities determined with four-ball wear tests

Four-ball wear tests were performed with commodity-grade SBO and commodity-grade triglyceride diester and monoester derivatives [3]. These tests were performed with stainless steel balls at 18 kg load at 75 °C, 1,200 rpm, and for 60 min, and the results are shown in Table 8.2. ZDDP (zinc dialkyl dithiophosphate) is wear additive and its use was evaluated at 1% with SBO and modified SBO samples. In four-ball wear tests, the lower the scar diameter, the higher is the lubricant's lubricity.

As shown in Table 8.2, nonmodified SBO had the highest scar diameter, which was reduced to about half with 1% ZDDP. The best performing modified oil was SBO dihexanoate, *without any ZDDP*, and its scar diameter was equal to or less than that obtained from a fully formulated SHC-134 gear oil and an SAE 10W-30 motor oil. The next highest performer was SBO monoisobutyrate, which when formulated with 1% ZDDP, had a scar diameter only slightly larger than the scar diameters obtained with the commercial oils. The scar diameters of four other modified oils, when modified with 1% ZDDP, were shown to be only slightly higher than those of the commercial oils. In summary, these test results show that certain commodity-grade SBO diesters and monoesters had significantly smaller wear scars than commodity-grade SBO, and their scar diameters were equal to or comparable to those of fully additized gear oil and commercial motor oil [3].

Table 8.2: Four-ball wear tests of commodity-grade SBO, diester, monoester, and commercial oils at 18 kg load.

Description:	SBO	Dihexanoate	Monoisobutyrate	SHC-134	SAE 10W-30
Wear scar (mm)	0.62	0.26	0.32	0.26	0.27
Wear scar + 1% ZDDP (mm)	0.32	0.28	0.28	–	–

The reduced scar diameters displayed by the behavior of SBO ester derivatives reflects enhanced lubricities provided by the added ester groups positioned along the triglyceride fatty acids. These extra ester groups presumably provide extra bonding with the metal surfaces, in addition to the bonding provided by the three triglyceride ester carbonyl groups.

Four-ball wear tests were also performed under 40 kg load, in which diesters and monoesters did not perform as well as commercial oils. However, addition of 1% ZDDP provided performances equal to or nearly equal to fully additized commercial oils. Continued research could evaluate the use of SBO diesters and monoester of grades other than commodity, to determine if suitable performance can be obtained in 40 kg four-ball wear tests without ZDDP.

8.5.2 EPDM seal swelling in the presence of lubricants and hydraulic fluids

Seal swelling is a problem in many lubricant and hydraulic fluid applications, and the seal material EPDM (ethylene propylene diene, M class, hardness) is commonly used in these applications.

EPDM pieces were weighed and immersed in commodity-grade-derived SBO diesters and monoesters for 24 h at 68 °C, thoroughly dried and then weighed again to determine any weight gains or losses [3]. These tests determined that 8.9% of commodity-grade SBO was absorbed in EPDM, while zero or nil quantities of the diesters SBO triglyceride dihexanoate, SBO triglyceride-bis-2-ethylhexanoate, and SBO triglyceride-bis-2-ethyl butyrate were absorbed in EPDM. Close to 1% of two commodity-grade SBO triglyceride monoesters were absorbed in EPDM.

In summary, these tests indicated that EPDM was essentially resistant to the commodity-grade SBO triglyceride diesters tested. Further testing would determine the magnitude of swelling over extended times to determine the potential long-term absorption of these SBO derivatives in EPDM. Nitrile rubber was essentially unchanged when exposed to both nonmodified and modified commodity-grade SBO.

8.6 Utility of SBO triglyceride diester and monoester lubricants

8.6.1 SBO triglyceride dihexanoates

Table 8.3: Viscosities of SBO triglyceride dihexanoate lubricants from various SBO grades.

Description	90% oleic	High oleic	Mid-oleic	Commod.	Low saturates	SAE 50
Viscosity (cSt)						
At 40 °C	210	272	349	398	601	222
At 100 °C	23.5	28.8	33.6	36.2	47.6	20.4
VI	138	141	137	134	132	107
Pour point	−23	−24	−21	−21	−21	−18

Data shown in Table 8.3 indicate that the ISO values [viscosity values in centistokes (cSt) at 40 °C] of SBO triglyceride dihexanoates steadily increase between 90% oleic and the low saturate SBO grade [1]. As given in Table 8.1, this viscosity shift is accompanied by a corresponding general reduction in the concentration of oleic acid (containing one double bond) and an increase in the combined concentrations of linoleic and linolenic acids (containing two and three double bonds, respectively). Thus, this compositional shift provides an increase in the number of triglyceride fatty acid double bonds as one moves from left to right in Table 8.1. Given that each double bond is functionalized by two hexanoate ester groups, the increase in ISO viscosity is primarily explained by the increased contribution of these dihexanoate ester groups to the molecular weights of triglyceride fatty acids as they undergo this compositional shift. However, increased amounts of saturated fatty acids also contribute to higher ISO viscosities.

It can be seen that the SBO triglyceride dihexanoate lubricant, prepared from 90% oleic acid with an ISO value of 210, is in the range of SAE 50 motor oil that is used in high-performance race cars.

Viscosity index (VI) is a measure of a fluid's viscosity change with change in temperature, and higher VIs are desirable since they undergo significantly lower viscosity changes with temperature changes. Note in Table 8.3 that the VIs for all SBO triglyceride dihexanoates are slightly higher than that of the commercial SAE 50 oil shown.

The higher viscosity SBO triglyceride dihexanoates, shown in Table 8.3 could be used as base oils for grease applications and gear oils. It has been found that lower molecular weight diesters have higher ISO viscosities than higher molecular weight diesters. Thus, the ISO viscosity of commodity-grade SBO triglyceride dihexanoate is 398 (pour point: – 21 °C) while the ISO viscosity of commodity-grade SBO triglyceride dipentanoate is 422 cSt (pour point: – 21 °C). The ability to readily achieve a variety of high-viscosity grease base oils presents the opportunity to use decreased amounts of expensive thickeners such as polyureas.

8.6.2 SBO triglyceride monohexanoates

Commodity-grade SBO was also converted to SBO triglyceride monohexanoate (ISO 136) and monoisobutyrate (ISO 222) that could be used for SAE 40 and SAE 50 motor oils, respectively. This comparison illustrates that monoester derivatives have lower ISO viscosities than comparable diester derivatives, when functionalized with the same ester due to their lower overall molecular weights.

8.7 Glycerin replacement (backbone modification) in SBO triglycerides with dialcohols and monoalcohols

One approach for overcoming high viscosities found in triglyceride derivatives is to hydrolyze a desired SBO-grade sample and re-esterify the recovered fatty acids with a dialcohol (diol) or a monoalcohol (mono-ol) [1]. Figure 8.3 shows the esterification of high oleic acid with the diol 1,2-propylene glycol (1,2 PG) to generate 1,2 PG dioleate that would be epoxidized and then acylated with a carboxylic acid anhydride of choice to generate the product 1,2 PG dioleate dicarboxylate. Exchanging the alcohol in an ester is referred to as changing the ester "backbone" or performing a "backbone" modification.

Another approach to obtain the same product would involve the transesterification of high-oleic SBO with methanol to prepare high-oleic methyl oleate (a specific soy biodiesel), which would be efficiently transesterified with 1,2 PG to produce 1,2 PG dioleate. This intermediate would also undergo the last two reactions shown in Figure 6.3 to produce 1,2 PG dioleate dicarboxylate.

8.7.1 Utility of 1,2 PG disoyate dihexanoates

Note that the ISO viscosities of the SBO 1,2 PG disoyate dihexanoate samples shown in Table 8.4 are appreciably lower than the ISO viscosities of the corresponding grade SBO triglyceride dihexanoate samples in Table 8.3 due to their lower molecular weights [1]. Also note that as one moves from left to right in Table 8.4, the ISO viscosities of different grades of SBOs increase significantly as the oleic acid concentration decreases, while the linoleic plus linolenic acid concentrations increase (see Table 8.1). This ISO viscosity shift represents another example of viscosity dependence on the "oleic acid/(linoleic plus linolenic acid)" ratio.

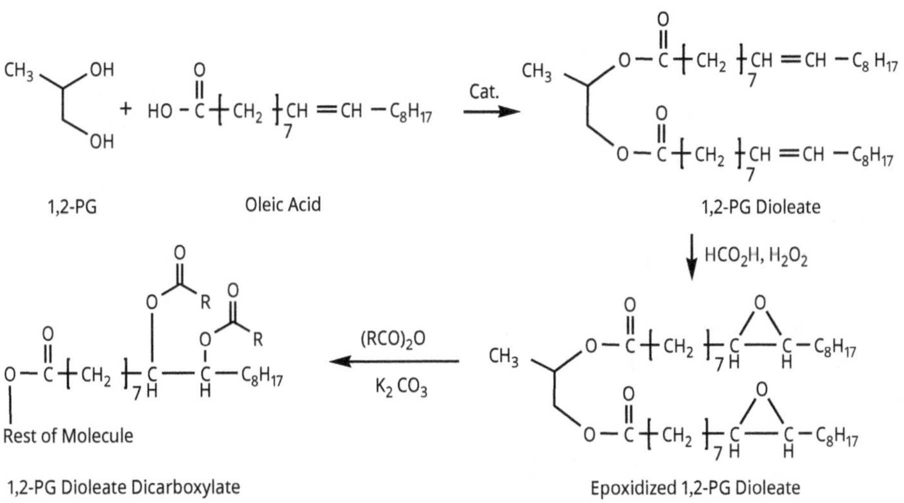

Figure 8.3: Re-esterification of fatty acids from SBO with 1,2 PG.

Table 8.4: Viscosity and pour point data for 1,2 PG disoyate dihexanoates for various SBO grades and select commercial motor oils.

Description	90% oleic	High oleic	Commodity	SAE 10W-30	SAE 10W-40	SAE 50
Viscosity (cSt)						
At 40 °C	78.0	99.3	232	69.7	95.4	222
At 100 °C	12.1	13.7	20.8	10.7	14	20.4
VI	153	139	105	141	151	107
Pour point (°C)	−34	−12	−12	-	−27	−18

The 1,2 PG disoyate dihexanoate sample from 90% oleic SBO, with an ISO viscosity of 78 cSt, is in the range of 10W-30 motor oil. The beneficial pour point of − 34 °C is due to the relatively low percentage of saturated fatty acids in this sample (nominally 8%).

The 1,2 PG disoyate dihexanoate sample from high oleic SBO has an ISO viscosity of 99.3 cSt, with a pour point of − 12 °C. This sample is in the range of an ISO 100 hydraulic fluid as well as 10W-40 motor oil. High oleic SBO has a nominal saturate level of 12%, which contributes to this sample's higher pour point. Another factor contributing to this high pour point is that the sum of linoleic and linolenic acids in high oleic SBO is 11%, while it is only 2% in 90% oleic SBO (whose 1,2 PG disoyate dihexanoate's pour point is − 34 °C, as described earlier).

The 1,2 PG disoyate dihexanoate from commodity-grade SBO, with an ISO viscosity of 232, is in fairly close range to SAE 50 motor oil. Contributing to its relatively high pour point of − 12 °C is the relatively high saturate content of 16%, while the sum of linoleic plus linolenic acid (61%) also contributes to this sample's high viscosity and pour points. Use of pour point depressants should also improve its low-temperature behavior.

A specialty 88% "oleic" oil (Emersol 233) containing 74% oleic, 14% palmitoleic (a 16-carbon, monounsaturated fatty acid like oleic acid), 4% linoleic, and 8% stearic acids was converted to 1,2 PG "Di-Emersol 233" dihexanoate. This product has an ISO viscosity of 97.2, with a VI of 149 and pour point of − 39 °C, and is thus in the range of an ISO 100 hydraulic oil and 10W-40 motor oil. The low pour point of − 39 °C is an attractive property for either use.

8.7.2 Utility of 2-butyl "oleate" diisobutyrate

Branched mono-ols, such as 2-butanol, can be esterified with different grades of soy fatty acids to produce diesters and monoesters with relatively low viscosities due to their relatively low overall molecular weights. Pioneer Plenish-brand SBO (oleic acid content 76%) was hydrolyzed to produce oleic acid that was esterified with 2-butanol, and this ester reacted with diisobutyric anhydride to generate 2-butyl "oleate" diisobutyrate. This product has an ISO viscosity of 25.0, with a VI of 109 and a pour point of <− 40 °C, so it is expected to be an effective low-temperature hydraulic fluid.

8.8 Biodegradable SBO diesters and monoesters and potential uses

Commodity-grade SBO triglyceride dihexanoate and SBO triglyceride-mixed mono (hexanoate/2-ethylhexanoate) were shown to be highly biodegradable when using ASTM 5,864 [3]. That these diverse diester and monoester samples demonstrated biodegradability strongly suggests that the range of SBO ester derivatives is also biodegradable and this property will help mitigate irresponsible dumping and related environmental pollution. Given their adaptability to fit various viscosity and low-temperature profiles, specific SBO-based esters described herein could also fit well in outdoor applications, including two-stroke engine oils, open-gear lubricants, rail flange oils and greases (to limit friction on straight and especially curved rails), rock-drilling fluids, and transformer fluids (with sufficiently high dielectric strength).

8.9 Controlling hydrolysis rates of ester-based lubricants and hydraulic fluids

An issue in ester-based motor oils is their potential hydrolysis due to contact with water produced in combustion-driven engines. It is interesting that hydrolysis of ester-based motor oils should be nonexistent in battery-powered car engines due to the absence of "water of combustion" in their engines.

It is known in general that the rates of ester hydrolysis are retarded by steric crowding around the ester carbonyl group provided by alkyl substituents' positioned alpha and beta to the alcohol hydroxyl functionality. Alcohols that provide some degree of steric crowding to reduce ester hydrolysis include the diols 1,2 PG, 2-methyl-1,3-propane diol, and the mono-ol, 2-butanol. Inspection of models of the triol 1,1,1-tris(hydroxymethyl) propane [also known as trimethylolpropane (TMP)] indicates that the ester carbonyl groups of TMP fatty acid esters, where each TMP alcohol group is esterified with a fatty acid, should experience significant steric hindrance toward hydrolysis.

Ester hydrolysis in hydraulic systems should be at a minimum since these are closed systems.

The hydrolytic stabilities of ester-based samples were obtained via ASTM D2619 by heating weighed samples in a tumbled vessel immersed in water heated at 93 °C for 48 h and then titrating with KOH, to determine released carboxylic acid. A summary of an increase in acid number of these samples as well as other test data is given in Table 8.5.

A passing score in the hydrolytic stability test is that not higher than 2.0 mg KOH/g is needed, so all substrates passed this test. It can be seen that replacing the glycerin backbone with 2-methyl-1,3-propane diol and TMP provided increased hydrolytic stability. The pendant methyl groups in 1,2 PG and 2-butanol fatty acid esters are also expected to provide some hydrolytic stability.

In the Noack test, engine oils require volatility losses of less than 15% for certification; hence, the above mentioned samples pass this test for motor oil use, given the above volatilization data.

The observed high flash points also appear to be reasonable for a number of uses.

Table 8.5: Backbone effects on hydrolytic stabilities, volatilities, and flash points.

Hydrolytic stability tests (ASTM D2619)		Noack % volatility ASTM 5800B	Flash point (°C) ASTM D92-05a
Lubricants: dihexanoates of commodity SBO	**Change in acid number**		
Triglyceride backbone	+1.34mg KOH/g	6.1*	298
2-Methyl-1,3-propane diol backbone	+0.65mg KOH/g	5.8	273
TMP backbone	+0.22mg KOH/g	6.3	243

*Penn state microoxidation test: % weight loss at 180 °C for 60min.

8.10 Viscosity and oxidation stability comparisons between nonmodified SBO and SBO diester and monoester derivatives

For comparison purposes, refined, bleached, and deodorized (RBD) commodity-grade samples of nonderivatized SBO were tested and determined to have an average ISO viscosity of 32.0, with an average VI of 228 [3]. These commodity-grade SBO samples and an 82% oleic SBO sample were tested by the Penn. state microoxidation test, and each sample demonstrated some deposit weight gain and appreciable evaporative loss, indicating that oxidative degradation processes were active in both commodity-grade SBO and relatively high oleic acid SBO, which indicates that it also has some oxidative vulnerability.

Importantly, the low ISO viscosities of these RBD SBO samples do not come close to matching the much higher viscosities of high-performance motor oils, gear oils, and base oils for grease applications that are described in Section 8.6 for specified SBO triglyceride diesters and monoesters. These RBD SBO viscosities are also much lower than the viscosities described in Section 8.7.1 for 1,2 PG disoyate dihexanoate derivatives that fit the viscosity requirements of 10W-30, 10W-40, SAE 50 motor oils, and an ISO 100 hydraulic fluid.

This comparative data supports the practice of converting specific grades of SBO to their diester and monoester derivatives to generate a range of usable and oxidatively stable SBO-derived lubricants and hydraulic fluids that are not available from direct use of nonmodified SBO.

8.11 Continued development and commercialization activities

This chapter describes various SBO modifications that overcome the oxidative vulnerability of SBO while simultaneously increasing the inherently low SBO viscosities and producing numerous applications with useful properties. Companies currently producing lubricants, motor oils, greases, and hydraulic fluids from petroleum sources should be highly motivated in further evaluating the potential and cost-effectiveness of SBO diesters and monoesters described in this chapter. Their interest would be mainly driven by the apparent shift away from fossil fuels due to climate change considerations that may significantly reduce the availability of petroleum-based starting materials.

Continued research could focus on preparing lubricants and hydraulic fluid candidates that more closely matched the viscosities of commercialized materials to evaluate performance evaluations.

Data presented in Section 8.5.1 shows that SBO triglyceride dihexanoate in 18 kg load four-ball wear tests without the use of the zinc-containing wear additive ZDDP provided scar diameters that were lower than those obtained from fully additized commercial lubricants. However, when this test was repeated at 40 kg load, this SBO derivative did not perform as well as these commercial lubricants. Continued research could determine whether a range of 1,2 PG disoyate dicarboxylates would pass the 40 kg load test to determine if metals containing wear additives were not needed, which would provide environmental and cost advantages while also providing systems more resistant to hydrolytic cleavage.

Research could also be continued in evaluating the same type of 1,2 PG diester derivatives to potentially obtain more effective seal-resistant lubricants and hydraulic fluids compared to previously tested SBO triglyceride diesters (see Section 8.5.2).

The US patents 8357643 and 9359572 will remain in effect until 2030 and 2033, respectively, thus allowing interested companies time to recoup continued research and development costs before their expiration.

8.12 Summary

The tendency of SBO to readily undergo oxidative cleavage and polymerization greatly inhibits SBO's direct use as a base oil for a variety of lubricants and hydraulic oils. Based on Noack and other test results, this liability is overcome by modification of SBO's double bonds by the addition of two ester groups to form "diesters" or by the addition of one ester group and a hydrogen atom across these double bonds to form monoesters [1, 3].

The attributes of diester and monoester derivatives across the range of SBO grades and compositions are as follows:

- A wide range of base oil viscosities are obtained from different grades of SBO that are characteristic of gear oils, greases, engine oils, and hydraulic fluids.
- Gear oils and greases maintain their triglyceride structures and higher viscosities.
- Engine oils and hydraulic fluids may require triglyceride backbone exchange with dialcohols such as 1,2 PG or monoalcohols such as 2-butanol to achieve the needed decreased viscosities and decreased pour points.
- The ability to readily achieve high-viscosity greases presents the opportunity to avoid the use of expensive thickeners such as polyureas.
- An SBO triglyceride diester and a mixed monoester were shown to be highly biodegradable, and this general property will help mitigate irresponsible dumping.
- Based on their biodegradability, the described modified lubricants fit well with outdoor applications such as two-stroke engine oils, open-gear lubricants, rail flange oils and greases, rock-drilling fluids, and transformer fluids (with sufficiently high dielectric strength).
- In ester-modified SBO, viscosities decrease with increases in oleic acid content and decrease in linoleic and linolenic acids, while viscosities increase with decreases in oleic acid concentration and increases in linoleic plus linolenic acids.
- Hydrolytic stability of diester-based lubricants and hydraulic fluids is increased by triglyceride backbone exchange with alcohols bearing substituents at their alpha and beta positions.
- Hydrolysis of diester and monoester lubricants would be minimal when used in electric motors due to the lack of "water of combustion" formed in these engines.
- Pour points move to the desired lower values with decreasing saturated fatty acid contents.
- Hydraulic fluid candidates that fit the high-volume ISO 100 market (pour point: −39 °C) and also the low-temperature ISO 22 market (pour point: <−40 °C) have been prepared.
- Four-ball wear-test scar diameters (18 kg load) of certain SBO triglyceride diester and monoesters are significantly smaller than that obtained from SBO (all without antiwear additives) and match those attained in commercial additized lubricants.
- Swelling of EPDM rubber by certain SBO diesters was significantly reduced (or was nil) compared to significant swelling in SBO, which has positive implications in overcoming seal swelling.
- Due to its reduction in double bonds, higher oleic SBO requires only about two-thirds of the reagents needed to produce diester or monoester derivatives compared to commodity-grade SBO.

References

[1] Benecke HP, Garbark DB, Vijayendran BR, Cafmeyer J Modified vegetable oil lubricants. US Patent 9359572 (issued on Jun. 7, 2016).

[2] Vijayendran BR Biobased motor oils are ready for primetime. Ind Biotech 2014;10:1–5.

[3] Benecke HP, Vijayendran BR, Cafmeyer J Lubricants derived from plant and animal oils and fats. US Patent 8357643 (issued on Jan. 22, 2013).

Claudia Creighton, Nguyen Duc Le, Huma Khan, Srinivas Nunna, Russell J. Varley*, Cai Li Song*

Chapter 9
An introduction to lignin-based carbon fiber

9.1 Background and history of carbon fiber

Carbon fiber possesses a unique combination of properties that sets it apart from most other fibers, such as high specific strength and modulus, low density, creep resistance, high thermal and electrical conductivity, excellent chemical resistance, and high temperature resistance in inert atmospheres [1–4]. Translating these properties into a carbon fiber composite therefore has enabled the development of lightweight, high-strength and stiffness structural materials and components that have revolutionized a wide range of industries such as transportation, sporting goods, defense, wind energy, pressure vessels and construction [2, 5–7].

Since the discovery of carbon fiber more than 140 years ago, carbon fiber properties have continued to improve and evolve as improved manufacturing methods have been developed, and the knowledge of the underlying materials chemistry of the precursor fiber has increased. The first carbon fiber was made by Joseph W. Swan and Thomas A. Edison in 1880 by meticulously carbonizing cotton threads and used as a filament for an incandescent bulb [8–10]. Unfortunately, soon after its invention, tungsten replaced carbon fiber as a bulb filament [11]. However, in 1909,W.R. Whitney continued the research and showcased an effective way of graphitizing cellulosic fibers for bulb filament applications [12]. After 50 years, Roger Bacon and a group of scientists at Union Carbide Corporation, Ohio, USA, extensively researched pathways to produce high-performance carbon fibers from cellulose-based rayon precursor [12–14]. Their research efforts led to the commercialization of high-modulus rayon-based carbon fibers and found applications in the defense industry sector [15]. While researchers based in the USA were exploring rayon-based carbon fibers, in 1961, Akio Shindo from the Government Industrial Research Institute, Osaka, invented polyacrylonitrile (PAN)-based carbon fibers [16–18]. Shortly after this invention, in 1967, scientists R. Moreton, W. William and W. Johnson from Royal Aircraft Establishment (RAE) developed proto-

*Corresponding author: **Russell J. Varley**, Carbon Nexus at the Institute for Frontier Materials, Deakin University, 75 Pidgons Rd, Waurn Ponds, VIC 3216, Australia, e-mail: russell.varley@deakin.edu.au

*Corresponding author: **Cai Li Song**, Petronas Research Sdn Bhd, Lot 3288 and 3289, Off Jalan Ayer Itam, Kawasan Institusi Bangi, 43000 Bangi, Selangor, Malaysia, e-mail: song.caili@petronas.com

Claudia Creighton, Nguyen Duc Le, Huma Khan, Srinivas Nunna, Carbon Nexus at the Institute for Frontier Materials, Deakin University, 75 Pidgons Rd, Waurn Ponds, VIC 3216, Australia

 | https://doi.org/10.1515/9783112212141-009

cols to produce high-strength and high-modulus carbon fibers from PAN-based precursor fibers with the application of tension during the thermal treatment process [19, 20]. Two years later, in 1969, carbon fibers were also produced from other sustainable precursors such as lignin, and this work was patented by Sugio Otani et al. [21], which was assigned to Nippon Kayaku Co Ltd. However, after few years of commercialization, in 1973, the production of lignin carbon fibers ceased due to its poor mechanical properties [22]. Nevertheless, due to their characteristic advantages, research on lignin-based carbon fibers has continued until today, primarily focusing upon improving its mechanical performance. In 1977, Leonard Sidney Singer from Union Carbide Corporation, New York, USA, invented high-modulus carbon fibers from mesophase pitch precursor [23]. Horikori et al. [24] from Sumitomo Chemical Company Ltd, Japan, published a patent in 1978, which describes the method to produce carbon fibers from polyethylene precursor source through sulfonation and subsequent carbonization process. As the history of making carbon fiber shows, over the past, scientists explored various precursor sources to produce carbon fibers; however, since the 1970s only PAN-based carbon fibers have seen an exponential market growth and remains by far the most commonly used precursor for carbon fiber today.

9.1.1 Market trends

In 2019, the global nameplate capacity of carbon fiber was 161,200 metric tons (MT). The carbon fiber market is dominated by PAN-based carbon fiber (representing almost 90% of the total carbon fiber production) [25]. By 2026, this figure is expected to increase to 180,400 MT [26]. In terms of value, this translates to roughly $3.5 billion [27]. The markets that are expected to drive the substantial increase in demand for carbon fiber over the next few years are primarily the wind, aerospace or defense, and the infrastructure and energy sectors (Table 7.1). In fact, the wind energy industry alone is already the largest consumer by volume of carbon fiber in the world, making up to 28,000 MT in 2021. Further driven by the rapid surge of onshore and offshore wind turbines, more so in China, this amount is expected to double to 54,000 MT in 5 years' time, with the USA being the largest supply of wind turbine blades. In contrast, in the aerospace industry, although the demand for composite-intensive aircrafts, such as the likes of Boeing 787 and Airbus A350, has suffered from the COVID-19 pandemic, it is likely to return to pre-COVID-19 levels by 2025, with a market capacity of around 29,100 MT expected. In the aerospace sector, significant weight reduction can be achieved through the use of carbon fiber in manufacturing. According to a research from Toray Industries, if all passenger cars and aircraft in Japan adopted carbon fiber to reduce weight and improve fuel efficiency, 22 million tons of CO_2 could be saved [28].

By far, due to the complicated carbon fiber manufacturing process requiring some of the world's most advanced production technology, the global carbon fiber manufacturers are still mainly concentrated in Japan and the USA. Some big players on the PAN-based carbon fiber market are Toray (Japan), Hexcel (USA), Mitsubishi Chemical Carbon Fiber & Composites (USA), Teijin Limited (Japan), SGL Carbon (Germany), and many more. Among all the carbon fiber manufacturers and suppliers, Toray dominates the carbon fiber supply chain with 54,870 MT of annual capacity, in both PAN-based regular tow and large tow carbon fiber [30]. Meanwhile, the pitch-based carbon fiber on the market started with Kureha (Japan), the first company in the world, to successfully develop an industrial manufacturing process for pitch-based carbon fiber in 1970, and has now been penetrated by companies like Cytec Industries, Inc. (USA), Hexcel Corporation (USA), and Nippon Graphite Corporation (Japan) [31, 32]. Tempering the figure of the estimated market capacity in the future is the knockdown effect, which is the potential carbon fiber shortfall between the name-plate capacity of existing carbon fiber facilities and the actual production capacity (125,150 MT) that is estimated to reach 55,250 MT in 2026, after taking into account all the recent expansion announcement by Zoltek (Toray Group, Japan), Hexcel (USA), Hyosung (South Korea), and DowAksa (a joint venture of DOW Chemical, USA, and AKSA, Turkey). Among the issues that hinder the growth and competitiveness of the carbon fiber industry to fill the shortfall, the most significant is the production costs that directly affect the price of carbon fiber. Besides, there is the issue with the cost of energy, primarily in the context of energy policy or strategy that would account for the energy savings contributed by the lightweight materials [29]. This deficit in production will start to show its sign as early as in 2023 or 2024, which could cause market disequilibrium, and subsequently, a surge in price of the carbon fiber [33].

Table 9.1: Summary of carbon fiber demand by its tow and geographic region (adopted with permission from [29]; Copyright 2022 USDOE/OSTI).

	CAGR 2012–2018*	2018 demand (tons)*	Fiber tow	Region of greatest CF demand (% of worldwide demand)
Wind energy	11.5%	25,070	24K–50K	Europe (68%)
Aerospace	13.0%	18,667	Small tow	Europe (46%)
Automotive	20.5%	9,863	Large tow (>24K)	Europe (62%)
Pressure vessel	21.9%	7,225	12K–50K	North America (60%)
Overall	12.5%	92,802	–	–

*Average of estimates by Red and Zimm (2012), Lucintel (2012), and Industry Experts (2013).

CAGR, compound annual growth rate.

The demand for carbon fiber composites has grown over the years due to the need for light-weighting that is required to meet the stringent emission controls set by regulatory authorities around the world [34]. In 2018, it was estimated that the composite market revenue will increase from US$26 to 41.4 billion by 2025 [34]. Aligning with the predicted progress trend, market reports in 2018 [35] and 2019 [36] show-cased an increase of 1.1% in global production capacity of carbon fibers in 2019 compared to 149.3 kilotons in 2018; at the same time, the compound annual growth rate of carbon fiber market demand from 2010 to 2018 is obtained as 11.45% [35]. The global demand for carbon fiber composites from 2010 to 2018 and the estimated values from 2019 to 2023 are shown in Figure 9.1 [36].

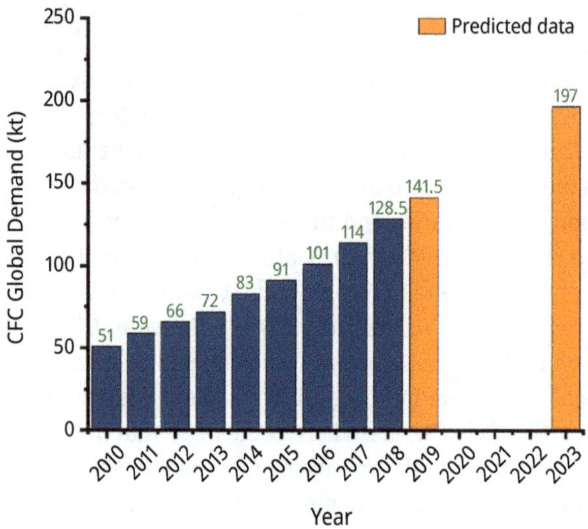

Figure 9.1: Carbon fiber composites' global demand from 2010 to 2018 and estimations from 2019 to 2023 [36].

9.1.2 Various grades of carbon fibers

A fiber with more than 92 wt% of carbon atoms is defined as carbon fiber [37]. The carbon content and its properties are strongly governed by the final carbonization temperature. Carbonizing fibers at temperatures up to 1,500°C results in approximately 95 wt% carbon, while further increasing the treatment temperature to 3,000°C leads to 99 wt% carbon content in fibers. Commonly, these fibers are described as graphite fibers and possess a highly aligned microstructure and corresponding high tensile modulus [37]. Carbon fibers with different combinations of mechanical properties can be manufactured by controlling the process parameters along the entire process chain, from the initial development of the precursor to carbonization. Various grades of carbon fibers that are on the commercial market are given in Table 9.2 and tend to be categorized according to their modulus. It is important to note, however, that these fibers are primarily produced from fossil fuel-derived precursors such as PAN and pitch in the case of ultra-high-modulus fiber.

Carbon fibers also come in various tow sizes, or number of individual fiber filaments, varying generally from 1k to 50k, although even higher tow sizes continue to be developed. With an increase in tow size, the cost per kg of carbon fibers reduces, but tends to come with higher variability in properties and hence performance [38, 39]. Larger tow sizes, such as 48k and 50k, are mainly used for industrial and non-aerospace applications, and the smaller tow sizes are generally used for aerospace applications and stricter requirements with respect to variability in fiber specifications and performance [39].

Table 9.2: Different grades of carbon fibers from PAN and pitch precursor types.

Carbon fiber grade	Tensile strength (GPa)	Tensile modulus (GPa)	Examples of commercial carbon fibers
Low modulus	1.0–2.5	50–229	XN-15-30S, XN-10-30S, XN-05-30S
Standard modulus	3.5–6.0	230–279	T300, T700S, PX35, IM6
Intermediate modulus	4.4–7.0	280–350	T800S, T1100S, IM9, IMS6L
High modulus	3.0–4.7	351–500	M46J, UMT430, HM63
Ultra-high modulus	2.5–4.0	501–950	M60J, UMT530, K13C2U, XN-80-60S

9.1.3 Carbon fiber precursors

Carbon fibers from fossil fuel-derived precursors such as PAN dominate the industry with more than 90% of all carbon fibers currently being produced from PAN. There are many reasons for this, but its high carbon yield of about 50–60% and the long history and experience of working with PAN enable a wide variety of properties to be manufactured [1, 40–42]. However, the high cost associated with the production of PAN precursor remains high, accounting for more than 50% of the carbon fiber production cost. To date, this has impeded its wider proliferation into high-volume industries such as the automotive industry and others that do not have the stringent requirements of the aerospace sector [43]. To enter broader utilization of carbon fiber composite material in the automotive industry, for example, it has been proposed that carbon fibers that cost less than US$15.4/kg whilst achieving a minimum tensile strength of 1.72 GPa and tensile modulus of 172 GPa [44] are required. In order to meet these requirements, potentially inexpensive alternative sustainable precursor materials such as lignin and cellulose are considered excellent candidates for carbon fiber production. While cellulose is the most widely available natural resource on the planet, lignin is the second most abundantly existing sustainable resource [45]. Cellulose is the most preferred bio-based precursor, in view of its availability and molecular orientation, but its poor carbon yield of only 10–30% after its carbonization inhibits its use in structural material applications [46]. Recently, Vocht et al. [47] developed novel stabilization protocols for cellulose precursor on lab scale to produce subsequent carbon fibers with a tensile strength of 2.8 GPa and a tensile modulus of 112 GPa. Results from this study provide an optimistic future for it to replace PAN-based fibers for specific mass market applications. Compared to cellulose, lignin possesses a high carbon content of 60–65%, which is extremely advantageous and making it a potentially ideal sustainable precursor if its processing challenges such as long stabilization times during manufacture can be overcome [46]. The present state of carbon fiber development from renewable precursor sources is still in its infancy at the research scale, despite ongoing efforts to scale up the manufacture of lignin and cellulose for fiber production. At the moment, the properties of the resultant carbon fibers, cellulose, lignin, and their blends are mostly inferior to carbon fibers produced from PAN and pitch-based precursor fibers. However, given the availability and potential cost of these bio-derived precursors if lignin carbon fibers with similar properties to PAN-based carbon fiber can be manufactured, the cost of carbon fiber could be cut in half [37]. Considering the economic and environmental benefits, current research is now refocusing on the development of bio-based precursors more explicitly using lignin as a primary source. The relative advantages and disadvantages of the most common alternative precursors and how they compare to PAN are given in Table 9.3.

Table 9.3: Benefits and limitations of using various renewable and nonrenewable precursors for the production of carbon fibers [37, 40, 44–46, 48–52].

Precursor	Advantages	Disadvantages
PAN	High tensile strength, good compression properties, and good carbon yield	High cost per kg and HCN emissions during thermal treatment
Pitch	High tensile modulus, high thermal conductivity, and high carbonization yield	Expensive and complex purification process and low tensile strength
Polyethylene	Low cost	Harsh chemicals required for the stabilization process and low mechanical properties
Cellulose	Low raw material cost and high molecular orientation	Low carbon yield and high carbon fiber production cost
Lignin	Inexpensive, high carbon content, thermally stable, high aromatic structure, and no HCN emission as by-product	Low tensile properties, disordered and complex microstructure, diversity of source, many purification steps required prior to spinning, and longer oxidation time

9.2 Synthesis and processing of lignin-based carbon fiber

9.2.1 Source, structure, properties, and extraction method

Lignin is a natural phenolic-based polymer that is abundant in the cell walls of plants, especially in land-based plants. It provides mechanical support for plant growth, as well as assists in the long-distance transport of water in xylem, resisting a variety of stresses and providing protection against pathogens [53, 54]. The chemical structure of native lignin comprises three *p*-hydroxycinnamyl alcohol monomers, referred to as monolignols. These are *p*-coumaryl alcohol (*p*-hydroxyphenyl, H unit), coniferyl alcohol (guaicyl, G unit), and sinapyl alcohol (syringyl, S unit), which are characterized by their degree of methoxyl substitution at the third and fifth carbon (C_3 and C_5) positions of the aromatic ring, that is, *p*-coumaryl alcohol has no methoxyl group, whereas coniferyl alcohol has a methoxyl group at C_3 position, and sinapyl alcohol has a methoxyl group at both C_3 and C_5 positions, respectively [55]. The most predominant inter-monomer linkage within lignin gives rise to its complex three-dimensional amorphous network structure called phenylpropane β-aryl ether (β-*O*-4). Other inter-unit linkages that have been identified to date are biphenyl and dibenzodioxocin (5–5), phenylcoumarin (β-5), 1,2-diaryl propane (β-1), phenylpropane α-aryl ether (α-*O*-4), diaryl ether (4-*O*-5) and β–β-linked structures (β–β) (Figure 9.2) [56, 57].

Lignin, being a highly branched and amorphous macromolecule, varies greatly in its composition depending on the plant species, maturity stage, and environmental conditions [50, 53]. In general, hardwood (dicotyledonous angiosperm) lignin consists principally of G and S units and traces of H units, whilst softwood (gymnosperm) lignin is made up mostly of G units with low levels of H and S units. By contrast, lignin from grass (monocot) has comparable levels of G and S units, and more H units than hardwood and softwood lignins [54]. For all biomass types, these monomers are linked without defined repeating units [49].

However, before lignin can be used to make a precursor fiber, let alone a carbon fiber, it needs to be extracted from the plant. Whilst in principle it can be extracted from any plant, in practice, viable sources are required such as jute, hemp, flax, pine, cotton, and wood pulp [59]. There are many different methods of extracting lignin from lignocellulosic feedstocks using a different chemical and physical processing routes, all of which play a critical role in controlling carbon yield, mechanical properties, and hence economic viability.

To produce a value-added product, lignin is first separated from the biomass, and is commonly referred to as “technical lignin,” which differs from native lignin in its physical properties, such as molecular structure and molecular weight, as well as chemical properties, such as solubility and hydrophobicity [60]. It is important to understand that the properties of the lignin are changed as a result of the processing

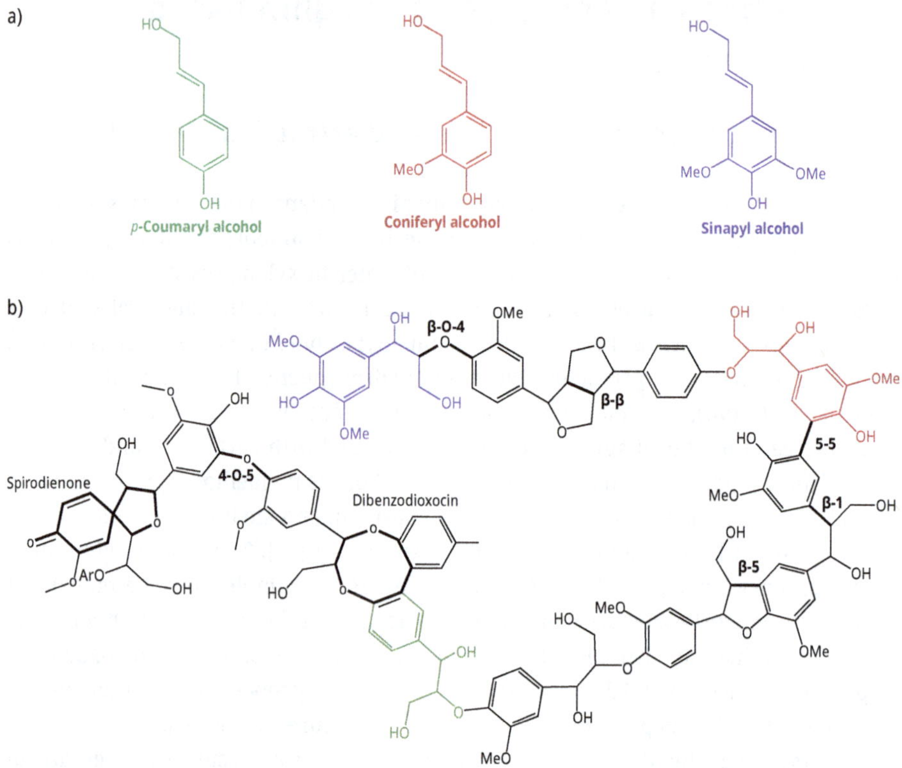

Figure 9.2: Molecular structures of (a) the monomers of lignin and (b) the lignin compound with the functional groups. Reprinted (adapted) with permission from [58]. Copyright 2022 American Chemical Society.

conditions and chemical reactions performed during the isolation procedure. Technical lignin is a common by-product of the paper and pulping industry, whereby the qualitative and quantitative yields of cellulose have been optimized in a series of delignification processes [61]. In these extraction media, lignin is progressively broken down into smaller molecular weight fragments via the cleavage of bonds [62]. At present, the chemical pulping industry uses primarily sulfite, kraft, or soda pulping process for separating lignin on an industrial scale from lignocellulosic feedstocks. Other popular fractionation methods include the organosolv treatment and steam explosion [63]. Some of these common technical lignins are described as follows.

9.2.1.1 Kraft lignin

Kraft pulping is the most dominant form of chemical pulping in the world producing about 130 million tons of pulp annually. In the kraft pulping process, lignocellulosic feedstock is digested at elevated temperature (145–170°C) and pressure in a blend solution of sodium sulfide (Na_2S) and sodium hydroxide (NaOH), also known as the “white liquor,” for a few hours. Under these conditions, the hydrosulfide and hydroxide anions react with the lignin polymers, by hydrolyzing the β-*O*-4 bond and fracturing the lignin polymer into smaller fragments, which readily dissolve in alkali solution, thus liberating the cellulose fibers initially bound with lignin. The dissolution of lignin in the spent pulping liquor gives its distinct darkish brown appearance, thereby giving rise to the name – the “black liquor” [64]. As high as 95% of the kraft lignin can effectively be recovered in the black liquor by acidification, with molecular mass ranging from 150 to 200,000 [65]; the side chain structure of the kraft lignin experiences substantial structural changes, comprising a reduction in the interunit ether linkage and a concomitant increase in C–C bonds, thiol groups, and phenolic hydroxyl groups [66]. Kraft lignin is hydrophobic and has a lower molecular mass than native lignin. Furthermore, it is almost insoluble in water and most solvents, but is highly soluble in alkaline media with pH above 11 [65]. In 2015, the production of kraft lignin reached about 630 ktons/year; however, it is still predominantly utilized in low-value applications, such as a combustion material for heat recovery, with little industrialization for high-value applications [67]. This limitation for high-value applications is largely attributed to the contamination of kraft lignin with carbohydrate molecules from hemicelluloses and fatty acids and the considerable amount of sulfur (1–3 wt%) covalently bound to kraft lignin in the form of thiols [65].

9.2.1.2 Lignosulfonate

Due to the limitations of kraft lignin, lignosulfonate is currently the main commercial source of lignin, accounting up to 90% of the total lignin market, with an annual global production of lignosulfonates reaching up to 1.8 million tons [65]. Lignosulfonate and kraft lignin have different properties, most notably its high solubility in water attributed obviously to the presence of the sulfonate group but also tends to have a similar or higher molecular weight to that of kraft lignin. The disadvantage of lignosulfonate is that it contains high ash content. The extraction of lignosulfonates involves the digestion of lignocellulosic feedstock at a high temperature ranging from 130 to 180 °C, in an aqueous solution of a sulfite (SO_3^{2-}) or bisulfite (HSO_3^-) salt of sodium, ammonium, magnesium, or calcium. In this process, the most important reaction is the sulfonation of the aliphatic chains of lignin that lends the lignosulfonate its key characteristics.

9.2.1.3 Soda lignin

Soda lignin is obtained from the soda pulping process. Briefly, soda pulping is similar to the kraft process, except that sodium sulfide is excluded from the cooking medium. Soda lignin displays properties similar to kraft lignin, but unlike kraft lignin and lignosulfonates, which are classified as sulfur-containing polymers that would give adverse environmental impact in their processing, soda lignin is sulfur free and has a higher purity, which makes its structure akin to native lignin when compared to other types of technical lignins [68]. Consequently, soda lignin is preferred in many applications such as tissue engineering, drug encapsulation, animal nutrition, and also in polymer synthesis such as phenolic resin whereby a high purity and biocompatibility of lignin are required [69]. That said, soda pulping has limited use on easily pulped materials, for example, straws and bagasse, and still poses a great challenge when it comes to the digestion of most hardwoods. Owing to that, soda lignin from the softwood source can contain a significant amount of silicate and nitrogen, along with higher sodium ion content, depending on the cooking procedure used.

9.2.1.4 Organosolv lignin

The organosolv pretreatment process, commonly known as organo-solvation, is used to extract lignin by mixing organic solvents with an inorganic acid catalyst to break the lignin bonds. The latter functions to facilitate the breakdown of the β-ether linkages. Studies show that the presence of a catalyst allows the removal of 91% of the lignin, while its absence reaches only 14% [70]. Another similar technique is a mild organosolv extraction with alcohol, which is carried out at a lower temperature of less than 150°C. Ethanol in particular is among the most popular alcohols in this process to obtain organosolv lignin. When ethanol is used as the solvent, it is believed that it can be incorporated in the β-aryl ether units via α-alkoxylation, resulting in the β-*O*-4-rich lignin. This solvent incorporation is a crucial step in preventing undesired condensation reactions by trapping the reactive benzylic cations, which is typically formed under acidic pretreatment process, more so in harsh acidic environments [71]. The mild extraction method has shown to be able to produce clean lignin with excellent retention of a high amount of β-aryl ether content, and contains very low carbohydrate, ash, and other non-lignin-derived impurities [71]. After all, there is still limited understanding of the actual relationship between the extraction method and on the lignin characteristics. The chemical structures of lignin and its properties have to be considered on a case-by-case basis as they are very much dependent on processing conditions.

9.2.2 Lignin fiber spinning and processing methods

Lignin, the second most abundant biopolymer after cellulose, has been investigated as a carbon fiber precursor since the 1960s primarily due to its vast abundance and bio-based origin [72, 73]. Successful use of lignin as an alternative precursor polymer is predicted to have a cost reduction of up to 50% in finished carbon fiber over conventional PAN polymer [43]. However, despite many studies published on investigating different routes for producing lignin-based carbon fibers [74-77], much further development is required to make lignin carbon fiber a commercial product.

Lignin-based precursor fibers have most frequently been produced using melt-spinning technique due to its affordability and scalability. Generally, to produce melt-spun lignin-based fibers, lignin powder is converted into pellets and extruded into fibers by heating them above the softening temperature (30–70°C higher than theT_g of lignin) [78], followed by cooling in cold air before collecting the fibers [79]. The melt-spinning technique offers several advantages such as no solvent is required for fiber spinning, high-throughput rate, and ease of the process. However, to process lignin fiber through melt-spinning, lignin should have a melting point well below its decomposition temperature, and its viscosity should remain stable during spinning [80]. A major drawback of melt-spinning lignin fibers is that it provides limited scope to improve the mechanical fiber properties as the fiber morphology is challenging to control. The fiber structure is governed by the removal of volatile components from lignin when it is heated above its glass transition temperature (T_g) during spinning, leading to the formation of defects and voids in the fibers [78]. Besides, depending on the source material, lignin may have varying low-molecular-weight fractions and sometimes no specific melting point due to its highly amorphous chemical structure. Lignin tends to decompose when heated at a temperature higher than 200–250°C [81], making the processing window for melt extrusion particularly narrow. The presence of low-molecular-weight lignin is also reported to cause fiber fusion during the stabilization process due to its low T_g [82]. Many studies use melt-spinning to produce lignin-based carbon fibers obtained from various sources and have generally reported relatively low mechanical properties compared to PAN and pitch-based carbon fibers. For example, when carbon fibers were produced from melt-spun unmodified pyrolytic lignin, macro-voids were observed in the formed carbon fibers, resulting in a low tensile strength of 370 MPa and a modulus of 36 GPa [83].

The alternative wet-spinning technique is the industrially preferred technique for current commercial carbon fiber production from PAN polymer [84]. It offers greater scalability potential and a highly controlled coagulation process that minimizes the formation of defects in the fibers during spinning, leading to high-quality precursor fiber. However, the extremely low molecular weight and amorphous structure of lignin make it challenging to be converted into fibers through wet-spinning as the spinning dope requires a certain shear behavior, which is generally achieved by blending with other linear, high-molecular-weight polymers such as PAN, polylactic acid (PLA), and polyvinyl alcohol (PVA) [85, 86]. A common variation of the wet-spinning method is the dry-jet or air-gap spinning, which utilizes a spinneret mounted vertically, at an air gap of about 10–200 mm from the coagulation medium. When the spinning dope falls freely under gravity due to the air gap, it promotes molecular orientation prior to coagulation, leading to improvement in mechanical properties of fibers.

The electrospinning technique is used to produce continuous fibers at micrometer to nanometer scale. This technique was first used to form lignin-based fibers in 2007 [87]. In this process, a lignin-spinning solution is introduced through a voltage-driven jet nozzle to form thin fiber mats, which are solidified and collected by a grounded collector [88]. Successful electrospinning of lignin-based fibers requires modification of lignin or blending with other polymers to prevent electrospray, which is often caused due to lignin's inability to form chain entanglements [89]. Since the electrospinning process results in fiber mats rather than single filaments, it remains a great challenge to separate and measure the properties of individual fibers [90]. Another significant drawback of the electrospinning process is the low scalability potential and low yield. Thus, fibers produced from the electrospinning method have been limited to applications in energy storage and filtration [91].

9.2.3 Blending of lignin with cellulose and other polymers

The challenges in the lignin fiber-spinning process arise from the complex structure of lignin, the lack of a reliable supply chain of raw lignin, and insufficient understanding of the conversion mechanism of lignin precursor into carbon fiber [92]. Blending with cellulose, which is another abundant biopolymer, is a very attractive solution to produce renewable fibers [93]. Lignin-cellulose fibers have been prepared by wet-spinning method, where cellulose was dissolved in a suitable ionic liquid and added to a lignin/DMSO solution [94]. Blending cellulose with lignin greatly enhances the carbon yield of fibers when compared to cellulose alone, where significant mass loss occurs due to the release of various pyrolysis gases during the carbonization process [95]. Cellulose also grants better molecular orientation to the cellulose-lignin blend [96]. However, inadequate interaction between cellulose and lignin often results in the leaching of lignin during coagulation. Moreover, high costs of ionic liquids, which are the preferred solvents for cellulose, along with the constraints on lignin quantities that can be blended with cellulose, are some of the limitations hindering the widespread use of cellulose-lignin fibers.

As previously mentioned, most of the commercial carbon fiber is manufactured using PAN, which is a linear polymer of acrylonitrile and prepared by solution or suspension polymerization with the addition of some suitable comonomers, which aid the spinnability and stabilization process and also improve the carbonization yield. On the other hand, lignin is a highly cross-linked aromatic network, which is amorphous in nature and has many hydrophilic hydroxy groups in its structure [97]. These differences in the basic structure make it difficult to adapt standard spinning processes to lignin; hence, it requires the establishment of innovative processes for lignin fiber spinning and its carbonization.

To overcome some of these limitations, lignin is often blended with PAN, which, thanks to its linear long-chain polymer structure, is likely to improve the processability of PAN/lignin blends. Softwood kraft lignin has been blended in equal ratios with PAN to produce wet-spun lignin-based carbon fibers with a tensile strength of 1.2 GPa and modulus of 148 GPa [74]. However, when blended with PAN, solution spinning of lignin presents a risk of uncontrolled leaching of lignin into the coagulation bath resulting in void formation and brittle fibers [98–100]. Other than PAN, lignin has been blended with other materials such as PVA [85], PLA [101], polyethylene oxide (PEO) [102], and graphene oxide [103] successfully to aid the spinnability and improve the mechanical properties of the fibers. While blending with various materials has shown to improve the mechanical properties, researchers are seeking solutions to overcome the intrinsic lack of well-defined molecular architecture and orientation in lignin that led to the poor mechanical properties of fibers based solely on lignin. The best carbon fiber properties reported in the literature have been achieved by altering the structure of a bio-oil-based pyrolyzed lignin, modifying it to create a more linear polymer, followed by melt-spinning into fiber. Furthermore, the authors reported the application of higher tension during stabilization and carbonization process, which resulted in carbon fibers with high mechanical properties with a tensile strength of 1.7 GPa and a modulus of 182 GPa [92].

9.2.4 Stabilization and carbonization of lignin-based precursor

9.2.4.1 Stabilization of lignin-based precursor

Stabilization is an important step in the production of carbon fibers to make precursors thermally stable enough to withstand high temperatures applied during the carbonization process [104, 105]. Various chemical and physical pre-stabilization treatment methods for lignin have been proposed to address the challenges during carbon fiber manufacturing, including the enhancement of the carbon yield, reduction of stabilization time, and prevention of fiber fusion.

9.2.4.1.1 Thermal stabilization of lignin-based precursor

For lignin-based carbon fiber production, thermal treatment is the most common method where the precursor is passed through a series of low- or moderate-temperature air ovens often between 200 and 350°C [106–108]. Under thermal treatment conditions, lignin performs a significant chemical change via decomposition, rearrangement, and cross-linking. During the process, the carbon content in the fiber increases while other elements are removed, resulting in the formation of a thermally stable structure, which prevents the destruction of fibers in the carbonization step. The thermal stabilization of lignin starts around 200°C beginning with the cleavage of phenol monomer and methoxy groups (–OCH_3), which links with others via C–O–C bonds [108, 109], releasing formaldehyde, methane, and water as volatile by-products [110–112]. At the same time, oxidation reaction introduces new carbonyl groups (ketones and aldehydes) to the molecular structure, which further promotes condensation and rearrangement reactions [113, 114]. The predominant challenge of thermal oxidative stabilization of lignin is the long duration since a slow heating rate is required to prevent fiber fusion. For example, Baker et al. [115] used different heating rates (0.05, 0.025, and 0.01°C/min) to stabilize organic purified lignin up to 250°C. Zhang et al. [116] performed stabilization of acylated softwood lignin in air with rate of 0.2°C/min to 220°C. Norberg et al. [117] used ultrafiltration softwood kraft lignin and achieved a higher heating rate of 4°C/min but still needed to hold for 30 min at 250°C.

9.2.4.1.2 Pre-stabilization treatment of lignin-based precursor

In addition to oxidative thermal stabilization, several physical and chemical methods have been studied to stabilize lignin fiber or speed up the stabilization. Physical agents (electron beam, plasma, UV, etc.), which can introduce cross-linking in a polymeric structure have been proposed to be effective as alternative methods for stabilization. Zhang [118] pretreated softwood kraft lignin by UV for 15 min, resulting in a significant reduction in stabilization time from 40 h to 4 h. Electron beam stabilization has been conducted by Seo et al. [119] on electrospun lignin-PAN fibers, leading to an increase of 20% in carbon yields accounting for the enhancement of cross-links in treated fiber. Plasma treatment and microwave are other possible methods tested with PAN fiber but have not been applied to lignin fiber yet [120].

Chemical stabilization is another method with the potential to increase the rate of stabilization of lignin. Since lignin is a polyphenolic resin, Lin et al. [121] pretreated polyethylene glycol/lignin fibers with hexamethylenetetramine and hydrochloric acid as a catalyst to improve cross-linking through methylene groups. These treated fibers were able to stabilize at a higher heating rate of 2°C/min, leading to reduced stabilization time. Guo et al. [122] used a boron nitride preceramic polymer as a cross-linking agent for lignin-phenol-formaldehyde resin blend. The blend was electrospun, and fibers were subsequently thermally stabilized at 250°C for 2h before carbonization. The treatment led to reduced stabilization time over the traditional method, and the obtained fibers retained their shape without fusion after being carbonized at 800°C for 2 h. Cho et al. [123] added 5 wt% nanocrystalline cellulose into lignin dope. The cellulose-added electrospun lignin fibers showed good thermal stability at a high heating rate of 10°C/min during carbonization up to 1,000°C, indicating the assistance of cellulose in preventing lignin fusion. Le et al. [106, 124] have significantly reduced the stabilization time by 30% and increased the char yield by 50% for cellulose-lignin-blended fibers by using different catalysts (phosphoric acid and boric acid). All the abovementioned works show a clear potential for future research in promoting the stabilization of lignin fibers via chemical treatment.

9.2.4.2 Carbonization of lignin-based stabilized fibers

Carbonization typically occurs at 400–1,600°C in an inert atmosphere, converting stabilized fiber into carbon fiber [120, 125]. During this step, all elements other than carbon, such as nitrogen, oxygen, and hydrogen, are removed from the structure [125, 126]. Carbon in lignin continues to condense and arrange to form polycyclic aromatic hydrocarbons [127, 128], expanding the number of aromatic rings in the molecular structure as temperature increases [129, 130] to form a carbon network. The carbon structure in lignin-based carbon fiber is particularly turbostratic (the sheets of carbon atoms are haphazardly folded, or crumpled, together) (Figure 9.3a), in that it contains more defects and is more disoriented than PAN-based carbon fiber [131]. Because of this poor orientation, lignin-derived carbon fiber has inferior mechanical properties to PAN-based carbon fiber, leading Baker et al. to suggest that higher performance lignin-based carbon fiber could be achieved by enhancing the graphitic orientation (Figures 7.3b and c).

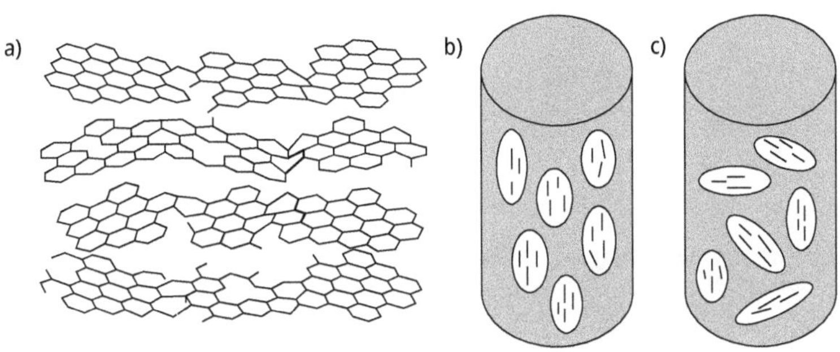

Figure 9.3: (a) Turbostratic carbon [108], (b) schematic representation of the orientation of graphite crystallites in PAN-based carbon fiber, and (c) lignin-based carbon fiber [131] (reprinted (adapted) with permission, copyright 2016 Wiley Periodicals, Inc.).

9.3 Mechanical properties of lignin-based carbon fibers

This turbostratic microstructure, lacking in graphitic structure with increased defects [108, 132], is the major reason for the reduced mechanical properties of lignin-based carbon fiber compared with PAN- and pitch-based carbon fibers. Some examples include that of Jin et al. who reported a tensile strength and modulus of 1.39 GPa and 98 GPa, respectively, using a fractionated-solvated softwood kraft lignin [133]. Zhang et al. made carbon fiber from acylated softwood lignin and reported tensile strength and modulus of 1.05 GPa and 52 GPa, respectively [116]. Blends of lignin and PAN have been shown to have better mechanical strength with tensile strength and modulus of 1.72 GPa and 230 GPa, respectively [134]. Cellulose and lignin blends (70:30) have been widely studied, including the work by Bengtsson, who reported tensile strength and modulus of 1.17 GPa and 77 GPa, respectively [135]. Blends of lignin with other polymers such as PEO, polyethylene terephthalate, and polypropylene have also been trialled somewhat successfully, but again display inferior mechanical properties due to the polymers' low thermal stability and resulting porous structure in the resulting carbon fiber [113, 136]. Despite this, Figure 9.4 summarizes the mechanical properties of carbon fibers made from lignin and lignin blends using the commercial Toray T300 carbon fiber as a benchmark, illustrating the general improvement blended systems compared with lignin only.

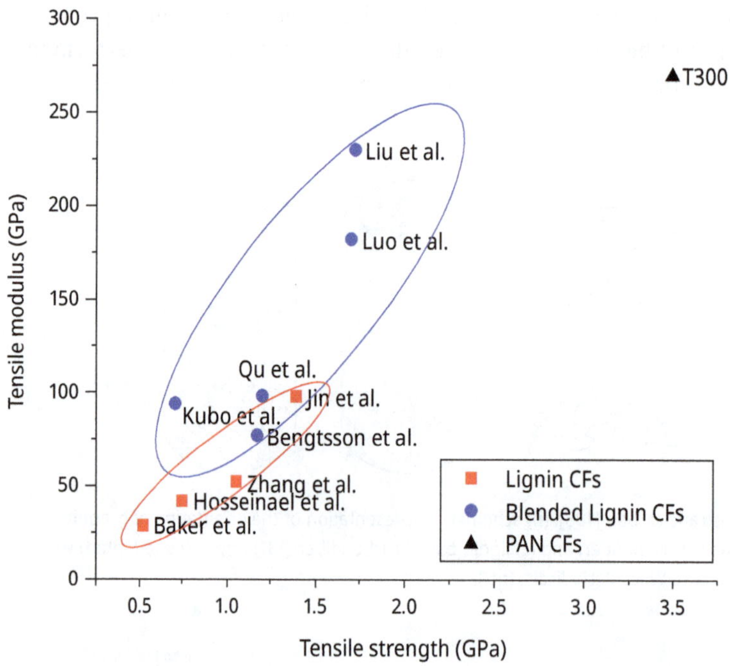

Figure 9.4: Mechanical properties of carbon fibers made from lignin and lignin blends in comparison with commercial Toray T300 carbon fiber.

Mechanical properties of lignin-based carbon fiber have been further improved via hot stretching during high-temperature graphitization above 2,000°C, which is commonly used to enhance the mechanical properties of cellulose-based carbon fiber

[120, 126]. The application of hot stretch is believed to refine the carbon structure by improving the graphitic levels and orientation of graphitic crystallites. However, the application of high-temperature graphitization of lignin-based carbon fiber may result in a lower carbon yield along with increased production cost [125]. In summary, the imperfect carbon structure of lignin-based carbon fiber is an open research question requiring further study to improve its mechanical properties.

9.4 Sustainable composites from bio-based carbon fiber

9.4.1 Applications and market demand

Lignin-based carbon fiber is an excellent and potentially inexpensive alternative to PAN-based carbon fiber particularly for general use composite materials. Unlike PAN-based carbon fiber where the manufacturing cost is highly dependent on oil price, with around 51% being precursor cost, 15% related to the oxidation process, and 23% related to the carbonization process [43, 137], the extraction of lignin from biomass waste or pulp mills comes at a much lower cost (Table 9.4). Due to the high cost of PAN, carbon fiber is rather expensive, limiting its use to high-value, low-volume products. These include aerospace structural components, the body and brake discs of Formula 1 racing cars, pressure vessels, and other defence-related products. The application of carbon fiber as structural composites is, in fact, dependent on its mechanical properties and cost. Replacing the PAN-based carbon fiber with lignin, in part or as a whole, offers the opportunity for it to be widely adopted in high-volume applications such as automotive, renewable energy, and construction applications. In recent years, automakers have started to use carbon fiber composites more widely, as add-on parts or whole chassis elements, in order to reduce weight, including the likes of Audi, Mercedes-Benz, BMW in their i3 and i8 models, as well as Toyota that pioneered in the carbon fiber composites for the tailgate in its Prius PHV [138]. However, for this progress to continue, further reductions in cost and volume are required. For every 10% reduction in vehicle weight, it has been calculated that this would translate into an increase in vehicle fuel economy of about 6%, with a concomitant reduction in emissions. To place the potential increase in fuel economy in perspective, body-in-white modeling indicates that more than 60% of the steel in a vehicle could be replaced with carbon fiber-reinforced polymers without impacting vehicle crashworthiness [139]. According to the market research by Mordor Intelligence, the automotive carbon fiber market was valued at US$20 billion in 2020, and it is expected to grow up to US$33 billion in 2026, attributed to the expected rapid growth of electric vehicles [140].

In addition to the excellent mechanical strength, durability, and levity, there is another important property that carbon fiber possesses – electrical conductivity. This

Table 9.4: Cost, CO_2,and energy usage comparison of PAN and lignin-based carbon fibers (adapted from [141]).

Material	Precursor cost	Production cost	Total cost	CO_2emissions	Total energy usage
Conventional PAN	\$11.1/kg	\$25.15/kg	\$36.25/kg	31 kg/kg	704 MJ/kg
Textile-grade PAN	\$4.4–13.2/kg	\$12.25–25.4/kg	\$16.65–38.6/kg	–	
Melt-spun PAN	\$6.3/kg	\$17.4/kg	\$23.7/kg	–	
Lignin	\$1.52/kg	\$6.27/kg	\$7.79/kg	24 kg/kg	554 MJ/kg

makes carbon fiber suitable to be used as an electrode in electrophysiological, electrochemical, biosensor and biofuel cell applications [142]. For example, traditionally, the negative electrode in lithium-ion battery is made up of copper, which makes up one-fourth of the total electrode weight but since the 1990s, carbon fiber has been employed as an alternative negative electrode in lithium-ion battery. This has the advantage of increasing efficiency by reducing the amount of non-active material in lithium-ion battery [143]. Lignin-derived carbon fiber is shown to be a promising precursor in several research studies. Some of the different forms studied includes fused electrospun carbon fibrous mats, fused lignin carbon fiber mats, or as micro-sized particles in a mixture with pitch. A high charge capacity of around 400–500 mAh/g and a good cyclic stability at different current rates have been obtained. In particular, carbon (nano)fiber spun from the electrospinning method is expected to improve the charge and mass transport at the cathode, through its exceptional through-plane and in-plane electrical conductivity, and high porosity after the carbonization step [144]. Moreover, the porosity can be controlled by adjusting the spinning parameters, which allows the design and synthesis of fiber of controlled morphology and activity. With further modification, enhanced improvement on the performance of the lignin-based carbon fiber electrode can be obtained. For example, after thermal annealing of the carbon fiber in the presence of urea, the charge capacity could be improved by around 20% [145]. While the carbon fiber tow is generally used as anode, when the surface of carbon fiber was coated with lithium oxide ($LiFePO_4$), it can function as a positive electrode [146]. A specific capacity of 100 mAh/g was recorded for this type of cathode, as good as commercial positive electrodes that have a theoretical capacity at about 150 mAh/g.

Nowadays, research is underway to create structural carbon fiber composites that double as energy storage system, that is, the structural batteries (or solid-state batteries) made of carbon fiber-based electrodes separated by a glass fiber separator in a structural battery electrolyte matrix material. These types of structural battery composites by far have only achieved an energy density of 24 Wh/kg but despite their inability to store as much energy as lithium-ion batteries, they are highly attractive for use in vehicles and other applications as "mass-less" energy storage because the weight of the battery vanishes when it becomes part of the load-bearing structure

[147]. The next-generation structural battery has fantastic potential in consumer technology. It could be quite possible within a few years to manufacture smartphones, laptops or electric bicycles that weigh half as much as those today and are much more compact. Apart from the extensive use of carbon fiber in batteries, it has also been demonstrated that carbon fiber, especially the electrospun carbon nanofibers (CNFs), can be manufactured into support for the platinum catalyst in the proton exchange membrane fuel cell. The incorporation of small catalyst particles into the CNF web is known as the microfibrous entrapped catalyst (MFEC) [148]. The major advantage of this is the high surface area to volume of the CNFs, high dispersion, and little catalyst aggregation, and consequently, the improved electrochemical active sites of the catalyst adhered on its surface as well as the synergism between the CNF and the immobilized catalytic species to achieve process intensification [144]. The particle size of MFECs is typically in the range of 100–200 μm, and the surface-area-to-volume ratio can reach up to 1×10^4 m^2/m^3 [148]. Depending on the catalytic species, alloys or bimetallic nanoparticles used to decorate the CNFs, either by electrospinning or doping in catalyst precursor, this leads to catalysts of various photo/electro/chemical catalytic properties. For example, CdS-TiO_2-doped CNFs display excellent photocatalytic activity for visible light hydrolytic dehydrogenation of ammonia borane; Au-TiO_2/CNFs for photodegradation of methyl orange, methylene blue and acid red 18 in wastewater under UV light irradiation [149]; Pd/CNFs for Heck's coupling reaction [150]; and last but not least, silver (Au)-doping CNFs in silver nitrate precursor could promote the reduction of 4-nitrophenol and benzylamine [151]. Noteworthy, the catalyst could be easily separated and recycled.

In short, lignin-based carbon fiber plays a paramount role in the transformation toward a more sustainable and affordable carbon fiber manufacturing. At the same time, there are challenges and issues ahead that need to be overcome. Undoubtedly, future works looking into a combined economical and ecological approach of improving the performance of the lignin-based carbon fiber would make them promising as the next generation of carbon fiber in wider applications.

References

[1] Morris EA, et al. Early development of multifilament polyacrylonitrile-derived structural hollow carbon fibers from a segmented arc spinneret. Carbon 2021;178:223–232.

[2] Zan Gao JZ, Rajabpour S, Joshi K, Kowalik M, Brendan Croom YS, Zhang L, Bumgardner C, Brown KR, Diana Burden JWK, Van Duin ACT, Zhigilei LV, Li X Graphene reinforced carbon fibers. Sci Adv 2020;6 (17):eaaz4191.

[3] Yang F, et al. Effect of amorphous carbon on the tensile behavior of polyacrylonitrile (PAN)-based carbon fibers. J Mater Sci 2019;54(11):8800–8813.

[4] Ge Y, et al. The effects of chemical reaction on the microstructure and mechanical properties of polyacrylonitrile (PAN) precursor fibers. J Mater Sci 2019;54(19):12592–12604.

[5] Groetsch T, et al. Gas emission study of the polyacrylonitrile-based continuous pilot-scale carbon fiber manufacturing process. Ind Eng Chem Res 2021;60(48):17379–17389.
[6] Nunna S, Setty M, Naebe M Formation of skin-core in carbon fibre processing: A defect or an effect?. Express Polym Lett 2019;13(2):146–158.
[7] Nunna S, et al. Time dependent structure and property evolution in fibres during continuous carbon fibre manufacturing. Materials 2019;12(7): p.1069.
[8] Dumanlı AG, Windle AH Carbon fibres from cellulosic precursors: A review. J Mater Sci 2012;47 (10):4236–4250.
[9] Edison TA Electric Lamp, USA: U.S.P. Office; Editor; 1880. 1–3.
[10] Swan JW Electric Lamp, U.S.P. Office; Editor; 1880. pp. 1–3.
[11] Park S-J Carbon fibers. In Robert Hull CJ, Osgood RM, Parisi J, T-y S, S-I U, Wang ZM, editors. Springer Series in Materials Science, Springer; 2015. 210, p.87.
[12] Ford CE, Mitchell CV Fibrous Graphite, U.S.P. Office; Editor; 1963.
[13] Tang MM, Bacon R Carbonization of cellulose fibers-I low temperature pyrolysis. Carbon 1964;2:211–220.
[14] Bacon R, Tang MM Carbonization of cellulose fibers-II physical property sudy. Carbon 1964;2:221–225.
[15] Gorss J High Performance Carbon Fibers. 2003.
[16] Cahn RW Graphitic carbon copies. Nature 1991;349:97.
[17] Swinbanks D MITI makes its move. Nature 1990;348:186.
[18] Nakamura O, et al. Study on the PAN carbon-fiber-innovation for modelling a successful R&D management. Synthesiology 2009;2(2):154–164.
[19] Moreton R, Watt W, Johnson W Carbon fibres of high strength and high breaking strain. Nature 1967;213(5077):690–691.
[20] Watt W, Johnson W New materials make their mark. Nature 1968;220:835.
[21] Otani S, et al. Method for Producing Carbonized Lignin Fiber, USA: U.S.P. Office; Editor; 1969. pp. 1–5.
[22] Souto F, Calado V, Pereira N Lignin-based carbon fiber: A current overview. Mater Res Express 2018;5(7):072001.
[23] Singer LS High Modulus, High Strength Carbon Fibers Produced From Mesophase Pitch, USA: U.S.P. Office; Editor; 1977. 1–25.
[24] Horikiri S, Iseki J, Minobe M Process for production of carbon fiber, U.S.P. Office; Editor. 1978.
[25] Soulis S, et al. Impact of alternative stabilization strategies for the production of pan-based carbon fibers with high performance. Fibers 2020;8(6):33.
[26] Composites world. Carbon fiber suppliers gear up for next-gen growth. 2020, [cited 2022 18 April]; Available from: https://www.compositesworld.com/articles/carbon-fiber-suppliers-gear-up-for-next-gen-growth.
[27] Fernández L Global market value of carbon fiber 2014–2026. 2021 [cited 2022 18 April]; Available from: https://www.statista.com/statistics/947941/global-market-value-carbon-fiber/.
[28] The Manufacturer. Creating a sustainable future for carbon fibre in manufacturing. 2021, [cited 2022 18 April]; Available from: https://www.themanufacturer.com/articles/creating-a-sustainable-future-carbon-fibre/.
[29] Das S, et al. Global Carbon Fiber Composites Supply Chain Competitiveness Analysis, United States; 2016.
[30] Toray Industries. I. *Production Capacity*. 2021 [cited 2022 18 April]; Available from: https://www.toray.com/global/ir/management/man_010.html.
[31] Ocean R Pitch-based carbon fiber Market Scope and overview, Strategy, Revenue, Opportunity, Business Segment Overview and Key Trends 2022–2030. 2022, [cited 2022 18 April]; Available from: https://www.taiwannews.com.tw/en/news/4408935.

[32] Nippon Graphite Fiber Co. L. What's Carbon fibers. 2022 [cited 2022 18 April]; Available from: https://www.ngfworld.com/en/fiber.html.
[33] Composites World. The outlook for carbon fiber supply and demand. 2021, [cited 2022 18 April]; Available from: https://www.compositesworld.com/articles/the-outlook-for-carbon-fiber-supply-and-demand.
[34] Zhang J, et al. Current status of carbon fibreand carbon fibrecomposites recycling. Compos Part B: Eng 2020;193.
[35] Witten E, et al. Composites Market Report 2018. 2018, AVK. p. 1–59.
[36] Sauer M *Composites Market Report 2019. 2019, AVK Carbon Composites e.V.* p. 1–11.
[37] Frank E, et al. Carbon fibers: Precursor systems, processing, structure, and properties. Angew Chem Int Ed 2014;53(21):5262–5298.
[38] Nunna S, et al. Development of a cost model for the production of carbon fibres. Heliyon 2019;5(10): e02698.
[39] Mirdehghan AS Fibrous polymeric composites. In Engineered Polymeric Fibrous Materials, 2021. pp. 1–58.
[40] Rahaman MSA, Ismail AF, Mustafa A A review of heat treatment on polyacrylonitrile fiber. Polym Degrad Stab 2007;92(8):1421–1432.
[41] Ghorpade RV, Lee S, Hong SC Effect of different itaconic acid contents of poly(acrylonitrile-co-itaconic acid)s on their carbonization behaviors at elevated temperatures. Polym Degrad Stab 2020;181.
[42] Dér A, et al. Modelling and analysis of the energy intensity in polyacrylonitrile (PAN) precursor and carbon fibre manufacturing. J Cleaner Prod 2021;303, p.127105.
[43] Mainka H, et al. Lignin – An alternative precursor for sustainable and cost-effective automotive carbon fiber. J Mater Res Technol 2015;4(3):283–296.
[44] Baker DA, Rials TG Recent advances in low-cost carbon fiber manufacture from lignin. J Appl Polym Sci 2013;130(2):713–728.
[45] Dai Z, et al. High-strength lignin-based carbon fibers via a low-energy method. RSC Adv 2018;8 (3):1218–1224.
[46] Bengtsson A, et al. Carbon fibers from lignin-cellulose precursors: Effect of stabilization conditions. ACS Sustain Chem Eng 2019;7(9):8440–8448.
[47] Vocht MP, et al. Preparation of cellulose-derived carbon fibers using a new reduced-pressure stabilization method. Ind Eng Chem Res 2022.
[48] Fang W, et al. Manufacture and application of Lignin-Based Carbon Fibers (LCFs) and Lignin-Based Carbon Nanofibers (LCNFs). Green Chem 2017;19(8):1794–1827.
[49] Poursorkhabi V, et al. Processing, carbonization, and characterization of lignin based electrospun carbon fibers: A review. Front Energy Res 2020;8, p.208.
[50] Yongjian X, et al. Current overview of carbon fiber: Toward green sustainable raw materials. Bioresources 2020;15(3):7234–7259.
[51] Bajpai P Raw materials and processes for the production of carbon fiber. Carbon Fiber 2021;13–50.
[52] Aldosari SM, Khan MA, Rahatekar S Manufacturing pitch and polyethylene blends-based fibres as potential carbon fibre precursors. Polymers 2021;13(9):1445.
[53] Kang X, et al. Lignin-polysaccharide interactions in plant secondary cell walls revealed by solid-state NMR. Nat Commun 2019;10(1):347.
[54] Lourenço A, et al. Lignin composition and structure differs between xylem, phloem and phellem in Quercus suber l. Front Plant Sci 2016;7:1612.
[55] Borges Gomes F J, et al. A review on lignin sources and uses. J Appl Biotechnol Bioeng 2020;7 (3):100–105.
[56] Del Río JC, et al. Lignin monomers from beyond the canonical monolignol biosynthetic pathway: Another brick in the wall. ACS Sustain Chem Eng 2020;8(13):4997–5012.

[57] Froass PM, Ragauskas AJ, Jiang J-E Chemical structure of residual lignin from kraft pulp. J Wood Chem Technol 1996;16(4):347–365.
[58] Zhang R, et al. Microbial ligninolysis: Toward a bottom-up approach for lignin upgrading. Biochemistry 2019;58(11):1501–1510.
[59] Watkins D, et al. Extraction and characterization of lignin from different biomass resources. J Mater Res Technol 2015;4(1):26–32.
[60] Ekielski A, Mishra PK Lignin for bioeconomy: The present and future role of technical lignin. Int J Mol Sci 2021;22(1):63.
[61] Li T, Takkellapati S The current and emerging sources of technical lignins and their applications. Biofuels Bioprod Biorefin 2018;12(5):756–787.
[62] Gratzl JS, Chen C-L Chemistry of pulping: Lignin reactions. In Lignin: Historical, Biological, and Materials Perspectives, American Chemical Society; 1999. pp. 392–421.
[63] McGrath JE, Hickner MA, Höfer R 10.01 – Introduction: Polymers for a sustainable environment and green energy. In Matyjaszewski K, Möller M. editors., Polymer Science: A Comprehensive Reference, Amsterdam: Elsevier; 2012. pp. 1–3.
[64] Cheremisinoff NP, Rosenfeld PE Chapter 6 – Sources of air emissions from pulp and paper mills. In Cheremisinoff NP, Rosenfeld PE. editors., Handbook of Pollution Prevention and Cleaner Production, Oxford: William Andrew Publishing; 2010. pp. 179–259.
[65] Gellerstedt G, Henriksson G Chapter 9 – Lignins: Major sources, structure and properties. In Belgacem MN, Gandini A. editors., Monomers, Polymers and Composites from Renewable Resources, Amsterdam: Elsevier; 2008. pp. 201–224.
[66] Fernández-Rodríguez J, et al. Chapter 7 – Lignin separation and fractionation by ultrafiltration. In Galanakis CM, editor. Separation of Functional Molecules in Food by Membrane Technology, Academic Press; 2019. pp. 229–265.
[67] Chen H 3 – Lignocellulose Biorefinery feedstock engineering. In: Chen H, editor. Lignocellulose Biorefinery Engineering. Woodhead Publishing; 2015. pp. 37–86.
[68] Vishtal AG, Kraslawski A Challenges in industrial applications of technical lignins. Bioresources 2011;6(3):2011.
[69] Witzler M, et al. Lignin-derived biomaterials for drug release and tissue engineering. Molecules, Basel, Switzerland: 2018. 23(8), pp. 1885.
[70] Fernández-Rodríguez J, De Diego-díaz B, Tapia-Martín ME Chapter 7 – Biomethanization of Agricultural Lignocellulosic wastes: Pretreatments. In Tyagi V, Aboudi K, editors. Clean Energy and Resources Recovery, Elsevier; 2021. pp. 155–202.
[71] Zijlstra DS, et al. Mild organosolv lignin extraction with alcohols: The importance of benzylic alkoxylation. ACS Sustain Chem Eng 2020;8(13):5119–5131.
[72] Kadla J, et al. Lignin-based carbon fibers for composite fiber applications. Carbon 2002;40 (15):2913–2920.
[73] Kubo S, Kadla J Lignin-based carbon fibers: Effect of synthetic polymer blending on fiber properties. J Polym Environ 2005;13(2):97–105.
[74] Jin J, Ogale AA Carbon fibers derived from wet-spinning of equi-component lignin/polyacrylonitrile blends. J Appl Polym Sci 2018;135(8):45903.
[75] Oroumei A, Fox B, Naebe M Thermal and rheological characteristics of bio-based carbon fibre precursor derived from low molecular weight organosolv lignin. ACS Sustain Chem Eng 2015;3 (4):758–769.
[76] Sudo K, Shimizu K A new carbon fiber from lignin. J Appl Polym Sci 1992;44(1):127–134.
[77] Johnson D, Tomizuka I, Watanabe O The fine structure of lignin-based carbon fibres. Carbon 1975;13 (4):321–325.
[78] Qu W, et al. Towards producing high-quality lignin-based carbon fibers: A review of crucial factors affecting lignin properties and conversion techniques. Int J Biol Macromol 2021;189:768–784.

[79] Nordström Y, et al. A new softening agent for melt spinning of softwood kraft lignin. J Appl Polym Sci 2013;129(3):1274–1279.

[80] Frank E, et al. Carbon fibers: Precursor systems, processing, structure, and properties. Angew Chem Int Ed 2014;53(21):5262–5298.

[81] Jin Y, et al. Lignin-based high-performance fibers by textile spinning techniques. Materials 2021;14 (12):3378.

[82] Qu W, et al. Potential of producing carbon fiber from biorefinery corn stover lignin with high ash content. J Appl Polym Sci 2018;135(4):45736.

[83] Qin W, Kadla JF Carbon fibers based on pyrolytic lignin. J Appl Polym Sci 2012;126(S1):E204–E213.

[84] Mataram A, et al. Characterization and mechanical properties of polyacrylonitrile/silica composite fibers prepared via dry-jet wet spinning process. Mater Lett 2010;64(17):1875–1878.

[85] Föllmer M, et al. Wet-spinning and carbonization of lignin-polyvinyl alcohol precursor fibers. Adv Sustainable Syst 2019;3(11):1900082.

[86] Shi X, et al. Stepwise fractionation extracted lignin for high strength lignin-based carbon fibers. New J Chem 2019;43(47):18868–18875.

[87] Lallave M, et al. Filled and hollow carbon nanofibers by coaxial electrospinning of Alcell lignin without binder polymers. Adv Mater 2007;19(23):4292–4296.

[88] Ma X, et al. Electrospun lignin-derived carbon nanofiber mats surface-decorated with MnO2 nanowhiskers as binder-free supercapacitor electrodes with high performance. J Power Sources 2016;325:541–548.

[89] Fang W, et al. Manufacture and application of lignin-based carbon fibers (LCFs) and lignin-based carbon nanofibers (LCNFs). Green Chem 2017;19:1794–1827.

[90] Poursorkhabi V, Mohanty AK, Misra M Electrospinning of aqueous lignin/poly (ethylene oxide) complexes. J Appl Polym Sci 2015;132(2):41260.

[91] Poursorkhabi V, et al. Processing, carbonization, and characterization of lignin based electrospun carbon fibers: A review. Front Energy Res 2020;8:208.

[92] Luo Y, et al. Enabling high-quality carbon fiber through transforming lignin into an orientable and melt-spinnable polymer. J Cleaner Prod 2021;307:127252.

[93] Olsson C, Sjöholm E, Reimann A Carbon fibres from precursors produced by dry-jet wet-spinning of kraft lignin blended with kraft pulps. Holzforschung 2017;71(4):275–283.

[94] Byrne N, et al. Enhanced stabilization of cellulose-lignin hybrid filaments for carbon fiber production. Cellulose 2018;25(1):723–733.

[95] Trogen M, et al. Cellulose-lignin composite fibres as precursors for carbon fibres. Part 1 – Manufacturing and properties of precursor fibres. Carbohydr Polym 2021;252:117133.

[96] Bengtsson A, et al. Carbon fibers from lignin–cellulose precursors: Effect of carbonization conditions. ACS Sustain Chem Eng 2020;8(17):6826–6833.

[97] Stewart D Lignin as a base material for materials applications: Chemistry, application and economics. Ind Crops Prod 2008;27(2):202–207.

[98] Ingildeev D, et al. Novel cellulose/polymer blend fibers obtained using ionic liquids. *Macromolecular Materials and Engineering*, 2012;297(6), pp.585-594.

[99] Song L, et al. Carbon fibers with low cost and uniform disordered structure derived from lignin/ polyacrylonitrile composite precursors. Fibers Polym 2021;22(1):240–248.

[100] Dong X, et al. Polyacrylonitrile/lignin sulfonate blend fiber for low-cost carbon fiber. RSC Adv 2015;5 (53):42259–42265.

[101] Wang Y, et al. Strong, ductile and biodegradable polylactic acid/lignin-containing cellulose nanofibril composites with improved thermal and barrier properties. Ind Crops Prod 2021;171:113898.

[102] Jayaramudu T, et al. Adhesion properties of poly(ethylene oxide)-lignin blend for nanocellulose composites. Compos Part B: Eng 2019;156:43–50.

[103] Föllmer M, et al. Structuration of lignin-graphene oxide based carbon materials through liquid crystallinity. Carbon 2019;149:297–306.
[104] Frank E, et al. Carbon fibers: Precursor systems, processing, structure, and properties. Angew Chem – Int Ed 2014;53:5262–5298.
[105] Le ND, et al. Understanding the influence of key parameters on the stabilisation of cellulose-lignin composite fibres. Cellulose 2021;28:911–919.
[106] Le N-D, et al. Chemically accelerated stabilization of a cellulose–lignin precursor as a route to high yield carbon fiber production. Biomacromolecules 2022;*23*(3), pp.839–846.
[107] Norberg I Carbon Fibres from Kraft Lignin, KTH Royal Institute of Technology; 2012. pp. 69.
[108] Ogale AA, Zhang M, Jin J Recent advances in carbon fibers derived from biobased precursors. J Appl Polym Sci 2016;133(45)43794.
[109] Le ND, et al. Effect of boric acid on the stabilisation of cellulose-lignin filaments as precursors for carbon fibres. Cellulose 2021;28:729–739.
[110] Braun JL, Holtman KM, Kadla JF Lignin-based carbon fibers: Oxidative thermostabilization of kraft lignin. Carbon 2005;43:385–394.
[111] Brodin I, et al. Oxidative stabilisation of kraft lignin for carbon fibre production. Holzforschung 2012;66(2):141–147.
[112] Beste A Reaxff study of the oxidation of lignin model compounds for the most common linkages in softwood in view of carbon fiber production. J Phys Chem A 2014;118:803–814.
[113] Kadla JF, et al. Lignin-based carbon fibers for composite fiber applications. Carbon 2002;40:2913–2920.
[114] Li Y, et al. Study on structure and thermal stability properties of lignin during thermostabilization and carbonization. Int J Biol Macromol 2013;62:663–669.
[115] Baker DA, Gallego NC, Baker FS On the characterization and spinning of an organic-purified lignin toward the manufacture of low-cost carbon fiber. J Appl Polym Sci 2012;124:227–234.
[116] Zhang M, Jin J, Ogale A Carbon fibers from UV-assisted stabilization of lignin-based precursors. Fibers 2015;3:184–196.
[117] Norberg I, et al. A new method for stabilizing softwood kraft lignin fibers for carbon fiber production. J Appl Polym Sci 2013;128:3824–3830.
[118] Zhang M Carbon Fibers Derived from Dry-Spinning of Modified Lignin Precursors. 2016;168.
[119] Seo DK, et al. Preparation and Characterization of the Carbon Nanofiber Mat Produced from Electrospun Pan/lignin Precursors by Electron Beam Irradiation. Rev Adv Mater Sci 2011;28(1):31–34.
[120] Choi D, Kil HS, Lee S Fabrication of low-cost carbon fibers using economical precursors and advanced processing technologies. Carbon 2019;142:610–649.
[121] Lin J, et al. Improvement of mechanical properties of softwood lignin-based carbon fibers. J Wood Chem Technol 2014;34:111–121.
[122] Guo Z, et al. The production of lignin-phenol-formaldehyde resin derived carbon fibers stabilized by BN preceramic polymer. Mater Lett 2015;142:49–51.
[123] Cho M, et al. Skipping oxidative thermal stabilization for lignin-based carbon nanofibers. ACS Sustain Chem Eng 2018;6:6434–6444.
[124] Le ND, et al. Cellulose-lignin composite fibers as precursors for carbon fibers: Part 2 – The impact of precursor properties on carbon fibers. Carbohydr Polym 2020;250:116918.
[125] Morgan P Carbon fibers and their composites. Carbon Fibers Compos, 2005.
[126] Al Aiti M, et al. On the morphology and structure formation of carbon fibers from polymer precursor systems. In Progress in Materials Science, Elsevier Ltd; 2018. pp. 477–551.
[127] Asmadi M, Kawamoto H, Saka S Thermal reactivities of catechols/pyrogallols and cresols/xylenols as lignin pyrolysis intermediates. J Anal Appl Pyrolysis 2011;92:76–87.
[128] Cao J, et al. Study on carbonization of lignin by TG-FTIR and high-temperature carbonization reactor. Fuel Process Technol 2013;106:41–47.

[129] Zhou H, et al. Polycyclic aromatic hydrocarbon formation from the pyrolysis/gasification of lignin at different reaction conditions. Energy Fuels 2014;28:6371–6379.

[130] Sharma RK, Hajaligol MR Effect of Pyrolysis conditions on the formation of Polycyclic Aromatic Hydrocarbons (PAHs) from Polyphenolic compounds. J Anal Appl Pyrolysis 2003;66(1–2):123–144.

[131] Baker FS Low Cost Carbon Fiber from Renewable Resources, U.S Department of energy; 2010.

[132] Sagues W, et al. Are lignin-derived carbon fibers graphitic enough?. Green Chem 2019;21(16), pp. 4253–4265.

[133] Jin J, et al. Carbon fibers derived from fractionated-solvated lignin precursors for enhanced mechanical performance. ACS Sustain Chem Eng 2018;6:14135–14142.

[134] Liu HC, et al. Processing, structure, and properties of lignin- and CNT-incorporated polyacrylonitrile-based carbon fibers. ACS Sustain Chem Eng 2015;3:1943–1954.

[135] Bengtsson A, et al. Carbon fibers from lignin-cellulose precursors: Effect of stabilization conditions. ACS Sustain Chem Eng 2019;7:8440–8448.

[136] Kubo S, Kadla JF Lignin-based carbon fibers: Effect of synthetic polymer blending on fiber properties. J Polym Environ 2005;13:97–105.

[137] Wang S, et al. Lignin-based carbon fibers: Formation, modification and potential applications. Green Energy Environ 2021;7(4), pp.578–605.

[138] Magazine J-A. Lightweighting and sustainability. June 2020 [cited 2022 18 April]; Sixth: [Available from: https://justauto.nridigital.com/just-auto_magazine_jun20/carbon_fibre_in_car_production_weighing_up_the_benefits.

[139] Ahmad H, et al. A review of carbon fiber materials in automotive industry. IOP Conf Ser Mater Sci Eng 2020;971(3):032011.

[140] Mordor Intelligence. Automotive Carbon Fiber Market – Growth, Trends, Covid-19 Impact, and Forecast (2022 – 2027). 2022, [cited 2022 18 April]; Available from: https://www.mordorintelligence.com/industry-reports/automotive-carbon-fiber-composites-market.

[141] Chen MC-W Commercial viability analysis of lignin based carbon fibre. 2014.

[142] Mohammadzadeh Kakhki R A review to recent developments in modification of carbon fiber electrodes. Arab J Chem 2019;12(7):1783–1794.

[143] Karl B, et al. Structural positive electrodes for multifunctional composite materials. In ICCM22 2019, Melbourne, VIC: Engineers Australia; 2019. pp. 2272–2278. In: ICCM22 2019, Melbourne, VIC: Engineers Australia; 2019. pp. 2272–78.

[144] Chan S, et al. Electrospun carbon nanofiber catalyst layers for polymer electrolyte membrane fuel cells: Structure and performance. J Power Sources 2018;392:239–250.

[145] Wang S-X, et al. Lignin-derived fused electrospun carbon fibrous mats as high performance anode materials for Lithium Ion batteries. ACS Appl Mater Interfaces 2013;5(23):12275–12282.

[146] Hagberg J, et al. Lithium iron phosphate coated carbon fiber electrodes for structural lithium ion batteries. Compos Sci Technol 2018;162:235–243.

[147] Asp LE, et al. A structural battery and its multifunctional performance. Adv Energy Sustainability Res 2021;2(3):2000093.

[148] Guomin X, Guan W, Sun X Filtration performance and application of activated carbon fiber enhanced microfibrous entrapped sorbent (ACF-MFES). Russ J Phys Chem A 2020;94(1):182–188.

[149] Gu Y, et al. One-step solvothermal synthesis of Au-TiO2 loaded electrospun carbon fibers to enhance photocatalytic activity. Vacuum 2016;130:1–6.

[150] Guo L, et al. Fabrication of palladium nanoparticles-loaded carbon nanofibers catalyst for the heck reaction. New J Chem 2013;37(12):4037–4044.

[151] Yu B, et al. Preparation of electrospun Ag/g-C_3N_4 loaded composite carbon nanofibers for catalytic applications. Mater Res Express 2017;4(1):015603.

Nur Liyana Ismail, Siti Fatihah Salleh, Jofry Othman, Sara Shahruddin*

Chapter 10 Bio-based surfactant: overview, trend, and future outlook

10.1 Introduction

Surfactants, an abbreviation of "surface-active agents," are compounds that reduce water-oil, liquid-gas, and solid-liquid or solid-gas medium surfaces and interfacial tension [1, 2]. The surface energy is reduced by the presence of hydrophilic and hydrophobic sections of the same surfactant molecule, owing to preferred interactions at surfaces and interfaces. In aqueous solution, surfactant molecules arrange themselves at the interface, where the hydrophobic part is in the air (or oil) and the hydrophilic part is in water, while at high concentration or concentrations above the critical micelle concentration (CMC), surfactant molecules self-assemble into micelles (Figure 10.1). Not only are they widely used as cleaning agents, but other beneficial properties such as foaming, emulsification, and particle suspension also make surfactants known for their wetting ability and effectiveness as emulsifiers and stabilizers. Due to this characteristic, surfactants are found in a variety of products that we use every day, including food, pharmaceuticals, toiletries, detergents, automotive fluids, paints, and coatings [2]. Surfactants have steadily grown in popularity since their debut in the early twentieth century, and they are now among the most widely used synthetic compounds on the planet [3, 4]. The dynamics of the surfactant market are driven by the cost, variety, and availability of hydrophobes as well as the cost and complexity of attaching or creating hydrophilic head groups.

Petrochemical and renewable sources are the two primary feedstock groups used in the manufacture of surfactants [5, 6]. The development of petrochemical processing led to the acquisition of hydrophobic structures of surfactant molecules through polymerization of alkenes such as ethylene or propylene. Although ethylene has been employed as a carbon chain building block, its increased applicability in industrial production resulted from the production of an intermediate or precursor, ethylene oxide [7]. Meanwhile, natural surfactants are usually derived from triglycerides found in vegetable oils or animal fats.

***Corresponding author: Sara Shahruddin**, Petronas Research Sdn. Bhd., Lot 3288 and 3289, Off Jalan Ayer Itam, Kawasan Institusi Bangi, 43000 Kajang, Selangor Darul Ehsan, Malaysia, e-mail: sarashahruddin@petronas.com

Nur Liyana Ismail, Siti Fatihah Salleh, Jofry Othman, Petronas Research Sdn. Bhd., Lot 3288 and 3289, Off Jalan Ayer Itam, Kawasan Institusi Bangi, 43000 Kajang, Selangor Darul Ehsan, Malaysia

 | https://doi.org/10.1515/9783112212141-010

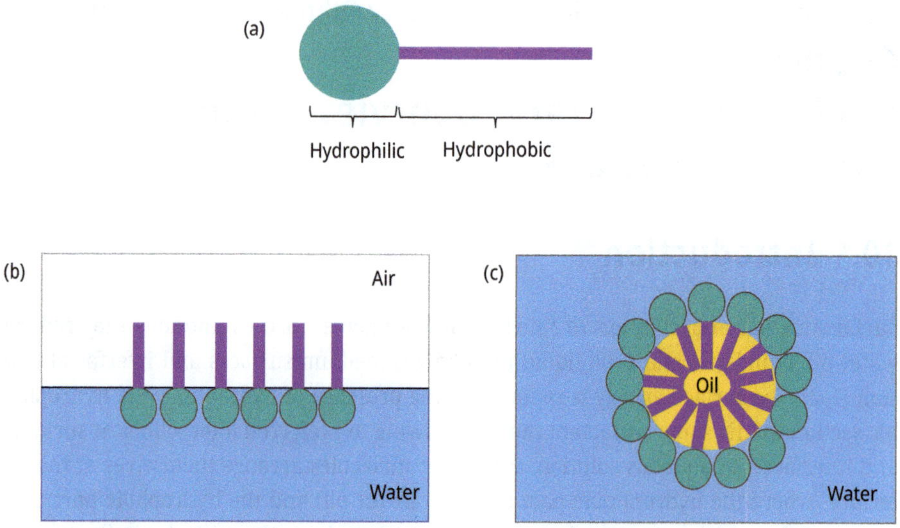

Figure 10.1: (a) Simplified surfactant molecule, (b) arrangement of surfactant monomers at the water surface, and (c) micelle formation above the critical micelle concentration (CMC).

The surfactant industry was focused on the saponification of oils and fats prior to petrochemical processing [8, 9]. Surfactants infiltrate water bodies after usage, where they can create issues if they remain for a long time, resulting in the build-up of potentially toxic or otherwise hazardous substances, causing significant environmental concerns [10–12]. Synthetic surfactant-related water contamination has increased in recent years because of their widespread usage in domestic, agricultural, and other cleaning activities. This occurrence has caused global concern, forcing establishment of a series of new rules governing their usage and disposal [13, 14].

In addition, experts relate the production of petrochemical-based surfactants to the high net output of CO_2, a greenhouse gas linked to climate change and global warming. By switching to renewable feedstock, this rate can be minimized. A previous study shows that by using renewable resources, instead of petrochemicals, for surfactant synthesis would cut CO_2 emissions by 37% in the EU [15]. Aside from environmental concerns and regulations, growing consumer awareness and market pressures have prompted considerable R&D into bio-based surfactants as potential substitutes for synthetic surfactants. Industry experts highlighted the importance of moving toward surfactant production from renewable sources, designing for biodegradation at the end of their life and improving surfactant efficiencies in use. Innovation in these areas will lessen the market's dependency on fossil-derived feedstocks and reduce waste creation, while maximizing the value of surfactants in their current applications.

The term "bio-based surfactant" refers to the surfactant produced by a chemical or enzymatic process that uses renewable substrates as raw materials [16, 17]. According to ISO/DIS 21680, the definition of bio-based surfactant is a surfactant wholly or partly derived from biomass (based on biogenic carbon) [18]. Most applications need further processing of bio-based feedstocks to incorporate functional groups that can give the surfactant's functional characteristics, resulting in a variety of anionic, cationic, nonionic, and amphoteric products. Many of these processes require the use of petroleum-based feedstocks or moieties that are not always environmentally friendly. The European Commission of Standardization has created categories for biosurfactants, including >95% completely bio-based, 50–94% majority bio-based, 5–49% minority bio-based, and 5% non-bio-based, to assist in analyzing the bio-based surfactants' sustainability criteria (Table 10.1) [19].

Table 10.1: Bio-based surfactant classes according to CEN/TS 17,035 [19].

Surfactant class	Bio-based carbon content, *X*% (m/m)
Wholly bio-based surfactant	≥95
Majority bio-based surfactant	$95 \geq X > 50$
Minority bio-based surfactant	$50 \geq X \geq 5$
Non-bio-based surfactant	$X < 5$

The hydrophobe, hydrophile, or both that derived from natural sources can be used in the production of bio-based surfactants. Plant oil fatty acids and animal fats are examples of natural hydrophobes, while glycerol, glucose, sucrose, and amino acids (aspartame, glutamic, lysine, arginine, alanine, and protein hydrolyzates) are the examples of natural hydrophiles. They can either be directly utilized from their original form or produced from complicated sources such as vegetable oil, sugarcane, sugar beets, and starch-producing crops. As for biosurfactants, they consist of hydrophilic sugar or peptide component and hydrophobic saturated or unsaturated fatty acid chains that are naturally produced by bacteria, yeast, and fungi. Hence, biosurfactant is classified as a wholly bio-based surfactant since all its raw materials are considered natural [20–22].

The hydrophobic part of bio-based and biosurfactant feedstock is mostly from fatty acyl groups. The fatty acyl groups are generally obtained from oilseeds in the form of triacylglycerol, but may also be derived from oleochemical by-products such as free fatty acid or phospholipids. Fatty acyl groups are generally utilized as lipophilic building blocks for surfactants in the form of free fatty acids or fatty acyl esters, which are produced via hydrolysis or alcoholysis of triacylglycerol [23, 24]. This fatty acyl group conjugates hydrophilic and lipophilic compounds via an ester bond. This bond makes the fatty acid-based surfactants suitable for foods, cosmetics, personal care, and pharmaceutical product applications, but not for laundry detergents, since the ester bonds are unstable. More stable bonds such as ether, amides, and carbonate bonds can be produced by converting the fatty acid groups to fatty alcohols, fatty amines, or fatty acid chloride [25–27].

Algae are another potential renewable source of fatty acids. It has been an active research area in recent years due to its potential for high oil production per acre and the ability to leverage nonarable soil [28–30]. Previously, Unilever partnered with Solazyme, a microalgae firm, in the aim of finding a palm-oil-free replacement for its soaps and surfactants. Solazyme used the advantage of their intellectual property in the areas of recombinant DNA expression in algae and algae bioprocessing to create oils with specific fatty acyl compositions [31]. Solazyme, later renamed as TerraVia, was acquired by Carbion in 2017 to focus on delivering innovative and high-value ingredients for food, personal care, and industrial applications [32]. Lignin has also been used as a feedstock in surfactant production due to its hydrophobic aromatic structure. Lignin-based surfactants are usually made by grafting hydrophilic groups or monomers onto the lignin to enhance its surface properties [33–35]. Extensive investigations are necessary to expedite the commercialization of lignin-based surfactants to the market since information on connecting performance and characteristics of lignin-based surfactants for their optimal usage is still lacking.

As for the hydrophilic part of biosurfactant, among the most significant renewable feedstocks are vegetable oils (for glycerol), sugarcane and sugar beets (for sucrose), and starch-producing crops such as maize, wheat, potato, and tapioca (for glucose) [4, 23, 36]. The use of glycerol as an alternative hydrophilic building block to replace ethylene oxide in the synthesis of nonionic surfactants is a feasible option. The major glycerol-based surfactants on the market are ester-based mono- and diglycerides, which are made by transesterifying triglycerides with excess glycerol and a base catalyst [4, 26, 37].

Carbohydrates such as sugar and sucrose are useful biorefinery feedstocks that make up as surfactant hydrophiles. Table 10.2 provides an extensive look at the three main classes of carbohydrate-based surfactants: sucrose esters, alkyl polyglucosides (APG), and sorbitan esters. The discovery of sucrose monoesters, or long-chain fatty acid esters, was one of the first major achievements of the Sugar Research Foundation, which led to their use as nonionic surfactants, food additives, and emulsifiers [38]. The global sucrose ester market amounted to $71.9 M in 2018 and is expected to reach $137.85 M by 2027 [39]. However, selectivity in the synthesis of these esters remains a challenge where acylation with a single fatty acid can yield a wide range of isomers with varying degrees of substitution [40]. One of the solutions to tackle the selectivity problem is by using enzyme catalysts such as lipases and proteases for regioselective sucrose ester production [41, 42]. Further improvement via lipase and protease protein engineering might increase the regioselectivity and yield of the catalyzed processes. The biotransformation of sucrose to sucrose esters, utilizing whole-cell fermentation methods, might also give a new path to sucrose-based surfactant production.

Besides sucrose esters, glucose can also react directly with fatty alcohol in a glycosidation process to produce APGs. APG is a nonionic surfactant class with growing production and popularity. Among carbohydrate-based surfactants, APG is the most popular one and is abundantly produced for wide range of applications due to its ex-

Table 10.2: Carbohydrate-based surfactants (alkyl polyglucosides, sorbitan esters, and sucrose esters) that are fully biodegradable.

Carbohydrate-based surfactants	Raw materials	Product characterization	Synthesis	Products	Properties	Primary functions	Application
Alkyl polyglucoside	– Head (hydrophilic): monosaccharides from starch, glucose syrup, glucose monohydrate, water-free glucose, D-glucose, D-xylose – Tail (hydrophobic group): fatty alcohols	– Degree of polymerization (no. of glucose attached to the head) – Alkyl chain length of the fatty alcohol, branching	– Fischer glycosation process: acetalization of glucose in molar excess fatty alcohol (acid catalyzed) – Solvents (optional): Dimethyl-formamide, dimethylsulfoxide	– Alkyl mono-, di-, tri-, and oligoglycosides as mixtures of α- and β-anomers.	– HLB range: 11–14 (relatively hydrophilic) – Critical micelle concentration decreases as alkyl chain increases	– Cleansing agent – Emulsifying agent – Wetting agent – Degreasing agent – Solubilizing agent – Hydrotope – Foaming agent	Medium-chain $C_{12/14}$ – Personal care: cosmetics, bath products, cleansers and wipes, oral care products – Homecare: surface cleaners, dishwashing agents, laundry detergents Short-chain $C_{8/10}$ (or branched C_8) – Hard surface cleaners – Agrochemicals (substitute alkyl phenol ethoxylate) – Industrial and institutional cleaning products Other applications – Oil fields – Admixtures for cement, concrete, and plaster

(continued)

Table 10.2 (continued)

Carbohydrate-based surfactants	Raw materials	Product characterization	Synthesis	Products	Properties	Primary functions	Application
Sucrose esters	– Head (hydrophilic group): disaccharides such as sucrose and lactose – Tail (hydrophobic group): fatty acids from vegetable oils such as lauric acid, stearic acid, palmitic acid, and oleic acid	– Degree of esterification (no. of acyl groups attached to sucrose) – Alkyl chain length of fatty acid – Degree of unsaturation of alkyl chain	– Direct esterification with fatty acids, or transesterification with fatty acid methyl ester (base catalyzed) – Usually carried out with emulsifier such as fatty acid soap and/or sucrose esters, and solvents	– Mono-, di-, tri-, tetra-, or penta-sucrose esters	– HLB range: 1–16 – Monoesters: very hydrophilic, HLB = 16 – Diesters and triesters: more lipophilic	– Emulsifying agent – Wetting agent – Solubilizing agent – Foaming agent	Limited application potential: cater to specific emulsification needs only for: – Food additives – Personal care – Detergent – Stabilizer – Pesticide – SE products are good candidates as insecticides and possibly antimicrobial agents
Sorbitan esters	– Head (hydrophilic group): dehydrated sugar alcohols (sorbitan) – Tail (hydrophobic group): fatty acids from vegetable oils such as lauric acid, stearic acid, palmitic acid, and oleic acid	– Degree of esterification – Alkyl chain length of fatty acid – Degree of unsaturation of alkyl chain	– Intramolecular dehydration of sorbitol (acid catalyzed) and subsequent esterification with fatty acids (base catalyzed)	– Mono-, di-, or tri-sorbitan ester	– HLB range: 4–8 (relatively lipophilic) – Surface tension increases as alkyl chain increases	– Emulsifying agent – Stabilizing agent	– Emulsifier – Pharmaceuticals – Foods – Cosmetic products – Pesticides – Coatings – Explosives

cellent dermatological properties, high stabilities toward oxidation and hydrolysis, as well as foaming and cleaning abilities [16]. APG's primary functions are cleansing, emulsifying, wetting, degreasing, solubilizing, hydrotope, and foaming. Besides its superior foaming performance, it also has an upper hand with respect to its use in the cosmetic and oil industries. Performance properties of APG can be adapted to meet market requirement via precise control of average degree of polymerization and the length of chain alcohols.

In addition, glucose may also be chemically converted to sorbitan to produce sorbitan esters. Like sucrose esters, sorbitan esters are esters of fatty acids, only that the polar head group is derived from sugar alcohols instead of sugar. Sorbitan esters are used as lipophilic nonionic surfactants in multitude of applications, including personal care, pharmaceutical, household, as well as food and beverages. Commercially, sorbitan esters are commonly referred to as spans such as Span 20 (sorbitan monolaurate), Span 40 (sorbitan monopalmitate), Span 60 (sorbitan monostearate), and Span 80 (sorbitan monooleate). Aside from sorbitant esters, other glucose-derived products are sorbitol, sorbitan, *N*-methyl glucamine, and *O*-methylglucoside, as well as the enzymatically converted products such as amino, lactic, and citric acids, all of which can be leveraged to produce surfactants (Figure 10.2) [4].

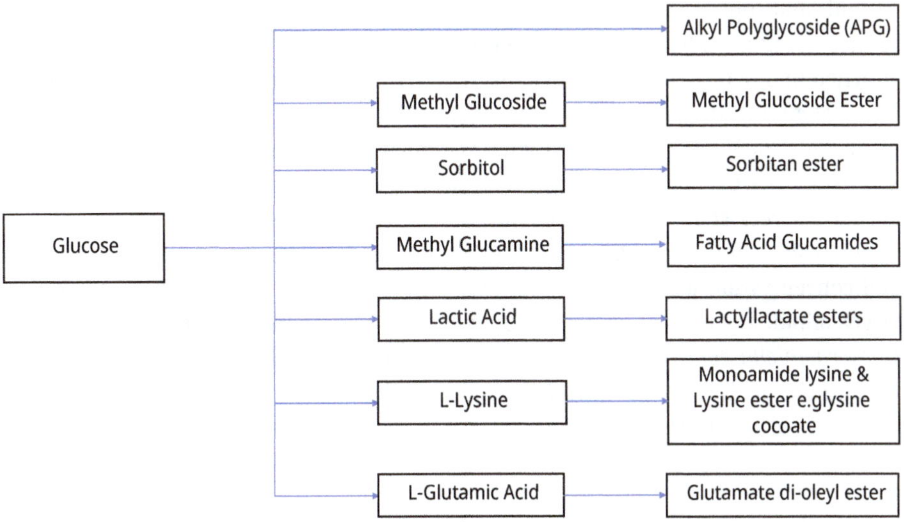

Figure 10.2: Simplified transformation pathway from glucose to several surfactant building blocks and surfactants.

Sugar-derived surfactants have a higher market demand than synthetic chemicals and surfactants due to their low toxicity, low cost, biodegradability, good cleaning and washing abilities, environmental compatibility, and high surface activity [43, 44]. However, if the demand for sugar surfactants grows in the long run, feedstock availability will become a concern. New methods that use bacteria and microorganisms to manufacture glucose are emerging; however, the issue of scalability has yet to be solved.

Aside from sugar-based surfactants, the development of new amino acid-based surfactants is on the rise, which may be influenced by advances in biotechnological amino acid synthesis. The two most produced amino acids in the market are L-glutamic acid and L-lysine. Besides that, alanine, aspartic acid, glycine, and arginine, as well as protein hydrolyzates are also used in the manufacture of some commercial surfactants [45–47]. In the meantime, sarcosine-based surfactants have been on the market for decades. Even though sarcosine is a naturally occurring molecule, it is mostly synthesized on a large scale by combining chloroacetic acid with *N*-methylamine [48–50]. Betaine, another naturally occurring molecule, is also synthesized in large scale using petrochemical-based trimethylamine and chloroacetic acid. Most betaine surfactants use an oleochemical hydrophobe precursor obtained from tropical oils as the bio-based component [51]. Glycine betaine is a promising biosurfactant that can be commercially extracted from brown algae and sugar beet molasses [52, 53].

Glycolipids are a type of complex carbohydrate that contains both a glycan and a lipid component. They are usually the main lipids of bacterial and fungal cell walls. In aqueous solution, glycolipids are amphiphilic substances that form stable micelles, and these molecules have the capacity to offer low interfacial tension [54, 55]. Rhamnolipids and sophorolipids are among the glycolipids that have been utilized the most as biosurfactants. Rhamnolipids are produced as one or two rhamnose sugar groups attached to one or two fatty acid chains by different bacterial species (i.e., *Pseudomonas aeruginosa*, *Pseudomonas chlororaphis*, and *Burkholderia pseudomallei*) [4, 56]. Aside from their favorable emulsifying, solubilizing, foaming, and antibacterial characteristics, the use of rhamnolipids is appealing due to their high production yields after relatively short incubation times [56]. Rhamnolipids are now available on a larger scale due to the optimized fermentation techniques and advanced extraction and concentration technologies. Sophorolipids, another extensively researched type of glycolipid, are biosynthesized by certain yeast strains like *Starmerella bombicola*, *Wickerhamiella domercqiae*, and *Candida batistae* from sophorose sugar and hydroxylated fatty acid. Sophorolipids are commercially used in dish and vegetable detergents, and in skin care formulations [57–60].

10.2 Recent progress in R&D and industrial production

Regulations on the environmental impact and hazardous chemicals are highly stringent, particularly in Europe and North America, which are the two largest markets for surfactants, especially in the home and personal care sectors. As a result, the surfactant industry is commencing to develop biosurfactants, which have lower levels of toxicity and a more environmentally friendly manufacturing process. Apart from complying with environmental regulations, the industry is seeing bio-based surfactants to achieve a sustainable competitive edge. The advent of biotechnology in the twenty-first century promoted the creation of novel bio-based and biosurfactants, along with their enhanced commercial and economic viability. Extensive and significant R&D has also enabled high-quality and high-functionality bio-based surfactant formulations to evolve from the lab-scale to niche applications to commercial-scale production. Some of the bio-based surfactants that are commercially available in the market, their main manufacturers, and their applications are listed in Table 10.3.

Table 10.3: Commercially available bio-based surfactants, their manufacturers, and their applications.

Bio-based surfactants	Selected manufacturers	Fields of applications
Anionic		
Lignosulfonate Methyl ester sulfonates Anionic derivatives of alkyl polyglucoside	Vanderbilt Minerals, LLC Huish Detergents, Inc., Lion Corp., Longkey, Stepan Cognis, Colonial Chemical	– Laundry – Food service and kitchen hygiene – Dishwashing – Hard surface cleaning – Institutional cleaning and sanitation – Vehicle and transportation care
Nonionic		
Fatty alcohol alkoxylate Fatty acid alkoxylate Alkyl polyglucoside Sorbitan ester Alkanoyl-*N*-methylglucamide Alkyl-ethoxylated mono- and diglycerides Polyglycerol esters	BASF, Dow BASF, Clariant, Croda Croda, Esterchem, Huntsman Akzo Nobel, BASF, Colonial Chemical, Dow, Huntsman BASF, Croda, Huntsman Clariant, Kao, Kerry Ingredients and Flavors BASF, Colonial Chemical, Hychem Corp., Kerry Ingredients and Flavors	– Dishwashing – Laundry – Hard surface cleaning – Food service and kitchen hygiene – Institutional cleaning and sanitation – Vehicle care – Personal care
Amphoteric		
Cocamidopropyl betaine Cocamidopropyl hydroxysultaine Lauryl hydroxysultaine	BASF, Colonial Chemical, Stepan Colonial Chemical, Stepan Colonial Chemical, Stepan	– Food and beverage processing – Personal care – Hard surface cleaning

Table 10.3 (continued)

Bio-based surfactants	Selected manufacturers	Fields of applications
Glycolipid		
Sophorolipid	BASF, Clariant, Ecover, Evonik, MG Intobio Co. Ltd., Saraya Co. Ltd., Soliance	– Personal care – Vegetable liquid wash – Dish washing
Rhamnolipid	AGAE Technologies, BASF, Biotensidon GmbH, Clariant, Evonik, GlycoSurf, Henkel, Jeneil Biotech Inc., Logos Technology Rhamnolipid Companies Inc., TeeGene Biotech	
Amino acid surfactants		
Sodium cocoyl glutamate	Ajinomoto Co. Inc., Stepan, Zschimmer and Schwarz	Personal care
Sodium methyl cocoyl taurate	Clariant	
α-Acyl glutamate and sarcosinate	Schill + Seilacher	

In the current development of novel surfactants, there is a growing trend of utilizing nontraditional naturally occurring branching hydrophobic chains [61–63]. Nonionic surfactants based on twin-tail glycerol have been synthesized, and they have good oil-in-water and water-in-oil emulsifying characteristics [64]. Other structural analogues of glycerol-based surfactants have recently been created by employing heterogeneous interfacial acidic catalysts to direct etherify glycerol and dodecanol. These surfactants have been shown to be comparable with commercially available surfactants in terms of physicochemical assessment and detergency ability [65]. Another class of amphiphilic compounds with a glycerol backbone is bio-based dialkyl glycerol ethers. These compounds have good solvosurfactant characteristics and can function as solubilizers for hydrophobic dyes in aqueous media [66].

Natural edible flavor vanillin is used to create a cleavable vanillin-based polyoxyethylene nonionic surfactant. Because it contains cleavable acetal bonds that break down quickly under acidic circumstances, this environmentally beneficial surfactant is totally biodegradable in nature. The surfactant's surface activity, wettability, emulsifying, and foaming properties are on par with nonylphenol ethoxylate surfactants, which are highly toxic to aquatic organisms and environment [67]. Several novel types of sustainable surfactants have been created in recent years by employing various types of terpenes, which are major components of essential oils derived from a variety of plants and flowers [68–71]. The terpenes were transformed to branched hydrophobic tail containing quaternary ammonium surfactants. Natural farnesol, a 15-carbon acyclic sesquiterpene alcohol found in neroli, lemon grass, tuberose, rose, citronella, and other plant species, was used to create a new form of terpene-based sustainable surfactant, which has demonstrated excellent surfactant performance [71]. Under the trade name ECO-SURF, Dow Chemical Co. is now offering a range of sustainable oilseed-based nonionic surfactants. These surfactants are claimed to have minimal aquatic toxicity and are biodegradable in nature, making them suitable candidates for paints and coatings, as well as home, industrial, and institutional cleansers and textiles [72].

TegraSurf, a sustainable water-based surfactant developed for energy, mining, agricultural, water treatment and other industrial applications, was released in July 2021 by Integrity BioChem, a technology-driven business producing next-generation biopolymers. TegraSurf is made of sustainable vegetal materials and is certified readily biodegradable by the OECD 301B guideline. After 90 days, it is no longer in the environment, making it safer and healthier for local population and allowing formulators to fulfill industry sustainability criteria [73]. BASF and Solazyme Inc. recently released Dehyton® AO 45, the first commercial microalgae-derived betaine surfactant made from microalgae oils as an alternative to conventional amidopropyl betaine surfactants [74]. Following the launch of sophorolipid-based surfactants in 2020, BASF formed an exclusive partnership with Holiferm Ltd. in the United Kingdom to focus on the development of glycolipids other than sophorolipids for personal and home care, as well as industrial uses [75].

Croda expanded its commercial-scale bio-based manufacturing capabilities and technology leadership in renewable raw materials by unveiling its 100% bio-based ethylene oxide production facility as an effort to make the world's products greener. Ethylene oxide is the key raw material used to produce surfactants. Croda's Atlas Point manufacturing plant in New Castle, Delaware, is the first of its type in the United States for the manufacture of 100% sustainable, 100% bio-based nonionic surfactants [76]. Ajinomoto is increasing to 60% of their global capacity for its Amisoft range of amino acid-based liquid surfactants by building a new plant in Brazil [77]. Sironix Renewables received $645,000 in investment from the University of Minnesota Discovery Capital Investment program and investors, as well as a $1.15 million grant from the US Department of Energy Advanced Manufacturing Office, to help them scale up their Eosix® production. The new renewable oleo-furan-based surfactants are 100% plant-based that offer unique and adjustable characteristics in a wide range of areas, including cleaning products, cleaners, cosmetics and personal care, agriculture and inks, and paint and coatings [78].

10.3 Industrial challenge on bio-based surfactant

This section covers the market performance, demand drivers, and growth prospects of biosurfactants. The market trend on bio-based and biosurfactants is discussed for the different geographic regions and in terms of changing market trends for biosurfactants in various application areas. Analysis of the industrial challenges of biosurfactants, which includes the growth-restraining factors and future opportunities, is provided.

10.3.1 The economy and market trend of bio-based surfactants

The worldwide surfactant industry, estimated to be worth $39 billion in 2019, is expected to expand at a rate of 2.6% per year over the following 5 years, reaching $46 billion in 2024. Surfactants are produced in a total of 17 million metric tons per year [79]. In the EU, of the 3 million metric tons of surfactants produced in 2019, roughly 50% were bio-based [80]. A market study by the Market Research Future [81] indicated that the global biosurfactant market value is around US $2.1 billion in 2020 and predicted it to reach US $2.8 billion by 2026, with a compound annual growth rate of over 5% from 2021 to 2026. The attractive performance of biosurfactants advances their high potential to substitute synthetic-based surfactants for drop-in applications and with unique properties that can overcome entry barriers for the emerging industrial areas.

Major types of biosurfactants such as sophorolipids, glycolipids, lipopeptides, polymeric biosurfactants, phospholipids, and fatty acids generally form the product demand application. Among them, sophorolipids provide the largest global market demand with detergents and industrial cleaning applications. The leading demand drivers for biosurfactants comprise a growing consumer preference, increasingly stringent regulatory requirements, and rising awareness toward eco-friendly alternatives. By being environmentally compatible and with low toxicity, many studies have considered biosurfactants as the next generation of industrial surfactants [82–84]. In terms of end-user applications, biosurfactants are finding usage in household detergents, industrial and institutional cleaners, cosmetics, and personal care within the major markets in Europe and North America [81]. Recently, they are gaining acceptance in the newer application areas such as in oil and gas as well as agricultural industries.

Furthermore, the increasing consumer awareness of biosurfactants' benefits and their wide range of application sectors form market drivers that increase their future growth potential. Higher growth of biosurfactants is seen in Asia-Pacific (APAC), especially in Southeast Asian countries, which have slightly different demand factors that involve the increasing purchasing power of mass consumers, growing concern on environmental issues, and the generation of harmful chemical by-products. In terms of APAC market segmentation, the major sales revenue for biosurfactants resides within the home care and personal care applications, as rising urbanization becomes the dominant factor for surfactant growth.

More importantly, a key growth enabler is in the innovative research on biosurfactants, especially when it can generate multifunctional and diversified products using renewable feedstock. This technological progress contributes to the desirable properties of biosurfactants to meet the changing consumer lifestyles in developing economies and consequently their increasing preference for usage in the end-user product formulation. As an example, within the home care detergent industry, the usage of biosurfactants as environmentally friendly products provides sustainable alternatives that are gaining a large market share [82, 85, 86].

The highest adoption of bio-based and biosurfactants is in Europe and North America, which dominates the bio-based surfactant market share in terms of revenue and volume. Increasingly stringent regulatory requirements enable a wider acceptance of biosurfactants in the place of synthetic surfactants. For example, the imposed government regulations such as CEN/TC-276 define the standards for surface-active agents and detergents to enhance the EU bio-based economy, and detergent regulation (EC) no. 648 requires surfactants used in detergents to be biodegradable under aerobic conditions as per OECD 301 test series. In addition, the COVID-19 pandemic results in a sharp increase in the bio-based surfactant product demand for household detergents, personal care, and industrial cleaners due to the rising trend for sanitation.

10.3.2 The industrial challenges for bio-based surfactant

Bio-based surfactants are synthesized via chemical reaction, which is usually carried out under harsh conditions. The use of hazardous solvents and toxic acid or base catalysts sometimes creates undesired waste or by-products that are detrimental to the environment. Enzymes have the potential to play a significant role in the production of numerous bio-based surfactants, although they are not currently used on a large basis. Enzymes provide several advantages over chemical processing, notably in terms of improving process sustainability. However, the main drawbacks of enzymes are their relatively higher price compared to chemical catalysts, as well as their slower reaction speeds. Since energy costs are expected to rise, the need for production sustainability (lower operating energy, less waste, and safer operating condition) is critical.

Despite the growing demand for bio-based surfactants, several challenges exist, which restrain their further market growth and wider adoption. General customers tend to choose cost-effective surfactants. At present, the high cost of bio-based surfactants is the biggest hurdle in meeting the requirement of price-sensitive Asian customers. Increased sustainability of biosurfactant alone, without significant higher performance, is not well accepted, as the usual consumers will not be willing to pay a "green" premium for bio-based products.

Higher complexity and low-efficiency microbial fermentation process in biosurfactant manufacturing contribute to the high production cost and expensive capital cost investment. For example, the average price of sophorolipids is US $34/kg as compared to sodium dodecyl sulfate and amino acid surfactants that are priced at US $1–4/kg [87]. High purification cost of bio-based surfactants is also a big issue, as it accounts up to 60% the production cost, but this can be minimized for the case of biosurfactant application in crude forms, such as in an industrial environment [88, 89]. However, for high-purity applications, improvement in downstream processing methods is needed to attain a competitive cost of production.

Furthermore, biosurfactant demand's reliance on the volatility and economic downturn of downstream end-user industries poses a serious risk to its long-term viability. Industries that are applicable for biosurfactant applications such as oil and gas, enhanced oil recovery, food industry, construction, textiles, paints, pharmaceuticals, and detergents are known to be susceptible to general macroeconomic performance. In addition, the COVID-19 pandemic further leads to disruption in the end-user industrial demand and sustainability concern on the raw material supply. Opportunity exists in developing a new technology solution that utilizes a low-cost raw material such as industrial wastes for biosurfactant production. However, this needs to consider the overall production impact factors that include the availability, stability, and variability of each component [88, 89, 90].

10.4 Future outlook and prospect

The development of bio-based surfactants from renewable feedstocks is an attractive alternative to fossil-based surfactants, with a significantly growing market attributed to their performance, biodegradability, biocompatibility, and nontoxicity [22, 33]. Additionally, advances in renewable technology, increased environmental concern, consumer awareness, and stringent regulatory requirements provide a continued push toward bio-based surfactant demand. Potential areas for use are growing fast and valuable outcomes depend on whether the bio-based surfactants can be customized for specific applications along with the ability to be produced at a price that will make them attractive alternatives to the fossil-based surfactants.

At present, fossil-based surfactants are less expensive than bio-based surfactants [4, 93, 94]. However, this trend will likely be changed in future, thereby increasing the prospects of bio-based surfactants. Feedstocks and how the bio-based surfactants are produced are the two key factors governing final product costs [4, 36, 95, 96]. To use renewable feedstock in the industry, they should be cost-effective, available in large quantities, and can effectively be converted to value-added surfactants [96]. The surfactant design requires careful selection of the hydrophile and hydrophobe pairs, so that they can be easily synthesized with minimum purification, and provides the desired properties for the intended application [4, 16, 93, 97]. Triglycerides, fatty acid methyl esters, fatty alcohols, fatty acids, and fatty amines are common examples of renewable hydrophobes used to produce bio-based surfactants. Meanwhile, sustainable hydrophilic head group can be designed using several molecules such as glycerol, carbohydrate feedstock such as sucrose, glucose, organic acids, and amino acids [4, 36, 95, 96].

Current production technologies are leveraged and further enhanced with green manufacturing principles to convert renewable feedstocks into valuable and new bio-based surfactants. Other than the aforementioned starting materials, alternative substrates such as agro-based industrial wastes or other suitable simple waste substrate is gaining a lot of research interest and can lead to significant cost reduction [98]. The use of biomass for surfactant manufacturing will help to further reduce greenhouse gas emissions. Besides that, as bio-based surfactants degrade, they only release back the quantitative amount of the carbon used by the plant to produce the surfactants [36]. Although the use of biomass appears to be a promising option, the lack of homogeneity and consistency in the feedstock is a serious issue that will result in inconsistency in the final products. Surfactant property and performance variations may result in unsatisfactory properties. As a result, waste material pretreatment is critical, and extensive testing will be required to minimize variations.

Besides feedstocks, catalyst design is crucial to ensure high selectivity of the processes by limiting the formation of by-products and to reduce the reaction times efficiently; with renewable materials, fundamental rethinking of catalyst development is necessary [99–101]. The other critical component toward achieving low product cost is the process required for optimal chemical conversion. Researchers are looking into equipment miniaturization such as continuous reactors to help reduce the raw material consumption and effluent production. Process intensification is another aspect that could help reduce the investment costs [100]. Research focusing on alternative or green solvents dedicated to the conversion of renewable feedstock to value-added products has led to several publications. Among those being researched include bio-based ionic liquids, deep eutectic solvents, bio-based solvents, CO_2-switchable solvents, and supercritical fluids [102–104].

Additionally, several operation and control factors provide important handles to minimize biosurfactant production costs. Batch cycle optimization on fermentation and purification process can reduce the idle time between batches and minimize chemical usage for equipment cleaning and energy use during sterilization. Productivity is the most important factor in the manufacturing economics of biosurfactant production at commercial scales [8]. Optimum batch sequencing campaign is critical to minimize start-up and shutdown frequency to lower the production downtime that improves productivity. Lastly, biosurfactant product development will need to fulfill time-consuming and expensive legislative requirements that restrain market growth [91], which will inevitably add a cost of compliance into the product development cost that is incurred by biosurfactant manufacturers. Other market entry requirements include biosurfactant products that are tested for long shelf life and the ability to maintain stable properties in the industrial environment [92].

10.5 Conclusions

In recent years, surfactant manufacturers have introduced a variety of new eco-friendly surfactant-based products to the market. Increased consumer awareness, combined with legislative demand for sustainable development, has resulted in the market introduction of numerous novel surfactant types based on renewable building blocks. Because of their improved biodegradability and low toxicity, these surfactants are a popular choice for innovative formulations in the industrial and consumer markets. However, the commercial production of bio-based surfactants faces numerous challenges that must be addressed in order for them to be economically viable, particularly the high production cost. As a result, low-cost biosurfactant production is critical for achieving an economically sustainable process and ensuring market continuity in the future. A balance should be struck between cost-effective formulations and efficient performance. Simultaneous design of bio-based surfactants for functional, economic, and environmental benefits will be required to ensure the long-term viability of business. Moving forward, continuous technological advancements using higher performance catalysts, optimized and integrated reaction processes using less expensive renewable feedstocks, and effective downstream purification are required in the future. It is hoped that these efforts will eventually lead to a greater use of bio-based surfactants, which will provide enormous benefits to consumers and the environment.

References

[1] Rosen MJ, Kunjappu JT. Surfactants and Interfacial Phenomena, 4th edition, Hoboken, New Jersey: John Wiley & Sons; 2012. ISBN 978-1-118-22902-6.

[2] Möbius D, Miller R, Fainerman VB. Surfactants: Chemistry, interfacial properties, applications, Volume 13 of Studies in Interface Science. ISSN 1383-7303.

[3] Traverso-Soto JM, González-Mazo E, Lara-Martín PA. Analysis of surfactants in environmental samples by chromatographic techniques, Chromatography – The most versatile method of chemical analysis, Leonardo de Azevedo Calderon, IntechOpen, doi: 10.5772/48475. Available from: https://www.intechopen.com/chapters/40379.

[4] Foley P, Beach ES, Zimmerman JB. Derivation and synthesis of renewable surfactants. Chem Soc Rev 2012;41:1499–1518.

[5] Jimoh AA. Lin J Biosurfactant: A new frontier for greener technology and environmental sustainability. Ecotoxicol Environ Saf 2019;184:109–607. https://doi.org/10.1016/j.ecoenv.2019.109607. PMid: 31505408.

[6] Taddese T, Anderson RL, Bray DJ, et al. Recent advances in particle-based simulation of surfactants. Curr Opin Colloid Interface Sci 2020;48:137–148. https://doi.org/10.1016/j.cocis.2020.04.001.

[7] Soler-illia G, Sanchez C. Interactions between poly(ethylene oxide)-based surfactants and transition metal alkoxides: Their role in the templated construction of mesostructured hybrid organic-inorganic composites. New J Chem 2000;24(7):493–499. https://doi.org/10.1039/b002518f.

[8] Zoller U, Sosis P, editors. Handbook of Detergents, Part F: Production, CRC Press; 2008 Nov 2008.

[9] Holmberg K. Natural surfactants. Curr Opin Colloid Interface Sci 2001;6:148–159.

[10] Deschênes L, Lafrance P, Villeneuve JP, Samson R. Adding sodium dodecyl sulfate and Pseudomonas aeruginosa UG2 biosurfactants inhibits polycyclic aromatic hydrocarbon biodegradation in a weathered creosote-contaminated soil. Appl Microbiol Biotechnol 1996;46 (5):638–646.
[11] Takeda K, Sasaoka H, Sasa K, Hirai H, Hachiya K, Moriyama Y. Size and mobility of sodium dodecyl sulfate – Bovine serum albumin complex as studied by dynamic light scattering and electrophoretic light scattering. J Colloid Interface Sci 1992;154(2):385–392.
[12] Rebello S, Asok AK, Mundayoor S, Jisha MS. Surfactants: Toxicity, remediation and green surfactants. Environ Chem Lett 2014;12(2):275–287.
[13] Siwayanan P, Bakar NA, Aziz R, Chelliapan S, Siwayanan P. Exploring Malaysian household consumers' acceptance towards eco-friendly laundry detergent powders. Asian Soc Sci 2015;11 (9):125–137.
[14] Farn RJ, editors.. Chemistry and Technology of Surfactants, John Wiley & Sons; 2008 Apr 15.
[15] Patel M. Surfactants based on renewable raw materials: Carbon dioxide reduction potential and policies and measures for the European Union. J Ind Ecol 2003;7(3-4):47–62.
[16] Hayes DG. Smith GA Bio-based surfactants: Overview and industrial state of the art. Bio Based Surfactants 2019;3–8.
[17] Kandasamy R, Rajasekaran M, Venkatesan SK, Uddin M. New trends in the biomanufacturing of green surfactants: Bio-based surfactants and biosurfactants. In: Next Gener Biomanuf Technol Am Chem Soc 2019;1329:243–260.
[18] ISO/DIS 21680(en) Surface active agents – Bio-based surfactants – Requirements and test methods {Internet]. 2020. Available from: https://www.iso.org/obp/ui/#iso:std:iso:21680:dis:ed-1:v1:en.
[19] Comite Europeen de Normalisation. CEN/TS 17035. In Surface Active Agents – Bio-Based Surfactants – Requirements and Test Methods, Brussels: CEN; 2017.pp. 2017.
[20] Md F. Biosurfactant: Production and application. J Pet Environ Biotechnol 2012;3(4):124.
[21] Van Renterghem L, Roelants SL, Baccile N, Uyttersprot K, Taelman MC, Everaert B, Mincke S, Ledegen S, Debrouwer S, Scholtens K, Stevens C. From lab to market: An integrated bioprocess design approach for new-to-nature biosurfactants produced by Starmerella Bombicola. Biotechnol Bioeng 2018;115(5):1195–1206.
[22] Moldes AB, Rodríguez-López L, Rincón-Fontán M, López-Prieto A, Vecino X, Cruz JM. Synthetic and bio-derived surfactants versus microbial biosurfactants in the cosmetic industry: An overview. Int J Mol Sci 2021;22(5):2371.
[23] Hayes DG, Solaiman DK, Ashby RD. Bio-Based Surfactants: Synthesis, Properties, and Applications, Elsevier; 2019 Apr 30.
[24] Salimon J, Salih N, Yousif E. Industrial development and applications of plant oils and their bio-based oleochemicals. Arab J Chem 2012;5(2):135–145.
[25] Kreutzer UR. Manufacture of fatty alcohols based on natural fats and oils. J Am Oil Chem Soc 1984;61(2):343–348.
[26] Biermann U, Bornscheuer U, Meier MA, Metzger JO, Schäfer HJ. Oils and fats as renewable raw materials in chemistry. Angew Chem Int Ed 2011;50(17):3854–3871.
[27] Giraldo L, Camargo G, Tirano J, Moreno-Pirajan JC. Synthesis of fatty alcohols from oil palm using a catalyst of Ni-Cu supported onto zeolite. E J Chem 2010;7(4):1138–1147.
[28] Jeon JY, Han Y, Kim YW, Lee YW, Hong S, Hwang IT. Feasibility of unsaturated fatty acid feedstocks as green alternatives in bio-oil refinery. Biofuels Bioprod Biorefin 2019;13(3):690–722.
[29] Hess SK, Lepetit B, Kroth PG, Mecking S. Production of chemicals from microalgae lipids–status and perspectives. Eur J Lipid Sci Technol 2018;120(1):1700152.
[30] De Luca M, Pappalardo I, Limongi AR, Viviano E, Radice RP, Todisco S, Martelli G, Infantino V, Vassallo A. Lipids from microalgae for cosmetic applications. Cosmetics 2021;8(2):52.

[31] Lerayer A. Solazyme:" unlocking the power of microalgae: A new source of sustainable and renewable oils". BMC Proc 2014 Oct;8(4):1–2. BioMed Central.
[32] Innovative microalgae specialist TerraVia acquired by Corbion [Internet]. 2017. Available from: https://www.corbion.com/media/624357/20170929-tvia-completion-eng-final.pdf.
[33] Alwadani N, Fatehi P. Synthetic and lignin-based surfactants: Challenges and opportunities. Carb Res Conv 2018;1(2):126–138.
[34] Schmidt BV, Molinari V, Esposito D, Tauer K, Antonietti M. Lignin-based polymeric surfactants for emulsion polymerization. Polymer 2017;112:418–426.
[35] Zhou M, Wang W, Yang D, Qiu X. Preparation of a new lignin-based anionic/cationic surfactant and its solution behaviour. RSC Adv 2015;5(4):2441–2448.
[36] Bhadani A, Kafle A, Ogura T, Akamatsu M, Sakai K, Sakai H, Abe M. Current perspective of sustainable surfactants based on renewable building blocks. Curr Opin Colloid Interface Sci 2020;45:124–135.
[37] Fan Z, Zhao Y, Preda F, Clacens JM, Shi H, Wang L, Feng X, De Campo F. Preparation of bio-based surfactants from glycerol and dodecanol by direct etherification. Green Chem 2015;17(2):882–892.
[38] Plat T, Linhardt RJ. Syntheses and applications of sucrose-based esters. J Surfact Deterg 2001;4 (4):415–421.
[39] Outlook for sucrose esters. Focus Surfactants 2020;2020(6):3.
[40] Queneau Y, Fitremann J, Trombotto S. The chemistry of unprotected sucrose: The selectivity issue. C R Chim 2004;7(2):177–188. t.
[41] Potier P, Bouchu A, Descotes G, Queneau Y. Lipase-catalysed selective synthesis of sucrose mixed diesters. Synthesis 2001;2001(03):0458–62.
[42] Cruces MA, Plou FJ, Ferrer M, Bernabé M, Ballesteros A. Improved synthesis of sucrose fatty acid monoesters. J Am Oil Chem Soc 2001;78(5):541–546.
[43] Jesus CF, Alves AA, Fiuza SM, Murtinho D, Antunes FE. Mini-review: Synthetic methods for the production of cationic sugar-based surfactants. J Mol Liq 2021;342:117389.
[44] Hill K, Rhode O. Sugar-based surfactants for consumer products and technical applications. Lipid/Fett 1999;101(1):25–33.
[45] Infante MR, Pérez L, Pinazo A, Clapés P, Morán MC, Angelet M, García MT, Vinardell MP. Amino acid-based surfactants. C R Chim 2004;7(6–7):583–592.
[46] Takehara M. Properties and applications of amino acid based surfactants. Colloids Surf 1989;38 (1):149–167.
[47] Bordes R, Holmberg K. Amino acid-based surfactants–do they deserve more attention?. Adv Colloid Interface Sci 2015;222:79–91.
[48] Baumann L. The preparation of sarcosine. J Biol Chem 1915;21(3):563–566.
[49] Cullum DC. The separation of sarcosine from methylaminediacetic acid. Analyst 1957;82 (977):589–591.
[50] Morris ED, Bost JC. In Kirk-Othmer Encyclopedia of Chemical Technology, John Wiley & Sons, Inc; 2002.
[51] Clendennen SK, Boaz NW. Betaine amphoteric surfactants – Synthesis, properties, and applications. In Bio-Based Surfactants, AOCS Press; 2019 Jan 1. pp. 447–469.
[52] Goursaud F, Berchel M, Guilbot J, Legros N, Lemiègre L, Marcilloux J, Plusquellec D, Benvegnu T. Glycine betaine as a renewable raw material to "greener" new cationic surfactants. Green Chem 2008;10(3):310–320.
[53] Nsimba ZF, Paquot M, Mvumbi LG, Deleu M. Glycine betaine surfactant derivatives: Synthesis methods and potentialities of use. Biotechnol Agron Soc Environ 2010;14(4):737–748.
[54] Bednarski W, Adamczak M, Tomasik J, Płaszczyk M. Application of oil refinery waste in the biosynthesis of glycolipids by yeast. Bioresour Technol 2004;95(1):15–18.

[55] Kamerling, J.P., Boons, G.J., Lee, Y.C., Suzuki, A., Taniguchi, N. and Voragen, A.G.J., 2007. Analysis of glycans, polysaccharide functional properties & biochemistry of glycoconjugate glycans, carbohydrate-mediated interactions. *Comprehensive Glycoscience, From Chemistry to Systems Biology, 2.*

[56] Irfan-Maqsood M, Seddiq-Shams M. Rhamnolipids: Well-characterized glycolipids with potential broad applicability as biosurfactants. Ind Biotechnol 2014;10(4):285–291.

[57] Van Bogaert IN, Zhang J, Soetaert W. Microbial synthesis of sophorolipids. Process Biochem 2011;46 (4):821–833.

[58] Hirata Y, Ryu M, Oda Y, Igarashi K, Nagatsuka A, Furuta T, Sugiura M. Novel characteristics of sophorolipids, yeast glycolipid biosurfactants, as biodegradable low-foaming surfactants. J Biosci Bioeng 2009;108(2):142–146.

[59] Develter DW, Lauryssen LM. Properties and industrial applications of sophorolipids. Eur J Lipid Sci Technol 2010;112(6):628–638.

[60] Roelants S, Solaiman DK, Ashby RD, Lodens S, Van Renterghem L, Soetaert W. Production and applications of sophorolipids. Bio Based Surfactants 2019;65–119.

[61] Iskandar WF, Salim M, Hashim R, Zahid NI. Stability of cubic phase and curvature tuning in the lyotropic system of branched chain galactose-based glycolipid by amphiphilic additives. Colloids Surf A Physicochem Eng Asp 2021;623:126697.

[62] Kim JH, Oh YR, Hwang J, Kang J, Kim H, Jang YA, Lee SS, Hwang SY, Park J, Eom GT. Valorization of waste-cooking oil into sophorolipids and application of their methyl hydroxyl branched fatty acid derivatives to produce engineering bioplastics. Waste Manag 2021;124:195–202.

[63] Elsoud MM. Classification and production of microbial surfactants. In Microbial Biosurfactants, Singapore: Springer; 2021. pp. 65–89.

[64] Zhang L, Zhang X, Zhang P, Zhang Z, Liu S, Han B. Efficient emulsifying properties of glycerol-based surfactant. Colloids Surf A Physicochem Eng Asp 2018;553:225–229.

[65] Fan, Z., Zhao, Y., Preda, F., Clacens, J.M., Shi, H., Wang, L., Feng, X. and De Campo, F., 2015. Preparation of bio-based surfactants from glycerol and dodecanol by direct etherification. Green Chemistry, 17(2), pp.882–892.

[66] Lebeuf, R., Illous, E., Dussenne, C., Molinier, V., Silva, E.D., Lemaire, M. and Aubry, J.M., 2016. Solvo-surfactant properties of dialkyl glycerol ethers: application as eco-friendly extractants of plant material through a novel hydrotropic cloud point extraction (HCPE) process. ACS Sustainable Chemistry & Engineering, 4(9), pp.4815–4823.

[67] Ding F, Zhou X, Wu Z, Xing Z. Synthesis of a cleavable vanillin-based polyoxyethylene surfactant and its pilot application in cotton fabric pretreatment. ACS Sustain Chem Eng 2019;7:5494–5500.

[68] Ogunkunle T, Fadairo A, Rasouli V, Ling K, Oladepo A, Chukwuma O, Ayoo J. Microbial-derived bio-surfactant using neem oil as substrate and its suitability for enhanced oil recovery. J Pet Explor Prod 2021;11(2):627–638.

[69] Ma J, Gao J, Wang H, Lyu B, Gao D. Dissymmetry gemini sulfosuccinate surfactant from vegetable oil: A kind of environmentally friendly fat liquoring agent in the leather industry. ACS Sustain Chem Eng 2017;5:10693–10701.

[70] Faßbach TA, Gaide T, Terhorst M, Behr A, Vorholt AJ. Renewable surfactants through the hydroaminomethylation of terpenes. ChemCatChem 2017;9:1359–1362.

[71] Bhadani A, Rane J, Veresmortean C, Banerjee S, John G. Bio-inspired surfactants capable of generating plant volatiles. Soft Matter 2015;11:3076–3082.

[72] ECOSURF™ SA surfactants- seed oil-based surfactants. Dow. 2008. (Internet) Available from: https://www.dow.com/documents/en-us/mark-prod-info/119/119-02222-01-ecosurf-sa-surfactants-seed-oilbased-surfactants.pdf.

[73] Integrity Biochem announces first-of-its-kind surfactant to improve industrial sustainability. Businesswire. 2021 (Internet). Available from: https://www.businesswire.com/news/home/

20210728005149/en/Integrity-BioChem-Announces-First-of-its-Kind-Surfactant-to-Improve-Industrial-Sustainability.

[74] BASF and solazyme launch the first commercial microalgae-derived betaine surfactant. BASF. 2015 (Internet) Available from: https://www.basf.com/us/en/media/news-releases/2015/07/P-US-15-137.html.

[75] How bio-based surfactants are turning the world green. Cosmetics & Toiletries. 2020 (Internet). Available from: https://www.cosmeticsandtoiletries.com/formulating/category/natural/How-Bio-Based-Surfactants-are-Turning-the-World-Green–570209271.html.

[76] BASF strengthens its position in bio-surfactants for personal care, home care and industrial formulators with two distinct partnerships. BASF. 2021. Available from: https://www.basf.com/my/en/media/news-releases/global/2021/03/p-21-148.html.

[77] Ajinomoto to expand amino acid-based surfactants. Specialty Chemicals Magazine. 2019 (Internet). Available from: https://www.specchemonline.com/index.php/ajinomoto-expand-amino-acid-based-surfactants.

[78] Sironix renewables closes seed round to scale production of its plant-based surfactant for detergents, cleaners & shampoos. GlobeNewswire. 2020 (Internet). Available from: https://www.globenewswire.com/news-release/2020/09/16/2094579/0/en/Sironix-Renewables-Closes-Seed-Round-to-Scale-Production-of-its-Plant-based-Surfactant-for-Detergents-Cleaners-Shampoos.html.

[79] Assessing the sustainability and performance of green surfactants. IHS Markit. 2020 (Internet). Available from: https://ihsmarkit.com/research-analysis/assessing-sustainability-and-performance-of-green-surfactants.html.

[80] Bio-based chemicals – a 2020 update. IEA Bioenergy Task 42 report. 2020 (Internet). Available from: https://www.ieabioenergy.com/wp-content/uploads/2020/02/Bio-based-chemicals-a-2020-update-final-200213.pdf.

[81] Market Research Future. Available at, https://www.https://www.reportlinker.com/p06081965/Biosurfactants-Market-Research-Report-by-Product-Type-by-Application-by-Region-Global-Forecast-to-Cumulative-Impact-of-COVID-19; 2021 [Accessed 2021-08–15].

[82] Nmp RES, Meira HM, Almeida FCA, et al. Natural surfactants and their applications for heavy oil removal in industry. Sep Purif Rev 2019;48(4): 267–281. doi: 10.1080/15422119.2018.1474477.

[83] Drakontis CE, Amin S. Biosurfactants: Formulations, properties, and applications. Curr Opin Colloid Interface Sci 2020;48:77–90. doi: 10.1016/j.cocis.2020.03.013.

[84] Naughton PJ, Marchant R, Naughton V, et al. Microbial biosurfactants: Current trends and applications in agricultural and biomedical industries. J Appl Microbiol 2019;127(1): 12–28. doi: 10.1111/jam.14243.

[85] Almeeida DG, Sores da de Silva RCF, Luna JM, et al. Biosurfactants: Promising molecules for petroleum biotechnology advances. Front Microbiol 2016;7:1718. doi: 10.3389/PMid:27843439.

[86] Silva RCFS, Almeida DG, Luna JM, et al. Applications of biosurfactants in the petroleum industry and the remediation of oil spills. Int J Mol Sci 2014;15(7): 12523–12542. doi: 10.3390/ijms150712523.

[87] Rodriguez-Lopez L, Rincon-Fontan M, Vecino X, et al. Extraction, separation and characterization of lipopeptides and phospholipids from corn steep water. Sep Purif Technol 2020; 248. doi: 10.1016/j.seppur.2020.117076.

[88] Roelants, S.L.K.W., Van Renterghem, L., Maes, K., Everaert, B., Redant, E., Vanlerberghe, B., Demaeseneire, S.L. and Soetaert, W., 2019. Microbial biosurfactants: from lab to market. Microbial biosurfactants and their environmental and industrial applications, pp.341-363.

[89] Singh P, Patil Y, Rale V. Biosurfactant production: Emerging trends and promising strategies. J Appl Microbiol 2019;126(1):2–13. doi: 10.1111/PMid:30066414.

[90] Santos DKF, Rufino RD, Luna JM, et al. Biosurfactants: Multifunctional biomolecules of the twenty-first century. Int J Mol Sci 2016;17:401. doi: 10.3390/ijms17030401.

[91] Natural Surfactants Market. Available at: https://www.marketsandmarkets.com/Market-Reports/natural-surfactant-market-25221394.html; 2018 [Accessed 2021-08–14].

[92] Souza EC, Vessoni-Penna TC, De Souza Oliveira RP. Biosurfactant-enhanced hydrocarbon bioremediation: An overview. Int Biodeterior Biodegrad 2014;89:88–94. doi: 10.1016/j.ibiod.2014.01.007.

[93] Geys R, Soetaert W, Van Bogaert I. Biotechnological opportunities in biosurfactant production. Curr Opin Biotechnol [Internet] 2014;30:66–72. Available from. http://dx.doi.org/10.1016/j.copbio.2014.06.002.

[94] Marchant R, Banat IM. Microbial biosurfactants: Challenges and opportunities for future exploitation. Trends Biotechnol 2012;30(11):558–565.

[95] Hayes DG. Fatty acids–based surfactants and their uses. In Fatty Acids [Internet], Elsevier Inc.; 2017. pp. 355–384. Available from http://dx.doi.org/10.1016/B978-0-12-809521-8.00013-1.

[96] Bhadani A, Iwabata K, Sakai K, Koura S, Sakai H, Abe M. Sustainable oleic and stearic acid based biodegradable surfactants. RSC Adv 2017;7(17):10433–10442.

[97] Farias CBB, Almeida FCG, Silva IA, Souza TC, Meira HM, Soares da Silva de RCF, et al. Production of green surfactants: Market prospects. Electron J Biotechnol [Internet]2021;51:28–39.Available from: https://doi.org/10.1016/j.ejbt.2021.02.002.

[98] Makkar RS, Cameotra SS, Banat IM. Advances in utilization of renewable substrates for biosurfactant production. AMB Express 2011;1(1):1–19.

[99] Marion P, Bernela B, Piccirilli A, Estrine B, Patouillard N, Guilbot J, et al. Sustainable chemistry: How to produce better and more from less?. Green Chem 2017;19(21):4973–4989.

[100] Perathoner S, Centi G. Science and technology roadmap on catalysis for Europe. 2016.

[101] Lange JP. Renewable feedstocks: The problem of catalyst deactivation and its mitigation. Angew Chem Int Ed 2015;54(45):13187–13197.

[102] Gu Y, Jérôme F. Bio-based solvents: An emerging generation of fluids for the design of eco-efficient processes in catalysis and organic chemistry. Chem Soc Rev 2013;42(24):9550–9570.

[103] Jessop PG, Mercer SM, Heldebrant DJ. CO_2-triggered switchable solvents, surfactants, and other materials. Energy Environ Sci 2012;5(6):7240–7253.

[104] Eckert CA, Knutson BL, Debenedetti PG. Supercritical fluids as solvents for chemical and materials processing. Nature 1996;383(6598):313–318.

Mohamad Fakhrul Ridhwan Samsudin, Nur Amalina Samsudin,
Wong Mee Kee, M. Syamzari B. Rafeen*

Chapter 11
Rethinking ethylene glycol: A historical and forward-looking perspective on sustainable production

Abstract: Since its discovery in 1859, ethylene glycol has evolved into a pivotal organic compound and chemical intermediate with diverse industrial applications. This chapter offers a retrospective and prospective analysis of ethylene glycol production, tracing its journey from early petrochemical synthesis routes to emerging sustainable alternatives. Key advances in synthesis methods, physicochemical properties, and performance characteristics are examined alongside the challenges that have shaped its industrial trajectory over recent decades. Special attention is given to current innovations aimed at improving environmental performance and enabling greener production pathways. Finally, the chapter highlights future opportunities and research directions that may redefine the role of ethylene glycol in a resource-constrained and sustainability-driven world. This chapter is intended to support researchers, innovators, and industry practitioners committed to reimagining chemical manufacturing for the future.

Keywords: ethylene glycol, empty fruit bunch, renewable, sustainable, alcohol

11.1 Introduction

Why ethylene glycol (EG)? Ethylene glycol is the simplest diol having direct application as an automotive antifreeze, coolant, and as a precursor or building block chemical for the production of plastics such as polyethylene terephthalate (PET) and production of polyester resins for fiber [1, 2]. Other applications of ethylene glycol are for film applications, defrosting and de-icing aircraft, heat-transfer solutions for coolant for gas compressors, heating, ventilating, and air-conditioning, and water-based formulations such as adhesives, latex paints, and asphalt emulsions, manufacture of capacitors and unsaturated polyester resins [3].

***Corresponding author: M. Syamzari B. Rafeen**, PETRONAS Research Sdn Bhd, Lot 3288 and 3289, Off Jalan Ayer Hitam, Kawasan Institusi Bangi, 43000 Kajang, Selangor, Malaysia,
e-mail: syamzari_rafeen@petronas.com.my
Mohamad Fakhrul Ridhwan Samsudin, Nur Amalina Samsudin, Wong Mee Kee, PETRONAS Research Sdn Bhd, Lot 3288 and 3289, Off Jalan Ayer Hitam, Kawasan Institusi Bangi, 43000 Kajang, Selangor, Malaysia

 | https://doi.org/10.1515/9783112212141-011

The birth of a molecule named "ethylene glycol" started with the work pioneered by Charles Adolphe Wurtz in 1859 where he treated 1,2-dibromoethane with silver acetate to produce ethylene glycol diacetate (see Figure 11.1) [4]. The intermediate product, ethylene glycol, diacetate was then hydrolyzed to ethylene glycol. The final product, ethylene glycol also known as 1,2-ethanediol ($HOCH_2CH_2OH$), possesses the properties of a colorless, practically odorless, low-viscosity, hygroscopic liquid of low volatility that is completely miscible with water and many organic liquids [3].

Figure 11.1: Charles Adolphe Wurtz (1817–1884), French chemist who discovered ethylene glycol.

Since its discovery in 1859, ethylene glycol was not produced industrially until World War I, when it was used as an intermediate for explosive (ethylene glycol dinitrate) substituting glycerol in explosive industries and as coolant [1]. The production of ethylene glycol back then was based on the chlorohydrin process developed also by Charles Adolphe Wurtz. The chlorohydrin process is the route for ethylene oxide production prior to further hydrolysis to produce ethylene oxide where ethylene oxide was prepared by elimination of hydrochloric acid from 2-chloroethanol (ethylene chlorohydrin) using potassium hydroxide solution [5, 6]. The chlorohydrin process enabled ethylene glycol industrial production in 1914, which marked the start of steady growth for ethylene glycol production [2, 3]. However, this process is no longer the intermediate pathway for ethylene glycol production due to pollution problems arising from the chlorine used being lost as calcium chloride and unwanted chlorine-containing by-products. Alternatively, direct oxidation was explored by Wurtz but not successful, until Lefort made a crucial discovery that the formation of ethylene oxide from ethylene and oxygen is possible via a catalytic process where metallic silver is used as the catalyst [1, 7].

Direct oxidation has advantages over chlorohydrin process due to the elimination of the need for large volume of chlorine and absence of chlorinated hydrocarbon by-products, simpler plant capex costs, and lower operating cost [2]. Direct oxidation technology can be divided into two categories depending on the source of the oxidizing agent, which are the air-based process and the oxygen-based process. Union Carbide Corp. was the first to commercialize the air-based direct oxidation process in 1937 whilst Shell Oil Co. was the first to commercialize oxygen-based system in 1958 [1, 8].

Lefort contribution on the direct oxidation for the production of ethylene oxide followed by hydrolysis of ethylene oxide to produce ethylene glycol became and remains the dominant commercial source of ethylene glycol production [5, 9]. Ethylene oxide hydrolysis can be performed with either acid or base catalyst system or uncatalyzed in neutral medium. In all ethylene hydrolysis processes, the main by-product is diethylene glycol (DEG) whereas higher glycols, i.e., triethylene glycol (TEG) and tetraethylene glycol are the remaining by-products [10, 11]. The large excess of water is used to minimize production of higher glycol homologues since production of higher homologues is inevitable because ethylene oxide reacts more quickly with ethylene glycols than with water. Typically, a 20-fold molar excess is used to maximize production of ethylene glycol [2]. It is noteworthy that the yield of ethylene glycol is solely dependent on the concentration of ethylene oxide in the ethylene oxide-water feed mixture, and remains unaffected by temperature, pressure, or contact time, as long as the reaction of ethylene oxide is substantially complete (>99%) [12]. In a typical process for ethylene oxide hydrolysis, large excess of water from hydrolysis is removed in a series of multiple-effect evaporators and ethylene glycol is refined using vacuum distillation [3].

Ethylene oxide hydrolysis is a well-established technology for the production of ethylene glycol and remains to be the commercial source of ethylene glycol. However, alternative synthesis routes are also being explored due to the constantly increasing demand for ethylene glycol with methods such as the syngas route, biomass-derived glycerol, cellulose conversion, formaldehyde, and its derivatives conversion, transesterification of ethylene carbonate, etc. [1].

From a commercial perspective, ethylene glycol (EG) is the highest-volume glycol product sold globally, surpassing other higher glycols such as DEG and TEG. According to a recent market outlook, the ethylene glycol market was valued at USD 37,743.8 million in 2023 and is projected to reach USD 75,303.1 million by 2034, growing at a compound annual growth rate (CAGR) of 6.5% from 2024 to 2034 [13]. This market growth is primarily driven by the use of ethylene glycol in the production of PET, which is widely used in packaging materials and synthetic fibers. The increasing global demand for environmentally friendly packaging and lightweight textiles has significantly boosted the consumption of ethylene glycol.

Ethylene glycol is commercially available in three main grades: fiber grade, industrial grade, and antifreeze grade [14]. Fiber-grade ethylene glycol typically contains at least 99.9 wt% monoethylene glycol (MEG) and is used in the production of high-performance materials such as polyester fibers and PET resins. Industrial-grade ethylene glycol, which generally contains ≥99.0 wt% MEG, is used in a broader range of chemical manufacturing and processing applications. Antifreeze-grade ethylene glycol, with a typical purity of ≥95.0 wt% MEG, is formulated for use in engine coolants and heat transfer fluids. Among these, fiber-grade MEG is subject to the most stringent quality specifications, particularly in terms of ultraviolet (UV) transmittance, as defined in standards such as ASTM E2470 [15]. These requirements are critical to ensure performance in applications sensitive to impurities and light transmission characteristics.

Additionally, ethylene glycol is a hygroscopic substance, meaning it readily absorbs moisture from the surrounding environment. This characteristic can significantly affect its purity if not properly handled [14]. Moreover, ethylene glycol can form complexes with dissolved oxygen upon exposure to air. This interaction can alter its UV absorption characteristics, causing the typical absorption peak at 180 nm to shift toward 220 nm due to the formation of ethylene glycol-oxygen complexes [15]. As such, to maintain the integrity of high-purity ethylene glycol, especially fiber-grade, proper storage and transportation protocols must be established. This includes the use of nitrogen sparging and nitrogen blanketing to prevent contamination by moisture and oxygen.

11.2 Synthesis of ethylene glycol

11.2.1 Hydration of ethylene oxide

The hydration of ethylene oxide (EO) at elevated temperatures is the most established method for the production of MEG. EO is obtained from the oxidation of petroleum-based ethylene with air or oxygen. MEG further reacts with EO to form DEG, TEG, and polymerized ethylene glycol in series. The aforesaid reactions are as follows:

$$\mathrm{EO} + \mathrm{H_2O} \rightarrow \mathrm{MEG} \tag{11.1}$$

$$\mathrm{EO} + \mathrm{MEG} \rightarrow \mathrm{DEG} \tag{11.2}$$

$$\mathrm{EO} + \mathrm{DEG} \rightarrow \mathrm{TEG} \tag{11.3}$$

By nature, EO has a higher tendency to react with MEG than water. The subsequent reactions that produce higher homologues are less desirable due to the economic value of the product [2].

In a typical thermal MEG production process scheme, EO reacts with water without catalysts at a temperature around 200 °C. The heavier glycols are typically suppressed by supplying excess water to the reaction. Increasing the molar ratio of water to EO has a significant effect on the selectivity of MEG. With mol ratio of water to EO above 20:1, 89% of MEG selectivity is achievable for non-catalyzed reaction [9]. About 8–10% of higher homologues, consisting of primarily DEG and TEG, are usually found in the reaction product. The excess water in the mixed EGs is removed through a series of evaporators. The dehydrated mixed EGs are then fed into distillation columns to separate the heavier EGs and purify MEG to the intended purity [16]. As the removal of water is an energy-intensive process, the higher water content that comes with the product increases the operation and capital expenses. Divided wall column, reactive distillation, and side reactor technology were among the techniques proposed to reduce the utility consumption for dehydration of MEG [17–19].

Considering the separation issues from excessive water and heavy EG, a modified two-step reaction is proposed by Mitsubishi Chemical Corporation (MCC) [9]. In the initial step (eq. (11.4)), ethylene is carbonated using carbon dioxide (CO_2) generated in the EO production process:

$$C_2H_4O + CO_2 \rightarrow C_3H_4O_3 \tag{11.4}$$

The ethylene carbonate is then hydrolyzed to form MEG with CO_2 as by-product:

$$C_3H_4O_3 + H_2 \rightarrow HOC_2H_4OH + CO_2 \tag{11.5}$$

The first step is catalyzed by phosphonium salt, which has demonstrated good stability and non-corrosive property. Attributing to its high catalytic activity for carbonation, phosphonium-based catalyst almost eliminates the possible dimerization reaction that produces DEG. Excellent EC conversion rate as high as 99.9999% is reported. Both carbonation and hydrolysis reactors can be operated at around 50 °C. A comparison of the conventional process and MCC's new MEG process is shown in Figure 11.2. Coupled with the low demand for water in the reactions, lower energy and capital investment are required in the purification section [9].

This technology is exclusively licensed to Shell under the name OMEGA process in 2002 [20]. The OMEGA process integrates Shell's high-performance EO process, which enables a plant-wide process intensification that produces more effluent and saves energy as well as minimizes capital cost. The technology is applied in commercial-scale plants ranging from 400 to 750 kton/year at various locations globally.

Another way to attain a higher MEG yield is by incorporating a catalyst in the reaction. Both homogeneous and heterogeneous catalysts are widely studied. Acids, soluble salts, cyclic amines, ion-exchange resins, zeolites, and ionic liquids (ILs) are among the catalysts tested to optimize selectivity, reduce the water requirement, and lower reaction temperature [21, 22]. Van Hal et al. [5] found that the pH of catalysts plays a central role in increasing the rate of hydration and maximizing MEG selectivity. Acid and bases of weaker strength are favored over their more concentrated counterparts, which promote oligomerization,, hence generating more undesired coproduct. Compared to homogeneous catalytic systems, heterogeneous catalysts have the advantages of ease of separation and fewer corrosion issues [21].

Recent development shows successful demonstration of tandem catalysis. [CoIII (salen)] encapsulated in nanocages of mesoporous silica (FDU-12) by Li et al. [23] showed superior activity and catalytic performance. Owing to the novel cooperative

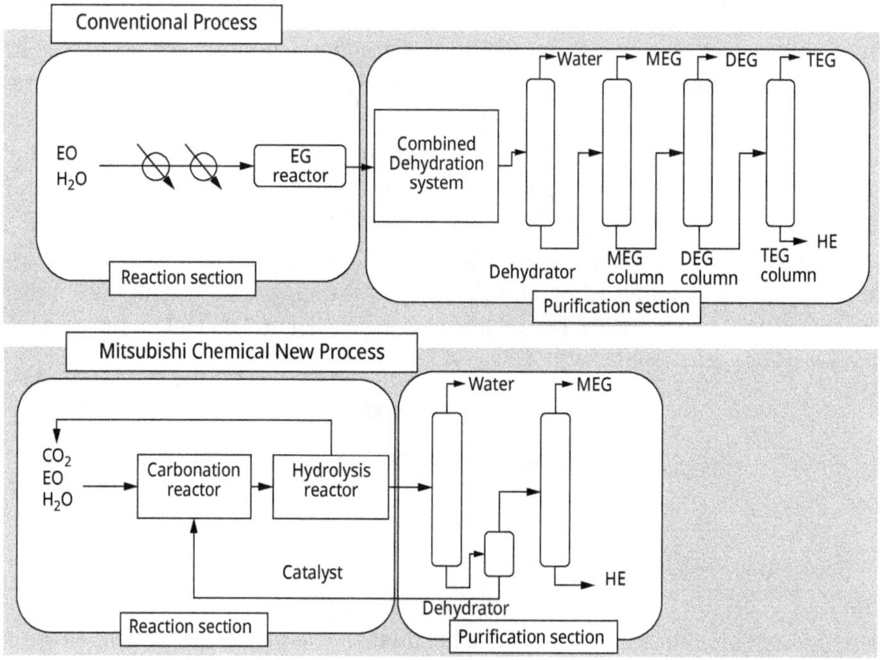

Figure 11.2: The conventional and new process of the ethylene glycol production (adapted from ref. [9] with permission).

activation effect, her work brought immobilization of homogeneous catalysts to the next level. Similar the mini reactor concept was applied by Dai et al. [24] to prepare a thermally stable Lewis acid (Sn-H-SSZ-13) confined in zeolite cages that is capable of pushing the EO conversion and selectivity to 99%. The zeolite cage is designed in such a way that it allows diffusion of EO, H_2O, and MEG through the pore windows, while the bigger molecules, such as DEG and TEG are retained. The narrow space in the cage restricts the collision of two or more MEG molecules, hence minimizing the formation of undesirable by-products. The diffusion control mechanism of zeolite framework during EO hydration is depicted in Figure 11.3. The high-performance catalytic process also has lower energy requirement because of the near ambient reaction temperature and more environmentally benign in terms of wastewater produced.

Despite the exceptional performance of these new generation catalysts, the synthesis of the catalysts is intricate in view of the complicated structure of the catalysts. It is challenging to synthesize these novel catalysts in bulk for industrial-scale application in a cost-effective manner. A simpler approach was proposed by Ding et al. [18], where hydrolysis of EO is assisted with CO_2 pressure using binary catalysts. A significant increase in MEG yield and selectivity are reported compared to the tests performed under the N_2 environment. One-pot two-step mechanism is elucidated, in which CO_2 participates in the reaction via the cycloaddition of EO to form ethylene carbonate intermediate. Subsequently, ethylene carbonate is hydrolyzed into MEG, while CO_2 is recovered. The process is catalyzed by IL and organic acid. About 97% selectivity is achieved at a low H_2O/EO ratio of 1.5:1 with CO_2 as a promoter and synergistic catalysis of IL and an organic base.

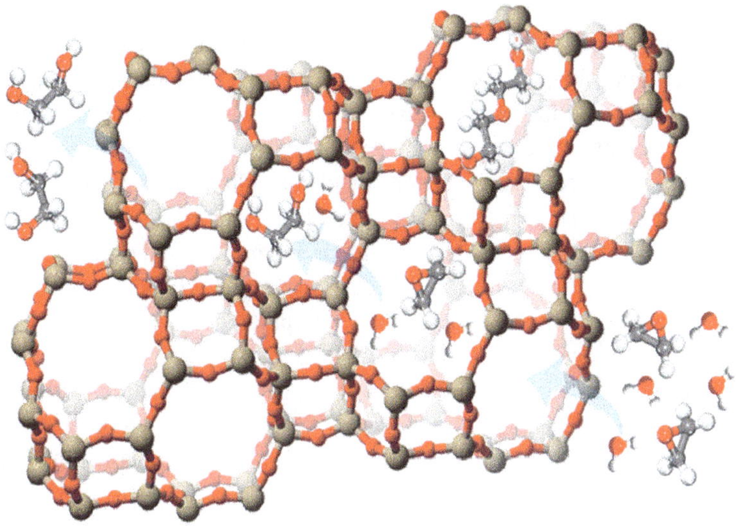

Figure 11.3: Schematic illustration of EO hydration to MEG via CHA zeolite (adapted from ref. [21] with permission).

11.2.2 Transesterification of ethylene carbonate and methanol

An alternative MEG synthesis method is transesterification of ethylene carbonate and methanol, in which MEG is co-produced with dimethyl carbonate (DMC). DMC is used in the automotive industry as a fuel additive and functions as carbonylation and methylation agents in chemical processes. Generally, ethylene carbonate is obtained from eq. (11.4). Then, ethylene glycol reacts with methanol in the presence of a catalyst as follows [25]:

$$C_3H_4O_3 + CH_3OH \rightarrow HOC_2H_4OH + OC(OCH_3)_2 \tag{11.6}$$

In the 1980s, a few patents related to the preparation of dialkyl carbonate via thermal catalytic reaction of glycol carbonate and alcohol were filed [26–28]. Glycols are collected as a by-product. A more refined process of transesterification of ethyl carbonate and methanol to produce MEG and DMC catalyzed by soluble salts of zirconium, titanium, and tin is patented by Texaco [27]. The reaction step (11.6) to co-synthesize MEG and DMC is equilibrium-limited. Continual withdrawal of liquid product is favored to enhance the conversion beyond its equilibrium limitation [29]. A reactive distillation design was described by Buysch et al. [30]. Both reactants are fed countercurrently into a gas-liquid contacting column. Ethylene carbonate is at the top, while methanol is in the vapor phase at the bottom. The heavier product, MEG is removed from the column as the bottom product. DMC can be continuously collected from the top of the column.

Knifton and Duranleau presented the outcome of six groups of homogeneous and heterogeneous catalysts, including tertiary amine, acidic resins, ammonium-exchanged zeolites, and phosphine polymer [31]. Base catalysts are reported to give better product yield and the catalytic activity improved with increasing temperature and base concentration. More than 98 mol% selectivities are achieved under mild operating conditions. Bhanage et al. [25] studied the use of layered clay minerals for the transesterification of ethylene carbonate and methanol. Smectites containing magnesium and/or nickel are highly selective for MEG and suppress unwanted oligomerization reactions. The catalytic performance is subjected to the amount of alkali incorporated.

Given the increased attention to IL due to its superior thermal stability, low vapor pressure, and tunability, a series of ILs were investigated for simultaneous production of MEG and DMC. All ILs demonstrate high selectivity of MEG with the highest selectivity of 99.9% recorded by EMImCl (1-ethyl-3-methylimidazolium with chloride anion). Ethylene carbonate conversion increase with decreasing size of the cation of the IL salt as well as increasing temperature and CO_2 pressure [32]. For the ease of post-reaction separation, researchers turn to immobilization of IL. Kim and colleagues proved that an IL grafted to commercial silica (RImX-CS) can be used for three cycles without significant activity deterioration based on a mole of DMC over mole of RImX ratio [33]. A longer alkyl chain length contributes to higher EC conversion because the bulkier alkyl chain extends the distance between anion and cation. The lower electrostatic bond increases the availability of anion. One-pot synthesis of MEG from supercritical CO_2, EO, and methanol was also attempted. Employing a catalyst system (potassium iodide supported on zinc oxide) that is active for both reaction steps ((11.4) and (11.6)), Chang et al. [34] successfully combined carbonation and esterification reactions as shown in Figure 11.4. EO conversion and MEG yield surpassed 99% and 57%, respectively, with by-products below 0.2%. Supercritical CO_2 plays a key role as a reactant and solvent in the reaction.

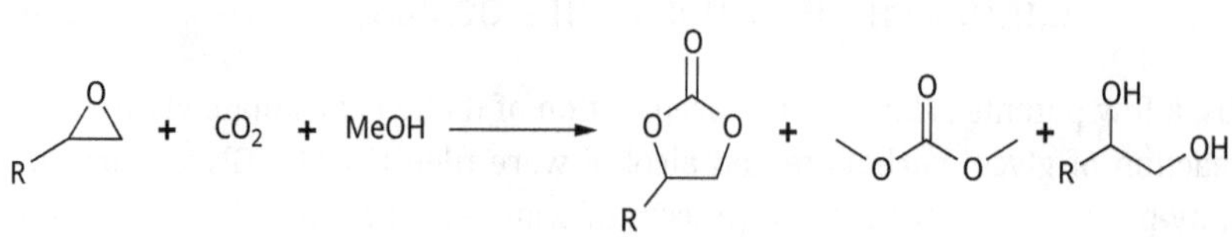

1 2 3 4

Figure 11.4: Schematic illustration of the one-pot synthesis process (adapted from ref. [31] with permission).

11.2.3 Biochemical process

In an effort to reduce dependence on petroleum, researchers have been exploring novel ways to produce EG from renewable resources. The valorization of biomass made up of cellulose, hemicellulose, and lignin received considerable attention in the past decade. Both cellulose and hemicellulose are composed of long chains of sugar. Upon breaking down this complex sugar, microorganisms, such as yeast, bacteria, and molds, can be used to convert the carbon-rich compounds into ethanol. The biochemical process proceeds slowly at low temperatures [35]. Subsequently, ethanol is dehydrated in the presence of phosphoric or activated alumina catalyst to form bioethylene [36]. Ethylene then goes through oxidation and hydration processes as explained earlier. Corn starch, sugarcane, and wheat starch are the major feedstock for global bio-ethanol production [37]. Owing to the abundance of sugarcane in Brazil, a commercial-scale bio-based EG plant with a capacity of 500 kton/year is available [36].

Lippits and Nieuwenhuys propose a one-pot conversion of bioethanol to ethylene oxide, which combines the dehydrogenation, dehydration, and oxidation reactions in one reactor [38]. With a prudent selection of catalysts and controlling the ethanol: O_2 ratio, the high selectivity of EO is achieved [39]. Taking into account the reaction kinetics and thermodynamics of separation, the feasibility of the direct conversion process was evaluated [40]. For higher plant throughput, the process design introduced diluted ethanol as solvent to enhance absorption and recycling of glycol to feed. Based on the process simulation, 90% conversion of bioethanol is achievable with 99% purity ethylene oxide product.

As an alternative to fermentation of sugar to bioethanol to obtain bio-based EG, scientists explored the metabolic pathways to synthesize EG at milder pressure and temperature. As there are currently no identified natural microbial pathways for direct production of EG, metabolic engineering approaches are tested to efficiently yield EG from lignocellulosic sugar by suppressing competing reactions and improving the activities of rate-limiting reactions. The following section elucidates the four synthetic biological routes for converting pentose and hexose into EG, which are Dahms pathway, xylulose-1-phosphate (X1P) pathway, ribulose-1-phosphate (R1P) pathway, and serine pathway.

The Dahms pathway, which enzymatically breaks down D-xylose, a five-carbon monosaccharide to obtain glycolaldehyde and pyruvate, was first introduced by Dahms [41]. Then, Liu et al. [42] demonstrated production of EG using engineered *Escherichia coli* via a four-step-modified Dahm's pathway. The main reactions along with by-product formation in Dahm's pathway are depicted in Figure 11.5. Following the successful biosynthesis of EG from the renewable carbon source by Liu et al. [42], a series of studies on metabolic interventions to enhance yield of EG via Dahm's pathway was conducted and the studies and outcome are compiled in Table 11.1.

Being an important metabolic process in human body, xylose is naturally converted to xylulose through an isomerization reaction. This reaction is essential for

Table 11.1: EG production from pentose and hexose in metabolic-engineered strains based on Dahm, X1P, R1P, and serine pathways.

Types of microorganism	Strain	Pathway	Titer (g/L)	Yield (g/g)	Yield (mol/mol)	Substrate	References
Bacteria	*E. coli* WL3110	Dahms	108.20	0.36	0.87	Xylose	[43]
	E. coli BL21	Dahms	72.00	0.40	0.97	Xylose	[44]
	E. coli W3110	Dahms	11.70	0.29	0.70	Xylose	[42]
	E. coli W3110	Dahms	7.72	0.39	0.95	Xylose	[45]
	E. coli MG1655	X1P	20.00	0.38	0.91	Xylose	[46]
	E. coli MG1655	R1P	40.00	0.35	0.85	Xylose	[47]
	E. coli MG1655	R1P	10.50	0.35	–	Arabinose + xylose	[47]
	C. glutamicum SL-1	R1P	5.80	0.31	0.75	Xylose	[48]
	H. bluephagenesis TDX9	R1P	0.91	0.14	–	Xylose	[49]
	E. coli MG1655	Serine	4.10	0.14	0.41	Glucose	[50]
	C. glutamicum Mdlc	Serine	3.50	0.09	0.25	Glucose	[51]
Yeast	*K. phaffii* JA122	Dahms	1.31			Xylose + glucose	[52]
	S. cerevisiae H4099	Dahms	0.01	–	–	Xylose + glucose	[53]
	S. cerevisiae yEG04	X1P	4.05	–	–	Xylose + glucose	[54]
	S. cerevisiae xks1	X1P	0.50	0.01	0.03	Xylose	[55]

sugar digestion and energy production in cells. Hence, this native reaction route competes with xylose oxidation in step (1a) to generate xylonate intermediate as illustrated in Figure 11.5. To encourage formation of xylonate over xylulose, the gene that is responsible for isomerization in xylose isomerase (*xylA*) is disrupted. With the elimination of *xylA*, reaction (1b) is blocked and metabolic flux is diverted towards Dahm's pathway. In the final step of the biotransformation of xylose to EG, glycolaldehyde can be converted to either EG or glycolic acid. Optimizing the reduction of glycolaldehyde to maximize EG production is crucial. After testing various endogenous glycolaldehyde reductases in *E. coli*, such as *yqhD*, *fucO*, and *yjgB*, at different expression levels, it was found that the overexpression of the yqhD gene had the most significant impact on achieving a higher titer of EG. In the effort to suppress accumulation of GA through step 4b to further improve EG yield, deletion of *aldA* gene was attempted with mixed results. A desirable increment of EG in final product was attained as per work performed by Wang et al. [44] and Pereira et al. [50], while Liu et al. [42] observed buildup of xylonate intermediate that impaired the growth of the engineered strain.

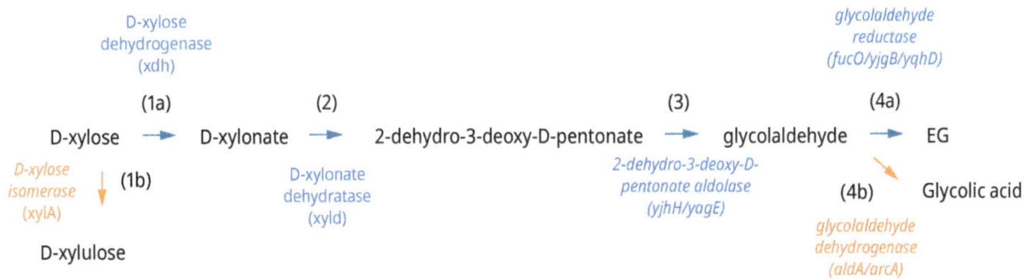

Figure 11.5: Schematic illustration of the one-pot synthesis process (adapted from ref. [42] with permission).

Yeast cells are often easier to cultivate, unlike bacteria, which may require more stringent growth conditions. Yeasts also generally have higher tolerance to harsher conditions and impurities, such as alcohols and organic acids, allowing for more robust fermentation processes in industry scale [56, 57]. Considering the advantages of yeast, *Saccharomyces cerevisiae,* is metabolically engineered to catabolize D-xylose for EG production. However, only 14 mg/L of EG titer was recorded due to the low enzymatic activity of *xyld* in *S. cerevisiae,* which resulted in excess accumulation D-xylonic [58]. Recently, a new xylonate dehydratase from bacterial species was integrated in the construction of an efficient *Komagataella phaffii* strain to boost the rate-limiting xylonate dehydration (Step 2) in Dahm's pathway. The natural existence of glucose units in lignocellulosic compound also prompted Carneiro et al. [52] to add the six-carbons-sugar as the substrate. Glucose may provide nutrient for microbial growth while xylose is metabolized to generate target chemical. The study demonstrated 30% improvement of EG production through co-fermentation of xylose and glycose but the EG yield was considerably lower than *E. coli.* Three constraints that cause bottlenecks in biotransformation of xylose to EG were identified: (1) *K. phaffii* promotes both glycolaldehyde reduction and oxidation reactions (i.e., step 4a and 4b in Figure 11.5) to concurrently produce EG and GA; (2) *K. phaffii* enables irreversible conversion of EG into GA; (3) oxidation of EG to form glycolaldehyde was observed in vitro in the presence of *K. phaffii* [52]. As such, the most effective metabolic interventions to maximize ethylene glycol (EG) production via the Dahms pathway are: (1) elimination of *xylA* to abolish loss of xylose through isomerization and (2) overexpression of the yqhD to drive glycolaldehyde reduction.

In 2015, a new synthetic pathway, namely X1P, was first investigated to coproduce EG and GA in *E. coli* cells. Following the natural pathway of wild-type *E. coli,* D-xylose goes through isomerization to obtain D-xylulose. The subsequent synthetic reaction sequence proceeded via phosphorylation of D-xylulose to D-xylulose-1-phosphate (D-xylulose-1P) and aldolytic cleavage of D-xylulose-1P to the two-carbon molecule, glyco-

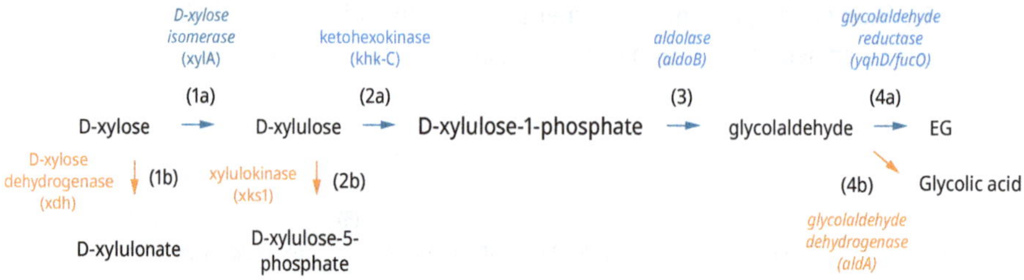

Figure 11.6: Steps in X1P pathway for EG production with competing reactions and respective enzymes (adapted from ref. [59] with permission).

laldehyde. Human ketohexokinase-C (*khk-C*) and aldolase-B (*aldoB*) genes were assembled to catalyze step 2 and step 3, respectively, as depicted in Figure 11.6 [59]. The bioconversion of pentose to EG was optimized by providing fully aerobic environment and deleting *aldA*, which takes part in glycolic acid by-product formation (step 4b in Figure 11.6) [46]. Being one of the preferred cell factories in industrial applications, Uranukul et al. [54] selected *S. cerevisiae* as yeast of choice to synthesize EG. The EG production in *S. cerevisiae* host was limited due to the lack of catalytic activity of phosphofructokinase in enhancing the rate controlling step of phosphorylation (step 2a in Figure 11.6). The process developed also requires continuous feeding of glucose as carbon source for cells' growth to prevent xylulose accumulation in fermentation broth. The highest EG production achieved through X1P route was 20 g/L with 0.91 mol EG/mol xylose reported by Alkim et al. [46].

The R1P route is the third pathway for xylose degradation aimed at the biosynthetic production of EG. This pathway was developed by Perreira et al. [50] by leveraging the native metabolic pathways of D-arabinose and L-lyxose in *E. coli*. This approach highlights the potential for utilizing existing metabolic routes to enhance the conversion efficiency of xylose. The five-step synthetic metabolic route is illustrated in Figure 11.7. R1P inherited reaction steps from both Dahm and X1P pathways. Instead of phosphorylation, D-xylulose is converted into D-ribulose (step 2a in Figure 11.7). The epimerization reaction is facilitated by D-tagatose epimerase (*dte*) enzyme. The endogenous fuculokinase (*fucK*) and d-ribulose1-phosphate aldolase (*fucA*) genes in *E. coli* are able to catalyze carbon flux in steps 3 and 4, respectively. The metabolic network was validated with a recombinant *E. coli* constructed by integrating a heterologous *dte* gene from *Pseudomonas cichorii*. Higher EG titer is attained with deactivation of *xks1*, *xylB*, and *aldA* genes encoding. The engineered strain had also shown that the valorization of L-arabinose, D-arabinose, and L-lyxose is feasible via R1P [50]. In more recent studies, *Corynebacterium glutamicum* and *Halomonas bluephagenesis* were investigated to obtain xylose-derived through R1P path EG [48, 49]. The EG product collected from both metabolic engineered hosts were comparatively lower than 10.5 g/L of EG in *E. coli* by Pereira et al. [50].

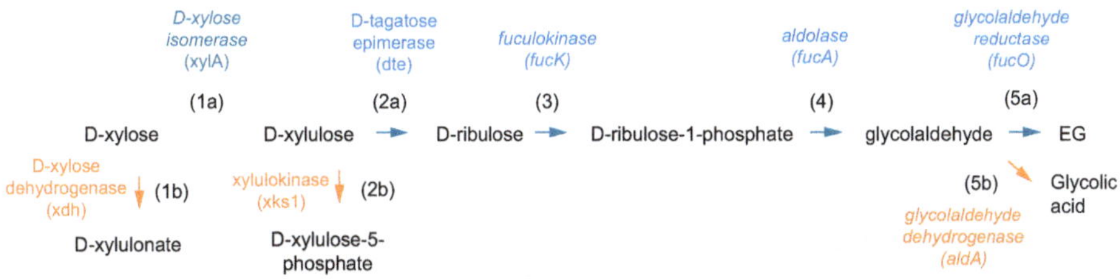

Figure 11.7: Steps in R1P pathway for EG production with competing reactions and respective enzymes (adapted from ref. [50] with permission).

Among the microorganisms experimented in three xylose consumption pathways, *E. coli* was shown to be the most successful engineered host for assimilation of D-xylose for EG production. The highest EG titer reached 108 g/L in *E. coli* through the Dahms pathway [43]. The yield of EG for these three routes is limited by the cleavage of five-carbon molecule to two-carbon glycolaldehyde and three-carbon pyuverate or dihydroxyacetone phosphate (DHAP) intermediates. Only glycolaldehyde is valorized to the desired EG product, while the pyuverate or dihydroxyacetone produced from pentose does not participate in any of the three pathways. Therefore, the stoichiometry of converting pentose to EG is not economically favorable. In view of the carbon economy, Pereira et al. [47] introduced the Serine pathway, which enables microbial conversion of three-carbon intermediates to generate EG.

The metabolic network of the serine pathway is illustrated in Figure 11.8. With glucose as starting material, serine is obtained from glycolysis of the monosaccharide, followed by oxidation and transamination reactions. Serine-to-EG can be implemented via two paths, either the decarboxylation to ethanolamine (step 5a) or deamination to hydroxypyruvate (step 5b). Both paths lead to formation of glycolaldehyde, a precursor to EG. The three competing reactions (steps 6c, 6c, and 7b) are repressed by removal of encoding serine dehydratase (*sdaA*), ethanolamine ammonia lyase (*eutBC*), and *aldA* in *E. coli* chromosomes. The modified *E. coli* demonstrated EG production of 4.1 g/L via the de novo Serine pathway [47]. Despite having the theoretical yield of 2 mol EG/mol glucose compared to 1 mol EG/mol xylose, yield of EG did not outperform Dham's pathway. Several metabolic modification strategies were also employed to optimize the Serine bioprocess in *C. glutamicum* as hosting microorganism but EG yield remains unsatisfactory for scale-up consideration [51]. This may be attributed to versatility of Serine as precursor to numerous metabolites, such as glycine, tryptophan, cysteine, and phospholipids. Besides, the major diversion of 3-phosphate glycerate (3P-glycerate) to central metabolism in step 1b as glycolysis for the cells' demand of energy and redox balance competes with the carbon flux to produce EG through step 1a.

The advancement of biotechnology in converting renewable sugar to EG builds a strong foundation in valorization the abundant lignocellulosic waste for chemical production, thereby promoting a circular economy. An effective metabolic engineering

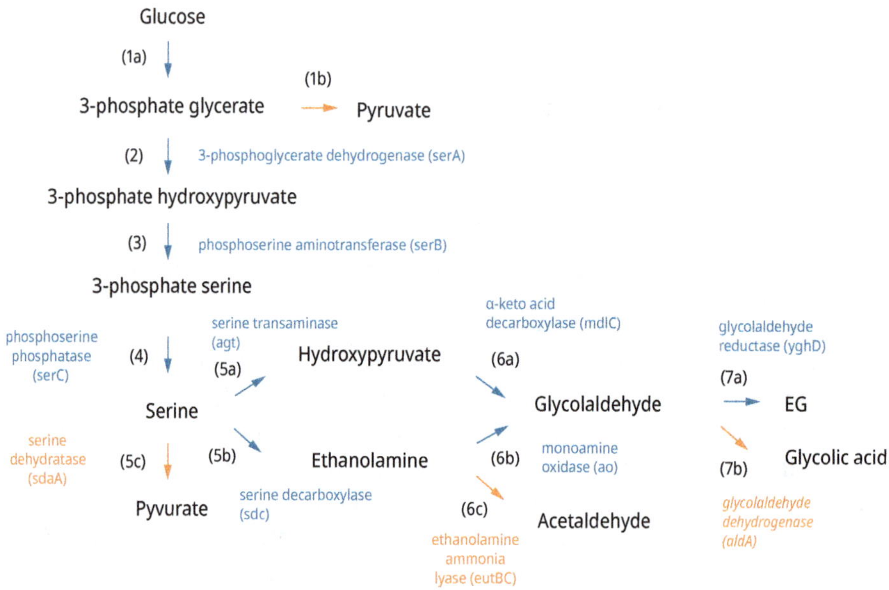

Figure 11.8: Steps in Serine pathway for EG production with competing reactions and respective enzymes (adapted from ref. [50] with permission).

approach is crucial for industrializing this bioconversion process. This involves the systematic modification of microbial pathways and resilient strain development that control the multiple parallel reactions and regulate metabolic flux for a balance of cell growth and EG formation. Further improving efficiency of the biosynthesis can be achieved through refinement of fermentation conditions. The downstream process, such as product recovery and purification shall be taken into account. Generally, fermentation broth contains high water content with complex composition; hence development of an energy efficient separation technique catalyzes commercialization of this microbial biochemical process.

11.2.4 Thermochemical process

Recently, catalyst specialists Haldor Topsoe and Braskem announced their collaboration to operate a demonstration plant employing a new sugar conversion technology [60]. The technology replaces the multistep sugar fermentation to bio-EG route with a simpler two-step process illustrated in Figure 11.9. Depolymerization of sugar was performed via pyrolysis, followed by catalytic hydrogenation of the C1–C3 oxygenates. In fact, the thermal cracking of sugar to obtain glycolaldehyde is documented in 2006 [61]. Glycolaldehyde is a versatile platform chemical that can be converted to EG and other numerous chemicals, including glycolic acid, methyl vinyl glycolate (MVG), and ethanol amines. Schandel et al. [62] studied hydrous thermolysis by spraying sugar solutions (glucose, fructose, sucrose, and xylose) into a fluidized bed of glass beads. Their works show that when glucose is subjected to high heat (525 °C), the sugar rings are broken giving yields as high as 73% of glycolaldehyde.

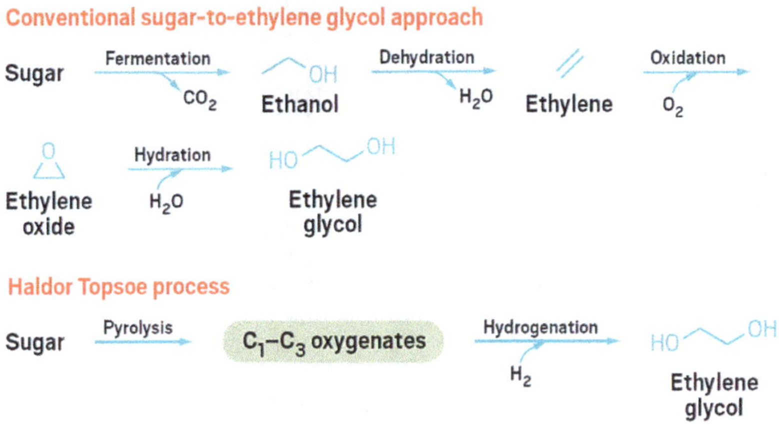

Figure 11.9: Comparison between conventional process and Haldor Topsoe process (adapted from ref. [60] with permission).

11.3 Properties of ethylene glycol

11.3.1 Physical properties

As mentioned earlier, ethylene glycol is the simplest diol that was developed back in 1859 by Charles-Adolphe Wurtz [1]. Since the discovery of this simplest diol, the development of ethylene glycol has started to bloom and received significant attention due to its multifunctional application. It is estimated that 70 million metric tons of ethylene glycol will be produced globally in 2025, which is a 40% increase in production in comparison to the year 2020 [20, 63]. Physical and chemical properties are detailed in Table 11.2.

Generally, ethylene glycol exists in a colorless and odorless form with a sweetish taste. It can be said that the physical properties of this ethylene glycol are standing between ethyl alcohol (ethanol) and glycerol. Despite the colorless properties, ethylene glycol possesses a specific gravity of 1.1135 g/cm^3 at 20 °C and it can be said that ethylene glycol is heavier than ethanol (0.787 g/cm^3) and lighter than glycerol (1.129 g/cm^3) [8]. Furthermore, this simplest glycol is known to be less viscous than glycerol (1.412 Pa.s) [2, 6]. In addition, the molecular weight of this diol is 62.07 g/mol and its

Table 11.2: Physical and chemical properties of ethylene glycol (adapted from ref. [1] with permission).

Property	Ethylene glycol
Molecular weight	62.07 g/mol
Physical state	Liquid
Odor	Odorless
Boiling point at 101.3 kPa	197.3 °C
Melting point	−12.9 °C
Auto-ignition temperature	427.0 °C
Critical temperature	446.9 °C
Flash point, closed cup (Pensky-Martens closed cup ASTM D93)	126.7 °C
Flash point, open cup (Cleveland Open Cup ASTM D92)	137.8 °C
Specific gravity at 20 °C	1.1135 g/cm^3
Solubility	Miscible with water
Solubility of water in ethylene glycol at 20 °C	100 wt%
Heat of combustion at 25 °C	−1,053.0 kJ/g mol
Heat of vaporization at 1 atm	53.2 kJ/g.mol
Critical specific volume	19.1 × 10^{-2} L/g mol
Lower explosive limit	3.2 vol%
Normal freezing point	−13.0 °C
Onset of initial decomposition	240.0 °C
Refractive index at 25 °C	143.0 × 10^{-2}
Surface tension at 25 °C	48.0 × 10^{-3} N/m
Upper explosive limit	53.0 vol%
Vapor density (air = 1)	2.1
Vapor pressure at 20 °C	7.5 Pa
Viscosity at 20 °C	19.8 × 10^{-3} Pa s

corresponding boiling point and melting point are 197.3 and −12.9 °C, respectively [1]. Table 11.2 summarizes the physical and chemical properties of the aforesaid diol.

11.3.2 Toxicity

Apart from being the simplest diols that are beneficial for multifarious applications, this diol possesses a toxicity behavior that is commonly available in many household and industrial products such as antifreeze, car wash fluids, detergents, cosmetics, paints, and vehicle brake fluids [2, 3]. Although ethylene glycol is odorless, exposure to this diol without proper personal protective equipment might cause significant morbidity and mortality to the user. The dangerous exposure can originate from two situations, either (i) accidental or (ii) intentional exposure. Both situations could cause the user to be treated with extensive medical treatments or minor treatment depending on the level of exposure.

As mentioned earlier, ethylene glycol has a sweet taste, which can attract children's attention while exploring their environment for unintentional consumption.

Thus, if this unfortunate situation occurs, the gastrointestinal tract will absorb the ethylene glycol upon ingestion and a series of alcohol and aldehyde dehydrogenase reactions, which consequently lead to glycolic acid [64, 65] occurs. These glycolic acids will cause the development of anion gap metabolic acidosis and further the development of acute renal injury. Thereunto, the consumption of ethylene glycol as small as 120 mL is known to be enough for killing an average-sized man [66]. Figure 11.10 delineated the schematic of the metabolic toxicity of the ethylene glycol accordingly.

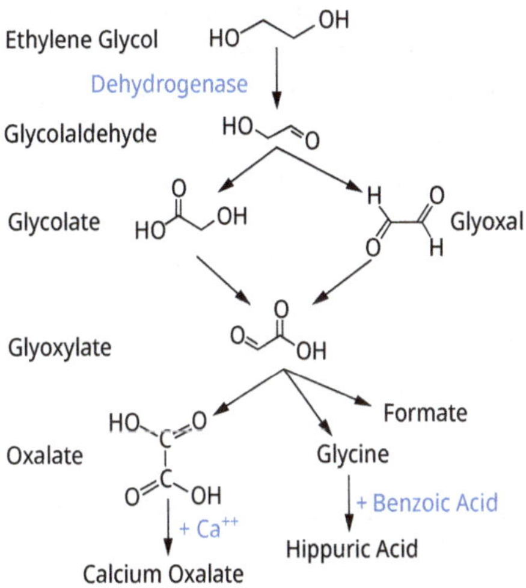

Figure 11.10: Toxicity pathways of ethylene glycol upon consumption (adapted from ref. [1] with permission).

11.4 Challenges and limitations

11.4.1 Production capacity

The traditional production of ethylene glycol (EG) relies heavily on crude oil, which limits the overall production capacity of ethylene glycol [67]. EG is predominantly produced from ethylene, which is derived from naphtha and ethane. Hence, the EG supply chain is susceptible to petrochemical market dynamics and crude oil price swings. This volatility may deter the investment in EG market, which in turn affects its supply chains. The regional imbalance of EG production, with its production being concentrated in Asia, the Middle East, and North America, further exposes the industry to geopolitical risks and supply disruptions. Despite the situation, little work has been done on the optimization of the system performance of EG production process [68].

The hydration of ethylene oxide is one of the most common routes to produce ethylene glycol and it is a highly energy-intensive process involving multiple stream recycling steps to achieve the required reaction conversion. Since the precursor for ethylene oxide is produced by the catalytic pyrolysis of petroleum, the rising demand for the ethylene glycol is in conflict with the diminishing supply of petroleum sources [69]. Kandasamy et al. [6] studied an alternative route to produce EG using glycerol waste from the biodiesel industry via hydrogenolysis. However, this method is still in the research and development phase and has not been tested at scale.

Environmental regulations impose constraints on EG production capacity. The conventional EG manufacturing process release substantial quantities of greenhouse gases, thereby raising significant environmental concerns. Stringent regulatory frameworks such as the European Union's REACH (Registration, Evaluation, Authorization, and Restrictions of Chemicals) have increased compliance costs for producers. Consequently, there has been a growing impetus toward the development of bio-based EG as a more sustainable alternative, although this approach encounters distinct technical and economic challenges. The transition to renewable feedstocks, particularly lignocellulosic biomass, necessitates elaborate pretreatment protocols to depolymerize cellulose and hemicellulose into fermentable sugars. These pretreatment processes are often typically energy-intensive, resulting in higher production costs and diminished conversion efficiency. In addition, the inherent variability in biomass composition, influenced by factors such as plant genotype, growth conditions, and harvesting methods complicates the standardization of production processes, thereby hindering the consistent product quality and yield.

The use of agricultural crops like corn and sugarcane for EG production competes with the food industry and raises public concerns regarding food security and land utilization. The seasonal variability in biomass feedstock availability also poses significant challenge to maintaining a consistent supply chain. Collectively, these constraints limit the scalability of bio-based EG production to bigger industrial levels.

In conclusion, multiple factors contribute to the limitations in EG production capacity. While notable progress has been made in developing bio-based EG technologies, their scalability and economic viability remain key barriers to displacing conventional petrochemical-based EG synthesis. Continued research and technological innovation are essential to surmount these challenges and establish sustainable EG production as a credible alternative to traditional fossil-derived methods.

11.4.2 Separation process for high-purity product

The purification of ethylene glycol to achieve high-purity products is a critical step in its production; however, the separation process presents inherent challenges and limitations that compromise both efficiency and scalability. One of the primary methods for EG purification is distillation, which is an energy-intensive technique due to EG's relatively high boiling point of 197 °C. To facilitate separation, additional reagents, such as benzene, may be employed during the distillation process [70]. These additives introduce environmental and health hazards, and their incorporation often complicates downstream purification protocols.

In conventional EG synthesis, the glycol is typically produced as part of a mixed alcohol stream. Significant separation challenges arise when isolating EG from these co-produced alcohols, particularly in achieving high yield and purity of the target compound. Moreover, EG and 1,2-butanediol (1,2-BDO) form an azeotropic mixture, as illustrated in Figure 11.11 [71]. Due to their closely aligned boiling points across the entire compositional range, separating these two alcohols is both technically demanding and energy-intensive. These purification constraints not only escalate operational expenditures but also impede the scalability of bio-based EG production platforms.

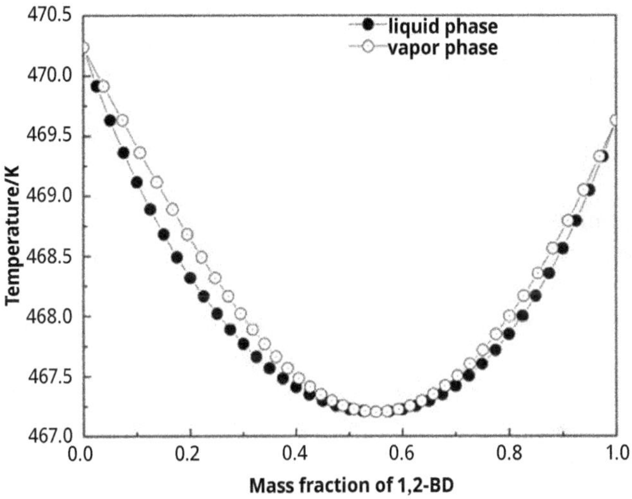

Figure 11.11: Binary phase diagram of ethylene glycol and 1,2-BDO (adapted from ref. [71] with permission).

In the coal-derived ethylene glycol industry, the distillation process comprises five columns: two methanol recovery columns, a light component removal column, a 1,2-BDO removal column, and an ethylene glycol refining column [68]. The limitations of this technology, however, are too glaring, which include high energy consumption and low recovery yield while the quality of the product does not necessarily meet the intended specifications [69, 71]. The excessive amount of energy required, combined with the large reflux ratio and tray number to remove other glycols from ethylene glycol also makes the traditional separation process come with a painfully high CAPEX and OPEX [72, 73].

In a study by Wang et al. [74], the technology of the aldehyde-assisted chemical looping separation process (CSLP) was applied for the purpose using three different aldehydes: (i) formaldehyde (FA), (ii) acetaldehyde (AA), and (iii) prolyl aldehyde (PA). With the objective to optimize the technoeconomic outcome of the process while maximizing production yield and minimizing the environmental impact, this study found that PA-assisted CSLP gives the yield of ethylene glycol with polyester-grade properties of 94.42% and an internal rate of return (IRR) of 137.28%. Nevertheless, such a promising process suffers from a high potential environmental impact due to liquid discharge. Meanwhile, AA-assisted CSLP has zero liquid discharge, while FA-assisted CSLP consumes large energy consumption and has high liquid discharge. These three CSLPs studied show that the separation of ethylene glycol from 1,2-BDO remains a challenge and is yet to be mitigated.

Significant technological advancements have been achieved in the field of EG production, notably by key industry players such as Shell, which introduced the OMEGA (only mono-ethylene glycol advantage) process. Through a two-step catalytic synthesis involving the formation of ethylene carbonate using a phosphonium halide catalyst, Shell's process achieves EG yields of 99–99.5%, eliminating the need for downstream purification [20]. The principal limitation of this technology lies in its substantial capital investment requirements, which may render it economically unviable for smaller-scale EG producers.

An alternative method proposed for EG purification is pervaporation, a membrane-based separation technique that offers both energy efficiency and environmental advantages [70]. Pervaporation operates by selectively separating components based on their permeation rates through a dense membrane. However, this approach remains unsuitable for industrial-scale deployment, as most membranes exhibit insufficient chemical stability and mechanical robustness under process conditions. Extensive research into membrane enhancement targeting on improving the selectivity and permeability of the membrane is essential in order to establish pervaporation as a viable substitute for conventional distillation-based EG purification.

Another promising method is melt crystallization, which separates EG from impurities based on differences in melting points [75]. This technique has the potential to yield high-purity EG without requiring elevated operating temperatures or chemical additives. However, melt crystallization is inherently complex and demands precise control over thermal and process parameters, which poses significant challenges for scale-up to industrial applications.

Each separation and purification method, be it conventional or emerging separation process, offers distinct advantages and limitations. Production facilities must place greater emphasis on this purification stage, as an efficient separation process is critical to ensuring the high product quality of ethylene glycol.

11.4.3 Release of hazardous gas

The production of ethylene glycol (EG) involves processes that release hazardous gases, posing threats to health, safety and environment. One of the emissions that raise major concern from EG production is ethylene oxide (EO), a volatile and highly reactive intermediate [76]. Given EO's carcinogenic nature, strict containment protocols and continuous emission monitoring are essential to mitigate exposure and ensure safe operational practices.

In coal-to-ethylene glycol (CtEG) processes, the production pathway involves the gasification of coal, which releases synthesis gas (syngas) [77]. This process releases substantial quantities of CO, a toxic gas at elevated concentrations. One of the foundational technologies for coal-based EG production was developed by Japan's Ube Industries, utilizing syngas in a multistep reaction sequence that includes the consumption and regeneration of methyl nitrite as a key intermediate [78]. Although coal-derived EG is economically attractive to polyester manufacturers, the process presents several challenges. The resulting EG often exhibits lower purity, and its supply reliability is hindered by the operational complexity of the plant. Moreover, a critical limitation of the Ube process is its dependence on methyl nitrite which is a hazardous and highly flammable compound.

Additionally, the combustion processes in the EG production can result in the emission of nitrogen oxides (NO_x) and sulfur oxides (SO_x), both of which contribute to atmospheric pollution and the formation of acid rain. A further environmental concern is the release of volatile organic compounds during EG manufacturing, which can undergo photochemical reactions in the atmosphere, resulting in the formation of ground-level ozone, which is a pollutant linked to adverse human health effects.

To reduce the hazardous gas emissions, the production plants need to implement effective gas treatment systems into their process. However, this is often seen as additional production cost and process complexity to the producers. Consequently, more stringent environmental regulations and safety protocols must be enforced through robust regulatory frameworks to address these environmental and public health challenges effectively.

11.5 Current and future opportunities

11.5.1 Advances in design, process simulation, and separation techniques

In the conventional distillation route, the ethylene glycol can be obtained at 95 wt% purity with other glycols from the top of the distillation column and can be sold as a qualified product at a relatively lower price. In addition, there is also a potential to explore further the possible applications of ethylene glycol with 95 wt% purity since its production consumes much less energy [71].

As such, Yu and Chein suggested an isothermal plug flow hydrogenation reactor and rearranging the reactors to further heighten the efficiency of the ethylene glycol production via a hydrogenation process [79]. In another related work published by Wei et al. [72], there are a few methodologies suggested to improve technology design related to this glycol production. This includes a CO coupling reactor (CCR) with four series functional reactors to overcome the technological difficulties related to the coal-based syngas.

On the other hand, Yang et al. [68] suggest that it is important to perform a simulation to optimize the product yield as well as the system performance of the ethylene glycol process derived from coal. They claimed that their simulated model is reasonable for such a process based on the increase in energy efficiency from 30.68% to 31.49%. Several key parameters were studied and optimized, which include the mole ratio between CO and methyl nitrate, the pressure, and the catalyst height. Alternatively, the ethylene glycol can be produced from xylitol by hydrogenolysis under heating in the presence of reducing agents and since xylitol is derived from biomass, this production route becomes favorable [80].

By studying the vapor-liquid equilibrium data of six binary systems consisting of ethylene glycol, 1,2-BDO, 2,3-BDO, and 1,2-propylene glycol (1,2 PG), Li et al. [71] simulated in Aspen Plus four separation processes to obtain polymer-grade ethylene glycol from the mixture components, comparing the total annual costs and profits for each. The four processes are: (1) conventional distillation, (2) azeotropic distillation, (3) distillation coupled with liquid-liquid extraction, and (4) reaction-assisted distillation. It is concluded that the economy depends mainly on the yield of the glycol, with reaction-assisted distillation being the current best separation method. This method utilizes a chemical reaction to enhance the separation process. The reaction-assisted distillation was proven to pose great potential in obtaining a high yield of ethylene glycol during separation [81].

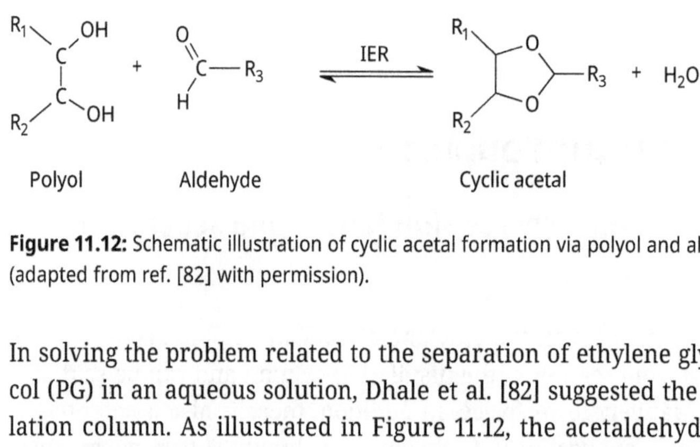

Figure 11.12: Schematic illustration of cyclic acetal formation via polyol and aldehyde reaction (adapted from ref. [82] with permission).

In solving the problem related to the separation of ethylene glycol and propylene glycol (PG) in an aqueous solution, Dhale et al. [82] suggested the use of a reactive distillation column. As illustrated in Figure 11.12, the acetaldehyde is used as a reactive agent and is combined with the polyols, subsequently forming a cyclic acetal. In light of this, the Amberlyst 15 cationic exchange resin is held in structured packing inside the column to catalyze the reaction. The glycols and the acetaldehyde are fed countercurrently into the column for mass transfer efficiency. This technology has been proven at a pilot scale with a 90% conversion rate achieved. Since this reaction is reversible, the downstream process involves the rapid process of hydrolysis of acetals to form high purity of ethylene glycol and propylene glycol, while the acetaldehyde is being recycled into the process. This method has an advantage over the hydration of water, which uses up a lot of energy; while solvent extraction although is theoretically feasible, yet the affinity of glycols towards water makes it difficult to find a suitable organic solvent to fulfil the separation requirement.

11.5.2 Emerging application of bio-based ethylene glycol

In the study carried out by India Glycols Limited (IGL), the comparison is made between the conventional production route of MEG and the production of bio-MEG from ethyl alcohol of sugarcane molasses-based on their life cycle assessment, respectively [83]. It is found that the production of 1 MT of MEG from petroleum emits 1,628 kilograms of CO_2. Meanwhile its bio-route counterpart only emits 1,221 kilograms of CO_2, a 25% reduction in the CO_2 emission over an MEG life cycle. This emission reduction is mainly contributed by the use of sugarcane molasses, which can be cultivated again, hence reducing the carbon footprint of the overall bio-MEG production.

The production of MEG from sugar has become more established with Braskem and Haldor Topsoe building their first demo-scale plant for the production of bio-MEG at the end of 2020 [84]. Although this technology is sustainable and has the potential to revolutionize the PET market, the source of feedstock tends to compete with the food production line. This is the same for other food-based feedstock such as corn, switchgrass, and wheat straw. Another challenge faced by bio-MEG production is that the cost and the price of bio-MEG are still higher compared to the fossil-based MEG. This wide price gap causes clients to take stock of the value of its use [85].

Bio-MEG can also be synthesized using the biotechnological route. The challenges of using this route predominantly lie in its yield, which does not meet the theoretical value expected while the production titers are too low for commercial-scale production [53]. Comparing the two main biotechnological routes, the metabolic engineering of *E. coli* shows a more promising result than the engineering yeast species.

11.5.3 Depolymerization of PET into ethylene glycol

Another interesting opportunity in this field is the depolymerization of PET. MEG is one of the main precursors in producing PET bottles. The extended period of time for PET to decompose naturally has caused the ever-rising environmental issue when these bottles end up in the landfills and in the oceans at the end of their use.

Realizing the opportunity to reverse the chemical process, a few companies including Eastman Chemical and Loop Industries have embarked on the pursuit to break down PET back into its building blocks, ethylene glycol (EG), and dimethyl terephthalate (DMT) via chemical processes such as methanolysis and hydrolysis [86].

These two monomers are then condensed again to produce recycled PET, which has properties similar to virgin PET and is suitable for packaging applications including in stringent-requirement industries such as the food industry. Although the mechanical recycling of PET bottles is already at a satisfying rate, this emerging depolymerization chemical recycling technique will tackle the PET fiber that faces degradation challenges when recycled mechanically, thus accelerating the recovery of environmental issues caused by used plastics.

Eastman Chemical and Loop Industries are already on their way to building plants using this technology [86, 87]. According to Loop Industries, the depolymerizing PET route to make new PET saves more than 2 kg of CO_2 per kilogram of PET relative to the conventional route of making PET from fossil fuel. Fueled by this motivation, the CEO of Loop Industries, Daniel Solomita envisions building one depolymerization of PET plant in every country to recycle polyester resins and fibers as he sees this technology to be an essential infrastructure to have. At the moment, Loop is still at its first project partnered with PET maker Indorama Ventures to process 40,000 tons per year of post-consumer PET in Spartanburg, South Carolina. This project is supported by PepsiCo, Coca-Cola, and Danone who made commitments to buy more than half of the plant's output.

Meanwhile, Eastman has also started its modest production of EG and DMT via glycolysis of PET last year [86]. The company will commence a larger plant in Kingsport, Tennessee worth $250 million with an increase in efficiency in its methanolysis technology by the end of 2022. This plant will break down 100,000 MTPA of post-consumer PET waste, including from PET packaging and carpet fiber, into EG and DMT. These monomers will then be made into specialty polyesters, one of the products being refillable water bottles. Eastman claims that resins production using recycled content will decrease greenhouse gas production by 20–30% as compared to fossil-fuel-based feedstocks.

11.5.4 Coal-to-ethylene glycol (CTEG) technology

Another emerging technology that has garnered much attention is coal-to-ethylene glycol, which has shown significant increase in the industrial use, which has the potential of replacing the conventional hydration of EO technology using this route. This is particularly in the case of EG producer in mainland China where the country is capitalizing their abundance resources of coal to produce value-added products such as EG [88]. Furthermore, the majority of this technology player resides within China particularly in remote areas that have a coal-rich region such as Inner Mongolia, Shaanxi, Ningxia, and Xianjiang. This development in the remote area reflects the national policies and agenda, which encourage the utilization of coal near mining sites to maximize the resource efficiency while valorizing the value-added products.

Coal-to-ethylene glycol (CTEG) is a multistep chemical process that produces MEG from coal-derived synthesis gas [89, 90]. The process typically involves several core units: an air separation unit (ASU), coal gasification (CG), sulfur recovery (SR), dimethyl oxalate synthesis (DMOS), ethylene glycol synthesis (EGS), and ethylene glycol refining (EGR). It begins with coal gasification under controlled temperature and pressure in the presence of oxygen and steam, generating synthesis gas (syngas) composed primarily of carbon monoxide (CO) and hydrogen (H_2). To optimize the CO/H_2 ratio for downstream reactions, the syngas undergoes a water-gas shift (WGS) reaction. Prior to this, acid gases such as carbon dioxide (CO_2) and sulfur dioxide (SO_2) must be removed to prevent catalyst poisoning and ensure process efficiency. The conditioned syngas is then fed into a carbonylation reactor, where CO reacts with methanol to form dimethyl oxalate (DMO) under specific operating conditions. In the subsequent hydrogenation step, DMO is selectively converted into MEG using catalysts. Finally, the crude MEG is purified through distillation to meet the desired product specifications.

Numerous CTEG plants have been established across China, and this number is projected to grow steadily in response to government incentives and localized support for coal conversion technologies [88]. The expansion of CTEG facilities, alongside conventional ethylene oxide-based MEG production, has significantly strengthened China's domestic MEG supply. As a result, this has led to a reduction in China's reliance on imports. Continued growth in this sector suggests that China will increasingly meet its internal demand through domestic output, with less dependence on foreign producers such as Saudi Arabia and the United States. This strategic shift enhances national control over MEG supply chains and mitigates exposure to global price fluctuations and trade uncertainties.

11.6 Conclusion

In conclusion, this chapter has discussed the two main synthesis routes for producing MEG, which are hydration of ethylene oxide and transesterification of ethylene carbonate. The advantages and disadvantages of each method were examined in detail. Several recent developments involving conventional, bio-based, and coal-derived MEG production pathways were reviewed, including the growing prominence of coal-to-ethylene glycol (CTEG) technology in China. The chapter also presented an overview of MEG's properties, including its toxicity profile, and highlighted current challenges in production capacity, high-purity separation, and mitigation of hazardous gas emissions. Finally, this chapter provides a comprehensive perspective on the current production landscape of MEG and outlines future opportunities that align with sustainable development and evolving industrial needs.

References

[1] Yue H, Zhao Y, Ma X, Gong J Ethylene Glycol: Properties, synthesis, and applications. Chem Soc Rev 2012;41:4218–4244.

[2] Rebsdat S, Mayer D Ethylene Glycol. Ullmanns Encycl Ind Chem 2012;13:531–546.

[3] Kirk RE, Othmer DF. Kirk-Othmer Encyclopedia of Chemical Technology, 4th edition, Wiley; 1995.

[4] Wurtz A Mémoire sur les glycols ou alcools diatomiques. Ann Chim Phys 1859;5(3):400–478.

[5] Van Hal J, Ledford JS, Zhang X Investigation of three types of catalysts for the hydration of ethylene oxide (EO) to monoethylene glycol (MEG). Catal Today 2007;123:310–315.

[6] Kandasamy S, Samudrala SP, Bhattacharya S. The route towards sustainable production of ethylene glycol from a renewable resource, biodiesel waste: A review. Catal Sci Technol 2019;9:567–577.

[7] Lefort TE. France Patent Pat 729952 (to Societe Francaise de Catalyse Generalise), 27 March 1931.

[8] Dye RF. Ethylene glycols technology. Korean J Chem Eng 2001;18:571–579.

[9] Kawabe K Development of highly selective process for mono-ethylene glycol production from ethylene oxide via ethylene carbonate using phosphonium salt catalyst. Catal Surv Asia 2010;14:111–115.

[10] Matar S, Hatch LF Hydrocarbon intermediates. Chem Petrochem Process 2001;29–48.

[11] Matar S, Hatch LF Chemicals based on ethylene. Chem Petrochem Process 2001;188–212.

[12] Shehata WM, Nady TG, Gad FK, Shoaib AM, Bhran AA A comparative study of mono Ethylene Glycol economic production via different techniques. Sci Rep 2024;14:1–19.

[13] Insights FM Ethylene Glycol market poised for significant growth: Demand to reach USD 75,303.1 million by 2034. Futur. Mark. Insights Glob. Consult. PVT, Ltd. (2024), [Online] https://www.globenewswire.com/news-release/2024/08. [Accessed 5 August 2025].

[14] Wong MK, Lock SSM, Chan YH, Yeoh SJ, Tan IS Towards sustainable production of bio-based Ethylene Glycol: Progress, perspective and challenges in catalytic conversion and purification. Chem Eng J 2023;468:143699.

[15] International A Standard Specification for Engine Coolant Grade Ethylene GlycolE2470-09; 2009

[16] Hv M, HV M, Wells GJ Enhancements in Ethylene Oxide/Ethylene Glycol manufacturing technology. 2021. [Online]. Available: https://catalysts.shell.com/en/ethylene-oxide-ethylene-glycol-production-white-paper. [Accessed 5 August 2025].

[17] Zhu F, Huang K, Wang S, Shan L, Zhu Q. Towards further internal heat integration in design of reactive distillation columns-part IV: Application to a high-purity ethylene glycol reactive distillation column. Chem Eng Sci 2009;64:3498–3509.

[18] Ding T, Zha J, Zhang D, Zhang J, Yuan H, Xia F, Gao G. CO_2 atmosphere enables efficient catalytic hydration of ethylene oxide by ionic liquids/organic bases at low water/epoxide ratios. Green Chem 2021;23:3386–3391.

[19] Yan L, Witt PM, Edgar TF, Baldea M. Static and dynamic intensification of water-ethylene glycol separation using a dividing wall column. Ind Eng Chem Res 2021;60:3027–3037.

[20] Harmsen J, Verkerk M. 15 shell OMEGA only monoethylene glycol advanced process. Process Intensif 2020;166–175.

[21] Zavelev DE, Tsodikov MV, Zhidomirov GM, Kozlovskii RA. Role of the surface hydroxyl groups of modified titanium oxide in catalytic ethylene oxide hydration. Kinet Catal 2011;52:659–671.

[22] Altiokka MR, Akyalçin S Kinetics of the hydration of ethylene oxide in the presence of heterogeneous catalyst. Ind Eng Chem Res 2009;48:10840–10844.

[23] Li B, Bai S, Wang X, Zhong M, Yang Q, Li C. Hydration of epoxides on [Co III(salen)] encapsulated in silica-based nanoreactors. Angew Chem Int Ed 2012;51:11517–11521.

[24] Dai W, Wang C, Tang B, Wu G, Guan N, Xie Z, Hunger M, Li L. Lewis acid catalysis confined in zeolite cages as a strategy for sustainable heterogeneous hydration of epoxides. ACS Catal 2016;6:2955–2964.

[25] Bhanage BM, Fujita SI, He Y, Ikushima Y, Shirai M, Torii K, Arai M. Concurrent synthesis of dimethyl carbonate and ethylene glycol via transesterification of ethylene carbonate and methanol using smectite catalysts containing Mg and/or Ni. Catal Lett 2002;83:137–141.

[26] Landscheidt H, Wolters E, Klausener A, Blank H, Birkenstock U Process for the preparation of dialkyl carbonates. Bayer Aktiengesellschaft Patent 1993;5(288):894.

[27] Knifton JF Process for co-synthesis of ethylene glycol and DMC. Texaco Patent 1987;4(661):609.

[28] Duranleau RG, Nieh ECY, Knifton JF Process for production of ethylene glycol and dimethyl carbonate. Texaco Patent 1986;4(691):041.

[29] Pacheco MA, Marshall CL Review of dimethyl carbonate (DMC) manufacture and its characteristics as a fuel additive. Energy Fuels 1997;11:2–29.

[30] Buysch H-J, Klausener A, Langer R, Mais F-J. "Process for the continuous preparation of Dialkyl Carbonates". Bayer Aktiengesellschaft Patent 1993;5(231):212.

[31] Knifton JF, Duranleau RG Ethylene glycol-dimethyl carbonate cogeneration. J Mol Catal 1991;67:389–399. doi: 10.1016/0304-5102(91)80051-4.

[32] Ju H-Y, Manju MD, Park D-W, Choe Y, Park S-W Performance of ionic liquid as catalysts in the synthesis of dimethyl carbonate from ethylene carbonate and methanol. React Kinet Catal Lett 2007;90(1):3–9.

[33] K-h K, D-W K, C-w K, Koh J-C, Park D-W Synthesis of dimethyl carbonate from transesterification of ethylene carbonate with methanol using immobilized ionic liquid on commercial silica. Korean J Chem Eng 2010;27(5):1441–1445.

[34] Chang Y, Jiang T, Han B, Liu Z, Wu W, Gao L, Li J, Gao H, Zhao G, Huang J One-pot synthesis of dimethyl carbonate and glycols from supercritical CO_2, ethylene oxide or propylene oxide, and methanol. Appl Catal A Gen 2004;263(2):179–186.

[35] Cherubini F The biorefinery concept: Using biomass instead of oil for producing energy and chemicals. Energy Convers Manage 2010;*51*(7), 1412–1421.

[36] Pang J, Zheng M, Sun R, Wang A, Wang X, Tao Z Synthesis of ethylene glycol and terephthalic acid from biomass for producing PET. Green Chem 2016;*18*(2), 342–359.

[37] Mohsenzadeh A, Zamani A, Taherzadeh MJ Bioethylene production from Ethanol: A Review and techno-economical evaluation. ChemBioEng Rev 2017;4:75–91.

[38] Lippits MJ, Nieuwenhuys BE Direct conversion of Ethanol into Ethylene Oxide on gold-based catalysts: Effect of CeOx and Li_2O addition on the selectivity. J Catal 2010;274:142–149.

[39] Lippits MJ, Nieuwenhuys BE Direct conversion of Ethanol into Ethylene Oxide on copper and silver nanoparticles: Effect of addition of CeOx and Li_2O. Catal Today 2010;154:127–132.

[40] Ripamonti D, Tripodi A, Conte F, Robbiano A, Ramis G, Rossetti I Feasibility study and process design of a direct route from Bioethanol to Ethylene Oxide. J Environ Chem Eng 2021;9:105969.

[41] Stephen Dahms A 3-Deoxy-D-Pentulosonic acid Aldolase and its role in a new pathway of D-Xylose degradation. Biochem Biophys Res Commun 1974;60:1433–1439.

[42] Liu H, Ramos KRM, Valdehuesa KNG, Nisola GM, Lee WK, Chung WJ Biosynthesis of Ethylene Glycol in Escherichia Coli. Appl Microbiol Biotechnol 2013;97:3409–3417.

[43] Chae TU, Choi SY, Ryu JY, Lee SY Production of Ethylene Glycol from Xylose by metabolically engineered Escherichia Coli. AIChE J 2018;64:4193–4200.

[44] Wang Y, Xian M, Feng X, Liu M, Zhao G Biosynthesis of Ethylene Glycol from D-Xylose in recombinant Escherichia Coli. Bioengineered 2018;9:233–241.

[45] Cabulong RB, Lee WK, Bañares AB, Ramos KRM, Nisola GM, Valdehuesa KNG, Chung WJ Engineering Escherichia Coli for Glycolic acid production from D-Xylose through the Dahms Pathway and Glyoxylate Bypass. Appl Microbiol Biotechnol 2018;102:2179–2189.

[46] Alkim C, Cam Y, Trichez D, Auriol C, Spina L, Vax A, Bartolo F, Besse P, François JM, Walther T Optimization of Ethylene Glycol Production from (d)-Xylose via a Synthetic Pathway Implemented in Escherichia Coli. Microb Cell Fact 2015;14:1–12.

[47] Pereira B, Li ZJ, De Mey M, Lim CG, Zhang H, Hoeltgen C, Stephanopoulos G Efficient utilization of Pentoses for bioproduction of the renewable two-carbon compounds Ethylene Glycol and Glycolate. Metab Eng 2016;34:80–87.

[48] Lee SS, Choi JI, Woo HM Bioconversion of Xylose to Ethylene Glycol and Glycolate in engineered Corynebacterium Glutamicum. ACS Omega 2019;4:21279–21287.

[49] Liu Y, Huo K, Tan B, He X, Wu Q, Li ZJ metabolic engineering of Halomonas Bluephagenesis for the production of Ethylene Glycol and Glycolate from Xylose. J Biotechnol 2024;396:36–40.

[50] Pereira B, Zhang H, De Mey M, Lim CG, Li ZJ, Stephanopoulos G Engineering a novel Biosynthetic pathway in Escherichia Coli for production of renewable Ethylene Glycol. Biotechnol Bioeng 2016;113:376–383.

[51] Chen Z, Huang J, Wu Y, Liu D Metabolic engineering of Corynebacterium Glutamicum for the de Novo production of Ethylene Glycol from Glucose. Metab Eng 2016;33:12–18.

[52] Carneiro CVGC, Trichez D, Bergmann JC, Reis VCB, Wagner N, Walther T, Almeida JRM de engineering Komagataella Phaffii for Ethylene Glycol production from Xylose. AMB Express 2024;*14*(1), 131.

[53] Salusjärvi L, Havukainen S, Koivistoinen O, Toivari M Biotechnological production of Glycolic Acid and Ethylene Glycol: Current state and perspectives. Appl Microbiol Biotechnol 2019;103:2525–2535.

[54] Uranukul B, Woolston BM, Fink GR, Stephanopoulos G Biosynthesis of Monoethylene Glycol in Saccharomyces Cerevisiae utilizing native Glycolytic Enzymes. Metab Eng 2019;51:20–31.

[55] Chomvong K, Bauer S, Benjamin DI, Li X, Nomura DK, Cate JHD Bypassing the Pentose Phosphate Pathway: Towards modular utilization of Xylose. PLoS One 2016;11:1–16.

[56] Gomes AMV, Carmo TS, Carvalho LS, Bahia FM, Parachin NS Comparison of Yeasts as Hosts for Recombinant Protein Production. Microorganisms 2018;6:1–23.

[57] Wang G, Huang M, Nielsen J Exploring the potential of Saccharomyces Cerevisiae for biopharmaceutical protein production. Curr Opin Biotechnol 2017;48:77–84.

[58] Salusjärvi L, Toivari M, Vehkomäki ML, Koivistoinen O, Mojzita D, Niemelä K, Penttilä M, Ruohonen L Production of Ethylene Glycol or Glycolic Acid from D-Xylose in Saccharomyces Cerevisiae. Appl Microbiol Biotechnol 2017;101:8151–8163.

[59] Cam Y, Alkim C, Trichez D, Trebosc V, Vax A, Bartolo F, Besse P, François JM, Walther T Engineering of a synthetic metabolic pathway for the assimilation of (d)-Xylose into value-added chemicals. ACS Synth Biol 2016;5:607–618.

[60] Tullo A New route planned to bio-based ethylene glycol. CEN Global Enterprise 2017;95(46):10.

[61] Majerski PA, Piskorz JK, Radlein DSAG United States Patent US7094932B2, 2006.

[62] Schandel CB, Høj M, Osmundsen CM, Jensen AD, Taarning E Thermal cracking of sugars for the production of glycolaldehyde and other small oxygenates. Chem Sus Chem 2020;13(4):688–692.

[63] Research TM Monoethylene Glycol Market to Expand at CAGR of 6.1% during Forecast Period, Notes TMR Study, (2022), [Online] https://www.prnewswire.com/news-releases/monoethyl.[Accessed 5 August 2025].

[64] Leth PM, Gregersen M Ethylene glycol poisoning. Forensic Sci Int 2005;155:179–184.

[65] Lovrić M, Granić P, Čubrilo-Turek M, Lalić Z, Sertić J Ethylene glycol poisoning. Forensic Sci Int 2007;170:213–215.

[66] Moore MM, Kanekar SG, Dhamija R Ethylene glycol toxicity: Chemistry, pathogenesis, and imaging. Radiol Case Rep 2008;3:122.

[67] Yang Q, Zhu S, Yang Q, Huang W, Yu P, Zhang D, Wang Z Comparative techno-economic analysis of oil-based and coal-based ethylene glycol processes. Energy Convers Manag 2019;198:111814.

[68] Yang Q, Zhang D, Zhou H, Zhang C Process simulation, analysis and optimization of a coal to Ethylene Glycol process. Energy 2018;155:521–534.

[69] Song H, Jin R, Kang M, Chen J Progress in synthesis of ethylene glycol through C1 chemical industry routes. Cuihua Xuebao/Chin J Catal 2013;34:1035–1050.

[70] Rostovtseva V, Faykov I, Pulyalina A A review of recent developments of pervaporation membranes for Ethylene Glycol purification. Membranes (Basel). 2022;*12*(3), 312.

[71] Li H, Zhao Z, Qin J, Wang R, Li X, Gao X Reversible reaction-assisted intensification process for separating the azeotropic mixture of ethanediol and 1,2-butanediol: Vapor-liquid equilibrium and economic evaluation. Ind Eng Chem Res 2018;57:5083–5092.

[72] Wei R, Yan C, Yang A, Shen W, Li J Improved process design and optimization of 200 kt/a ethylene glycol production using coal-based syngas. Chem Eng Res Des 2018;132:551–563.

[73] Chen YC, Hung SK, Lee HY, Chien IL Energy-saving designs for separation of a close-boiling 1,2-propanediol and ethylene glycol mixture. Ind Eng Chem Res 2015;54:3828–3843.

[74] Wang R, Na J, Li X, Li H, Gu S, Gao X Techno-economic and environmental evaluation of aldehyde-assisted chemical looping separation technology in coal-based and biomass-based ethylene glycol purification. J Clean Prod 2021;313:127675.

[75] Wang T, Dong J Ethylene Glycol purification by melt crystallization: Removal of 2-Methoxyethanol impurity. Chinese J Chem Eng 2021;37:39–45.

[76] Haque ME, Tripathi N, Palanki S, Xu Q, Nigam KDP Plant-wide modeling and economic analysis of Monoethylene Glycol production. Processes 2022;*10*(9), 1755.

[77] Wang J, Liu D, Sun J Life cycle energy consumption, environmental impact, and costing assessment of coal to Ethylene Glycol processes via Dimethyl Oxalate and Formaldehyde. Environ Sci Pollut Res 2023;30:31141–31156.

[78] Jean-François Tremblay. Polyester made from coal? china is betting on it. CEN Glob Enterp 2019;97:22–24.

[79] Yu BY, Chien IL Design and optimization of Dimethyl Oxalate (DMO) hydrogenation process to produce Ethylene Glycol (EG). Chem Eng Res Des 2017;121:173–190.

[80] Delgado Arcaño Y, Valmaña García OD, Mandelli D, Carvalho WA, Magalhães Pontes LA Xylitol: A review on the progress and challenges of its production by chemical route. Catal Today 2020;344:2–14.

[81] Li H, Huang W, Li X, Gao X Application of the aldolization reaction in separating the mixture of ethylene glycol and 1,2-butanediol: Thermodynamics and new separation process. Ind Eng Chem Res 2016;55:9994–10003.

[82] Dhale AD, Myrant LK, Chopade SP, Jackson JE, Miller DJ Propylene glycol and ethylene glycol recovery from aqueous solution via reactive distillation. Chem Eng Sci 2004;59:2881–2890.

[83] India Glycols Limited. Reduction of greenhouse gas emissions via use of chemical products. ICCA 2016.

[84] Braskem. Newsletter carbon neutral circular economy. Braskem, 4 February 2021. [Online]. Available: https://www.braskem.com/newsletter-carbon-neutral-detalhe-en/first-demonstration-scale-of-renewable-bio-meg-production. [Accessed 5 August 2025].

[85] Blog GC. What future for coal-to-meg and bio-meg in a low-cost brent world? Technon Orbichem, 11 June 2020. [Online]. Available: https://greenchemicalsblog.com/2020/06/11/what-future-for-coal-to-meg-and-bio-meg-in-a-low-cost-brent-world/.[Accessed 5 August 2025].

[86] Tullo A Eastman will build a $250 million plastics recycling plant. Chemical & Engineering News, 1 FEBRUARY 2021. [Online]. Available: https://cen.acs.org/environment/recycling/Eastman-build-250-million-plastics/99/web/2021/02.

[87] Tullo A Industry's plastic recycling bet. Chemical & Engineering News, 12 OCTOBER 2020 [Online]. Available: https://pubs.acs.org/doi/10.1021/cen-09839-cover

[88] Qiu C China MEG Demand Stable; Oversupply to Worsen on New Capacities; 2020

[89] Liu C, Zhao L, Zhu S, Shen Y, Yu J, Yang Q Advanced exergy analysis and optimization of a Coal to Ethylene Glycol (CtEG) process. Energy 2023;282:128790.

[90] Li D, Zhao L, Zha Z, Ma H, Wan B, Liu G Study on the influence of coal on Coal-to-Ethylene Glycol process considering the integration of reactor, distillation column, and heat exchanger network. J Clean Prod 2024;463:142715.

Daniel Garbark

Chapter 12 Soybean oil-based surfactants

12.1 Surfactants background and advantages of high oleic soybean oil (HOSO)

Surfactants are defined as “surface-active agents” [1]. These ingredients, commonly used in detergents, have unique properties that allow the component to be compatible with water and at the same time be attracted to organic materials such as grease and dirt. Surfactants are used in a variety of applications, with the largest market being detergents. These materials perform a balancing act of being attracted to the grease and dirt (lipophilic), while also being attracted to water (hydrophilic). Water alone will not adequately clean the laundry. The lipophilic tail is attracted to the grease and dirt, while the hydrophilic head group is attracted to the water. The surfactant suspends the grease and dirt in the water so that it can be rinsed away. Therefore, the surfactant performs the primary cleaning function for a laundry detergent.

Surfactants have changed significantly over the years. The earliest known detergents contain fatty acid soaps. These soaps helped in lowering the surface tension of water, allowing for appropriate cleaning of hands and clothing. However, with the advent of laundry machines, the fatty acid soaps were found to leave deposits on clothing. This led to the development of linear alkylbenzene sulfonate (LAS). With improved water miscibility, LAS was a superior choice to traditional soaps. Surfactant development has continued to increase efficacy and selectivity, leading to the development of nonionic surfactants and later cationic and amphoteric varieties. Overall, there are four categories of surfactants: anionic, nonionic, cationic, and amphoteric.

Surfactants developmental history consists of various feedstocks, including both petroleum and bio-based. For improved water miscibility, surfactants focused on shorter chain lengths such as those based on lauric acid (C_{12}). Even though many were produced from petroleum sources, a shift to bio-based options led to use of coconut fatty acids and palm kernel fatty acids in both purified and crude mixtures. While these bio-based feedstocks show promising performance, there is no US production. This led to increased consideration of feedstocks such as soybean oil (SBO). SBO is the largest vegetable oil harvested in the United States and with the advent of soy fatty acid varieties, some characteristics could be exploited for the production of equivalent performing surfactants versus coconut fatty acid derivations. This chapter presents recent work done in this area at Battelle [2, 3]. With the introduction of high

Daniel Garbark, formerly Lead Researcher at Battelle, Columbus, Ohio USA,
e-mail: dangarbark@yahoo.com

 | https://doi.org/10.1515/9783112212141-012

oleic soybean oil (HOSO), many reactions can be performed without interference from unwanted side reactions as seen previously with commodity SBO. For example, when reacting epoxidized SBO with hindered or low reactivity hydroxyls, significant portions of tetrahydrofuranyl (THF) ethers and cross-linked fatty acids can be formed. This is shown in Figure 12.1.

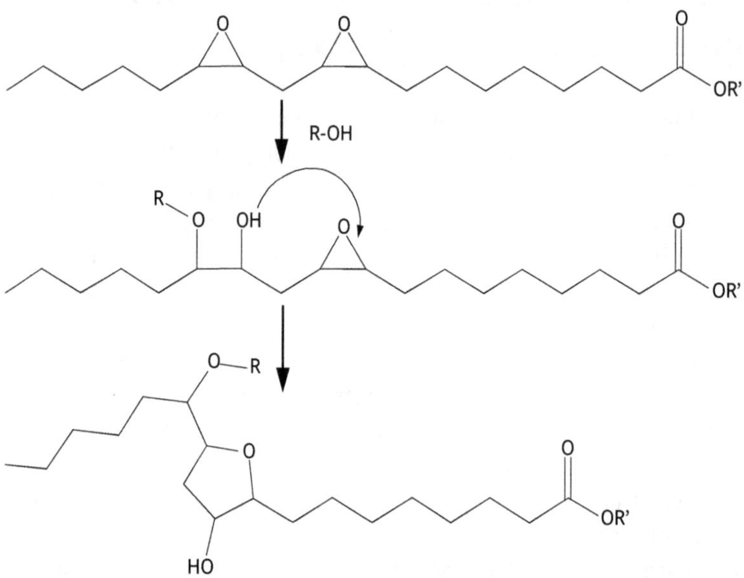

Figure 12.1: Reaction of epoxidized linoleate with hindered or low-activity alcohols.

The various levels of fatty acids in SBO can be found in Chapter 22. The reason for the THF ethers is due to the higher levels of linoleic and linolenic fatty acids (53wt% and 8wt%, respectively) in commodity SBO, which contain greater levels of doubly allylic olefins. In using epoxidized HOSO, the same reaction occurs slowly and due to the lower amount of linoleic and linolenic acids (10wt% and 1wt%, respectively), the potential for THF ethers is significantly reduced. This allows for the use of various moieties, when using HOSO as a building block to a variety of surfactants. Commodity SBO can be used in cases where the epoxide is reacted with highly active primary hydroxyls, such as those from alkyl alcohols, polyether alcohols, and polyether alcohol monoalkyl ethers.

Bio-based poly-acids have a unique opportunity for use in a variety of applications such as chelation in laundry detergent builders. This can be seen in the use of trisodium citrate as a water softener in various detergent formulations. These softeners, which bind metal cations, also work to maintain detergent stability, help keep soil from redepositing during washing, and improve the cleaning efficiency of the detergent formulation. Traditionally, materials used as detergent builders have included sodium triphosphate and ethylenediaminetetraacetic acid (EDTA). Regulations have pushed for reduction of phosphates used in detergents, which resulted in greater interest in identifying and developing environmentally friendly alternatives. One group of surfactants produced from HOSO consists of active structural sites similar to EDTA, but with a fatty acid inserted in the middle. This allows for dual functionality, as the water softener maintains surfactant behavior through the remaining aliphatic tail.

Through designed chemical structure, epoxidized HOSO allows for an array of functionalities, leading to a targeted structure to stain determination such as soy lactate performing improved make-up removal. The main improvement over the older fatty acid soaps is by having a hydrophilic group in the middle of the chain, one that creates greater water miscibility while still maintaining surface activity through an aliphatic C_8–C_9 chain. In the cases where commodity SBO is used, the aliphatic tail is shortened to C_5–C_6. With the ability to vary side chains at the original olefin site, along with functional groups at the original fatty acid group, the hydrophilic-lipophilic balance (HLB) of the molecule can be more readily controlled. Selection of functional groups and reactants also leads to high bio-based content and good biodegradability. In some cases, soy-based candidates were produced using hydroformylation and hydrogenation as an alternative to epoxidation. To date, 84 soy-based surfactant structures have been produced as potential replacements to those currently used in the marketplace. A few of these structures are described in this chapter, with additional details drawn from recent patent filings on bio-based surfactants derived from HOSO [12]. These patents highlight methods for producing surfactants with at least 50% or up to 100% bio-based content, emphasizing applications in laundry detergents, cleaners, and oilfield uses, while ensuring low toxicity, persistence, and by-products.

12.2 Anionic surfactants

Anionic surfactants are the largest group of surfactants, accounting for approximately 40% of the world's surfactant production [2]. They contain functional groups such as sulfonates, sulfates, carboxylates, or phosphates at the structural head. These products exhibit superior wetting and emulsifying properties and tend to be higher-foaming materials. Fatty acid soaps and alkylbenzene sulfonates fall into this category. Fatty acid soaps, the most common of surfactants, typically have a HLB of approximately 15, which means they function more as a cleaning agent than as a surfactant. Alkylbenzene sulfonates are petroleum-based and do not biodegrade under anaerobic conditions.

Anionic surfactants include structures such as sodium lauryl sulfate (SLS), sodium dodecylbenzene sulfonate (SDBS), and alkyl-ether sulfates such as sodium laureth sulfate. Other anionic surfactants consist of fluorinated materials such as perfluorooctanesulfonate and perfluorobutanesulfonate. However, fluorinated surfactants have come under scrutiny due to their resilience in decomposition being sometimes described as forever chemicals.

The HOSO-based anionic functional variations were produced by reacting epoxidized HOSO with a hydroxy acid such as glycolic, lactic, malic, and citric acids. Their production was performed to increase hydrophilicity through increased carboxylate salt groups, while maintaining a stable ether bond with the hydroxy acid. The reaction scheme for the HOSO-lactate is shown in Figure 12.2. The methyl ester of epoxidized HOSO is used for the reaction, as it decreases the amount of side reactions and the final intermediate viscosity. The ethyl or methyl ester of the hydroxy acid is used to protect the epoxide group from reacting more favorably with the carboxylic acid instead of the hydroxyl group. The final step not shown is the hydrolysis with sodium or potassium hydroxide and the removal of methanol and/or ethanol.

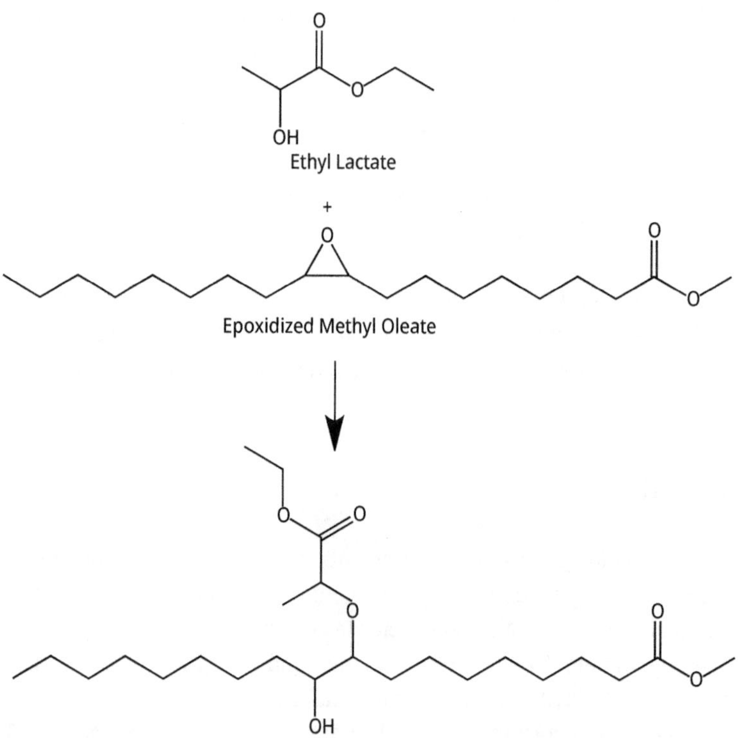

Figure 12.2: Reaction of ethyl lactate with epoxidized methyl HOSO.

An alternative reaction scheme allows for greater reaction speed to produce similar products more efficiently. The hydroxy acid esters can first be reacted through alkoxylation, preferably with ethylene or propylene oxide, to increase the reactivity of the hydroxy acid with the epoxidized SBO. The final structure produced then contains a polyalkoxyl ether chain between the hydroxy acid and the fatty acid.

When substituting the hydroxy acid-soy derivatives for SDBS, we observed that all formulations performed poorly in stain removal testing (ATSM D4265). Stain removal testing is performed in triplicate by utilizing reflectance of stain standards, before and after one laundry cycle. What was observed was that the citrate-soy candidate improved the removal of the stains from coffee, EMPA 112 (cocoa), EMPA 116 (blood, milk, and ink), and red wine. The lactate-soy candidate significantly improved the removal of makeup stains. We later found that by substituting the hydroxy acid-soy candidates for the nonionic portion of the standard formulation allowed for great improvements in overall stain removal.

In using similar reaction schemes (ring opening of epoxide, followed by secondary reactions), other anionic soy-based derivatives can be formed. For example, the epoxide group can easily be reacted with any polyethylene glycol (PEG) group. After the initial ring opening and purifying of the intermediate, the remaining primary hydroxyl group on the PEG ether has greater reactivity than the secondary and hindered hydroxyl at the original olefin site. Through reaction of the intermediate with chlorosulfonic acid, followed by purification and saponification, a structure containing both the sulfonate salt and carboxyl salt is produced. The final structure is shown in Figure 12.3.

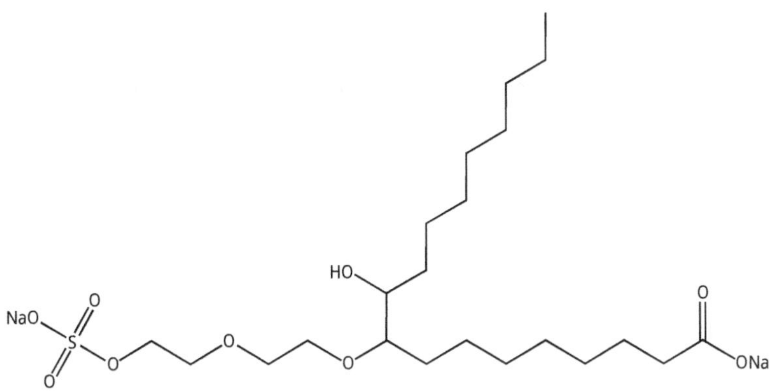

Figure 12.3: Candidate 16.

Other soy-based aromatic sulfonate structures were also developed; however, many were produced with low yields and were not evaluated further. Another candidate was produced through alternative chemistry versus epoxidation. Candidate 17, seen in Figure 12.4, was produced by reacting hydroformylated and hydrogenated methyl HOSO with chlorosulfonic acid, followed by saponification.

Evaluations were performed for laundry applications using ASTM D4265, where results of ±10% are statistically equivalent. When substituting Candidate 16 for SDBS in a standard formulation (Laundry Formulation 1), it was found that performance was acceptable when replacing 50% of the SDBS. When replacing 100% of the SDBS, the formulations were no longer in the 10% range. The results are given in Table 12.1.

The standard was held at 100%, while the other mixtures were seen performing below 100%. A direct comparison to SLS was also recorded. It was found during this research that sulfonates and sulfates were the greatest factor in attacking grass stains. However, we were surprised at the improvement in grass removal, when substituting at 50% with SDBS versus the 100% substitution. This result showed that while we could not replace the SDBS by 100%, there may be a blend levels that are advantageous to overall performance.

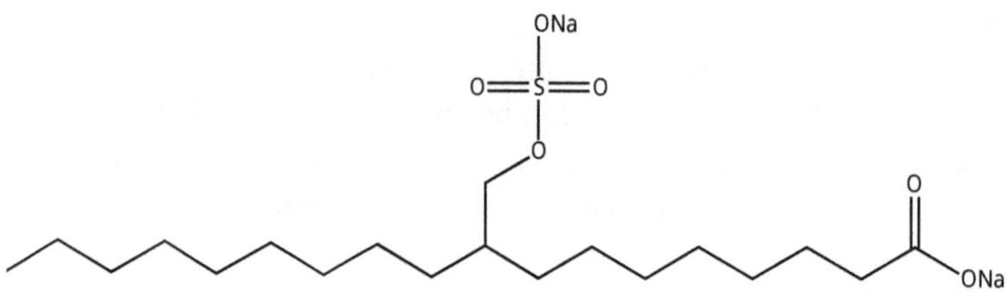

Figure 12.4: Candidate 17.

Table 12.1: Stain removal results when replacing SDBS.

Soil	Substrate	Standard (120 mL)	Candidate 16C 50% SDBS replacement (120 mL)	Candidate 16C 100% SDBS replacement (120 mL)	SLS bio-standard (normalized) (120 mL)	Sig. dif.
Clay	Cotton	58.8	53.8	58.6	60.8	6.9
Coffee	Cotton	51.2	51.5	44.8	42.2	8.6
Dust sebum	Cotton	56.3	44.1	40.4	44.7	5.6
EMPA 101 (olive oil)	Cotton	8.4	8.6	8.4	8.9	2.4
EMPA 112 (cocoa)	Cotton	7.1	4.4	11.1	9.3	5.9
EMPA 116 (blood, milk, ink)	Cotton	22.1	18.8	18.4	32.4	2.2
Grass	Cotton	59.1	72.8	48.3	33.9	6.2
Make up	Cotton	34.9	25.4	18.1	20.8	7.8
Red wine	Cotton	34.9	38.7	36.1	46.1	5.6
Spaghetti	Cotton	75.6	72	74.5	85.2	7.8
Overall soil		418.4	390.1	358.7	384.2	59.0
% of best		Best	93%	86%	92%	

Recent patent developments [12] further expand on anionic candidates, such as those comprising at least 70 wt% to 99 wt% oleic acid or other oleate, comprising a hydroformyl sulfonate based on functionalized epoxidized soybean oil. These include high oleic soybean oil comprising a hydroxyalkane sulfonate.

12.3 Nonionic surfactants

Nonionic surfactants are the second largest surfactant group manufactured globally by volume, at about 35% [4]. Fatty alcohol ethoxylates, alkyl phenol ethoxylates, fatty acid esters, and alkylpolyglucosides fall into this category. The hydrophilic groups are covalently bonded to the aliphatic tail, allowing for controlled miscibility and dispersibility. This class of surfactants tends to be less sensitive to water hardness and pH changes but do suffer in applications with temperature changes.

The main nonionic surfactants in laundry applications are based on ethoxylates of dodecanol (lauryl alcohol). As stated earlier, palm kernel and coconut fatty acids are rich in lauric acid (C_{12}), which is converted to alcohol through hydrogenation of the carboxylic acid. That hydroxyl is then alkoxylated at various chain lengths, depending on the application needs. Similar reaction processes have been used with stearyl (C_{18}) and cetyl alcohol (C_{16}). These longer-chain alcohols require greater amounts of ethoxylate groups to achieve water miscibility versus the lauric derivatives.

For nonionic surfactant replacement, we utilized laundry formulation 1, which contained a nonionic surfactant based on C_{12} alcohol with approximately 15 ethoxylate groups. For this reason, Tergitol® 15-S-15 was used. As mentioned above, our anionics were found to be good replacements for the nonionic portions of the formulation. We also found in other formulation evaluations that the stain correlating to ethoxylate groups was dust sebum removal; also known as ring around the collar. Our initial candidates for replacing nonionics were the epoxidized HOSO that was ring-opened with triethylene glycol, PEG 200, or PEG 400, followed by saponification of the triglyceride. Candidate 4 (the result of ring-opening HOSO with triethylene glycol, followed by saponification) can be seen in Figure 12.5. The benefits to producing through this reaction scheme are that no alkoxylation is being performed, therefore removing the concern for dioxane formation. What was found in laundry testing was that there wasn't significant improvement when going from three ethoxylates to approximately eight ethoxylates. The three ethoxylates soy product is water-miscible, so no increase in the amount of ethoxylate groups is needed to achieve solubility.

Later, other derivatives were also produced where alcohol amines were reacted with the intermediate to amidify versus the saponification route. An example of this can be found in Figure 12.6. The benefit is that the material will be less affected by formulation pH changes. However, it was not evaluated in laundry applications and in

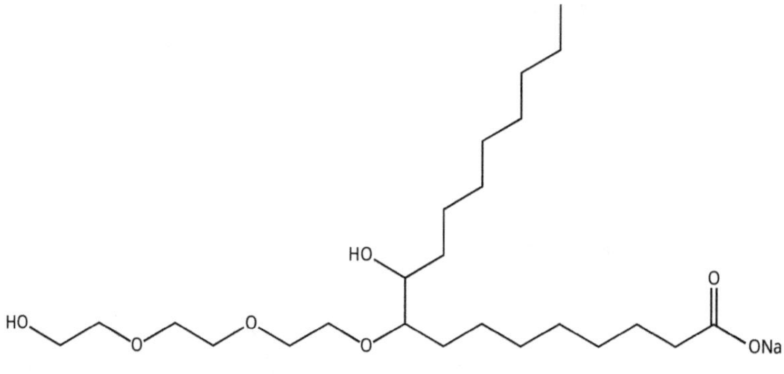

Figure 12.5: Candidate 4.

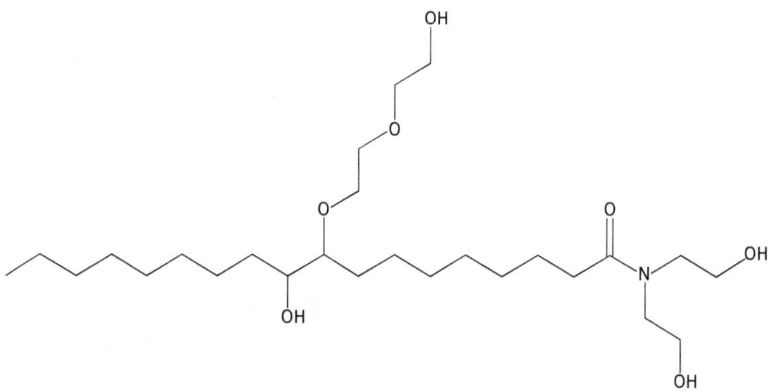

Figure 12.6: Candidate 52.

hard surface cleaning applications. A limited number of glucosides based on this structure were produced but tested in other applications such as fracking fluid additive.

A similar candidate (Candidate 22E) was found to perform well in hard surface cleaning formulations. Candidate 22E is produced by ring opening the epoxidized HOSO with ethanol, followed by amidifying with diethanolamine (DEA). The results are presented and discussed in the "Hard surface" section. The benefit was that the candidate achieved statistical equivalence in stain removal as a drop-in replacement for cocamide DEA. Cocamide DEA is the result of amidification of coconut fatty acids with DEA, which is rich in C12 fatty acid.

Patent advancements [12] introduce additional nonionic structures, such as those derived from epoxidized high oleic soybean oil reacted with a hydroxyalkyl group-containing reduced heteroatom, yielding intermediates like Candidate 25 Intermediates 1 and 2. These feature a hydroxyl group and a carbonyl or amide, providing enhanced performance in general purpose cleaning, with bio-based content exceeding 95%. For instance, a nonionic candidate comprising a polyalkylene glycol could be methoxy, ethoxy, propoxy, or other, maintaining low foam and high soil removal.

12.4 Cationic surfactants

Cationic surfactants typically have excellent antibacterial properties, provide good corrosion protection, and can be good demulsifiers [5]. Quaternary ammonium compounds and pyridines fall into this category, and the surfactants tend to be pH dependent. They also consist of primary, secondary, or tertiary amines. Some examples of these types of surfactants are cetrimonium bromide and cetylpyridinium chloride. A soy-based example of this is our Candidate 12 (see Figure 12.7). This compound is produced by reacting epoxidized HOSO, in the presence of excess ethylene diamine in an autoclave. The product can be achieved with or without catalyst. Candidate 12 was found to nonbiodegradable due to the antimicrobial nature of the structure, killing the microbes of ready biodegradation testing.

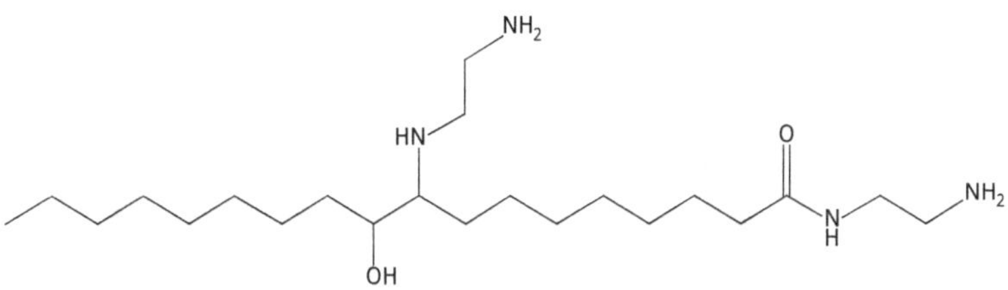

Figure 12.7: Candidate 12.

Work at Battelle has found that this class of surfactants works well in oilfield applications. Testing results can be found later in this chapter. Another soy cationic structure that performed well in oilfield testing is shown in Figure 12.8. This compound started with either stearyl alcohol treated with epichlorohydrin (as seen below), epoxidized SBO, or epoxidized methyl soyate. The soy-based products improve oil recovery, while in some cases, require less surfactant than the concentration currently used in the market.

Candidate 12 was found to improve oil recovery in imbibition tests, reaching up to 30% OOIP. Recent patents [12] also describe cationic variants where the amine was functionalized to a quaternary ammonium salt. These products contain highly antimicrobial characteristics.

12.5 Amphoteric surfactants

Finally, the last type of surfactants is amphoteric surfactants, also known as zwitterionic surfactants. These surfactants can behave as a cation or anion, depending on pH. They consist of structures such as phosphatides and lauroamphoacetates, have low reactivity with other materials, and are increasingly used in personal care products such as baby shampoo and facial cleanser. The cationic portion can be based on primary,

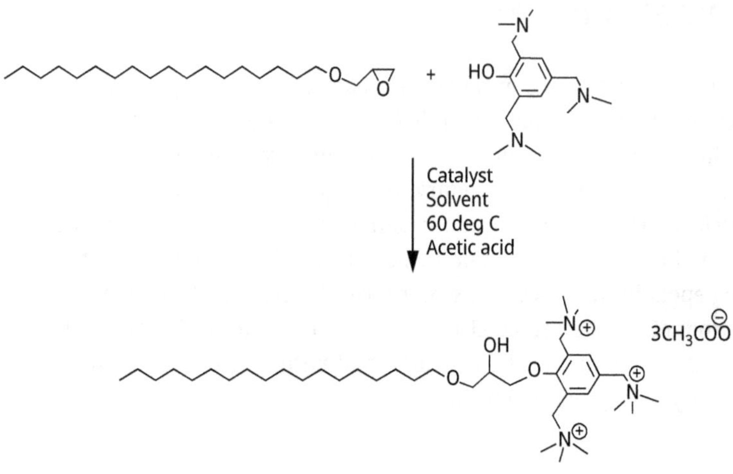

Figure 12.8: Reaction scheme for cationic soy surfactant [6].

secondary, or tertiary amines or quaternary ammonium cations. The anionic part can vary and include sulfonates such as hydroxysultaines and cocamidopropyl betaine.

As mentioned earlier, utilizing soy as the substrate to replace coconut fatty acid derivatives requires epoxidation and ring opening or hydroformylation and hydrogenation to the alcohol. The reaction scheme in Figure 12.9 is an example of that used to create a soy-based betaine (Candidate 43).

Two benefits were found in the soy betaines. The first benefit was that we were able to directly replace SDBS in standard laundry formulation 1, with statistically equivalent performance. The stain removal results are shown in Table 12.2.

The second benefit was found in our laundry standard formulation 2. In this formulation, both commodity and HOSO-based betaines performed well as drop-in replacements for cocamidopropyl betaine. The results are shown in Table 12.3.

Candidates 43 and 48 were HOSO and commodity soy-based, respectively. Both are produced by first reacting the epoxides with ethyl cellosolve, followed by the same chemistry used to produce cocamidopropyl betaine and used as a weight-to-weight drop-in replacement. The Soy Mix 1 was based on laundry formulation 1 and consisted of a blend of soy surfactants, including a soy-based builder.

Patent extensions [12] detail amphoteric candidates like Candidate 25, an amphoteric replacement for cocamide DEA in vinyl tile and wallboard cleaning (hard surface cleaning), showing 102.3% of best performance in low-solids formulations. Structures include a C_{18} carbon chain with a zwitterionic moiety, such as a sulfonate or carboxylate, paired with amine, achieving high stability and low foam.

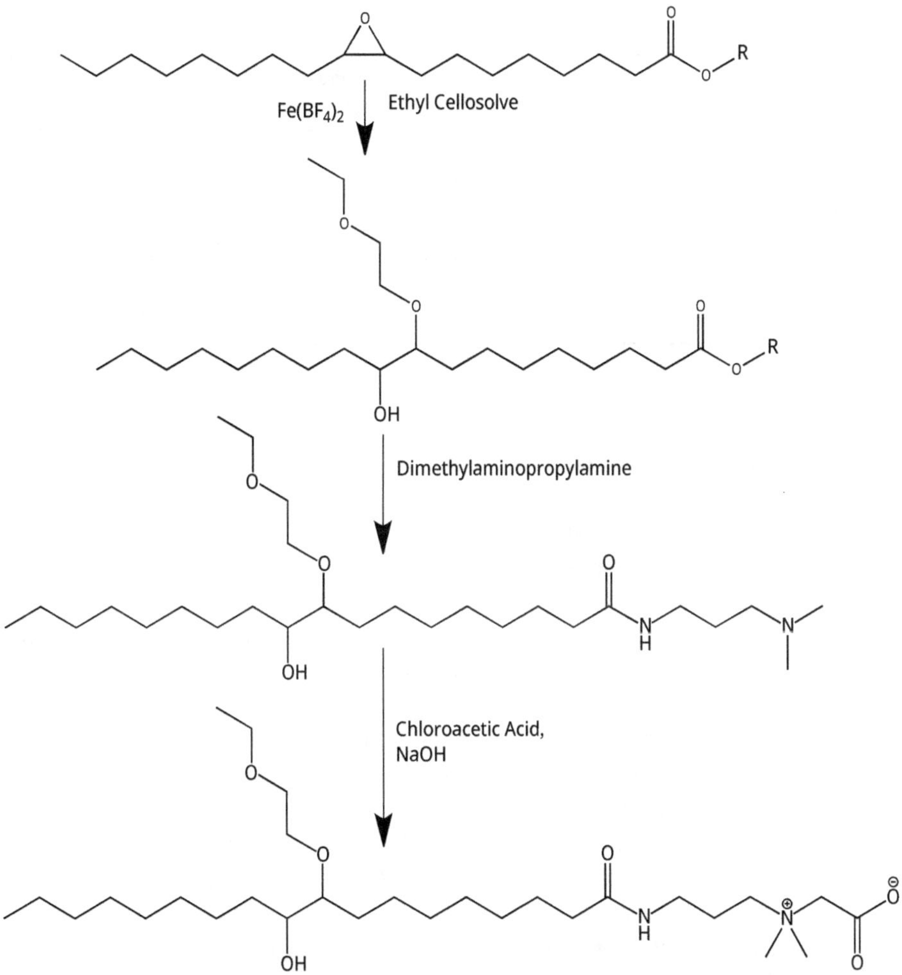

Figure 12.9: Reaction scheme for the production of Candidate 43.

Table 12.2: Replacement of SDBS with soy-betaine.

Soil	Substrate	Standard (120 mL)	Candidate 25, 100% replacement of SDBS (120 mL)	Sig. dif.
Clay	Cotton	56	50.5	3.1
Coffee	Cotton	42.5	43.7	2.6
Dust sebum	Cotton	64.3	62.1	6.2
EMPA 101 (olive oil)	Cotton	8.7	11.1	1.7
EMPA 112 (cocoa)	Cotton	1.1	2	5.2
EMPA 116 (blood, milk, ink)	Cotton	12.8	10.3	1.7

Table 12.2 (continued)

Soil	Substrate	Standard (120 mL)	Candidate 25, 100% replacement of SDBS (120 mL)	Sig. dif.
Grass	Cotton	47	42.6	1.6
Make up	Cotton	59.2	63.7	6.5
Red wine	Cotton	31.5	29.8	0.9
Spaghetti	Cotton	85	79.2	3.5
Overall soil Removal totals		408.1	395	
% of best		Best	97%	

12.6 Laundry formulations

Two formulations were used when evaluating the soy surfactants in laundry applications. The first was a very simple formulation shown in Table 12.4. The formulation consists of sodium carbonate as builder, Tergitol™ 15-S-15 as the nonionic surfactant, SDBS as the anionic surfactant, and sodium xylene sulfonate (SXS) as the hydrotrope, to keep everything compatible. This formulation was 39% solids.

The second formulation, seen in Table 12.5, was derived from patent WO2007/068390. This formulation was 18.7% solids and utilized SLS, potassium cocoate, Tergitol™ 15-S-7 and 15-S-3, cocoglucoside, and cocamidopropyl betaine.

Patent-integrated formulations [12] include variations with 100% SDBS replacement using soy betaines, showing 97% of best in overall soil removal, and pilot-scale tests at 30 gal, with Candidate 43 achieving 104% performance against off-the-shelf products. Later formulations based solely on soybean-based surfactants, performed better than the best fully formulated off-the-shelf standard by 108% in warm water evaluation. When evaluating in cold water (70 F), the full soy-mix with added enzymes performed statistically equivalent to the off-the-shelf standard. Finally, a full soy mixture was found to improve softening of fabric at just-below-moderate level. This is due to the C_{18} nature of soy-based surfactants.

Table 12.3: Standards and soy formulations versus off-the-shelf standards.

Soils on cotton	**Off-shelf tide**	**Off-shelf 7th generation**	**Off-shelf Purex**	**Standard 1**	**Standard 2**	**Soy mix 1**	**Candidate 43 (30 gal scale)**	**Candidate 48**	**Sig. dif.**
Clay	63.8	60.0	53.7	61.6	68	57.6	71.8	69.0	3.2
Soils on cotton	Off-shelf tide	Off-shelf seventh generation	Off-shelf Purex	Standard 1	Standard 2	Soy mix 1	Candidate 43 (30 gal scale)	Candidate 48	Sig. Dif.
Clay	63.8	60.0	53.7	61.6	68	57.6	71.8	69.0	3.2
Coffee	54.7	51.4	46.0	52.8	48.3	64.1	48.1	49.5	3.9
Dust sebum	63.9	60.1	53.8	61.7	59.4	60.6	69.2	65.5	5.3
EMPA 101 (olive oil)	8.7	8.2	7.3	8.4	9.7	9.9	8.7	7.8	0.9
EMPA 112 (cocoa)	2.6	2.4	2.2	2.5	3.8	11	5.1	6.0	3.2
EMPA 116 (blood, milk, ink)	8.8	8.3	7.4	8.5	7.7	15.6	10.4	10.6	0.9
Grass	55.8	52.5	47.0	53.9	47.2	45.4	59.9	53.6	4
Make up	59.6	56.0	50.1	57.5	71.3	54.4	61.1	55.2	9.8
Red wine	30.8	28.9	25.9	29.7	26.9	39.1	27.5	25.5	2.4
Spaghetti	86.0	80.9	72.3	83	82.9	84.7	80.6	78.7	2
Overall soil	434.7	408.8	365.7	419.6	425.2	442.4	442.4	421.4	
Removal totals									
% of best	102%	96%	86%	99%	100%	104%	104%	99%	

(continued)

Table 12.3 (continued)

Soils on cotton	Off-shelf tide	Off-shelf 7th generation	Off-shelf Purex	Standard 1	Standard 2	Soy mix 1	Candidate 43 (30 gal scale)	Candidate 48	Sig. dif.
Coffee	54.7	51.4	46.0	52.8	48.3	64.1	48.1	49.5	3.9
Dust sebum	63.9	60.1	53.8	61.7	59.4	60.6	69.2	65.5	5.3
EMPA 101 (olive oil)	8.7	8.2	7.3	8.4	9.7	9.9	8.7	7.8	0.9
EMPA 112 (cocoa)	2.6	2.4	2.2	2.5	3.8	11	5.1	6.0	3.2
EMPA 116 (blood, milk, ink)	8.8	8.3	7.4	8.5	7.7	15.6	10.4	10.6	0.9
Grass	55.8	52.5	47.0	53.9	47.2	45.4	59.9	53.6	4
Make-up	59.6	56.0	50.1	57.5	71.3	54.4	61.1	55.2	9.8
Red wine	30.8	28.9	25.9	29.7	26.9	39.1	27.5	25.5	2.4
Spaghetti	86.0	80.9	72.3	83	82.9	84.7	80.6	78.7	2
Overall soil	434.7	408.8	365.7	419.6	425.2	442.4	442.4	421.4	
Removal totals									
% of best	102%	96%	86%	99%	100%	104%	104%	99%	

Table 12.4: Laundry standard formulation 1.

Ingredient	Function	Standard formulation 1 mass (g)
Water	Diluent	305
Sodium xylene sulfonate	Hydrotrope	50
Sodium dodecylbenzene sulfonate	Anionic surfactant	50
Tergitol™ 15-S-15	Nonionic surfactant	90
Sodium carbonate	Builder	5
Total		500

Table 12.5: Laundry standard formulation 2.

Ingredient	Function	Standard formulation 1 mass (g)	Standard formulation 2 mass (g)
Water	Diluent	305	243.8
Sodium xylene sulfonate	Hydrotrope	50	
Sodium dodecylbenzene sulfonate	Anionic surfactant	50	
Tergitol™ 15-S-15	Nonionic surfactant	90	
Sodium carbonate	Builder	5	
Total		500	300
Sodium lauryl sulfate	Anionic surfactant		15.3
Potassium cocoate	Anionic surfactant		6
Tergitol™ 15-S-7	Nonionic surfactant		11.7
Coco glucoside	Nonionic surfactant		9
Cocamidopropylbetaine	Amphoteric surfactant		13.2
Tergitol™ 15-S-3	Nonionic surfactant		1

12.7 Other applications

12.7.1 Hard surface cleaning

A wide variety of functionalities are used in hard surface cleaning surfactants. The cleaners are used to remove dust, dirt, and stains. They typically consist of nonionic surfactants but can also include anionic and amphoteric surfactants. One hard surface surfactant is that of cocamide DEA. The concern with cocamide DEA is that it is listed in California Prop 65 as a known cancer-causing agent.

We initially produced candidates for the main purpose of replacing cocamide DEA in a standard formulation. The soy candidate 22, mentioned earlier, was created to contain chemistry similar to that of cocamide DEA. The formulations 1 and 2 evaluated are shown in Table 12.6. Formulation 1 was a high builder formulation that contained 30% solids, while Formulation 2 was only 5% solids. In formulation 2, a builder based on soybean flour was utilized.

Table 12.6: Standard formulations for hard surface cleaning.

Ingredient	wt%.	Function
Formulation 1		
Tetrapotassium pyrophosphate	20	Builder
Cocamide DEA	8	Surfactant
Sodium xylene sulfonate	2	Hydrotrope
Water	70	Diluent
Formulation 2		
Tergitol™ 15-S-15	2	Surfactant
Sodium xylene sulfonate	2	Hydrotrope
Soybuilder 18	1	Builder
Water	95	Diluent

The evaluations are given in Table 12.7(a). For Formulation 1, Candidate 22 (ethanol ring-opened diethanolamide) and Candidate 1 (soy-citrate) were considered statistically equivalent to the cocamide DEA standard. Candidate 4 (soy-triethylene glycol) and Candidate 13 (soy-EDTA) were just outside the 10% window. While Candidate 22 performed well, the other soy candidates are interesting, as they don't contain the diethanolamide associated with the Prop 65 concern.

For the low solids' formulation (Table 12.7(b)), Candidates 25 (soy-betaine) and 1 both performed exceptionally well. However, more testing is required as less ethoxylate groups are typically used for the nonionic surfactant component in hard surface cleaners.

Table 12.7a: Stain removal results for hard surface cleaning.

Soil	Type	Substrate	Cocamide DEA	Can. 22	Can. 1	Can. 4	Can. 13	Sig. dif.
(b)								
ASTM D4488-A3	Grime	Vinyl tile	76.5	74.2	66.7	63.9	72.5	7.2
ASTM D4488-A5	General purpose	Vinyl tile	89.6	89.3	83.6	85.1	83.7	4.9
ASTM D4488-A2	Kitchen grease	Wallboard	35.6	31.5	31.2	28.3	19.8	3.4
Overall soil			201.7	195	181.5	177.3	176	

Table 12.7b: Stain removal results for hard surface cleaning.

Soil	**Type**	**Substrate**	**Cocamide DEA**	**Can. 22**	**Can. 1**	**Can. 4**	**Can. 13**	**Sig. dif.**
Removal totals								
% of best			100%	96.70%	90.00%	87.90%	87.30%	
(b)								
Soil	Type	Substrate	Tergitol™ 15-S-15	Can. 25	Can. 1	Can. 4	Sig. dif.	
ASTM D4488-A3	Grime	Vinyl tile	33.7	39.9	22.5	6.5	3.2	
ASTM D4488-A5	General purpose	Vinyl tile	23.6	22.1	21	23	7.3	
ASTM D4488-A2	Kitchen grease	Wallboard	6.7	3.5	20.1	3.6	8.6	
Overall soil			64.0	65.5	63.6	33.1		
Removal totals								
% of best			100%	102.30%	99.40%	51.70%		

12.7.2 Oilfield application testing

An imbibition test is a comparison of the imbibition potential of water and oil into a rock. The wettability of the rock is determined by which phase imbibes more [7]. Many factors lead to good performance in oil removal from rock deposits. We initially performed a rough evaluation to determine candidates that were effective in causing separation between the oil-water phases. Candidates 10, 12, and 13 demonstrated superior phase separation in oil-water mixtures Figure 12.10.

We next focused on the overall imbibition testing and contact angles of the best candidates. Imbibition results are shown in Figure 12.11. Candidate 12 was the best performer, reaching 31% original oil in place (OOIP), which correlates to 72% improvement in oil recovery.

Contact angle with two oil basins were evaluated: Turner and Niobrara. The Turner results are shown in Figure 12.12. The unique property was the diminished contact angle at the 1 gpt (gallon per trillion) level.

Figure 12.10: Miscibility testing of heavy oil and water.

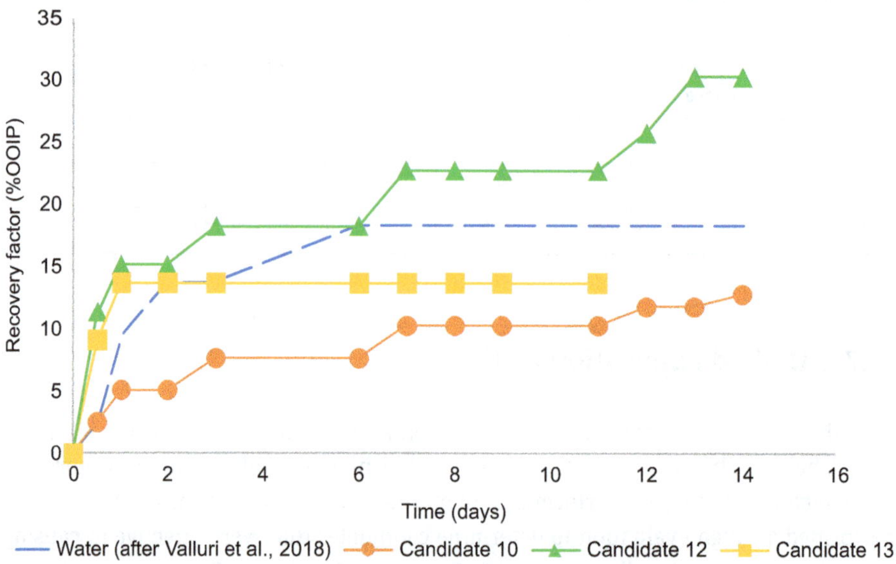

Figure 12.11: Imbibition testing.

12.7.3 Adjuvant-leaf wetting agents

Crop adjuvants are materials that are added to pesticide formulations to enhance their effectiveness and properties of application. Work focused on evaluating multiple formulations where soybean-based surfactants replaced 100% of the commodity surfactants. The soy materials performed best in the property of leaf wetting. In adjuvant formulations, a wetting agent is added into the spray mixture to reduce the surface tension of spray droplets, which prevents droplets from beading up on the leaf's waxy surface. Therefore, the wetting agent lowers spray droplet surface tension, which enables the spray to increase the leaf area covered by each droplet. This can be seen by

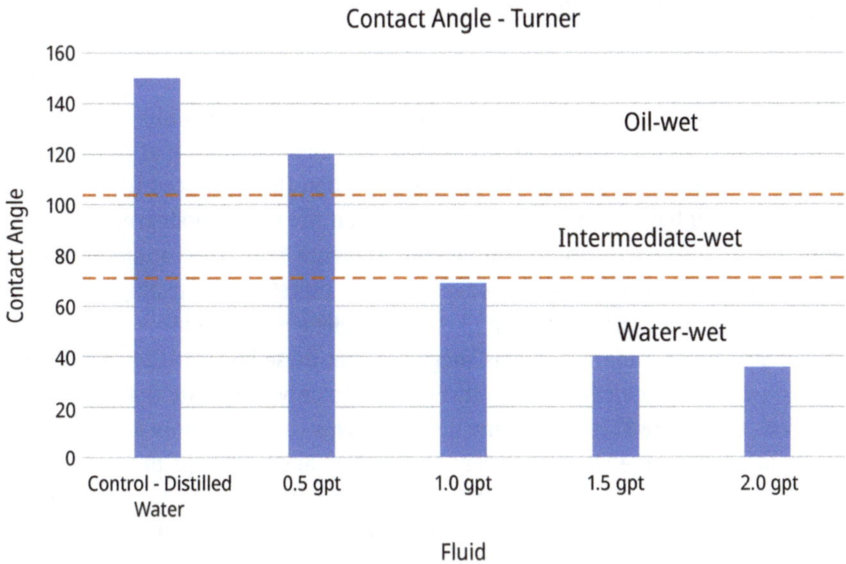

Figure 12.12: Contact angle with Turner basin oil.

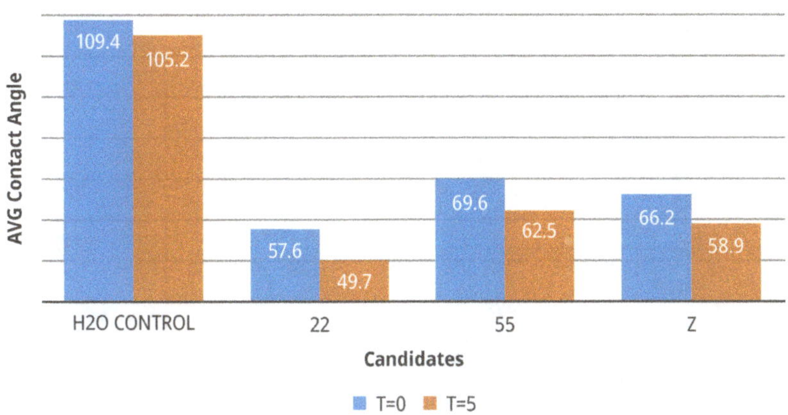

Figure 12.13: Average contact angle on wax paper versus time.

the differences in contact angle on wax paper at times 0 and 5 min in Figure 12.13. It was found that Candidate 22E (ethanol ring-opened epoxidized HOSO DEA) performed best in this application.

12.8 Biodegradation testing

OECD 310 is an aerobic biodegradation test that predominantly measures ready biodegradability by the evaluation of CO_2 in sealed vessels. In the OECD 310 test method, the CO_2 evolution resulting from the ultimate aerobic biodegradation of the test substance is determined by measuring the inorganic carbon (IC) produced in sealed test bottles [8]. Four soy-based candidates were evaluated for ready biodegradation testing. Candidate 1 (soy-citrate) results can be seen in Figure 12.14. The candidate is very close to the 10-day window quickly achieves the 60% threshold for testing. However, Candidate 2 (soy-lactate) and Candidate 25 (soy-diethylene glycol betaine) reached levels of about 50% biodegradation. It was suggested that these be run at lower concentrations and will likely achieve the 60% threshold. The only candidate that didn't perform well was Candidate 12 (soy-ethylene diamine). This candidate only achieved 10% biodegradation in the time window observed. The free amine groups present in Candidate 12 tend to kill microbes needed for breaking down the surfactant. While this opens up the possibility of preservatives, it does not perform well in this testing.

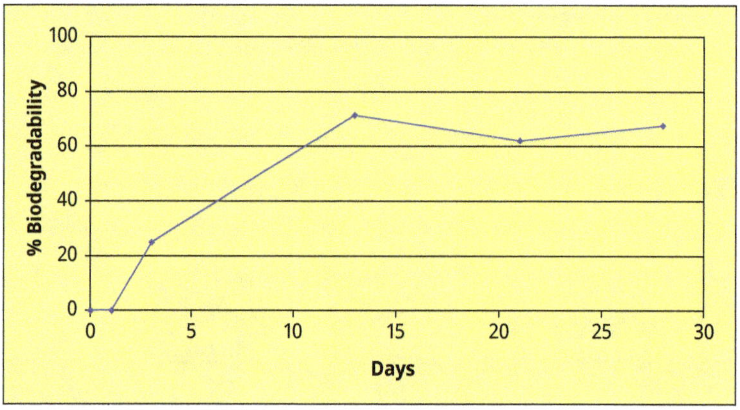

Figure 12.14: OECD 310 test results of Candidate 1.

12.9 Surface tension testing

Surface tension is the tendency of liquid surfaces at rest to shrink into the minimum surface area possible. Surface tension (dyn/cm) is what allows objects with a higher density than water such as razor blades and insects (e.g., water striders) to float on a water surface without becoming even partly submerged [9]. Surface tension is a good gauge for surfactant performance in cleaning applications, as lower surface tensions typically lead to better rinsing of stains on a substrate. Table 12.8 shows that many of the soy surfactants are in the same range as those used and accepted in the market.

When switching to soy-betaine-based candidates, we were able to drop the surface tension to below the level of 30 dyn/cm.

Table 12.8: Comparison of soy surfactants versus standards.

Sample	ST	Standard	ST
Water	71.2	SDBS	34
Candidate 1	32.5	Pluronic® L64	37
Candidate 2	36.6	Tergitol™ 15-S-3	
Candidate 4	34.5	Tergitol™ 15-S-7	
Candidate 5	33.6	Tergitol™ 15-S-15	36
Candidate 10	32.0	Tergitol™ 15-S-40	45
Candidate 11	36.5		
Candidate 12	32.4		
Candidate 13	35.2		
Candidate 14	32.0		
Candidate 15	37.3		
Candidate 16	39.8		
Candidate 25	28.2		
Candidate 29	26.8		
Candidate 32	27.1		
Candidate 39	27.1		
Candidate 43	25.5		

12.10 Critical micelle concentration

In colloidal and surface chemistry, the critical micelle concentration (CMC) is defined as the concentration of surfactants above which micelles form and all additional surfactants added to the system will form micelles [10]. The micelles can pack in various ways but are typically spherical. We determined the CMC values (Table 12.9) of select candidates and found that many were in the expected 0.01–0.5 range. For CMC testing, the lower the value the better.

12.11 Surfactant foam testing

The Ross-Miles method is used for measuring the foamability of surfactant solutions and the stability of the foam produced, which is based on height measurement. It is named after the authors John Ross and Gilbert D. Miles [11]. The foam test determines a foam profile over time for various functionalities of surfactants. Some select results are shown in Figure 12.15. Many of our anionic surfactant candidates followed a similar pattern to the standard nonionic Tergitol™ surfactant. Candidate 16C is interesting,

Table 12.9: CMC results of select candidates.

From plots	Candidate 1	Candidate 2	Candidate 3	Candidate 4	Candidate 5	Candidate 10	Candidate 11	Candidate 12	Candidate 13	Candidate 14	Candidate 15	Candidate 16
CMC approximate	wt%	wt%	wt%	wt%	wt%	wt%	wt%	wt%	wt%	wt%	wt%	wt%
Concentration	0.1	0.1	0.01	0.5	Weak transition	0.05	0.1	0.01	0.1	0.05	0.1	0.01
1% Turbidity *NTU	1,234.8	184.5	187.3	200.2	65.0	26.1	138.2	466.1	153.0	105.8	351.2	731.0

*NTU approximation by UV-vis

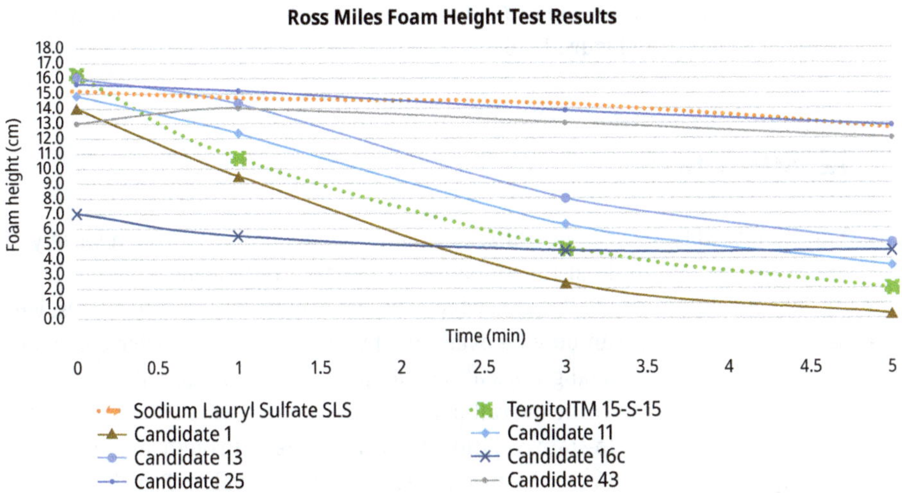

Figure 12.15: Ross-Miles foam height test results.

as it is a sulfonate that holds its foam height over extended time but starts at half the height of SLS at the same concentration. This potentially has implications where low foam is needed such as high efficiency laundry applications. The soy-betaine candidates contained similar foam patterns to SLS.

Having achieved good Ross-Miles foam height results, the surfactants were next analyzed for bubble concentration versus time on a Kruss analyzer. In the case of soy-surfactants, it was found that even though the foam height remained constant, the

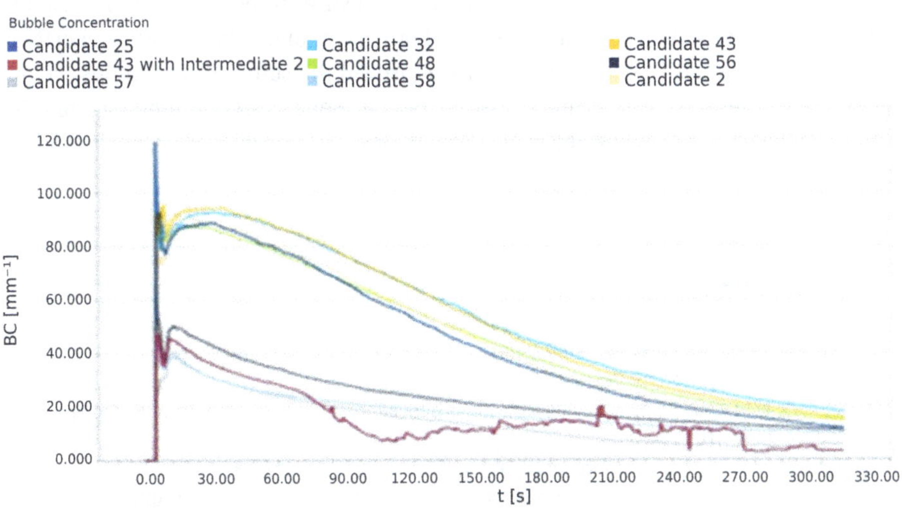

Figure 12.16: Bubble concentration versus time test results.

bubble concentration reduces over time by merging and forming larger bubble sizes. This behavior is shown in Figure 12.16.

12.12 Summary

HOSO and commodity SBO can be utilized as platform chemicals to create a variety of surfactants with a multitude of functionalities. C18-based fatty acids have not traditionally been used as surfactants, as they typically have lower water miscibilities, versus shorter chains, which can only be overcome by blending with shorter chain surfactants. Insertion of functional groups at the olefin site not only alters the molecules miscibility but also can affect the overall performance of the surfactant. This effect is shown through structure-to-stain relationships, identified through evaluation in cleaning applications and further by properties such as CMC value, Ross-Miles foam height testing, and determination of lowering of surface tension. Examples of structure-to-stain relationships are lactate to make-up, sulfonates/sulfates to grass, PEGs to dust sebum, and citrate to coffee, cocoa, blood, milk, ink, and red wine. A fully soy-based laundry formulation has been shown to perform equivalently to off-the-shelf fully formulated standards. The properties, such as foam height and surface tension, of many of the soy surfactants are in-line with standards currently in the marketplace made from predominately C_{12} chains.

The ability to perform weight for weight replacement of coconut fatty acid-based surfactants has become increasingly important, as supply line concerns increase. The United States has no domestic source of C_{12} fatty acids, other than petroleum derivatives. These soy surfactants are now ready for industrial trials, showing promise for the future of sustainable green materials in both warm and cold conditions. Further, it was found that the C_{18}-based surfactants improved the softness score to just below moderate softness. This platform chemistry was developed through funding from the United Soybean Board, while soy-chelant technology was developed through funding from Indiana Soybean Alliance.

References

[1] Surfactant. Merriam-Webster.com Dictionary, Merriam-Webster, https://www.merriam-webster.com/dictionary/surfactant. Accessed 27 Jun. 2022.

[2] Garbark, et al. Laundry builders and surfactants derived from bio-based hydroxyacids and epoxides, US Patent: pp. 11168284.

[3] Garbark, et al. Bio-based surfactants, US Patent Application 20220089983.

[4] Kumar A. Global Markets for Surfactant Chemicals and Materials, BCC Research; 2018.

[5] M CP-B, Fernández-Amado M, López-Mahía P, Muniategui-Lorenzo S, Prada-Rodríguez D. Chapter 11 – Surfactants In Cosmetics: Regulatory Aspects and Analytical Methods. In: Salvador A, Chisvert A, editors. Analysis of Cosmetic Products, 2nd ed., Elsevier; 2018. pp. 249–87. ISBN 9780444635082. https://doi.org/10.1016/B978-0-444-63508-2.00011-4.

[6] Lalgudi, et al. Bio-based surfactants, cationic surfactants comprising an ether link, US Patent Application; 20210139409.

[7] Imbibition. Oilfield Glossary, Schlumberger, https://glossary.oilfield.slb.com/Terms/i/imbibition.aspx#:~:text=An%20imbibition%20test%20is%20a%20comparison%20of%20the,which%20phase%20imbibes%20more.%20See%3A%20drainage%2C%20oil-wet%2C%20water-wet Accessed.

[8] OECD 310 – Ready biodegradation – CO2. Situ Biosciences, https://www.situbiosciences.com/biodegradation/oecd-310-ready-biodegradation-co2/#:~:text=OECD%20310%20is%20an%20aerobic%20biodegradation%20test%20that,inorganic%20carbon%20%28IC%29%20produced%20in%

[9] Surface tension – an overview | sciencedirect topics. www.sciencedirect.com. Retrieved 2021-12–30.

[10] IUPAC. Compendium of chemical terminology, 2nd ed. (the “Gold Book”) (1997). Online corrected version: (2006–) “critical micelle concentration”. doi: 10.1351/goldbook.C01395.

[11] Ross J, Miles GD. An apparatus for comparison of foaming properties of soaps and detergents. Oil Soap 1941;99–102. https://doi.org/10.1007/BF02545418.

[12] Garbark, et al. Bio-based surfactants, US Patent Application Publication US; 2025. pp. 0257030. A1, Aug. 14, 2025..

Rachel Yoho*, Michelle Young, Aditi Hoffman

Chapter 13
An innovative microbial future: bioconversion products of microbial electrochemical systems

13.1 Introduction

There are few things in life more valuable than creating something from nothing. Creating something of value from invaluable materials is a rare and special human endeavor. Human history has a rich tradition of seeking to create value-added products from raw materials with little to no monetary value or usefulness. Medieval alchemy sought to turn less valuable metals into gold. Creating wine from water is a popular story in one major world religion. These examples illustrate the desire to do something more than other types of creation that we may encounter every day. Cooking, sculpting, painting, and many forms of creation shape outcomes of beauty and wonder (and valuable end products) from raw materials.

So, the idea of creating something valuable from nearly useless materials is not new. Yet, there are few activities as remarkable and versatile as microbial systems to create a wide range of products or facilitate such a vast diversity of processes. Microbial systems clean up wastewater [1], remediate metals [2], remove hazardous organic compounds [3], synthesize many products from nanomaterials to hydrocarbons [4, 5], recover phosphorous [6], and so much more. The possibilities and applications of a number of designs and systems seem nearly limitless.

When considering the challenges facing humanity, the intertwined climate and energy crises come to mind as the pressing issue, as well as the drivers of several other global pressing needs. The climate crisis elicits a number of potential emotional responses and shapes actions and decision-making [7]. These emotional and intellectual responses have led to numerous innovations in the research sector over the last couple of decades. Barriers to progress stem largely from the perceptions of [8] and implementation of [9] renewable energy technologies rather than the development of the technologies themselves.

***Corresponding author: Rachel A. Yoho,** (formerly) George Mason University, Fairfax, Virginia, USA, (currently) College of Southern Nevada, Las Vegas, Nevada, e-mail: rachel.yoho@csn.edu
Michelle Young, Carollo Engineers, Inc., 390 Interlocken Crescent, Suite 800, Broomfield, CO 80021, USA (did not contribute to second edition)
Aditi Hoffman, Department of Civil Engineering, Kansas State University, 2118 Fiedler Hall, 1701C Platt Street, Manhattan, KS 66502, USA, e-mail: aditip@ksu.edu

 | https://doi.org/10.1515/9783112212141-013

13.1.1 Some recent trends

The advancement of renewable energy technologies, alongside the development of methods for creating valuable products from wastes, bridges the gap in terms of addressing needs, adding clear value, driving sustainability, and demonstrating practical application. Microbial electrochemical systems are one such area of versatile application.

The last two and a half decades have seen the development and advancement of microbial electrochemistry, particularly with practical applications. These practical applications [10] range from energy production [11], wastewater treatment and resource recovery [12], niche energy availability for remote sensing [13], and removal of environmental contaminants [14]. The microbial research community grew in prominence particularly after the potential engineering applications were realized, specifically, the waste-to-energy processes [15]. In these early examples, microorganisms oxidized organic compounds in wastewater to generate energy. This work had important connections not only to decreasing the costs of wastewater treatment but also more broadly to climate needs and the energy crisis [15].

The technologies, broadly, demonstrate a large versatility in terms of application and product output. The designs, materials, and microorganisms then also vary widely. In this work, we go backward, highlighting the products generated and back to specific details about the design. Finally, we wrap up the chapter with a brief overview of a new frontier in microbial technologies and research applications: generative artificial intelligence.

13.1.2 Microbial electrochemical cells

Broadly, the microbial electrochemical cells (MXCs) represent a unique class of technologies relying on microorganisms and spanning wastewater treatment, bioremediation, and energy generation through multiple bioprocesses [16]. Important distinctions in this space are the acronyms occasionally representing different time periods in the development of the field, advancements, and applications. Here, we will use the general term of MXC as the overarching term for a microorganism-based system that generates many different potential products. Other terms that appear in this space are microbial fuel cell (MFC), typically focused on energy output [15] and microbial electrolysis cell (MEC), typically requiring energy input to produce valuable products including hydrogen [17]. Certain applications have their own acronyms as well including the microbial desalination cell (MDC) [18] and plant applications (PMFCs) [19].

Beyond the importance of the related acronyms, other key identifying factors are whether the reaction of interest occurs on the bioanode or the biocathode, the biological processes, and any other applicable living organisms. For example, these include biocathode bioremediation [20, 21], product synthesis [22, 23], and fermentation [24], among others [10].

13.2 Bioanodes

Generally understood as the growth of a film of some sort of living microorganism on the surface of an electrode, the bioanode presents a number of important benefits and challenges to MXCs. Overall, microbial electrochemical systems necessitate the microbial colonization of an anode in order to generate the electrical current [25]. These are the best-known biofilms in the field, yet many unknowns remain. Importantly, MXCs are taking advantage of the "enzymatic machineries" in the varied microorganisms to facilitate reactions not possible in nonbiological engineered systems [15]. The factors that define and shape the formation of bioanodes are still poorly understood [25], even considering their deep history and importance in the field.

The performance of the bioanode shapes the overall performance of the MXC [26]. The many factors affecting the performance of the bioanode are well outside of the scope of this chapter; however, the bioanode itself shows promise in adding opportunity in the field. For example, capacitive electrodes in the anode compartment of an MXC create a situation in which a capacitive bioanode can form. This type of integrated system allows the system to both generate and store renewable energy [27]. Unlike solar panels that work only during the day or wind generation facilitates that have production only when the wind blows, capacitive opportunities for MXCs make the nonfossil fuel energy generation more like nuclear energy – in that it would be less dependent on external or weather conditions – but without the scale of the hazardous waste disposal challenges.

While we certainly do not propose that MXC systems are at the scalable point to replace nuclear energy generation, the similarities in advantage and potential lack of serious waste disposal concerns can be compelling. Improving the capacitance of bioanodes presents an opportunity for improving wastewater treatment, even as this research area continues to develop [28].

The poor understanding of biofilm formation on anodes presents an important challenge in the scale-up of systems [25]. Enrichment processes typically occur in batch systems over multiple cycles until current generation is stable [25, 29, 30].

Early challenges to MXC systems were the use of various mediators in the system. Early iterations of these systems relied on soluble mediator compounds to facilitate electron transfer [31]. These mediators were generally toxic and impossible to use in open systems [31], including potassium ferricyanide [32], anthraquinone 2, 6-disulfonic acid, cobalt sepulchrate [33], thionine [34], neutral red [35], and azure A [36]. Challenges today include the continued need to improve performance of specific applications of MXC systems [37], as well as the field's ongoing needs for demonstrating the ability to scale up through system stability, overall efficiency, and economic feasibility [38].

13.3 Anodophilic microorganisms

Microorganisms with the ability to use a solid material external to the cell as a final electron acceptor in the respiratory process generally are considered to be anodophilic microorganisms. These organisms typically oxidize some sort of organic material (the "food source") in a liquid medium or liquid waste stream and transfer the electrons to the final, external electron acceptor [39].

The mechanisms through which these organisms transfer the electrons outside of their cells vary among microbial species and are also highly contested in the field. Cytochromes, particularly the outer membrane *c*-type cytochromes, are considered as key components of these organisms' ability to respire a final extracellular electron acceptor [40]. Others in the field assert the "metal-like conductivity" of the microbial nanowires as pili [41]. This group further espouses the metal-like conductivity as due to the overlapping π–π orbitals of aromatic amino acids within the nanowires themselves [41].

More widely studied are the specific cytochrome mechanisms, for example, with multiple pathways of electron transport observed depending on the anode potential for *Geobacter sulfurreducens* [42] and *Geoalkalibacter ferrihydriticus* [43]. Specific cytochromes appear essential to potential-dependent extracellular electron transport. For example, *imcH* in *G. sulfurreducens* is an inner membrane cytochrome identified to be required for high redox potential extracellular electron acceptors [44]. Other work indicates "electron conduits" each with several key components: a periplasmic multiheme *c*-type cytochrome, an integral outer membrane anchor, and outer membrane redox lipoproteins [45]. This further supports the necessity of cytochrome-dependent pathways for extracellular electron transport, particularly depending on the applied potential of the anode [42].

Importantly, real-life biofilms are unlikely to be pure cultures. The reality of the extracellular electron transport is much more likely to be some mix among the two proposed mechanisms. Naturally, real life is wildly complex. Mixed culture extracellular electron transport in larger systems and in the environment likely have these complexities and many more.

Here, we highlight several important organisms to this type of work, notably *Geobacter* and *Shewanella* as well as a few related extremophiles with particularly compelling and niche applications.

13.4 *Geobacter*

Among mixed culture systems, members of the *Geobacter* genus are essential in both generating current and decreasing charge transfer resistance in the anode biofilms [25, 46, 47].

Extracellular electron transport is a complex process, with notable advancements in the field occurring only in the last 10–15 years. Essential to these advancements were the study of pure culture biofilms rather than mixed cultures dominated by *Geobacter* or another exoelectrogenic species.

The family, Geobacteraceae, and genus *Geobacter* are deltaproteobacteria considered to be ubiquitous in many environmental settings including soil, sediments, and similar subsurface environments. In these environments, metal reduction is a typical practice [48]. This family, identified just over 30 years ago, provides researchers and practitioners with essential services such as degrading hydrocarbon contaminants, reduction of radionucleotides, toxic metals, and nitrogen. Therefore, these organisms appear as frequently studied and applied components of several bioapplications: bioremediation, bioenergy, and bioelectronics [48].

Among the *Geobacter* species, the most well-known and well-studied is G. sulfurreducens. The first identified, however, was *G. metallireducens*, following investigations of sediments from the Potomac River in the eastern portion of the United States [49]. In the over three decades since, *G. sulfurreducens* emerged as the more studied organism, particularly as the first to have a sequenced genome and more understood genetic system [50]. For *Geobacter*, the outer membrane OmcE and OmcS outer membrane *c*-type cytochromes are understood to mediate the extracellular electron transfer mechanisms [40].

Electroactive biofilm development on anodes depends on the energy obtained from electron transfer to the electrode [51]. Enrichment with *Geobacter* species accelerates electrochemical reactions in MFCs [51]. Acetate-fed pre-enrichment of anodic communities enhances *Geobacter* abundance in biofilms, thereby improving coulombic efficiency in MFCs [52]. Recent studies have shown that overexpression of nanowire proteins in *G. sulfurreducens* like PilA, OmcZ, OmcS, and OmcT (see earlier discussion about mechanisms) decrease the internal charge transfer resistance and improve the biofilm formation on anodes, thereby enhancing the current production [53]. The *Geobacter*-dominated biofilm species are in fact so resilient, that they have been observed to generate current densities despite a lack of anode potential in single chamber MFCs providing reduced operational costs and simplifying the system [54].

13.5 *Shewanella*

Aside from the rich history of *Geobacter*, the other most well-known organism in this field is *Shewanella*. Overall, the genus has more than 50 known species across different types of marine environments and ecosystems [55]. These are aquatic microbes found worldwide [56], much like *Geobacter*. What sets these organisms apart is their diversity in respiratory pathways and ability to survive, and thrive, at low temperatures [48].

The biotechnology applications of *Shewanella* are diverse. From energy generation through biocatalysis to remediation of chlorinated compounds, radionucleotides, and pollutants to the environment, *Shewanella* presents a complex physiology [48]. Interestingly, beyond the heavy metal remediation and MXC applications, members of this same genus are becoming more well-known as human pathogens, particularly from occupational or marine exposure [55]. These human illnesses typically are skin infections, middle ear inflammation and infections, and blood infections (bacteremia) [55]. Research indicates that the human infections are limited to *Shewanella algae* [55], so risk in biotechnology research likely is negligible.

Shewanella oneidensis stands out among the genus for research and biotechnology applications. For *S. oneidensis* MR-1, the MtrC, and OmcA are understood to be the *c*-type cytochromes involved in extracellular electron transport [40]. Overall, multiheme cytochromes exist in a number of phylogenetically diverse microorganisms and enable long-range electron transfer (tens of nanometers) [57]. These abilities have important global-scale impacts in the biogeochemical cycling of N, S, and Fe [57]. More specifically, *S. oneidensis* is well known to be able to use ferric iron [Fe(III)], a popular research application choice, as an external final electron acceptor [58, 59].

Interestingly, flavins are also found to have a role in extracellular electron transport for *Shewanella*. In a study of *S. oneidensis* MR-1 and MR-4 grown on poised anodes, the intact biofilms revealed redox-active molecules inside the biofilms. These were riboflavin and riboflavin-5'-phosphate, with riboflavin the dominant molecule for established biofilms. Removal of the biofilm from the electrode also indicated that riboflavins remained on the surface of the electrode as well as could adhere to other materials of interest in these systems like Fe(III) and Mn(IV) oxy(hydro)oxides [60]. Other molecules of interest are water-soluble molecules secreted by the organism. For example, *S. oneidensis* secretes water-soluble molecules acting as electron shuttles or Fe(III) complexing ligands, quinones, and quinone moieties are observed to accelerate Fe(III) reduction by functioning as mediators, and extracellular appendages ("nanowires") have also been observed [58]. By comparison, *Geobacter* are not known to produce any electron shuttles [22, 31].

Shewanella as a genus overall is unique and varied, particularly within this context of interest to biotechnology applications and bioproduct development. Much research has been done and is ongoing on key species, particularly *S. oneidensis*.

Shewanella species are notable for their unique ability to transfer electrons generated during intracellular metabolism to external solid electron acceptors, including manganese and iron oxides, as well as electrodes. This mechanism, called extracellular electron transfer (EET), enables *Shewanella* to contribute to biogeochemical cycles and presents potential for microbial electrochemical technologies [61]. EETs can be classified based on the direction of electron flow into outward and inward transfer based EET. However, limited substrate metabolism and bidirectional EET have hindered practical application of *Shewanella*, despite having unique physiological traits with immense diverse potential. Recent studies have indicated that genetic engineering has made it possible to rewire native metabolism in *Shewanella* cells to achieve specific goals like production of biofuels, chemicals, or environmental remediation agents [62, 63].

Shewanella-based biosensors are known for their ability to detect Shewanella's EET ability, on-site and real-time monitoring abilities [64]. Organic nutrients like biological oxygen demand and volatile fatty acids can be used by Shewanella as electron donors for catabolism, which generates EET currents to external electrodes. Similarly, toxic pollutants can compromise the current generated [65, 66]. Beyond biosensor development, machine learning is emerging as a valuable tool in the metabolic engineering of *Shewanella* strains. While its application in this context remains at an early stage, successes in other microbial chassis demonstrate significant potential for optimizing metabolic pathways [67, 68].

13.6 Other species

Other relatives of *Geobacter* have gained prominence in pure culture studies of anode biofilms. In particular, these have been the "extremophiles" and are all members of the δ-proteobacteria [10] such as *G. ferrihydriticus* [69, 70] and *Geoalkalibacter subterraneus* [69, 71]. Even outside of the δ-proteobacteria, other organisms are able to respire anodes. *Thermincola ferriacetica* is an example of this from the class *Firmicutes* [10, 72–74]. As the name implies, *Thermincola ferriacetica* is best known for its thermophilic existence and ability to respire an anode while under high temperatures.

13.7 Biocathodes

As compared to early and ongoing work in the field that grows biofilms on anodes, the biocathode has emerged as an interesting area of research. As the name implies, the biocathode has a living biofilm growing on the cathode of the system. The systems then are microbial-cathode MFCs, or generally, biocathodes [75], or can broadly be called biocathode MXCs.

The "traditional" MFC or MXC have the biological anode and abiotic cathode and occasionally require the use of a catalyst or electron mediator. This increases costs and limits application possibilities [76]. Biocathodes specifically are seen as a way to overcome these limitations [76].

Biocathodes are versatile, with oxygen as an electron acceptor, or possibly even the use of contaminants as electron acceptors [75]. This allows for several different applied processes for nutrient removal or bioremediation and can be co-occurring with biological production of electricity. These systems can be applied to convert electricity into reduced products [75]. Overall, the classification of the biocathode depends on which final electron acceptor is used in the reaction, including aerobic biocathodes (oxygen) and anaerobic biocathodes (e.g., nitrate or sulfate) [76]. The general reactions for aerobic biocathodes have the electron mediators (e.g., iron and manganese) reduced by the cathode in an abiotic process then reoxidized by the microbes [76]. Aerobic biocathodes can facilitate an oxygen reduction reaction with water as a product [77]. The anaerobic biocathodes reduce the final electron acceptors through a process of accepting electrons from the cathode via microbial metabolism [61]. In a fascinating application of both engineering and microbiology, it is possible to have an MXC with both a bioanode and biocathode, even without requiring the use of chemical catalysts [78].

13.8 MXCs for products

Living systems are typically more complex to understand and control than other systems; however, this has been addressed in major areas such as wastewater treatment. The living fuel-cell systems have important advantages over other methods of energy generation. Fuel cells, whether biological or not, avoid the inefficiencies of combustion engines, as oxidation and reduction can occur on the electrode surfaces [31]. The inclusion of the biological aspects of the highly efficient fuel-cell system incorporates the advantages of microorganisms to break down organic compounds and, generally, produce electricity [31, 79]. Of the dozen or so processes identified for the microbial electrochemical technologies, the "X" in MXC dilutes out the meaning into many different technical goals [15].

13.8.1 Production and recovery of valuable products

In this work, we highlight several notable products or activities of microbial systems: electricity, hydrogen, hydrogen peroxide (H_2O_2), ammonium (NH_4^+), volatile fatty acids (VFAs), and heavy metals. We chose these specific processes, as they represent examples of widely studied or niche applications, great value-adds to specific systems, or opportunities specifically available with microbial systems.

13.8.1.1 Electricity

The most well-known product of the MXC system is electricity. From the early days in the field, the opportunities for niche applications to produce energy appear to have sustained the general interest in the technology as a whole. For electricity generation from biological systems, many details are shared in earlier sections, including on *Geobacter*, *Shewanella*, and bioanodes.

Multiple fields of study are involved in bioproduct generation from MXC systems. Starting from the microbiological areas of interest for a new type of biological process to engineering applications, research areas are abundant. The realization of the ability of microorganisms to perform redox reactions directly on electrodes, specifically without the use of chemical mediators, provides unique research areas in both scientific and engineering fields [15]. Anecdotally, high current densities (~11 A/m^2) and ~70% Coulombic efficiencies were observed in *Geobacter*-enriched butyrate (an important fermentation product) oxidizing communities [80]. Importantly, this study indicated that mixing complementary cultures outperformed simply enriching cultures. Basically, the butyrate-oxidizing community combined with a *Geobacter*-rich mixed culture created a new mixed community outperforming other research studies [76].

While this may be seen to further complicate the research field, the applications are profound. Carefully creating the physical systems and development of the biological communities may continue to enhance the field, including increased current densities and high Coulombic efficiencies.

13.8.1.2 Biogreen hydrogen

Although vehicle technology has moved on from past interest in hydrogen fuel cells, this particular potential bioproduct is still of interest from the MXC system. The microbial bioelectrochemical system research areas generally progressed from MFC to MXCs. The MFC systems are highlighted by the possible application in wastewater treatment and energy generation but have more recently expanded in one way through the use of biocathodes [81]. The systems and processes are so varied [76] that it is difficult to summarize well in this context.

Several technologies that can be used to produce hydrogen from a biological approach: biophotolysis, photo fermentation, dark fermentation, and hybrid biohydrogen production via electrochemical methods [82]. Several different types of organisms can be used to facilitate the reactions in these different systems, from cyanobacteria and green microalgae to different types of bacteria [81].

Research applications to produce hydrogen vary widely. In the biocathode application, hydrogen can be produced by the microorganisms. A study with biocathode and bioanode application revealed approximately 20% hydrogen recovery at the cathode, but experienced losses after 1,600 h of operation due to calcium phosphate buildup on the cathode [78]. In an MEC application, a mixed culture of anodic syntrophic fermentative and anode respiring bacteria produced hydrogen from a synthetic milk wastewater source with the best current density (150 A/m^3) as compared to glycerol and starch synthetic wastewater sources [83]. Excellent hydrogen recoveries

(~90%) and production (0.94 m^3/m^3/day) observed experimentally [83] provide great promise for the field.

One of the biggest obstacles for large scale implementations of MECs has been their dependence on external electrical power supply. Even though an MEC requires a minimum theoretical voltage supply of 0.135 V, the actual voltage for HER (hydrogen evolution reaction) is much higher [84]. Different studies have identified approaches to overcome this limitation by applying and controlling external voltage using real-time and intermittent modes, which help optimize performance and energy consumption in MECs [85]. Some other approaches to reduce this overpotential have been by modifying membrane materials, nonconventional electrode modification and high surface area electrodes using granular activated carbon, bio-char, etc., using photo-anodes and photo-catalysts [86–90]. Focus has also been on making self-sustaining MECs by integrating them with carbon-neutral renewable energy sources like solar, photovoltaic, photoelectrode microbial (MFCs), thermoelectric, hydroelectric, osmotic energy, and others, making them a net-zero CO_2 emission process for sustainable hydrogen production [85]. A review study [84] mentions the use of photocatalysts in MECs to study TiO_2 as a photocatalyst resulting in a 15–20% increase in columbic efficiency. However, biocathodes can be an efficient alternative for expensive metal catalysts. They have a unique capability of self-catalyzing the HER. Both types of biocathodes enzyme-based (hydrogenase) and microbial biocathodes have recently been explored [91–93].

13.8.1.3 Hydrogen peroxide

H_2O_2 is a valuable industrial chemical and oxidizing agent with a wide range of uses ranging from a bleaching agent for textiles and paper, a propellant, and an oxidant in chemical manufacturing [94]. Approximately 5% of the U.S annual H_2O_2 usage is in the water and wastewater treatment industry [95]. This typically is in low concentrations for advanced oxidation coupled with either Fenton reactions or UV treatment for the removal of complex organic chemicals and emerging contaminants [94, 96, 97], and for disinfection [98–102]. H_2O_2 is also used in advanced oxidation processes to remove biological products or contaminants that contribute to taste and odor issues [100, 103, 104]. This makes it an important value-added bioproduct, particularly due to the importance of addressing taste and odor for the consumer. H_2O_2 is utilized at wastewater treatment plants (WWTPs) for hydrogen sulfide removal in scrubbing towers to reduce odor emissions [105, 106].

Electrolysis of H_2O_2 has been commercially available for more than 100 years but is economically disadvantageous due to large energy requirements [94]. During electrolysis, H_2O_2 is produced through oxygen reduction at the cathode:

$$O_2 + 2H_2O + 2e^- \rightarrow H_2O_2 + 2OH^- \quad (E^{o\prime} = +0.424\,V_{Ag/AgCl})$$

One potentially energy-neutral application of MXC technology is cathodic electrosynthesis of H_2O_2 using electricity produced by anodophilic microorganisms, using microbial peroxide-producing cells (MPPCs). One advantage of H_2O_2 production using MPPCs is that it requires simple catalysts like carbon rather than Pt or Au for electricity and H_2 production.

Researchers have produced H_2O_2 in MPPCs at concentrations suitable for tertiary water and wastewater treatment with little-to-low energy input (30–50 mg H_2O_2/L for disinfection and 55–20 mg H_2O_2/L for micropollutant removal) when fed with acetate- or lactate-based synthetic wastewater at the laboratory scale [97, 107]. Rozendal et al. [108] produced 3.9 g H_2O_2/L-day and concentrations as high as 1.3 g/L with energy input of 0.93 W.h/g H_2O_2. To maximize H_2O_2 production, Modin and Fukushi [109] produced 9 g/L H_2O_2 at a rate of 11.1 g/L-day while applying a high amount of energy at 3.0 W.h/g H_2O_2. Griffin et al. [110] produced 4.6 g H_2O_2/L at a production rate of 4.6 g/L-day with 1.6 W.h/g H_2O_2 energy input.

Young et al. [111, 112] demonstrated that H_2O_2 production and concentration are a trade-off on most MPPC: reactors produced as high as 3.1 g H_2O_2/L and a rate of 0.76 g H_2O_2/L at a 4 h hydraulic retention time (HRT), but the net production rate of H_2O_2 was higher 1.02 g H_2O_2/L-day at a 1-h HRT with an effluent concentration of 1 g/L. The decrease in concentration is attributed to degradation of H_2O_2 in the reactor through reaction with the cathode catalyst.

A few pilot-scale MPPC studies are available. Sim et al. [113] operated at 110 L MPPC with acetate as the anode media and tap water as the catholyte. The reactor achieved 98 mg H_2O_2/L production with low H_2O_2 conversion efficiency (7.2%). Conversion efficiency was likely impacted by the low conductivity of tap water, which did not provide adequate anions or cations for transfer to the anode chamber to maintain charge neutrality.

13.8.1.4 Ammonia

Ammonia (NH_3) is an important part of the food-energy-water nexus. The NH_3 used in fertilizers is produced via the energy-intensive Haber-Bosch process, which requires 28–30 GJ of energy per ton of NH_3 produced [114] and 949 m^3 of natural gas as a substrate [115]. At the same time, urine and domestic wastewater are high in nitrogen that, when naturally hydrolyzed, becomes inorganic NH_3. It is estimated that total Kjeldahl nitrogen (TKN), which is the sum of organic N and NH_3, is about 50 mg/L in domestic wastewater, and humans produce about 12 g TKN per person per day [106, 116]. Thus, there is ample opportunity to use wastewater and urine treatment as potential sources of NH_3.

One potential method of separation from waste streams is MXCs, where the focus at the anode is used for COD removal and the cathode for NH_3 recovery using an MFC, MEC, or MDC. In the MXC, COD is utilized by anodophilic microorganisms to produce electricity, releasing NH_4^+ during substrate utilization and current that is transferred to the cathode [117]. To maintain charge neutrality within the cell, NH_4^+ can transfer to the cathode chamber when a cation exchange membrane (CEM) is used to separate the anode and cathode chambers. When internal resistances are low and the current high, a recovery system can operate at an MFC producing electricity, while recovery systems with low current production operate at an MEC.

Many researchers have focused on NH_3 removal using MECs and MFCs through promotion of nitrification and denitrification processes in the anode chamber and using a biocathode [118–121]. Other researchers have focused on maximizing NH_3 transport to and off-gassing at the cathode chamber. Villano et al. [122] operated an MEC with a methane (CH_4)-producing biocathode captured a modest 33 mg/day of NH_3 (50% of daily input at the cathode) in the cathode chamber at a reasonable 8.3 h HRT in the cathode chamber. Off-gassing of NH_3 was not observed mostly due to the pH being 8.2, well below the pK_a of 9.35 [123]. Kuntke et al. [124] used MFCs with up to 3 g/L NH_4^+ in the influent to isolate 124 mg NH_4^+/day of at the cathode while producing 0.23 W/m^2 of energy. Qin and He [125] used an MFC coupled with forward osmosis to capture and concentrate 81% of NH_4^+ supplied to the anode chamber. Aeration of the cathode chamber encouraged stripping of NH_3 from the catholyte, with two-thirds of NH_3 off-gassing from the catholyte.

Significant research has been dedicated to NH_3 from urine. Using real urine at 4 g NH_4^+/L, Kuntke et al. [126] recovered 3.29 g/m^2/day with 40% off-gassing as NH_3 while producing 0.05 W/m^2 of energy. Kuntke et al. [127, 128] used different concentration methods to recover NH_3 at rates as high as 162 g/m^2/day with 2.3 W.h/gN recovered. In the only scaled-up reactor, Zamora et al. [129] operated an MEC with 2.5 L anode and cathode volumes using real urine to recover NH_4^+. Ninety-two percent of influent NH_4^+ migrated from the anode to the cathode chamber, of which 31% was recovered.

In microbial desalination cells (MDCs), the anode and cathode chambers are separated by a center chamber that acts as a desalination chamber. The anode and desalination chambers are separated by anion exchange membrane (AEM), and the desalination and cathode chambers are separated by a CEM. Zhang and Angelidaki [130, 131] applied a lab-scale-modified submerged MDC for in situ NH_3 recovery in an anaerobic digester to help prevent NH_3 inhibition of methanogenesis. The MDC recovered an average of 80 gN/m^2/day and achieved a maximum power density of 0.71 W/m^2, making the cell a net energy producer.

Other NH_4^+-rich waste streams have also been explored for MFC and MEC applications including pig slurry [132], landfill leachate [133], and livestock water [134, 135]. Most notably, Sotres et al. [132] used an MEC to recover up to 25.5 g/m^2/day from piggery waste with 0.4 V applied voltage. In the same study, an MFC was used to recover 7.2 g/m^2/day while producing 0.46 W/m^2. Sotres et al. [132] also saw increased NH_4^+

recovery rates (2.3–7.2 g/m^2/day) with increasing N and organic loading rates that correlated with increased current production.

13.8.1.5 Volatile fatty acids

Volatile fatty acids (VFAs) are valuable commodities used in a variety of industries including chemical manufacturing, food manufacturing, consumer and personal care products, and bioplastics [136]. Short-to-medium chain VFAs are classified as carboxylic acids with two to six linearly chained carbons with a carboxyl group at one end. During typical anaerobic digestion, carbohydrates, proteins, and lipids are hydrolyzed to amino acids, monosaccharides, and larger fatty acids [106, 137]. These hydrolysis products can be fermented to VFAs by acetogenic and acidogenic fermenters. In a typical anaerobic digester, the presence of methanogenic *Archaea* would convert the VFAs to CH_4; however, VFAs can accumulate to high concentrations when methanogenesis is inhibited [137]. One major focus is VFA recovery from fermentation of organic-rich waste streams like municipal sludge, food waste, fats, oils, and grease (FOG), and other municipal solid wastes [136, 138, 139].

MXCs have demonstrated CO_2 valorization to VFAs using two different methods: as a separation process for high-strength VFA waste streams in systems similar to desalination processes and as a production source and recovery system using microbial electrosynthesis (MES) cells and biocathodes [140–142]. These MXC technologies lend themselves well to VFA separation since membranes and cathode-operating potential can potentially be tailored to select for specific VFAs [143]. For example, production of acetate from CO_2 requires a theoretical potential of −0.28 V.

13.8.1.5.1 Source separation via desalination cells

Similar to NH_4^+ recovery, MDCs are effective at isolating VFAs in a middle chamber. Zhang and Angelidaki [140] used a system similar to an MDC to isolate VFAs from a fermentation broth consisting of acetate, propionate, and butyrate. The microbial bipolar electrodialysis cell was fed to fermentation broth to dialysis chambers while the anode and cathode were operated using domestic wastewater and NaCl solution, respectively. The system achieved 98% VFA recovery from the fermentation broth with 1.2 V applied potential during a 94-h retention time.

13.8.1.5.2 Production via microbial electrolysis and biocathodes

Biocathode systems that produce VFAs are based on conversion of CO_2 valorization to carboxylic acids [142]. A variety of pure and mixed acetogenic biocathode cultures have demonstrated the ability to reduce CO_2 to VFAs with the addition of electrons from either external electricity sources or anodophilic bacteria in the anode chamber [144].

Acetate has been a major focus of MES technologies. Acetate recoveries have ranged 0.02–1,330 g/m^2/day at applied current densities ranging from 0.01 to 200 A/m^2 [145–148]. Generally, increased supplied current to the cathode resulted in higher acetate production rates. Most notably, Jourdin et al. [146, 147] used abiotic anodes to supply electricity to several layers of macroporous carbon felt cathode stacked together and inoculated with a mixed culture. They achieved 1,330 g/m^2/day acetate production at neutral pH (6.7) and with 200 A/m^2 of applied current. More importantly, Jourdin et al. reported >99% current recovery to acetate. Yu et al. [145] used the thermophilic bacteria *Moorella thermoautotrophica* to produce acetate and formate up to 2.9 and 3.5 g/m^2/day, respectively, with about 1 A/m^2 of added current density. Roy et al. [149] utilized unpurified CO_2 from a brewery as a substrate source for MES performance. The CO_2 was 97.9% pure and supplied to a mixed-culture biocathode producing up to 66 g/m^2/day of acetate while producing small amounts of formate and butyrate and 84% conversion of electrons to VFAs.

Other VFAs have also been recovered via MES, primarily caproate [150–152], butyrate [150, 151, 153–155], and formate [145, 149]. These VFAs were generally produced simultaneously with acetate [145, 149, 151, 154, 156]. Production rates for different VFAs ranged from 0.04 g/L/day to 5.7 g/L/day for butyrate, 0.95–2.4 g/L/day for caproate, and 0.36 g/L/day for formate [156, 157]. Jourdin et al. [124] manipulated cathode HRT and CO_2 loadings to favor butyrate and caproate production. At a 14-day HRT and 173 L/day CO_2 loading, Jourdin et al. [150] observed chain elongation of ethanol with acetate to produce 106 and 64 g/m^2/day of *n*-butyrate to *n*-caproate, respectively. Batlle-Vilanova et al. [155] selectively produced butyrate over other products by maintaining a pH around 5 and H_2 partial pressures >1 atm.

13.8.1.6 Heavy metals

Microbial electrochemical cells have significant promise for the remediation and recovery of different heavy metals. As shown in Figure 13.1, many of the heavy metals that are common contaminants in water and acid mine drainage have positive redox potentials, making them ideal for recovery at cathode, although some anode processes exist [158]. Table 13.1 summarizes select works on heavy metal recovery from MXCs, particularly at the lab scale. No pilot-scale studies have been performed using these technologies, largely due to limitations in membrane materials and mass transfer limitation due to electrode spacing and reactor volume. Unless otherwise noted in the figure, all MXCs were dual chamber in design, with either a CEM or AEM separating the anode and cathode chambers and performed as batch experiments.

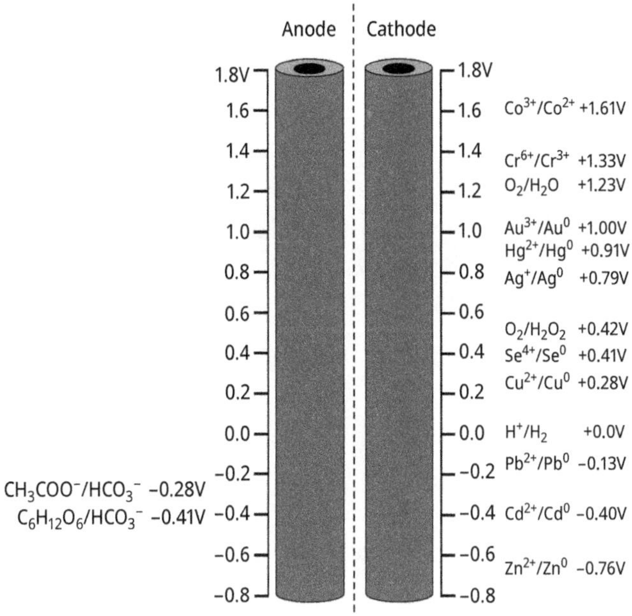

Figure 13.1: Redox potentials for common products recovered in microbial electrochemical cells (adapted from Nancharaiah et al. [158]).

13.8.1.6.1 Chromium(VI)

Chromium is a common groundwater contaminant and common industrial chemical in the metallurgy, electroplating, leathering, dyes, and corrosion control industries [158]. The two most common forms are hexavalent (Cr(VI)), which is highly soluble at all pHs and toxic, mutagenic, and carcinogenic, and trivalent (Cr(III)), which is significantly less toxic than Cr(VI) and tends to precipitate to chromium hydroxide ($Cr(OH)_3$) under moderately acidic to alkaline conditions. The reduction reaction of aqueous Cr (VI) has a highly favorable redox potential at the cathode that is even more favorable than O_2 reduction ($E^{o\prime} = +1.23\ V_{SHE}$), making Cr(VI) reduction an electricity-producing process [159]:

$$Cr_2O_7^{2-} + 6e^- + 14H^+ \rightarrow 2Cr_3^+ + 7H_2O \qquad (E^{o\prime} = +1.33\,V_{SHE})$$

Complete $Cr_2O_7^{2-}$ was generally achievable at concentrations less than 150 mg/L while using simple carbon-based catalysts. Wang et al. [159] were the first to demonstrate Cr(VI) reduction from contaminated catholyte in a batch cathode chamber using a graphite rod, achieving 100% remediation of 100 mg/L of $Cr_2O_7^{2-}$ at pH 2 in 150 h. Li et al. [160] coupled Cr(VI) reduction at the cathode with light irritation to achieve 97% Cr(VI) reduction. Li et al. [161] achieved 100% conversion of 80 mg/L $Cr_2O_7^{2-}$ to Cr(III) while producing 700 mW/m^2. In the same study, Li et al. [161] performed scanning

Table 13.1: Select metals recovery works using MXCs.

Authors	Cathode setup	Membrane type	Cathode HRT (h)	Anolyte/ buffer	Catholyte	Maximum % contaminant remediation and reduction rate	pH of cathode	Current density/ power density at maximum removal	Columbic efficiency
Chromium(VI)									
Wang et al. [159]	Graphite plate	PEM	150	20 mM acetate	100 mg/L $Cr_2O_7^{2-}$	100% 0.67 g/m^3/h	2.0	0.34 A/m^2 and $R = 1{,}000\ \Omega$ 155 mW/m^2	59%
Li et al. [160]	Rutile	PEM	26	27 mM acetate	26 mg/L $K_2\ Cr_2O_7$	97%	2.0	0.3 A/m^2	
Yi et al. [161]	Carbon cloth	PEM	72	13 mM acetate	80 mg/L $Cr_2O_7^{2-}$	100%	2.0	1.6 A/m^2 and 700 mW/m^2	
Liu et al. [162]	Carbon cloth with polypyrrole MnO_2 catalyst	PEM	32	9 mM acetate	50 mg/L $Cr_2O_7^{2-}$	100%	2.0	3.7 A/m^2 and 1,429 mW/m^2	
Gupta et al. [163]	Carbon nanofibers with Al-Ni catalyst	PEM		12 mM acetate and 100 mM sucrose	200 ppm $Cr_2O_7^{2-}$	85%	2.0	7.86 A/m^2 and 1,530 mW/m^2	93%
Zhang et al. [164]	Carbon rod	CEM	4	12 mM acetate	50 mg/L Cr(VI)	68%	2.0	0.55 A/m^2 and 3.1 W/m^2	69%

Cobalt									
Huang et al. [165]	MFC: Graphite felt MEC: Carbon rod	CEM	MFC: 6 MEC: 6	17 mM acetate	MFC: 10 mM HCl MEC: 250 mg/L Co(II)		MFC: 1.0 MEC: 6.0		MFC: 41% MEC: 75%
Jiang et al. [166]	Graphite felt	CEM	6.2	17 mM acetate	847 μm Co(II)	92%		13.7 A/m^2 and 0.2 V applied voltage	
Huang et al. [167]	GAC and Pt-coated carbon felt	AEM	12 d	18 mM acetate	40–60 mg/L $CoSO_4$	95%	8.4	8.2 W/m^3	22%
Saad et al. [168]	Stainless steel screens	CEM	6	12 mM acetate	50 ppm	99%	7.4–8.4	3.2 mA and 1.8 V applied voltage	
Huang et al. [169]	Graphite felt	CEM	6	17 mM acetate	0.5 mM Co(II)	4.7 g/L/h	5.6	13.3 A/m^3 and 1.5 W/m^3	
Shen et al. [170]	Graphite felt	CEM		–	40 mg/L Co(II) and 50 mg/L Cu(II)	5.3 g/L/h	5.5	4.8 W/m^3	54%
Copper									
Shen et al. [170]	Graphite felt	CEM		–	40 mg/L Co(II) and 1,000 mg/L Cu(II)	83 g/L/h	2.9	26 W/m^3	55%
ter Heijne et al. [171]	Graphite foil	Bipolar	Anaerobic: 7d Aerobic: 6d	20 mM acetate	1 g/L Cu(II)	>99.9%	3	Anaerobic: 0.43 W/m^2 Aerobic: 0.80 W/m^2	
Tao et al. [172]	Graphite felt	PEM	264	28 mM glucose	200 mg/L Cu(II)	96%	4.7	0.11 W/m^2	

(continued)

Table 13.1 (continued)

Authors	Cathode setup	Membrane type	Cathode HRT (h)	Anolyte/ buffer	Catholyte	Maximum % contaminant remediation and reduction rate	pH of cathode	Current density/ power density at maximum removal	Columbic efficiency
Wu et al. [173]	Carbon cloth with Pt catalyst	*N/A	28	16 mM acetate	12.5 mg/L Cu(II)	98.3%	6	10.2 W/m^3	30%
Modin et al. [174]	Titanium wire	AEM	2	20 mM acetate	0.8 g/L Cu(II), 0.4 g/L Pb(II), 0.8 g/L Cd(II), and 0.3 g/L Zn(II)	84.3%	< 2	0.1–7.0 A/m^2 and not applied	77%
Zhang et al. [164]	Carbon rod	CEM	4	12 mM acetate	50 mg/L Cu(II)	60%	2.0	0.55 A/m^2 and 3.2 W/m^2	54%
Ai et al. [175]	Carbon cloth	*N/A	24	20 mM acetate				–	–
Cadmium(II)									
Modin et al. [174]	Titanium wire	AEM	20–24	20 mM acetate	0.8 g/L Cu(II), 0.4 g/L Pb(II), 0.8 g/L Cd(II), and 0.3 g/L Zn(II)	62%	<2	0.1–6.0 A/m^2 and 0.51 V applied	5.3%
Choi et al. [176]	Carbon cloth	AEM	60	12.2 mM acetate	50–200 mg/L Cd(II)	90–94%	6	–	
Abourached et al. [177]	Carbon cloth	*N/A	30	60 mM acetate	200 μM Cd(II) and 400 μM Zn(II)	90%	–	3.6 W/m^2	34%
Colantonio et al. [178]	Stainless steel mesh	*N/A	48	8.3 mM acetate	10 mg/L Cd(II)	95%		0.42–0.45 mA and 0.6 V applied	34–37%

Ai et al. [175]	Carbon cloth	*N/A	24	20 mM acetate	132 mg/L Cd(II)	>99.9%	7.1	–	–
Zhang et al. [164]	Titanium sheet	CEM	4	12 mM acetate	50 mg/L Cr(VI), 50 mg/L Cu(II), and 50 mg/L Cd(II)	29%	2.0	0.4 A/m^2 and N/A**	28%
Silver									
Choi and Cui [179]	Carbon cloth	AEM	8	12 mM acetate	200 mg/L Ag(I)	>99%	2.0	3.8 A/m^2 and 2.6 W/m^2	
Tao et al. [180]	Graphite plate	PEM	5.5	21 mM acetate	1 mM $AgNO_3$	>89%	2.0	–	
Tao et al. [172]	Graphite plate	PEM	12	21 mM acetate	1 mM $[AgS_2O_3]^-$	84%	6.3	–	
Wang et al. [181]	Graphite	Bipolar	21	27 mM acetate	1.6 g/L Ag^+	>99%	9.2	–	
Lim et al. [182]	Graphite felt	AEM		10 mM acetate	4,000 ppm Ag^+	92%	7	4.25 A/m^2 and 1.9 W/m^2	
Kamperidis et al. [183]	Graphite cloth with Pt	PEM	24	8.3 mM glucose	50 mg/L	>93%	2 and 7	248 W/m^2	
Gold									
Choi and Hu [184]	Carbon cloth	CEM	12	12 mM acetate	200 mg/L Au(III)	99.8%	2.0	2.1 A/m ^{2} and 1.9 W/m^2	80%

(continued)

Table 13.1 (continued)

Authors	Cathode setup	Membrane type	Cathode HRT (h)	Anolyte/ buffer	Catholyte	Maximum % contaminant remediation and reduction rate	pH of cathode	Current density/ power density at maximum removal	Columbic efficiency
Lead(II)									
Modin et al. [174]	Titanium wire	AEM	20–24	20 mM acetate	0.8 g/L Cu(II), 0.4 g/L Pb(II), 0.8 g/L Cd(II), and 0.3 g/L Zn(II)	48%	<2	0.1–5.1 A/m^2 and 0.34 V applied	5.3%
Selenite									
Catal et al. [185]	Carbon cloth with Pt	*N/A	72	1 mM glucose	200 mg/L Se^{2-}	99%	–	1,000 W/m^2	38%
Zinc(II)									
Modin et al. [174]	Titanium wire	AEM	20–24	20 mM acetate	0.8 g/L Cu(II), 0.4 g/L Pb(II), 0.8 g/L Cd(II), and 0.3 g/L Zn(II)	44%	< 2	16.8 24.7 A/m^2 and 1.5 V applied	1.2%
Abourached et al. [177]	Carbon cloth	*N/A	50	60 mM acetate	200 μM Cd(II) and 400 μM Zn(II)	97%	–	–	34%
Mercury									
Wang et al. [186]	Carbon paper	AEM	8	17 mM acetate	100 mg/L Hg(II)	99.5%	2	1.44 A/m^2 and 0.4 W/m^2	4%

*Single-chamber MXC.

**Powered by MFCs with no direct energy production.

electron microscopy (SEM) on the carbon cloth cathode and showed that the chromium was precipitating on the cathode.

Other researchers have explored catalysts attached to carbon electrodes to improve Cr(VI) reduction speeds and energy production. Alumina-nickel nanoparticles were dispersed on a carbon nanofiber cathode to serve as catalysts, increasing the Cr(VI) reduction rate to 2.13 g/m^3/h [163]. Deposition of the nanoparticles improved the general electroconductivity of the cathode, improving power density to 1,530 mW/m^2. However, the system was only about to remove 85% of the 200 ppm present as Cr (VI). Liu et al. [162] achieved 100% remediation of 50 mg/L $Cr_2O_7^{2-}$ in less than 32 h using a carbon cloth coated with a polypyrrole-MnO_2 catalyst. Results were repeatable over more than six cycles, with intermediate removal of Cr(III) from the cathode using a 100 mM NaOH wash. This is the first study to indicate that the Cr(III) could easily be recovered from the cathode surface and the cathode reused.

13.8.1.6.2 Cobalt(II)

Recovery of Co(II) has become an increasingly important sustainability process due to its presence in Li-ion batteries in the form of $LiCoO_2$. The $LiCoO_2$ present in batteries must first be isolated from the cathode materials and then the Co(III) leached into aqueous solution for recovery using MECs. Co(III) can be reduced to Co(II):

$$LiCoO_2 + 4H + e^- \rightarrow Co^{2+}(aq) + Li(aq) + 2H_2O \quad (E^{o\prime} = +1.61\ V_{SHE})$$

and Co(II) reduced to Co(0) through

$$pH \le 4.0{:}Co^{2+}(aq) + 2e^- \rightarrow Co^0(s) \qquad (E^{o\prime} = \pm 0.232\ V_{SHE})$$

$$pH > 4.0{:}Co^{2+}(aq) + 2H_2O + 4e^- \rightarrow Co^0(s) + 2OH^- + H_2(g) \qquad (E^{o\prime} = \pm 0.232\ V_{SHE})$$

Since the energy produced during Co(III) is more than sufficient to power Co(II) reduction to Co(O), Huang et al. [165] developed a combined MFC–MEC system that first releases Co(II) from $LiCoO_2$ at an MFC cathode and then reduces Co(II) at an MEC cathode. The systems were relatively small, ranging up to 25 mL in anode and cathode chamber volumes. In the MFC, cobalt was leached from 1 g/L of $LiCoO_2$ loaded on the cathode at a rate of 46 mg/L/h. In the MEC, Co(II) was reduced to Co(0) at 7 mg/L/h. More importantly, the entire system was run without external energy input. Jiang et al. [166] recovered Co(0) while producing H_2 gas using an MEC. Peak Co(II) recovery of 92% occurred at 0.5 V applied voltage and 13.2 A/m^2, after which H_2 production increased with the increased power being added to the system. SEM and XRD confirmed the presence of crystalline Co(0) structures on the cathode surface. Huang et al. [167] increased surface area at the cathode by utilizing GAC touching the cathode surface, Co(0) deposition to the GAC surface. Co(0) was effectively removed in that form from the cathode and GAC surfaces using a stripping solution. Saad et al. [187] utilized multiple-stacked stainless steel screens as the cathode to increase surface area and conductivity. With an applied voltage of 1.8 V, the system remediated >99% of 50 ppm Co (II). The porous nature of the screen improved mass transport by reducing the diffusion layer between the bacteria and bulk liquid.

Researchers have used biocathode systems to make Co(II) remediation energy positive [170, 169]. Huang et al. [165] recovered 4.6 mg Co(II) /L/h while producing 1.5 W/m^3 of power. Shen et al. [170] performed concomitant Co(II) and Cu(II) recovery 5.3 mg Co(II) /L/h using a biocathode. For this important contaminant, these biobased systems provide a promising step to address environmental issues.

13.8.1.6.3 Copper(II)

Copper is a metal that is essential for cell metabolism but is also a broad concern [188], with long-term exposure causing toxicity and organ failure. Copper waste sites are associated with mining, metals, and electronic manufacturing, and some pesticides. MFC remediation of Cu(II) is an emerging recovery technique due to copper's positive redox potential, making it an excellent candidate for recovery on the cathode:

$$Cu^{2+}(aq) + 2e^- \rightarrow Cu^0(s) \quad \left(E^{o\prime} = +0.28\ V_{SHE}\right)$$

There have been limited uses of MFCs for Cu recovery, but all researchers reported producing pure Cu(0) on the cathode surface [170–172]. These studies also reported Cu(II) was reduced to pure Cu(0), not CuO or CuO_2, and was deposited on the cathode surface.

Ter Heijne et al. [171] performed one of the first studies on Cu(II) remediation in a cathode chamber, finding little difference in results under aerobic and anaerobic conditions. More than 99.9% removal was achieved from 1 g/L Cu(II) initial concentration while producing 0.43 and 0.8 W/m^2 under anaerobic and aerobic conditions, respectively.

Lower power densities were as effective at removing Cu(II): Tao et al. [172] was able to achieve >96% Cu(II) recovery as Cu(0) with 0.11 W/m^2 power density. Shen et al. [170] used an MFC to produce electricity for simultaneous Cu(II) and Co(II) removal on a MEC biocathode consisting primarily of *Proteobacteria* and *Firmicutes.* There were increased removal rates of Cu(II) (83 mg/L/h) and Co(II) (6.4 mg/L/h) with increased influent Cu(II) concentration up to 1,000 mg Cu(II)/L.

Single-chamber MFCs have also been used to recover Cu(II) [173, 175]. The MFC reduced low concentrations <12.5 mg/L Cu(II) with >98% efficacy [173]. Consistent with the microbial community data from Shen et al. [170], Cu was consumed by a *Proteobacteria* and *Firmicutes* biocathode and reduced to Cu_2O or elemental copper. Another study performed simultaneous removal of Cu(II) and Cd(II) in simulated acid mine drainage, achieving >99.9% recovery of Cu mostly as Cu(0) with trace amounts of $Cu_{gl}O$ [175]. The advantage of a single chamber system is the absence of a membrane, which is more costly and increases Ohmic resistance in the system.

13.8.1.6.4 Cadmium(II)

Cadmium(II) is a carcinogenic and toxic metal highly regulated in waterways. Cd(II) is generally used in electroplating, pigments, and battery manufacturing. Traditional activated sludge processes are ineffective at removing it from wastewater. Cd(II) reduction to Cd(0) has a low redox potential:

$$Cd^{2+} + 2e^- \rightarrow Cd^0(s) \qquad (E^{o\prime} = -0.40\ V_{SHE})$$

Because of the low redox potential, Cd(II) is often recovered at a bioanode of a MEC using Cd(II) as a substrate for biomass growth like acetate or at the cathode with power input from an outside source. Reduction to solid Cd is dominant at a low pH (≤2). At higher pH, Cd can be precipitated as cadmium hydroxide ($Cd(OH)_2$) or cadmium carbonate ($CdCO_3$).

Cd(II) recovery research has focused on both single- and dual-chamber MECs, with many researchers coupling MEC Cu(II) recovery at a bioanode with an MFC to power the system. Choi et al. [176] used two dual-chamber MXCs to compensate for the energy required to remediate Cd(II): they used a Cr(VI)-reducing MFC to power cathodic Cd(II) removal from a separate MEC. Choi et al. [176] obtained 90–94% Cd(II) recovery as Cd(0) while entirely powered by the Cr(VI) MFC. Zhang et al. [164] used Cr(IV) and Cu(II)-recovery MFCs in series and parallel to power a Cd(II)-recovery MEC. Cd(II) recovery was highest (29%) when powered in parallel connection by both MFCs. Colantonio and Kim [178] used a single-chamber MEC to achieve 95% Cd (II) removal in 48 h when applying 0.6 V potential. Another single-chamber MXC used a novel anodophilic bacteria, *Pseudomonas* sp. E8, to achieve >99.9% cathodic removal of Cd(II) and Cu(II) in simulated acid mine drainage [175]. SEM, XRD, and EDXS confirmed that the Cd(II) was reduced to Cd(0) at the cathode surface.

Other researchers explored concomitant removal of Cd with other metal contaminants. Modin et al. [174] performed cathodic Cd recovery from fly ash contaminated with copper, lead, cadmium, and zinc in a dual-chamber MEC. The applied potential of the system was varied to remove each contaminant independently. For Cd(II), 62% of Cd was recovered as Cd(0) with 0.51 V of applied potential. Abourached et al. [177] used a single-chamber MEC to recover Cd(II). The MEC obtained 90% removal at 200 μM Cd(II) but exhibited signs of decreased voltage likely due to microorganism toxicity at 300 μM Cd(II), indicating the potential need for a membrane separator to prevent inhibition of anode microorganisms.

13.8.1.6.5 Silver(I) and gold(III)

Precious metals like silver and gold are used in a variety of applications including jewelry, acid mine drainage, electronics manufacturing, and dental applications. Silver and gold(III) are also common in wastewater streams [189]. Both metals have highly favorable redox potentials for capture at a cathode. Gold is generally present in solution as $AuCl_4^-$ and can be converted to solid gold via:

$$AuCl_4^- + 3e^- \rightarrow Au(s) + 4Cl^- \quad (E^{o\prime} = +1.00\ V_{SHE})$$

Silver can be present as multiple soluble compounds including Ag^+, $Ag(NH_3)^{2+}$, and $[AgS_2O_3]^-$. While these compounds have slightly less favorable redox potentials versus gold, they are generally more omnipresent, providing more opportunities for capture:

$$Ag^+(aq) + e^- \rightarrow Ag^0(s) \quad (E^{o\prime} = +0.80\ V_{SHE})$$

$$Ag(NH_3)^{2+}(aq) + e^- \rightarrow Ag^0(s) + 2NH_3(aq) \quad (E^{o\prime} = +0.37\ V_{SHE})$$

$$[AgS_2O_3]^-(aq) + e^- \rightarrow Ag^0(s) + S_2O_3^{2-}(aq) \quad (E^{o\prime} = +0.25\ V_{SHE})$$

Limited research has been performed with gold, with 99.8% removal in 12 h with an initial solution concentration of 200 mg Au(III)/L and limited power production (1.9 W/m^2) [184]. SEM confirmed that Au(0) deposited on the cathode as clusters of crystals.

Conversely, significant research has pursued silver recovery using MFCs. Silver removal in MFCs has consistently exhibited high removal efficiencies of high initial concentration Ag(I) waste streams [179–183]. Initial research by Tao et al. [180] and Choi and Cui [179] demonstrated >95% removal of 1.6 g/L $Ag(NH_3)^{2+}$ and 200 mg/L $AgNO_3$, respectively, in less than a day through electrochemical deposition of Ag(0) crystals on the cathodes. Interestingly, removal of the Ag ions resulted in the cathode chamber pH increasing from 2.0 to about 7.0, making the system more sustainable for general water treatment [179, 180].

Other researchers have demonstrated >99% removal of Ag in slightly alkaline cathodic conditions. Yun-Hai et al. [190] achieved >99% removal of Ag at high initial concentrations (1.6 g $Ag(NH_3)_2^+$/L).

13.8.1.6.6 Other metals

As summarized in Table 13.1, limited research has been performed in the removal of other metals, including several divalent metals such as zinc(II) [174], lead(II) [174], and mercury(II) [186] as well as selenite [185] with relatively positive results. Since lead and zinc have negative redox potentials, they require an MEC configuration for remediation. Conversely, the positive redox potentials of mercury and selenite mean their remediation has the potential of being energy-generating. These metals were usually investigated as part of an amalgam of heavy metals, making each specific metal difficult to remove.

13.9 Challenges to MXC implementation

MXCs provide several opportunities for valuable product recovery with power generation, including electricity production, H_2O_2 production, and heavy metals recovery. Regardless of the product produced, implementation of valuable product recovery in

MXCs has faced many challenges. Scaling up MXCs for practical application in wastewater treatment and heavy metal recovery requires building robust systems at large capacities. To obtain power or produce H_2 at levels sufficient to power a water resource recovery facility, multiple MXCs will be required to be operated either in parallel or series depending upon the desired outcome [191]. Several groups have explored stacking systems to increase overall voltage and current output but have produced energy far below expectations [192–194]. Stacking systems will require complex control systems to prevent voltage reversal to less active cells [195].

Materials provide a particular challenge for MXC implementation. The separator membrane in a dual-chambered MXC must be able to withstand different head pressures, pHs, and chemicals which can deteriorate the membrane's integrity [111, 112]. Aerobic cathodes can require expensive catalysts like Pt to effectively produce electricity [76, 78] and a gas diffusion layer that is hydrophobic to retain the catholyte while allowing for sufficient transfer of O_2 to the cathode chamber. For metals recovery, the membrane and cathode require materials resistant to the acids often required to recover the metals.

In addition, H_2O_2, NH_4^+, and heavy metals recovery research have not discerned if the bacteria on anodes or biocathodes will be affected or inhibited by transfer of ions across a membrane. For H_2O_2 produced at pH >11.35, the chemical deprotonates to HO_2^-, which can transport from the cathode to the anode chamber across a membrane and potentially destroy the anode biofilm. Ki et al. [196] determined that fungi grow at the anode and consume destructive HO_2^- before it can contact the anode biofilm. It is unclear if the transport of chemicals across the membrane can cause antimicrobial-resistant gene expression at the anode chamber. Work performed by Chen et al. [197] found that *Geobacter* can biodegrade oxytetracycline without gene expression.

However, possibly the most inhibitory part of MXC manufacturing is cost. One of the most popular AEM and CEM manufacturers charges \$200/m^2 of membrane [198]. Ge and He [199] reported that over 60% of MXC costs were associated with the membrane. Precious metal catalysts for energy generation are extremely expensive.

Overall, biocathodes are seen as one of the "low-hanging fruit" in the field. It is an area ripe for continued development. Bioprospecting in this is abundant and remains well open to new and important developments. Also, biocathodes have some of the easier challenges in these fields. The focus areas for the MXC, MEC, and similar acronym systems certainly have shifted in the decades since the early identification of *Geobacter* in sediments [49]. As the focus areas continue to evolve, signs point to greater focus on specific bioproduct and bioremediation applications.

13.10 Artificial intelligence and microbial electrochemical systems

With the expansion of artificial intelligence (AI) technologies in the last few years, applications across many, varying fields are becoming prevalent. In just a few years, AI has quickly revolutionized many aspects of labor and automation [200], teaching and learning [201], human interactions and carryover effects [202, 203], many aspects of STEM-related research [204], text analysis [205], and many other areas. This pace of innovation presents both many challenges and many opportunities.

Mathematical modeling for bioelectrochemical systems – in general – is an established practice (e.g., for overviews, see [206, 207]) within the field predating the widespread use of AI by many years. Today, AI presents an opportunity for studying these systems, with particular advantages in application because of the many unknowns in such systems [208], such as substrate characteristics, microorganisms and diversity in mixed cultures, the materials for electrodes, and operating conditions for the systems [90]. The main factors (and influencing factors) involved in a general bioelectrochemical system are the microorganisms (community composition and dominant species), substrates and products (profiles of each and the headspace), and reactor performance (for example, temperature, pH, current, voltage) [209]. Today, the focus of AI transformation and innovation lies largely in the areas of efficiency, predictive capabilities, and controlling processes [210]. These new research directions can be useful in performance enhancement based on the ability to identify patterns and provide new perspectives on data analysis [209] or modeling outputs [211]. A key example of the application of AI to these systems is optimizing system operation conditions [212] to maximize desired results, such as electricity production or contaminant removal [90], as well as modeling and applying AI to optimize the biological processes themselves [90]. In predictions, AI can be useful for assessments of the microbial community, substrates and products, and, as mentioned, overall performance [209].

We note that the full specifics, methodologies, and procedural details of applying AI technologies to MXC systems lie significantly outside the scope of this chapter. Further, applications of AI and machine learning advancements can also be found as part of the frontiers in research for many of the individual sub-areas of MXC research (e.g., the sub-topics of this chapter). The general process for AI modeling in these systems involves having experimental data, importing that to the AI model, performing system testing and updating the model, and then, being able to provide predictions on the system (as outputs of the AI model) [209]. We note recent work in wastewater treatment [210, 213, 214], nitrogen and sulfur removal [215], anaerobic processes [216], electricity generation [90], hydrogen production [212], biosensors [217, 218], many aspects of pollution reduction and bioremediation [219], and many more topics. Again, due to the scope of this particular chapter and the rate of production and volume of current research on AI and machine learning integration, we refer readers to seeking current review articles and state-of-the-art research combining these topics – AI and MXCs – such as Mousazadehgavan et al.'s recent work [210] and the scoping of AI applications, alongside directions and modeling processes detailed by Li et al [209]. We are encouraged by the ways in which AI may be applied to facilitate continued innovation and application across MXC technologies.

13.11 Summary

Microbial electrochemical systems demonstrate incredible versatility. From unique niche applications to product generation to materials recovery, these systems have a short but rich research history and promising future. These systems can perform value-added services from recovering rapidly depleting materials from waste streams to electricity generation to bioremediation. The future of both research areas and practical applications for microbial electrochemical systems and microbial electrochemical technologies is bright. As the climate crisis continues to escalate, advancements are needed across multiple sectors. The application of artificial intelligence-based technologies may provide further advancements and innovations at a rapid pace across many of the sub-topics discussed in this chapter. However, as discussed in Chapter 29, the application of artificial intelligence has other potential "costs" in research. Overall, the bioproducts produced from these microbial systems may not power the world, but they will have important, niche applications. The demand for harnessing microbial abilities to either recover or create valuable products seems to be a promising direction for a more sustainable future.

References

[1] Daims H, Taylor MW, Wagner M. Wastewater treatment: A model system for microbial ecology. Trends Biotechnol 2006;24(11):483–489.
[2] Roane TM, Pepper IL, Miller RM. Microbial remediation of metals. Biotechnol Res Ser 1996;6:312–340.
[3] Kobayashi H, Rittmann BE. Microbial removal of hazardous organic compounds. Environ Sci Technol 1982;16(3):170A–183A.
[4] Gericke M, Pinches A. Microbial production of gold nanoparticles. Gold Bull 2006;39(1):22–28.
[5] Ladygina N, Dedyukhina EG, Vainshtein MB. A review on microbial synthesis of hydrocarbons. Process Biochem 2006;41(5):1001–1014.
[6] Rittmann BE, Mayer B, Westerhoff P, Edwards M. Capturing the lost phosphorus. Chemosphere [Internet] 2011;84(6):846–853. Available from: http://dx.doi.org/10.1016/j.chemosphere.2011.02.001.
[7] Davidson DJ, Kecinski M. Emotional pathways to climate change responses. Wiley Interdiscip Rev Clim Change 2022;13(2):1–19.

[8] Sovacool BK. The cultural barriers to renewable energy and energy efficiency in the United States. Technol Soc [Internet] 2009;31(4):365–373. Available from: http://dx.doi.org/10.1016/j.techsoc.2009.10.009.

[9] Susskind L, Chun J, Gant A, Hodgkins C, Cohen J, Lohmar S. Sources of opposition to renewable energy projects in the United States. Energy Policy 2022 February;165:112922. Available from: https://doi.org/10.1016/j.enpol.2022.112922.

[10] Yoho R. Energy and the environment: Electrochemistry of electron transport pathways in anode-respiring bacteria and energy technology and climate change in science textbooks, Arizona State University; 2016.

[11] Rabaey K, Verstraete W Microbial fuel cells: Novel biotechnology for energy generation. Trends Biotechnol 2005;23(6):291–298.

[12] Munoz-Cupa C, Hu Y, Xu C, Bassi A. An overview of microbial fuel cell usage in wastewater treatment, resource recovery and energy production. Sci Total Environ [Internet] 2021;754:142429. Available from: https://doi.org/10.1016/j.scitotenv.2020.142429.

[13] Ewing T, Ha PT, Babauta JT, Tang NT, Heo D, Beyenal H. Scale-up of sediment microbial fuel cells. J Power Sources [Internet] 2014;272:311–319. Available from: http://dx.doi.org/10.1016/j.jpowsour.2014.08.070.

[14] Vijay A, Khandelwal A, Chhabra M, Vincent T. Microbial fuel cell for simultaneous removal of uranium (VI) and nitrate. Chem Eng J [Internet] 2020;388:January, 124157. Available from: https://doi.org/10.1016/j.cej.2020.124157.

[15] Torres CI. On the importance of identifying, characterizing, and predicting fundamental phenomena towards microbial electrochemistry applications. Curr Opin Biotechnol 2014;27:107–114.

[16] Yoho R, Popat S, Fabregat-Santiago F, Giménez S, Heijne A, Torres C. Electrochemical impedance spectroscopy as a powerful analytical tool for the study of microbial electrochemical cells. In: Biofilms Bioelectrochem Syst 2015;249–280.

[17] Hasany M, Mardanpour MM, Yaghmaei S cells: A review. Int J Hydrogen Energy [Internet] 2016;41(3):1477–1493. Available from: http://dx.doi.org/10.1016/j.ijhydene.2015.10.097.

[18] Cao X, Huang X, Liang P, Xiao K, Zhou Y, Zhang X, et al. A new method for water desalination using microbial desalination cells. Environ Sci Technol 2009;43(18):7148–7152.

[19] Deng H, Chen Z, Zhao F. Energy from plants and microorganisms: Progress in plant-microbial fuel cells. ChemSusChem 2012;5(6):1006–1011.

[20] Aulenta F, Reale P, Canosa A, Rossetti S, Panero S, Majone M. Characterization of an electro-active biocathode capable of dechlorinating trichloroethene and cis-dichloroethene to ethene. Biosens Bioelectron [Internet] 2010;25(7):1796–1802. Available from: http://dx.doi.org/10.1016/j.bios.2009.12.033.

[21] Tandukar M, Pavlostathis SG. Co-digestion of municipal sludge and external organic wastes for enhanced biogas production under realistic plant constraints. Water Res [Internet] 2015;87:432–445. Available from: http://dx.doi.org/10.1016/j.watres.2015.04.031.

[22] Nevin KP, Lovley DR. Lack of production of electron-shuttling compounds or solubilization of Fe(III) during reduction of insoluble Fe(III) oxide by geobacter metallireducens. Appl Environ Microbiol 2000;66(5):2248–2251.

[23] Rabaey K, Rozendal RA. Microbial electrosynthesis – Revisiting the electrical route for microbial production. Nat Rev Microbiol 2010;8(10):706–716.

[24] Dennis PG, Harnisch F, Yeoh K, Tyson GW. Dynamics of cathode-associated microbial communities and metabolite profiles in a glycerol-fed bioelectrochemical system. Appl Environ Microbiol 2013;79(13):4008–4014.

[25] Ortiz-Medina JF, Call DF. Electrochemical and microbiological characterization of bioanode communities exhibiting variable levels of startup activity. Front Energy Res 2019 September;7:1–11.

[26] Pham TH, Aelterman P, Verstraete W. Bioanode performance in bioelectrochemical systems: Recent improvements and prospects. Trends Biotechnol 2009;27(3):168–178.

[27] Deeke A, Thja S, Hamelers HVM, Buisman CJN Capacitive bioanodes enable renewable energy storage in microbial fuel cells. Environ Sci Technol 2012;46(6):3554–3560.

[28] Xu H, Yang XL, Liu HP, Song HL, Li T Enhancing anode intrinsic capacitance improves the performance of microbial fuel cells in wastewater treatment. *J Environ Manage* 2025;*390*:126251.

[29] Hutchinson AJ, Tokash JC, Logan BE. Analysis of carbon fiber brush loading in anodes on startup and performance of microbial fuel cells. J Power Sources 2011;196(22):9213–9219.

[30] Paitier A, Godain A, Lyon D, Haddour N, Vogel TM, Monier JM. Microbial fuel cell anodic microbial population dynamics during MFC start-up. Biosens Bioelectron [Internet] 2017;92:357–363. October 2016 Available from: http://dx.doi.org/10.1016/j.bios.2016.10.096.

[31] Bond DR, Lovley DR. Electricity production by *Geobacter sulfurreducens* attached to electrodes. Appl Environ Microbiol 2003;69(3):1548–1555.

[32] Emde R, Swain A, Schink B Anaerobic oxidation of glycerol by *Escherichia coli* in an amperometric poised-potential culture system. Appl Microbiol Biotechnol 1989;32(2):170–175.

[33] Emde R, Schink B. Enhanced propionate formation by propionibacterium freudenreichii subsp. freudenreichii in a three-electrode amperometric culture system. Appl Environ Microbiol 1990;56 (9):2771–2776.

[34] Kim N, Choi Y, Jung S, Kim S. Effect of initial carbon sources on the performance of microbial fuel cells containing proteus vulgaris. Biotechnol Bioeng 2000;70(1):109–114.

[35] Park DH, Zeikus JG. Electricity generation in microbial fuel cells using neutral red as an electronophore. Appl Environ Microbiol 2000;66(4):1292–1297.

[36] Choi Y, Jooyoung S, Seunho J, Sunghyun K. Optimization of the performance of microbial fuel cells containing alkalophilic bacillus sp. J Microbiol Biotechnol 2001;11(5):863–869.

[37] Zhao C, Song Y, Chen H, Chen H, Li Y, Lei A, Wu Q, Zhu L. Improving the performance of microbial fuel cell stacks via capacitive-hydrogel bioanodes. *Int J Hydrogen Energy* 2025;*97*:708–717.

[38] Boto ST, Cristiani L, Rosenbaum MA. Biochemical production with microbial bioelectrochemical systems. *Curr Opin Biotechnol* 2025;*93*:103291.

[39] Pinto RP, Srinivasan B, Manuel MF. Tartakovsky B.A two-population bio-electrochemical model of a microbial fuel cell. Bioresour Technol [Internet] 2010;101(14):5256–5265. Available from: http://dx.doi.org/10.1016/j.biortech.2010.01.122.

[40] Shi L, Richardson DJ, Wang Z, Kerisit SN, Rosso KM, Zachara JM, et al. The roles of outer membrane cytochromes of *Shewanella* and *Geobacter* in extracellular electron transfer. Environ Microbiol Rep 2009;1(4):220–227.

[41] Malvankar NS, Lovley DR. Microbial nanowires for bioenergy applications. Curr Opin Biotechnol 2014;27:88–95.

[42] Yoho RA, Popat SC, Torres CI. Dynamic potential-dependent electron transport pathway shifts in anode biofilms of *Geobacter sulfurreducens*. ChemSusChem 2014;7(12):3413–3419.

[43] Yoho RA, Popat SC, Rago L, Guisasola A, Torres CI. Anode biofilms of *Geoalkalibacter ferrihydriticus* exhibit electrochemical signatures of multiple electron transport pathways. Langmuir 2015;31 (45):12552–12559.

[44] Levar CE, Chan CH, Mehta-Kolte MG, Bond DR. An inner membrane cytochrome required only for reduction of high redox potential extracellular electron acceptors. Mbio 2014;5(6).

[45] Otero J et al. 2018–Identification of different putative outer membrane electron conduits necessary for Fe (III) citrate, Fe (III) oxide, Mn (IV) oxide, or electrode reduction by *Geobacter sulfurreducens*.pdf.

[46] Marsili E, Sun J, Bond DR. Voltammetry and growth physiology of *Geobacter sulfurreducens* biofilms as a function of growth stage and imposed electrode potential. Electroanalysis 2010;22(7–8):865–874.

[47] ter Heijne A, Schaetzle O, Gimenez S, Navarro L, Hamelers B, Fabregat-Santiago F. Analysis of bio-anode performance through electrochemical impedance spectroscopy. Bioelectrochemistry [Internet] 2015;106:64–72. Available from: http://dx.doi.org/10.1016/j.bioelechem.2015.04.002.

[48] Reguera G, Kashefi K. The electrifying physiology of geobacter bacteria, 30 years on. Adv Microb Physiol 2019;74:1–96.

[49] Lovley DR, Phillips EJP. Rapid assay for microbially reducible ferric iron in aquatic sediments. Appl Environ Microbiol 1987;53(7):1536–1540.

[50] Tabares M, Dulay H, Reguera G. *Geobacter sulfurreducens*. Trends Microbiol [Internet] 2020;28 (4):327–328. Available from: https://doi.org/10.1016/j.tim.2019.11.004.

[51] Derosa J, Garrido-Barros P, Li M, Peters JC. Use of a PCET mediator enables a Ni-HER electrocatalyst to act as a hydride delivery agent. *J Am Chem Soc 144*(43):20118–20125.

[52] Christgen B, Spurr M, Milner EM, Izadi P, McCann C, Yu E, Curtis T, Scott K, Head IM. Does pre-enrichment of anodes with acetate to select for Geobacter spp. enhance performance of microbial fuel cells when switched to more complex substrates?. *Frontiers in Microbiology* 2023;*14*:1199286.

[53] Wang Z, Hu Y, Dong Y, Shi L, Jiang Y. Enhancing electrical outputs of the fuel cells with Geobacter sulferreducens by overexpressing nanowire proteins. *Microbial Biotechnology 16*(3):534–545.

[54] Yadav RS, He W, Liang D, Li C, Yu Y, Feng Y. Enrichment of Geobacter on Anode Biofilms from domestic wastewater without posing anode potential in microbial electrochemical cells. *Microbiol Res 15*(3):1859–1869.

[55] Janda JM, Abbott SL. The genus *Shewanella*: From the briny depths below to human pathogen. Crit Rev Microbiol 2014;40(4):293–312.

[56] Hau HH, Gralnick JA. Ecology and biotechnology of the genus *Shewanella*. Annu Rev Microbiol 2007;61:237–258.

[57] Breuer M, Rosso KM, Blumberger J, Butt JN. Multi-haem cytochromes in *Shewanella* oneidensis MR-1: Structures, functions and opportunities. J R Soc Interface 2015;12(102).

[58] Shi L, Rosso KM, Clarke TA, Richardson DJ, Zachara JM, Fredrickson JK. Molecular underpinnings of Fe(III) oxide reduction by *Shewanella oneidensis* MR-1. Front Microbiol 2012;3(Feb):1–10.

[59] Myers CR, Nealson KH. Respiration-linked proton translocation couples to anaerobic reduction of manganese(iv) and iron(iii) in *Shewanella putrefaciens* MR-1. J Bacteriol 1990;172(11):6232–6238.

[60] Marsili E, Baron DB, Shikhare ID, Coursolle D, Gralnick JA, Bond DR. *Shewanella* secretes flavins that mediate extracellular electron transfer. Proc Natl Acad Sci U S A 2008;105(10):3968–3973.

[61] Lovley DR, Holmes DE. Electromicrobiology: The ecophysiology of phylogenetically diverse electroactive microorganisms. *Nature Reviews Microbiology 20*(1):5–19.

[62] Li F, Wang L, Liu C, Wu D, Song H. Engineering exoelectrogens by synthetic biology strategies. *Curr Opin Electrochem* 2018;*10*:37–45.

[63] Zhang J, Li F, Liu D, Liu Q, Song H. Engineering extracellular electron transfer pathways of electroactive microorganisms by synthetic biology for energy and chemicals production. *Chem Soc Rev 53*(3):1375–1446.

[64] Prévoteau A, Rabaey K. Electroactive biofilms for sensing: Reflections and perspectives. *ACS Sens 2*(8):1072–1085.

[65] Jiang Y, Yang X, Liang P, Liu P, Huang X. Microbial fuel cell sensors for water quality early warning systems: Fundamentals, signal resolution, optimization and future challenges. *Renew Sustain Energy Rev* 2018;*81*:292–305.

[66] Sonawane JM, Ezugwu CI, Ghosh PC. Microbial fuel cell-based biological oxygen demand sensors for monitoring wastewater: State-of-the-art and practical applications. *ACS Sens 5*(8):2297–2316.

[67] Kim GB, Kim WJ, Kim HU, Lee SY. Machine learning applications in systems metabolic engineering. *Curr Opin Biotechnol* 2020;*64*:1–9.

[68] Patra P, Disha BR, Kundu P, Das M, Ghosh A. Recent advances in machine learning applications in metabolic engineering. *Biotechnol Adv* 2023;*62*:108069.

[69] Badalamenti JP, Krajmalnik-Brown R, Torres CI. Generation of high current densities by pure cultures of anode-respiring Geoalkalibacter spp. under alkaline and saline conditions in microbial electrochemical cells. Mbio 2013;4(3).

[70] Zavarzina DG, Kolganova TV, Boulygina ES, Kostrikina NA, Tourova TP, Zavarzin GA. *Geoalkalibacter ferrihydriticus* gen. nov. sp. nov., the first alkaliphilic representative of the family geobacteraceae, isolated from a soda lake. Microbiol 2006;75(6):673–682.

[71] Greene A, Patel B, Yacob S. Geoalkalibacter subterraneus sp. nov., an anaerobic fe (iii)-and mn (iv)-reducing bacterium from a petroleum reservoir, and emended descriptions of the family desulfuromonadaceae and the genus geoalkalibacter. Int J Syst Evol Microbiol 2009;59(4):781–785.

[72] Parameswaran P, Bry T, Popat SC, Lusk BG, Rittmann BE, Torres CI. Kinetic, electrochemical, and microscopic characterization of the thermophilic, anode-respiring bacterium thermincola ferriacetica. Environ Sci Technol 2013;47(9):4934–4940.

[73] Marshall CW, May HD. Electrochemical evidence of direct electrode reduction by a thermophilic Gram-positive bacterium, thermincola ferriacetica. Energy Environ Sci 2009;2(6):699–705.

[74] Zavarzina DG, Sokolova TG, Tourova TP, Chernyh NA, Kostrikina NA, Bonch-Osmolovskaya EA. Thermincola ferriacetica sp. nov., a new anaerobic, thermophilic, facultatively chemolithoautotrophic bacterium capable of dissimilatory Fe(iii) reduction. Extremophiles 2007;11 (1):1–7.

[75] Huang L, Regan JM, Quan X. Electron transfer mechanisms, new applications, and performance of biocathode microbial fuel cells. Bioresour Technol 2011;102(1):316–323.

[76] He Z, Angenent LT. Application of bacterial biocathodes in microbial fuel cells. Electroanalysis 2006;18(19–20):2009–2015.

[77] Milner EM, Popescu D, Curtis T, Head IM, Scott K, Yu EH. Microbial fuel cells with highly active aerobic biocathodes. *Journal of Power Sources* 2016;*324*:8–16.

[78] Jeremiasse AW, Hamelers HVM, Buisman CJN. Microbial electrolysis cell with a microbial biocathode. Bioelectrochemistry 2010;78(1):39–43.

[79] Wingard LB, Shaw CH, Castner JF. Bioelectrochemical fuel cells. Enzyme Microb Technol 1982;4 (3):137–142.

[80] Miceli JF, Garcia-Peña I, Parameswaran P, Torres CI, Krajmalnik-Brown R. Combining microbial cultures for efficient production of electricity from butyrate in a microbial electrochemical cell. Bioresour Technol 2014;169:169–174.

[81] Harnisch F, Schröder U. From MFC to MXC: Chemical and biological cathodes and their potential for microbial bioelectrochemical systems. Chem Soc Rev 2010;39(11):4433–4448.

[82] Azwar MY, Hussain MA, Abdul-Wahab AK. Development of biohydrogen production by photobiological, fermentation and electrochemical processes: A review. Renew Sustain Energy Rev 2014;31:158–173. Available from: http://dx.doi.org/10.1016/j.rser.2013.11.022.

[83] Montpart N, Rago L, Baeza JA, Guisasola A. Hydrogen production in single chamber microbial electrolysis cells with different complex substrates. Water Res 2015;68:601–615.

[84] Gautam R, Nayak JK, Ress NV, Steinberger-Wilckens R, Ghosh UK. Bio-hydrogen production through microbial electrolysis cell: Structural components and influencing factors. *Chem Eng J* 2023;*455*:140535.

[85] Phan QHH, Phan TP, Nguyen PKT. Recent advances in powering microbial electrolysis cells for sustainable hydrogen production. *Int J Hydrogen Energy* 2025;*147*:150034.

[86] Ki D, Popat SC, Torres CI. Reduced overpotentials in microbial electrolysis cells through improved design, operation, and electrochemical characterization. *Chem Eng J* 2016;*287*:181–188.

[87] Lavender MB, Pang S, Liu D, Jourdin L, Ter Heijne A. Reduced overpotential of methane-producing biocathodes: Effect of current and electrode storage capacity. *Bioresource Technology* 2022;*347*:126650.

[88] Pophali A, Singh S, Verma N. Simultaneous hydrogen generation and COD reduction in a photoanode-based microbial electrolysis cell. *Int J Hydrogen Energy 45*(48):25985–25995.

[89] Park SG, Rajesh PP, Sim YU, Jadhav DA, Noori MT, Kim DH, Al-Qaradawi SY, Yang E, Jang JK, Chae KJ. Addressing scale-up challenges and enhancement in performance of hydrogen-producing microbial electrolysis cell through electrode modifications. *Energy Rep* 2022;*8*:2726–2746.

[90] Mohammadi F, Vakili-Nezhaad GR, Al-Rawahi N, Gholipour S. Microbial electrochemical systems for bioelectricity generation: Current state and future directions. *Results Eng* 2023;*20*:101619.

[91] Sun Y, He J, Yang G, Sun G, Sage V. A review of the enhancement of bio-hydrogen generation by chemicals addition. *Catalysts 9*(4):353.

[92] Jafary T, Daud WRW, Ghasemi M, Kim BH, Jahim JM, Ismail M, Lim SS Biocathode in microbial electrolysis cell; present status and future prospects. *Renew Sustain Energy Rev* 2015;*47*:23–33.

[93] Wang L, Long F, Liang D, Xiao X, Liu H Hydrogen production from lignocellulosic hydrolysate in an up-scaled microbial electrolysis cell with stacked bio-electrodes. *Bioresource Technology* 2021;*320*:124314.

[94] Campos-Martin JM, Blanco-Brieva G.. Fierro JLG.Hydrogen peroxide synthesis: An outlook beyond the anthraquinone process. Angew Chem Int Ed 2006;45(42):6962–6984. Available from:: http://dx.doi.org/10.1002/anie.200503779.

[95] FMC Corporation. Vision. 2012.

[96] De la Cruz N, Giménez J, Esplugas S, Grandjean D, De Alencastro LF, Pulgarin C. Degradation of 32 emergent contaminants by UV and neutral photo-fenton in domestic wastewater effluent previously treated by activated sludge. Water Res 2012;46(6):1947–1957.

[97] Yang W, Zhou H, Cicek N. Treatment of organic micropollutants in water and wastewater by UV-based processes: A literature review. Crit Rev Environ Sci Technol 2014;44(13):1443–1476.

[98] Glaze WH, Kang JW, Chapin DH. The chemistry of water treatment processes involving ozone, hydrogen peroxide and ultraviolet radiation. Ozone Sci Eng 1987;9(4):335–352. Available from:: http://dx.doi.org.ezproxy1.lib.asu.edu/10.1080/01919518708552148.

[99] Wagner M, Brumelis D, Gehr R. Disinfection of wastewater by hydrogen peroxide or peracetic acid: Development of procedures for measurement of residual disinfectant and application to a physicochemically treated municipal effluent. Water Environ Res 2002;33–50.

[100] Ksibi M. Chemical oxidation with hydrogen peroxide for domestic wastewater treatment. Chem Eng J 2006 Autumn;119(2–3):161–165. Available from: http://www.sciencedirect.com.ezproxy1.lib.asu.edu/science/article/pii/S1385894706001331.

[101] Kruithof JC, Kamp PC, Martijn BJ. UV/H_2O_2 treatment: A practical solution for organic contaminant control and primary disinfection. Ozone Sci Eng 2007;29(4):273–280. Available from: http://search.ebscohost.com/login.aspx?direct=true&db=eih&AN=26099550&site=ehost-live.

[102] Snyder S, Lei H, Wert E, Westerhoff P, Yoon Y. Removal of edcs and pharmaceuticals in drinking water. Water Environ Res Found 2008.

[103] Acero JL, Von Gunten U. Influence of carbonate on the ozone/hydrogen peroxide based advanced oxidation process for drinking water treatment. Ozone Sci Eng 2000;22(3):305–328. Available from: http://dx.doi.org.ezproxy1.lib.asu.edu/10.1080/01919510008547213.

[104] Drogui P, Elmaleh S, Rumeau M, Bernard C, Rambaud A. Oxidising and disinfecting by hydrogen peroxide produced in a two-electrode cell. Water Res 2001;35(13):3235–3241.

[105] Charron I, Feliers C, Couvert A, Laplanche A, Patria L, Requieme B. Use of hydrogen peroxide in scrubbing towers for odor removal in wastewater treatment plants. Water Sci Technol 2004;50(4):267–274.

[106] Metcalf & Eddy Inc. Wastewater Engineering: Treatment and Reuse, 5th edition, New York: McGraw-Hill Education; 2014.

[107] Rajala-Mustonen RL, Heinonen-Tanski H. Effect of advanced oxidation processes on inactivation of coliphages. Water Sci Technol 1995;31(5):131–134.

[108] Rozendal RA, Leone E, Keller J, Rabaey K. Efficient hydrogen peroxide generation from organic matter in a bioelectrochemical system. Electrochem Commun 2009 Sep;11(9):1752–1755.

[109] Modin O, Fukushi K. Production of high concentrations of H_2O_2 in a bioelectrochemical reactor fed with real municipal wastewater. Environ Technol 2013;34(19):2737–2742.

[110] Griffin J, Taw E, Gosavi A, Thornburg NE, Pramanda I, Lee HS., et al. Hybrid approach for selective sulfoxidation via bioelectrochemically derived hydrogen peroxide over a niobium(v)-silica catalyst. ACS Sustain Chem Eng 2018;6(6):7880–7889.

[111] Young MN, Links MJ, Popat SC, Rittmann BE, Torres CI. Tailoring microbial electrochemical cells for production of hydrogen peroxide at high concentrations and efficiencies. ChemSusChem 2016;9(23).

[112] Young MN, Chowdhury N, Garver E, Evans PJ, Popat SC, Rittmann BE., et al. Understanding the impact of operational conditions on performance of microbial peroxide producing cells. J Power Sources 2017;356.

[113] Sim J, Reid R, Hussain A, An J, Lee HS. Hydrogen peroxide production in a pilot-scale microbial electrolysis cell. Biotechnol Rep 2018;19:e00276. Available from: https://doi.org/10.1016/j.btre.2018.e00276.

[114] Kirova-Yordanova Z. Exergy analysis of industrial ammonia synthesis. Energy 2004;2912–2915. SPEC. ISS.:2373–84.

[115] Fowler D, Coyle M, Skiba U, Sutton MA, Cape JN, Reis S, et al. The global nitrogen cycle in the twenty-first century. Philos Trans R Soc B Biol Sci 2013;368:1621.

[116] Randall DG, Naidoo V. Urine: The liquid gold of wastewater. J Environ Chem Eng 2018;6 (2):2627–2635.

[117] Kelly PT, He Z Nutrients removal and recovery in bioelectrochemical systems: A review. Bioresour Technol 2014;153:351–360. Available from: http://dx.doi.org/10.1016/j.biortech.2013.12.046.

[118] Clauwaert P, Rabaey K, Aelterman P, De Schamphelaire L, Pham TH, Boeckx P, et al. Biological denitrification in microbial fuel cells. Environ Sci Technol 2007;41(9):3354–3360.

[119] You SJ, Ren NQ, Zhao QL, Kiely PD, Wang JY, Yang FL, et al. Improving phosphate buffer-free cathode performance of microbial fuel cell based on biological nitrification. Biosens Bioelectron 2009;24(12):3698–3701.

[120] Virdis B, Rabaey K, Yuan Z, Keller J Microbial fuel cells for simultaneous carbon and nitrogen removal. Water Res 2008;42(12):3013–3024.

[121] Virdis B, Rabaey K, Rozendal RA, Yuan Z, Keller J. Simultaneous nitrification, denitrification and carbon removal in microbial fuel cells. Water Res 2010;44(9):2970–2980.

[122] Villano M, Scardala S, Aulenta F, Majone M. Carbon and nitrogen removal and enhanced methane production in a microbial electrolysis cell. Bioresour Technol 2013;130:366–371.

[123] Snoeyink VL, Jenkins D. Water Chemistry, New York: John Wiley & Sons, Inc.; 1980.

[124] Kuntke P, Geleji M, Bruning H, Zeeman G, Hamelers HVM, Buisman CJN. Effects of ammonium concentration and charge exchange on ammonium recovery from high strength wastewater using a microbial fuel cell. Bioresour Technol 2011;102(6):4376–4382. Available from: http://dx.doi.org/10.1016/j.biortech.2010.12.085.

[125] Qin M, He Z. Self-supplied ammonium bicarbonate draw solute for achieving wastewater treatment and recovery in a microbial electrolysis cell-forward osmosis-coupled system. Environ Sci Technol Lett 2014;1(10):437–441.

[126] Kuntke P, Śmiech KM, Bruning H, Zeeman G, Saakes M, Sleutels THJA, et al. Ammonium recovery and energy production from urine by a microbial fuel cell. Water Res 2012;46(8):2627–2636.

[127] Kuntke P, Zamora P, Saakes M, Buisman CJN, Hamelers HVM. Gas-permeable hydrophobic tubular membranes for ammonia recovery in bio-electrochemical systems. Environ Sci Water Res Technol 2016;2(2):261–265.

[128] Kuntke P, THJA S, Saakes M, Buisman CJN. Hydrogen production and ammonium recovery from urine by a microbial electrolysis cell. Int J Hydrogen Energy 2014;39(10):4771–4778. Available from: http://dx.doi.org/10.1016/j.ijhydene.2013.10.089.

[129] Zamora P, Georgieva T, ter Heijne A, Sleutels THJA, Jeremiasse AW, Saakes M, et al. Ammonia recovery from urine in a scaled-up microbial electrolysis cell. J Power Sources 2017;356:491–499. Available from. http://dx.doi.org/10.1016/j.jpowsour.2017.02.089.

[130] Zhang Y, Angelidaki I. Submersible microbial desalination cell for simultaneous ammonia recovery and electricity production from anaerobic reactors containing high levels of ammonia. Bioresour Technol 2015;177:233–239.

[131] Zhang Y, Angelidaki I. Counteracting ammonia inhibition during anaerobic digestion by recovery using submersible microbial desalination cell. Biotechnol Bioeng 2015;112(7):1478–1482.

[132] Sotres A, Cerrillo M, Viñas M, Bonmatí A. Nitrogen recovery from pig slurry in a two-chambered bioelectrochemical system. Bioresour Technol 2015;194:373–382. Available from: http://dx.doi.org/10.1016/j.biortech.2015.07.036.

[133] Qin M, Molitor H, Brazil B, Novak JT, He Z. Recovery of nitrogen and water from landfill leachate by a microbial electrolysis cell-forward osmosis system. Bioresour Technol 2016;200:485–492.

[134] Qin M, Hynes EA, Abu-Reesh IM, He Z. Ammonium removal from synthetic wastewater promoted by current generation and water flux in an osmotic microbial fuel cell. J Clean Prod 2017;149:856–862. Available from: http://dx.doi.org/10.1016/j.jclepro.2017.02.169.

[135] Zou S, Qin M, Moreau Y, He Z. Nutrient-energy-water recovery from synthetic sidestream centrate using a microbial electrolysis cell – Forward osmosis hybrid system. J Clean Prod 2017;154:16–25. Available from: http://dx.doi.org/10.1016/j.jclepro.2017.03.199.

[136] Veluswamy GK, Shah K, Ball AS, Guwy AJ, Dinsdale RM. A techno-economic case for volatile fatty acid production for increased sustainability in the wastewater treatment industry. Environ Sci Water Res Technol 2021;7(5):927–941.

[137] Rittmann BE, McCarty PL. Environmental Biotechnology: Principles and Applications, 2nd edition, New York: McGraw-Hill; 2020.

[138] Dahiya S, Sarkar O, Swamy YV, Venkata Mohan S. Acidogenic fermentation of food waste for volatile fatty acid production with co-generation of biohydrogen. Bioresour Technol 2015;182:103–113. Available from: http://dx.doi.org/10.1016/j.biortech.2015.01.007.

[139] Puyol D, Batstone DJ, Hülsen T, Astals S, Peces M, Krömer JO. Resource recovery from wastewater by biological technologies: Opportunities, challenges, and prospects. Front Microbiol 2017;7 (Jan):1–23.

[140] Zhang Y, Angelidaki I. Bioelectrochemical recovery of waste-derived volatile fatty acids and production of hydrogen and alkali. Water Res [Internet] 2015;81:188–195. Available from: http://dx.doi.org/10.1016/j.watres.2015.05.058.

[141] Liu B, Kleinsteuber S, Centler F, Harms H, Sträuber H. Competition between butyrate fermenters and chain-elongating bacteria limits the efficiency of medium-chain carboxylate production. Front Microbiol 2020;11(March):1–13.

[142] Jiang Y, May HD, Lu L, Liang P, Huang X, Ren ZJ. Carbon dioxide and organic waste valorization by microbial electrosynthesis and electro-fermentation. Water Res 2019;149:42–55.

[143] Aktij SA, Zirehpour A, Mollahosseini A, Taherzadeh MJ, Tiraferri A, Rahimpour A. Feasibility of membrane processes for the recovery and purification of bio-based volatile fatty acids: A comprehensive review. J Ind Eng Chem 2020;81:24–40. Available from: https://doi.org/10.1016/j.jiec.2019.09.009.

[144] May HD, Evans PJ, LaBelle EV. The bioelectrosynthesis of acetate. Curr Opin Biotechnol [Internet] 2016;42:225–233. Available from: http://dx.doi.org/10.1016/j.copbio.2016.09.004.

[145] Yu L, Yuan Y, Tang J, Zhou S. Thermophilic moorella thermoautotrophica-immobilized cathode enhanced microbial electrosynthesis of acetate and formate from CO_2. Bioelectrochemistry [Internet] 2017;117:23–28. Available from: http://dx.doi.org/10.1016/j.bioelechem.2017.05.001.

[146] Jourdin L, Grieger T, Monetti J, Flexer V, Freguia S, Lu Y, et al. High acetic acid production rate obtained by microbial electrosynthesis from carbon dioxide. Environ Sci Technol 2015;49(22):13566–13574.

[147] Jourdin L, Freguia S, Flexer V, High-rate KJB.. Co_2-based microbial electrosynthesis closer to practical implementation through improved electrode design and operating conditions. Environ Sci Technol 2016;50(4):1982–1989.

[148] LaBelle EV, May HD. Energy efficiency and productivity enhancement of microbial electrosynthesis of acetate. Front Microbiol 2017;8(May):1–9.

[149] Roy M, Yadav R, Chiranjeevi P, Patil SA. Direct utilization of industrial carbon dioxide with low impurities for acetate production via microbial electrosynthesis. Bioresour Technol 2021;320(PA):124289. Available from: https://doi.org/10.1016/j.biortech.2020.124289.

[150] Jourdin L, Winkelhorst M, Rawls B, Buisman CJ, Strik DP. Enhanced selectivity to butyrate and caproate above acetate in continuous bioelectrochemical chain elongation from CO_2-steering with CO_2 loading rate and hydraulic retention time. *Bioresource Technology Reports 7*:1000284.

[151] Vassilev I, Hernandez PA, Batlle-Vilanova P, Freguia S, Krömer JO, Keller J, et al. Microbial electrosynthesis of isobutyric, butyric, caproic acids, and corresponding alcohols from carbon dioxide. ACS Sustain Chem Eng 2018;6(7):8485–8493.

[152] Jiang Y, Chu N, Qian DK, Jianxiong Zeng. R Microbial electrochemical stimulation of caproate production from ethanol and carbon dioxide. Bioresour Technol 2020;295(September 2019):122266. Available from: https://doi.org/10.1016/j.biortech.2019.122266.

[153] Ganigué R, Puig S, Batlle-Vilanova P, Balaguer MD, Colprim J. Microbial electrosynthesis of butyrate from carbon dioxide. Chem Commun 2015;51(15):3235–3238.

[154] Raes SMT, Jourdin L, Buisman CJN, DPBTB S. Continuous long-term bioelectrochemical chain elongation to butyrate. Chemelectrochem 2017;4(2):386–395.

[155] Batlle-Vilanova P, Ganigué R, Ramió-Pujol S, Bañeras L, Jiménez G, Hidalgo M, et al. Microbial electrosynthesis of butyrate from carbon dioxide: Production and extraction. Bioelectrochemistry 2017;117:57–64.

[156] Jourdin L, Raes SMT, Buisman CJN, Dpbtb S. Critical biofilm growth throughout unmodified carbon felts allows continuous bioelectrochemical chain elongation from CO_2 up to caproate at high current density. Front Energy Res 2018;6(Mar):1–15.

[157] Dessì P, Rovira-Alsina L, Sánchez C, Dinesh GK, Tong W, Chatterjee P, et al. Microbial electrosynthesis: Towards sustainable biorefineries for production of green chemicals from CO2 emissions. Biotechnol Adv 2021;46(December I):2020.

[158] Nancharaiah YV, Venkata Mohan S, Lens PNL. Metals removal and recovery in bioelectrochemical systems: A review. Bioresour Technol 2015;195:102–114.

[159] Wang G, Huang L, Zhang Y. Cathodic reduction of hexavalent chromium [Cr(vi)] coupled with electricity generation in microbial fuel cells. Biotechnol Lett 2008;30(11):1959–1966.

[160] Li Y, Lu A, Ding H, Jin S, Yan Y, Wang C, et al. Cr(vi) reduction at rutile-catalyzed cathode in microbial fuel cells. Electrochem Commun 2009;11(7):1496–1499. Available from. http://dx.doi.org/10.1016/j.elecom.2009.05.039.

[161] Li M, Zhou S, Xu Y, Liu Z, Ma F, Zhi L, et al. Simultaneous Cr(vi) reduction and bioelectricity generation in a dual chamber microbial fuel cell. Chem Eng J 2018;334(November 2017):1621–1629. Available from:. https://doi.org/10.1016/j.cej.2017.11.144.

[162] Liu X, Yin W, Liu X, Zhao X. Enhanced Cr reduction and bioelectricity production in microbial fuel cells using polypyrrole-coated MnO_2 on carbon cloth. Environ Chem Lett 2020;18(2):517–525.

[163] Gupta S, Yadav A, Verma N. Simultaneous Cr(vi) reduction and bioelectricity generation using microbial fuel cell based on alumina-nickel nanoparticles-dispersed carbon nanofiber electrode. Chem Eng J 2017;307:729–738. Available from: http://dx.doi.org/10.1016/j.cej.2016.08.130.

[164] Zhang Y, Yu L, Wu D, Huang L, Zhou P, Quan X, et al. Dependency of simultaneous Cr(vi), Cu(ii) and Cd(ii) reduction on the cathodes of microbial electrolysis cells self-driven by microbial fuel cells. J Power Sources 2015;273:1103–1113.

[165] Huang L, Yao B, Wu D, Quan X. Complete cobalt recovery from lithium cobalt oxide in self-driven microbial fuel cell – Microbial electrolysis cell systems. J Power Sources 2014;259:54–64. Available from: http://dx.doi.org/10.1016/j.jpowsour.2014.02.061.

[166] Jiang L, Huang L, Sun Y. Recovery of flakey cobalt from aqueous Co(ii) with simultaneous hydrogen production in microbial electrolysis cells. Int J Hydrogen Energy 2014;39(2):654–663.

[167] Huang T, Song D, Liu L, Zhang S. Cobalt recovery from the stripping solution of spent lithium-ion battery by a three-dimensional microbial fuel cell. Sep Purif Technol 2019;215(January):51–61. Available from: https://doi.org/10.1016/j.seppur.2019.01.002.

[168] Saad DR, Alismaeel ZT, Abbar AH. Cobalt removal from simulated wastewaters using a novel flow-by fixed bed bio-electrochemical reactor. Chem Eng Process – Process Intensif [Internet] 2020;156 (May):108097. Available from: https://doi.org/10.1016/j.cep.2020.108097.

[169] Huang L, Liu Y, Yu L, Quan X, Chen G. A new clean approach for production of cobalt dihydroxide from aqueous Co(ii) using oxygen-reducing biocathode microbial fuel cells. J Cleaner Prod 2015;86:441–446.

[170] Shen J, Sun Y, Huang L, Yang J. Microbial electrolysis cells with biocathodes and driven by microbial fuel cells for simultaneous enhanced Co(ii) and Cu(ii) removal. Front Environ Sci Eng 2015;9 (6):1084–1095.

[171] Heijne AT, Liu F, van der W R, Weijma J, Buisman CJN, Hamelers HVM. Copper recovery combined with electricity production in a microbial fuel cell. Environ Sci Technol 2010;44(11):4376–4381.

[172] Tao HC, Liang M, Li W, Zhang LJ, Ni JR, Wu WM. Removal of copper from aqueous solution by electrodeposition in cathode chamber of microbial fuel cell. J Hazard Mater 2011;189(1–2):186–192. Available from: http://dx.doi.org/10.1016/j.jhazmat.2011.02.018.

[173] Wu Y, Zhao X, Jin M, Li Y, Li S, Kong F, et al. Copper removal and microbial community analysis in single-chamber microbial fuel cell. Bioresour Technol 2018;253(November 2017):372–377. Available from. https://doi.org/10.1016/j.biortech.2018.01.046.

[174] Modin O, Wang X, Wu X, Rauch S, Fedje KK. Bioelectrochemical recovery of Cu, Pb, Cd, and Zn from dilute solutions. J Hazard Mater 2012;235–236:291–297.

[175] Ai C, Hou S, Yan Z, Zheng X, Amanze C, Chai L, et al. Recovery of metals from acid mine drainage by bioelectrochemical system inoculated with a novel exoelectrogen, pseudomonas sp.E8. Microorganisms 2020;8(1).

[176] Choi C, Hu N, Lim B. Cadmium recovery by coupling double microbial fuel cells. Bioresour Technol 2014;170:361–369.

[177] Abourached C, Catal T, Liu H. Efficacy of single-chamber microbial fuel cells for removal of cadmium and zinc with simultaneous electricity production. Water Res 2014;51:228–233.

[178] Colantonio N, Kim Y. Cadmium (II) removal mechanisms in microbial electrolysis cells. J Hazard Mater 2016;311:134–141. Available from: http://dx.doi.org/10.1016/j.jhazmat.2016.02.062.

[179] Choi C, Cui Y. Recovery of silver from wastewater coupled with power generation using a microbial fuel cell. Bioresour Technol 2012;107:522–525.

[180] Tao HC, Gao ZY, Ding H, Xu N, Wu WM. Recovery of silver from silver(i)-containing solutions in bioelectrochemical reactors. Bioresour Technol 2012;111:92–97.

[181] Yh W, BS W, Pan B, Chen QY, Yan W. Electricity production from a bio-electrochemical cell for silver recovery in alkaline media. *Appl Energy* 2013 Dec 1;112:1337–1341.

[182] Lim BS, Lu H, Choi C, Liu ZX. Recovery of silver metal and electric power generation using a microbial fuel cell. Desalin water treat 2015;54(13):3675–3681.

[183] Kamperidis T, Tremouli A, Remoundaki E, Lyberatos G. Silver recovery from wastewater, simulating the chemical extract originating from a PV panel using microbial fuel cell technology. Waste Biomass Valorization 2022;7(0123456789):Available from: https://doi.org/10.1007/s12649-022-01793-y.

[184] Choi C, Hu N. The modeling of gold recovery from tetrachloroaurate wastewater using a microbial fuel cell. Bioresour Technol 2013;133:589–598.

[185] Catal T, Bermek H, Liu H. Removal of selenite from wastewater using microbial fuel cells. Biotechnol Lett 2009;31(8):1211–1216.

[186] Wang Z, Lim B, Choi C. Removal of Hg^{2+} as an electron acceptor coupled with power generation using a microbial fuel cell. Bioresour Technol 2011;102(10):6304–6307. Available from: http://dx.doi.org/10.1016/j.biortech.2011.02.027.

[187] Saad DR, Alismaeel ZT, Abbar AH. Cobalt removal from simulated wastewaters using a novel flow-by fixed bed bio-electrochemical reactor. Chem Eng Process – Process Intensif [Internet] 2020;156 (August):108097. Available from: https://doi.org/10.1016/j.cep.2020.108097.

[188] Gaetke LM, Chow CK. Copper toxicity, oxidative stress, and antioxidant nutrients. *Toxicology* 2003 Jul 15;189(1–2).

[189] Westerhoff P, Lee S, Yang Y, Gordon GW, Hristovski K, Halden RU, et al. Characterization, recovery opportunities, and valuation of metals in municipal sludges from U.S. wastewater treatment plants nationwide. Environ Sci Technol 2015;49(16):9479–9488.

[190] Yun-Hai W, Bai-Shi W, Bin P, Qing-Yun C, Wei Y. Electricity production from a bio-electrochemical cell for silver recovery in alkaline media. Appl Energy 2013;112:1337–1341.

[191] Gajda I, Greenman J, Ieropoulos IA. Recent advancements in real-world microbial fuel cell applications. Curr Opin Electrochem [Internet] 2018;11:78–83. Available from: https://doi.org/10.1016/j.coelec.2018.09.006.

[192] Kim T, Kang S, Kim HW, Paek Y, Sung JH, Kim YH, et al. Assessment of organic removal in series- and parallel-connected microbial fuel cell stacks. Biotechnol Bioprocess Eng 2017;22(6):739–747.

[193] Yang W, Li J, Ye D, Zhang L, Zhu X, Liao Q. A hybrid microbial fuel cell stack based on single and double chamber microbial fuel cells for self-sustaining ph control. J Power Sources 2016;306:685–691. Available from: http://dx.doi.org/10.1016/j.jpowsour.2015.12.073.

[194] Oh SE, Logan BE. Voltage reversal during microbial fuel cell stack operation. J Power Sources 2007;167(1):11–17.

[195] Kim B, Mohan SV, Fapyane D, Chang IS. Controlling voltage reversal in microbial fuel cells. Trends Biotechnol 2020;38(6):667–678. Available from: https://doi.org/10.1016/j.tibtech.2019.12.007.

[196] Ki DW. Anaerobic Conversion of Primary Sludge to Resources in Microbial Electrochemical Cells, Arizona State University; 2016.

[197] Chen J, Yang Y, Liu Y, Tang M, Wang R, Tian Y, et al. Bacterial community shift and antibiotics resistant genes analysis in response to biodegradation of oxytetracycline in dual graphene modified bioelectrode microbial fuel cell. Bioresour Technol 2019;276(January):236–243. Available from. https://doi.org/10.1016/j.biortech. 2019.01.006.

[198] Membranes Interenational Inc. MI. Price Li [cited 2022 Jun 4]. Available from: https://ionexchangemembranes.com/price-list/st [Internet]. 2022.

[199] Ge Z, He Z Long-term performance of a 200 liter modularized microbial fuel cell system treating municipal wastewater: Treatment, energy, and cost. Environ Sci Water Res Technol 2016;2 (2):274–281.

[200] Frank MR, Autor D, Bessen JE, Brynjolfsson E, Cebrian M, Deming DJ, Feldman M, Groh M, Lobo J, Moro E, Wang D. Toward understanding the impact of artificial intelligence on labor. *Proc Natl Acad Sci U S A 116*(14):6531–6539.

[201] Yoho R. No, let's not go back to handwritten activities: Inclusive teaching strategies in the context of ChatGPT. *Natl Teach Learn Forum* 2023October;32(6).

[202] Hohenstein J, Kizilcec RF, DiFranzo D, Aghajari Z, Mieczkowski H, Levy K, Naaman M, Hancock J, Jung MF. Artificial intelligence in communication impacts language and social relationships. *Sci Rep* *13*(1):5487.

[203] Guingrich RE, Graziano MS. Ascribing consciousness to artificial intelligence: Human-AI interaction and its carry-over effects on human-human interaction. *Frontiers in Psychology* 2024;*15*:1322781.

[204] Musib M, Wang F, Tarselli MA, Yoho R, Yu KH, Andrés RM, Greenwald NF, Pan X, Lee CH, Zhang J, Dutton-Regester K. Artificial intelligence in research. *Science 357*(6346):28–30.

[205] Sripathi KN, Moscarella RA, Steele M, Yoho R, You H, Prevost LB, Urban-Lurain M, Merrill J, Haudek KC. Machine learning mixed methods text analysis: An illustration from automated scoring models of student writing in biology education. *J Mix Methods Res 18*(1):48–70.

[206] Luo S, Sun H, Ping Q, Jin R, He Z. A review of modeling bioelectrochemical systems: Engineering and statistical aspects. *Energies 9*(2):111.

[207] Gadkari S, Gu S, Sadhukhan J. Towards automated design of bioelectrochemical systems: A comprehensive review of mathematical models. *Chem Eng J* 2018;*343*:303–316.

[208] Gadkari S. Application of artificial intelligence methods for the optimization and control of bioelectrochemical systems, In *Scaling Up of Microbial Electrochemical Systems*, Elsevier; pp. 437–455.

[209] Li C, Guo D, Dang Y, Sun D, Li P. Application of artificial intelligence-based methods in bioelectrochemical systems: Recent progress and future perspectives. *J Environ Manage* 2023;*344*:118502.

[210] Mousazadehgavan M, Hajalifard Z, Basirifard M, Afsharghoochani S, Mirkhalafi S, Kabdaşlı I, Hashim K, Nakouti I. The critical role of artificial intelligence in optimizing electrochemical processes for water and wastewater remediation: A state-of-the-art review. *ACS EST Water* 2025.

[211] Garg A, Vijayaraghavan V, Mahapatra SS, Tai K, Wong CH. Performance evaluation of microbial fuel cell by artificial intelligence methods. *Expert Syst Appl 41*(4):1389–1399.

[212] Fathy A, Rezk H, Yousri D, Alharbi AG, Alshammari S, Hassan YB. Maximizing bio-hydrogen production from an innovative microbial electrolysis cell using artificial intelligence. *Sustainability 15*(4):3730.

[213] Senthil Rathi B, Senthil Kumar P, Sanjay S, Prem Kumar M, Rangasamy G. Artificial intelligence integration in conventional wastewater treatment techniques: Techno-economic evaluation, recent progress and its future direction. *Int J Environ Sci Technol 22*(1):633–658.

[214] Gupta A, Verma A, Katiyar K. Synergistic potential of microbial communities and Artificial Intelligence in strengthening sustainable wastewater treatment solutions, In *Microbial Niche Nexus Sustaining Environmental Biological Wastewater and Water-Energy-Environment Nexus*, Cham: Springer Nature Switzerland; pp. 225–249.

[215] Mahmood Q, Gazder U, Cai J, Khan IA, Hung YT. Explainable artificial intelligence analysis of simultaneous anaerobic Nitrogen and Sulfur removal in anaerobic Sulfide Oxidation bioreactor. *Water 17*(13):1880.

[216] Jiang J, Lin Q, Guan X, Zhou S, Zhong S, Xiang X, Wang Z, Chen G, Guo G. Using machine learning and GPT Models To enhance electrochemical pretreatment of Anaerobic Cofermentation: Prediction, early warning, and biomarker identification. *ACS ES&T Engineering 5*(5):1149–1159.

[217] Akkaş T, Reshadsedghi M, Şen M, Kılıç V, Horzum N. The role of artificial intelligence in advancing Biosensor technology: Past, present, and future perspectives. *Advanced Materials* 2025;2504796.

[218] Bocan A, Siavash Moakhar R, Del Real Mata C, Petkun M, De Iure-Grimmel T, Yedire SG, Shieh H, Khorrami Jahromi A, Mahshid SS, Mahshid S. Machine-learning-aided advanced electrochemical biosensors. *Advanced Materials* 2025;2417520.

[219] Trivedi P, Kumar A, Gupta N, Patel CN. Artificial intelligence and machine learning in microbial degradation of pollutants and toxins, In *Microbial Metabolomics: Recent Developments, Challenges and Future Opportunities*, Singapore: Springer Nature Singapore; pp. 377–400.

Section III: **Synthetic biology**

Beth Bannerman, Sunil Chandran

Chapter 14
Clean chemistry: powered by biology

Editor's note: This chapter provides a compelling look at the startup process and technology development for a bioproducts company. This chapter is retained from the first edition of the book to highlight the opportunity and early stages of creative development.

14.1 Introduction

"By far, nature is the best chemist of all time." This is a statement by Nobel Laureate Frances Arnold, who conducted the first directed evolution of enzymes – proteins that catalyze chemical reactions. The uses of her results include more environmentally friendly manufacturing of chemical substances, such as pharmaceuticals, and the production of renewable fuels.

Nature, as we know, has been catalyzing chemical reactions for 4.5 billion years. It is indisputable that organisms have adapted and evolved during the history of life on Earth.

Natural chemical reactions and biological evolution, the process by which species of life (plants and animals) adapt over time, in response to their changing environment, have created enormous diversity of life. There is a variety of evidence for evolution – one is molecular biology. DNA and the genetic code reflect the shared ancestry of life.

We see evidence in fossils too. From the shape of a human skeleton and skull, we can see how humans have evolved over time and differently, by geography. What is not as easy to see, but no less clear, are the imprints of microscopic organisms left in rocks about 3.7 billion years old [1].

Although the biotechnology developed by Amyris Inc. is the most advanced in the world, it is simply science powered by nature. Amyris leverages nature's own catalytic conversion process (fermentation) to produce molecules that already exist in nature.

That is how we define clean manufacturing. In this chapter, we will describe the science in more detail, the near limitless opportunity of its application and benefits to nature as well as the economy. We will also describe our own evolution as a company through mid-2022 (the time of this writing) and share the lessons we have learned along the way.

Beth Bannerman, Formerly of Amyris, Inc., San Francisco, CA, USA
Sunil Chandran, Amyris, Inc., San Francisco, CA, USA, Amyris Company, e-mail: info@amyris.com

 | https://doi.org/10.1515/9783112212141-014

14.2 The founders

The year was 2002. In a laboratory at the University of California, Berkeley, four postdoctoral fellows, along with the professor of Chemical and Biomolecular Engineering, were discovering new ways to apply engineering principles to biological systems.

Working with single-cell organisms such as bacteria or yeast, known for their catalytic capabilities to ferment beer, bread, and wine, they discovered how to transform yeast into a "living factory" that could produce almost any compound that exists in the natural world through fermentation. It was a major scientific breakthrough with enormous potential to revolutionize medicine and every other commercial field, with a dependence on natural compounds. Especially for natural compounds that were rare or difficult and expensive to manufacture by conventional chemical routes, this clean manufacturing process was significantly faster [2] and less costly. In addition, the natural compounds produced using this new method were pure and bioidentical to those that existed in nature. The process was sustainable and less destructive to the environment.

Professor Jay Keasling had already been widely known as a pioneer in synthetic biology. Together with Vince Martin, Jack Newman, Kinkead Reiling, and Neil Renninger [3], these scientists recognized the potential of their discovery and in 2003, started Amyris Biotechnologies to change the world.

Galley sidebar:
The name Amyris (plant name *Amyris balsamifera*) was chosen by Jack Newman's wife. An avid gardener, she had been taken by the plant's use as a substitute for real sandalwood. The promise of substituting nature-for-nature fit well with the researchers' work to use biology, to develop an alternative source for Artemisinin. The scientists' passion for their work was also neatly captured in the word Amyris, a homophone for Amorous.

With a company name chosen and the realization that their technology could be applied to almost any chemical compound to address a range of global problems, their first question was where to begin?

14.3 Purpose-driven from the start

That same year, the Africa Malaria report[1] was published by the World Health Organization (WHO) and UNICEF, which said that malaria remained "the single biggest cause of death of young children in Africa and one of the most important threats to the health of the pregnant women and their newborns."

1 http://apps.who.int/iris/bitstream/handle/10665/67869/WHO_CDS_MAL_2003.1093.pdf;jsessionid=63282432317F59B29A64066179B73771?sequence=1.

According to Renninger, "We saw this as a great application of the technology. If you can make one compound, you've built most of the technology required to make any one of those compounds. This was a great way to launch the company."

The team published a paper in *Nature Biotechnology* describing how they created a pathway by inserting genes into *Escherichia coli*, a common bacteria. The findings demonstrated how biotechnology could help advance the United Nations' Millennium Development Goals [4], aligned to reversing global poverty, hunger, and disease affecting billions of people. The UN called for reducing the deaths from malaria, a preventable mosquito-borne infectious disease and WHO was recommending artemisinin-based combination therapies (ACTs) as an effective malaria treatment.

The challenge was the availability of the key ingredient, artemisinin, a natural compound traditionally found in the extract of *Artemisia annua*, or the sweet wormwood plant. While ACTs are the most effective treatment, *A. annua* crop yields can be unpredictable from year to year, causing price volatility, ranging from $400 to $1,200 per kg [5], and impacting the overall availability of this life-saving drug.

The next year, in 2004, the Bill and Melinda Gates Foundation [6] granted $42.6 million to the Institute for OneWorld Health (a nonprofit pharmaceutical company) to create new approaches to developing a more affordable cure for malaria. Amyris partnered with OneWorld Health and UC Berkeley to develop a scalable, stable, and cost-effective production process for artemisinin. This funding set Amyris on the path to engineer its first molecule, proving its capability as a biotech pioneer on a global stage, and paving the way to a dependable and secure supply of antimalarial treatments produced in a sustainable and scalable way.

Amyris successfully engineered a microbe to produce artemisinic acid, a molecule that can be converted to artemisinin, allowing for on-demand, clean manufacturing to meet the global need for artemisinin and ACTs. Amyris demonstrated that its technology not only demonstrated how modern manufacturing could stabilize costs and supply but also save lives. Amyris went on to license its artemisinic acid production strain and process to Sanofi via OneWorld Health and later, Amyris was recognized with a United Nations Global Citizen Award [7] based on its contributions toward several aspects of the UN's Millennium Development Goals, including the production of artemisinic acid through clean manufacturing, powered by biology. The demand and production for artemisia farmers, as shown in Figure 14.1, is also an important consideration.

By mid-2005, after receiving the Gates grant, Amyris was exploring new applications and chemical compounds, with a connection to their earlier work. Also at this time, the team was seeking new investment from Silicon Valley venture capitalists, who happened to be particularly interested in global warming and renewable fuels. Amyris had identified a few interesting fuel candidates belonging to the isoprenoid class of molecules (also called terpenoids), a source of complex hydrocarbons, which are the dominant component of crude oil, gasoline, and diesel fuel.

This would prove to be an important turning point for the company.

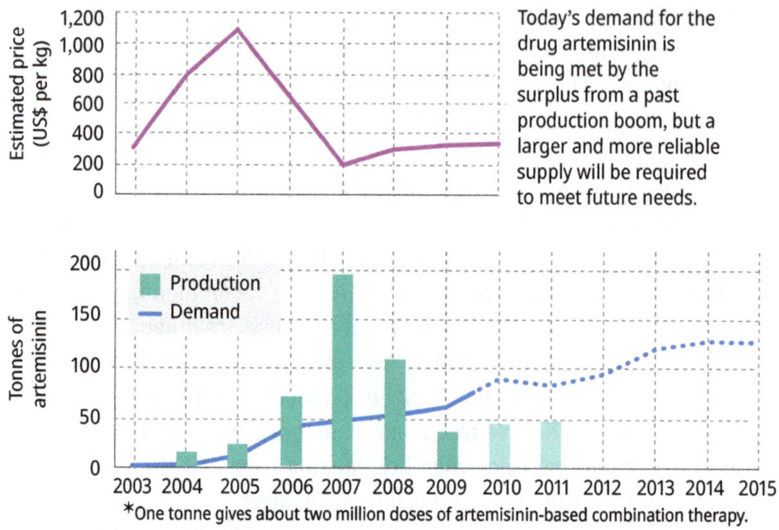

Figure 14.1: Boom and bust for artemisia farmers.
Source: *iGem Academy*.

14.4 Purpose-driven: clean air

Fossil fuel use is one of the biggest sources of greenhouse gas emissions that contribute to global warming. The decade ahead (2010s) would become known for its climate change consequences, human-driven extreme weather events, including heatwaves, floods, droughts, and wildfires, impacting millions of people around the world, from Russia to Pakistan and East Africa. Global temperatures soared to new records and by 2019, global levels of carbon dioxide in the atmosphere reached 415 parts per million for the first time in human history.

Amyris was positioned well with technology to help address this global problem. With demonstrated capability in the pharmaceutical sector, Amyris had also secured $20 million [8] in Series A funding from Khosla Ventures, Kleiner Perkins Caufield & Byers, and Texas Pacific Group Ventures and, at the same time, in 2007, brought on a new Chief Executive Officer, John Melo, formerly President of US fuel operations at the oil firm, BP.

According to Amyris Board Director Geoff Duyk, Amyris was realistic in terms of what it would take to fully realize the value of its technology. "Unless you are vertically integrated, too much of the rent flows away from you," he said. "Amyris recognized the need to be responsible for both building and deploying." Driven by the belief that being vertically integrated would capture the most value over time, Amyris prepared itself for what would be a major change for the company in the years to come.

In the next few years, Amyris would partner with one of the largest sugar and ethanol distributors in Brazil and form a joint venture with an ethanol-producing mill to build a production plant for renewable fuels. By early 2009, against the backdrop of a global financial collapse, Amyris' renewable diesel biology was technically viable, the product was of the highest quality but production costs were not yet competitive. The company had grown to 200 employees in Emeryville, CA, with a team of 40 people working in Brazil. Manufacturing operations had been established in partnership with Brazilian producers of sugarcane – the feedstock for Amyris' fermentation and the essential element to its clean manufacturing.

Innovation in the global biofuels market came in several waves: the first from corn, which then raised concerns about food availability and questions about the amount of energy that corn produces, being lower than the amount of energy it takes to produce; the second wave of biofuel production was from biomass such as food crop waste, eliminating the main issues of the first generation; and the third generation used algae, more efficient than previous generations. Amyris participated in the fourth wave of biofuels, which used biomass that captured CO_2 even more efficiently [9]. Global production of biofuels was up, but the cost of petroleum had swung wildly between 2000 and 2010; from $30/barrel in 2000 to $133/barrel in July 2008 and back down to $39/barrel in early 2009 [10]. Ultimately, the economics of production could not compete with the variability of the oil market and so Amyris began exploring extensions of its science and technology platform, beyond biofuels.

In 2010, Amyris became a public company, selling its stock in an initial public offering [11]. By the end of 2012, Amyris was producing its first commercial molecule, farnesene, for specialty chemicals and renewable fuels. Branded by Amyris as Biofene, this terpenoid, which is found in many plants, including green apples, can also be converted to other compounds, including ingredients used for flavors, fragrances, and cosmetics. In 2014, Amyris earned the distinction of the Presidential Green Chemistry Award from the Environmental Protection Agency [12] for its clean manufacturing approach to farnesane, a breakthrough renewable hydrocarbon for use as diesel and jet fuel: Initial studies showed that farnesane can reduce greenhouse gas emissions by as much as 80% when compared to fossil jet fuels [13].

By this time, Amyris had reached another crossroads. Its science had so much optionality that the challenge once again was to evaluate and choose the best next step. The company had increasingly become interested in supplying ingredients to consumer product companies because their products used many petroleum-based ingredients that could be replaced by farnesene alternatives.

The company's growth plan would rely on partnerships, which take time to develop. Said two Amyris executives at the time,

> We don't plan to market the consumer products ourselves, so we are working with partners; however, it might take these companies some time to research and launch potential applications. As a result, it is hard to predict the volume and pricing of the chemical products at this point. It's

> more than convincing our customers. Our [business] customers convince their consumers, that the new formulation is better.

Because the production pipeline at the time was singularly focused on farnesene, a precursor to other chemical compounds, there was minimal impact on their manufacturing strategy and a wide array of opportunity for end uses. To determine the best end-use for farnesene, Amyris looked at three primary criteria: technical fit, financial impact, and regulatory approval and timelines. Amyris recognized that the first company with commercially viable products would lead the market and the perception of the biotech industry. So, Amyris invested further in R&D with new robotics, machine learning, and artificial intelligence. The company developed its own programming code, Genotype Sequencing Language. Amyris soon began optimizing the speed and costs of developing new yeast strains, efficient living factories, to get the most conversion from sugars from the fermentation process to the final product. By now, there were more chemical compounds for specialty ingredients in active development and testing.

Three major insights influenced the direction Amyris took next with its science: (1) Amyris was leading the biotech sector with its ability to commercialize products by converting sugarcane syrup into specific chemical compounds, providing the company with optionality to disrupt virtually any industry, particularly consumer goods; (2) by moving up the value chain, from ingredient supplier to consumer goods producer, Amyris could realize a greater percentage of the end products' profitability, and (3) Amyris could potentially gain 10–30 times better capital efficiency [14] by producing specialty ingredients (lower volume, higher margin), compared to biofuel or other commodity ingredients (high volume, low margin).

Galley sidebar:
Commodity chemicals and renewable fuels generated less than $1 of revenue for every kilogram of product and less than $0.2 per kilogram in gross margins, while specialty chemicals generated $39 per kilogram of revenue and $25 per kilogram of gross margins [15].

14.5 The business of Lab-to-Market™

These insights marked an important pivot in Amyris' evolution. Amyris could now begin positioning itself with broader value: Amyris is disrupting conventional production systems that rely on destructive and nonsustainable practices and replace them with highly scalable, clean manufacturing powered by biology. Its science and technology platform enables a scientist to precisely engineer microbes like *Saccharomyces cerevisiae* (brewer's yeast) to ferment sugar and produce a variety of bioidentical molecules, which serve as natural ingredients that consumers use on a day-to-day basis. Using its technology platform, Amyris accesses better performing natural ingredients at lower cost, and then produces them at scale in a way that is far better for the environment, by not depleting natural resources.

A microbe like *Saccharomyces cerevisiae*, which has evolved for billions of years, can be further modified to consume sugar to produce a target molecule. These microbes can be seen as hardware and the genetic code as software to turn the hardware into tiny living factories that go on to make ingredients used in skincare, fragrances, flavors, health, and nutritional supplements, in addition to biofuels. Coding is complex enough in computer science, let alone in living organisms. Amyris accelerates the cycle of design, test, learn, and iterate with every microbe it programs.

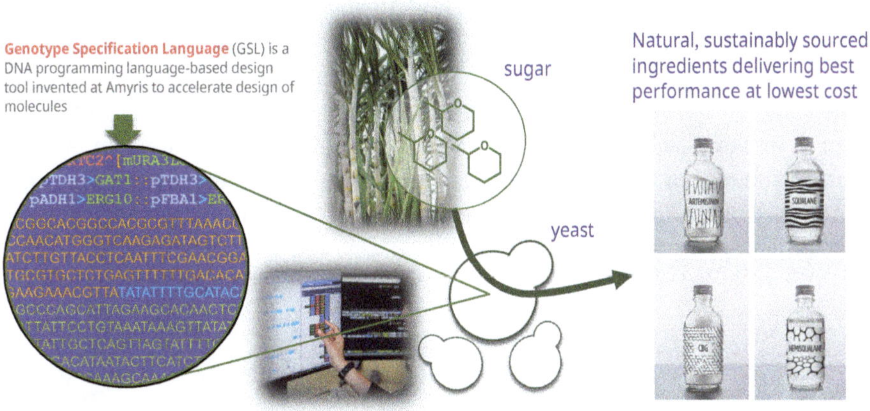

Figure 14.2: We program cells to create high-performing sustainable ingredients.
Source: *Amyris.*

Amyris improves the predictability and repeatability of its success using its proprietary Lab-to-Market™ technology platform. This platform has evolved to continuously learn from previous iterations via highly optimized and automated molecular biology, and analytical and process development tools, combined with machine learning algorithms and statistical models, enabling the collection of larger amounts of quality data. A fully integrated, end-to-end, R&D, process development, and manufacturing solution can disrupt how companies traditionally approach this problem. This integration means that Amyris starts every project with the end in mind. Before design on the first strain begins, the team already has a good understanding of its manufacturing requirements, cost targets, and project timeline.

Since its start, Amyris has grown its business designing, developing, and producing clean chemistry. By shifting to sustainably sourced, clean manufacturing, Amyris is advancing its mission to accelerate the world's transition to sustainable consumption. A key to its success is the prioritization of opportunities. Amyris has applied its technology to solve supply chain volatility and make rare molecules from nature abundant, while also being sustainably sourced. This aligns with a great consumer demand trend of our time and the need for companies to lower costs and access reliable supply. The flexibility of its technology platform has enabled Amyris to build an expansive product portfolio of molecules.

What Amyris can make with this platform is almost unbounded in diversity. Already, Amyris has successfully engineered strains that have produced greater than 250 molecules through fermentation from over 20 different biological pathways. These 250 molecules unlock thousands more: not only can Amyris access upstream biochemical intermediates, but every time the company optimizes a pathway, it can also produce hundred more related compounds. Moreover, Amyris can use simple chemistry to further diversify every scaffold. All told, Amyris can access hundreds of thousands of potential targets, or over half of all small molecule diversity found in nature. See Figure 14.2.

This technology is at the core of Amyris' value – the ability to make molecules and the ability to scale molecules –13 to date – to the metric ton or even kiloton scale, at a speed unparalleled in the sector to date. With each successful scale-up, Amyris leverages its know-how to optimize its next success in pathway engineering and analytics to measure the molecules, developing processes for fermentation and purification, and running unit operations at scale.

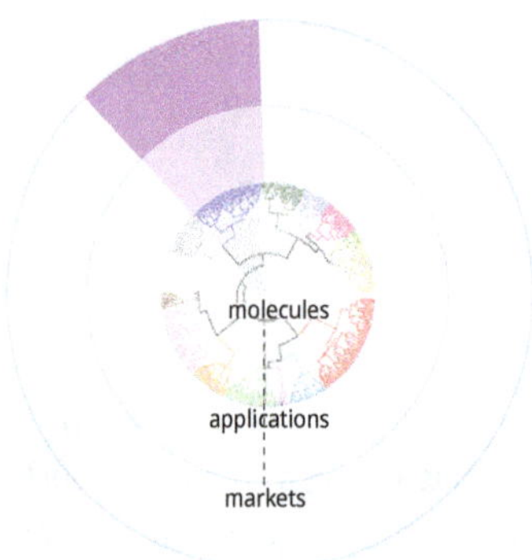

Ingredient	Application	Market
Farnesene	Polymers	High Performance Materials
Vitamins	Nutrition Ingredient	Human and Animal Health
Squalane	Emollient	Clean Beauty
Hemisqualane	Silicone Replacement	Clean Beauty/Personal Care
Squalene	Immunostimulant	Pharma

Figure 14.3: A single class of molecules creates significant commercial value.
Source: *Amyris.*

Technology on its own is not as impactful as its application. In the realm of personal care, Amyris has commercialized effective formulations, which are on the market today – as borne out by extensive clinical study and benchmarking. Like Amyris' core technology, this know-how builds on itself and becomes exponentially stronger over time as Amyris enhances its knowledge base. In applications where Amyris is not an expert, it forms partnerships with market leaders to stay focused on only the most relevant and urgent applications and to drive adoption for accelerated scale and impact.

The real value and impact are realized once Amyris' ingredients reach consumers at a massive scale. Amyris has leveraged economies of scale across its formulation, manufacturing, distribution, digital infrastructure, marketing, and consumer experience capabilities to grow brands that will reach new audiences. And this again is why Amyris selects top partners in their respective markets.

One example of how value shows up through Amyris' capabilities of scaling ingredients, identifying applications, and getting them to market is a single molecule known as farnesene. Farnesene has given rise to billions of dollars of value to the company through a variety of applications that can access a wide range of end markets. Derivatives of farnesene include vitamin E, squalane, hemisqualane, and squalene, proving how combining fermentation and chemistry amplifies value. As with all its molecules, Amyris provides a No Compromise™ solution: best performance in its category, for the best cost, using a sustainable process. See Figure 14.3.

When choosing the next target molecule to make, Amyris applies three filters: market, technology, and efficacy.

Ideally, there is a large total addressable market for a target molecule made from animals or plants within a challenged supply chain that is subject to variable availability, price, and quality. There may be further value driven for our brands and partnerships through offering a distinctive advantage in cost and performance.

Amyris also considers the technology investment that would be required. How much existing experience in R&D, process development, and manufacturing can be leveraged? What strains, processes, and capital expenditures are already in place? If little, does the market opportunity justify new investment?

Efficacy is critical to Amyris. First and foremost, the molecule must be safe for its application. Second, the target molecule must have best-in-class activity in its respective applications. Finally, there must be complementary and synergistic functions to ingredients in Amyris' existing portfolio to capture added value from the portfolio of ingredients available for formulation.

Over the years, Amyris has been continuously innovating and investing in its technology platform. Its first commercial product took nearly 40 months to develop from strain to pilot plant run. And today, Amyris averages less than 12 months. The cost of product development has dropped by 90%, time to market has reduced by 80%, and Amyris has been able to do this by increasing the bandwidth of its R&D pipeline by 500% since 2012, while increasing operating expenditures by only 20%.

14.6 Purpose-driven: clean beauty

One notable chemical compound that Amyris developed from farnesene was squalane. For decades, squalene had been harvested from sharks' liver and then converted via a hydrogenation process into squalane. Livers of an estimated 3,000 sharks are required to produce just under 1 ton of squalane. Up to 2.7 million deep-sea sharks [16] a year were killed to meet the global demand for squalane in the cosmetic industry alone. After environmentalists pushed for the cosmetic industry to stop using shark liver oil, scientists turned to olives, a plant-based source of squalene.

The undisputable skin health benefits [17] of squalane are the reason for its longstanding global appeal:

- Provides superior moisturization to the skin longer, compared to other oils
  - Moisture is vital for healthy skin. Your skin produces oil naturally to support moisture, strength, and elasticity. When applied to the skin, squalane mimics the natural oils your skin produces and because of these biomimetic qualities, squalane has been proven to have exceptional moisturizing properties. Lightweight and nongreasy, squalane is efficiently absorbed into the skin, without leaving an oily residue, and works quickly to replenish moisture in the skin, evenly, helping to balance the skin's natural oil production.
- Lowers transepidermal water loss (TEWL) by 13.9%
  - Moisturized skin also looks and feels firmer and reduces the appearances of fine lines and wrinkles. Studies have shown that squalane has better absorption than other moisturizers like Jojoba and Argan oils, with greater short-term moisturization, a 51% immediate improvement in moisturization and continued moisturization for 24 h. See Figure 14.4. Squalane has also been proven to reduce TEWL by 13.9%, keeping that moisture locked in.

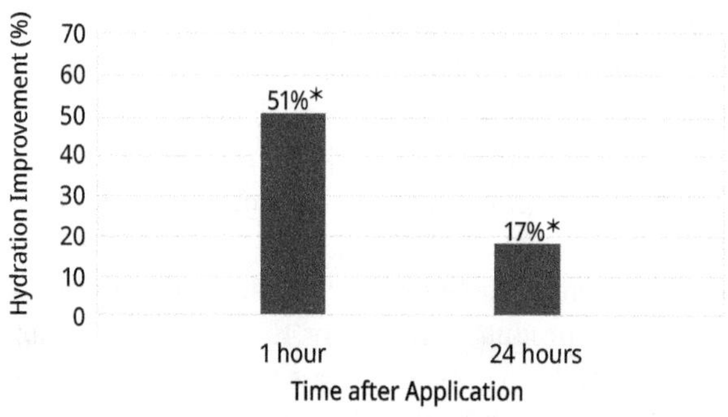

Figure 14.4: Moisturizes skin.
Source: *Amyris.*

- Reduces skin roughness by 28%
  - A deficiency in essential skin lipids can result in the skin feeling tight and itchy with a dull appearance and rough or even cracked texture. A compromised skin barrier further exacerbates dryness, as moisture easily escapes from the skin. Regular use of squalane replenishes the skin's lipid content and prevents moisture loss due to its outstanding occlusive emollient properties.

- Promotes healthier skin with diverse microbiome
  - The microbiome is an ecosystem of bacteria living on the skin's surface that keeps it healthy. The microbiome aids in wound healing, limits exposure to allergens, minimizes oxidative damage, keeps the skin plump and moist, and even protects it from harmful UV rays. Studies have shown that younger-looking skin can be achieved by a healthy and well-balanced skin microbiome. A microbiome that's out of balance can leave skin vulnerable to concerns such as rosacea, atopic dermattitis, psoriasis, acne, eczema, and inflammation. Clinical studies have shown that squalane not only supports a healthy skin microbiome but also increases microbiome diversity.
- Promotes cell turnover by 34%
  - Skin cell turnover is the continuous process of shedding dead skin cells and replacing them with younger cells. This process is essential to healthy skin as dead, dull skin cells are shed to reveal a fresh, radiant complexion. The longer the cell turnover cycle takes, the more dead cells are left to build up on the surface of the skin. This proliferation of cellular debris can make skin appear dry and dull, clog the pores, leading to breakouts, make lines and wrinkles appear more pronounced, and make age spots and hyperpigmentation appear darker. Healthy cell turnover is key to smoother, more even-toned, healthy, vibrant skin. Regular use of squalane has been shown to promote cell turnover by 34%, which in turn improves the texture and tone of the skin. See Figure 14.5.

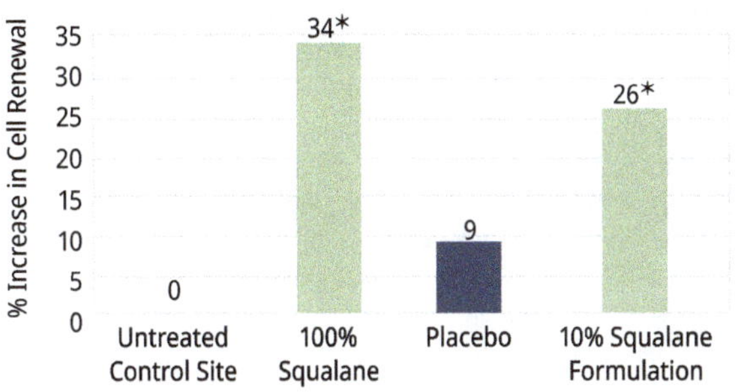

Figure 14.5: Improves cell turnover.
Source: *Amyris.*

- Repairs and reinforces the skin barrier
  - A healthy barrier is critical to normal skin function, but everything from environmental changes and stress, to the use of harsh soaps and overexfoliation, can compromise it. When your barrier is compromised, your skin is not able to perform its two primary functions: inhibit TEWL and protect itself from environmental pollutants. Squalane's emollient properties support and

enhance the skin barrier. Your skin barrier is also slightly acidic. This acidity (the acid mantle) is crucial to a healthy skin barrier. With a pH range of 5.5–6.5, squalane protects and maintains the skin's acid mantle and promotes optimal barrier function.

- Suitable for all skin types
  - We have all heard that oils can cause breakouts but not all oils are created equal. Squalane is an oil that mimics squalene, an essential component of SSL. When skin is dry, as a result of a lack of lipids, environmental factors, harsh cleansers, stress, or even a hormonal imbalance, this can lead to an overproduction of sebum to compensate, both congesting the skin and leaving it feeling excessively oily. Squalane is ideal for all skin types, even oily skin and sensitive skin, and all ages from birth up. Squalane is a lightweight, colorless, odorless, anti-inflammatory oil that is noncomedogenic, meaning it will not clog your pores or leave a greasy residue, while giving skin the moisture it needs. Squalane is nonirritating up to 100% concentration and is safe for use on all skin from birth onward.
- Supercharges the skincare routine
  - Because squalane is so readily absorbed in the skin, it can also enhance absorption of other active ingredients in your skin care products. Squalane is the most effective emollient for delivering CBD compared to other traditional emollients.

Galley sidebar:
What is Amyris squalane? Amryis squalane, branded as Neossance™, is a plant-derived, sustainable, ethical, and renewable version of nature's most effective emollient, squalene. In humans, squalene is one of the main, naturally occurring components of our skin surface lipids (SSL), a complex mixture of lipids that moisturize the skin and protect it from the environment. Newborn babies have the greatest concentration of squalene in their skin, but the reserve begins to drop suddenly between 30 and 40 years. Because of squalene's superior function as an occlusive emollient, both moisturizing and preventing TEWL, it has long been used in skincare products as a moisturizer.

In 1910, Japanese scientist Mitsumaro Tsjuiimoto discovered squalene in shark liver oil – specifically, sharks from the family *Squalidae* – and so, for many years, sharks were killed for their livers and the squalene extracted for use in cosmetic products. This unethical practice contributed to many shark species becoming endangered, and had a devastating environmental impact. Thanks to advances in the field of biotechnology, Amyris has pioneered a process through which to create a clean, sustainable, and ethical squalene alternative known as squalane: a biomimetic lipid that uses the power of science to mimic nature with improved purity, delivery, and efficacy.

In an alternative process, reliant on 600× more land and the harvesting of olive trees, squalene is extracted from olive pulp, skin, and pits and then converted via a hydrogenation process into squalane. Olive-derived squalane however is priced higher than shark-derived squalane and has a lower purity (squalane from olive oil wastes can be contaminated with by-products from processing – including plant waxes, free fatty acids, phytosterols, and neutralization by-products), with a characteristic odor and light yellow color.

Furthermore, both natural sources – shark and olive – presented a structural mismatch between the organization and the resource needs of the market, leading to cycles of shortages and higher prices. Shark liver oil and the unsaponifiable fraction of olive oil are by-products of the food industry; hence the amounts available varied, depending on many environmental factors over which squalane manufacturers had no control.

Seeing the need for a truly renewable, sustainable, and innovative solution, Amyris pioneered the technology to produce squalane from sustainable sugarcane.

Amyris squalane comes from 100% sustainably and ethically grown and harvested, renewable sugarcane, which is then biofermented to create the purest squalane available. Lightweight, odorless, colorless, antibacterial, noncomedogenic, and suitable for all skin types, squalane is nature's most powerful emollient.

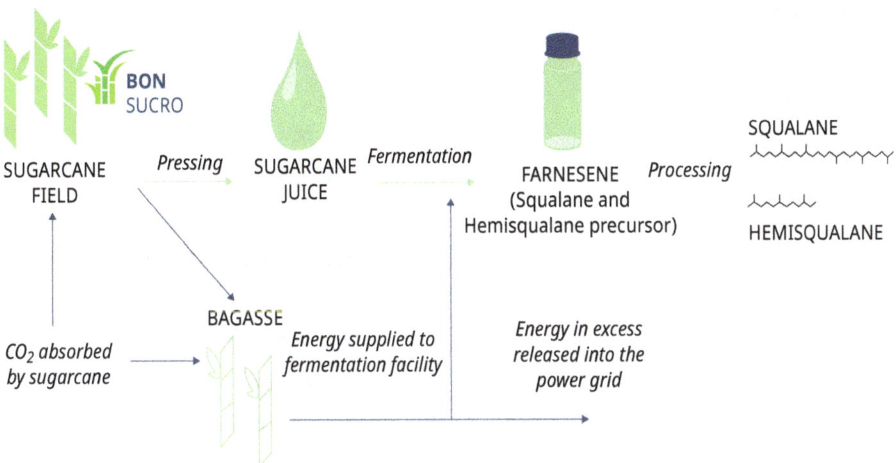

Figure 14.6: Sugarcane to squalene and the hemisqualene process.
Source: *Amyris.*

The Amyris squalane clean manufacturing process starts in the sugarcane fields of Brazil. The sugarcane is certified by Bonsucro, a nonprofit that promotes sustainable sugarcane production. Amyris believes that independent certification is important to ensure sustainability at our ingredient's source. See Figure 14.6.

Amyris' sustainably and ethically grown and harvested non-GMO sugarcane crops are located well outside the threatened Amazon region and do not contribute to deforestation of the rainforest nor compete with food agriculture. The sugarcane sourced by Amyris is a fast growing, rapidly renewable crop that requires minimal irrigation due to abundant rainfall in Brazil. The sugarcane is harvested and crushed to extract the juice from the plant that contains that sugar. This sugary juice is then boiled, until thickened, and spun through a centrifuge to remove any impurities. This process is done utilizing sustainable production practices, including the use of cogeneration systems to supply energy. Through cogeneration, the by-products of converting sugarcane into sugar are used to produce energy to power Amyris' plant.

The cane juice is then fermented with Amyris' engineered yeast to produce β-farnesene, the chemical precursor to squalane. The resulting β-farnesene undergoes a distillation and hydrogenation process to become pure squalane. Amyris' process is more efficient than the olive oil squalane process from seed to market, requiring four less manufacturing steps. Neossance™ squalane has 20 times less impurities than olive oil, ensuring safe, high-quality products.

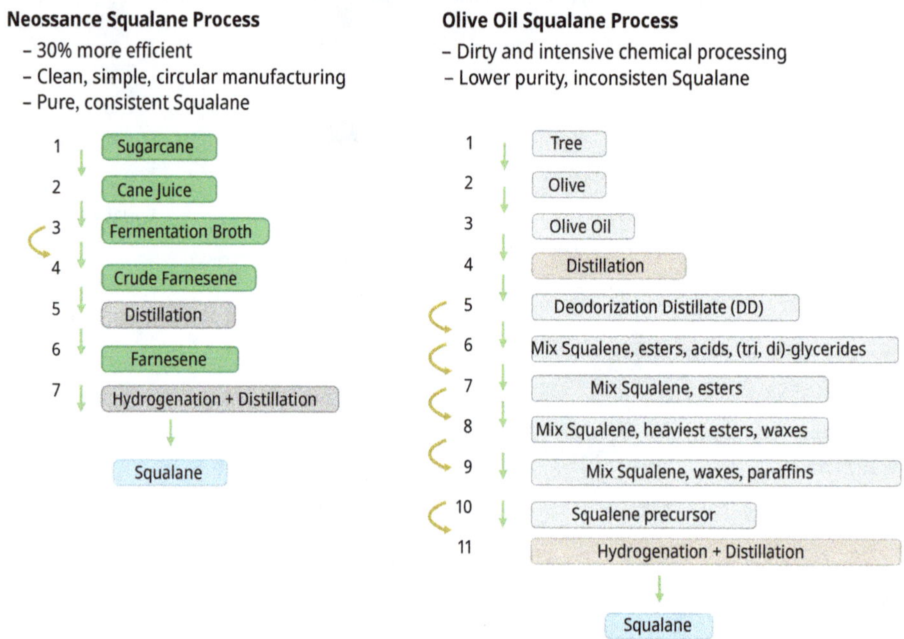

Figure 14.7: Neossance squalene process and olive oil squalene process.
Source: *Amyris.*

Sugarcane-derived Amyris squalane is the most chemically pure variety of squalane available with 94–97% purity. Squalane has the same chemical structure, regardless of its origin. However, there are issues with sustainability and purity, depending on the source. For example, olive-oil-derived squalane can be contaminated with by-products from processing, including plant waxes, free fatty acids, phytosterols, and neutralization by-products, and has a characteristic odor and light yellow color. See Figure 14.7 and 14.8.

These impurities can range anywhere from 6 to 18%, depending on the manufacturing process, including:
- Isosqualane (3–5%)
- Monocyclosqualane (1–3%)
- Hemisqualane (C15; 0–1%)
- Sesquisqualane (C45; 0–1%) Control

PROPERTIES	FIRST GENERATION SHARK **Squalane**	SECOND GENERATION OLIVE **Squalane**	THIRD GENERATION NEOSSANCE **Squalane (sugar)**
C30 Purity	Pure	Less Pure	Less Pure
Initial Odor Intensity	80% In-spec	45% In-spec	100% In-spec
Saturation	Medium	Low	Medium
Volatile Impurities	Some Impurities	More Impurities	Fewest Impurities
Sustainability	Not Sustainable	Not Sustainable	Most Sustainable
Consistency	Inconsistent	Inconsistent	Always Available

Figure 14.8: Properties of first-, second-, and third-generation squalene.
Source: *Amyris.*

Galley sidebar:
What role does squalane play in skin health? Our skin is our largest organ and the stratum corneum (SC) is the skin's outermost layer, acting both as the barrier against external aggressors like bacteria, pollution, and UV radiation, while simultaneously preventing water loss (TEWL) to keep skin hydrated. The SC layer of the skin is also where SSL, a film that is part of the epidermal barrier, is produced. SSL is a layer of lipids, ceramides, and fatty acids, with squalene accounting for around 13% of these lipids. Studies have shown that as we age, the SC displays a 30% reduction in total lipid content. The importance of skin lipids in the maintenance of skin homeostasis (healthy, balanced skin) is well-documented, and the disturbance of these lipids is linked to conditions including dry skin, rough surface texture, tightness, dullness, and volume loss; skin diseases such as acne, atopic dermatitis, and psoriasis; and the development of chronic, low-grade inflammation. Fortunately, the topical application of lipids can combat these conditions, improve barrier repair, and reduce TEWL. In other words, replenishing our lipid levels with topical applications of squalane has multiple benefits for skin health.

While squalene is one of the most important human skin cell lipids due to its multiple double bonds, it is unstable, very sensitive to air, and is considered a fragile compound for most practical uses. In squalane, these double bonds have been eliminated, making it more stable, with a total absence of toxicity, being odorless and colorless, with excellent sensorial properties, and offering a longer efficacy. If we use the analogy that skin is made up of "bricks" (protein from skin cells) and "mortar" (lipids), the lipid layer is what holds the walls together and keeps the skin strong, hydrated, and soft. The composition, size, and structure of squalane makes for the perfect "mortar," diffusing into the spaces between cells and providing moisture and filling in any breaches in the skin barrier.

Because Amyris controls custody of its ingredients – from Lab-to-Market™– the biotech company can confidently claim that its squalane is 100% plant-based and sustainable.

In February 2010, Amyris entered into a partnership with Soliance, a provider of ingredients to the French cosmetic industry, to sell large quantities of its squalane to the cosmetic industry. In 2011, Amyris also entered into a multiyear agreement to provide large quantities of squalane to Nikko Chemicals Co. Ltd. for distribution in the Japanese market.

In Amyris' 2014 Annual Report, Melo said, "Our business model has two key sources of revenue: sales of our renewable products and inflows from product development collaborations. Our business model is simple – we partner with leading companies to solve their supply, performance, cost, and sustainability problems. Our partners provide market insight and market access, which helps us to prioritize molecules for development as well as an effective channel access to the market."

Said Melo at the time, "In order to effectively scale our business, we need to be in a production environment. Only then will we learn what isn't working at larger volumes. We will learn from being in the market with real products." This statement would play out later when Amyris expanded its business into new markets in the consumer sector, flavors and fragrances, beauty, and health and wellness.

In 2015, Melo shared his plan to begin selling specialty, formulated products directly to industrial and consumer markets by establishing Amyris' own consumer brands in markets where the company already had the best performing molecule and a significant cost advantage, "Our objective for adding these new points of access is to accelerate our market reach and deliver positive impact on the planet while generating as much as 10 times the revenue and three to five times the margin dollars for each liter of product we produce."

The same year, Amyris announced a partnership with Squalane Natural Health, a privately held personal care products company based in the Netherlands, to enter in agreements for the production and marketing of squalane. Amyris branded its squalane ingredient, Neossance™, and sold it through global distributors to large beauty companies around the world, and formulated in a range of products from antiaging skincare to suncare and deodorant.

Galley sidebar:
Amyris squalane is:
- 100% naturally derived
- USDA Certified Bio-based Product
- Approved by Ecocert
- Microbiome friendly
- My Green Lab Platinum Certified
- EWG Skin Deep Rated
- Bonsucro Certified

In 2016, starting with Biossance™, Amyris launched its own in-house consumer brands championing their squalane. Clean beauty skincare brand, Biossance™ started as an experiment with a soft internal launch before officially launching on the Home Shopping Network. The following year, the brand launched in Sephora and then expanded beyond the USA to Canada and Brazil. Since its launch, Biossance has grown 300% year over year, with global reach across five continents, earning over 50 global awards and attracting millions of consumers to its direct-to-consumer site, Biossance.com, which now offers more than 20 clean beauty products. The success of Biossance is a testament to the quality performance of the squalane ingredient and demonstrated an impressive consumer response, which was not lost on major consumer products companies looking for a sustainable alternative to their shark and olive squalane.

Biossance was launched as a clean beauty brand just as the term "natural" began to lose its label luster due to consumer confusion around its definition and the greenwashing by many brands. Clean beauty focuses on ingredients that are safe and nontoxic as well as manufacturing processes that are sustainable and socially responsible. Amyris' clean manufacturing and commitment to safe, effective, and sustainable ingredients propelled Biossance to lead the clean beauty trend. Positioned perfectly for the megatrend of healthy living from the outside in and inside out, Biossance launched the Clean Academy, a beauty industry first, connecting directly with consumers to educate about the ingredients, formulations, and offering guidance on healthy skin care routines.

At the time of this writing, the Amyris family is made up of nine consumer brands, with three using squalane formulations: Biossance, Pipette™, and Rose Inc.™, with more consumer brand launches planned in 2022 and beyond.

As with Biossance, squalane is the hero ingredient powering the Pipette line of baby and mother care products. Amyris launched Pipette in 2019, earning certifications by EWG Verified, National Eczema Association, Leaping Bunny and approvals by pediatricians, Good Housekeeping, Parents magazine, Allure, and WWD. Distributed internationally, Pipette products have seen a 900% growth in points in distribution and are now sold at Target, Ulta, Walmart, and Walgreens. Free of sulfates, silicone, mineral oil and petrolatum, the clean formulation and manufacturing have a special appeal to new parents.

In August 2021, Amyris launched Rose Inc., a color cosmetics brand with founder Rosie Huntington-Whiteley, entrepreneur and super model. Applying Amyris' technology and unique, biodesigned, and fermented ingredients, including squalane to color was a logical next step. Today, the brand is sold globally, including at Sephora and direct-to-consumer, and offers 14 products, including color, skincare, and skincare tools.

At the same time, Amyris also announced the launch of JVN™, a brand of clean hair products created for all hair types, with founder Jonathan Van Ness, hair stylist, TV personality, and *New York Times* best-selling author. Free of silicone and sulfates, each of the 10 products addresses the top hair health and styling concerns; they are color-safe, cruelty-free, and vegan. Already an award winner, including from Vogue, the JVN line is sold internationally, including at Sephora and direct-to-consumer. The retail launch orders exceeded the first full year of Biossance revenues, an indication that clean manufacturing was gaining traction with consumers. See Figure 14.9.

Figure 14.9: Examples of beauty products.
Source: *Amyris.*

14.6.1 Amyris family of award-winning consumer brands

The JVN product line is formulated with a signature ingredient, hemisqualane, a derivative of farnesene, created by Amyris through its fermentation, distillation, and hydrogenation process. Amyris leveraged its Lab-to-Market™ technology platform and clean manufacturing to create hemisqualane, to provide a clean, sustainable alternative to silicones and existing silicone replacements. Amyris' process is more efficient and more sustainable than leading alternatives to silicone, which are either petrochemically derived or derived from nonsustainable plant sources like palm oil, a major driver of deforestation of some of the world's most biodiverse forests, destroying the habitat of already endangered species. Hemisqualane is also biodegradable and does not harm aquatic life. Hemisqualane is the best performing, natural alternative to silicones in haircare. Unlike silicones used in rinse-off hair products, hemisqualane penetrates hair, nourishing it, strengthening it, and preventing color degradation.

The performance results of Amyris hemisqualane and its comparative benefits to nonsustainable hair ingredients are quickly winning loyal customers.

- Protects and repairs damaged hair
  - Hemisqualane improves your hair over time by creating healthier, stronger strands. It penetrates the hair's core better to deeply repair, protect, and smooth all types and textures.
  - 40% reduction in hair cortex damage after one treatment
  - 15% damage reduction from bleaching after one treatment, seals cuticles
- Makes hair easier to comb, wet and dry
  - 30% reduction in combing energy (WET)
  - 27% reduction in combing energy (DRY)
- Improves hair elasticity
  - 7% increase in elongation Improves elasticity
  - strengthens the hair, prevents breakage
- Maintains color
  - 30% increase in color maintenance
- Reduces frizz
  - In clinical studies, hemisqualane has shown a significant, instant reduction in frizz. Hemisqualane keeps frizz at bay, long after applying, and improves hair health over time.
  - 29% reduction in frizz after application

Galley sidebar:
What is hemisqualane?
Hemisqualane is a plant-derived, clean, sustainable, biodegradable silicone alternative. Amyris developed hemisqualane as a safe, sustainable solution to growing consumer concerns about silicone toxicity and their impact on the environment. Cyclomethicones (D4 and D5) have been essential compo-

nents of personal care and beauty for decades but have been classified as a toxic "Substance of Very High Concern" by the European Union (EU).

Silicones are also not biodegradable, leading to concerns about their environmental impact on waterways, oceans, and accumulation inside of wildlife. Amyris leverages its Lab-to-Market™ technology to create a clean, sustainable, and biodegradable hemisqualane, the most safe, sustainable silicone alternative available. Lightweight, odorless, transparent, nongreasy noncomedogenic, and suitable for all skin and hair types, hemisqualane is the best-performing natural silicone replacement available.

Silicones have been essential components of personal care and beauty for decades, as they deliver even coverage, spreadability, and a light, dry feel in antiperspirant, sun care, color cosmetics, and makeup removal applications. In haircare, silicones support combability, thermal and color protection, antifrizz, and elongation features. In recent years, however, there has been a growing concern about silicone toxicity and its impact on the environment. Environment Canada assessments concluded that cyclotetrasiloxane and cylcopentasiloxane – also known as D4 and D5 – are toxic, persistent, and have the potential to bioaccumulate in aquatic organisms, as they are not biodegradable.

The EU classifies D4 as an endocrine disruptor, based on evidence that it interferes with human hormone function, and a possible reproductive toxicant that may impair human fertility. In laboratory experiments, exposure to high doses of D5 has been shown to cause uterine tumors and harm to the reproductive and immune systems. D5 can also influence neurotransmitters in the nervous system. Structurally similar to D4 and D5, cyclohexasiloxane (or D6) is persistent and has the potential to bioaccumulate.

Environment Canada's assessment of D6 concluded that this third siloxane is not entering the environment in a quantity or concentration that endangers human health or the environment, but noted significant data gaps concerning its toxicity. Common silicones in cosmetics are: dimethicone, cyclomethicone, cyclohexasiloxane, cetearyl methicone, and cyclopentasiloxane. As a result, cyclomethicones are restricted by the EU and have been banned by a host of clean beauty brands and retailers.

Hemisqualane is sustainably sourced from renewable sugarcane. Amyris' clean manufacturing process reduces CO_2 emission compared to traditional petrochemical processes and respects biodiversity. Hemisqualane is also biodegradable and does not harm aquatic life. Hemisqualane is superior to silicones in skincare in haircare. In addition to concerns about the impact of silicone on health and the environment, there are additional drawbacks to the use of silicones in skincare and haircare.

While D5 is well-known for its ability to remove long-wear and waterproof makeup, its occlusive properties can act as a barrier, trapping oil, dirt, dead skin cells, and other comedogenic substances, thereby increasing the chance of clogged pores and potentially leading to acne. Hemisqualane is noncomedogenic.

In hair products, silicones create a seal that keeps hair hydrated from within, but this protective layer can block other nourishing ingredients from penetrating into the hair follicle. Over time, silicone can also build up on hair, resulting in a dry feel and dull appearance. Unlike silicones, hemisqualane penetrates the hair shaft for maximum nourishment and protection.

Both consumer and expert evaluations demonstrate that hemisqualane provides excellent sensorial and spreading properties and is ideal for replacing silicones in color cosmetics, suncare, haircare, and more. It also has makeup removal benefits and is a top replacement for isohexadecane, amodimethicone, and D5.

14.7 Purpose-driven: health and wellness

From its earliest days, Amyris was focused on health and wellness. Years after its success with artemisinin, Amyris partnered with a pharmaceutical intermediate Nenter & Co. to convert farnesene to vitamin E for distribution to the global nutraceuticals market [18]. This fat-soluble antioxidant vitamin is important for a strong immune system and healthy skin and eyes [19]. Human consumption of vitamin E supplements makes up a fraction of the market. Nearly 90% of the vitamin E produced goes into animal feed to promote health and immunity. Amyris' clean manufacturing lowered the traditional manufacturing cost by as much as 35% by simplifying the process and reducing the number of required process steps. In 2019, Amyris sold its vitamin E royalties to DSM for $57 million.

Just as Amyris was exiting the vitamin E business, it was entering the sugar substitute market, which today is projected to reach $20.6 billion globally by 2025, recording a CAGR of 4.5% during the forecast period [20].

In 2018, Amyris launched its newest compound, Rebaudioside M (Reb M), which is the active ingredient in the stevia leaf and found only at such extremely low concentrations (less than 0.1%), which makes it very difficult to isolate the molecule using traditional extraction and harvesting processes. Using its fermentation and clean manufacturing process, Amyris successfully produces pure Reb M at scale, without reliance on leaf extraction or the associated challenges. Amyris Reb M sweetener, which received FDA GRAS (Generally Regarded as Safe) designation, is 350×–500× sweeter than sugar, depending on the application, from baking to beverages.

Requiring only miniscule amounts of Reb M in formulations, this pure, high-potency sweetener tastes like sugar without the lingering and bitter aftertaste normally associated with stevia. It also has a zero glycemic index, making it suitable for people living with diabetes. One in three Americans live with prediabetes, with minority and low income communities most impacted. Seven in ten deaths are caused by illness related to sugar intake. In the USA alone, $3 trillion is spent annually on healthcare costs related to diabetes, obesity, and heart disease.

In 2021, Amyris signed a partnership valued at $100 million with Ingredion, a market leader in the global distribution of sweeteners. Its global customer reach, formulation capabilities, and commercial teams accelerate the availability and adoption of the fermentation-based, zero-calorie sweetener. Ingredion is also a minority owner of Amyris' Barra Bonita manufacturing facility that comes fully online in 2022. The partnership underscores Amyris' recognized leadership in synthetic biology in successfully developing, scaling, and commercializing pure ingredients that are sustainably sourced and produced by fermentation.

"We are excited by the addition of this breakthrough ingredient to our sugar reduction portfolio, which complements our PureCircle stevia product line and will allow us to provide our customers with the broadest selection of nature-based, high-intensity sweeteners on the market," said Jim Zallie, Ingredion's president and chief executive officer. "Our partnership with Amyris positions Ingredion well to meet our customers' increasing demand for quality ingredients that will drive transformational change in the food industry."

According to Ingredion, which offers Reb M made from all three production methods, Reb M from fermentation and bioconversion outperformed Reb M extracted from the leaf across all four key sustainability metrics (GHG emissions, land use, water scarcity, and culmulative energy demand), with the fermentation-based approach (which doesn't involve stevia plants at all) generating the lowest environmental impact by some degree. Ingredion went on to say that fermentation-based Reb M is the cheapest option, in addition to having the lowest environmental footprint [21]. See Figure 14.10.

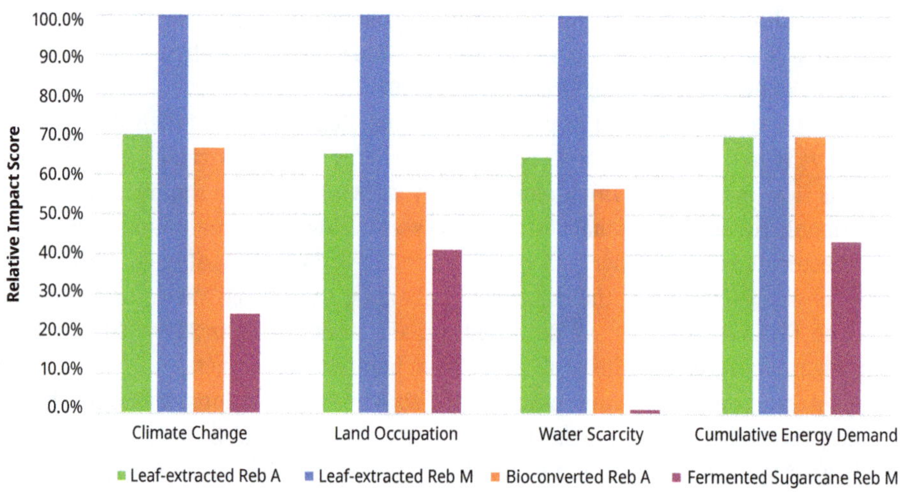

Figure 14.10: Relative impact score and climate change, land occupation, water scarcity, and cumulative energy demand for leaf-extracted Reb A, Leaf-extracted Reb M, Bioconverted Reb A, and Fermented Sugarcane Reb M.
Source: *Ingredion.*

Amyris also launched Purecane™, its own consumer brand of no-calorie sweetener, powered by the hero ingredient, Reb M. The brand was named the No. New Release on Amazon.com, consistently earning 4+ star reviews for its products, including zero-calorie packets and variety of zero-calorie, culinary, and baking sweeteners.

One of the best known, clean manufactured ingredients by Amyris is squalane, as discussed earlier. However, Amyris also produces squalene. While squalane is a stable, noncomedogenic lipid compound that is better suited for cosmetics, squalene is an active ingredient in an important class of adjuvants (including influenza vaccines), which are used in vaccines to increase the immune response, and help the vaccine to work more effectively. Squalene is traditionally harvested from deep-sea shark livers, which has the potential to devastate shark populations and contribute to long-term environmental issues. With funding from the National Institutes of Health (NIH), Amyris developed a pharmaceutical-grade squalene adjuvant that is a sustainable and scalable alternative to shark-derived adjuvants (vaccine boosters). Amyris is in conversation with leading pharma partners to commercialize the production and use of its squalene vaccine adjuvant. In 2022, Amyris entered into a joint venture with ImmunityBio to accelerate the commercialization of a next-generation Covid-19 vaccine.

Amyris and ImmunityBio combine important vaccine technology and manufacturing capabilities in the joint venture. Upon completion of successful human trials and regulatory approval, the joint venture's goal is to start delivering the second-generation vaccine as soon as is practically possible with the goal of delivering immunity for Covid-19 and access to underserved parts of the world, where current vaccine technology is challenged due to cost and supply chain limitations.

"We are pleased to combine our expertise in human trials, T-Cell technology, and our access to RNA manufacturing capacity with the Amyris and Infectious Disease Research Institute's (IDRI [22]) RNA technology platform and Amyris' adjuvant technology," said Patrick Soon-Shiong, M.D., Executive Chairman and Global Chief Scientific and Medical Officer at ImmunityBio.

14.8 Partnerships are key

Partnering is key to Amyris' strategy. As a recognized leader in the biotechnology sector, Amyris has made deliberate choices to partner with industry leaders to maximize value from its technology and to demonstrate its ability to scale and commercialize ingredients faster, at a better cost point and with greater purity. Amyris has earned its leadership position in biotechnology from its repeated success at leveraging its proprietary Lab-to-Market™ technology platform to commercialize fermentation-derived ingredients.

Galley sidebar:
"We will meet the growing demand for clean ingredients, not by depleting nature but through clean chemistry, working with nature. Clean chemistry will continue to become a bigger component in applications for everyday consumer products, as consumers continue to demand more natural and sustainable solutions. Our future growth is about more molecules, into more products and used at higher rates in each application. We're able to offer a reliable, high-quality supply of molecules that are lower cost and sustainable. It's the better choice in every way." Amyris President & CEO John Melo

Amyris first partnered with global leader Firmenich in 2011 to produce three bio-based flavor and fragrance ingredients and then later added new partnerships with Givaudan and International Flavors & Fragrances – two of the largest and best known flavor and fragrances houses in world. Ten years later, in a $50 million deal, Amyris licensed rights to DSM to assume the supply of farnesene to Givaudan for the production and sale of a single specialty ingredient. Later the same year, Amyris sold its flavors and fragrances (F&F) bio-based intermediates business to DSM for $150 million. DSM has been an important strategic partner to Amyris over the years. In 2017, Amyris sold its manufacturing facility (Brotas 1) to DSM and accelerated its own plans for the construction of a new, state-of-the-art manufacturing plant for specialty products.

In April 2022, Amyris announced the commissioning of its newest, state-of-the-art fermentation plant in Barra Bonita, Brazil, with abundant availability of sustainable raw materials (sugarcane), securing production capacity for its rich pipeline of sustainable and bio-based compounds. The new manufacturing site at Barra Bonita is large: 185,000 m^2 land area; 1,500 pieces of process equipment; 50,000 m of process piping; and with a 4,000-ton steel structure.

Brazil has always been important to Amyris and will continue to play an important role in its future. Its new manufacturing plant is located next to the Raizen Barra Bonita sugar mill, which is the world's second largest sugar mill, at the heartland of Brazilian sugarcane production, in São Paulo State. Amyris has access to Barra Bonita's economic and sustainable sugarcane, the feedstock for Amyris' specialty fermentation process.

Amyris' feedstock supply partner, Raizen, is the largest sugarcane producer in the world. They have led the way in ensuring that their Barra Bonita sugarcane is Bonsucro-certified. This certification ensures that our feedstock is ethically and sustainably produced and harvested under the strictest guidelines to ensure societal contributions and benefits. Amyris is the only biotechnology company that also has received this certification, an important distinction for Amyris' commercial customers and distribution partners, because it provides traceability for sustainable supply chains.

The Barra Bonita plant aids Amyris in its efforts to achieve a zero-carbon footprint in fermentation production. The plant's electric substation is colocated with Raizen's cogeneration plant. As of the time of this writing, Amyris' electrical plant converts the fiber waste stream from the sugarcane production into energy. And Amyris is exploring future innovations to reduce emissions, including the use of sustainable biogas, processed from the vinasse waste stream. There are already plans underway to use biogas as sustainable fuel for its steam.

Amyris did not lose sight of the opportunity to positively impact other social aspects of Barra Bonita's community. The company prioritized local labor since the start of construction, with nearly 80% of the workforce on the construction site from the local Barra Bonita region. Amyris has also partnered with technical and professional schools for the training of skilled operators at its highly automated plant.

The new manufacturing plant has a proprietary design, with a process tower two times taller than Amyris' former Brotas plant, which allows Amyris to reduce the amount of energy required in production, and to take full advantage of gravity with the vertical fermentation process.

The clean manufacturing process starts at the top where the sugarcane syrup enters. Sterilizing and enriching the sugarcane into several variations of fermentation feedstock happen at the next level. Each feedstock is optimized to precise specifications for Amyris' different compounds. The materials go through three stages of fermentation: inoculation, seed growth, and then full-scale fermentation. The controls needed to achieve precise fermentation conditions for feeding, cooling, and production are located at the center of the tower. At the base, Amyris controls the harvesting of its fermentation broth before moving it to the purification stage.

The tower clusters five fermentation "mini factories," capable of concurrently producing any combination of Amyris' 13 different ingredients. Each line, which consists of two fermentation tanks, operates independently of the other. In total, these lines can produce around 6,000 tons per year of fermented ingredients sourced sustainably.

Amyris continues to invest in transformative technologies that benefit the environment and that are biology-based. As one of the earliest synthetic biology companies to successfully reach full-scale manufacturing with fermentation technology, Amyris has a unique perspective on what it takes to scale up fermentation and downstream processing technology. As a fundamental principle, Amyris prioritizes scaling up as quickly as possible to gain competitive business advantages. By reaching industrially relevant scales faster, Amyris can produce larger samples quickly, sample customers early and often for feedback, initiate any regulatory processes early on, and quickly assess how the technology is performing in manufacturing facilities.

Individually, any of those factors is compelling enough to rapidly push to industrial scale, and when leveraged holistically, Amyris has an unmatched view into manufacturing operations and market potential. From strain engineering to process development to full-scale deployment, Amyris has achieved impressive manufacturing milestones, with 13 sustainable products in market and almost two decades of experience in 17 full-scale production facilities across seven countries, in addition to its new specialty precision fermentation plant in Barra Bonita.

With every successful fermentation run at industrial scale, Amyris gains another opportunity to closely examine sustainability metrics – like CO_2 footprint and water usage – to identify areas for optimization. As a result, Amyris continuously refines its life cycle analysis (LCA), receiving more precise feedback at every stage from inception to final product shipping to waste disposal. This feedback informs where the team should direct technology development efforts to improve the sustainability of Amyris products.

Through this understanding, Amyris can also better estimate the sustainability impact for new product opportunities being explored and thus make better choices for new technology development and deployment.

Clean manufacturing at full-scale capacity has allowed Amyris to deliver on its commitment to people and the planet – to Make Good. No Compromise™. As the company continues to expand its reach to even more consumers around the world, the

Amyris LCA engine will only grow stronger, enabling the highest level of quality and performance for the world's sustainable ingredients.

Amyris is delivering clean products that today's consumers demand, while building a robust supply chain that allows the company to increase its clean manufacturing capacity. Amyris has invested in advanced, specialty fermentation capability. With its own manufacturing capability, Amyris gains more control over its ability to grow sustainably as a company. Amyris has learned that it is impossible to have real impact on sustainability and the health of people and the planet, without clean manufacturing at scale. Because Amyris believes that fermentation is the future of chemistry, we have a responsibility to clean up the world's chemistry in the fastest way possible.

Galley sidebar:
Amyris has commercialized 13 biobased ingredients.

Artemisinin is a key ingredient used in antimalarial drugs, and ACTs are the first-line treatment recommended by the WHO. Artemisinin is traditionally found in the extract of *Artemisia annua* or the sweet wormwood plant. While ACTs are the most effective treatment, Artemisia crop yields can be unpredictable from year to year, causing price volatility of the plant-derived molecule and impacting the overall availability of this life-saving drug. Amyris successfully engineered yeast to produce artemisinic acid, a molecule that can be converted to artemisinin at a significantly lower cost and without the risks of climate-related crop variability. Using synthetic biology, this approach allows for "on demand" clean manufacturing to meet the global need for artemisinin and ACTs.

Biofene, or *trans*-β-farnesene, is a sustainable and versatile ingredient, produced via fermentation and synthetic biology. Biofene is a precursor molecule that, with a bit of clean chemistry, can be converted to other natural ingredients that Amyris produces. Applications of Amyris' Biofene include performance materials, adhesives, fragrances, surfactants, stabilizers, oligomers and polymers, resins, emulsifiers, vitamin precursors, crop protection, and foams, coatings, and sealants.

Biosilica™ (silica) is used widely in personal care and cosmetics and is traditionally sourced from nonrenewable sand dredging, which requires significant energy consumption and emits large amounts of CO_2. Amyris' sustainable and renewable Biosilica™ is derived from discarded sugarcane ashes and has similar or better performance characteristics compared to industry benchmarks. Applications suited for Biosilica™ are foundations, creams, lotions, and powders, as a sustainable and better performing alternative to extractive silica and microplastics. Biosilica™ has unique sphericity, high oil absorption, particle size, and oil/water absorption ratio. These characteristics offer premium benefits such as elegant texture, sebum control, mattification, and antiaging effects, without drying the skin. Biosilica™ is also 100% biogenic (produced by nature), homogenous, soft to the skin, and has a silky application without blurring characteristics.

Bisabolol is used in personal care products such as lotions, sunscreens, and face/eye creams. It has been a popular personal care ingredient for hundreds of years because of its skin-healing antimicrobial, anti-inflammatory, and antiirritant properties. Bisabolol is traditionally sourced from either the German chamomile (*Matricaria recutita*) or more commonly, the Brazilian Candeia tree. The challenge with botanical sources comes down to availability, purity, and sustainability. When extracted from German Chamomile, the amount of bisabolol produced can vary greatly. It also requires a significant amount of plant material, making it cost-prohibitive. The more common botanical source is from the endangered Brazilian Candeia tree. On average, it takes 12 years to grow a new tree, and a metric ton of tree material yields only about 7 kg of bisabolol. The limited availability of the Candeia tree also leads to significant price instability and fluctuations in product supply. Growing efforts to preserve Brazilian biodiversity have also led to innovation to identify alternative sources for bisabolol.

CBG (cannabigerol) is the precursor from which all other cannabinoids are synthesized and has exciting therapeutic potential to address consumer skincare concerns. To date, cannabinoids have been extracted directly from hemp plants, which require significant land usage, are difficult to scale, and often have regulatory risks related to THC. Using biotechnology and sugarcane fermentation, Amyris produces clean CBG, without the hemp plant. This process guarantees no THC in the final product and, therefore, carries no regulatory risk or challenges related to THC. Additionally, the entire process ensures environmental and social responsibility, starting with its Bonsucro-certified sugarcane feedstock in the fermentation process. Studies demonstrate CBG provides benefits to address common skincare conditions like dry skin, irritation, and redness. This fermented ingredient gives consumer goods companies a new option to formulate a completely new variety of oils, creams, and lotions with clean CBG.

Neossance™ **hemisqualane** is the preferred sustainable and naturally derived alternative to silicone by top global brands. As a sustainable ingredient in virtually unlimited supply, it provides an equivalent sensory and functional performance to unhealthy, banned silicones like isohexadecane and cyclomethicone in skincare, haircare, suncare, makeup removal, and color cosmetics. Certain types of silicones can build up on hair that block out moisture and subsequently prevent the hair from receiving moisture. Over time, silicones can make the hair appear dull and can also result in weaker hair due to lack of moisture. Hemisqualane is ideal for conditioners, treatments, and styling products as a sustainable and high-performing silicone replacement for hair care. Hemisqualane moisturizes and repairs the hair shaft at the root and from the inside out. It delivers color and thermal production, improved manageability, reduced frizz, and reduced greasiness without drying the scalp.

Manool is a key ingredient used to make woody, amber notes in the fragrance industry. Traditionally, manool is sourced from fallen *Manoao* pine trees, a native New Zealand species. *Manoao* pine trees are an endangered species and currently, the harvesting of fallen *Manoao* pines is restricted to one company to preserve New Zealand's natural biodiversity. As a result, manool is only available in small quantities from the traditional agricultural source and long-term availability can be difficult to predict. Amyris has developed a renewable solution to help protect the Manoao pine species as well as ensure a reliable supply of manool for the future.

Patchouli oil, native to Southeast Asia, has been traded for centuries along the Silk Road. It is extracted from the leaves of *Pogostemon cablin*, an exotic mint, and is one of the largest fragrance crops globally. Patchouli oil is used around the world, most commonly in fragrances and personal care products, with production estimated at 1,200 tons per year. Currently, the patchouli plant is one of the largest, but most variable fragrance crops in the world. In its botanical state, the fragrance is inconsistent and can have burnt or rubbery undertones, due to extraction methods and variability in region and environmental conditions. Amyris' fermentation-produced patchouli-type oil provides a cleaner, more consistent fragrance profile and is not subject to fluctuations in cost or availability. Because of the clean manufacturing, Amyris' patchouli-type oil doesn't have the harsh notes, typical of patchouli extracted from agricultural sources, offering a unique, elevated woody note. For one hectare of land, traditional plant-based sources for patchouli yield less than 0.15 metric tons, whereas Amyris' fermentation-based approach yields 0.73 metric tons. Further, the land that patchouli grows on can be continuously cultivated on a single plot of land for only 3–4 years before soils are depleted and unsuitable for patchouli production; in comparison sugarcane – Amyris' feedstock – can be cultivated on the same plot of land for up to 200 years. Amyris' patchouli-type oil helps meet the growing demand from formulators interested in this fragrance and offers higher purity, stabilized pricing, and a reliable supply chain source. Due to the elevated, clean scent of Amyris' patchouli-type oil, it has also led to innovative product applications and can be found in fine fragrances, haircare, body care, soap, candles, laundry detergents, fabric softeners, room sprays, and more.

Purecane™ Reb M is a zero-calorie sweetener – pure, sustainable with a zero glycemic index, and is certified non-GMO. Reb M offers formulators a high-purity, high-potency sweetener, with a taste profile that mimics sucrose with no lingering aftertaste. Reb M is bioidentical to stevia's sweetest molecule, present in the leaves at only very low concentrations (less than 0.1%), making it difficult to isolate using traditional extraction and harvesting processes. But Amyris creates the pure Reb M molecule through fermentation, by engineering yeast and fermenting sugarcane, so none of the challenges that exist with leaf purification exist with Amyris' clean chemistry.

Santalols are the active components of sandalwood oil that gives it its signature woody aroma. Sandalwood has been deeply tied to Indian culture over generations and has traditionally been harvested with great respect and a mind toward sustainability – a sandalwood tree must grow for 30 years to reach full maturity before its essential oil can be harvested. But over the course of decades, between population growth and globalization, the trees have been overharvested and the supply has dwindled. As a result, pricing of sandalwood has skyrocketed, it has become inaccessible, and today, it is a threatened species.

Sclareol is a key component in plants ranging from clary sage to key limes that makes aromas linger longer in products. Sclareol has come to replace a fragrance material called ambergris, which is derived from sperm whale secretion and is difficult to find and cost prohibitive – its key chemical constituent is ambroxide. As a result of the limited supply available from sperm whales, the industry now uses an alternative to ambroxide: sclareol, commonly derived from clary sage plants, a flowering herb native to Europe.

Squalane. Amyris' Neossance™ squalane is a high-quality and versatile emollient used in a wide variety of applications such as skincare, sun care, color cosmetics, makeup removal, and deodorants. Squalane has proven skin therapy benefits in formulations, including an increase in skin firmness and the reduction in the appearance of fine lines and wrinkles. Squalane boosts skin health, with moisturizing, antiaging, and brightening skin improvements from a natural ingredient that is ethically and sustainably sourced. Squalane is the more stable, hydrogenated version of squalene, a compound we are born with as babies but naturally diminishes as we age. Traditionally, beauty companies have relied on shark livers for this ingredient, threatening shark populations with long-lasting consequences to our ocean's biodiversity. An alternate traditional source, olive oil, carries dependencies on weather, crop yields, and environmental supply chain risks. Neossance™ squalane offers purity and quality, helping brands formulate consistent and high-performing products, without causing environmental harm.

Squalene is a natural oil that is the active component in an important class of vaccine adjuvants. Adjuvants are a key ingredient used in vaccines to increase the immune response and ultimately help the vaccine work more effectively. Adjuvants thus allow less of the immunogen (the microbe or protein that the vaccine is targeting) to be used when manufacturing vaccines. Squalene has known properties as an effective carrier system for immunogens. Traditionally, squalene is harvested from deep-sea shark livers, which has the potential to devastate shark populations and have long-lasting ecological consequences. Amyris' pharmaceutical-grade squalene is molecularly identical to the traditional shark source and can also provide a sustainable, traceable, and reliable source of a critical vaccine ingredient.

References

[1] Dodd MS, et al. Evidence for early life in Earth's oldest hydrothermal vent precipitates. Nature 2017;543:60–64.

[2] Interlandi J. Newsweek 2008.

[3] Magat C Stanford University Graduate School of Business Case Study on Amyris, 2009.

[4] Amyris News Release, Globe newswire, (2016). Specter M. The New Yorker, 2009.

[5] Cohen JM, Singh I, O'Brien ME.. Predicting global fund grant disbursements for procurement of artemisinin-based combination therapies. Malar J 2008;7:200. doi: 10.1186/1475-2875-7-200. PMID: 18831742; PMCID: PMC2570684.

[6] Sanders R. UC Berkeley News 2004.

[7] Amyris News Release, Globe newswire, 2015.

[8] Gormley B. *Wall Street Journal* 2016.

[9] Villas-Boas JM.. Berkeley Haas Case Series, Amyris 2019.

[10] Villas-Boas JM. Berkeley Haas Case Series, Amyris, 2019.

[11] Nasdaq.com, 2010.

[12] Environmental Protection Agency, 2014.

[13] Amyris Press Release, Globenewswire, 2014.

[14] Villas-Boas JM.. Berkeley Haas Case Series, Amyris 2019.

[15] Amyris Inc. Q2 2018 Earnings Call, 2018.

[16] Ducos L, et al. Bloom Association, 2015.

[17] McPhee D, et al, Cosmetics & Toiletries magazine. 2014.

[18] Amyris News Release, Globe newswire, 2017.

[19] National Institutes of Health, Office of Diary Supplements, 2021.

[20] Sugar Substitutes Market Report, MarketsandMarkets, 2020.

[21] Navigator F.. Which Reb M production method is best for the environment? Ingredion LCA probes stevia sustainability metrics. August 2022.

[22] AAHI Press release. IDRI now AAHI, Businesswire, 2022.

Crispinus Omumasaba, Alain A. Vertés, F. Blaine Metting*,
Hideaki Yukawa*

Chapter 15 Hydrogen bacteria: technology development and deployment

15.1 Introduction

A decades-long international debate on how to tackle global climate change has failed to elicit a response equal to the daunting task of stemming the uninterrupted rise in greenhouse gas (GHG) emissions [1, 2]. Carbon dioxide (CO_2), the largest component of GHG emissions now accounts for about 65% of global warming and consequent climate change for which the overwhelming scientific consensus blames human activity [3, 4]. The emissions come primarily from burning fossil fuels to produce vital energy for heating, electricity and transportation, and to manufacture everyday items like plastics, paints, fertilizers, and even cosmetics. In line with recent trends, global energy demand grew in 2024, but at a 2.2% rate that was double the 1.3% average growth over the previous 10 years. Global CO_2 emissions in 2024 likewise increased 0.8–0.9% on the previous year to 37.8 billion tons (GT CO_2eq) [5].

Adoption of the United Nations Framework Convention on Climate Change (UNFCC) in 1992, the Kyoto Protocol in 1997, and the Paris Agreement in 2015 collectively heightened awareness and concern about climate change and intensified focus on CO_2 management [6–8]; G20 economies implemented carbon emission taxes ranging from \$3 to \$60 per ton to incentivize CO_2 capture from industrial processes via legally binding carbon budgets. But global carbon credits markets alone have proved inadequate as a response to rising CO_2 emissions for 30 years [9, 10]. Unchecked emissions strain remaining carbon budgets, with some estimates suggesting the 1.5 °C budget will be surpassed within the next 5 years – and may indeed have been already. Climate model scenarios inform that keeping global temperature rise to 1.5 °C above pre-industrial levels by 2050 is going to take more than GHG emissions control; deliberate carbon dioxide removal (CDR) – taking CO_2 out of the atmosphere through nature-based and technological solutions – is inescapable [11]. CDR technologies play a crucial role in supporting the transition of heavy industries toward a decarbonized

***Corresponding author: F. Blaine Metting**, Utilization of Carbon Dioxide Institute Co. Ltd., 2-3-1 Daiba, Minato-ku, Tokyo 135-0091, Japan
***Corresponding author: Hideaki Yukawa**, Utilization of Carbon Dioxide Institute Co. Ltd., 2-3-1 Daiba, Minato-ku, Tokyo 135-0091, Japan, e-mail: hyukawa@co2.co.jp
Crispinus Omumasaba, Alain A. Vertés, Utilization of Carbon Dioxide Institute Co. Ltd., 2-3-1 Daiba, Minato-ku, Tokyo 135-0091, Japan

 | https://doi.org/10.1515/9783112212141-015

future and could facilitate net-negative emissions scenarios. Indeed, CDR credits have emerged as a critical component of the carbon market landscape in recent years [12]. CDR methods within the bioenergy with carbon capture and storage/utilization (BECCS/U) group of negative emissions technologies (NETs) have spawned technologies for CO_2 conversion with potential to grow the capacity of engineered systems to utilize CO_2 in alignment with circular economy principles of converting waste CO_2 into valuable products [13] (Figure 15.1).

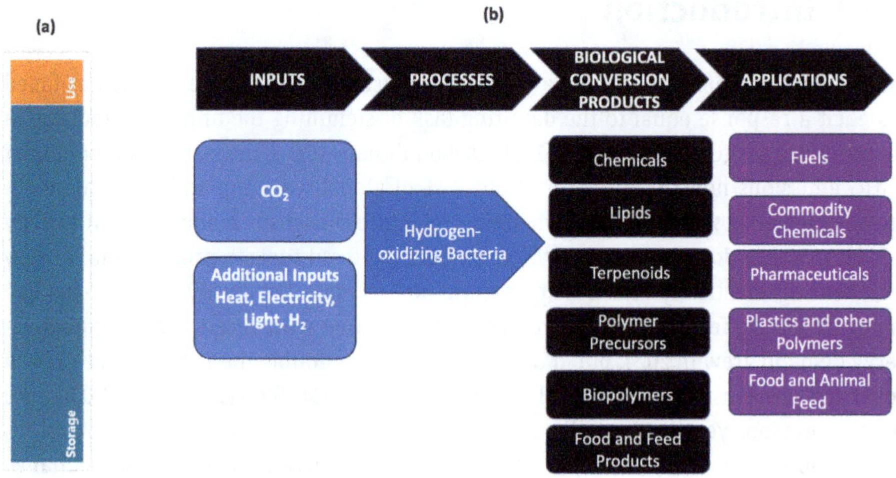

Figure 15.1: Bioenergy with carbon capture and storage (BECCS) relies on nature to remove CO_2 from the atmosphere for use elsewhere. (a) Most of the resulting CO_2 emissions are then stored underground. The small fraction used as inputs for biochemicals and industrial products can be enhanced with HOB-based technology [220].

New opportunities to utilize CO_2 in the development of products and services are capturing the attention of governments, industry, and the investment community. Climate change mitigation is the primary driver for this increased interest, but other factors, including technology leadership and supporting a circular economy are gaining in importance [14]. A single technology with potential for resource recovery, waste treatment, and sustainable food production taps the ability of hydrogen-oxidizing bacteria (HOBs) to readily oxidize hydrogen (H_2) or reduce protons (H^+), using the energy derived to "fix" CO_2. But our understanding of biological CO_2 fixation in general and HOBs in particular is poor, particularly regarding HOBs ecological roles and their interactions within microbial communities. While HOBs are known to utilize H_2 for energy and are found in diverse environments, including soil and water, their specific contributions to ecosystems and responses to environmental changes are not fully understood. New strains, substrates, and cultivation conditions on the one hand, and further development of existing strains on the other, could help turn CO_2 into a mainstay feedstock for the bio-based industry.

Chemosynthetic microorganisms derive nutrition from inorganic substances without sunlight, relying instead on the oxidation of sulfur (S), H_2, hydrogen sulfide (H_2S), and methane (CH_4) for their energy needs. In atmospheric chemosynthesis, the consumption of ubiquitous trace levels of atmospheric gases (H_2, CO, and CO_2) provides the energy and carbon needs for bacteria to literally "live on air." It is a globally distributed phenomenon with significant implications for the global carbon cycle and bacterial survival within environmental reservoirs. Its discovery has elicited a palpable shift in the perception of HOBs [15]. Traditionally considered a specialized functional group benefiting from the energy potential of H_2 for mixotrophic growth or persistence, HOBs are now earning their due recognition for their ability to utilize not just H_2 but multiple trace gases [16]. HOBs play heretofore noncanonical roles in various ecosystem services such as carbon cycling [17], plant growth promotion [18], and primary production [19]. The new recognition warrants new experimental approaches to the development of HOB industrial strains.

Tokyo-based Utilization of Carbon Dioxide Institute Co. Ltd. (UCDI) has stayed at the forefront of research into CO_2-centered biotechnologies since its founding in 2015. UCDI works with the non-model, moderately thermophilic HOB, *Hydrogenophilus thermoluteolus* [20]. The UCDI strain's uncharacteristically fast growth on $H_2/CO_2/O_2$ gas mixtures allowed UCDI to lay the foundation of an intellectual property (IP) portfolio based on modifications of the microbe's central metabolic pathways to produce ethanol, amino acids, and organic acids. But the central metabolic pathways constitute only one of four recognized modules of HOB metabolism [21]. For UCDI strain *H. thermoluteolus*, genetic engineering tools to facilitate easier access to the remaining modules are in active development with lessons from progress made with in the development of related HOB strains [22, 23]. Firsthand experience with the strain informed the company's pursuit of technological footholds in three different fields: CO_2 to protein/food, CO_2 to biofuels (sustainable aviation fuel), and CO_2 to chemicals. This chapter picks up from where HOBs were introduced as important cogs in a carbon-neutral, bio-based industry of tomorrow [24]. This time around, recent technology deployments featuring HOB Microbial Cell Factory (MCF) chassis for CO_2 valorization are noted and discussed in light of new recognition of HOB characteristics that recent scientific literature reveals. Developing and implementing a recalcitrant HOB MCF chassis oozing evidence of exceptional industrial characteristics and profound effects on circular economy contributes to the significant efforts worldwide focused on identifying and developing non-model microorganisms with excellent characteristics for biomanufacturing.

15.2 State of the art of CO_2 utilization

15.2.1 CO_2 utilization is a small fraction of CDR

CO_2 utilization as a process in which one or more economically valuable products are produced using CO_2, the source of the CO_2 notwithstanding, is the subject of increasing interest from the scientific community [25]. To be utilized, the CO_2 must be captured from the atmosphere via land-based measures or Direct Air Capture (DAC) [26, 27]. Land-based CDR measures in general and BECCS in particular are critical to achieving the long-term mitigation goals of the Paris Agreement. BECCS traditionally involved cultivating energy crops, burning them as fuel to produce bioenergy, and then capturing the CO_2 emitted during combustion for long-term storage. The technique thus combines energy production with CDR, and comes with an estimated mitigation potential of 0.4–11.3 GT CO_2 eq/year [28].

Capture of CO_2 from ambient air was commercialized in the 1950s before the advent of captured CO_2 as a feedstock for hydrocarbon fuels in the 1960s [29]. Large-scale capture of CO_2 as a tool for managing climate risk, now commonly referred to as DAC, was first explored in the 1990s [30]. The first DAC paper with commercial engineering cost breakdown reported levelized costs of \$94 to \$232 per ton CO_2 from the atmosphere [31]. And direct air capture of CO_2 from the atmosphere has gained significant attention for achieving negative emissions [32, 33]. New developments combining CO_2 capture from flue gas and ambient air abound even though atmospheric CO_2 capture remains very challenging as is determining how to efficiently and economically utilize the collected carbon [34].

More than 2 GT CO_2 eq/year are removed from Earth's atmosphere but only about 0.1% to around 2.3 million tons per year (MT CO_2 eq/year) – is removed using new technologies. The new technologies have been projected to account for 11.75 MT CO_2 eq/year by 2025 even as the world will need to remove another 7–9 GT CO_2 eq/year by mid-century from the atmosphere if the 1.5 °C Paris Agreement target is to be met [35].

15.2.2 Prospects for future growth

Biological processes that tap energy stored in CO_2 and other low value carbon sources are self-renewing and their chemical energy storage largely compatible with existing industrial and commercial infrastructure. CO_2 utilization involves a suite of chemical and biological processes that transform the carbon in CO_2 into other forms, such as mineral carbonates, alcohols, hydrocarbons, polymers, or elemental carbon materials. Such processes allow CO_2 to replace fossil fuels to enable sustainable production of chemicals and materials and support durable storage of CO_2 in long-lived products. Thus, CO_2 serves as a feedstock for any carbon-based product [36].

Nature is replete with microorganisms that grow on $CO_2/H_2/O_2$ feedstreams via mechanisms that are not yet entirely understood. They include HOBs isolated from diverse ecological niches [20, 37–40]: obligate chemolithoautotrophic growth is particularly well-suited to extreme environments such as burning coal piles from which *Streptomyces thermoautotrophicus* was isolated, hot springs where the moderately thermophilic *Hydrogenobacter thermophilus* originated, and dynamic environments of tidal zone sediment from where *Thiomicrorhabdus marina* was isolated [41–43]. These aerobic HOBs are distinct from hydrogen-oxidizing acetogens that do oxidize H_2 but without autotrophic CO_2 fixation, placing them at the periphery of the present topic of discussion [44].

There are more than 25 different utilization pathways that could materialize over the next decade or two to create building materials, chemicals, fuels, proteins, etc. from captured CO_2. Several CO_2 utilization pathways where the technology is still developing should be monitored going forward given potentially sizable markets. These include green methanol to olefins, dimethyl ether, formic acid, polymers, and CO_2-derived proteins, and CO_2 to ethanol [45]. CO_2 utilization is enabled by economic and noneconomic policies that develop the sector's value for sustainable product formation and long-term carbon storage. Though CO_2 conversion technologies are operating today in some jurisdictions, most areas of the product and process landscape still require investment in research, development, and demonstration. Every jurisdiction must assess the infrastructure, market, policy, and research needs that enable CO_2 utilization to serve the needs of circular carbon economies and durable carbon storage [46].

15.3 Metabolic modules of hydrogen-oxidizing bacteria

15.3.1 About hydrogen oxidizing bacteria (HOB)

Diverse aerobic bacteria oxidize the three most abundant reduced trace gases in the atmosphere (H_2, CO, and CH_4). Together, they account for the net consumption of approximately 75% of H_2, 15% of CO, and 4% of CH_4 lost from the atmosphere each year. Thus, they perform a major but often overlooked ecosystem-regulating service by mitigating emissions of climate-active and toxic gases [47]. HOBs grow chemolithoautotrophically by using H_2 and CO_2 as their sole energy, electron, and carbon sources, whereas carbon monoxide-oxidizing carboxydotrophs can grow chemolithoautotrophically by using CO as the sole energy, electron, and carbon source. Methane-oxidizing bacteria use CH_4 as their sole source of energy, electrons and, often, carbon [48].

Chemolithoautotrophy describes a lifestyle where microorganisms derive their nourishment from chemical sources, inorganic electron donors, and inorganic carbon sources [16]. HOBs comprise a polyphyletic group (i.e., organisms with the same traits but different evolutionary origins) of facultative chemolithoautotrophs traditionally defined by the ability to utilize H_2 as an electron donor to reduce a terminal electron acceptor. Bacteria and archaea, and even the rare eukaryote [49], can oxidize H_2 using oxygen (O_2), carbon dioxide (CO_2), sulfate (SO_4^{2-}), ferric ion (Fe^{3+}), nitrate (NO_3^-), or bicarbonate (HCO_3^-) as terminal electron acceptors. Chemolithoautotrophy is not limited to HOBs. Sulfur-oxidizing bacteria such as *Hydrogenophilus thiooxidans* [50], ammonia-oxidizing microorganisms such as *Nitrosospira multiformis* [51], iron-oxidizing bacteria such as *Mariprofundus ferrooxydans* [52], and carbon monoxide-oxidizing bacteria such as *Mycobacterium smegmatis* [53] oxidize an inorganic compound other than H_2 for their chemolithotrophic growth. Regardless, the energy generated by the oxidation of inorganic compound is used to sustain bacterial basal metabolic processes, which can result in beneficial outcomes such as biomass synthesis for microbial protein production [54], carbon capture for CDR [21], nitrogen fixation for improved agricultural yields [55], and chemical oxygen demand (COD) reduction for wastewater treatment [56].

The first HOB isolated exhibited chemolithoautotrophic growth supported by CO_2 and H_2 as carbon and energy sources, respectively [57]. Most aerobic trace-gas-fermenting bacteria isolated since display facultative chemolithoautotrophic or mixotrophic growth, i.e., the capacity of heterotrophic growth on organic substances but also being capable of chemolithoautotrophic proliferation relying on the oxidation of gas fractions. The first report of “Knallgas bacteria” as potential sources of single-cell protein drew interest from NASA and catapulted these bacteria into the limelight for their ability to fix CO_2 in the absence of light [58, 59]. This non-phototrophic CO_2 fixation, often referred to as dark CO_2 fixation, is carried out by both chemolithoautotrophic and heterotrophic microbes via several metabolic pathways and reactions [60].

The metabolic diversity of aerobic HOBs allows the bacteria to colonize diverse natural environments and, with a little nudging, engineered systems [61]. Many are well-adapted to substrate limitation owing to their high affinity for hydrogen and capacity to use alternative energy sources. The metabolic pathways underlying the reputed versatility of HOBs have been grouped into four modules: a nitrogen fixation module, an energy and redox generation module, a carbon assimilation module, and a product synthesis module [21]. Considering the wide distribution of sulfate-reducing and denitrifying HOBs, to name but two types, modules other than these four could well be functional.

15.3.2 Energy and redox generation module: hydrogenases, sox enzyme system

Hydrogen gas (H_2) is an important source of reductant utilized by a wide range of bacteria: the hydrogenase enzymes that catalyze hydrogen oxidation are present in photosynthetic bacteria (*Rhodobacter*), nitrogen fixers (*Azotobacter*, *Rhizobium*), cyanobacteria (*Synechocystis*), strict anaerobes (*Clostridia*), and *Salmonella* and *Escherichia coli* species [62]. Many microorganisms use H_2 as an electron donor to support aerobic respiration, anaerobic respiration, and carbon fixation. Others produce H_2 to dispose reducing equivalents through fermentative and photobiological processes [63]. All H_2-consuming and -producing bacteria catalyze the reversible oxidation of H_2 using the complex metalloenzymes. As the vehicles by which cells access the energy stored in H_2, understanding the principles of H_2 production and utilization by hydrogenases is important both for basic research and possible applications.

Hydrogenases are generally classified by the structure of their catalytic site into [NiFe]-hydrogenases (hydrogen sensing and consumption), [FeFe]-hydrogenases (hydrogen-evolving), or [Fe]-hydrogenases (methanogenesis) [64, 65]. *H. thermoluteolus* genome encodes three [NiFe]-hydrogenases to the four of model HOB *Cupriavidus necator* H16. Membrane-bound [NiFe]-hydrogenase (MBH) and a soluble NAD^+-reducing [NiFe]-hydrogenase (SH) generate cellular energy and reducing power [66–68]. Regulatory [NiFe]-hydrogenase (RH) triggers hydrogenase gene expression via an atypical two-component regulatory system when H_2 is present in the environment. The actinobacterial-like [NiFe]-hydrogenase (AH) of *C. necator* H16 is missing from the *H. thermoluteolus* genome [69, 70]. Examination of 44 bacterial phyla genomes in which genes coding for hydrogenases were detected, 36 (>80%) harbored [NiFe] hydrogenases, the most widely distributed type of hydrogenase [71].

Many [NiFe]-hydrogenases are highly active and readily catalyze both hydrogen consumption (uptake hydrogenases) and production [72–74]. A notable exception is an important class of membrane-bound O_2-tolerant [NiFe] hydrogenases that only catalyze H_2 oxidation (the uptake reaction) yet possess an active site that is structurally very similar to that of classical, reversible [NiFe] hydrogenases [75]. The existence of hydrogenases has been recognized since the 1930s but their huge potential as biocatalysts in many emerging technologies in the field of energy has not come to pass [76]. According to the affinity of their hydrogenases to bind H_2, HOBs have also been classified into high-affinity (can consume H_2 at sub-atmospheric concentrations) and low-affinity (rapid uptake of elevated concentrations of H_2) subgroups [77]. But it is widely accepted now that HOBs do not form two distinct groups, but rather a continuum according to the affinity spectrum of their [NiFe]-hydrogenase toward H_2 [78]. [NiFe]-hydrogenases have received considerable attention as hydrogen-based catalysts for fuel cells and H_2 sensors, but most characterized [NiFe]-hydrogenases are inhibited by O_2, limiting their activity and stability [79]. Nevertheless, the excellent selectivity of hydrogenases leads to many interesting possibilities for biotechnology driven by H_2 as a clean reductant [80].

Reduced inorganic sulfur compounds are common effluents of textile manufacturing, dye and heavy water productions, tanning and paper industries, coal processing, and sulfidic waste management systems [81, 82]. These compounds can be harmful to the environment and human health. Therefore, remedial measures to lessen their levels in industrial waste are often necessary to comply with environmental regulations [83]. Biological oxidation by sulfur-oxidizing bacteria (SOB) is a low maintenance and operationally inexpensive method that has drawn immense research interest recently [82]. But the sulfur-oxidizing ability has been reported in a diverse range of bacteria.

Chemolithotrophic bacteria can oxidize various sulfur species for energy and electrons using reduced inorganic sulfur compounds such as sulfite (SO_3^{2-}), thiosulfate ($S_2O_3^{2-}$), hydrogen sulfide (H_2S) as energy sources. The best-studied pathway of bacterial sulfur-chemolithotrophy involves direct oxidation of thiosulfate to sulfate (SO_4^{2-}) without any free intermediate. The pathway, mediated by SoxXAYZBCD, is the exclusive mechanism of thiosulfate oxidation in facultatively chemolithotrophic α-proteobacteria and the intact system exists on the *H. thermoluteolus* genome [84]. Sulfide (S_2^{-}) and sulfate are the most abundant and significant components of sulfur cycles worldwide; thiosulfate also constitutes a key junction in the network of sulfur-species transformations in a number of ecosystems including hot spring waters from where *H. thermoluteolus* was isolated [20]. Taxonomically diverse microorganisms oxidize or reduce thiosulfate to meet their bioenergetic requirements. Oxidation of thiosulfate as a chemolithotrophic substrate is widespread in bacteria and *Archaea*.

The metabolic pathways of *H. thermoluteolus*, as reconstructed using the KEGG automatic annotation server [85], reveal a complete set of genes for the tricarboxylic acid cycle (TCA), the Calvin-Benson-Bassham cycle (CBB), oxidative phosphorylation (PPPox), sulfr oxidation (*sox*), nitrite (NO_2^-) reduction to nitrous oxide (N_2O) (*nirS-norBC*) and nitrite ammonification (*nirBD*) [50]. The nitrous oxide reductase gene (*nosZ*) harbors a transposon insertion that would prevent native *H. thermoluteolus* from denitrifying nitrite (NO_2^-) to N_2. Genes encoding an uptake hydrogenase (*hup*) are located approximately 30 kb between hydrogenase maturation proteins (*hyp*) and bidirectional hydrogenase (*hox*) in *H. thermoluteolus*.

During autotrophic metabolism, all energy and reducing power required for the metabolic activity of HOBs derive from oxidation of H_2. The reversible interconversion of molecular H_2 and protons is one of the most ancient microbial metabolic reactions. First, membrane-bound hydrogenase (MBH) binds molecular H_2 and converts it to protons and electrons. The free energy generated thereof is used to generate a sodium motive force (smf) across the biological membrane that powers ATP synthesis using proton pumps and cytochrome oxidase (cyt). H_2 that diffuses into the cell binds to soluble hydrogenase (SH) for H_2-driven nicotinamide and flavin cofactor regeneration [86]. The energy and reducing power generated by H_2 oxidation also drive the CBB cycle for CO_2 fixation [87]. CBB synthesizes acetyl-CoA, a key intermediate linking energy production, biosynthesis, and gene regulation [88]. The CBB pathway and hydrogenases of *R. eutropha* were engineered for improved activity, in one of only a few cases where the efficiency of CO_2 assimilation pathway was improved [89].

15.3.3 Carbon assimilation module: RuBisCO, other carboxylases

CO_2, an important carbon feedstock for a future green economy, is the most prevalent form of elemental carbon in the atmosphere. The biological conversion of CO_2 into biomass is critical for the global carbon cycle, and key to establishing a sustainable (bio) economy, in which CO_2 serves as the main carbon source. Biological carbon fixation requires several enzymes to turn CO_2 into biomass. Most CO_2-fixing organisms use "carbon-fixation" routes, in which CO_2 is first incorporated into an acceptor molecule, subsequently reduced and directly converted into a multicarbon compound through the same pathway [90]. The acceptor molecule of most consequence is ribulose 1,5-bisphosphate (RuBP) carboxylase/oxygenase (RubisCO, EC 4.1.1.39). Alternatively, CO_2 can be assimilated through a "reduction-first" strategy [91], in which CO_2 is initially converted into a reduced C1-compound (e.g., formate or carbon monoxide), which is condensed into multicarbon compounds in a subsequent phase through downstream reactions. Chemolithoautotrophic bacteria fix CO_2 using eight known pathways, with the CBB cycle being quantitatively the most significant [92]. CBB and other "carbon-fixation" routes are less energy-efficient than the reductive citric acid cycle (rTCA) and the Wood Ljungdahl pathway (WLP), two "reduction-first" routes. It is inconvenient that the two are restricted to microaerophilic and anaerobic microorganisms, making them unavailable to photosynthetic organisms and other aerobic bioproduction hosts.

More than 99% of global carbon fixation is catalyzed by RuBisCO, the central enzyme of the CBB cycle and most abundant protein in the biosphere [93]. RuBisCO can be found in all domains of life, including autotrophic bacteria where it carboxylates ribulose-1,5-bisphosphate (RuBP) into two molecules of 3-phosphoglyceric acid (PGA). RuBisCO is a notoriously inefficient catalyst, suffering from low turnover rates, a low affinity for its CO_2 substrate, and a competitive inhibition by O_2 as an alternative substrate. Elevated CO_2 concentration increases the enzyme's carboxylation rate and competitively inhibits the oxygenase reaction [94]. RuBisCO is also inhibited by sugar-phosphate inhibitors [95]. It has evolved into multiple enzymatic forms with a range of kinetic properties, and coevolved with CO_2 concentrating mechanisms. The *cbb* operons of facultative autotrophs are more strictly regulated than those of obligate autotrophs. *cbbR*, the master regulator for microbial carbon dioxide fixation is usually located directly upstream of the *cbb* operon that it regulates but in the opposite orientation [96]. Exceptions to this orientation abound, including in *Hydrogenophilus thermoluteolus* [97]. Facultative autotrophs encode one or two gene clusters dedicated to carbon assimilation to the three or more of obligate autotrophs [98].

Due to its central role in the global carbon cycle, RuBisCO has been the subject of intense research for over 50 years. Significant effort has been invested into better understanding its structure-function details, but rational design approaches have largely failed to provide catalysis enhancing solutions [99]. Modern directed evolution tools are a promising conduit to improving RuBisCO [100]. Efforts to improve the catalytic efficiency of RuBisCO should facilitate a better understanding of biological CO_2 fixation mechanisms and help unlock untapped reserves of sustainable energy and value-added chemicals [101, 102].

RuBisCO is but one of many CO_2 acceptor molecules available for carbon fixation by chemolithoautotrophs. Heterotrophs as well can utilize enzymes of assimilatory carboxylations such as PEP carboxykinase, isocitrate dehydrogenase, malic enzyme, 2-oxoacid synthase, and crotonyl-CoA carboxylase/reductase as CO_2 acceptors [103, 104]. HCO_3^- acceptors such as PEP carboxylase, pyruvate carboxylase, acetyl-CoA/propionyl -CoA carboxylase, and methylcrotonyl-CoA carboxylase round out the arsenal of acceptor molecules available for carboxylation-dependent carbon capture. Formate dehydrogenase (FDH) is an acceptor molecule of "reduction-first" carbon capture, the best-known pathway of which is the energy-efficient reductive acetyl-CoA pathway commonly known as the Wood-Ljungdahl pathway – the only carbon fixation pathway that produces ATP rather than consumes it [91]. Its end-product, acetyl-CoA, is the same end-product of "reduction-first" pathways, which can support higher biomass yield than their purely carboxylation counterparts. From a biotechnological perspective, high biomass yield could be translated into high product yield, making the "reduction-first" pathways very interesting for the bio-based industry. Considering that photorespiration remains the principal drawback to agriculture and other biotechnologies that rely on primary producers such as plants and algae, chemolithoautotrophic HOBs represent the best option to realize high yield bioprocessing [105]. It is instructive that harvested HOBs are typically richer in protein content (50–83% dry cell weight) than plants or algae [106].

The CBB cycle has evolved for optimal synthesis of C3 metabolites but not for the production of the C2 building block acetyl-CoA [107]. Carbon lost in producing acetyl-CoA from stepwise decarboxylation of C3 sugar limits the maximum carbon yield of the CO_2 fixation process but avails crucial nodes for anaplerotic carboxylation reactions catalyzed by PEP carboxylase (*ppc*) and malic enzyme (*maeB*) to replenish the TCA cycle. The same nodes are available to biosynthetic carboxylation pathways such as those leading to fatty acid and polyketide biosynthesis. Moreover, the nodes anchor heterologous pathways leading to other valuable chemicals (Figure 15.2).

Success in constructing heterologous pathways starting from the pyruvate node and a few starting from the PEP node is reflected in the number of patent applications for pyruvate- and aspartate-derived amino and organic acids [24]. Interest in leveraging these off-ramps parallels those in cyanobacteria in which the carbon flux in two prolific strains has been redirected to produce chemicals via a heterologous pathway centered on the CBB cycle [108]. Robust patenting activity in the nascent HOB-based biomanufacturing from CO_2 space is testament to the improving feasibility of HOB as industrial strains.

Efficiency of the CBB cycle is limited primarily by the O_2 sensitivity of RuBisCO. Research has explored circumventing this limitation through bioprospecting nature for RuBisCO variants with better characteristics [109]: protein engineering [100], directed evolution [110], and augmenting the CBB cycle with synthetic pathways insensitive to O_2 [111, 112]. CO_2 fixation by HOB is comparatively more resilient to O_2 owing to the presence of multiple hydrogenases that not only catalyze the oxidation of H_2 to form $2e^-$ and $2[H^+]$ but also support cell growth and survival [71, 113, 114]. It is noteworthy that the genomes of HOB reveal a carbonic anhydrase (CA) enzyme which should function to generate HCO_3^- ions from CO_2 gas bubbled into reaction medium. Not much has been reported about the impact of this enzyme on the observed efficacy of CO_2 fixation by HOB. Regardless, the conversion efficiency of CO_2 to acetyl-CoA is no doubt improved by anaplerotic carboxylation pathways and may contribute to the fast growth of HOB cells. The essential amino acid content of the biomass compares favorably with the best alternative proteins emerging on the market.

15.3.4 Product synthesis module: TCA, anaplerotic node, β-oxidation, PHB pathway

Metabolic flexibility in aerobic HOBs enhances cell growth and survival when nutrients are imbalanced or limiting. This phenomenon has been exploited in the production of intracellular storage compounds (such as glycogen or polyhydroxyalkanoates (PHAs)) under conditions of nutrient limitation. Model HOB strain *Cupriavidus necator* H16 can grow autotrophically on H_2 and CO_2 via CBB or formate. When nutrients are abundant, acetyl-CoA enters the TCA cycle for biomass synthesis (Figure 15.2). The TCA intermediate α-ketoglutarate (α-KG) plays a crucial role in peptide and amino acid synthesis and coordinates carbon and nitrogen utilization and lipid synthesis [115]. α-KG finds broad applications in the pharmaceutical, food, and chemical industries, and its industrial production is shifting from chemical-based processes utilizing toxic reagents to biomass-based feedstocks utilizing heterogeneous catalyst [116, 117]. Incorporation of ammonia into the α-KG carbon skeleton produces the essential amino acid precursor glutamate [118]. A transamination reaction between glutamate and a keto-acid produces an amino acid. HOBs accumulate protein levels above 70% on dry weight basis [106]; *H. thermoluteolus* protein content exceeds 80% on dry weight basis, making it an excellent single-cell protein (SCP) source and anchoring the CO_2 to protein/food business at UCDI.

If the concentration of nitrogen, oxygen, or phosphorus in the medium becomes limiting, HOB cells convert acetyl-CoA into polyhydroxybutyrate (PHB) to store excess energy and carbon sources in response to environmental stresses. Should the nutrient limitation resolve, PHB can be converted back to acetyl-CoA and enter the TCA cycle

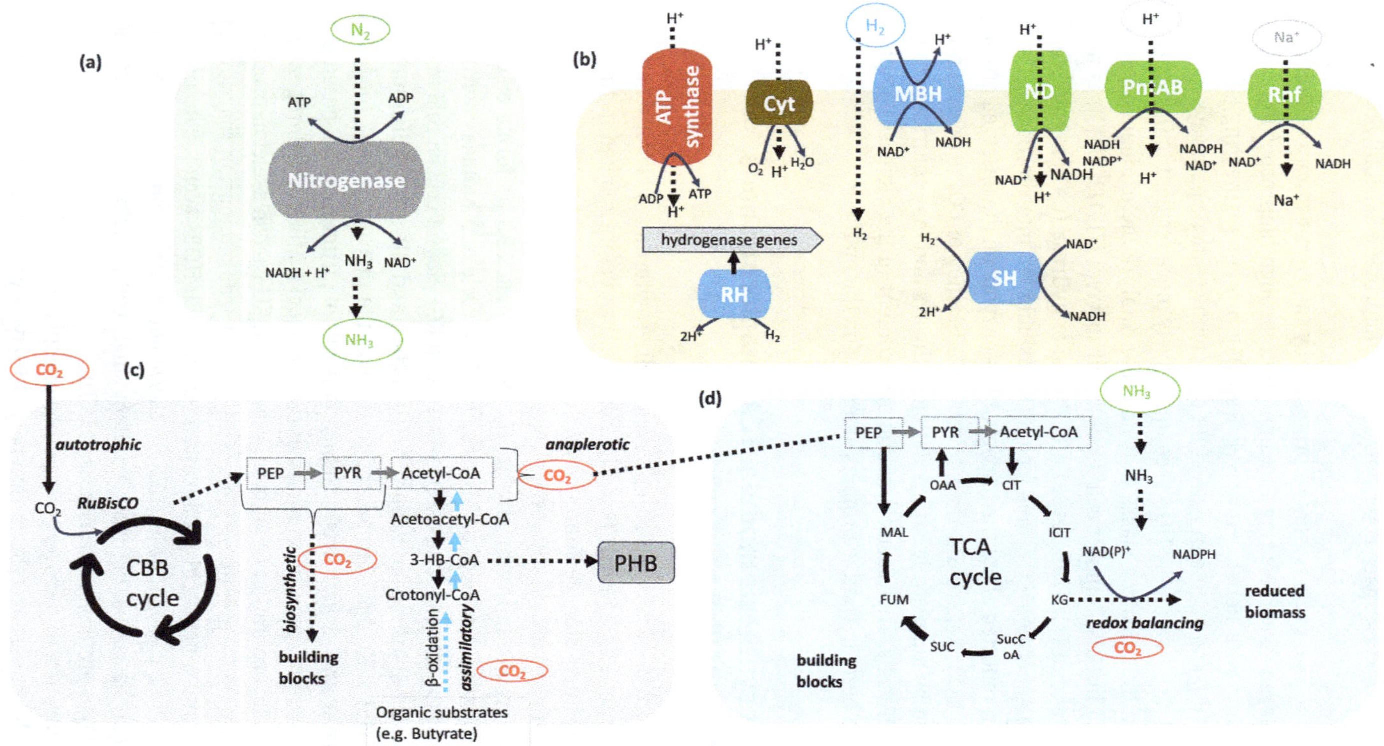

Figure 15.2: Schematic representation of the four recognized modules of HOB: (a) N_2 fixation (green background), (b) energy and reducing power generation (gold background), (c) carbon assimilation (blue-grey background), and (d) product synthesis (blue background). *H. thermoluteolus* does not encode nitrogenase but does encode a full set of *sox* genes for sulfite reduction. Abbreviations: MBH, membrane bound hydrogenase; SH, soluble hydrogenase; RH, regulatory hydrogenase; Cyt, cytochrome oxidase; ND, NADH dehydrogenase; PntAB, transhydrogenase; Rnf, Rhodobacter nitrogenase factor; Nfn,; PEP, phosphenolpyruvate; PYR, pyruvate; 3HB-CoA, 3-hydroxybutyrate-CoA; CIT, citrate; ICIT, isocitrate; KG, α-ketoglutarate; SucCoA, succinyl-CoA; SUC, succinate, FUM, fumarate; MAL, malate; OAA, oxaloacetate.

to synthesize biomass [119]. Mesophilic *Cupriavidus necator* is a proficient producer of PHB with excellent protein synthesis capabilities. Moderately thermophilic UCDI strain *H. thermoluteolus*, a facultatively chemolithoautotrophic hydrogen- and thiosulfate-oxidizing bacterium able to grow autotrophically, heterotrophically, and mixotrophically likewise yields biomass with a high protein content [120]. Heterotrophic growth characteristics vary marginally among *Hydrogenophilus* species, and growth on thiosulfate is confined to *H. thermoluteolus*. Mixotrophic growth with a number of organic substrates in an atmosphere containing hydrogen, oxygen, and carbon dioxide was reported for *Alcaligenes eutrophus, Pseudomonas facilis, Paracoccus denitrificans* strain "Vogt", *Mycobacterium gordonae, Arthrobacter* 11/*x*, and *H. islandicus* [40, 121].

Microbial protein is by far the most common product of companies exploiting HOBs at the moment, with great strides made toward bringing to market HOB-derived microbial protein products. Great strides have also been made in the synthesis of value-added products such as oils, biochemicals, bioplastics, and biofuels from a variety of low value substrates. Finland-based Solar Foods Oy manufactures a microbial protein it calls Solein using *Xanthobacter* sp. SF1. Solein received a novel food regulatory approval that allows for the sale of food products containing Solein in Singapore in September 2022. In February 2024, the company declared a self-affirmed GRAS in readiness for Solein's commercialization and entry into the US market. It awaits a Novel Food license approval from the European Union. Belgium-based Avecom has developed several SCPs from low-value substrates and UK-based Deep Branch Biotechnology and Denmark-based Aerbio produce microbial protein with a tailored amino acid profile using CO_2 from flue gas. US-based Superbrewed Foods produces a postbiotic cultured microbial protein using *Clostridium tyrobutyricum*. The company secured FDA-GRAS notification in March 2024 and European Union EFSA QPS list in January 2024. As HOBs begin to permeate contemporary food and chemicals industries, they contribute to value creation and provide new jobs and livelihoods based on the nascent CO_2 industry. The combined market for meat, eggs, dairy, and seafood products is projected to grow sevenfold from $35 billion to reach at least $290 billion by 2035 as growing consumer sentiment drives unparalleled growth in plant-, microorganism-, and animal cell-based alternatives [122].

The food industry currently lacks a non-animal-based protein that can offer all the benefits of animal protein. RuBisCO is naturally abundant and offers nutritional benefits superior to other plant-based proteins. Regarded as a nonallergenic protein that possesses numerous organoleptic, functional, and physical properties that are highly promising for the food industry, RuBisCO as a protein source is held back chiefly by the lack of efficient methods for large-scale purification and subsequent utilization in the food sector. RuBisCO as a food has been said to enhance the secretion of serotonin (happiness hormone) and melatonin (sleep regulation hormone) and elevate tryptophan levels in the human brain [123, 124]. Plants are the preferred choice of RuBisCO at present but the advent of the microbial protein wave could make HOB as a viable source of the protein soon.

15.3.5 Nitrogen fixation module: nitrogenase

Some bacteria have the remarkable capacity to fix atmospheric nitrogen to ammonia under ambient conditions, a reaction only mimicked on an industrial scale by a chemical process that requires high temperatures, elevated pressure, and special catalysts. The ability of microorganisms to use nitrogen gas as the sole nitrogen source incurs physiological penalties because the process is oxygen-sensitive and energy-dependent [125]. But it could prove advantageous over the microorganisms' dependence on reactive nitrogen sources for their protein needs. Autotrophic nitrogen-fixing hydrogen-oxidizing bacteria (NF-HOB) use the energy derived from H_2 oxidation to drive nitrogen fixation, on top of driving carbon fixation. With their ability to directly utilize CO_2 and N_2, NF-HOBs offer a promising alternative for food and feed applications worth investigating further [55, 126].

NF-HOB, also called "H-C-N microorganisms" include members from the following genera: *Azospirillum* (*A. lipoferum*), *Ancylobacter* (*A. aquaticus*, *A. vacuolatus*), *Xanthobacter* (*X. autotrophicus*, *X. flavus*, *X. variabilis*), *Bradyrhizobium* (*B. japonicum*), *Azohydromonas* (*A. lata*), and *Derxia* (*D. gummosa*) [127–130]. *X. autotrophicus* and *Azotobacter vinelandii* can synthesize ammonia from atmospheric N_2 and fix it into biomass [131, 132]. As the ATP and NADH involved in nitrogen fixation must come from H_2 oxidation, it follows that a greater amount of H_2 must be oxidized to satisfy the energy demand for N_2 fixation [127]. The generated ammonia (NH_3) can combine with the TCA intermediate α-KG to serve as a nitrogen source for amino acid and protein synthesis. Any excess NH_3 is excreted, especially when biomass synthesis is inhibited. Uptake of this NH_3 provides nitrogen for other biological or chemical processes. HOBs with N_2-fixation capacity are common in the rhizosphere microbial community [133].

Suppressed production of microbial protein weighs against development of NF-HOB strains for SCP. But the fact that NF-HOB can produce microbial protein sustainably by directly using N_2 as a nitrogen source without extracting reactive nitrogen sources from wastewater and industrial nitrogen fertilizers is worth exploring [134]. The separation of microbial protein production from centrally synthesized Haber-Bosch nitrogen and higher CO_2 fixation potential are important merits for pursuing the development of improved NF-HOB strains.

However, many nitrogen sources, including ammonia (NH_3), nitrate (NO_3^-), dinitrogen (N_2), amino acids, and other nitrogen-containing organic compounds can be exploited by HOBs. Nitrogen sources other than NH_3 are first converted to NH_3, often with some expenditure of electrons and ATP [135]. Therefore, NH_3 is still considered the optimal nitrogen source for HOBs.

15.4 High-affinity hydrogenases as biocatalysts for energy conversion

Hydrogen gas (H_2) has been highly touted as a revolutionary fuel for 50 years and a prime contender for the next generation of energy, yet usage is confined to oil refining and fertilizer production [136]. When utilized in fuel cells or combusted cleanly, H_2 produces only water, eliminating greenhouse gas (GHG) and pollutant emissions such as CO_2, SO_x, and NO_x. Hydrogen is versatile – with many ways to produce H_2, e.g., electrolysis, thermal conversion (natural gas/coal conversion and carbon capture, utilization, and storage), photoelectrochemical and biological production; and many ways to use H_2, e.g., from transportation and grid stabilization to industrial and chemical processes. The versatility of hydrogen avails opportunities in large-scale, long-duration energy storage, and the decarbonization of industry. As nations strive toward net-zero carbon targets by 2050, H_2 integration into the energy mix has become urgent, with the launch of roadmaps by many countries prioritizing H_2 technologies [5, 137].

But hydrogen has serious drawbacks in terms of storage and transportation; it takes up too much volume, rendering it impractical to store. Compressed H_2 gas tanks and liquified H_2 tanks are the conventional methods for storing and transporting hydrogen, but the compression is expensive and essentially turns hydrogen storage tanks into high-pressure explosives [138]. Various alternatives to store hydrogen have been evaluated, including liquid hydrogen carriers (LHCs), i.e., molecules that store hydrogen in chemical bonds in the liquid phase. They provide an approach to store hydrogen with a smaller volumetric footprint. Standout LHCs include ammonia (NH_3) [139], liquid hydrogen (LH_2) [140], liquid organic hydrogen carriers (LOHCs) [141], and methanol (CH_3OH) [142]. NH_3 and methylcyclohexane (MCH) are the leading contenders, stable for long durations and can use existing infrastructures to both store and transport hydrogen at large scales. But both require production of H_2 at high temperatures, with consequent CO_2 emissions, while NH_3 is difficult to recycle and MCH requires precious metal catalysts [143]. The search for better LHCs continues to draw intense research interest [144–146]. Hydrogen storage alternatively employs solid-state hydrogen carriers (SHC), which exploit the ability of hydrogen to stick on the surface or move into the bulk of the storage materials. SHCs, in particular metal hydrides (MH), are a commercially viable alternative to compressed gas or liquid hydrogen storage solutions that offer promise in terms of operating conditions and volumetric energy density [147]. But all current carriers have their own weaknesses, and a shift away from conventional chemical hydrogen carriers is attracting research interest [143].

Biological H_2 production is a promising alternative to industrial H_2 synthesis from fossil fuel. It is mediated by hydrogenases, structurally complicated enzymes requiring specialized metal cofactors for both catalysis and electron transfer. They have received considerable attention in recent decades for application as biocatalysts for both biohydrogen production and H_2 oxidation as an energy source in enzymatic fuel cells [148]. [NiFe]-hydrogenases, the preferred hydrogenases for industrial applications, are gaining in importance as catalysts in fuel cells, which utilize H_2 to produce electrical current. They are a promising alternative to the platinum-based catalysts that currently predominate in fuel cells due to the abundance of nickel and iron, and the resistance of some of the enzymes to inhibition by gases, including carbon monoxide (CO), which can rapidly poison platinum-based catalysts used in proton membrane fuel cells (PMFCs) [79, 149, 150]. Hydrogenase enzyme-modified electrodes as compared with platinum electrodes, when exposed to CO, quickly regain catalytic activity once the CO gas is removed. The H_2 oxidation activity of hydrogenases has been exploited in electrochemical fuel cells, which are independent of the traditional platinum catalysts [150]. Furthermore, hydrogenases have been successfully used in H_2-driven cofactor regeneration systems to recycle cofactors such as NAD(H) and NADP(H), which are required for many oxidoreductases of commercial interest [86, 151, 152]. These biotechnological applications of hydrogenases are, however, still in their infancy.

The majority of characterized [NiFe]-hydrogenases are inhibited by oxygen (O_2), limiting their activity and stability. Recent studies show that bacteria and archaea produce diverse [NiFe]-hydrogenases that function in oxic environments and often support aerobic respiration. Noted O_2-tolerant hydrogenases from *Cupriavidus necator* [153], *Hydrogenophilus thermoluteolus* [68], *Bradyrhizobium japonicum* [154], *Hydrogenovibrio marinus* [66], *Aquifex aeolicus* [155], *Salmonella enterica* serovar *Typhimurium* [156], and *Mycobacterium smegmatis* [157] support aerobic hydrogenotrophic and mixotrophic growth of their respective hosts. Their high affinity for H_2 makes these enzymes promising candidates for the development of enzyme-based fuel cells (EBFCs), which utilize H_2 at low concentrations and in impure gas mixtures [79]. Hydrogenases allow microorganisms to use H_2 as an energy source by yielding low-potential electrons for respiration and carbon fixation. Other microbes produce H_2 as a means of disposing of excess reductant during fermentation, as a terminal electron-acceptor, or to maintain redox balance during photosynthesis. Additionally, some hydrogenases are capable of bifurcation, coupling the oxidation of H_2 to the simultaneous reduction in a high and low-potential substrate, or confurcation via the reverse process.

Synthesis of [FeFe]-hydrogenases is generally more successful than that of other classes of hydrogenases [158]. The most efficient microbial enzymes for catalytic H_2 conversion, [FeFe]-hydrogenases have not yet been implemented into industrial applications of an economic scale, but much research to allow this to happen has demonstrated scalable [FeFe]-hydrogenase production, allowing operation of these biocatalyst systems without the need for specialist infrastructure such as anaerobic glove boxes [159, 160]. The game-changing discovery of CbA5H [FeFe]-hydrogenase, a soluble hydrogenase (SH) from *Clostridium beijerinckii* SM10, was the first description of a hydrogenase that can withstand the oxidative action of oxygen by switching to an inactive (H_{inact}) protected form in an aerobic environment [161]. CbA5H is also stable in high concentrations of acetone and acetonitrile and withstands intermediate concentrations of dimethylsulfoxide (DMSO), ethanol, and methanol [162]. The exchange of a methionine for a glutamate residue leads to an increased resistance of the hydrogenase to dioxygen.

15.5 Artificial autotrophy attracting scientific interest

Fossil fuels – coal, oil, and natural gas – account for more than 81.5% of the world's primary energy consumption in 2024 [163]. These are finite resources the extraction of which becomes more challenging as resources are depleted. Their existence can be traced back to autotrophic CO_2 fixation, the primary route through which organic carbon enters the biosphere, and a key step in the biogeochemical carbon cycle. But burning of fossil fuels for energy releases CO_2, of which the concentration in the atmosphere has increased gradually since the Industrial Revolution [164]. Natural CO_2 fixation assimilates more than 380 GT CO_2 per year and plays an essential role in the global carbon cycle as a major carbon sink but has not kept pace with human-made CO_2 emissions [165, 166]. The CBB pathway accounts for more than 90% of all biological CO_2 fixations despite being a notoriously inefficient natural carboxylation process. It is therefore tempting to speculate that even a small increase of CBB performance could have vast downstream effects, assimilating even greater amounts of atmospheric CO_2 from the atmosphere. But rational design approaches have largely failed to enhance CBB catalysis, and it remains debatable whether a more efficient CBB would mean faster growing autotrophs with improved CO_2 fixation [99].

Biological carbon sequestration is the better option for CO_2 fixation owing to its mild reaction conditions and eco-friendly approach [167, 168]. It is supposed that CO_2 fixation might be more economic and efficient once sustainable bioprocesses producing biofuel and chemicals directly from CO_2 are realized. In this respect, autotrophic microorganisms make the most suitable cell factories. But for autotrophic microorganisms to convert CO_2 into value-added products effectively, their slow growth, low productivity, and inferior energy conversion efficiency must be overcome. Most autotrophs fix CO_2 via the CBB cycle in three phases – carboxylation, reduction, and ribulose 1,5-bisphosphate (RuBP) regeneration [169]. RuBisCO fixes more than 250 GT CO_2 from the atmosphere per year but it is notoriously inefficient, with a slow carboxylation turnover rate of 1–10/s compounded to a low affinity for CO_2 [104, 170].

Attempts to increase the efficiency of the CBB cycle aimed at improving autotrophic growth and consequent product synthesis or enhancing CO_2 utilization involve modifying catalytic properties of RuBisCO by protein engineering [100, 101, 171], introducing more efficient photorespiration pathways [87, 172], and introducing CO_2 concentration mechanisms to minimize photorespiration and increase the carboxylating activity of RuBisCO [173]. The efforts have resulted in some minor improvements in performance of autotrophs employing the CBB cycle but replacement of the cycle and accompanying enzyme RuBisCO by potentially more efficient alternatives seems like the more promising path to take [174, 175].

Natural autotrophs can produce organic compounds from inorganic compounds and heterotrophs cannot, but heterotrophs are proven efficient biofuel and chemical production chasses [176, 177]. Heterotrophs also boast more advanced genetic toolboxes that stretch the microorganisms' inherent advantages even further. Synthetic biology already enables the engineering of novel functions and *de novo* metabolic networks in heterotrophic microorganisms for industrial biotechnology applications. It has allowed significant progress in artificial pathway design from advocacy to practice in a short period of time but there remains a huge scope for exploration of pathway diversity, operational efficiencies, and host suitability. Synthetic one-carbon assimilation pathways can offer faster and more efficient alternatives to their native counterparts, enabling enhanced bio-based production of fuels and chemicals from one-carbon substrates [178, 179]. To date, six naturally occurring carbon assimilation pathways with different physiological properties, adaptation to environmental conditions, and bioenergetic trade-offs have been validated through either directly enzymatic or phenotypic tests or genomic analysis followed by experimental tests [102, 180, 181]. The in vivo feasibility of eight other CO_2 fixation pathways has been confirmed [92, 182]. In total, more than 30 new-to-nature CO_2-fixation pathways with improved capture and conversion of CO_2 that outcompete natural pathways have been designed theoretically but only a handful have been implemented in heterotrophic microorganisms to produce biofuels and chemicals from CO_2 [183, 184]. It has been established that RuBisCO and phosphoribulokinase (PRK) expressed in *Escherichia coli, Saccharomyces cerevisiae, Methylobacterium extorquens, Kluyveromyces marxianus,* and *Pichia pastoris* gains the heterotrophs the ability to fix CO_2. Engineering synthetic CBB cycle in the hemiautotrophs allows for autotrophic growth on CO_2 [185]. The resultant hemiautotrophs produce biomass from CO_2 but derive energy and reducing power from organic compounds [186]. This total dependence on sugars and other multicarbon organic compounds for cellular biomass is a notable weakness of hemiautotrophy.

Conversion of hemiautotrophs into fully functional artificial autotrophs is a daunting process involving replacement of substrate transport systems, partial deletion of the sugar metabolic pathway, integration of systems for CO_2 fixation and central metabolism, supply of reducing power, and adaption of the strain to grow under this rewired metabolism [182, 187]. An *E. coli* strain deriving reducing power and energy from a one-carbon formate energy module was successfully constructed [188]. In the final analysis, a set of just three mutations of genes involved in glycolysis (*pgi*, encoding phosphoglycerate isomerase), central-carbon regulation (*crp*, encoding cAMP receptor protein), and RNA transcription (*rpoB*, encoding the β subunit of RNA polymerase) promoted autotrophy, highlighting the malleability of host *E. coli* metabolism [189]. Advancing synthetic biology is opening the way to novel carbon assimilation pathways and artificial autotrophic strains that can be used in low carbon biomanufacturing. Interest in converting heterotrophic organisms into facultative autotrophs is expected to have a positive impact on Earth's ecosystem.

Established model organisms such as *Escherichia coli, Bacillus subtilis, Saccharomyces cerevisiae,* and *Corynebacterium glutamicum* are extensively exploited in the commercial production of amino acids, enzymes or insulin, for example. But given the large variation of essential requirements between bioprocesses, these model heterotrophs often lack some of the essential requirements for a given bioprocess, such as high growth rates, easy handling in diverse bioprocesses, robustness against harsh process conditions, and tolerance toward high substrate and product concentrations [190]. New microbial chassis strains with better-tailored traits, essential to fulfil the needs of a sustainable bio-based industry, are now more accessible following the collective progress in synthetic biology, metabolic engineering, and genome reduction. Autotrophic microorganisms are a feasible sustainable/green production platform for commercially essential products such as biofuels, commodity chemicals, and biopolymers. With access to a maturing synthetic biology toolkit [22], *Cupriavidus necator* metabolism has been redesigned to assimilate formate not through the native CBB pathway but through the more efficient synthetic reverse glycine pathway [179, 191]. *C. necator* is one of a few biotechnologically relevant hosts that can grow on formate [192]. The resultant growth is faster and more efficient.

15.6 Demand for CO_2 as a feedstock is rising

Conversion of captured CO_2 to valuable products and fuels such as methanol, urea, and plastics is a mature technology involving high-temperature, high-pressure catalysis [193]. Such production and use of methanol alone results in about 165 MT CO_2 eq/year of emissions or about 0.3% of the world's total [194]. 12.5 MT CO_2 eq/year is emitted from urea production via the Bosch-Meiser process, making it one of the largest emitters in the chemical industry. UK-based Econic Technologies (www.econic-technologies.com), a pioneer in the transformation of CO_2 into high performance products, estimates that if 30% of all polyols (used as cross-linking agents in plastics production) were made from CO_2, this would save 90 MT CO_2 eq/year from the atmosphere. What's more, because CO_2 is so cheap – less than $100 a ton, compared to $2,000 for propylene oxide (the standard raw material for plastics production) – it would save a manufacturer $10 million a year for a plant with a 50 kiloton annual output [195]. Of the current global demand of around 230 MT CO_2 eq/year, the fertilizer industry alone consumes 125 MT CO_2 eq/year for urea manufacturing. Oil and gas producers use around 70–80 MT CO_2 eq/year for enhanced oil recovery (EOR). CO_2 is also an integral part of food and beverage production, cooling, water treatment and agriculture, but the demand is relatively small. Additionally, CO_2 is utilized in the chemicals industry in Sabatier synthesis (methane production), Fischer-Tropsch synthesis (liquid hydrocarbon production), hydrogenation to methanol, and hydrogenation to formic acid. The global demand for CO_2 as a raw material is forecast to grow from 634 MT CO_2 eq/year in 2030 to 6,076 MT CO_2 eq/year by 2050 [196].

Approaches to make conventional chemical conversion of CO_2 more sustainable focus primarily on utilizing captured CO_2 with renewable energy sources and other inputs: green urea from green ammonia, green methanol through biomass gasification to create biomethanol, or by synthesizing e-methanol from green hydrogen, and green plastics from biomass sources (bio-based plastics) processes like microbial fermentation or chemical recycling to reduce reliance on fossil fuels. But electrochemical reduction processes that produce e-chemicals for green hydrogen or green ammonia are not yet competitive with fossil resources, and so technological advancements (e.g., for catalysts) and legislative adjustments, if enacted as currently anticipated, could enable the commercialization of CO_2 reduction processes in the near future [197].

Microbial cell factories offer an eco-friendly alternative for transforming CO_2 into commercially valuable products because of their reduced carbon impact and milder operating conditions compared to conventional industrial procedures. Biological conversion promotes carbon circularity, reduces reliance on fossil resources, and helps mitigate climate change. Applications range from industrial processes like manufacturing and enhanced oil recovery to the creation of new materials and fuels, driving a potential new "industrial revolution" toward a carbon-neutral society. Biofuels and bioproducts that have been produced in microbes from biomass-derived sugars can also be produced directly from CO_2. Significant improvements in titer, rate, and yield are required to establish this pathway for biofuels and bioproducts. Also, strategies where intermediates are produced biologically from CO_2 (ethanol and isobutanol) and then converted to final products by catalysis need to be considered alongside complete biological conversion.

Acetogens, a broad group of strictly anaerobic bacteria that produce acetate from C1 substrates including CO_2, often using H_2 as an energy source via the Wood-Ljungdahl pathway, are synonymous with the acetone-butanol-ethanol (ABE) fermentation. ABE fermentation has regained interest due to the environmental problems generated by fossil fuels and the need to modify production chains including considerations of sustainability and circular economy [198]. ABE production traditionally depended on starchy biomass, but efforts are being made to shift to lignocellulosic biomass instead [199]. The inevitable transition toward C1 substrates began with chemoautotrophic gas-fermenting bacteria including *Clostridium autoethanogenum* [200], *Clostridium ljungdahlii* [201], and *Acetobacterium woodii* [202]. The currently prevailing iteration of C1-based third generation (3G) biomanufacturing aimed ultimately at converting atmospheric CO_2 and renewable energy sources, such as waste streams and solar power, into valuable products. CO_2-based biochemicals production achieves net-negative greenhouse gas emissions, ranging from −2.43 kg CO_{2eq} to −0.38 kg CO_{2eq} per kg of biochemical, primarily due to the permanent sequestration of biogenic carbon within the final products [203]. However, existing industrial infrastructure and market demands impose constraints on microbes, bioprocesses, and products that require careful consideration to ensure technical and economic success of gas fermentation technologies at scale.

A CO_2-based biochemicals industry evolving toward established production methods in the chemical industry ultimately aims to convert atmospheric CO_2 to fuels and chemicals using renewable energy [44, 204]. Where biomass is the desired end-product, 3G biomanufacturing makes it trivial to recover the biomass without fundamental alterations to the manufacturing process itself [205]. Consequently, the CO_2-based bioindustry can be said to be organized in value-added chains from inputs to applications via specific intermediates (Figure 15.1). CO_2-containing feedstream gases are still largely of fossil fuel origin, but the use of gas streams from renewable materials is on the rise owing to increased efficiencies in the biomass gasification process [206].

Gaseous $CO_2/H_2/O_2$ feedstreams as substrates have advantages of global availability, noncompetitiveness with food or feed, predictable pricing through well-proven market mechanisms, and full metabolic utility. The "food vs. fuel" debate intensified a while ago in view of the need to meet the caloric and nutritional requirements of a human population projected to approach or exceed 10 billion within three decades [207]. Moreover, the use of HOBs results in few impurities derived from nonreactive carbon, good stability, reduced risk of microbial contamination, and accessibility from waste streams. The gaseous $CO_2/H_2/O_2$ feedstream as a substrate is an important enabler for circular economy considerations.

15.7 Opportunities, challenges, and future directions

There have been new entrants in the CO_2 as a feedstock for microbial fermentation space over the last two years even though the HOB strains favored have stayed the same. Colipi (colipi.com), based in Germany, is one such new entrant, doubling the number of companies working on *H. thermouteolus*. Colipi is focused on biomanufacturing of oils, and a patent they have filed describes a two-step process for HOB cultivation [208]. A query for patent records containing the phrase "hydrogen oxidizing bacteria" returned about 100 different patent applicants, 10 of which were credited with three or more patent records. *Cupriavidus necator* H16 continues to draw the most interest from the scientific community with its confirmed ability to produce several biochemicals. Already in use for the industrial production of PHAs, *C. necator* and powder derived from its biomass have undergone toxicological tests to clear regulatory hurdles before entering the market for microbial protein [209]. The strain has been successfully genetically engineered, and great progress has been made on synthetic biology toolkit and synthetic biology applications that can produce an efficient cell factory for converting CO_2 to alcohols, fatty acids, and terpenoids and other value-added products [22]. It has been utilized as model to study H_2 metabolism and the biochemical basis of [NiFe]-hydrogenase maturation, activation, and catalysis and CO_2-based biotechnology in general.

Few other HOB strains have received anywhere near that much attention for industrial-scale production of biochemicals. *Rhodococcus opacus* is a gram-positive bacterium with exciting potential for lignin and industrial waste valorization, lipid production for nutrition, biofuel, and commodity chemicals [210]. *Xanthobacter autotrophicus,* a gram-negative rod-shaped bacterium, which produces considerable amounts of the food dye and antioxidant zeaxanthin and has shown potential remediation of groundwater, is making waves in the microbial protein space [211-213]. *Paracoccus denitrificans,* like *H. thermoluteolus* and *Hydrogenobacter thermophilus,* can derive energy from thiosulfate oxidation and are capable biodenitrifying bacteria that can play a vital role in the remediation of nitrogen-contaminated water [120, 215]. *Azohydromonas* (*A. lata* and *A. australica*) are also renowned PHA-producing strains that can utilize wastewater from the agricultural industry as feedstock [216]. *Herbaspirillum, Sulfuricurvum,* and *Azonexus* have drawn interest for their bioremediation of contaminated soils and wastewater treatment [217].

In 2022, the Japanese government launched a ¥150-trillion 10-year Green Transformation (GX) initiative to transform her entire economic and social system from an economy, society, and industrial structure dependent on fossil fuels to "structures driven by clean energy." The overarching aim of the initiative is to drive economic growth and development through emissions mitigation. Included in the initiative is a commitment of ¥4 trillion to build a CCUS value chain and capture 120–240 MT CO_2 eq/year by 2050. Also included is a carbon levy targeting fossil fuel importers, and an emissions trading scheme (GX-ETS) [218]. Equivalent GX investment promotion by other governments include the EU's €1-trillion 10-year public private investment of 2020, the USA's US$369-billion 10-year Inflation Reduction Bill of 2022, Germany's €50-billion 2-year government support of 2020, France's €30-billion 2-year government support of 2020, and the UK's £26-billion 8-year government support of 2020. These commitments demonstrate a collective desire to pursue policies that will decarbonize leading economies. The question is how to transition away from a CO_2-emissions-producing economy without slowing down current social development. A rethink and redesign for energy systems to gradually increase the market share of renewables at the same time as larger amounts of CO_2 are recycled back into the carbon cycle is inescapable.

To reduce global reliance on fossil fuels, renewable energy must be brought online as a matter of urgency. Moving from fossil fuels to renewable energy resources is driven by urgency to alleviate the impact of energy shortages and environmental problems. As technologies for the utilization of renewable energy, such as photovoltaic and wind power, have advanced and the price of renewable electricity has continued to drop, the era of competitive renewable energy is approaching. But difficulties in integrating renewable energy into the power grid dictate that the energy be converted and stored into appropriate forms, including heat or chemicals such as hydrogen, syngas, and synthetic fuels. The most technologically advanced of these, power-to -hydrogen ($P2H_2$) aligns perfectly with HOB-based bioprocesses, supplying the hydrogen needed to drive the bioprocesses. Several laboratory-scale electromicrobial setups have demonstrated production of chemicals using HOBs. Significant challenges exist in fundamental research, and industrial deployment is not feasible at present.

High-value bulk chemicals (e.g., CO, HCOOH, C_2H_4, and CH_3CH_2OH) are now heavily dependent on fossil fuels and belong to energy-intensive industries because of the high-temperature and high-pressure processes involved. Sustainable bulk-chemical production is possible with the right biotechnology (feedstock, conversion, and downstream recovery), but industrial implementation requires a range of new technologies and further research. The cost, volume, and energy density of alternative fuels are concerns that can be addressed by better technologies to improve cost-efficiencies. Bulk chemicals, which can be produced using HOB-based bioprocesses, are prime targets as a market-entry strategy. The challenge lies in the slow growth, low productivity, and inferior energy conversion efficiency of current HOB strains. A few strains, notably *Hydrogenophilus thermoluteolus,* show atypically fast growth that, if elucidated, could yield a richer pool of fast-growing HOB strains. More advanced genetic engineering tools for HOB strains are in rapid development. Low productivity and energy conversion efficiencies inferior to heterotrophic strains are challenges that are likely to take longer to resolve.

Current biofuels are susceptible to fluctuations in oil prices, with the resulting extreme variability of biorefinery profit margins a major hurdle to large-scale deployment of biomass as an industrial feedstock. Biodiesel and bioethanol, the two primary biofuels, suffer from dependence on a narrow range of agricultural feedstocks as well as a limited variety of production processes. Carbon-based synthetic biofuels, which can be produced from any type of biomass could address these limitations if suitable technological progress could be realized. HOB-based bioprocesses can enter this domain through synthetic biofuels manufactured via biochemical conversion processes from “defossilized” CO_2 sources such as point source capture from industrial processes and direct capture from air. Biochemical and computational analyses have uncovered a large diversity of natural and synthetic CO_2 fixation pathways, revealing pathways that are more ATP-efficient and with faster kinetics than the ubiquitous CBB cycle. The challenge has been the paucity of successful implementations of the pathways in heterotrophic or autotrophic cell factories until the recent successes with artificial heterotrophy [188]. Future progress in new pathway implementations will benefit from developing synthetic biology approaches involving rational pathway design, enzyme engineering and modularization, in combination with indispensable laboratory evolution approaches.

Plastics are ubiquitous in contemporary society; about 99% are produced using fossil fuels as feedstock materials. These fossil-based plastics show the strongest production growth of all bulk materials over the last decade. The plastics industry's current growth trajectory is exponential, and plastic production is expected to double or triple by 2050. The rapidly increasing production of plastics and the continued reliance on fossil fuels for production have contributed to numerous environmental problems and health harms. The ever-increasing use of these fossil-based materials is responsible for emissions of 2.24 GT CO_2 eq/year, corresponding to roughly 5.3% of global GHG emissions. Their disposal exerts undue stress on the environment with up to 8 million tons of plastic waste collecting annually in the world's oceans with particularly detrimental effects of microplastics on aquatic life and, by extension, human health. A recently described chemical method to repurpose polyethylene terephthalate (PET) plastic waste to capture CO_2 holds the twin promise of tackling the plastic waste problem and reducing GHG emissions. It helps that the captured CO_2 is transforming into a resource for biomanufacturing. Bioplastics, manufactured from bio-based polymers, can have a lower carbon footprint and exhibit advantageous materials properties than fossil-based plastics; moreover, they can be compatible with existing recycling streams and some offer biodegradation as an end-of-life (EOL) scenario if performed in controlled or predictable environments. There is growing interest in using CO_2 as a feedstock for plastics production, but the environmental impact of CO_2-derived plastic polymers is not well understood. Only a few CO_2-derived plastic polymers are being produced and not on a commercial scale.

The use of enzymes to generate H_2, instead of using rare metal catalysts, is an exciting area of study in modern biochemistry and biotechnology, as well as biocatalysis driven by sustainable H_2. It is inspired by the ability of some bacteria to use hydrogenase enzymes to selectively activate H_2, thereby obtaining electrons or hydride ions, depending on the type of redox partner. The enzymes make use of abundant metals (iron and sometimes nickel), in contrast to the often-expensive synthetic catalysts that rely on rarer elements such as ruthenium or platinum. [NiFe]-hydrogenases catalyze both H_2 oxidation and H_2 evolution at ambient temperature, with no release of CO_2 and no requirement for precious metals. The enzymes can extract and store electrons or hydride ions from H_2 and use them when needed [143]. Typically, [NiFe]-hydrogenases are preferred for biotechnological applications due to the better O_2-tolerance and therefore simplified enzyme handling. A major weakness of the hydrogenases is their reliance on unstable FeS clusters for electrons. The oxygen tolerance of FeS clusters is crucial for their function in various biological processes. Fortification of FeS clusters has demonstrated a marked improvement in air stability and is emerging as an area of research interest.

Scholarly publications on HOBs in bioremediation are few and oscillate around the topics of orthophosphate removal from ground water, HOBs enrichment as biofertilizers, and treatment of nitrate-containing wastewater. The number of patent records is similarly small and is dominated by the Japanese company, Sekisui Chemical. HOBs reduce orthophosphate to ultralow concentrations and reduce perchlorate (ClO_4^-) using a variety of electron donors including organic carbon compounds, hydrogen, iron, and reduced sulfur compounds to the benign products chloride and oxygen. To the extent that HOBs allow for treatment technologies that can remove perchlorate from drinking water without introducing organic chemicals that stimulate bacterial growth in water distribution systems, they warrant greater research attention. HOBs that accept several electron donors in addition to H_2 are particularly interesting in co-cultures as a means to extend the potential range of useful biotechnological processes. Anaerobic HOBs are particularly interesting because anaerobiosis is typically less challenging to achieve at large industrial scale.

References

[1] Dabholkar U. Environmental perspective to the year 2000 and beyond: A framework for world development. *Environ Conserv* 1989;16:49–64.

[2] CO_2 emissions since Industrial Revolution. *Statista*, https://www.statista.com/statistics/264699/worldwide-co2-emissions/ (accessed 8 September 2025).

[3] Ritchie H, Rosado P, Roser M. CO_2 and Greenhouse Gas Emissions. *Our World in Data*, https://ourworldindata.org/co2-and-greenhouse-gas-emissions (2023, accessed 11 September 2025).

[4] Lynas M, Houlton BZ, Perry S. Greater than 99% consensus on human caused climate change in the peer-reviewed scientific literature. *Environ Res Lett* 2021;16:114005.

[5] IEA. Global Energy Review 2025, 2025, https://www.iea.org/reports/global-energy-review-2025.

[6] United Nations Framework Convention on Climate Change, https://unfccc.int/files/essential_background/background_publications_htmlpdf/application/pdf/conveng.pdf.

[7] Kyoto Protocol to the United Nations Framework Convention on Climate Change, https://unfccc.int/resource/docs/convkp/kpeng.pdf (accessed 9 September 2025).

[8] Paris Agreement: Sustainable Development Knowledge Platform, https://sustainabledevelopment.un.org/frameworks/parisagreement (accessed 9 September 2025).

[9] Wahyudi H, Leny SM.. Higher carbon tax rates more effective in reducing emissions in G20 countries?. *J Environ Earth Sci* 24 December 2024;7:Epub ahead of print. doi: 10.30564/jees.v7i1.7256.

[10] Stoddard I, Anderson K, Capstick S, et al. Three decades of climate mitigation: Why haven't we bent the global emissions curve?. *Annu Rev Environ Resour* 2021;46:653–689.

[11] Lück S. Scientific literature on carbon dioxide removal revealed as much larger through AI-enhanced systematic mapping. *Nat Commun*.

[12] Johnstone I, Fuss S, Walsh N, et al. Carbon markets for carbon dioxide removal. *Clim Policy* 2025;1–8.

[13] Chauvy R, De Weireld G. CO_2 Utilization Technologies in Europe: A Short Review, December 2020, Epub ahead of print, doi: 10.1002/ente.202000627.

[14] IEA. *Putting CO_2 to Use: Creating value from emissions*, International Energy Agency.

[15] Ray S, Jin J-O, Choi I., et al. Recent trends of biotechnological production of polyhydroxyalkanoates from C1 carbon sources. *Front Bioeng Biotechnol* 2023;10:907500.

[16] Baril X, Constant P. Life history strategies to study the succession of H2-oxidizing bacteria and their ecological role along H_2 concentration gradients in upland soils. *Lett Appl Microbiol* 2025;78:ovaf061.

[17] Piché-Choquette S, Constant P. Molecular hydrogen, a neglected key driver of soil biogeochemical processes. *Appl Environ Microbiol* 2019;85:e02418-18.

[18] Maimaiti J, Zhang Y, Yang J, et al. Isolation and characterization of hydrogen-oxidizing bacteria induced following exposure of soil to hydrogen gas and their impact on plant growth. *Environ Microbiol* 2007;9:435–444.

[19] Pander B, Mortimer Z, Woods C, et al. Hydrogen oxidising bacteria for production of single-cell protein and other food and feed ingredients. *Eng Biol* 2020;4:21–24.
[20] Goto E, Kodama T, Minoda Y. Isolation and culture conditions of thermophilic hydrogen bacteria. *Agric Biol Chem* 1977;41:685–690.
[21] Li R, Jiang Y, Luo K, et al. Hydrogen-oxidizing bacteria: A promising contributor to environmental sustainability and resource recycling, March 2025, Epub ahead of print. doi: 10.1016/j.greenca.2025.02.004.
[22] Pan H, Wang J, Wu H, et al. Synthetic biology toolkit for engineering Cupriviadus necator H16 as a platform for CO2 valorization. *Biotechnol Biofuels* 2021;14:212.
[23] Soltysiak MPM, Ory ALH, Lee AD, et al. XanthoMoClo—A robust modular cloning genetic Toolkit for the Genera *Xanthobacter* and *Roseixanthobacter*. *ACS Synth Biol* 2025;14:1173–1190.
[24] Yukawa H, Metting FB, Omumasaba C, et al. Chapter 13 Hydrogen bacteria: A developing platform for the sustainable carbon dioxide-based bioindustry. In: Vijayendran BR, edtors.. *BioProducts*, De Gruyter; pp. 243–256.
[25] *Sackler Forum: Dealing with carbon dioxide at scale*, May 2018.
[26] Hirata A, Ohashi H, Hasegawa T, et al. The choice of land-based climate change mitigation measures influences future global biodiversity loss. *Commun Earth Environ* 2024;5:259.
[27] Li G, Yao J. Direct Air Capture (DAC) for achieving net-zero CO2 emissions: Advances, applications, and challenges. *Eng* 2024;5:1298–1336.
[28] Dooley K, Pelz S, Norton A. Understanding land-based carbon dioxide removal in the context of the Rio Conventions. *One Earth* 2024;7:1501–1514.
[29] Beller M, Steinberg M. *Liquid fuel synthesis using nuclear power in a mobile energy depot system*, BNL-955; pp. 4580336.
[30] Grimes P, Lackner K, Ziock H-J, et al. Carbon Dioxide Extraction From Air: Is It An Option?.
[31] Keith DW, Holmes G, St. Angelo D, et al. A process for capturing CO2 from the atmosphere. *Joule* 2018;2:1573–1594.
[32] Wang Y, Qu L, Ding H, et al. Distributed direct air capture of carbon dioxide by synergistic water harvesting. *Nat Commun* 2024;15:9745.
[33] Schäppi R, Rutz D, Dähler F, et al. Drop-in fuels from sunlight and air. *Nature* 2022;601:63–68.
[34] Zhang R, Xie Z, Ge Q, et al. Recent advancements in integrating CO2 capture from flue gas and ambient air with thermal catalytic conversion for efficient CO2 utilization. *J CO2 Util* 2024;89:102973.
[35] Smith SM, Geden O, Gidden MJ, et al. *The State of Carbon Dioxide Removal – 2nd Edition*, The State of Carbon Dioxide Removal; 2024,https://osf.io/f85qj/. accessed 23 May 2024.
[36] Zeitler E. Carbon Dioxide conversion: Pathway to Defossilize carbon-based materials and durably store Carbon. *Clim Energy* 2025;41:1–8.
[37] Ehsani E, Dumolin C, Arends JBA, et al. Enriched hydrogen-oxidizing microbiomes show a high diversity of co-existing hydrogen-oxidizing bacteria. *Appl Microbiol Biotechnol* 2019;103:8241–8253.
[38] Nakagawa S, Takai K. Deep-sea vent chemoautotrophs: Diversity, biochemistry and ecological significance: Chemoautotrophy in deep-sea vents. *FEMS Microbiol Ecol* 2008;65:1–14.
[39] Sass K, Güllert S, Streit WR, et al. A hydrogen-oxidizing bacterium enriched from the open ocean resembling a symbiont. *Environ Microbiol Rep* 2020;12:396–405.
[40] Vésteinsdóttir H, Reynisdóttir DB, Örlygsson J. Hydrogenophilus islandicus sp. nov., a thermophilic hydrogen-oxidizing bacterium isolated from an Icelandic hot spring. *Int J Syst Evol Microbiol* 2011;61:290–294.
[41] Gadkari D, Schricker K, Acker G, et al. *Streptomyces thermoautotrophicus* sp. nov., a Thermophilic CO- and H_2 -Oxidizing Obligate Chemolithoautotroph. *Appl Environ Microbiol* 1990;56:3727–3734.
[42] Kawasumi T, Igarashi Y, Kodama T, et al. Hydrogenobacter thermophilus gen. nov., sp. nov., an extremely Thermophilic, Aerobic, Hydrogen-Oxidizing Bacterium. *Int J Syst Bacteriol* 1984;34:5–10.

[43] Tan X-Y, Liu X-J, Li Z, et al. Thiomicrorhabdus marina sp.nov., an obligate chemolithoautotroph isolated from tidal zone sediment, and genome insight into the genus Thiomicrorhabdus. *Front Mar Sci* 12 April 2023;10:Epub ahead of print. doi: 10.3389/fmars.2023.1144912.

[44] Takors R, Kopf M, Mampel J, et al. Using gas mixtures of CO, CO_2 and H_2 as microbial substrates: The do's and don'ts of successful technology transfer from laboratory to production scale. *Microb Biotechnol* 2018;11:606–625.

[45] Carbon capture and utilization as a decarbonization lever. *OGCI*, https://www.ogci.com/wp-content/uploads/2024/05/CCU-Report-vf.pdf (accessed 19 September 2025).

[46] Liguori R, Faraco V. Biological processes for advancing lignocellulosic waste biorefinery by advocating circular economy. *Bioresour Technol* 2016;215:13–20.

[47] Greening C, Grinter R. Microbial oxidation of atmospheric trace gases. *Nat Rev Microbiol* 2022;20:513–528.

[48] King GM, Weber CF. Distribution, diversity and ecology of aerobic CO-oxidizing bacteria. *Nat Rev Microbiol* 2007;5:107–118.

[49] Aguilera-Campos KI, Stairs CW. Hydrogen metabolism: A eukaryote taps into the electron sink. *Curr Biol* 2022;32:R49–R51.

[50] Kawai S, Ishikawa M, Hanada S, et al. Hydrogenophilus thiooxidans sp. nov., a moderately thermophilic chemotrophic bacterium unable to grow on hydrogen gas, isolated from hot spring microbial mats. *Int J Syst Evol Microbiol* 2022;72:005355.

[51] Qin N, Li L, Wan X, et al. Increased CO_2 fixation enables high carbon-yield production of 3-hydroxypropionic acid in yeast. *Nat Commun* 2024;15:1591.

[52] Emerson D, Fleming EJ, McBeth JM. Iron-Oxidizing Bacteria: An environmental and genomic perspective. *Annu Rev Microbiol* 2010;64:561–583.

[53] Cordero PRF, Grinter R, Hards K, et al. Two uptake hydrogenases differentially interact with the aerobic respiratory chain during mycobacterial growth and persistence. *J Biol Chem* 2019;294:18980–18991.

[54] Angenent SC, Schuttinga JH, Van Efferen MFH, et al. Hydrogen Oxidizing Bacteria as novel protein source for human consumption: An overview. *Open Microbiol J* 2022;16:e187428582207270.

[55] Wang H. Enhancement strategies for the microbial protein production of nitrogen-fixing hydrogen-oxidizing bacterial community, 2024. 2024.08.07.607053.

[56] Chen Y-Z, Zhang L-J, Ding L-Y, et al. Sustainable treatment of nitrate-containing wastewater by an autotrophic hydrogen-oxidizing bacterium. *Environ Sci Ecotechnology* 2022;9:100146.

[57] Repaske R. Characteristics of hydrogen bacteria. *Biotechnol Bioeng* 1966;8:217–235.

[58] Schlegel HG, Lafferty R. Growth of 'Knallgas' Bacteria (Hydrogenomonas) using direct Electrolysis of the culture medium. *Nature* 1965;205:308–309.

[59] Mateles RI, Baruah JN, Tannenbaum SR. Growth of a Thermophilic Bacterium on hydrocarbons: A new source of single-cell protein. *Science* 1967;157:1322–1323.

[60] Miltner A, Kopinke F-D, Kindler R, et al. Non-phototrophic CO_2 fixation by soil microorganisms. *Plant Soil* 2005;269:193–203.

[61] Ray AE, Zhang E, Terauds A, et al. Soil microbiomes with the genetic capacity for atmospheric Chemosynthesis are widespread across the poles and are associated with moisture, Carbon, and Nitrogen limitation. *Front Microbiol* 2020;11:1936.

[62] Dubini A. How bacteria get energy from hydrogen: A genetic analysis of periplasmic hydrogen oxidation in Escherichia coli. *Int J Hydrog Energy* 2002;27:1413–1420.

[63] Frielingsdorf S, Pinske C, Valetti F, et al. Editorial: Hydrogenase: Structure, function, maturation, and application. *Front Microbiol* 2023;14: https://www.frontiersin.org/articles/10.3389/fmicb.2023.1284540, accessed 22 November 2023.

[64] Vignais PM, Occurrence BB.. Classification, and biological function of Hydrogenases: An overview. *Chem Rev* 2007;107:4206–4272.

[65] Adam N, Perner M. Novel hydrogenases from deep-sea hydrothermal vent metagenomes identified by a recently developed activity-based screen. *ISME J* 2018;12:1225–1236.

[66] Shomura Y, Taketa M, Nakashima H, et al. Structural basis of the redox switches in the NAD+-reducing soluble [NiFe]-hydrogenase. *Science* 2017;357:928–932.

[67] Kulka-Peschke CJ, Schulz A-C, Lorent C, et al. Reversible Glutamate Coordination to High-Valent Nickel Protects the Active Site of a [NiFe] Hydrogenase from Oxygen. *J Am Chem Soc* 2022;144:17022–17032.

[68] Kumar R, Escorcia AM, Stein M. Fully Oxidized state of the oxygen-tolerant [NiFe] Hydrogenase from *Hydrogenophilus thermoluteolus* SH: A quantum mechanics cluster and quantum mechanics/ molecular mechanics study. *Inorg Chem* 2025;acs.inorgchem.5c00503.

[69] Schäfer C, Friedrich B, Novel LO.. Oxygen-Insensitive Group 5 [NiFe]-Hydrogenase in Ralstonia eutropha. *Appl Environ Microbiol* 2013;79:5137–5145.

[70] Kanehisa M, Goto S. KEGG: Kyoto encyclopedia of genes and genomes. *Nucleic Acids Res* 2000;28:27–30.

[71] Greening C, Biswas A, Carere CR, et al. Genomic and metagenomic surveys of hydrogenase distribution indicate H_2 is a widely utilised energy source for microbial growth and survival. *ISME J* 2016;10:761–777.

[72] Radan A, Milčić N, Sudar M, et al. Hydrogenases – Types, sources, properties, and the potential for their application. *Kem U Ind* 2025;74:33–42.

[73] Cha M, Park M, Kim S. Biological routes for biohydrogen production: A clean and carbon-free fuel. *Biotechnol J* 2025;20:e70074.

[74] Xuan J, He L, Wen W, et al. Hydrogenase and Nitrogenase: Key catalysts in Biohydrogen production. *Molecules* 2023;28:1392.

[75] Volbeda A, Amara P, Darnault C, et al. X-ray crystallographic and computational studies of the O2-tolerant [NiFe]-hydrogenase 1 from Escherichia coli. *Proc Natl Acad Sci U S A* 2012;109:5305–5310.

[76] Jugder B-E, Welch J, Aguey-Zinsou K-F, et al. Fundamentals and electrochemical applications of [Ni–Fe]-uptake hydrogenases. *RSC Adv* 2013;3:8142.

[77] Häring V, Conrad R. Demonstration of two different H2-oxidizing activities in soil using an H_2 consumption and a tritium exchange assay. *Biol Fertil Soils* 1994;17:125–128.

[78] Constant P, Chowdhury SP, Pratscher J, et al. Streptomycetes contributing to atmospheric molecular hydrogen soil uptake are widespread and encode a putative high-affinity [NiFe]-hydrogenase. *Environ Microbiol* 2010;12:821–829.

[79] Greening C, Kropp A, Vincent KA, et al. Developing high-affinity, oxygen-insensitive [NiFe]-hydrogenases as biocatalysts for energy conversion. *Biochem Soc Trans* 2023;51:1921–1933.

[80] Sokolova D, Vincent KA. Exploiting hydrogenases for biocatalytic hydrogenations. *Chem Commun* 2024;60:13667–13677.

[81] Pokorna D, Zabranska J Sulfur-oxidizing bacteria in environmental technology. *Biotechnol Adv* 2015;33:1246–1259.

[82] Pirieh P, Naeimpoor F. Oxic/micro-oxic/anoxic growth of chemolithoautotrophic sulfur oxidizing bacteria: Model reconstruction and probing shifts in energy metabolism by flux balance analysis. *Environ Res* 2025;283:122126.

[83] Nguyen PM, Do PT, Pham YB, et al. Roles, mechanism of action, and potential applications of sulfur-oxidizing bacteria for environmental bioremediation. *Sci Total Environ* 2022;852:158203.

[84] Rameez MJ, Pyne P, Mandal S, et al. Two pathways for thiosulfate oxidation in the alphaproteobacterial chemolithotroph Paracoccus thiocyanatus SST. *Microbiol Res* 2020;230:126345.

[85] Moriya Y, Itoh M, Okuda S, et al. KAAS: An automatic genome annotation and pathway reconstruction server. *Nucleic Acids Res* 2007;35:W182–W185.

[86] Ratzka J, Lauterbach L, Lenz O, et al. Systematic evaluation of the dihydrogen-oxidising and NAD+-reducing soluble [NiFe]-hydrogenase from Ralstonia eutropha H16 as a cofactor regeneration catalyst. *Biocatal Biotransformation* 2011;29:246–252.

[87] Yu J. Fixation of carbon dioxide by a hydrogen-oxidizing bacterium for value-added products. *World J Microbiol Biotechnol* 2018;34:89.

[88] Wang Y, Yang H, Geerts C, et al. The multiple facets of acetyl-CoA metabolism: Energetics, biosynthesis, regulation, acylation and inborn errors. *Mol Genet Metab* 2023;138:106966.

[89] Li Z, Xin X, Xiong B, et al. Engineering the Calvin–Benson–Bassham cycle and hydrogen utilization pathway of Ralstonia eutropha for improved autotrophic growth and polyhydroxybutyrate production. *Microb Cell Factories* 2020;19:228.

[90] Bar-Even A, Noor E, Milo R. A survey of carbon fixation pathways through a quantitative lens. *J Exp Bot* 2012;63:2325–2342.

[91] Cotton CA, Edlich-Muth C, Bar-Even A. Reinforcing carbon fixation: CO2 reduction replacing and supporting carboxylation. *Curr Opin Biotechnol* 2018;49:49–56.

[92] Berg IA. Ecological aspects of the distribution of different Autotrophic CO $_2$ fixation pathways. *Appl Environ Microbiol* 2011;77:1925–1936.

[93] Bar-On YM, Milo R. The global mass and average rate of rubisco. *Proc Natl Acad Sci U S A* 2019;116:4738–4743.

[94] Drake BG, Gonzalez-Meler MA, Long SP. More efficient plants: A consequence of rising atmospheric CO2?. *Annu Rev Plant Physiol Plant Mol Biol* 1997;48:609–639.

[95] Orr DJ, Robijns AKJ, Baker CR, et al. Dynamics of Rubisco regulation by sugar phosphate derivatives and their phosphatases. *J Exp Bot* 2023;74:581–590.

[96] Dangel AW, Tabita FR. CbbR, The master regulator for Microbial Carbon Dioxide fixation. *J Bacteriol* 2015;197:3488–3498.

[97] Terazono K, Hayashi NR, Igarashi Y. CbbR, a LysR-type transcriptional regulator from Hydrogenophilus thermoluteolus, binds two cbb promoter regions, *FEMS Microbiol Lett.*

[98] Esparza M, Cárdenas JP, Bowien B, et al. Genes and pathways for CO2 fixation in the obligate, chemolithoautotrophic acidophile, Acidithiobacillus ferrooxidans, Carbon fixation in A. ferrooxidans. *BMC Microbiol* 2010;10:229.

[99] Prywes N, Phillips NR, Tuck OT, et al. Rubisco function, evolution, and engineering. *Annu Rev Biochem* 2023;92:385–410.

[100] Zhao L, Cai Z, Li Y, et al. Engineering Rubisco to enhance CO2 utilization. *Synth Syst Biotechnol* 2024;9:55–68.

[101] Carmo-Silva E, Sharwood RE. Rubisco and its regulation – Major advances to improve carbon assimilation and productivity. *J Exp Bot* 2023;74:507–509.

[102] Cocon KD, Luis P. The potential of RuBisCO in CO_2 capture and utilization. *Prog Energy Combust Sci* 2024;105:101184.

[103] Alonso-Sáez L, Galand PE, Casamayor EO, et al. High bicarbonate assimilation in the dark by Arctic bacteria. *ISME J* 2010;4:1581–1590.

[104] Erb TJ. Carboxylases in natural and synthetic microbial pathways. *Appl Environ Microbiol* 2011;77:8466–8477.

[105] Hasegawa S, Suda M, Uematsu K, et al. Engineering of Corynebacterium glutamicum for High-Yield l -Valine Production under Oxygen Deprivation Conditions. *Appl Environ Microbiol* 2013;79:1250–1257.

[106] Matassa S, Boon N, Pikaar I, et al. Microbial protein: Future sustainable food supply route with low environmental footprint. *Microb Biotechnol* 2016;9:568–575.

[107] K-h T, Yj T, Blankenship RE. Carbon metabolic pathways in phototrophic bacteria and their broader evolutionary implications. *Front Microbiol* 1 August 2011;2:Epub ahead of print. doi: 10.3389/fmicb.2011.00165.

[108] Zhang A, Carroll AL, Atsumi S. Carbon recycling by cyanobacteria: Improving CO2 fixation through chemical production. *FEMS Microbiol Lett* 2017;364:fnx165.

[109] Zhang J, Liu G, Carvajal AI, et al. Discovery of a readily heterologously expressed Rubisco from the deep sea with potential for CO_2 capture. *Bioresour Bioprocess* 2021;8:86.

[110] Zhou Y, Whitney S. Directed evolution of an improved Rubisco; In Vitro Analyses to Decipher fact from fiction. *Int J Mol Sci* 2019;20:5019.
[111] Yu H, Li X, Duchoud F, et al. Augmenting the Calvin–Benson–Bassham cycle by a synthetic malyl-CoA-glycerate carbon fixation pathway. *Nat Commun* 2018;9:2008.
[112] Lu K-J, Hsu C-W, Jane W-N, et al. Dual-cycle CO_2 fixation enhances growth and lipid synthesis in *Arabidopsis thaliana*. *Science* 2025;389:eadp3528.
[113] Islam ZF, Welsh C, Bayly K, et al. A widely distributed hydrogenase oxidises atmospheric H2 during bacterial growth. *ISME J* 2020;14:2649–2658.
[114] Wilde E, Schlegel HG. Oxygen tolerance of strictly aerobic hydrogen-oxidizing bacteria. *Antonie Van Leeuwenhoek* 1982;48:131–143.
[115] Doucette CD, Schwab DJ, Wingreen NS, et al. α-Ketoglutarate coordinates carbon and nitrogen utilization via enzyme I inhibition. *Nat Chem Biol* 2011;7:894–901.
[116] Hong C-B, Hua W, Liu L, et al. Sustainable synthesis of α-ketoglutaric and methanetriacetic acids from biomass feedstocks. *Nat Commun* 2025;16:1245.
[117] Chiang C-J, Lin J-H, Ta D-T, et al. Metabolic Engineering of Escherichia coli for Overproduction of Alpha-Ketoglutarate Using Crude Glycerol. *J Agric Food Chem* 2025;73:18346–18352.
[118] Hertz L, Chen Y. Integration between Glycolysis and Glutamate-Glutamine Cycle Flux May Explain Preferential Glycolytic Increase during Brain Activation, Requiring Glutamate. *Front Integr Neurosci* 2017;11:18.
[119] Sun S, Ding Y, Liu M, et al. Comparison of Glucose, Acetate and Ethanol as Carbon Resource for Production of Poly(3-Hydroxybutyrate) and Other Acetyl-CoA Derivatives. *Front Bioeng Biotechnol* 2020;8:833.
[120] Miyake D, Ichiki S, Tanabe M, et al. Thiosulfate oxidation by a moderately thermophilic hydrogen-oxidizing bacterium, Hydrogenophilus thermoluteolus. *Arch Microbiol* 2007;188:199–204.
[121] Bowien B, Schlegel HG. Physiology and biochemistry of Aerobic Hydrogen-Oxidizing Bacteria. *Annu Rev Microbiol* 1981;35:405–452.
[122] Witte B, Obloj P, Koktenturk S, et al. Food for thought: The protein transformation. *Ind Biotechnol* 2021;17:125–133.
[123] Pearce FG, Brunke JE. Is now the time for a Rubiscuit or Ruburger? Increased interest in Rubisco as a food protein. *J Exp Bot* 2023;74:627–637.
[124] Nawaz MA, Kasote DM, Ullah N, et al. RuBisCO: A sustainable protein ingredient for plant-based foods. *Front Sustain Food Syst* 13 May 2024;8:Epub ahead of print. doi: 10.3389/fsufs.2024.1389309.
[125] Dixon R, Kahn D. Genetic regulation of biological nitrogen fixation. *Nat Rev Microbiol* 2004;2:621–631.
[126] Zhang Y-X, Gao J-Y, Wang -M-M, et al. Enrichment of nitrogen-fixing hydrogen-oxidizing bacteria community for efficient microbial protein production in airlift reactor. *Bioresour Technol* 2025;428:132443.
[127] Hu X, Kerckhof F-M, Ghesquière J, et al. Microbial protein out of thin air: Fixation of nitrogen gas by an autotrophic Hydrogen-Oxidizing Bacterial enrichment. *Environ Sci Technol* 2020;54:3609–3617.
[128] Lu S, Guan X, Liu C. Electricity-powered artificial root nodule. *Nat Commun* 2020;11:1505.
[129] Malik KA, Schlegel HG. Chemolithoautotrophic growth of bacteria able to grow under N2-fixing conditions. *FEMS Microbiol Lett* 1981;11:63–67.
[130] Scott WT, Van Mastrigt O, Block DE, et al. Nitrogenous compound utilization and production of volatile organic compounds among commercial Wine Yeasts highlight strain-specific Metabolic diversity. *Microbiol Spectr* 2021;9:e0048521.
[131] Liu C, Sakimoto KK, Colón BC, et al. Ambient nitrogen reduction cycle using a hybrid inorganic-biological system. *Proc Natl Acad Sci U S A* 2017;114:6450–6455.
[132] Ortiz-Marquez JCF, Do Nascimento M, Curatti L. Metabolic engineering of ammonium release for nitrogen-fixing multispecies microbial cell-factories. *Metab Eng* 2014;23:154–164.

[133] Ling N, Wang T, Kuzyakov Y. Rhizosphere bacteriome structure and functions. *Nat Commun* 2022;13:836.

[134] Bertasini D, Binati RL, Bolzonella D, et al. Single Cell Proteins production from food processing effluents and digestate. *Chemosphere* 2022;296:134076.

[135] Leigh JA, Dodsworth JA. Nitrogen regulation in bacteria and archaea. *Annu Rev Microbiol* 2007;61:349–377.

[136] Johnson N, Liebreich M, Kammen DM, et al. Realistic roles for hydrogen in the future energy transition. *Nat Rev Clean Technol* 2025;1:351–371.

[137] Iida S, Sakata K. Hydrogen technologies and developments in Japan. *Clean Energy* 2019;3:105–113.

[138] Usman MR. Hydrogen storage methods: Review and current status. *Renew Sustain Energy Rev* 2022;167:112743.

[139] Madheswaran DK, Vengatesan S, Jegadheeswaran S, et al. Ammonia as a hydrogen carrier: A comprehensive analysis of electrolysis efficiency and its potential in sustainable energy systems. *Renew Sustain Energy Rev* 2025;221:115915.

[140] Schiaroli A, Claussner L, Campari A, et al. A comprehensive review on liquid hydrogen transfer operations and safety considerations for mobile applications. *Int J Hydrog Energy* 2025;107:164–182.

[141] Chilunda MD, Talipov SA, Farooq HMU, et al. Electrochemical cycling of liquid organic hydrogen carriers as a sustainable approach for hydrogen storage and transportation. *ACS Sustain Chem Eng* 2025;13:1174–1195.

[142] Spatolisano E, Restelli F, Matichecchia A, et al. Assessing opportunities and weaknesses of green hydrogen transport via LOHC through a detailed techno-economic analysis. *Int J Hydrog Energy* 2024;52:703–717.

[143] Ogo S, Takahashi C, Yatabe T, et al. Hydrogen energy system that uses [NiFe] Hydrogenase functional models and works at 20–40 °C. *ACS Sustain Chem Eng* 2025;13:1051–1055.

[144] Kaiwa Y, Kobayashi K, Kataoka M, et al. Polymers for reversible hydrogen storage inspired by Electrode-active materials in organic batteries. *Int J Soc Mater Eng Resour* 2022;25:1–9.

[145] Tang J, Xie R, Pishva P, et al. Recent progress and perspectives of liquid organic hydrogen carrier electrochemistry for energy applications. *J Mater Chem A* 2024;12:15580–15591.

[146] Ngasse Moudio ND, Bian X-Q, Chinamo DS. Liquid hydrogen carriers for clean energy systems: A critical review of chemical hydrogen storage strategies. *Fuel* 2026;404:136329.

[147] Altaf M, Demirci UB, Haldar AK. Review of solid-state hydrogen storage: Materials categorisation, recent developments, challenges and industrial perspectives. *Energy Rep* 2025;13:5746–5772.

[148] Fan Q, Neubauer P, Lenz O, et al. Heterologous hydrogenase overproduction systems for biotechnology – an overview. *Int J Mol Sci* 2020;21:5890.

[149] Lojou E. Hydrogenases as catalysts for fuel cells: Strategies for efficient immobilization at electrode interfaces. *Electrochimica Acta* 2011;56:10385–10397.

[150] Chenevier P, Mugherli L, Darbe S, et al. Hydrogenase enzymes: Application in biofuel cells and inspiration for the design of noble-metal free catalysts for H2 oxidation. *Comptes Rendus Chim* 2012;16:491–505.

[151] Mertens R, Greiner L, Van den Ban ECD, et al. Practical applications of hydrogenase I from *Pyrococcus furiosus* for NADPH generation and regeneration. *J Mol Catal B Enzym* 2003;24–25:39–52.

[152] Lauterbach L, Lenz O, Vincent KA. H_2 -driven cofactor regeneration with NAD(P) + -reducing hydrogenases. *FEBS J* 2013;280:3058–3068.

[153] Radu V, Frielingsdorf S, Evans SD, et al. Enhanced Oxygen-Tolerance of the full Heterotrimeric Membrane-Bound [NiFe]-Hydrogenase of *Ralstonia eutropha*. *J Am Chem Soc* 2014;136:8512–8515.

[154] Stults LW, Moshiri F, Maier RJ. Aerobic purification of hydrogenase from Rhizobium japonicum by affinity chromatography. *J Bacteriol* 1986;166:795–800.

[155] Giudici-orticoni M, Leger C. Hyperthermostable and oxygen resistant hydrogenases from a hyperthermophilic bacterium Aquifex aeolicus: Physicochemical properties. *Int J Hydrog Energy* 1 January 2006;Epub ahead of print. doi: 10.1016/J.IJHYDENE.2006.06.007.

[156] Parkin A, Bowman L, Roessler MM, et al. How *Salmonella* oxidises H_2 under aerobic conditions. *FEBS Lett* 2012;586:536–544.

[157] Grinter R, Leung PM, Wijeyewickrema LC, et al. Protease-associated import systems are widespread in Gram-negative bacteria. *PLoS Genet* 2019;15:e1008435.

[158] Webb S, Milton RD. Towards scalable Electroenzymatic Hydrogen production with [FeFe]-Hydrogenase. *ChemElectroChem* 2025;12:e202400700.

[159] Kleinhaus J T, Wittkamp F, Yadav S, et al. [FeFe]-Hydrogenases: Maturation and reactivity of enzymatic systems and overview of biomimetic models. *Chem Soc Rev* 2021;50:1668–1784.

[160] Cleary SE, Hall SJ, Galan-Bataller R, et al. Scalable Bioreactor Production of an O2-Protected [FeFe]-Hydrogenase Enables Simple Aerobic Handling for Clean Chemical Synthesis. *ChemCatChem* 2024;16: e202400193.

[161] Rutz A, Das CK, Fasano A, et al. Increasing the O_2 Resistance of the [FeFe]-Hydrogenase CbA5H through Enhanced Protein Flexibility. *ACS Catal* 2023;13:856–865.

[162] Gerbaulet M, Hemschemeier A, Happe T. The O2-stable [FeFe]-hydrogenase CbA5H reveals high resilience against organic solvents. *Catal Sci Technol* 2024;14:7093–7106.

[163] *Statistical Review of World Energy*, 74th edn., London: Energy Institute; 2025, https://www.energyinst.org/statistical-review.

[164] Sorimachi K Innovative method for CO_2 fixation and storage. *Sci Rep* 2022;12:1694.

[165] Liu X, Li L, Zhao G, et al. Optimization strategies for CO_2 biological fixation. *Biotechnol Adv* 2024;73:108364.

[166] Gong F, Zhu H, Zhang Y, et al. Biological carbon fixation: From natural to synthetic. *J CO2 Util* 2018;28:221–227.

[167] Zhao T, Li Y, Zhang Y. Biological carbon fixation: A thermodynamic perspective. *Green Chem* 2021;23:7852–7864.

[168] Liang X, Duan Y, Su Y, et al. Carbon capture by biological methods. *Camb Prisms Carbon Technol* 2025;1:e4.

[169] Hudson EP The Calvin Benson cycle in bacteria: New insights from systems biology. *Semin Cell Dev Biol* 29 March 2023;Epub ahead of print. doi: 10.1016/j.semcdb.2023.03.007.

[170] Flamholz AI, Prywes N, Moran U, et al. Revisiting trade-offs between Rubisco kinetic parameters. *Biochemistry* 2019;58:3365–3376.

[171] Galmés J, Conesa MÀ, Díaz-Espejo A, et al. Rubisco catalytic properties optimized for present and future climatic conditions. *Plant Sci* 2014;226:61–70.

[172] Smith EN, Van Aalst M, Weber APM, et al. Alternatives to photorespiration: A system-level analysis reveals mechanisms of enhanced plant productivity. *Sci Adv* 2025;11:eadt9287.

[173] Hennacy JH, Jonikas MC Prospects for engineering biophysical CO2 concentrating mechanisms into land plants to enhance yields. *Annu Rev Plant Biol* 2020;71:461–485.

[174] Claassens NJ A warm welcome for alternative CO_2 fixation pathways in microbial biotechnology. *Microb Biotechnol* 2017;10:31–34.

[175] Matsumura H, Shiomi K, Yamamoto A, et al. Hybrid Rubisco with complete replacement of Rice Rubisco small subunits by Sorghum Counterparts Confers C4 plant-like high catalytic activity. *Mol Plant* 2020;13:1570–1581.

[176] Liu X, Luo H, Yu D, et al. Synthetic biology promotes the capture of CO2 to produce fatty acid derivatives in microbial cell factories. *Bioresour Bioprocess* 2022;9:124.

[177] Yan X, He Q, Geng B, et al. Microbial Cell Factories in the Bioeconomy Era: From Discovery to Creation. *BioDesign Res* 2024;6:0052.

[178] Claassens NJ, Bordanaba-Florit G, Cotton CAR, et al. Replacing the Calvin cycle with the reductive glycine pathway in Cupriavidus necator. *Metab Eng* 2020;62:30–41.

[179] Dronsella B, Orsi E, Schulz-Mirbach H, et al. One-carbon fixation via the synthetic reductive glycine pathway exceeds yield of the Calvin cycle. *Nat Microbiol* 2025;10:646–653.

[180] Wurtzel ET, Vickers CE, Hanson AD, et al. Revolutionizing agriculture with synthetic biology. *Nat Plants* 2019;5:1207–1210.

[181] Taha A, Patón M, Rodríguez J. Bioenergetic trade-offs can reveal the path to superior microbial CO_2 fixation pathways. *mSystems* 2025;10:e01274.

[182] Liang B, Zhao Y, Yang J. Recent advances in developing artificial autotrophic microorganism for reinforcing CO_2 fixation. *Front Microbiol* 2020;11:592631.

[183] Luo S, Diehl C, He H, et al. Construction and modular implementation of the THETA cycle for synthetic CO2 fixation. *Nat Catal* 2023;6:1228–1240.

[184] Angermayr SA, Gorchs Rovira A, Hellingwerf KJ. Metabolic engineering of cyanobacteria for the synthesis of commodity products. *Trends Biotechnol* 2015;33:352–361.

[185] Feng J, Ma D, Gao S, et al. Recent advances in engineering heterotrophic microorganisms for reinforcing CO_2 fixation based on Calvin–Benson–Bassham cycle. *ACS Sustain Chem Eng* 2023;11:9509–9522.

[186] Antonovsky N, Gleizer S, Noor E, et al. Sugar Synthesis from CO2 in Escherichia coli. *Cell* 2016;166:115–125.

[187] Puiggené Ò, Favoino G, Federici F, et al. Seven critical challenges in synthetic one-carbon assimilation and their potential solutions. *FEMS Microbiol Rev* 2025;49:fuaf011.

[188] Gleizer S, Ben-Nissan R, Bar-On YM, et al. Conversion of Escherichia coli to Generate All Biomass Carbon from CO_2. *Cell* 2019;179:1255-1263.e12.

[189] Ben Nissan R, Milshtein E, Pahl V, et al. Autotrophic growth of Escherichia coli is achieved by a small number of genetic changes. *eLife* 2024;12:RP88793.

[190] Ravagnan G, Schmid J Promising non-model microbial cell factories obtained by genome reduction. *Front Bioeng Biotechnol* 5 August 2024;12:Epub ahead of print. doi: 10.3389/fbioe.2024.1427248.

[191] Claassens NJ, Scarinci G, Fischer A, et al. Phosphoglycolate salvage in a chemolithoautotroph using the Calvin cycle. *Proc Natl Acad Sci* 2020;117:22452–22461.

[192] Calvey CH, Sànchez I, Nogué V, AM W, et al. Improving growth of Cupriavidus necator H16 on formate using adaptive laboratory evolution-informed engineering. *Metab Eng* 2023;75:78–90.

[193] Biswal T, Shadangi KP, Sarangi PK, et al. Conversion of carbon dioxide to methanol: A comprehensive review. *Chemosphere* 2022;298:134299.

[194] De Jong M, Bunse M, Hamelinck C. *Methanol carbon footprint and certification*, Amsterdam: International Methanol Producers and Consumers Association IMPCA; 8 September 2025. https://impca.eu/wp-content/uploads/2024/06/GU_IMPCA_Methanol-product-carbon-footprint-and-certification.pdf. (accessed.

[195] Cormier Z Turning carbon emissions into plastic. *BBC Earth*, https://www.bbcearth.com/news/turning-carbon-emissions-into-plastic (accessed 11 September 2025).

[196] Galimova T, Ram M, Bogdanov D, et al. Global demand analysis for carbon dioxide as raw material from key industrial sources and direct air capture to produce renewable electricity-based fuels and chemicals. *J Clean Prod* 2022;373:133920.

[197] Bachleitner S, Ata Ö, Mattanovich D. The potential of CO_2-based production cycles in biotechnology to fight the climate crisis. *Nat Commun* 2023;14:6978.

[198] Oehlenschläger K, Hengsbach J-N, Volkmar M, et al. From pre-culture to solvent: Current trends in Clostridium acetobutylicum cultivation. *Appl Microbiol Biotechnol* 2025;109:47.

[199] Serrano-Echeverry VA, Guerrero-Fajardo CA, Castro-Tibabisco KT. A Review of Biobutanol: Eco-Friendly Fuel of the Future – History, Current Advances, and Trends. *Fuels* 2025;6:55.

[200] Humphreys CM, McLean S, Schatschneider S, et al. Whole genome sequence and manual annotation of Clostridium autoethanogenum, an industrially relevant bacterium. *BMC Genomics* 2015;16:1085.

[201] Köpke M, Held C, Hujer S, et al. Clostridium ljungdahlii represents a microbial production platform based on syngas. *Proc Natl Acad Sci* 2010;107:13087–13092.

[202] Moon J, Dönig J, Kramer S, et al. Formate metabolism in the acetogenic bacterium *Acetobacterium woodii*. *Environ Microbiol* 2021;23:4214–4227.

[203] Sagani A, Anagnostara IM, Gennitsaris S, et al. Performance evaluation of biogenic CO2-based renewable chemicals: A holistic life cycle assessment and multi-criteria approach. *Biofuels Bioprod Biorefining* 2025;19:1219–1237.

[204] Mohan SV, Butti SK, Amulya K, et al. Waste biorefinery: A new paradigm for a sustainable bioelectro economy. *Trends Biotechnol* 2016;34:852–855.

[205] Liu Z, Wang K, Chen Y, et al. Third-generation biorefineries as the means to produce fuels and chemicals from CO_2. *Nat Catal* 2020;3:274–288.

[206] Sher F, Hameed S, Smječanin Omerbegović N, et al. Cutting-edge biomass gasification technologies for renewable energy generation and achieving net zero emissions. *Energy Convers Manag* 2025;323:119213.

[207] Ezeh AC, Bongaarts J, Mberu B Global population trends and policy options. *Lancet Lond Engl* 2012;380:142–148.

[208] Arbter P, Utesch T, Zeng A-P. *Process for biotechnological production of a bioproduct.* US20250163480A1, https://patents.google.com/patent/US20250163480A1/en?oq=US+2025%2f0163480+A1 (2025, accessed 19 September 2025).

[209] Modica V, Glávits R, Murbach TS, et al. Toxicological evaluation of protein powder derived from Cupriavidus necator. *J Appl Toxicol* 2023;43:887–912.

[210] Castro AR, Rocha I, Alves MM, et al. Rhodococcus opacus B4: A promising bacterium for production of biofuels and biobased chemicals. *AMB Express* 2016;6:35.

[211] Choi B, Glávits R, Murbach TS, et al. 90-day oral toxicity study in rats of a protein-rich powder derived from Xanthobacter sp. SoF1. *J Appl Toxicol* n/a; doi: 10.1002/jat.4663.

[212] Klinzing K, Aabrandt Søndergaard I, Chirom T, et al. In vitro genotoxicological evaluation of protein-rich powder derived from *Xanthobacter* sp *SoF1*. *J Appl Toxicol* 2024;jat.4621.

[213] Meusel M, Rehm H-J. Biodegradation of dichloroacetic acid by freely suspended and adsorptive immobilized Xanthobacter autotrophicus GJ10 in soil. *Appl Microbiol Biotechnol* 1993;40:Epub ahead of print October. doi: 10.1007/BF00170446.

[214] Miyake D, Ichiki S, Tanabe M, et al. Thiosulfate oxidation by a moderately thermophilic hydrogen-oxidizing bacterium, Hydrogenophilus thermoluteolus. *Arch Microbiol* 2007;188:199–204.

[215] Sano R, Kameya M, Wakai S, et al. Thiosulfate Oxidation by a Thermo-Neutrophilic Hydrogen-Oxidizing Bacterium, Hydrogenobacter thermophilus. *Biosci Biotechnol Biochem* 2010;74:892–894.

[216] EggPlant S.r.l. – Siryo SpA, https://www.siryo.it/en/portfolio/eggplant/ (accessed 15 September 2025).

[217] Venkatachalam J, Mohan H, Seralathan -K-K Significance of Herbaspirillum sp. in biodegradation and biodetoxification of herbicides, pesticides, hydrocarbons and heavy metals – A review. *Environ Res* 2023;239:117367.

[218] Japan GR. *Overview of Japan's Green Transformation (GX)*, GR Japan; January 2023, https://grjapan.com/sites/default/files/content/articles/files/gr_japan_overview_of_gx_plans_january_2023.pdf. accessed 12 February 2024.

[219] Karali N, Khanna N, Shah N Climate Impact of Primary Plastic Production, 2024, https://escholarship.org/uc/item/6cc1g99q.

[220] Biniek K, Henderson K, Rogers M, Santoni G. Driving CO_2 emissions to zero (and beyond) with carbon capture, use, and storage. McKinsey Quarterly 2020.

Vikas Kumar*, Anisa Mitra

Chapter 16
Renewable waste feedstocks for a sustainable aquaculture industry

16.1 Introduction

Aquaculture holds great promise for global economies, environment, and food security. The projected rise in global population has led to an increase in aquaculture operations [1]. However, commercial aquaculture development faces several obstacles and is alleged to be extremely wasteful, harmful to the environment, and unable to produce healthy outcomes [2]. Intensive fisheries depend heavily on artificial feed requirements, which raise production costs [3]. "Aquafeed 2.0" refers to the recent evolution of aquafeed, which was primarily fish meal (FM)- and fish oil (FO)-based before becoming primarily terrestrial-based [4]. In today's food production system, resources are limited, and the cost of energy, particularly from fossil fuels, continues to rise, putting additional strain on the achievement of the sustainable development goals of the aquaculture industry. As a result, in recent years, the concept of "regenerative" aquaculture has emphasized "precision aquaculture," "sustainable feeds," and "circular economy" [5]. In recent years, the circular bioeconomy has been used to produce ingredients for "Aquafeed 3.0" to improve aquaculture's sustainability in terms of reducing its environmental footprint. To address this critical bottleneck, the aquaculture industry must embrace circular economy (CE) to reduce environmental stress [6] rather than the traditional, linear economy approach of "take-make-dispose" [7]. CE principles encourage valorization or the value of biomass cascading down the value chain, by reducing waste, increasing efficiency, and supporting more sustainable systems. The concept of CE involves greater diversification and resilience by using renewable natural resources, with the goal of achieving sustainable ecological conservation and food production. By reusing products or coproducts in aquaculture, waste is reduced, and the value extracted from natural resources is increased. A reduction in feed waste due to improved fish nutrition can increase financial sustainability [8]. The modern-day aquafeed 3.0 is in consonance with the circular bioeconomy approach by the use of more nutrient-dense ingredients, and makes our domesticated carnivorous species much more omnivorous. Aquaculture needs to create novel tech-

***Corresponding author: Vikas Kumar**, Department of Animal, Veterinary and Food Sciences, Aquaculture Research Institute, University of Idaho, Moscow, ID 83844-2330, USA, e-mail: vikaskumar@uidho.edu
Anisa Mitra, Department of Zoology, Sundarban Hazi Desarat College, Pathankhali, South 24 Parganas 743611, West Bengal, India

 | https://doi.org/10.1515/9783112212141-016

niques where production and profitability are at par with reduced environmental impact and carbon footprint [4, 9]. Amidst a changing climate, consistently tumultuous economies, and rapidly changing social dynamics to decrease reliance on pricey imported feeds, there is a threat to the availability, affordability, environmental viability, and social acceptability of raw materials for aquafeed, which are still being tested to find the best formulation. The promotion of alternate ingredients as aquafeed can help to achieve sustainable development by profit maximization and livelihood opportunities [10]. In this chapter, we focus on the alternate ingredients, mainly the by-products/coproducts from nonconventional sources, to illustrate their importance and efficiency as aquafeed.

16.2 Waste coproducts

16.2.1 Distiller's dried grains with solubles

In recent decades, the growing demand for ethanol as a fuel additive and the decreasing reliance on fossil fuels have resulted in a dramatic increase in ethanol production from a variety of cereal grains [11], following ethanol removal from cereal grains via a series of particulate coproducts, known as distiller's dried grains with solubles (DDGS) (Figure 16.1) [12, 13]. In comparison to other traditional alternative protein sources, DDGS are widely available and reasonably priced. DDGS has sufficient amount of protein (usually between 250 and 450 g/kg, depending on the type) (Table 16.1) [14, 15] and can act as an alternate protein source as aquafeed. However, a few essential amino acids (EAAs) are also lacking in DDGS. All the same, by including these compounds in the feed formulation, these deficiencies can be corrected. The incorporation of DDGS into trout diets is further complicated by the fact that the quality and nutritional composition of DDGS grain sources and processing plants can differ significantly [14, 16]. But the primary factor limiting higher levels of DDGS incorporation in fish feed is high-structural fiber content. However, study has proved that fractionated DDGS products can be tolerated to a higher degree than nonfractionated DDGS and have been used successfully up to 300 g/kg in rainbow trout diets, without EAA supplementation or adverse effects on growth performance. Previous researchers demonstrated the importance of DDGS in rainbow trout [17], channel catfish [Lim et al., 2009; 18], Tilapia [19, 20], and sunshine bass [21]. Zhou et al. [22] proposed that growth in channel catfish juveniles that are fed diets containing 35–40% DDGS was comparable to fish fed a standard commercial manufactured formulation. Dietary inclusion of 30% DDGS in Nile tilapia improved feed consumption, feed efficiency, and growth [23] and incorporation of DDGS up to 30% in Mrigal [24] and Channel catfish [18] resulted in higher growth and feed conversion. A mixture of DDGS and soybean meal can be incorporated up to 49.5% in yellow perch (*Perca flavescens*) to increase growth rate [25]. DDGS can be incorporated in the diet of *Oreochromis niloticus* at 20% level without affecting growth performance and body composition [26]. In rainbow trout (*Oncorhynchus mykiss*), DDGS and high-protein DDGs were used to determine the efficiency of these two coproducts. DDGS, substituting 50% (DDGS50 diet) of the plant protein, resulted in higher feed intake and weight gain (WG) and lower feed conversion ratio (FCR) in trout fed, while feeding the DDGS100 diet resulted in a lower FCR compared with the control and the DDGS50 diets. DDGS100-containing diets tended to increase energy compared to fish fed with control and DDGS50 diets, hence high energy diet led to lower fish performance. However, the energy digestibility of rainbow trout fed with HPDDG100 and HPDDG50 diets was higher than control. The relative weight of the distal intestine, intestinal enzyme activity, or plasma metabolites was unaffected by either the DDGS or the HPDDG diets. When rainbow trout diets were substituted for the typical plant ingredient-based diets, it was demonstrated that both DGGS and HPDDG were suitable for energy and protein sources up to the level tested [27]. Significantly improved per-hectare profit and benefit-cost ratios were used to illustrate the economic benefit of high DDGS inclusion in common carp feeds. It was found that traditional cereal-based semi-intensive carp farming performed worse in terms of production, nutrient use, and economic performance than when DDGS-based compound feeds were combined with the most effective use of pond food [28]. Another experiment using high-protein distillers dried grains (HP-DDG) with 25% of FM in *Pangasius hypopthalmus* is acceptable for positive growth and feed utilization [29] (Table 16.2).

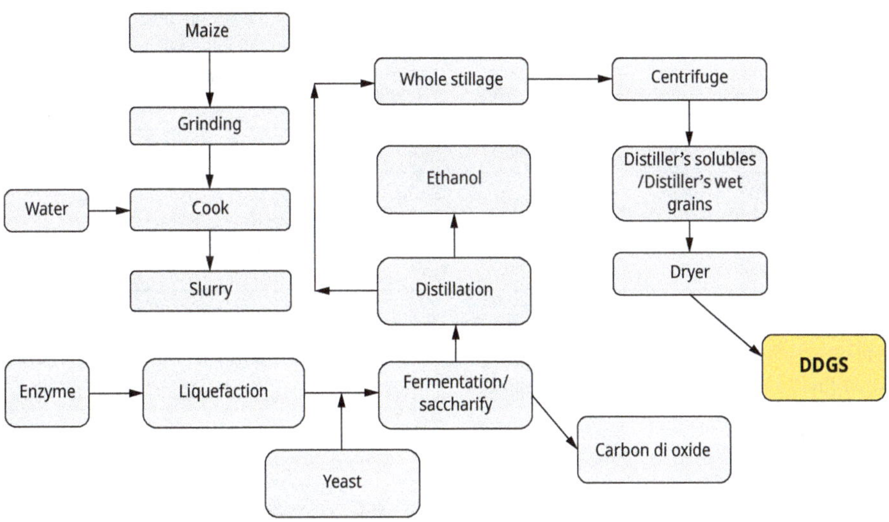

Figure 16.1: Process to produce maize distillers dried grains with soluble (DDGS) using dry-grind technology (modified after [33, 34]).

Table 16.1: Comparison of dietary component of DDGS from different sources.

Dietary component	DDGS source		
	Maize[a,b] (mg/kg)	Wheat[c] (mg/kg)	Sorghum grain[d] (mg/kg)
Dry matter	910 (870–920)	930 (920–940)	870–890
Crude protein	280 (260–310)	510 (390–590)	89–92
Fat	100 (90–117)	37 (29–50)	27–30
Crude fiber	110 (102–130)	77 (62–92)	21–24
Ash	52 (45–67)	50 (47–53)	14

[a][35],[b][36],[c][37], [d][129].

Table 16.2: Effects of feeding DDGS to Pacific white leg shrimp and fish.

Pacific white leg shrimp

Body weight (initial – final, g)	DDGS %	Ingredients replaced	Trial duration (days)	Fish meal (%)	Supplemental lysine (%)	Optimum DDGS%	References
0.49–7.2	0–40	Soyabean meal and corn starch	56	6	0–0.13	40	Rhodes et al.
0.99–6.1	0–30	Fish meal, soybean meal, and wheat flour	56	0		Partial replacement of soybean meal with DDGS in diets containing no fish meal reduced growth performance	[38]
0.04–16.3	10	Sorghum	126	0	None	10	[39]
0.45–25	0–10	Sorghum, fish meal	53	0	None	Up to 10	[40]

Fish

Body weight (initial-final, g)	DDGS (%)	Ingredients replaced	Trial duration (days)	Optimum DDGS%	References
Cyprinus carpio Juvenile					
45.4 ± 0.6 to 497.9 ± 25.7	40	Plant ingredients	155 days	40	[41]
Pangasius hypophthalmus	0%, 25%, 50%, and 75% HP-DDG	Fish meal	14 weeks	25% of fish meal by HP-DDG	[6]

To evaluate the suitability of corn DDGS as a protein source for European catfish (*Silurus glanis*), an in vivo digestibility assessment was done to determine apparent digestibility coefficients (ADCs) for protein, lipid, phosphorus, and amino acids available in DDGS and growth, nutrient utilization, and metabolism in juveniles. ADC of dry matter, crude protein, crude fat, and phosphorus of DDGS ingredient for European catfish juveniles were found to be 49.42%, 73.39%, 77.38%, and 87.98%, respectively. The study concluded that 30% DDGS can be included in the diet of European catfish without compromising the growth performance and nutrient utilization [30]. Thus, it is evident that the use of protein-enriched DDGS or high-protein distiller's grains shows promise but the lack of data and familiarity with the feed ingredient, misconceptions about DDGS [31], and large-scale production are still the main constraints [32].

16.2.2 Beer industry coproduct

Brewer's spent grains (BSG) are the main by-coproducts (80–85% of total) of beer manufacturing industries [42] and widely available throughout the world [43]. Brewer's yeast is rich in protein and has 30% EAAs of total protein [44], vitamins, and minerals [45], and other bioactive compounds such as β-glucans, mannan oligosaccharides, vitamins, minerals, and nucleic acids [46]. After beer production, the different left-over forms of brewer's waste are of potential environmental hazard [47], and their safe disposal is a major environmental concern for the industry. But it is a good source of nondegradable protein and water-soluble vitamins [48]. Proximate analysis suggests that it contains 43.51% crude protein, 1.05% lipids, and 8.33% total ash [49, 50]. Moreover, in recent past, several attempts were carried out to exploit its beneficial use in aquafeed [44, 51–53]. Several authors [54–56] cited this by-product as a potential alternative to FM in the feed for fish [17, 51, 57–60] and crustaceans [61] due to its high content in protein and fiber as well as lipids, minerals, and vitamins [62].

Jayant et al. [59] made an effort to develop an environmentally friendly feed formulation in *Pangasianodon hypophthalmus* fingerlings, formulated with 0%, 25%, 50%, 75%, or 100% BSGs, as a replacement for soybean meal. The BSG contributed 0%, 24.26%, 47.10%, 69.51%, and 91.26% of the total protein in the diets. The study suggested that BSGs can serve as an alternate protein source with 50% level substitution of soybean meal without any adverse effect on growth, nutrient utilization, and feed conversion.

A trial was conducted to analyze the effect of the inclusion of yeast and spent grain obtained from breweries in feeds for rainbow trout (*Oncorhynchus mykiss*). The study showed that the combination of 20% yeast and 15% spent grain in the feed, formulated with only 15% inclusion of FM, showed 87–89% protein digestibility and increased growth [63]. Another experiment showed that partial replacement of FM by inclusion of up to 30% brewers' spent yeast and 15% spent grain in isoproteic and isolipidic diets for gilthead sea bream (*Sparus aurata*) gave similar results in terms of growth, food conversion, and fillet final composition for a feed with FM as the main protein source, and showed a protein digestibility of 89–95% [64]. Brewery waste (brewer's grains) was used at four different levels (10%, 20%, 30%, and 40% w/w) replacing rice bran in fish diet under a semi-intensive culture system and its impact on the growth of catla, *Catla catla*; rohu, *Labeo rohita*; and mrigal, *Cirrhina mrigala*, was studied. A better growth performance was attributed in *C. catla* and *L. rohita* fed on a diet containing 30% brewery waste in the feed, whereas *C. mrigala*, fed on a diet containing brewery waste at the above-mentioned levels, showed poorer growth than the control.

16.2.3 Soymeal coproduct-processed soybean meal (EnzoMeal™)

Soybean products and derivatives thereof are considered good alternatives to FM. There are many advantages in using soybean compared to FM, such as competitive market price, stable supply, and available protein and balanced amino acids, among others. Despite the obvious advantages of using soybeans, it still contains indigestible components such as nonstarch polysaccharides and other carbohydrates. Moreover, it has many antinutritional compounds and factors that restrict nutrient digestion and absorption. These compounds remain a hindrance to using the full potential of soybean products and derivatives as FM replacement in the diets of several aquatic animals, especially fish [65–69]. Hence, through the efforts of soybean producers, conventional soybean meal has undergone several processes to improve and exploit the potential of soybean meal. EnzoMeal™, an improved soybean meal developed by the Ohio Soybean Council, Worthington, Ohio, USA, and Battelle, contains higher protein (20%) and lower antinutritional factors (50%) (Table 16.3). Table 16.4 shows the comparative amino acid profiles of EnzoMeal™ and conventional soybean meal.

The product EnzoMeal™ has been tested as an aquafeed in several aquatic animals. When used as an FM replacement for diets of yellow perch (*Perca flavescens*), a level of 50% replacement resulted in better growth and feed performance [70] compared to 100% FM replacement using conventional soybean meal and EnzoMeal™. The product used contained 580 g/kg crude protein and low levels of antinutritional factors compared to conventional soybean meal. However, like other plant-based proteins, EAA supplementation was implemented for EnzoMeal™-based diets. On the other hand, when EnzoMeal™ was used as an FM replacement for diets of Pacific white shrimp (*Litopenaeus vannamei*), the growth performance and product quality of the shrimp remained the same [71]. The study is notable as it effectively showed that shrimps could sustainably use diets with low levels of animal-based proteins, when combined with alternative aquafeed ingredients such as EnzoMeal™. Last, when EnzoMeal™ was used as a long-term FM replacement for diets of rainbow trout (*Oncorhynchus mykiss*), growth and feed performance enhancement was achieved. In addition, the higher the inclusion levels of EnzoMeal™, the greater is the performance-reaching levels, as seen in rainbow trout fed with conventional soybean meal. This effect has been linked to lesser GH/IGF axis system disruptions, which resulted in better nutrient usage. The IGFBP-1b1 expression in the hepatic tissues was upregulated in rainbow trout fed with high amounts of conventional soybean meal. This particularly resulted in growth inhibition. No adverse effects have been observed in rainbow trout fed with EnzoMeal™ [72]. Hence, modification of conventional soybean meal, particularly to increase protein contents and to lower antinutritional factors, may pave the way for more effective alternative FM replacements.

Table 16.3: Comparative nutritive value of soybean meal vs. modified soybean meal/EnzoMeal™.

	Commercial SBM	EnzoMeal™	EM benefits over SBM
Protein (%)	45.9	56	22% increase
Carbohydrate (%)	39.85	27.5	45% decrease
Oligosaccharide (%)	15	<0.05	~100% removal
Phytic acid (%)	1.52	0.88	42% decreased

Table 16.4: Amino acid profile (%) of commercial soybean meal and modified soybean meal.

	Soybean meal	EnzoMeal
Essential amino acids		
Lysine	2.07	2.4
Methionine	0.64	0.83
Histidine	1.22	1.41
Threonine	1.84	2.25
Tryptophan	0.71	0.84
Leucine	3.69	4.32
Isoleucine	2.19	2.65
Phenylalanine	2.41	2.81
Valine	2.34	2.77
Nonessential amino acids		
Tyrosine	1.56	1.89
Cysteine	0.62	0.72
Arginine	3.36	3.64
Alanine	2.03	2.48
Aspartic acid	5.3	6.39
Glutamic acid	8.66	9.94
Glycine	1.98	2.4
Ornithine	0.03	–
Proline	2.41	2.77
Serine	2.3	2.86

16.2.4 Poultry coproduct meal

The most common poultry-based ingredients used in feed formulations consist of ground-rendered clean parts of the carcasses of slaughtered poultry [73, 74]. Poultry and livestock by-products such as poultry meat flour, feather, and blood meal (BM) are important sources of protein for use in feeding fish. Poultry-based meal (PBM) emerged as one of the most promising alternative ingredients as a relatively cheap source of protein compared with FM, and available in large quantities throughout the year, especially in poultry-producing regions such as Asia. They contain essential fatty acids, vitamins, and minerals and are virtually free of antinutrients, unlike plant proteins [75, 76]. However, like other animal-based proteins, variation in compositional quality is common and the manufacturing characteristics also vary based on the products (Table 14.5) [77–81]. However, advancements in processing technologies can counteract these challenges [82]. In addition, laws and regulations in the selection of raw materials), sanitary transportation, and handling play a vital role in safeguarding the product quality to stimulate its use in aquaculture feeds [73, 79, 83]. Studies have proved that PBM can successfully replace FM in different dietary inclusion levels for numerous finfish and shellfish species, although a recent study using meta-analysis reported that no significant effect on growth was detected with PBM-supplemented diets, compared with control diets with 100% FM across a wide range of studies, species, and environmental conditions. Similar trends were observed for freshwater fish, marine water fish, and crustaceans for final weight. Significantly higher FCR was detected in fish, compared with crustaceans, and in freshwater fish, compared with marine fish. In both freshwater and marine fish, effects varied across different species. Certain factors, such as nutritional quality (specifically amino acids or fatty acids), palatability, and digestibility of PBM, could also mediate the influence of PBM supplements on the growth and FCR of fish [84].

16.2.5 Blood meal

Blood meal (BM) is considered a proteinaceous concentrate [35]. But the protein digestibility of BM manufactured using different techniques has been shown to differ significantly [83, 85]. Blood proteins appear to be especially sensitive to heat damage, and the drying technique used can have a very significant effect on digestibility of BM. Bureau et al. [83] observed that the digestibility of crude protein in spray-dried blood products was significantly higher than that of roto-plate dried, steam-tube dried, and ring-dried BMs. A study suggests that BM can be a very good source of bioavailable amino acids. Spray- and ring-dried BMs are widely used in salmonid feeds due to their very high digestibility and consistent quality (Table 16.5). Good performance of fish has been observed for fish fed diets containing approximately 8–20% BM, in conjunction with high (more than 20%) FM levels [86, 87]. A study was conducted by [88] to investigate the feasibility of utilizing BM in diets for feeding Nile tilapia fingerlings and determine the optimum level for the BM. An average percentage Weight gain (WG) was 165% and 157% in fish that were fed diets containing 10% and 15% BM. BM also improved food conversion ratios (FCR), which were 2.06 and 2.16 for the 10% and 15% BM treatments, respectively, as compared to 2.33 for the control diet. Moreover, the addition of BM to the basic diet reduced the cost of the diet substantially. The study showed that the 10% BM treatment gave better performance than the 15% BM treatment. This probably relates to the known effect of BM on palatability of diets, particularly at high inclusion levels. Hence, it is recommended to use only 10% of BM in fish diets. In another feeding trial, diets were formulated to replace 0% (control), 9.80%, 19.61%, and 29.41% of FM with chicken hemoglobin powder in Largemouth bass. However, the fish-specific growth rate was significantly reduced when diet replacement level was up to 19.61%, which may be related to the feed intake and ADC of protein and amino acids. But the feed efficiency and protein efficiency ratio were not significantly decreased until the replacement level, beyond 29.41% [89].

16.2.6 Feather meal

Feather meal (FeM) is produced by processing the feathers obtained after slaughtering poultry. FeM is a low-quality protein source as it lacks EAAs found in many livestock species, including lysine, methionine, histidine, and tryptophan [90]. But, keratin being the primary component of feather proteins (80–100%), is poorly digestible when raw [91] Table 16.5. Hence it is hydrolyzed before it can be used as a valuable source of protein in feed [92; 93]. Bureau et al. [83] estimated the digestibility of crude protein of FeMs found to vary between 77% and 86%. A comparable ADC value was reported by [94] for a FeM fed to rainbow trout. [95] estimated that the digestibility of crude protein of FeM was about 79% for rockfish (*Sebastes schlegeli*). A recent study showed that the addition of chicken FeM, fermented with *Bacillus subtilis*, could increase the activity of proteolytic and lipolytic activity and nutrient digestibility of silver pompano [96]. Studies reported that 15–25% FeM could be included in the diet of rainbow trout [97, 98] and 15% FeM (90% crude protein), replacing FM [99], could be included in the diet of Chinook salmon, without effect on growth and feed efficiency. Bishop et al. [100] reported that FeM can replace 66% of FM and meat and bone meal (MBM) (9.9% of the total diet) without any significant effect on the growth of *Oreochromis niloticus* fry, although FeM, as a sole source of animal protein in fishfeed, appears to result in a decrease in WG and the associated growth parameters. A 16-week feeding trial on red hybrid tilapia (*Oreochromis* sp.) showed that up to 15% of FeM can be included into the fish diet with good growth performance and FCR [101].

16.2.7 Meat and bone meal

The regulatory definition of MBM is the rendered product from mammalian tissues, including bone, and must contain a minimum of 4% phosphorus, with calcium level not exceeding 2.2 times the actual phosphorus level. MBM can be used to design aquafeed to provide protein, amino acids, phosphorus, calcium, energy, and other nutrients (Tables 16.5 and 16.6), although regulations and directives vary worldwide and from country to country [102]. The degree of protein digestibility in MBM appears to vary based on the source and processing. For MBM fed to red drum, [103] and [104] observed protein digestibility of approximately 74–79% (*Sciaenops ocellatus*). Allan et al. [105] reported lower digestibility values for MBM fed to silver perch (*Bidyanus bidyanus*]. Many freshwater and marine fish species were found to have high digestive ability of meat meal [106, 107]. The maximum level of MBM in rainbow trout feeds could be up to 24% MBM or about 25% of the total digestible protein, and significant amounts of MBM could be consumed by Mozambique tilapia and gilthead seabream without having an adverse effect on their performance [98]. In a trial, MBM (53% CP, 15% CL) was used to replace FM in the diets of juvenile gilthead seabream (*Sparus aurata*). Three extruded experimental diets (45% CP; 20% CL) were created, with 0, 50, and 75% of the protein coming from MBM (diets MBM0; MBM50; MBM75). Overall, without affecting growth performance, feed utilization, or nutrient retention, MBM protein can replace up to 50% of FM protein in diets for young gilthead seabream. In contrast to feed intake, growth performance and feed efficiency were lower with MBM75 diet, but similar with MBM0 and MBM50 diets. Except for crude lipid and energy content, which were lower with the MBM75 diet, total body composition was unaffected by diet composition. The composition of the diet had no impact on the retention of protein or EAAs, but the MBM75 diet had a lower retention of energy. Economic efficiency showed that diets containing MBM reduced production costs, with the MBM diets achieving the lowest economic conversion ratio and the MBM50 diet achieving the highest economic profit [108]. The study's findings showed that replacing 50% of the FM with MBM had no negative effects on the growth and survival rate of tilapia (*O. niloticus*) fry and made fry production more affordable [109]. In a 90-day feeding trial, the impact of switching from FM to MBM in diets for juvenile *Pseudobagrus ussuriensis* was assessed. Six diets were developed, with MBM replacing FM at 0 (S0), 200 (S20), 400 (S40), 600 (S60), 800 (S80), and 1,000 g/kg (S100), respectively. The findings revealed no discernible difference in WG between fish fed with S0, S20, and S40 diets. However, when MBM was used in place of 600, 800, and 1,000 g/kg FM protein, WG was significantly reduced. The S80 and S100 diets had significantly lower ADCs for protein and dry matter than the other diets. According to the study's findings, the ideal dietary MBM replacement level was 34.3% [110]. In juvenile *Macrobrachium nipponense*, with 15% or 50% of the FM protein replaced by either MBM or PBM [MBM(15), MBM(50)] demonstrated that adding MBM to diets in place of FM had no effect on *M. nipponense's* ability to grow, though PBM(15) had a significantly higher specific growth rate than the other groups. However, compared to other groups, shrimp fed with MBM(15) diet had significantly higher survival rates [111].

Table 16.5: Crude protein (CP) content and apparent digestibility coefficients (ADC) of CP and gross energy (GE) of rendered animal protein ingredients of various origins (modified after [66]).

Ingredients along with their processing conditions	CP (%)	ADC of CP (%)	ADC of GE (%)
Feather meals			
Steam hydrolysis 30 min at 276 kPa, disc dryer	75	81	80
Steam hydrolysis 448 min at 276 kPa, disc dryer	82	81	78
Steam hydrolysis 40 min at 276 kPa, ring dryer	76	81	76
Steam hydrolysis 40 min at 276 kPa, steam tube dryer	75	87	80
Meat and bone meals			
125–135 °C, 20–30 min, 17–34 kPa	57	83	68
Same as above but air classification of final product to reduce ash content	55	87	73
133 °C, 30–40 min, 54 kPa	50	88	82
128, 20–30 min, 17–34 kPa	48	87	76
132–138 °C, 60 min	50	88	82
127–132 °C, 25 min	54	89	83
Poultry by-product meals			
138 °C, 30 min	65	87	77
127–132 °C, 30–40 min, 54 kPa	63	91	87
Blood meals			
Steam coagulated blood rotoplate dryer	83	82	82
Steam coagulated blood ring dryer	84	88	88
Whole blood spray-dryer	83	96	92
Blood cells, spray-dryer	86	96	93
Blood plasma, spray-dryer	71	99	99
Steam coagulated blood steam tube dryer	91	84	79
Whole blood, spray-dryer	82	97	94
Steam coagulated blood steam ring-dryer	86	85	86

16.2.8 Rendered animal fats

Rendered animal fats, which have been used in fish feeds for decades due to their low cost and widespread availability (Table 16.7), could be an interesting alternative to FO. In the last 10 years, the price of inedible animal fats has dropped by 40–50%, to around $0.30/kg, hence FO replaced by rendered fats could reduce the cost of aquafeed significantly. However, due to a variety of factors, including poor digestibility, variable quality, effects on growth, and product quality, and more recently, concern over disease transmission, the use of these products in aquafeeds has historically been restricted or even avoided [112]. These worries are less important today, according to a growing body of research, and rendered animal fats can be beneficial components of fish feed. Studies have shown that most fish species can effectively use rendered animal fats as long as the diet contains enough mono- or polyunsaturated fatty acids to ensure proper digestion of saturated fatty acids and the essential fatty acid requirements are met. The evidence suggests that rendered animal fats at levels corresponding to 30–40% of total lipids have no negative effects on growth performance, feed efficiency, or product quality in most fish species studied. Overall, using these low-cost alternatives to FO could result in significant immediate savings. With the increasing use of this FO alternate, research is needed to support expanded applications of rendered fat from various potential sources in aquafeeds [120]. The only remaining barrier to the widespread use of animal fats in aquaculture feeds is the widespread fear of bovine spongiform encephalopathy and other transmissible spongiform encephalopathies, which is still not confirmed [121].

Table 16.6: Proximate analysis of meat and bone meals [113–118], amino acid composition [116, 119], and mineral content, reported in [115, 116, 118].

Characteristic	Typical values
Dry matter (%)	92.9–100
Crude protein (%)	49.5–59.4
Ether extract (%)	8.9–16.0
Crude fiber (%)	1–5.13
Ash	20.7–52.9%
Moisture content	1.9–5.7%
pH	5.89–7.19
Protein solubility	2.2–7.2%
Particle size	25.6–800 μm
Color	Light tan brown to dark brown

Amino acids	%CP
Alanine	9.19
Arginine	6.98–7.06
Aspartic acid	4.25
Cystine	0.94–1.01
Glutamine	5.40
Glutamic acid	7.79
Glycine	16.67
Histidine	1.89–2.29
Isoleucine	2.76–3.69
Leucine	6.13–6.85
Lysine	5.18–6.08
Methionine	1.40–2.12
Phenylalanine	3.36–3.56
Proline	11.3
Serine	4.00
Threonine	3.27–4.65
Tryptophan	0.58–0.75
Tyrosine	2.79
Valine	4.20–4.29

Mineral content	
Ca	10.60–13.50%
P	4.73–6.50%
Mg	0.24–1.20%
K	1.02–1.56%
Na	0.71–0.78%
Cl	0.44–0.80%
S	0.39%
Cu	1.5–10 mg/kg
Fe	500–602 mg/kg
Mn	12.3–22 mg/kg
Zn	94 mg/kg
Mo	2.7 mg/kg

Table 16.7: Types and characteristics of rendered fats (modified after [120, 121]).

Types of rendered fat	Characteristics
Feed-grade animal fat	Derived from by-products of different species; a ratio of unsaturated beef and pork unsaturated to saturated (U to S) of between 1:1 and 1.6:1, depending on material used
Poultry fat	Fat from poultry by-products; U to S ratio of 2:1, specific gravity of 0.91 at 15–15.6 °C; light brown liquid/pale brown solid; gross energy (kcal/kg) of 9,000

Table 16.7 (continued)

Types of rendered fat	Characteristics
Choice white grease	Derived mainly from the rendering of pork tissue; U to S ratio is 1:1; certain blends of beef, pork, and poultry fat are sold as choice white grease; specific gravity is 0.84 at 15–15.6 °C; yellow liquid/pale brown solid; gross energy (kcal/kg) of 9,350
Tallow	Derived from rendered beef tissue, but may contain other animal fats; U to S ratio ranges from 0.9 to 1.6:1; specific gravity ranges from at 0.89–0.91 at 15–15.6 °C; color: <3 red (AOCS Wesson); gross energy is 9,020 kcal/kg
Yellow grease	Mainly restaurant grease but can contain dead stock fat and/or dark color, high FFA, and high MIU fat from any type of rendering operation; specific gravity is 0.92 at 15–15.6 °C; color: med-brown liquid/brown solid; gross energy is 9,372 kcal/kg
Blended animal and vegetable fat	Blended animal fat, vegetable oil, acidulated vegetable oil, soapstock, and/or restaurant grease.

16.3 Single-cell protein

Single-cell protein (SCP) products can be prepared from different microbial sources, including microalgae, yeast and other fungi, and bacteria (Figure 16.2). They are the sources of almost all EAAs, which may not be available in plant derivatives. The nutrient varies based on the source of the SCPs. The source from microbial cells contains 70–80% amino-nitrogen of the total nitrogen. Algae contains a rich amount of fat and various vitamins like A, B, C, D, and E, with β-carotene, tocopherols, and vitamin B. Bacterial SCP is high in protein (around 80% of total dry weight) and certain essential amino acids. *Bacillus* species have carotenoid pigments with antioxidant properties [122]. SCP meals are the promising alternate aquafeed, showing positive feeding trial data on salmon, trout, and shrimp, with modulation of intestinal microbiota, enhancement of innate immunity, and strengthened resistance against stress [123] (Table 16.8). But the main constraint of SCP production in mass scale is the processing price of the product, finding new strains, and developing new technologies [124]. SCPs can be produced in a range of environments using various technologies such as aerobic, anaerobic, and gas bioreactors. Though SPC might have a comparable protein content to FM, it has some major drawbacks including presence of toxins, contains antinutritional compounds like phytic acid, and unbalanced aminoacidic proportions, especially lysine and methionine, compared to FM. It is also less palatable, is difficul to digest, or has challenges to scale up the production process [123, 125].

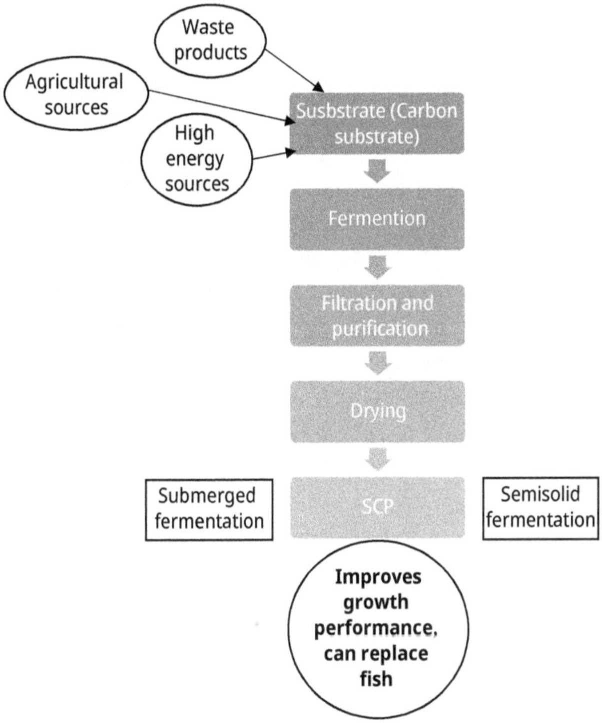

Figure 16.2: Production process of SCP.

Table 16.8: Advantages of SCP used as aquafeed.

– Recycling diverse types of wastes such as those from agriculture, urban, and food; can provide carbon source for SCP production
– Environmentally friendly and does not require large expanse of land
– Production is not seasonal or climate-dependent
– High-protein content, with good amino acid profile, low fat content, and good carbohydrate content, with high content of digestible proteins
– Large quantity of biomass can be produced in a comparatively shorter duration due to high rate of multiplication
– Modulate intestinal microbiota, enhance innate immunity, and strengthen resistance against stress

16.4 Summary

Sustainable aquaculture requires aquafeeds that go beyond the utilization of FM and other standard protein replacements. The promotion of alternate ingredients as aquafeed can help to achieve sustainable development by profit maximization and livelihood opportunities. Various alternative ingredients, mainly waste coproducts or byproducts from different industries such as breweries, soymeal processing, poultry and meat processing, and even SCPs, are promising alternates in partially and/or completely replacing FM from aquafeeds. They are proven as nutrient-dense, digestible, and economical aquafeeds. Value-adding efforts and additional processing steps can improve protein content or reduce antinutrients that seem to increase the viability of these alternatives, though there is a threat to the availability, affordability, environmental viability, and social acceptability of raw materials for these renewable waste feedstocks that are still being tested to find the best formulation.

References

[1] Bostock J, McAndrew B, Richards R, Jauncey K, Telfer T, Lorenzen K, Little D, et al. Aquaculture: Global status and trends. Philos Trans R Soc B Biol Sci 2010;365(1554):2897–2912.

[2] Macarthur ELLEN, Heading HEADING. How the circular economy tackles climate change. Ellen MacArthur Found 2019;1:1–71.

[3] Hertrampf JW, Piedad-Pascual F. Handbook on Ingredients for Aquaculture Feeds, Springer Dordrecht: Springer Science & Business Media; 2003.

[4] Cottrell RS, Metian M, Froehlich HE, Blanchard JL, Jacobsen NS, McIntyre PB, Nash KL, et al. Time to rethink trophic levels in aquaculture policy. Rev Aquacult 2021;13(3):1583–1593.

[5] Centre for Tropical and subtropical aquaculture. Recycling feed byproducts shows promise for aquaculture nutrition. 2022.

[6] Foundation EM. Delivering the Circular Economy: A Toolkit for Policymakers, Ellen MacArthur Foundation; 2015.

[7] Kusumowardani N, Tjahjono B. Circular economy adoption in the aquafeed manufacturing industry. Procedia CIRP 2020;90:43–48.

[8] Ahmad AL, Chin JY, Harun MHZM, Low SC. Environmental impacts and imperative technologies towards sustainable treatment of aquaculture wastewater: A review. J Water Process Eng 2022;46:102553.

[9] Colombo S, Turchini G. 'Aquafeed 3.0': Creating a more resilient aquaculture industry with a circular bioeconomy framework. Rev Aquacult 2021;13. doi: 10.1111/raq.12567.

[10] Mitra A. Thought of alternate aquafeed: Conundrum in aquaculture sustainability?. In Proceedings of the Zoological Society, 2020. Vol. 74. pp. 1–18.

[11] Liu K, Han J. Changes in mineral concentrations and phosphorus profile during dry-grind processing of corn into ethanol. Bioresour Technol 2011;102(3):3110–3118.

[12] Bothast RJ, Schlicher MA. Biotechnological processes for conversion of corn into ethanol. Appl Microbiol Biotechnol 2005;67(1):19–25.

[13] Welker TL, Lim C, Barrows FT, Liu K. Use of distiller's Dried Grains with Solubles (DDGS) in rainbow trout feeds. Anim Feed Sci Technol 2014;195:47–57.

[14] Ortín N, Waldo G, Yu P. Nutrient variation and availability of wheat DDGS, corn DDGS and blend DDGS from bioethanol plants. J Sci Food Agric 2009;89(10):1754–1761.

[15] Widyaratne GP, Zijlstra RT. Nutritional value of wheat and corn distiller's dried grain with solubles: Digestibility and digestible contents of energy, amino acids and phosphorus, nutrient excretion and growth performance of grower-finisher pigs. Can J Anim Sci 2007;87(1):103–114.

[16] Liu SX, Singh M, Inglett G. Effect of incorporation of distillers' dried grain with solubles (DDGS) on quality of cornbread. LWT-Food Science and Technology 2011;44(3):713–718.

[17] Cheng ZJ, Hardy RW. Effects of microbial phytase supplementation in corn distiller's dried grain with solubles on nutrient digestibility and growth performance of rainbow trout, oncorhynchus mykiss. J Appl Aquacult 2004;15(3–4) 83-100.

[18] Li MH, Robinson EH, Oberle DF, Lucas PM. Effects of various corn distillers by-products on growth, feed efficiency, and body composition of channel catfish, Ictalurus punctatus. Aquacult Nutr 2010;16 (2):188–193.

[19] Wu YV, Rosati R, Sessa DJ, Brown P. Utilization of protein-rich ethanol co-products from corn in tilapia feed. J Am Oil Chem Soc 1994;71(9):1041–1043.

[20] Shelby RA, Lim C, Yildrim-Aksoy M, Klesius PH (2008). Effect of distillers dried grains with solubles-incorporated diets on growth, immune function and disease resistance in Nile tilapia (Oreochromis niloticus L.).

[21] Thompson KR, Rawles SD, Metts LS, Smith RG, Wimsatt A, Gannam AL, . . . Webster CD. Digestibility of dry matter, protein, lipid, and organic matter of two fish meals, two poultry by-product meals, soybean meal, and distiller's dried grains with solubles in practical diets for Sunshine bass, Morone chrysops× M. saxatilis. J World Aquacult Soc 2008;39(3):352–363.

[22] Zhou P, Davis DA, Lim C, Yildirim-Aksoy M, Paz P, Luke AR. Pond demonstration of production diets using high levels of distiller's dried grains with solubles with or without lysine supplementation for channel catfish. N Am J Aquacult 2010;72(4):361–367.

[23] Lim C, Garcia JC, Yildirim-Aksoy M, Klesius PH, Shoemaker CA, Evans JJ. Growth response and resistance to streptococcus iniae of Nile tilapia, oreochromis niloticus, fed diets containing distiller's dried grains with solubles. J World Aquacult Soc 2007;38(2):231–237.

[24] Paul BN, Das S, Giri SS, Chattopadhay DN, Mandai RN, Pandey BK, Chakraborti PP. Performance of Cirrhinus mrigala fingerlings on feeding dried distillers grain soluble. Indian Journal of Animal Research 2012;46(3):272–275.

[25] Schaeffer TW, Brown ML, Rosentrater KA. Effects of dietary distillers dried grains with solubles and soybean meal on extruded pellet characteristics and growth responses of juvenile yellow perch. N Am J Aquacult 2011;73(3):270–278.

[26] Schaeffer TW, Brown ML, Rosentrater KA, Muthukumarappan K. Utilization of diets containing graded levels of ethanol production co-products by Nile tilapia. J Anim Physiol Anim Nutr 2010;94 (6):e348–e354.

[27] Øverland M, Krogdahl Å, Shurson G, Skrede A, Denstadli V. Evaluation of distiller's dried grains with solubles (DDGS) and high protein distiller's dried grains (HPDDG) in diets for rainbow trout (*Oncorhynchus mykiss*). Aquacult 2013;416:201–208.

[28] Sándor ZJ, Révész N, Varga D, Tóth F, Ardó L, Gyalog G. Nutritional and economic benefits of using DDGS (distiller'dried grains soluble) as feed ingredient in common carp semi-intensive pond culture. Aquacult Rep 2021;21:100819.

[29] Allam BW, Khalil HS, Mansour AT, Srour TM, Omar EA, Nour AAM. Impact of substitution of fish meal by high protein distillers dried grains on growth performance, plasma protein and economic benefit of striped catfish (*Pangasianodon hypophthalmus*). Aquacult 2020;517:734792.

[30] Sándor ZJ, Révész N, Lefler KK, Čolović R, Banjac V, Kumar S. Potential of corn distiller's dried grains with solubles (DDGS) in the diet of European catfish (Silurus glanis). Aquacult Rep 2021;20:100653.

[31] U.S. Grains Council. DDGS – Production and Exports, 2020. https://grains.org/buying-selling/ddgs/.

[32] Welker TL, Lim C, Barrows FT, Liu K. Use of distiller' s dried grains with solubles (DDGS) in rainbow trout feeds. Anim Feed Sci Technol 2014;195:47–57. doi: 10.1016/j.anifeedsci.2014.05.011.
[33] Gyan WR, Yang QH, Tan B, Xiaohui D, Chi S, Liu H, Zhang S. Effects of replacing fishmeal with dietary dried distillers grains with solubles on growth, serum biochemical indices, antioxidative functions, and disease resistance for Litopenaeus vannamei juveniles. Aquacult Rep 2021;21:100821.
[34] Pal K, Maji C, Tudu B. J Food Anim Sci 2020;1:117–120. Journal of Food and Animal Sciences, 1, 117–120.
[35] Feedstuffs, 2011.
[36] Spiehs MJ, Whitney MH, Shurson GC. Nutrient database for distiller's dried grains with solubles produced from new ethanol plants in Minnesota and South Dakota. J Anim Sci 2002;80 (10):2639–2645.
[37] Nrc. Nutrient Requirements of Swine, 10th edition, Washington DC: National Academy Press; 1998.
[38] Cummins VC, Webster CD, Thompson KR, Velasquez A. Replacement of fish meal with soybean meal, alone or in combination with distiller's dried grains with solubles in practical diets for pacific white shrimp, Litopenaeus vannamei, grown in a clear-water culture system. J World Aquacult Soc 2013;44(6):775–785.
[39] Sookying D, Davis DA. Pond production of Pacific white shrimp (Litopenaeus vannamei) fed high levels of soybean meal in various combinations. Aquaculture 2011;319(1–2):141–149.
[40] Roy LA, Bordinhon A, Sookying D, Davis DA, Brown TW, Whitis GN. Demonstration of alternative feeds for the Pacific white shrimp, Litopenaeus vannamei, reared in low salinity waters of west Alabama. DoAquaculture research 2009;40(4):496–503.
[41] Sándor ZJ, Révész N, Varga D, Tóth F, Ardó L, Gyalog G. Nutritional and economic benefits of using DDGS (distiller'dried grains soluble) as feed ingredient in common carp semi-intensive pond culture. Aquacult Rep 2021;21:100819.
[42] Tang D, Yin G, He Y, Hu S, Li B, Li L, Liang H, Borthakur D. Recovery of protein from brewer's spent grain by ultrafiltration. Biochem Eng J 2009;48:1–5.
[43] Essien J, Udotong I. Amino acid profile of biodegraded brewers spent grains (BSG). J Appl Sci Environ Manage 2010;12(1):12. doi: 10.4314/jasem.v12i1.55582. ISSN: 1119–8362.
[44] Hassan MA, Aftabuddin M, Meena DK, Mishal P, Gupta SD. Effective utilization of distiller's grain soluble-an agro-industrial waste in the feed of cage-reared minor carp *labeo bata* in a tropical reservoir. India Environ Sci Pollut Res 2016;23:1e6.
[45] Ovie SO, Eze SS. Utilization of saccharomyces cerevisiae in the partial replacement of fishmeal in clarias gariepinus diets. Int J Adv Agric Res 2014;2:83–88.
[46] Implvo F, Pinho O, Vieira E, Tavarela JG. Brewer's saccharomyces yeast biomass: Characteristics and potential applications. Trends Food Sci Techno 2010;21(2):77–84.
[47] Hang YD, Splitstoesser DF, Woodams EE. Utilization of brewery spent grain liquor by Aspergillus niger. Appl Microbiol 1975;30(5):879–880.
[48] Taylor J, Taylor JRN. Some potential applications for brewers spent grains the protein-rich co-products, from sorghum lager beer brewing. In 12th scientific and technical convention, 2009.
[49] Singh P, Paul BN, Rana GC, Mandal RN, Chakrabarti PP, Giri SS. Evaluation of breweries waste in the feed of catla catla fingerlings. Indian J Anim Nutr 2014;31(2):187–191.
[50] Singh P, Paul BN, Giri SS. Potentiality of new feed ingredients for aquaculture: A review. Agric Rev 2018;39(4).
[51] Kaur VI, Saxena PK. Incorporation of brewery waste in supplementary feed and its impact on growth in some carps. Bioresour Technol 2004;91(1):101–104.
[52] Levic J, Djuragic O, Sredanovic S. Use of new feed from brewery by-products for breeding layers. Rom Biotechnol Lett 2010;15:5559–5565.

[53] Zerai DB, Fitzsimmons KM, Collier RJ, Duff GC. Evaluation of brewer's waste as partial replacement of fish meal protein in nile tilapia, oreochromis niloticus, diets. J World Aquacult Soc 2008;39 (4):556–564.

[54] Oliva-Teles A, Gonçalves P. Partial replacement of fishmeal by brewers yeast (*Saccaromyces cerevisae*) in diets for sea bass *(Dicentrarchus labrax*) juveniles. Aquacult 2001;202(3–4).

[55] Ozório ROA, Portz L, Borghesi R, Cyrino JEP. Effects of dietary yeast (*Saccharomyces cerevisia*) supplementation in practical diets of tilapia (*Oreochromis niloticus*). Animals 2012;2(1):16–24.

[56] Hauptman BS, Barrows FT, Block SS, Gibson Gaylord T, Paterson JA, Rawles SD, Sealey WM. Evaluation of grain distillers dried yeast as a fish meal substitute in practical-type diets of juvenile rainbow trout, *Oncorhynchus mykiss*. Aquacult 2014;432:7–14.

[57] Campos INÊ, Matos ELI, Aragão CLÁ, Pintado M, Valente LMP. Apparent digestibility coefficients of processed agro-food by-products in European seabass (*Dicentrarchus labrax*) juveniles. Aquacult Nutr 2018;24(4):1274–1286.

[58] Huige NJ. Brewery byproducts and effluents. In Priest FG, Stewart GG, Handbook of Brewing, 2nd edition, CRC Press; 2006. ISBN 9780824726577.

[59] Jayant M, Hassan MA, Srivastava PP, Meena DK, Kumar P, Kumar A, Wagde MS. Brewer's spent grains (BSGs) as feedstuff for striped catfish, *Pangasianodon hypophthalmus* fingerlings: An approach to transform waste into wealth. J Clean Prod 2018;199:716–722.

[60] Yamamoto T, Marcouli PA, Unuma T, Akiyama T. Utilization of malt protein flour in fingerling rainbow trout diets. Fish Sci 1994;60(4):455–460.

[61] Muzinic LA, Thompson KR, Morris A, Webster CD, Rouse DB, Manomaitis L. Partial and total replacement of fish meal with soybean meal and brewer's grains with yeast in practical diets for Australian red claw crayfish *Cherax quadricarinatus*. Aquacult 2004;230(1–4) 359-76.

[62] Mussatto SI, Dragone G, Roberto IC. Brewers' spent grain: Generation, characteristics and potential applications. J Cereal Sci 2006;43(1):1–14.

[63] Estevez A, Padrell L, Iñarra B, Orive M, San Martin D. Brewery by-products (yeast and spent grain) as protein sources in rainbow trout (*Oncorhynchus mykiss*) feeds. Front Mar Sci 2022.

[64] Estevez A, Padrell L, Iñarra B, Orive M, San Martin D. Brewery by-products (yeast and spent grain) as protein sources in gilthead seabream (*Sparus aurata*) feeds. Aquacult 2021;543:736921.

[65] Hernández MD, Martinez FJ, Jover M, García B. Effects of partial replacement of fish meal by soybean meal in sharp snout seabream (*Diplodus puntazzo*) diet. Aquacult 2007;263:159–167.

[66] Kumar V, Makkar HPS, Becker K. Detoxified *Jatropha curcas* kernel meal as a dietary protein source: Growth performance, nutrient utilization and digestive enzymes in common carp (*Cyprinus carpio* L.) fingerlings. Aquacult Nutr 2011a 17:313–326.

[67] Kumar V, Makkar HPS, Becker K. Nutritional, physiological and hematological responses in rainbow trout (*Oncorhynchus mykiss*) juveniles fed detoxified *Jatropha curcas* kernel meal. Aquacult Nutr 2011b 17:451–467.

[68] Kumar V, Makkar HPS, Becker K. Evaluations of the nutritional value of *Jatropha curcas* protein isolate in common carp (*Cyprinus carpio* L.). J Anim Physiol Anim Nutr 2012a 96:1030–1043.

[69] Kumar V, Sinha AK, Makkar HPS, De Boeck G, Becker K. Phytate and phytase in fish nutrition: A rev. J Anim Physiol Anim Nutr 2012b 96:335–364.

[70] Kumar V, Wang H-P, Lalgudi R, Cain R, McGraw B, Rosentrater KA. processed soybean meal as an alternative protein source for yellow perch (*Perca flavescens)* feed. Aquacult Nutr 2020;25 (4):917–931.

[71] Hulefeld R, Habte-Tsion H-M, Lalgudi R, Cain R, McGraw B, Tidwell JH, Kumar V. Nutritional evaluation of an improved soybean meal as a fishmeal replacer in the diet of pacific white shrimp, *Litopenaeus vannamei*. Aquacult Res 2018;49(4):1414–1422.

[72] Kumar V, Lee S, Cleveland B, Romano N, Lalgudi R, Rubio M, McGraw B, Hardy R. Comparative evaluation of processed soybean meal (enzomealtm) vs. regular soybean meal as a fishmeal

replacement in diets of rainbow trout (*Oncorhynchus mykiss*): Effects on growth performance and growth-related genes. Aquacult 2019;516:734652.

[73] Cruz-Suárez LE, Nieto-López M, Guajardo-Barbosa C, Tapia-Salazar M, Scholz U, Ricque-Marie D. Replacement of fish meal with poultry by-product meal in practical diets for *Litopenaeus vannamei*, and digestibility of the tested ingredients and diets. Aquacult 2007;272(1–4) 466-76.

[74] Dong FM, Hardy RW, Haard NF, Barrows FT, Rasco BA, Fairgrieve WT, Forster IP. Chemical composition and protein digestibility of poultry by-product meals for salmonid diets. Aquacult 1993;116(2–3) 149-58.

[75] Abdul-Halim HH, Aliyu-Paiko M, Hashim R. Partial replacement of fish meal with poultry by-product meal in diets for snakehead, *Channa striata* (bloch, 1793), fingerlings. J World Aquacult Soc 2014;45 (2):233–241.

[76] Gunben EM, Senoo S, Yong A, Shapawi R. High potential of poultry by-product meal as a main protein source in the formulated feeds for a commonly cultured grouper in Malaysia (*Epinephelus fuscoguttatus*). Sains Malays 2014;43(3):399–405.

[77] Davis DA, Arnold CR. Replacement of fish meal in practical diets for the pacific white shrimp, *Litopenaeus vannamei*. Aquacult 2000;185(3–4) 291-98.

[78] Dawson MR, Shah Alam M, Watanabe WO, Carroll PM, Seaton PJ. Evaluation of poultry by-product meal as an alternative to fish meal in the diet of juvenile black sea bass reared in a recirculating aquaculture system. N Am J Aquacult 2018;80(1):74–87.

[79] de Yta A G, Davis DA, Rouse DB, Ghanawi J, Saoud IP. Evaluation of practical diets containing various terrestrial protein sources on survival and growth parameters of redclaw crayfish *Cherax quadricarinatus*. Aquacult Res 2012;43(1):84–90.

[80] Robinson EH, Li MH, Manning BB. Evaluation of corn gluten feed as a dietary ingredient for pond-raised channel catfish *Ictalurus punctatus*. J World Aquacult Soc 2001;32(1):68–71.

[81] Tacon AGJ, Hasan MR, Subasinghe RP. FAO. Use of Fishery Resources as Feed Inputs to Aquaculture Development: Trends and Policy Implications. No. 1018, Rome: Food and Agriculture Organization of the United Nations; 2006.

[82] Najafabadi HJ, Moghaddam HN, Pourreza J, Shahroudi FE, Golian A. Determination of chemical composition, mineral contents, and protein quality of poultry by-product meal. Int J Poult Sci 2007;6 (12):875–882.

[83] Bureau DP, Harris AM, Cho CY. Apparent digestibility of rendered animal protein ingredients for rainbow trout (*Oncorhynchus mykiss*). Aquacult 1999;180(3–4):345-58.

[84] Galkanda-Arachchige HSC, Wilson AE, Davis DA. Success of fishmeal replacement through poultry by-product meal in aquaculture feed formulations: A meta-analysis. Rev Aquacult 2020;12 (3):1624–1636.

[85] Cho CY, Slinger SJ, Bayley HS. Bioenergetics of salmonid fishes: Energy intake, expenditure and productivity. Comp Biochem Physiol Part B 1982;73(1):25–41.

[86] Abery NW, Gunasekera RM, De Silva SS. Growth and nutrient utilization of murray cod *Maccullochella peelii* peelii (Mitchell) fingerlings fed diets with varying levels of soybean meal and blood meal. Aquacult Res 2002;33(4):279–289.

[87] Luzier JM, Summerfelt RC, Ketola HG. Partial replacement of fish meal with spray-dried blood powder to reduce phosphorus concentrations in diets for juvenile rainbow trout, *Oncorhynchus mykiss* (Walbaum) 1, Aquacult Res; 1995, Vol. 26, 8, pp./ 577–587.

[88] Obey, et al. 2019.

[89] Ding G, Li S, Wang A, Chen N. Effect of chicken haemoglobin powder on growth, feed utilization, immunity and haematological index of largemouth bass (*Micropterus salmoides*). Aquac Fish 2020;5 (4):187–192.

[90] Crawshaw R. Co-product Feeds in Europe: Animal Feeds Derived from Industrial Processing. Lulu. comz 2019.

[91] Moran ET Jr, Summers JD, Slinger SJ. Keratins as sources of protein for the growing chick: 2. hog hair, a valuable source of protein with appropriate processing and amino acid balance. Poult Sci 1967;46(2):456–465.

[92] El Boushy AR, Van der Poel AFB, Walraven OED. Feather meal – A biological waste: Its processing and utilization as a feedstuff for poultry. Biol Wastes 1990;32(1):39–74.

[93] Papadopoulos MC. Processed chicken feathers as feedstuff for poultry and swine. A review. Agric Wastes 1985;14(4):275–290.

[94] Sugiura SH, Babbitt JK, DONG FM, et al. Utilization of fish and animal by-product meals in low-pollution feeds for rainbow trout Oncorhynchus mykiss (Waulbaum). Aquacult Res 2000;31:585–593.

[95] Lee SM. Apparent digestibility coefficients of various feed ingredients for juvenile and grower rockfish (Sebastes schlegeli). Aquaculture 2002;207(1–2):79–95.

[96] Adelina A, Feliatra F, Siregar YI, Putra I. Suharman I Use of chicken feather meal fermented with *Bacillus Subtilis* in diets to increase the digestive enzymes activity and nutrient digestibility of silver pompano *Trachinotus Blochii* (Lacepede, 1801). F1000 Res 2021;10:25. doi: 10.12688/f1000research.26834.2. PMID: 33868644; PMCID: PMC8030120.

[97] Henrichfreise B. Bewertung von aufgeschlossenem Getreide und hydrolysiertem Federmehl in der Ernahrung von Regenbogenforellenfuttern, Diss. Landwirtsch Fak. Rheinische Friedrich-Wilhelms-Univ, Bonn: 1989.

[98] Bureau DP, Harris AM, Bevan DJ, Simmons LA, Azevedo PA, Cho CY. Feather meals and meat and bone meals from different origins as protein sources in rainbow trout (*Oncorhynchus mykiss*) diets. Aquacult 2000;181(3–4):281–291.

[99] Fowler LG. Feather meal as a dietary protein source during Parr-Smolt transformation in fall chinook salmon. Aquacult 1990;89(3–4):301–314.

[100] Bishop CD, Angus RA, Watts SA. The use of feather meal as a replacement for fish meal in the diet of Oreochromis niloticus fry. Bioresour Technol 1995;54(3):291–295.

[101] Yong ST, Mardhati M, Farahiyah IJ, Noraini S, Wong HK. Replacement of fishmeal in feather meal-based diet and its effects on tilapia growth performance and on water quality parameters. J Trop Agric Food Sci 2018;46(1):47–55.

[102] Ockerman HW, Basu L. Production and consumption of fermented meat products. In Toldrá F, Hui YH, Astiasaran I, Sebranek J, Talon R, editors. Handbook of Fermented Meat and Poultry, 2nd ed., 2014. pp. 7–11.

[103] McGoogan BB, Reigh RC. Apparent digestibility of selected ingredients in red drum (*Sciaenops ocellatus*) diets. Aquacult 1996;141(3–4):233-44.

[104] Gaylord TG, Dm G III. Determination of digestibility coefficients of various feedstuffs for red drum (*Sciaenops ocellatus*). Aquacult 1996;139(3–4):303-14.

[105] Allan GL, Parkinson S, Booth MA, Stone DAJ, Rowland SJ, Frances J, Warner-Smith R. Replacement of fish meal in diets for Australian silver perch, *Bidyanus bidyanus*: I. Digestibility of alternative ingredients. Aquacult 2000;186(3–4):293–310.

[106] Gomes EF, Rema P, Kaushik SJ. Replacement of fish meal by plant proteins in the diet of rainbow trout (*Oncorhynchus mykiss*): Digestibility and growth performance. Aquacult 1995;130(2–3):177–186.

[107] Watanabe T, Verakunpiriya V, Watanabe K, Viswanath K, Satoh S. Feeding of rainbow trout with non-fish meal diets. Fish sci 1998;63(2):258–266.

[108] Moutinho S, Peres H, Serra C, Martínez-Llorens S, Tomás-Vidal A, Jover-Cerdá M, Oliva-Teles A. Meat and bone meal as partial replacement of fishmeal in diets for gilthead sea bream (*Sparus aurata*) juveniles: Diets digestibility, digestive function, and microbiota modulation. Aquacult 2017;479:721–731.

[109] Hasan MT, Khalil I, Mohammed S, Kashem M, Hashem S, Mazumder S. Substitution of fish meal by meat and bone meal for the preparation of tilapia fry feed. Int J Anim Fish Sci 2012;5:464–469.

[110] Wang Y, Tao S, Liao Y, Lian X, Luo C, Zhang Y, Yang C, Cui C, Yang J, Yang Y. Partial fishmeal replacement by mussel meal or meat and bone meal in low-fishmeal diets for juvenile ussuri catfish (*Pseudobagrus ussuriensis*): Growth, digestibility, antioxidant capacity and IGF-I gene expression. Aquacult Nutr 2020;26(3):727–736.

[111] Yang Y, Xie S, Cui Y, Lei W, Zhu X, Yang Y, Yu Y. Effect of replacement of dietary fish meal by meat and bone meal and poultry by-product meal on growth and feed utilization of gibel carp, *Carassius auratus* gibelio. Aquacult Nutr 2004;10(5):289–294.

[112] Bureau DP. Rendered products in fish aquaculture feeds. Essential rendering, Alexandria, Virginia, USA: National Renderers Association; 2006, 179–184.

[113] Garcia R, Phillips J. Physical distribution and characteristics of meat & bone meal protein. J Sci Food Agric 2009;89. doi: 10.1002/jsfa.3453.

[114] Howie SA, Calsamiglia S, Stern MD. Variation in ruminal degradation and intestinal digestion of animal byproduct proteins. Anim Feed Sci Technol 1996;63(1–4):1-7.

[115] National Research Council and Canadian Department of Agriculture, 1971.

[116] National Research Council, 2001

[117] Nengas I, Alexis MN, Davies SJ, Petichakis G. Investigation to determine digestibility coefficients of various raw materials in diets for gilthead sea bream, *Sparus auratus* l. Aquacult Res 1995;26 (3):185–194.

[118] Preston RL. Feed Composition Table, Minneapolis, MN: Penton Media; 2014, 2014. pp. 18–26.

[119] Li YL, McAllister LTA, Beauchemin KA, He ML, McKinnon JJ, Yang WZ. Substitution of wheat dried distillers grains with solubles for barley grain or barley silage in feedlot cattle diets: Intake, digestibility, and ruminal fermentation. J Anim Sci 2011;89(8):2491–2501.

[120] Trushenski JT, Lochmann RT. Potential, implications and solutions regarding the use of rendered animal fats in aquafeeds. Am J Anim Vet Sci 2009;4:4.

[121] Bureau DP, Gibson J, El-Mowafi A. Use of animal fats in aquaculture feeds, Avances en Nutrición Acuicola; 2002.

[122] Bharti V, Pandey PK, Koushlesh SK. Single cell proteins: A novel approach in aquaculture systems. World Aquaculture 2014;45(4):62–63.

[123] Pereira AG, Fraga-Corral M, Garcia-Oliveira P, Otero P, Soria-Lopez A, Cassani L, Cao H, Xiao J, Prieto MA, Simal-Gandara J. Single-cell proteins obtained by circular economy intended as a feed ingredient in aquaculture. Foods 2022;11:2831.

[124] Jones SW, Karpol A, Friedman S, Maru BT, Tracy BP. Recent advances in single cell protein use as a feed ingredient in aquaculture. Current opinion in biotechnology 2020;61:189–197.

[125] FAO. The Growth of Single-cell Protein in Aquafeed, Bangkok: 2022.

Section IV: **Some commercial bioproducts**

Dan Derr, Huu Tuan Tran

Chapter 17 Microbially produced green surfactant bioproducts

17.1 Introduction

17.1.1 Surfactants

Surfactants [1] are chemicals that have an affinity for surfaces. Hence, the name is a portmanteau for **surf**ace **act**ive **agent**. Surfactants are composed of a polar, hydrophilic head group, often depicted as a ball, and a nonpolar, hydrophobic, usually oleophilic tail (see Figure 17.1).

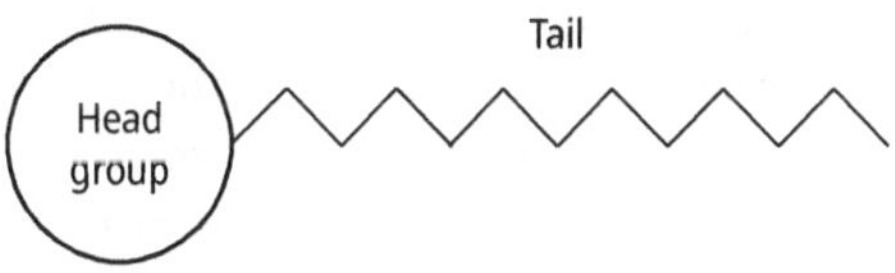

Figure 17.1: A commonly used cartoon model of a surfactant molecule.

Figure 17.1 is an appropriate model for a small-molecule surfactant, but polymers can also be surfactants, though they will not be discussed in detail here. Nearly all well-characterized microbial biosurfactants are small molecules.

Briefly, surfactants' preference for residing at surfaces comes from the fact that their polar and nonpolar portions are connected but spatially separated from each other. If a surfactant is put in water, it is driven to the surface where the head group can be just inside the water phase, and the tail sticks out into the air, or any other nonpolar phase that is in contact with the water.

Surfactants are an important commercial product in modern society but have also been used for at least 4,500 years by humanity. In fact, the saponification reaction using oil as a feedstock and producing soap is described as the first-recorded chemical reaction [2], at least the earliest found to date. Soap is the archetype of surfactants.

The global surfactant market is currently valued at approximately USD 40 billion per year [3], reflecting its critical role in both consumer products and industrial applications. Surfactants are amphiphilic compounds that reduce surface and interfacial tension, enabling processes such as emulsification, dispersion, foaming, and detergency [4]. Because of these versatile physicochemical properties, they are widely em-

Dan Derr, Integrity BioChem
Huu Tuan Tran, Kansas State University, e-mail: huutuan@ksu.edu

 | https://doi.org/10.1515/9783112212141-017

ployed across diverse sectors. In the consumer market, surfactants are essential components of home cleaning and laundry products, dishwashing liquids, and a broad range of personal care items such as shampoos, soaps, and cosmetics [5]. In agriculture, they serve as wetting agents, emulsifiers, and adjuvants that improve the efficiency of pesticides and herbicides [6]. Within the energy sector, surfactants play a role in enhanced oil recovery, biofuel production, and drilling fluids. They are also indispensable in paints, coatings, textiles, food processing, pharmaceuticals, and manufacturing industries, where they function as stabilizers, dispersants, and processing aids [7]. This breadth of applications underscores the economic significance of surfactants as well as the growing demand for sustainable, bio-based alternatives to conventional petroleum-derived products [8].

17.1.2 Microbial biosurfactants

Here, biosurfactants are defined as surfactants that are what are known as "natural products" by chemistry professionals; in other words, they are produced in natural systems. Specifically, the focus here will be on microbial biosurfactants. Microbial biosurfactants are produced biosynthetically by microbes. Typically, microbes do not produce a pure compound biosurfactant; instead, they produce a mixture of compounds that differ slightly in the length and unsaturation level of the tail, and in at least one case, the makeup of the head group. The exact structural details of each mixture will not be discussed at great length, here.

In addition to microbial biosurfactants, there are also plant-based biosurfactants, saponins being a well-studied example [9], but those will also not be discussed here.

There was a period in the twentieth century where surfactants, like many other commonly used chemicals, were dominated by synthetic products [2]. But over the last couple of decades, industry has shifted to using biological sources for synthetic surfactants, producing what are known as "bio-based" products. Natural products such as microbial biosurfactants are an extension of this trend. Microbial biosurfactants currently have a miniscule share of the overall surfactants market, but the share is growing, and new products are now being introduced regularly, as will be discussed in more detail below.

It is notable that the first surfactant, soap, was – and is – a bio-based chemical: a natural oil is processed using synthetic chemistry – in this case, simple alkaline hydrolysis chemistry, but synthetic nonetheless – to produce a useful product. Thus, the arc of surfactants history started with a bio-based solution several thousand years ago, traveled through a synthetic phase around the twentieth century, has reverted in the direction of bio-based, again, and is now moving toward natural products. That is not to say that synthetic surfactants will be totally displaced. Synthetic surfactants will likely remain a large slice of the market pie, as there are many applications where synthetic surfactants are quantifiably the best choice when all three legs of the sustainability stool – economic, social, and environmental – are considered. But microbial biosurfactants will almost certainly continue to increase their market share for the foreseeable future.

17.1.3 Classes of surfactants

Without going into too much detail – whole books are written on this topic [10] – surfactants are often broken down into four classes:

1. Anionic – the head group has a negative charge
2. Nonionic – the entire molecule is neutral
3. Cationic – the head group has a positive charge
4. Zwitterionic or amphoteric – the head group carries both a positive and a negative charge

Microbial biosurfactants are predominantly anionic or nonionic in nature, reflecting the chemical composition of their hydrophilic head groups [11]. Anionic biosurfactants carry a net negative charge, often due to the presence of carboxyl, phosphate, or sulfate functional groups [12]. Examples include rhamnolipids (produced by *Pseudomonas aeruginosa*), which are glycolipids with rhamnose sugar moieties, and sophorolipids (produced by *Candida bombicola*), which contain sophorose linked to hydroxylated fatty acids [13]. In contrast, nonionic biosurfactants lack a net charge on their head groups, relying instead on neutral sugar or polyol structures for hydrophilicity [13, 14]. Examples include mannosylerythritol lipids (MELs) and trehalose lipids, which are produced by various yeasts and actinomycetes [15]. They tend to be good cleansers, emulsifiers, and sometimes have antimicrobial properties, making them attractive in personal and home care. Many biosurfactants are also being considered in applications with uncontrolled release, such as agriculture, wastewater treatment, remediation, and energy production because of their attractive environmental footprint.

Multiple biosurfactants will be nonionic in acidic conditions and anionic under less acidic conditions, though some are also nonionic at all pH they are stable at.

17.1.4 History

The scientific history of microbial biosurfactants begins around the middle of the twentieth century. The first use of the term biosurfactant appears to occur in 1960 [16], referring to rhamnolipids. Rhamnolipids had their structure elucidated in 1949 [17]. Another class of compounds that is now considered a biosurfactant, trehalose lipids, was first reported in 1933 [18]. In the 1960s, the term biosurfactant appears to be used in the academic literature a handful of times and becomes more common as the decades pass.

Fatty acids, the original soaps developed by human civilizations, can be produced microbially as what are known as free fatty acids (FFA). They would then be microbial biosurfactants. However, it is unlikely this is a viable production route, as soap is easily and cost-effectively obtained from saponification of natural oils. As mentioned above, this makes them a bio-based surfactant. FFA will not be considered any further here, as there is no incentive to make them as microbial biosurfactants.

Another surfactant that can be produced microbially is phospholipids. However, phospholipids are readily obtained as a bioproduct during the production of vegetable oils. As they are produced today, phospholipids would be considered plant-derived biosurfactants, and as with FFA, there is no incentive to produce them via fermentation. Phospholipids will not be further considered here.

17.2 Microbial biosurfactants

Most microbial biosurfactants that are known to be commercially available today are glycolipids, in which Desai and Banat describe as "carbohydrates in combination with long-chain aliphatic acids or hydroxy aliphatic acids," in their seminal 1997 review [19]. Put more simply, glycolipids are a combination of sugar (glyco) and fat (lipid).

17.2.1 Sophorolipids

Sophorolipids (SL) were the first biosurfactant to be commercialized in industrial quantities and they lead the microbial biosurfactant market today. SL are glycolipids.

SL were first reported and described in 1961 by Gorin et al. [20]. The first mention that they are surfactants was found by Inoue and Ito in 1982 [21]. One of the reasons SL have been successfully commercialized is the productivities, and titers of SL fermentations have been demonstrated at very high levels. For example, in 2004, Pekin et al. [22] showed 400 g/L titers, with productivities approaching 1 g/L/h. The companies producing SL commercially have exceeded these productivity numbers, but it can be difficult to document commercial progress, as companies often keep production metrics as trade secrets.

SL contain a sophorose sugar head group and a long-chain hydroxy fatty acid tail, as depicted in Figure 17.2. The carboxylic acid at the end of the tail can form a lactone with one of the sugar hydroxyl groups, as illustrated in the figure.

As produced using modern production organisms, for example, *Candida bombicola*, SL tend to be richer in the lactone form than the ring-opened form, though this varies with fermentation conditions [23].

In the lactone form, SL are nonionic surfactants at all pH they are stable at and have low water solubility. Alkali conditions can hydrolyze the lactone, producing the ring-

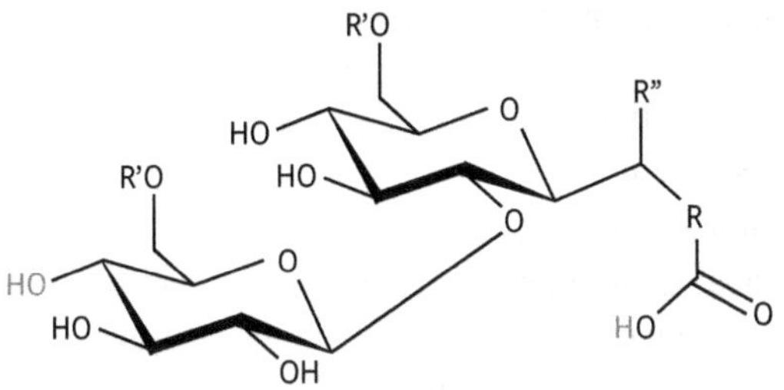

Figure 17.2: A general structure for SL. R = a hydrocarbon, often unsaturated, ~15 carbons in length, R′ = (C = O)CH_3 or H, and R″ = CH_3 or H. The gray OH on the terminal glucose and H on the carboxylic acid are commonly dehydrated to form a lactone ring in the SL produced biocatalytically.

opened form, which is a nonionic surfactant at low pH, and an anionic surfactant at neutral and alkaline pH. Again, at low pH in the nonionic form, ring-opened SL has low water solubility. But the anionic surfactant at higher pH is highly water soluble [23].

17.2.1.1 Commercial status

Evonik is almost certainly the largest volume producer of SL. They first began publicizing SL products around 2015 [24], and REWOFERM SL ONE [25] was commercialized in 2016 [26]. Evonik has indicated in the past that they produce SL at what they call their Fermas site in Slovakia. There has been conflicting information released on the fermentation capacity at this site over the years, but it is likely on the order of 1,000 m^3, and growing (see 17.2.2 section), as of 2022.

Givaudan purchased Soliance in 2014 [27]. One of the products Soliance had developed was an SL product called Sopholiance S, which continues to be marketed by Givaudan as of early 2022.

Also in 2014, DSM partnered with Synthezyme, which was a company commercializing sophorolipid technology at the time [28]. However, in this case, no DSM or Synthezyme SL products could be found still marketed in 2022.

Holiferm is a biosurfactant company founded in 2018 that announced plans in 2021 to build a 1,100 tons/year SL plant [29] in Wallasey, UK [30]. At the time of this writing, they market a product line for personal care applications called HoliSurf and one for cleaning applications called HoneySurf. They use a proprietary continuous separation approach to improve productivities and titers compared to batch production.

BASF displayed an interest in SL technology in 2021, with an investment in Allied Carbon Solutions [31]. At the same time, they announced an agreement with Holiferm to jointly develop glycolipids for a variety of end uses. Allied Carbon Solutions is a company headquartered in Japan that focuses on SL. The tradename for their SL product is ACS-Sophor.

Locus has been pursuing SL for several years. In 2022, their performance ingredients unit announced an exclusive agreement with Dow for SL as home and personal care ingredients [32]. At in-cosmetics in 2022, Dow announced EcoSense SL-60 HL and HA products for personal care applications [33].

It is difficult to capture all the products that contain SL, but a few highlights are worth noting. Saraya, a Japan-based global cleaning products company, had products as early as 2014 that contained SL [34]. Ecover has a number of homecare products they market in Europe that contain SL, and SL are now being used more and more in personal care products worldwide.

17.2.2 Rhamnolipids

Rhamnolipids (RL), also a glycolipid, have recently entered the market as an ingredient in consumer care applications.

The structure of rhamnolipids was first elucidated in 1949 by Jarvis and Johnson [17]. Rhamnolipids typically have one or two rhamnose sugar units as the headgroup and two medium chain 3-hydroxy fatty acids as the tail; see Figure 17.3.

Jarvis and Johnson attributed the first isolation of rhamnolipids to Bergstrom, Theorell, and Davide in 1947, or possibly Birch-Hirschfeld in 1935. *Pseudomonas aeruginosa* was the originally produced organism. It then took more than a decade for anyone to report on their surfactant properties, which as mentioned in the introduction, appears to have occurred in 1960 [16]. Productivities have now been improved to commercializable levels [35]. While *Pseudomonas* is the most prolific natural producer, many genetically modified organisms (GMOs) have also been developed that biosynthesize RL with commercializable metrics [23].

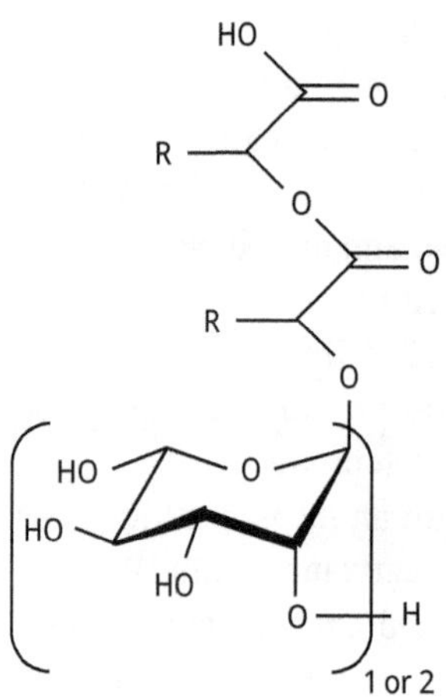

Figure 17.3: RL are made up of one or two rhamnose sugar units attached to two medium-chain length 3-hydroxy fatty acids.

17.2.2.1 Commercial status

Jeneil Biotech marketed an early commercial product that contained RL, a biofungicide called Zonix. They obtained EPA approval for this product in 2004 [36]. Stepan recently began distributing Zonix and it is still available today [37]. Jeneil continues to market RL for agriculture, bioremediation, consumer care, and antimicrobial applications on their website [38].

Evonik, with their market-leading SL position, announced their first RL product, RHEANCE ONE, at in-cosmetics Global in April of 2018 for use in personal care and cosmetics [39]. Around the New Year of 2022, Evonik announced a "three-digit million-euro" capital investment in a rhamnolipid production facility [40]. The new facility will be at their Fermas site in Slovakia, mentioned above in the SL section. That level of investment will likely result in thousands of tons, or even tens of thousands of tons, of production capacity for microbial biosurfactants. Evonik released REWOFORM RL 100 in early 2022 [41].

In 2020, Stepan acquired NatSurFact from Logos Technologies [42]. (For full disclosure, note that this author worked on NatSurFact at Logos for several years.) Early in 2021, Stepan announced the acquisition of a commercial fermentation production plant in Lake Providence, LA, originally constructed by Myriant for succinic acid production. Stepan indicated they will use it for microbial biosurfactant production [43]. As noted above, they also distribute Zonix Biofungicide, an RL-based product.

The agreement between Holiferm and BASF mentioned above in the SL section also applies to RL [44].

Advanced BioCatalytics announced in early 2021 that they are marketing BioSS RL for oil & gas, cleaning, agriculture, and wastewater markets [45].

Without going into any further detail, a number of companies indicated an interest in RL sometime in the last few decades, but appear to have stopped pursuing them. Either there is no longer any evidence that they are working with RL, or their commercial momentum appears to have stalled.

Unlike for SL, above, RL are early enough in their commercial deployment that only a couple of products have been announced that contain them. In 2019, Unilever announced they were marketing a hand dishwashing soap in Chile called Quix that contained Evonik's RHEANCE ONE. Additionally, a niche personal care cleanser direct marketed by Booni Doon contains NatSurFact RL [46]. MarkNature, an online store that sells products suitable for farming, gardening, food and beverages, and laboratories, markets RL ingredients in small lot sizes, as well [47].

17.2.3 Mannosylerythritol lipids

It appears that the earliest elucidation of mannosylerythritol (MEL) structures was in 1983 by Kawashima, Nakahara, Oogaki, and Tabuchi [48]. The MEL portion of the molecule is the hydrophilic head group, and they have two ~ 6–10 carbon hydrophobic tails attached via ester linkages to the mannose, as shown in Figure 17.4. Kawashima et al. [48] used a *Candida sp.* to produce their MEL, but other yeasts such as *Pseudozyma* sp. and *Ustilago* sp. are also known as producers [23].

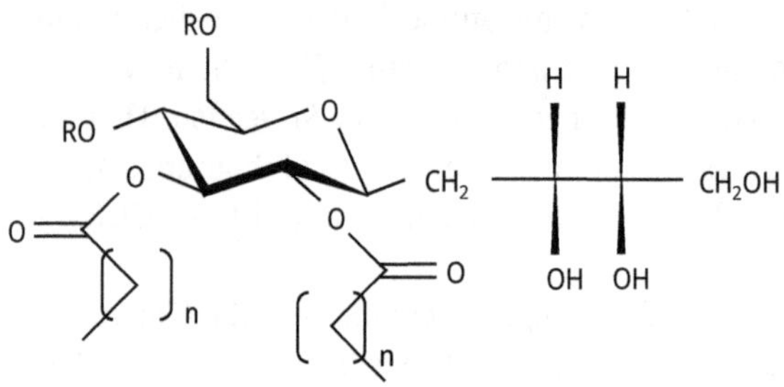

Figure 17.4: MEL are made up of a mannosylerythritol unit attached to two medium-chain-length fatty acids. R's are H or (C = O)CH_3 and n = 6–10.

17.2.3.1 Commercial status

MEL lipids were reported to be available from Toyobo in 2019 [25]. However, Toyobo's website only lists their MEL product, Ceramela, as available in 1–5 kg lot sizes [49].

Advanced BioCatalytics announced in early 2022 the opening of a pilot plant that will be used to produce commercial quantities of MEL in the United States [50].

17.2.4 Surfactin

In contrast to most of the flycolipid microbial biosurfactants discussed here, surfactin is a lipopeptide or a combination of a hydrocarbon tail (lipo) with a cyclic, small protein head group (peptide). Surfactin was first isolated, characterized, and immediately classified as a surfactant in 1968 by Arima et al. [51]. As shown in Figure 17.5, the tail of the most common congener is 15 carbons long and attaches to the peptides through an α-hydroxy carboxylic acid. The head group of this congener has seven amino acid residues, including an aspartic acid. The aspartic acid provides a free carboxylic acid to allow the same pH behavior, as described several times above: at highly acidic pH, it will be a nonionic surfactant, and at neutral and alkaline pH, an anionic surfactant. This congener is produced using *Bacillus subtillis* as the producing organism.

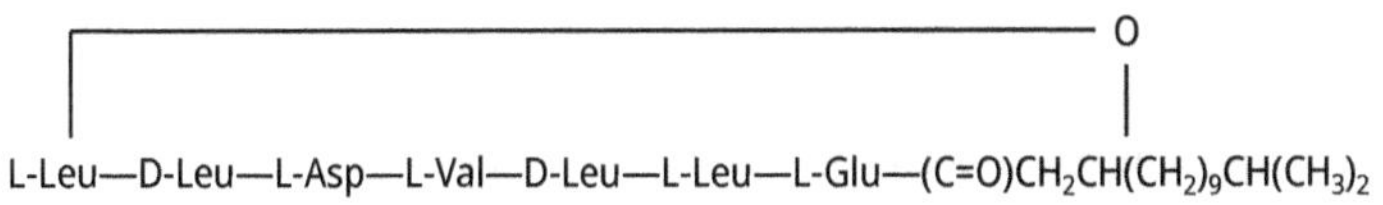

Figure 17.5: A common congener of surfactin. The amino acids make up the head group, and the hydrocarbon chain is the tail.

17.2.4.1 Commercial status

In 2014, Kaneka announced they had succeeded in mass-producing surfactin [52]. They currently market it as a cosmetics ingredient and to undisclosed other applications. Sabo is also involved in a partnership with Kaneka that includes surfactin [53].

17.2.4.2 Other lipopeptides

There are other lipopeptides that might operate as surfactants, for example, iturin, serrawettin (a.k.a. serratamolide), and lichenysin, though this has not been investigated to this author's knowledge. These materials appear to be of more interest as actives than surfactants.

17.2.5 Future options

There are a number of microbial biosurfactants that have been discussed in the academic literature or are used for applications other than as surfactants that will likely be commercialized in the near to not-too-distant future.

17.2.5.1 Cellobiose lipids

CL are a glycolipid that could be introduced in the next few years. The earliest mention that could be found of CL occurred in the mid-1980s and they were immediately classified as biosurfactants [54]. They have a cellobiose – a glucose dimer – headgroup with acyl esterification of some hydroxy groups and a ~ 16-carbon tail, with a terminal α-hydroxy carboxylic acid. With a pK_a of ~ 3.5, CL will be nonionic at highly acidic pH and anionic at neutral and alkaline pH.

An interesting aspect of CL with regard to commercialization is that they are produced by the same organisms as MEL. Thus, the near-term commercial providers of MEL mentioned above will have institutional knowledge about their producing organism, which may allow them to commercialize CL, as well.

17.2.5.2 Trehalose lipids

Trehalose lipids (TL) are a glycolipid that are likely further back in the pipeline, even though they discovered natural product earliest. They were first reported by Anderson and Newman in 1933 [18]. They appear to be first classified as biosurfactants around 1980 [55]. The headgroup of TL is a trehalose moiety, which is another glucose dimer. The tail is composed of two comparatively long-chain (>18 carbons) fatty acids. They have been found to be produced by a variety of organisms, including *Mycobacteria, Corynebacteria, Pseudomonas, Arthrobacter, Rhodococcus,* and *Micrococcus.* One potential issue of trehalse lipids is they are a component of the cell wall in many producing species, which would make them more difficult to isolate in downstream processing than excreted products, like all those listed above [23]. CL are not known to be commercially available, today.

17.2.5.3 Flavolipids

Even further back in the pipeline is flavolipids, which have a unique combination of head group and tails. Badour et al. [56] describe flavolipids as having a headgroup made up of a citric acid and two cadaverine moieties and two moderate-length hydrocarbon tails. These are also referred to in the literature as siderolipids, but flavolipids appear to be the more common usage. They were first reported and classified as biosurfactants in 2004 by Bodour et al. [56]. Titers in the initial report are 0.1 g/L or less, but as the authors note, could likely be improved with optimization. No production optimization studies have been found in the literature. In fact, the amount of interest in flavolipids appears to be low. What is of interest here is that as recently as 20 years ago, new microbial biosurfactants were still being discovered. The implication is that there are likely more out there, and one of them, including flavolipids, could be commercializable.

17.2.5.4 Polymer biosurfactants

Microbially produced polymers that have surfactant properties can be difficult to characterize. Examples include polysaccharide–protein–fatty acid complexes and lipopolysaccharide-protein complexes. Because they are poorly characterized and are not marketed as surfactants, they are not discussed here in any more detail, with one exception, emulsan.

Emulsan is a well-characterized polymer that has repeating hydrocarbon chains pendant from a glycopolymer backbone. There are also carboxylic acid moieties every three units on the glycopolymer backbone. The pK_a of the carboxylic acid is reported to be 3 [57], meaning that the polymer chain will have some anionic character at pH as low as

1.5. At pH above ~5, it would be expected that the carboxylic acids are nearly 100% deprotonated, producing one negative charge for every three sugar units at pH >5.

While emulsan is a well-characterized microbial biosurfactant, it is not known to be commercially available today. That could change in the future.

17.3 Perspective

It appears that in the last few years, after decades of expectation, microbial biosurfactants are finally becoming available in large commercial quantities. Specifically, SL have been available for the last several years, and RL are entering the market at the time of this writing, in 2022. There is also a healthy pipeline of other biosurfactants that can grow their current small market share or be introduced in the future.

Notably, the large capital investments by Evonik and Stepan – and partnerships announced by BASF and Dow – are a sign that global leaders in the surfactant market believe there is a future in microbial biosurfactants.

Because microbial biosurfactants are effectively new ingredients, they will likely open up new end-uses or improve the functionality of current products that use surfactants. Additionally, because of their positive ecological profile, they can lower the environmental impact of the industry.

The next generation of biosurfactant research and application is expected to be strongly shaped by advances in machine learning (ML) and data-driven approaches. While microbial surfactants already hold great promise as biodegradable, low-toxic, and sustainable alternatives to synthetic surfactants, their industrial adoption has been hindered by challenges in strain selection, process optimization, yield improvement, and cost-effectiveness [58]. ML models trained on genomic, metagenomic, and metabolomic datasets offer transformative opportunities to address these limitations [59]. Also, ML-based tools (e.g., deep learning and Bayesian optimization) can predict the influence of pH, temperature, nutrient ratios, and aeration on biosurfactant yield and quality [60].

Also, integrating ML with techno-economic analysis and life cycle assessment will help evaluate the environmental and economic impacts of large-scale biosurfactant production [61, 62]. This can guide industries toward circular bioeconomy models, emphasizing waste-to-value approaches such as using agricultural or industrial residues as feedstocks.

This chapter helps look to the future of microbial biosurfactants with bubbly enthusiasm!

References

[1] Myers D. Surfactant Science and Technology, John Wiley & Sons, Inc.; 2006.
[2] Verbeek H. Historical review in surfactants. In Consumer Products: Theory, Technology and Application., Springer-Verlag; 1987. pp. 1–4.
[3] Sakac N, Markovic D, Sarkanj B, Madunic-Cacic D, Hajdek K, Smoljan B, Jozanovic M. Direct potentiometric study of cationic and nonionic surfactants in disinfectants and personal care products by new surfactant sensor based on 1,3-dihexadecyl-1h-benzo[d]imidazole-3ium. Molecules 2021;26:1366–1381. doi: 10.3390/molecules26051366.
[4] Letcher RJ, Chu S, Smyth S-AJJOHM.. Side-chain fluorinated polymer surfactants in biosolids from wastewater treatment plants, 2020. 388, pp. 122044.
[5] Zargar AN, Srivastava P.. Biosurfactants: Sustainable alternatives to chemical surfactants. In Industrial Applications of Biosurfactants and Microorganisms, Elsevier; 2024. pp. 425–436.
[6] Silva MDGC, et al, Biosurfactants: Promising biomolecules for agricultural applications, 2024. 16, 1, pp. 449.
[7] Veeramanoharan A, Kim S-CJRA.. A comprehensive review on sustainable surfactants from CNSL: Chemistry, key applications and research perspectives, 2024. 14, 35, pp. 25429–25471.
[8] Das R, et al, A comprehensive review on the creation, description, and utilization of surfactants containing multiple hydroxyl groups, 2024. 2024, 1, pp. 6120535.
[9] Moghimipour E, Saponin HS.. Properties, methods of evaluation and applications. Annu Res Rev Biol 2015;5(3):207–220. doi: 10.9734/ARRB/2015/11674.
[10] Porter M. Handbook of Surfactants, 2nd edition, Blackie Academic & Professional; 1991.
[11] Thakur V, et al. Recent progress in microbial biosurfactants production strategies: Applications, technological bottlenecks, and future outlook. 2024;408:131211.
[12] Dhuldhaj UP, et al. Application of Microbial Biosurfactants and Factors Needed for their Production: An Overview, 2025. pp. 1–25.
[13] Solomon AA, Vishnu DJRIE.. Integrated strategies for biosurfactant production and scale-up: Advances in fermentation engineering and computational modelling, 2025. pp. 106853.
[14] Ou-douaou A, et al. Advances in microbial biosurfactants for mineral flotation: Towards sustainable phosphate ore beneficiation, 2025. 76, pp. 108267.
[15] Bueno-Mancebo J, et al. Preliminary study on novel sustainable production of mannosylerythritol lipids by solid-state fermentation, 2025. 38, pp. 104144.
[16] Anderson K. Folding of outer membrane proteins in the anionic biosurfactant rhamnolipid. Febs Lett 1960;588(10):1955.
[17] Jarvis F, Johnson M. A glyco-lipid produced by Pseudomonas aeruginosa. J Am Chem Soc 1949;71 (12):4124–4126.
[18] Anderson R, Neman M. The chemistry of the lipids of tubercle bacilli: Xxxiii. isolation of trehalose from the acetone-soluble fat of the human tubercle baillus. J Biol Chem 1933;101(2):499–504.
[19] Desai J, Banat I. Microbial production of surfactants and their commercial potential. Microbiol Mol Biol Rev 1997;61(1):47–64.
[20] Gorin P, Spencer J, Tulloch A. Hydroxy fatty acid glycosides of sophorose from torulopsis magnoliae. Can J Chem 1961;39(4):846–855.
[21] Inoue S, Ito S. Sophorolipids from torulopsis bombicola as microbial surfactants in alkane fermentations. Biotechnol Lett 1982;4(1):3–8.
[22] Pekin G, Vardar-Sukan F, Kosaric N. Production of sophorolipids from Candida bombicola ATCC 22214 using Turkish Corn oil and honey. Eng Life Sci 2005;5(4):357–362. doi: 10.1002/elsc.200520086.
[23] Soberon-Chavez G, editor. Biosurfactants from Genes to Applications, Springer; 2011.
[24] McCoy M. American cleaning institute's 2015 conference. In Chem Eng News, 2015.

[25] Salek K, Euston S. Sustainable microbial biosurfactants and bioemulsifiers for commercial exploitation. Process Biochem 2019;85:143–155. 101016/j.procbio.2019.06.027.
[26] Tullo A. Evonik industries is c&en's company of the year for 2016. Chem Eng News 2017.
[27] Reisch M. Givaudan will buy ingredients maker. Chem Eng News 2104.
[28] Bomgardner M. Biobased chemicals advance. Chem Eng News 2014.
[29] Culliney K. UK biotech startup holiferm eyes €8m for commercial biosurfactant upscale. Cosmet Des Eur 2021.
[30] Hatmaker J. Biosurfactant company signs for 90,000 sq. ft. Wallasey industrial, Place North West; January, 10, 2022.
[31] Bettenhausen C. BASF invests in biosurfactants. Chem Eng News 2021.
[32] Bettenhause C. Dow signs deal with locus performance ingredients for sophorolipids. Chem Eng News 2022.
[33] Happi https://www.happi.com/live-from-shows/in-cosmetics/2022-04-05/dow-introduces-high-performing-and-sustainable-technologies (accessed 2022–06–02).
[34] Kulakovskaya E, Kuakovskaya T. Prospects of practical application of Sophorolipids, Cellobios Lipids, and Mels. In Kulakovskaya E, Kuakovskaya T, editors. Extracellular Glycolipids of Yeasts, Academic Press; 2014. pp. 75–83. doi: 10.1016/B978-0-12-420069-2.00006-6.
[35] Lohitharn N.. Enhanced Production of Rhmanolipids Using at Least Two Carbon Sources., US 11 142 782 B2, 2021.
[36] US Environmental Protection Agency Office of Pesticide Programs. Biopesticides registration action document hamnolipid biosurfactant, March 23 2004.
[37] Growing Produce. https://www.growingproduce.com/fruits/whats-new-for-2022-in-biological crop protection-products/ (accessed 2022–06–02).
[38] Biotech J. Inc Product Offerings., https://www.jeneilbiotech.com/biosurfactants. accessed 2022–06–02.
[39] Stuart E. Evonik to present innovations at in-cosmetics global 2018. Glob Cosmet Ind.
[40] Bettenhausen C. Evonik invests in rhamnolipid biosurfactants. Chem Eng News 2022.
[41] Happi. https://www.happi.com/contents/view_breaking-news/2022-03-17/evonik-launches-rewoferm-rl-100 (accessed 2022–06–02).
[42] Bettenhausen C Stepan buys biosurfactant business from logos. Chem Eng News 2020.
[43] Tullo A. Stepan makes a pair of acquisitions. Chem Eng News 2021.
[44] Bettenhausen C. Switching to sustainable surfactants. Chem Eng News 2022.
[45] PR Newswire. https://www.prnewswire.com/news-releases/advanced-biocatalytics-is-pleased-to-announce-the-commercial-release-of-bioss-rl-our-new-rhamnolipid-based-surfactant-and-part-of-our-fermactants-product-line-301293551.html (accessed 2022–06–02).
[46] Bettenhausen C. Rhamnolipids rise as a green surfactant. Chem Eng News 2020.
[47] MarkNature. https://www.marknature.com/products/rhamnolipid-biosurfactantandhttps://www.marknature.com/products/purify-rhamnolipid?pr_prod_strat=copurchase&pr_rec_id=8d88a6c91&pr_rec_pid=7216096149679&pr_ref_pid=7030679896239&pr_seq=uniform (accessed 2022–06–02).
[48] Kawashima H, Nakahara T, Oogaki M, Tabuchi T. Extracellular production of a mannosylerythritol lipid by a mutant of Candida sp. from n-alkanes and triacylglycerols. J Ferment Technol 1983;61 (2):143–149.
[49] Toyobo Cosmetic Ingredients Department. https://www.toyobo-global.com/products/cosme/cate gory/ceramela/index.html) (accessed 2022–06–02).
[50] PR Newswire. https://www.prnewswire.com/news-releases/abc-opens-commercial-mel-biosurfactant-production-in-irvine-ca-301549977.html (accessed 2022–06–02).

[51] Arima K, Kakinuma A, Tanura G. Surfactin, a crystalline peptide lipid surfactant produced by Bacillus subtilis: Isolation, characterization and its inhibition of fibrin clot formation. Biochem Biophys Res Commun 1968;31(3):488–494.

[52] Org. PHYS. https://phys.org/news/2014-10-drastic-reduction-synthetic-surfactants-power.html (accessed on 2022–06–02).

[53] COSSMA. https://www.cossma.com/business/article/sabo-partners-with-kaneka-34277.html (accessed 2022–06–02).

[54] Frautz B, Lang S, Wagner F. Formation of cellobiose lipids by growing and resting cells of Ustilago maydis. Biotechnol Lett 1986;8:757–762. doi: 10.1007/BF01020817.

[55] Cooper D, Zajic J. Surface-active compounds from microorganisms. Adv Appl Microbiol 1980;26:229–253. doi: 10.1016/S0065-2164(08)70335-6.

[56] Bodour A, Guerrero-Barajas C, Jiorle B, Malcomson M, Paull A, Somogyi A, Trinh L, Bates R, Maier R. Structure and characterization of lavolipids, a novel class of biosurfactants produced by flavobacterium sp. Strain MTN11. Appl Environ Microbiol 2004;70:114–120. doi: 10.1128/AEM.70.1.114-120.2004.

[57] Gutnick D. The emulsan polymer: Perspectives on a microbial capsule as an industrial product. Biopolymers 1987;26(S0):S223–S240. doi: 10.1002/bip.360260020.

[58] Sharma N, et al, A comprehensive review on microbial production and significant applications of multifunctional biomolecules: Biosurfactants, 2025. 36, 2, pp. 26.

[59] Pandit NK, Meena SSJW, Valorization B.. Exploring sustainable biosurfactant production through waste valorization. Emerg Res Trends Ind Appl 2025;1–35.

[60] Vares CA, et al, Machine Learning Strategies for Forecasting Mannosylerythritol Lipid Production Through Fermentation: A Proof-of-Concept, 2025. 15, 7, pp. 3709.

[61] Noll P, et al. Limits for sustainable biosurfactant production: Techno-economic and environmental assessment of a rhamnolipid production process, 2024. 25, pp. 101767.

[62] Bippus L, et al. Life cycle assessment for early-stage process optimization of microbial biosurfactant production using kinetic models – A case study on mannosylerythritol lipids (MEL), 2024. 12, pp. 1347452.

Jacyr Quadros

Chapter 18 Epoxidation of vegetable oils and their use in plasticized polyvinyl chloride (PVC) applications

This updated chapter covers the chemistry of the epoxidation of vegetable oils, focusing on the chemistry of epoxidized soybean oil (ESO) and its derivatives. In this, we also present a review of the use of these materials as plasticizers in flexible polyvinyl chloride (PVC) applications.

18.1 Chemistry of epoxidation

Epoxidation is a commonly used reaction to promote the modification of vegetable oils to generate chemicals that are useful in various applications, such as flexible PVC, polyurethanes, coatings, etc.

18.1.1 Epoxidation mechanism

The epoxidation of vegetable oils has been known for a long time. In the first experiments, the main reagent was peracetic acid [1], but the use of formic acid and the "in situ" formation of performic acid [2] have been used industrially as the preferred method due to its better conversion and improved safety.

The epoxidation is described as a two-phase reaction [3] in Figure 18.1.

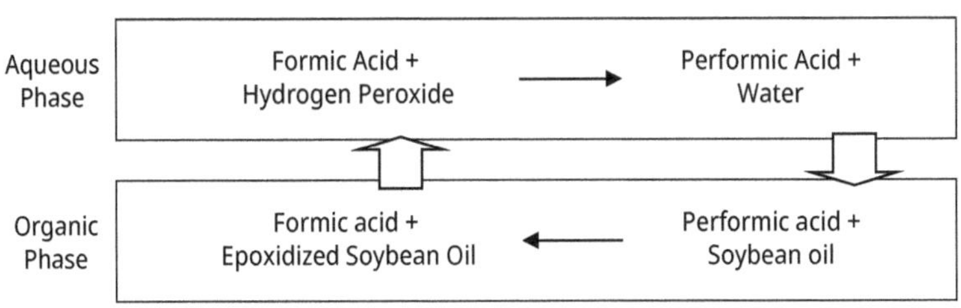

Figure 18.1: Two-phase reaction for epoxidation.

Jacyr Quadros, Innoleics Serv. e Cons. Ltda., São Paulo, Brazil, e-mail: jquadros@innoleics.com

 | https://doi.org/10.1515/9783112212141-018

These reactions may be represented in Figure 18.2. A single example of a molecule found in soybean oil (linolenic triglyceride) has been chosen for simplification.

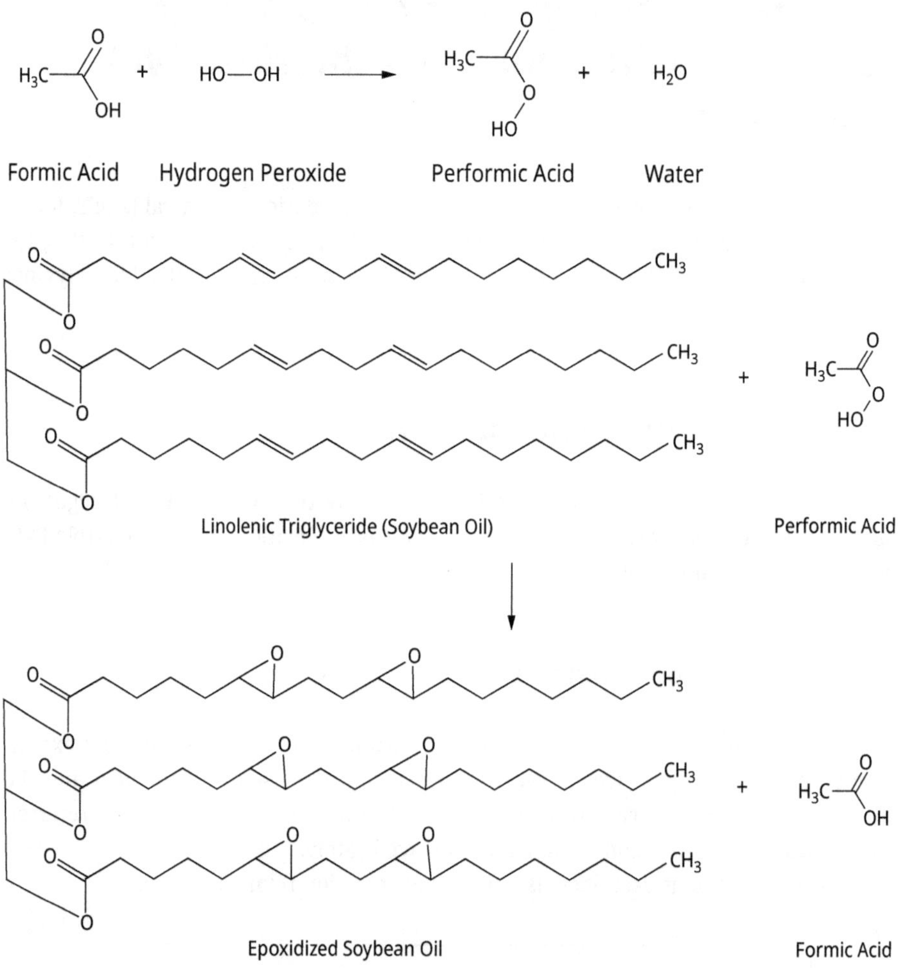

Figure 18.2: First part of epoxidation reaction representations.

Mass transfer between aqueous and organic phases is a limiting factor for the apparent kinetics of the reaction. Thus, a significant level of agitation and breakdown of droplets is critical.

At the same time, concurrent reactions that promote undesired ring-opening happen, as indicated in Figure 18.3. These reactions are intensified by the presence of acids or excess hydrogen peroxide.

The main application for epoxidized soybean oil and derivatives is plasticizers for PVC and the oxirane index is directly related to the product's compatibility with

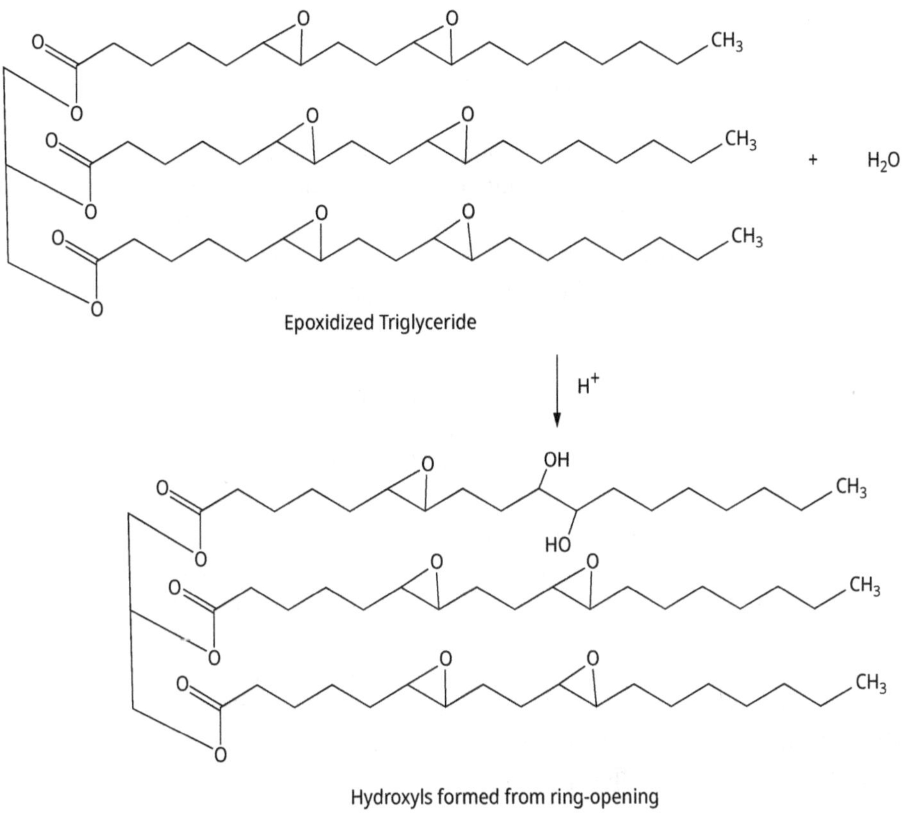

Figure 18.3: Second part of epoxidation reaction representations.

the resin. The oxirane index, ASTM D 1652 method, measures the weight percent of oxirane ring oxygen atoms in a given molecule. It indicates the conversion of the epoxidation process. Double bonds are undesirable, as they are inversely related to the long-term stability of the epoxidized molecules [4]. A complete epoxidation of the double bonds while minimizing the ring-opening reaction is fundamental to maximizing plasticizer compatibility and permanence.

The epoxidation reaction is highly exothermic [5], and controlling the temperature is one of the key considerations in the commercial manufacture of epoxidized vegetable oils.

18.1.2 Industrial experience

The traditional industrial process relies on a controlled addition of hydrogen peroxide to the premixed formic and soybean oil. Intense agitation is fundamental to ensure good mass transfer between phases and maximize heat transfer between reaction media and cooling devices.

The ingredients are most commonly refined soybean oil, formic acid at 85% purity, and hydrogen peroxide (ideally greater than 60% in water). Higher concentrations of hydrogen peroxide allow better conversion and higher reaction rates. Some producers preheat the mix before starting the addition of hydrogen peroxide, but that is not necessary, as the reaction is highly exothermic even at room temperature. The reaction times vary from 7 to 18 h, depending on reactor size and heat exchange capacity.

The intensification of the heat removal through increased agitation and greater exchange surface areas allows a faster addition of the peroxide, resulting in improved conversion and a reduced reaction time [5].

Once the reaction has achieved its optimum conversion, where the maximum oxirane index is obtained, it is important to stop the reaction; otherwise the ring-opening reactions will continue and decrease the oxirane index. Stopping the reaction can be done by adding water to dilute the hydrogen peroxide, cooling the reaction media, or ensuring quick phase separation.

After the reaction is finished, the next step is phase separation, which is typically done by decanting, normally requiring from 12 to 24 h depending on vessel volume, or by centrifuging.

Once the phases are separated, the residual volatiles – water, peroxide, and formic acid – are removed by flash distillation followed by stripping in vacuum. Temperatures should ideally be maintained below 120 °C, a general rule to avoid additional ring opening during these final processes.

18.1.3 Latest academic developments (state of the art)

The most promising area seems to be process intensification, such as micro-reactors [6], microemulsion [7], new catalysts [8], or addition of all reactants at once [5]. Enzymatic catalysis [9] has also been studied, demonstrating significant reduction of undesired ring-opening reactions but with significant increases in reaction times and process complexity.

In recent years, researchers have leveraged advances in process intensification and the development of new simulation tools to evaluate continuous processes [10, 11], and the kinetics of epoxidizing the methyl esters of vegetable oils [12].

It is important to note that epoxidation does not necessarily yield a material that will be suitable as a chemical intermediate. The total oxirane content, the remaining unreacted double bonds, and the formation of hydroxyls impacts directly on the performance of the final product in various applications.

Different oils can be epoxidized, such as canola, corn, sunflower, etc. The resulting materials are dependent on the available double bonds. Generally, it is expected that the oils with the highest contents of unsaturation will yield more compatible materials, which attain superior performance in most applications.

18.1.4 The relevance of soy-based solutions

Soy has been the vegetable oil of choice for epoxidation, as a combination of various factors: First, its composition provides an appropriate level of unsaturation to obtain materials that are technically viable, especially in the plasticizers segment. Second, soybean oil is a commodity with reasonable pricing, providing final products with competitive costs. Third, soybean oil is produced on a very large scale, in two major global regions (South America and North America), and chemicals represent less than 1% of total demand share, providing supply security. Fourth, soybean demand is majorly affected by the demand for protein (soybean meal), and any increase in the demand for soybean oil would have a cost reduction impact on food, since most of the soy-dependent food derivatives is associated with the soybean meal price, as demonstrated by a study at Purdue University [13].

It is important to keep in mind that there are several initiatives to evaluate alternative sources of triglycerides that would provide even better viability and sustainability features to bio-based epoxides [14].

18.2 Transesterification

Transesterification is one of the preferred methods to further modify epoxidized vegetable oils to obtain useful chemicals by breaking down the triglycerides with alcohols. These materials can be used as primary plasticizers for PVC and offer improved performance as intermediates for other polymers, as lubricants, or even as additives for agricultural chemicals.

18.2.1 Other esterification technologies

There are several technologies available to split the triglycerides of vegetable oils. But transesterification has proven to be the most effective in terms of conversion, process simplicity, and cost.

18.2.2 Transesterification mechanism

Transesterification is a reaction where triglycerides react with a mono-alcohol to form fatty acid esters. The mono-alcohol normally ranges from one (methanol) to nine (isononanol) carbons.

The reaction can be carried out with base or acid catalysts, but base catalysis is preferred, such as sodium methoxide. In the case of the transesterification of epoxidized oils, acid catalysts could damage the oxirane rings, as discussed previously in this chapter.

For base catalysis, low humidity and low acidity of all reactants are fundamental to avoid poor conversion and undesirable soap formation.

The three-step reaction is represented in Figure 18.4.

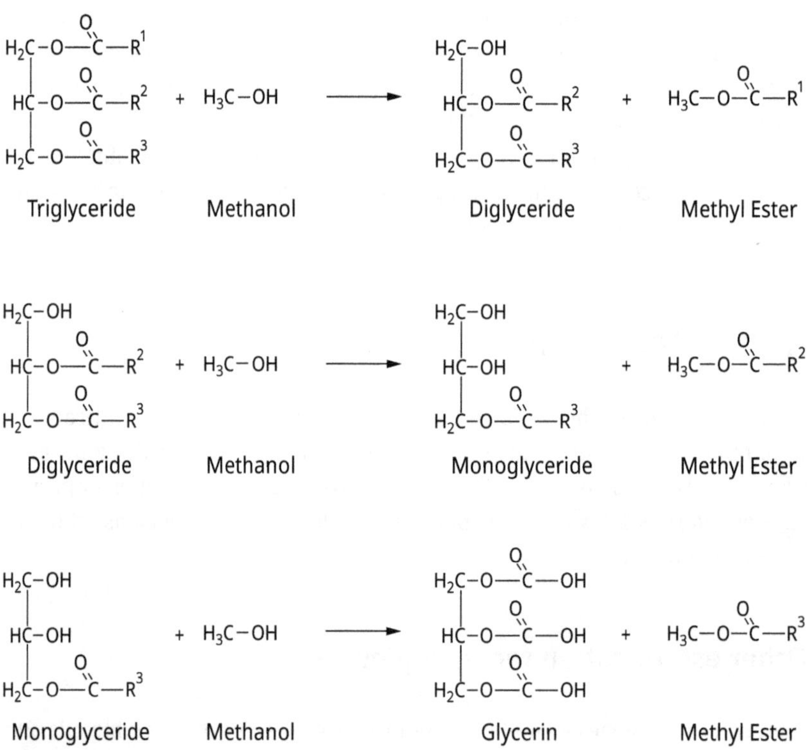

Figure 18.4: Three-step reaction for transesterification from triglyceride and mono-alcohol to glycerine and methyl ester.

18.2.3 Equilibrium

Transesterification is a reversible equilibrium reaction. This equilibrium is dependent on reaction time and initial ratios of reactants.

The reaction is intensified by agitation, as methanol and glycerine have low solubility in the glycerides and esters. When carried out at 60 °C, excess methanol, and intense agitation, such as a stirred reactor with high shear or ultrasound cavitation, the reaction achieves equilibrium in a few minutes, and the reaction media changes from a light yellow to a darker, reddish tone. After setting, the glycerine sets to the bottom and can be separated. From the top layer is then removed the excess alcohol by distillation, which can be reutilized. After the distillation, the final product is stored.

Should the reaction be interrupted earlier than at equilibrium or should there be a lower ratio of alcohol to triglycerides, the reaction will produce a combination of esters and tri-, di-, and monoglycerides. The proportions of these molecular species depend on the reaction conditions. The chart in Figure 18.5 shows a representation of the conversion over time for an excess ratio of methanol.

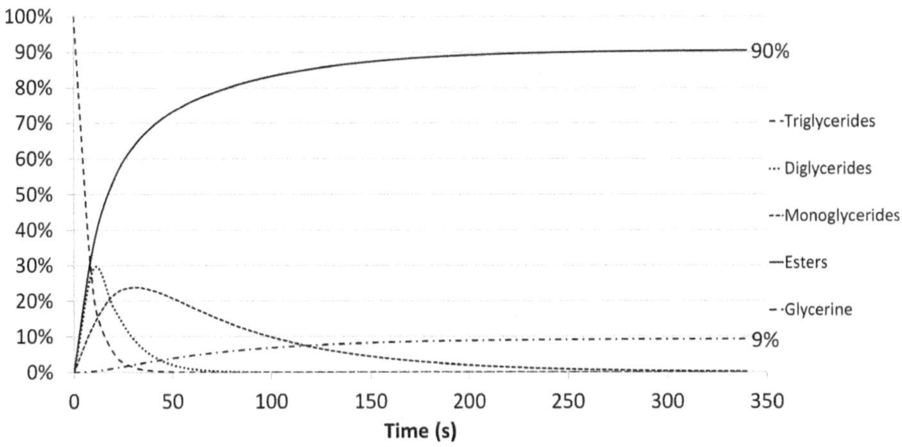

Figure 18.5: Percentage over time for triglycerides, diglycerides, monoglycerides, esters, and glycerine with excess methanol.

18.3 Acylation

Acylation or acetylation is a common reaction in the chemistry of vegetable oil and animal fat-based raw materials. The conversion of available hydroxyl groups into acetyl esters is instrumental in modifying molecules to achieve desired performances and behavior, as well as improving overall stability. In the case of PVC applications, hydroxyls are known to cause compatibility problems.

18.3.1 Acylation mechanism

The reaction is represented in Figure 18.6, with the acylation of a diglyceride as an example:

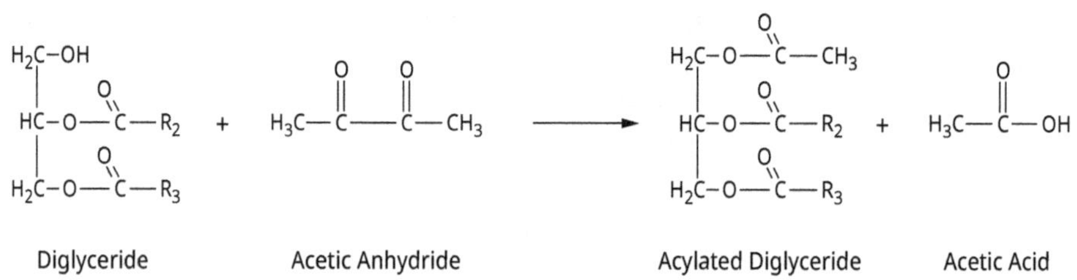

Figure 18.6: Acylation of a diglyceride.

18.3.2 Industrial experience

The most commonly applied method of acylation, widely adopted in the industry, is the reaction of available hydroxyl groups with acetic anhydride carried out without the need for a catalyst.

After the reaction is concluded, the removal of excess acetic anhydride and residual acetic acid is done by distillation. The removal of these acids is fundamental to ensuring the final material's quality and long-term stability.

Once all higher volatility components are properly removed, the product is finished and stored.

18.4 Bio-based plasticizers

18.4.1 Bio-plasticizers gain momentum

The flexible PVC industry has an interesting opportunity to increase the contents of bio-based materials, promoting a reduction of the carbon footprint in the supply chain. Many segments, such as automotive, footwear, flooring, and synthetic leather, are introducing low carbon or even negative carbon footprint materials as part of their regular product lines.

This movement presents a great opportunity for PVC formulators, as bio-based plasticizers can be used to achieve these goals at a competitive cost.

18.4.2 Bio-based plasticizer generations

Epoxidized materials have been used as plasticizers for PVC since the early stages of flexible PVC technology. Plasticizers are normally high molecular weight esters with solubility parameters within an ideal range, promoting plasticity and flexibility to the PVC resin, reducing hardness.

The first vegetable-based PVC plasticizers used were simple epoxidized oils, such as soybean oil and linseed oil. These materials, considered the first generation of epoxy plasticizers, have presented an acceptable performance in some applications. Nevertheless, due to cost and performance limitations when compared to phthalate plasticizers, epoxidized plasticizers were relegated to a secondary role in the business. Epoxidized soybean oil (ESO) had been used typically as a co-stabilizer, applied at low concentrations, below 2%, in PVC compounds, or at most as a secondary plasticizer. Although presenting superior performance, epoxidized linseed oil was commercially limited due to its higher cost. Despite these limitations, epoxy derivatives represented the largest volume of noncyclic plasticizers in the 1970s [4].

It was only in the 2000s that epoxy derivatives presented a significant penetration in the PVC plasticizers market, mainly due to increased interest in sustainability and increased petroleum-based materials costs. At the same time, ESO producers significantly improved product quality, enabling its use at higher concentrations.

As a consequence of these favorable market conditions, other epoxy derivatives also started to gain momentum, especially epoxidized esters of soybean oil, due to a good combination of cost and performance. With improved compatibility, epoxidized esters of lower alcohols (methyl, ethyl, butyl, amyl, and 2-ethyl-hexyl) [15] entered the market as primary plasticizers, establishing what can be considered the second generation of epoxy derivatives. There were still a few limitations when compared to traditional phthalate plasticizers.

More recently, additional developments have introduced even better performing materials, such as epoxidized acylated mono- and diglycerides of soybean oil [16].

These third-generation materials are replacing existing petroleum-based plasticizers with little performance deficits.

As a separate group, specialty bio-based materials such as acylated castor oil esters [17] present superior performance but are limited due to their higher cost.

18.4.3 Performance comparisons

Vegetable-based plasticizers can be considered primary plasticizers for flexible PVC applications. As these are molecules with different structures and functionalities when compared to traditional general-purpose plasticizers such as di-isononyl phthalate (DINP) or dioctyl terephthalate (DOTP), there are naturally some performance differences. In most applications, these differences are either not significant or adequately mitigated by formulation adjustments or special additives.

The chart, Figure 18.7, shows some of the key parameters observed by formulators in the flexible PVC industry. A third-generation vegetable plasticizer with a 750-Da molecular weight was chosen for comparison with DOTP and di(2-ethylhexyl) phthalate (DEHP or DOP).

The high efficiency of the vegetable-based plasticizer is readily noticeable. The efficiency is similar to that of DOP, but with a considerably lower volatility. Also, a typical characteristic is noted in the mechanical properties: higher tensile strength while also having a higher elongation at break, which can be observed in final applications, where materials formulated with vegetable-based epoxy plasticizers present a higher resilience.

In addition to these characteristics, compounds made with epoxy plasticizers also present a considerably superior performance in thermal stability, as shown in the picture below (see Figure 18.8), where methyl epoxy soyate is compared to DOP and di-isobutyl phthalate (DIBP). Because of this inherent thermal stability property, PVC compounded with vegetable oil plasticizers requires less potentially harmful and toxic metal soap stabilizers used to ensure PVC heat stability. Such thermal stabilizers are also quite expensive, adding to the overall cost of compounded PVC formulations.

Another characteristic of epoxy plasticizers is viscosity increase in paste PVC applications, which require adjustments or blends with other plasticizers. This is expected as epoxy materials are inherently more efficient and cause the resin particles to swell faster, increasing viscosity over time. The chart in Figure 18.9 shows this effect, comparing traditional plasticizers, including dioctyl adipate (DOA) with octyl epoxy soyate, methyl epoxy soyate, and a 470-Da third-generation vegetable epoxy plasticizer.

Adjustments can be made, depending on the formulation, with a combination of different PVC resins that will have lower viscosity aging or by including lower viscosity plasticizers.

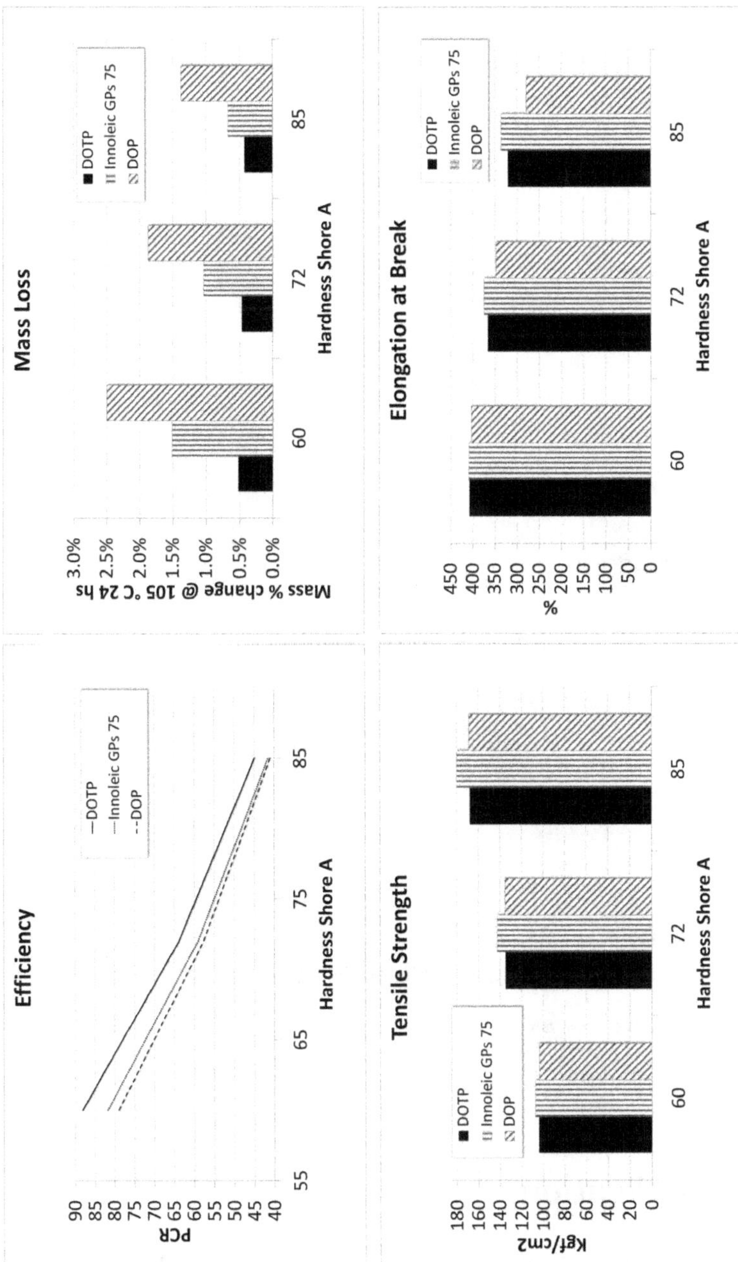

Figure 18.7: Efficiency, mass loss, tensile strength, and elongation at break data as key parameters in flexible PVC.

Figure 18.8: Metrastat thermal stability comparison of DOP, DIBP, and MES.

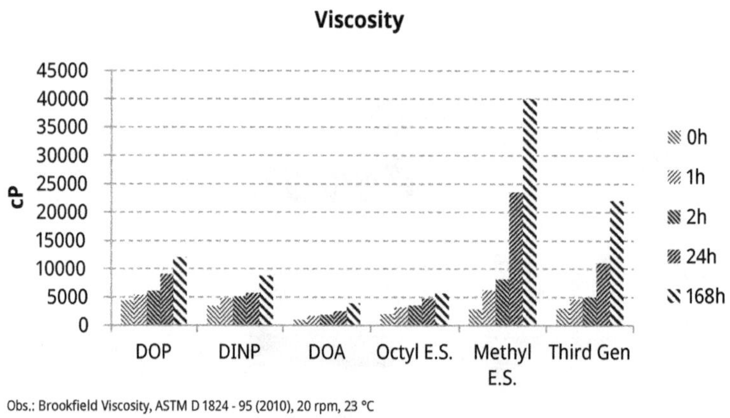

Figure 18.9: Comparison of plasticizers.

18.5 Other applications

Although not as significant as PVC plasticizers, epoxidized vegetable oils are also commercially successful in some other applications. Among these applications, the most notable is the use of epoxy materials as raw materials to produce polyols. The oil is fully epoxidized and the polyol is obtained by the subsequent ring opening of the epoxy rings [18]. These bio-based polyols are primarily used in polyurethane foams, fully or partially replacing fossil-based materials.

The use of epoxidized esters or oils in coatings is also relevant, replacing mostly the lower alcohol phthalates, such as di-butyl phthalate and di-isoamyl phthalate.

There are some emerging areas of applications, such as 3D printing [19, 20], where epoxidized materials are replacing traditional fossil-based polymeric systems with promising results and adequate competitiveness.

18.6 Challenges for growth and wider adoption

18.6.1 Key areas for technical improvements

Epoxy plasticizers present some areas of improvement to enable higher growth and wider adoption.

One of the key sensitivities is higher water extraction when compared to petroleum-based materials. This is an important performance aspect, especially for outdoors and intense water-exposed applications, such as pool liners, hoses, roofing, and awnings. Since this is intrinsic to the polar nature of the epoxidized molecules, additives may play an important role in mitigating this effect. Hydrophobic components, such as silicone oils, have been considered and can be effective, but are limited due to the low compatibility with the PVC matrix. Another solution is to create physical barriers, layering different compositions to minimize the direct contact of water with the bio-based plasticizers.

Another area of concern is hydrolysis [21]. This is a chemical reaction where water breaks the oxirane rings forming hydroxyls, turning the product incompatible with the PVC matrix. The consequence is an accelerated exudation of the plasticizer, generating undesirable surface defects, such as tack or oil build. Although critical in accelerated weathering tests, this reaction is a concern mostly in applications where articles are exposed to high humidity or moisture and high temperatures. In addition, proper processing of the epoxidized material to minimize acidity also significantly reduces the hydrolysis reaction rates.

Finally, a surface effect known as blooming [22] is also a sensitive area for epoxidized materials. This effect is noted in articles with darker colors and glossy finish. Occasionally, when submitted to low temperatures (−5 to 0 °C), a white bloom becomes visible on the surface. If heated to >30 °C, the effect disappears completely. There is no conclusive study of the mechanism that generates this problem, but it could be associated with the presence of saturated fatty acid esters, which have a very high melting point. These molecules, present even at very low levels on the surface of the PVC, not detectable visibly, might nucleate and crystallize, creating the visible whitish bloom. When heated, these molecules return to their liquid state and are not visible. This effect is not easy to replicate in laboratory conditions, as the rate of temperature drop and increase seems to be a key factor for its appearance. High oleic soybean oil, with reduced fatty acid saturates through genetic modification, could potentially mitigate this effect, but the cost of this emerging technology is still a limiting factor.

18.6.2 Marketing and regulatory opportunities

Governments and regulatory agencies are promoting the reduction of carbon emissions through programs and regulations, creating new opportunities for bio-based materials. Companies are now faced with sustainability and emission reduction goals, forcing them to seek process improvements and new materials.

A study [23] published in 2014 shows that the industry was not ready to adopt chemicals from renewable sources. The respondents indicated that the barriers to adopting bio-based materials were, in order of importance: high cost, little market demand, inferior performance, and credibility of sustainability claims. Considering the current scenario for bio-based plasticizers, cost has significantly been reduced; the demand is increasing due to the environmental pressure; performance has improved significantly with the new generations of plasticizers; and sustainable claims can be certified through reliable eco-labels.

A study presented at PVC 2021 [24] concluded that companies with higher awareness about eco-labels have a higher perception of demand and premium value for renewable plasticizers. Also, there was a direct correlation between company size and its perception of demand for bio-based materials.

Younger consumers are also significantly more impacted by sustainability than older ones [25]. In response to this demand, companies are introducing new product lines directed at this public segment that is considerably more likely to change its purchasing behavior toward more sustainable products.

References

[1] Findley TW, Swern D, Scanlan JT Epoxidation of unsaturated fatty materials with peracetic acid in glacial acetic acid solution. J Am Chem Soc 1945;6(3):412–414.

[2] Niederhauser WD, Koroly JE Process for the epoxidation of esters of oleic and linoleic acids. US2485160, USA; 1949.

[3] Santacesaria E, et al. Biphasic model describing soybean oil epoxidation with H2O2 in continuous reactors. Ind Eng Chem 2011;51(26):8760–8767.

[4] Sears J. K, Darby JR. The Technology of Plasticizers, s.l: Wiley; 1982.

[5] de Quadros JV Jr, Giudici R Epoxidation of soybean oil XE "soybean oil" at maximum heat removal and single addition of all reactants. Chem Eng Process Intensif 2016;100:87–93.

[6] Novel process window for the soybean oil XE "soybean oil" epoxidation-Simulation of high-p, T processing microreactors. In: Cortese B, De Croon MHJM, Hessel V. 8th European Congress on Chemical Engineering (ECCE-8), Berlin, Germany: s.n; 2011. pp. 42.

[7] Epoxidation of Soybean Oil in a Microemulsion-Assisted Environment. Rethwisch D et al. Cincinnati, OH, USA: s.n., 2005. The 2005 Annual Meeting.

[8] Wai PT, Jiang P, Shen Y, Zhang P, Gu Q, Leng Y Catalytic developments in the epoxidation of vegetable oils and the analysis methods of epoxidized products. RSC Adv 2019;9(65):38119–38136.

[9] Vlček T, Petrović ZS. Optimization of the chemoenzymatic epoxidation of soybean oil XE "soybean oil, 2006. Vol.83, 3, pp. 247–252.

[10] Olivieri GV, de Quadros JV Jr, Giudici R Evaluation of potential designs for a continuous epoxidation reactor. Ind Eng Chem Res 2021;60(39):14099–14112.

[11] Cogliano T, Turco R, Di Serio M, Salmi T, Tesser R, Russo V. Epoxidation of vegetable oils via the Prilezhaev reaction method: A review of the transition from batch to continuous processes. Ind Eng Chem Res 2024;26:63.

[12] Meng Y, Ferreira EN, Maria Pontes Silva Ricardo N, Kebir N, Leveneur S. Epoxidation showdown: Unveiling the kinetics of vegetable oils vs Their Methyl Ester counterparts. Ind Eng Chem Res 2024;44:63.

[13] Lusk JL Food and Fuel: Modeling Food System Wide Impacts of Increase in Demand for Soybean Oil , 2022.

[14] Towards a biorefinery processing Sargassum seaweed: Techno-economic assessment of alginate and fucoidan production through SuperPro Designer® Designer® process simulation. Ramírez-Partida AE, García-Cayuela T, Amador-Castro LF, Alper HS, Carrillo-Nieves D In: Environmental Technology & Innovation, 2024. 34, 103587.

[15] Benecke HP, Vijayendran B, Elhard JD. Plasticizers derived from vegetable oils. US6797753B2, USA; January 31, 2002.

[16] Quadros J Jr., Carvalho JAD. Plasticized PVC composition. US 8,623,947, USA; 01 07, 2014.

[17] Galanakis CM, editor Bio-based products and industries, s.l.: Elsevier; 2020. pp. 390.

[18] Yufeng M, PhD, et al. Recent advances in vegetable oil based fine chemicals and polymers, s.l.: Emerald Publishing; 2024. ISSN, pp. 2049–1220.

[19] Thorbole A, Bodhak C, Sahu P, Gupta RK. Green ink revolution: Pioneering eco-friendly 3D printing with vegetable oil-derived inks. Polym Eng Sci 2024;64:12.

[20] Alarcon RT, Porcarello M, Cellai A, Schmitt CC, Sangermano M. Advancing green additive manufacturing: Epoxidized vegetable oils for room-temperature Photopolymerization 3D printing. Polymer 2025;128868.

[21] Guo Y, Hardesty JH, Mannari VM, Massingill JL Hydrolysis of epoxidized soybean oil XE "soybean oil" in the presence of phosphoric acid. J Am Oil Chem Soc 2007;84(10):929–935.

[22] Lauridsen CB, Hansen LW, Brock-Nannestad T, Bendix J, Simonsen KP A study of stearyl alcohol bloom on Dan Hill PVC dolls and the influence of temperature. Stud Conserv 2017;62(8):8 445–455.

[23] Special Chem. Polymer-Additives. [Online] February 1, 2014. [Cited: April 7, 2022.] https://polymer-additives.specialchem.com/tech-library/article/unquestionable-verdict-for-newer-bio-based-chemicals-or-additives-no-thanks-too-expensive.

[24] Eco-Labels, Applicability and Impact on the Flexible PVC Sector, Quadros JQ, Garcia-Tousa DO-L. s. n.; 2021.

[25] Nielsen. Global consumers seek companies that care about environment issues. Nielsen. [Online] 2018. https://www.nielsen.com/us/en/insights/article/2018/global-consumers-seek-companies-that-care-about-environmental-issues/.

Gunnar Lynum, Steven Lynum

Chapter 19 The positive impact bio-solutions have on the global construction industry

Addressing the COP26 Conference's preliminary opening on a Sunday afternoon, Abdulla Shahid, the President of the UN General Assembly called for an honest look at where things stand: "We have had decades to argue the facts about climate change, about the power of renewables, about the fine details of monitoring or cost-sharing. Yet, we have still failed to act with the conviction and determination required" [1]. In this chapter, SMD Products Company Inc. would like to share how bio-based solutions can have a positive impact on the global construction industry. As a small private enterprise, it is sometimes easier to impact change, thanks to public policy initiatives, as smaller enterprises are sometimes missing the large bureaucratic red tape that sometimes blocks innovation.

19.1 Introduction

The goal of this chapter is to walk the reader through one of the greatest environmental challenges we face as a civilization. Modern day teaching focuses on the environmental impact of several industries, which include the sectors of energy, transportation, mining, manufacturing, and so on. But one that is often overlooked is the one that impacts every person on this earth, namely the construction industry. Without buildings or modern-day structures, one can only fathom how the world would have evolved. It is believed that about 1.8 million years ago humans may have lived in what may have been the first "house" in Tanzania in Olduvai Gorge, and started heading down the path, which is now often referred to as environmental destruction, which in turn has been analyzed to exhaustion [1]. If one is interested in archeology, there are many sources that are highlighted at the end of this chapter. For avoidance of doubt, the authors of this chapter want to be clear to the reader that we are capitalists, realists, and not tenured environmentalists. We believe in free markets, the modernization of the world, and know this can be done in a sustainable and environmentally friendly manner, thanks to bio-solutions that have evolved in recent years. It is important to set the stage and understand the global impact that industrialization has

Gunnar Lynum, Steven Lynum, SMD Products Company, Inc. and Strategic Market Development, LLC P.O. Box 1634, Manchester, MO 63021, e-mail: customerservice@smdproductscompany.com, www.smdproductscompany.com

 | https://doi.org/10.1515/9783112212141-019

had on the environment. It is also important to recognize the commonsense solutions that have occurred and continue to occur, thanks to the unified efforts of several industrialized nations that are trying their utmost to protect the world for future generations, as most recently observed at COP26 in Glasgow, Scotland [2]. We will explore, propose, and share the new way of doing things. It was Peter F. Drucker who stated, "If you want something new, you have to stop doing something old." Let us now proceed and look at something new, but first, by understanding how we got here and why we are proposing something new, for example using bio-based solutions (see images of a soybean field in Figure 19.1 and a bio-based product label in Figure 19.2).

Figure 19.1: Soybean field in the USA, Autumn 2021 (courtesy of Gunnar Lynum).

19.1.1 The history of construction (i.e., concrete)

The first evidence of the use of a concrete-like material being used dates to roughly 3,000 BC. The Egyptian pyramids are a perfect example of how a material, unlike modern-day concrete, was used to build solid structures. It can be argued that there must have been other structures built prior to the pyramids using a concrete-like substance, but for interest of getting to the end of our story, let us start with the pyramids of Egypt as our jump-off point when trying to understand the impact that construction has had on all of us. It is believed that the Egyptians' definition of concrete was a unique mixture of gypsum, lime, mud, and straw that was used to bind the large blocks of stone that make up the pyramids. Nearly 2,500 years later, it is clear that the Romans improved upon the work of the Egyptians and formed a structure like the

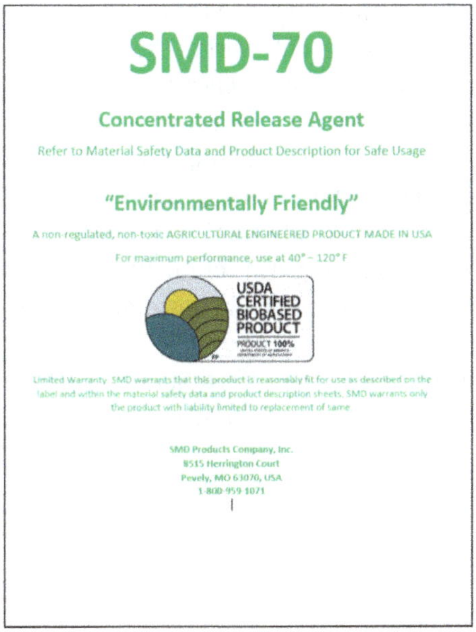

Figure 19.2: SMD-70 product label (courtesy of SMD Products Company, Inc.).

Coliseum in Rome, which was built under the instruction of Emperor Vespasian, and completed during the time of Emperor Titus in AD 80 [3]. A fantastic marvel, the Amphitheatre, able to hold between 50,000 and 80,000 spectators, is sometimes called one of the Seven New Wonders of the World [4].

Between the Roman Empire and the Middle Ages, it gets a little murky as the formulation of concrete was believed to have been lost until it was found again in 1414 in old manuscripts [5]. The invention of early concrete is credited to John Smeaton. A perfect account of this historically significant fact goes to the Canadian authors at Giatec Scientific, who share the story on their website as follows:

> It wasn't until 1756 that the technology took a big leap forward when John Smeaton discovered a more modern method for producing hydraulic lime for cement. He used limestone containing clay that was fired until it turned into clinker, which was then ground into powder. He used this material in the historic rebuilding of the Eddystone Lighthouse in Cornwall, England in 1793.

Today's concrete is made from what is called Portland cement. Just over 30 years after Smeaton famously used his first modern day concrete in the Eddystone Lighthouse; Joseph Aspdin invented a mixture, which he made by burning grounded chalk and finely crushed clay in a lime kiln until carbon dioxide evaporated, producing a strong cement, later named "Portland Cement." The inventor – Joseph Aspidin – is credited with inventing Portland cement in 1867 [3]. And, the idea of reinforcing concrete, the modern day use of rebar, was done by Joseph Monier through his 1867 patent of the concept that one can reinforce concrete using an iron mesh.

From the industrial revolution onward, the advancements and the use of concrete cannot be understated. It has become a part of every person's world – from streets to small structures to high-rise buildings to major dams. Without concrete, many things we take for granted would not have been possible. Can one imagine what large cities would look like if the world's first concrete high-rise in Cincinnati, Ohio had never been built in 1903? The Ingalls Building, a modest 16-story building is credited to be the world's first reinforced concrete skyscraper. This architectural first, listed in the National Register of Historic Places, has now been renovated, as of the timing of this publication, and is now operated as a Courtyard Marriott Hotel in downtown Cincinnati [6]. What if the Hoover Dam had not been built in 1936? At over 221 m in height, it is still one of the largest hydroelectrical stations in the United States [7]. These achievements have helped society, but at what cost? Our next section shall discuss the environmental footprint the concrete industry has created.

19.1.2 The environmental footprint issue

Now that we understand the colorful and detailed history of the industry, let us step back and understand what environmental footprint has been created, and what opportunity there is for sustainable bio-based materials to be used to reduce this environmental impact on the future. It is interesting to note, as we just learned, the first high-rise was built in the Great State of Ohio, and one of the solutions we plan to introduce is also grown in the same great state of Ohio. But, before we find the solution, let us first understand the problem.

In 2019, the National Academy of Sciences brought attention to a paper that was published by the University of Rochester, which took a broad look at the problem [8]. It was cited that carbon dioxide (CO_2) from the cement industry accounts for up to 8% of the global CO_2 emissions. The industry has its challenges, giving off up to 2.8 GT/year of environmentally unfriendly CO_2 gasses – CO_2 emissions due to the decomposition of $CaCO_3$ to CaO and the use of fossil fuels in that process. This is a problem, but one that will be solved in the coming years, as there are several well-known efforts in developing countries to produce a form of "green cement" that offers lower overall emissions during production, primarily due to the effort of taking industrial waste and using it as a raw material in the cement. So, it is safe to say that the core cement industry is already looking at various solutions in order to reduce its ecological footprint. One subsegment of the construction industry, which has not yet been addressed, is the often overlooked environmental impact of "form-release agent," sometimes called "form oil." At a gathering of industry leaders in Washington, DC, at the 2021 AFCC Biobased Economy Conference, we were able to hear the industry challenges and highlight what sustainable bio-based products can do to help the industry lower its environmental footprint [9].

The University of Cambridge dictionary defines environmental footprint as: the effect that a person, company, activity, etc. has on the environment [10]. For the purposes of this chapter, we will focus on the use of form oil and other release agents. It is said that form oil or form release touches nearly every concrete project being executed. So, what is a concrete form release? It is a chemical that is applied to prevent the adhesion of freshly placed concrete to the forming surface. The forming surface is usually plywood, steel, metal, silicon, or some combination of these materials. The form release market is segmented into various uses – PU molding, rubber molding, concrete molding, die-Casting, plastics moldings, and others – and is dominated by industry powerhouses such as Chem-Trend L.P. (US), Henkel AG & Co. KGaA (Germany), Croda International Plc (UK), LANXESS Group (Germany), Shin-Etsu Chemical Co., Ltd. (Japan), Daikin Industries Ltd. (Japan), Michelman, Inc. (US), Marbocote Limited (UK), McGee Industries, Inc. (US), Miller-Stephenson, Inc. (US), TAG Chemicals GmbH (Germany), and LORD Corporation (US) [11]. The concrete form release market is the third largest segment in the greater concrete release agent market, which has suppliers that are more from the petroleum industry, like Shell and Chevron. Here the goal of a good concrete form release is to create a barrier between the fresh concrete and the form; thus, highly toxic and high VOC content oils are sometimes used from these suppliers as they are easy to apply in spray form. The environmental challenge is that this form release, which acts as a barrier runs off the form during the process and off the concrete after the process. One can only imagine what this concrete form release runoff does to the local ecosystem when building a dam or bridge, which is close to local water sources. It is increasingly evident that demand for biodegradable form release agents will become even greater.

The US Government attempted to limit the VOC (volatile organic compounds) emissions from architectural coatings under Section 183(e) of the Clean Air Act in 1999. The new regulation set form-release VOC content to 450 grams per liter or 3.8 pounds per gallon for newly manufactured products. In the Act, form releases were classified as architectural coatings, but only applied to the manufacturers and not the users. These new regulations forced more processing steps in the creation of petroleum-based form-release agents, and for this reason, also increased the environmental footprint of these products. Though the intention of the regulation was noble, the large loophole continued to exist [12]. The usage of petroleum-based form releases causes strong smells due to the VOC content, which is still allowed, and skin irritations due to the petroleum content in the form release. These minor discomforts are easily avoidable by implementing a system whereby bio-based products are used, as they do not contain VOCs. The raw materials in biodegradable solutions have been shown to reduce skin-irritation, which continues to plague workers who handle form release daily.

19.2 The global initiative: UCOP26/Paris Accord/Bio-Preferred

As noted at the beginning of this chapter, the most recent global initiatives for the reduction of environmental damage on a broad basis has been led by the Paris Accord and by, what was recently called, COP26 held in Scotland. Addressing the Conference's preliminary opening, Abdulla Shahid, the President of the UN General Assembly called for an honest look at where things stand: "We have had decades to argue the facts about climate change, about the power of renewables, about the fine details of monitoring or cost-sharing. Yet, we have still failed to act with the conviction and determination required" [13]. For this reason, it became clear that certain governments and international agencies would need to step up and take action into their own hands. This movement has created the untiring demand for bio-based solutions that can not only contribute to reducing greenhouse gases, but also reduce the pollution of the waterways. One example of a roadmap for a sustainable future was presented by the Tokyo Organizing Committee of the Olympic and Paralympic Games [14]; see Figure 19.3 for a conceptual map of global sustainability efforts and entities shaping the 2020 Olympic Games. In parallel, the United States Department of Agriculture (USDA) was also working for the promotion of bio-based solutions to be used by federal contractors on infrastructure projects in the United States. This program run by the USDA is called the Bio-Preferred program; see Chapter 1 for more details, and involves third-party testing of products for their compliance to very rigid criteria, in which those that pass are invited to attach the USDA Bio-Preferred label on their product [15]. Activities by entities such as the International Olympic Committee, United Nations, and the US Government help stimulate industry to come up with creative solutions to complex problems. But, as noted in the beginning of our chapter, the fine balance between regulation and capitalism should be maintained to stimulate innovation.

Bio-based solutions can produce better concrete products, be sustainable, easier to handle, and cost effective.

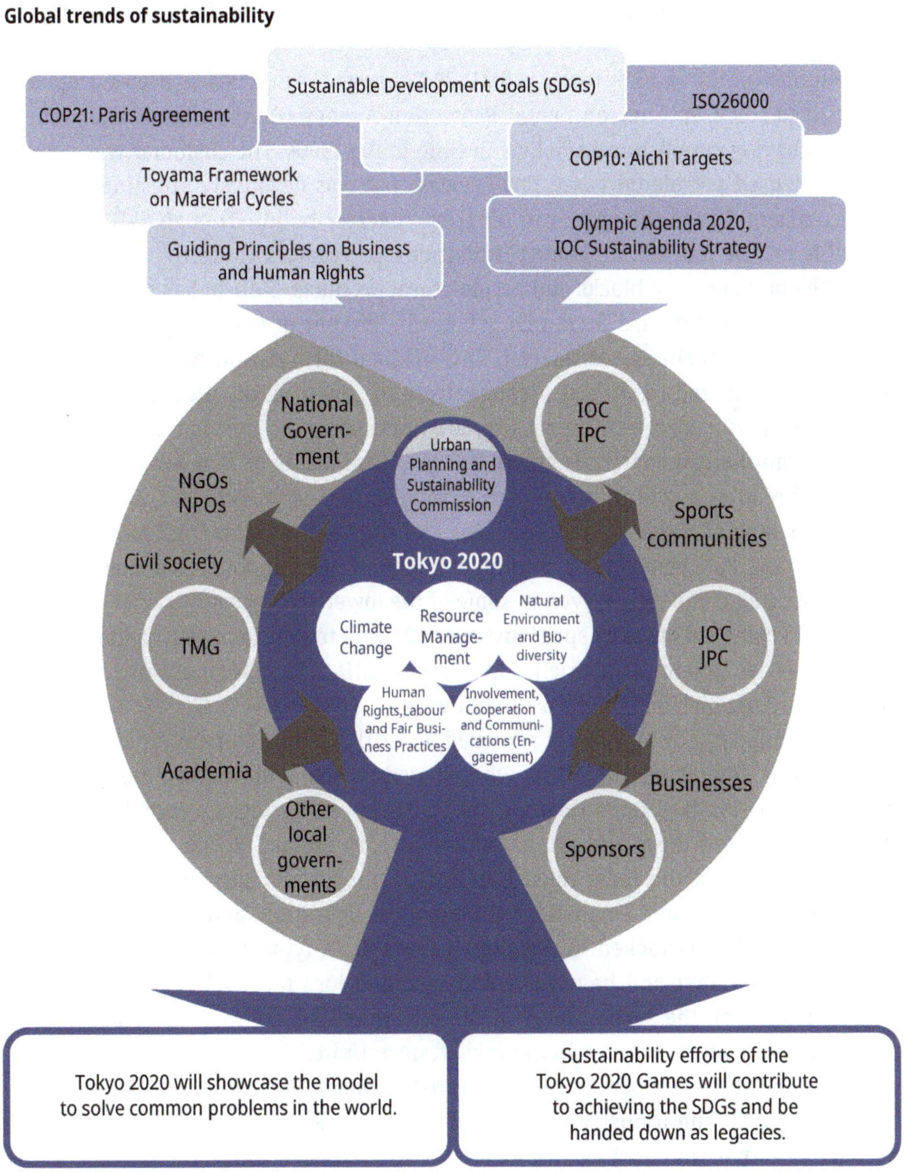

Figure 19.3: *Sustainable Sourcing Code* (3rd edition) – public handout to potential material suppliers from welcome packet of the organizing committee, January 2016.

19.3 Introduction to SMD-70 concrete form release

One solution on the market today is SMD-70 Concrete Form Release. It was originally designed to solve an Environmental Protection Agency (EPA) violation for a client who was having runoff into Lake Erie in Ohio in the 1990s. The customer was using a commonly used petroleum-based form release but was negatively affecting the local fish ecosystem due to the large run off from concrete builds. Though SMD-10 is still available on the market, and several boutique pre-casters continue to say SMD-10 makes the best concrete blocks and architectural products, SMD-10 has the same logistical footprint as standard form oils, when the amount of energy spent to ship the product to the job site is considered. This original formulation has now been perfected and stabilized to ship as a concentrate to construction sites throughout the world as SMD-70 concrete form release. It is interesting to note that SMD-70, though derived from the raw materials grown in the United States of America, is commonly shipped abroad as a concentrate due to its environmental friendliness, ability to support the making of higher quality concrete, and smaller environmental footprint, as it is shipped as a 20% concentrate, eliminating 80% of the shipping costs, thereby making this product competitive, even in some of the lower income countries in Southeast Asia. SMD Products Company spent just over 10 years to develop SMD-70, after having seen the market potential experienced by selling SMD-10 to several of North Americas largest construction companies for several years.

The complexity of logistics is removed as SMD-70 ships globally in standard 55-gallon drums. One drum of SMD-70, when mixed on site, will produce the equivalent of 5 drums of standard form release. The ability to reduce logistic costs by not shipping water around the world has created a positive impact on the overall environmental footprint of the form release industry. One 20-foot shipping container can hold over 40 drums on a single-floor shipment configuration, and a grand total of 78 drums when double-stacked in the same container. A typical customer will buy 40 drums in a container and have 2.5-gallon cans and jugs for smaller projects, which may be loaded on the second level of the container. All the raw materials used in SMD-70 are regionally sourced in the mid-western United States of America, from the original soybean oil component to the proprietary blend of chemicals used in the creation of SMD-70. The industry was misled by some legacy manufacturers who conducted a negative campaign against form releases that have a mixture of water, and it has taken years for SMD-70 to catch on as "the form release" of choice for several job sites. The image below shows why the legacy manufacturers pushed hard to prevent SMD-70 from becoming the new standard on the market. SMD-70 quickly proved to major contractors that it provided a better release.

The unique concept of having water as the delivery agent is the key to the product. This is the secret sauce of SMD-70. Unlike traditional petroleum-derived release agents who use petroleum as the delivery agent that evaporates quickly, leaving the fatty acid (lubricant) on the form material, SMD-70 does something previously un-heard-of: SMD-70 is blended with water that remains as an emulsion. This means the fatty acid molecule is encapsulated by water. The water is used to deliver the product as a fine mist, but never allowing the water to enter the molecule. This is the key point and leads to better concrete with fewer bug holes. Please see Figure 19.4. Though users have been confused by the legacy suppliers to believe that when you add water, the product becomes a water-based product, this is not the case with SMD-70 as the water is only used as the delivery mechanism. For this reason, SMD-70 was the product of choice, used in the building of the Tokyo Olympic Stadium and the Olympic Village. Please see Figure 19.5, which highlights the Tokyo 2020 Olympics project (courtesy of Saitama Ryokoku).

Figure 19.4: Concrete product using traditional release agent on left and using SMD-70 on the right (courtesy of SMD Products Company, Inc.).

During the 2021 AFCC Biobased Economy Conference, the effective use of bio-based form releases was discussed in detail. The consensus of the panel discussion was that one does not need to give up concrete strength or the quality of the concrete product when using the latest technologies that are on the market. Sophisticated engineering companies are using artificial intelligence to determine the concrete strength using very powerful image-based algorithms that rely on deep learning to find areas of weakness in concrete samples. Figure 19.6 shows one such solution from Aiforia that is applying its innovative use of artificial intelligence (AI) using image recognition, combined with deep learning algorithms, to the oncology space to potentially solve

Figure 19.5: Report of document sharing construction project details (courtesy of Saitama Ryokoku).

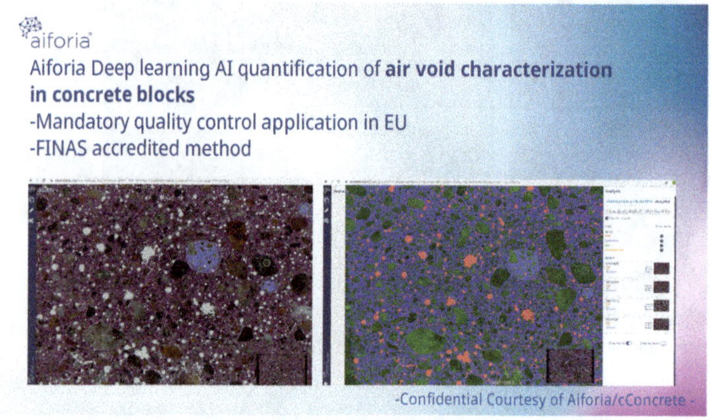

Figure 19.6: Color image of how AI is applied in the concrete industry (image courtesy of Aiforia).

some of the greatest challenges at some of the world's largest medical institutions. The process is also being used in the areas of sustainable construction and sustainable fish farming. We believe the use of AI, especially the type of AI deployed by Aiforia will continue to help advance the construction industry.

The structural integrity of concrete is becoming more of a recent talking point as several recent mishaps have occurred with infrastructure that has started decaying. Western nations are now looking into perfecting the process to make better concrete. As we know, there are seven steps for making concrete: (1) Make cement; (2) set forms; (3) apply form release; (4) place rebar; (5) pour concrete; (6) let concrete set; and (7) cure the concrete. It is now becoming more evident that the timing to apply the rebar and the process of curing concrete play important roles in the process. Historically, rebar was placed and then sprayed with form release, as the form release would quickly evaporate and lose its lubricating properties. SMD-70 has been shown to be the ideal solution for this timing problem, as it can be applied and will not evaporate (only the delivery agent, "the water" evaporates), allowing for more time to place the rebar with no loss of lubricating features. Please see Figure 19.7, where one can observe the proper placement of rebar. We observed this first hand when we noticed and observed the use of SMD-70 by some of the suppliers to Singapore's world-renowned HDB flat construction, where torrential rains and tropical heat would sometimes wash away legacy form releases, when SMD-70 would remain in its useful state even after some of the harshest weather patterns had taken their toll on the construction.

Figure 19.7: Typical image of rebar being placed in form prior to concrete pour (photo courtesy of SMD Products Company, Inc.).

Research out of Japan is now being validated by the Japanese Government for use in major infrastructure projects where we are seeing the curing benefits of a derivative of SMD-70. A nationally recognized university in Japan has conducted testing, which showed that the application of SMD-70 in its concentrated form will create a better environment for curing the concrete when compared to the environmentally damaging methods of water bath, plastic, and heating. This was validated, and can be seen in the Figure 19.8, where it is shown that under the comprehensive strength test, air permeability test, and neutralization acceleration test, there is no worse outcome when compared to traditional curing methods. Though there is still work to be done, it seems clear that using a concentrated version of SMD-70 may solve some of the environmental footprint issues that occur downstream after the concrete is poured. This is an evolving project that may open up new uses for SMD-70, here again showing that bio-based solutions can help not only with the release part of the project but the downstream curing part as well.

Examination	Curing Method (When Compared to Whats Commercially Available))			
	SMD-70 100%	SMD-70 75%	SMD-70 50%	SMD-70 33%
Comprehensive Strength Test	Effective	Effective	Effective	Effective
Rebound Hammer Test	No consistent effect was observed during this review			
Air Permeability Test	Effective	**Superior Effect**	**Superior Effect**	**Not Very Effective**
Neutralization Acceleration Test	**Superior Effect**	**Superior Effect**	Effective	**Not Very Effective**

Figure 19.8: Diagram of test results carried out in collaboration with academia. Test data may be received upon request to SMD Products Company, Inc. (courtesy of Saitama Ryokoku & University Professor).

19.3.1 Tokyo Olympics and Major Projects

As noted in the previous section, SMD-70 was found to be the best solution for the building of the Tokyo Olympic Stadium and the Athlete Village. The global concern regarding concrete strength and concrete quality, combined with the demand for an environmentally friendly solution, created the unique environment whereby SMD-70 became the only viable candidate. Thanks to the partnership of one of the largest construction companies in Japan with the exclusive partner in Japan for SMD-70, the company can proudly say it took part in the behind the scenes work of building a successful 2020 Tokyo Olympic games.

In parallel to the Tokyo Olympic infrastructure building, the positive publicity helped SMD-70 become the obvious choice for Japan's largest dam, built in the North of the Tokyo Kanto-region. SMD-70, being a bio-based product, is fully biodegradable, and allows it to be used in complex waterways, where the ecological system is of such great concern when building this 87-meter-tall dam. Images can be seen in Figure 19.9 and on the publicly available website set up for the public to review this important project.

Figure 19.9: Japan's largest dam project in an environmentally important area, north of Tokyo (photo's courtesy of Saitama Ryokoku, artist rendition courtesy of the publicly available project website).

19.4 Conclusion

As we started in Tanzania, went to Ohio, and finished with SMD-70 from the heartland of the United States of America, it is quite apparent: bio-based products are not the compromise, but the solution. The aspect of forming better concrete-finished product with less need for touch-up will reduce the overall project labor costs. The ability to ship as a concentrate reduces the logistics costs by 80%. The opportunity to potentially offer stronger concrete, thanks to the use of SMD-70 as a curing agent, is under investigation, but preliminary results have led to the Japanese National Highway Board (NEITS) granting the SMD-70 derivative with recognition for use on core infrastructure projects. All these benefits alone justify the immediate use of SMD-70, the bio-based solution for the concrete industry. On top of all these obvious benefits, when one considers the fact that the solution is sustainable, biodegradable, and environmentally friendly, it is clear that sometimes the best and easiest solution to tackle a problem is right in front of you and not as complicated as one may have thought. Over the coming years, we can all look forward to observing the benefits that SMD-70 and other bio-based materials will have on the global construction industry. The authors would like to encourage governments and institutions to create a free and open space for products like SMD-70 to be used, not because they are from a small business, but because they are the best solutions for the business.

In conclusion, the authors would like to share some highlights and closing points from the presentation given at the 2021 AFCC Biobased Economy Conference in Figure 19.10. We stand committed to help provide the best solution for all stakeholders with no compromises.

Figure 19.10: Sample slides showing global utility of SMD-70 from the 2021 AFCC Biobased Economy Conference (September 2021, courtesy of SMD Products Company).

References

[1] de la Torre I. The OGAP Core Team. 2022, May 27. https://www.olduvai-gorge.org/.

[2] Worth K. COP26 opens in Glasgow with calls for ambitious solutions to tackle climate emergency. 2022, May 27. https://news.un.org/en/story/2021/11/1104522.

[3] Day M. Everything about concrete. 2022, May 27). https://www.everything-about-concrete.com/.

[4] Weber B. About new 7 wonders. 2022, May 27. https://world.new7wonders.com/.

[5] Scientific G.. 2017, July 28. The history of Concrete. https://www.giatecscientific.com/education/the-history-of-concrete/.

[6] Hospitality Net. Courtyard by Marriott Cincinnati Downtown. 2022, May 30. https://www.hospitalitynet.org/announcement/41006628/courtyard-by-marriott-cincinnati-downtown.html.

[7] Engineer G.. The history of hoover dam. 2020, June 30. https://www.geoengineer.org/news/history-of-the-hoover-dam.

[8] Ellis LD., et al. Toward electrochemical synthesis of cement – An electrolyzer-based process for decarbonating $CaCO_3$ while producing useful gas streams, 2019.

[9] Alternative Fuels & Chemicals Coalition. 2022, May 30. https://www.altfuelchem.org/2021-afcc-workshop-sessions.

[10] University of Cambridge. 2022, May 30. https://www.cam.ac.uk/.

[11] Markets & Markets. Precast Market Research Report, March 2021.

[12] Environmental Protection Agency Fact Sheet. 2022, May 30. https://www.epa.gov/stationary-sources-air-pollution/fact-sheets-consumer-and-commercial-product-section-183e-voc-rules.

[13] Worth K. COP26 opens in Glasgow with calls for ambitious solutions to tackle climate emergency. 2022, May 27. https://news.un.org/en/story/2021/11/1104522.

[14] United States Department of Agriculture, Bio-Preferred website. 2022, June 25. https://www.biopreferred.gov/BioPreferred/.

[15] Tokyo 2020 Website. 2022, March 31.https://www.tokyo2020.jp.

Jonathan Cristiani*, Grace Dearnley, Justin Distler, Andrew Doerflinger, Vincent Mazzoni

Chapter 20 Green hydrogen production, use, and project delivery

20.1 Introduction

As interest in bioproducts rises with the progress of the clean energy transition, hydrogen has become an increasingly attractive fuel, feedstock, and energy carrier due to its compatibility with both traditional and decarbonized processes and systems. Hydrogen is a fundamental building block of the natural world that provides a diverse array of sources for production of molecular hydrogen, such as biomass, waste, fossil resources, and water-splitting technologies. Hailed for its flexibility, hydrogen can be used in a range of markets such as transportation fuels, energy storage, and chemical precursors.

Hydrogen is valued for its high energy content on a mass basis, which makes it favorable as a fuel for both aerospace and terrestrial transportation. "Green" hydrogen, produced via electrolysis and integrated with renewable energy resources, has gained significant attention in the sustainability movement for its potential as a decarbonization tool in historically carbon-intensive markets. It shows potential for use as feedstock to reduce carbon intensity in the "hard-to-abate" cement, steel, chemical and petrochemical industries, as a zero-carbon fuel in fuel cell electric vehicles, and to support the production of synthetic fuels and chemical products.

The broad field of hydrogen production and use has wide implications for the bioproducts industry, particularly in the marriage of renewable net-zero carbon resources such as biomass with renewable electricity. Despite the challenges associated with transportation and storage, low-carbon hydrogen can be viewed as a powerful energy carrier and feedstock in the successful implementation of integrated biorefineries. It is also highly complementary in less apparent ways, such as the decarbonization of the multitude of transportation systems associated with biomass feedstock and bio-

Acknowledgments: Contributors and reviewers: Jacob Clanton, Jack Donoghue, Michael Clifford, Michael Goff, and Frank Jakob.

***Corresponding author: Jonathan Cristiani**, Black & Veatch Corporation, 11401 Lamar Avenue, Overland Park, KS 66211, USA, e-mail: CristianiJM@BV.com
Grace Dearnley, Justin Distler, Andrew Doerflinger, Vincent Mazzoni, Black & Veatch Corporation, 11401 Lamar Avenue, Overland Park, KS 66211, USA

 | https://doi.org/10.1515/9783112212141-020

product logistics. This chapter provides an overview of the many hydrogen production and utilization alternatives, as well as a discussion of hydrogen infrastructure projects and their synergies with the bioproducts sector.

20.2 Hydrogen production

Numerous processes have been developed to produce hydrogen, and all are used today with varying levels of scale and technological maturity. Each of these hydrogen production processes relies upon different technologies and reaction pathways. Determining a best-fit process for any given hydrogen production operation requires evaluation of a range of factors, from affordability and spacing constraints to technological shortfalls and advantages. Each production type holds a unique role in the overall hydrogen production market.

20.2.1 Electrolysis

Electrolysis is the process of splitting water into hydrogen and oxygen using electricity in an electrochemical cell. Electrolyzers come in a variety of capacities and chemistries, but the fundamental concept remains the same. Electrolyzers, like fuel cells and batteries, have electrodes (i.e., anodes and cathodes) separated by an electrolyte. The combination of electrodes and electrolyte – the stack – varies by the type of chemical reactions taking place. Electrolysis is considered "green" when the electricity consumed is provided by renewable energy resource(s). Here, four types of electrolyzers are examined: proton exchange membrane (PEM), alkaline water electrolysis (AWE), anion exchange membrane (AEM), and solid oxide electrolysis (SOE).

PEM electrolyzers exchange a proton (H^+ion) through the polymer electrolyte between the electrodes. In a PEM electrolyzer, water is split into oxygen and hydrogen, with the hydrogen ions traveling from the anode to the cathode and exiting out of the cathode side of the stack. Oxygen exits out of the anode side of the stack. Electrocatalysts help lower the activation energy required for the splitting of water. Although high electrolyzer capital cost is a barrier to adoption, recent research and development initiatives have optimized the catalytic activity of the cell while minimizing the amount of expensive electrocatalysts, thereby lowering the cost of PEM electrolyzers [1].

AWE electrolyzers fundamentally function similarly to PEM electrolyzers; however, the hydroxide anion (OH^-) ion transported in the electrolyte and travels from the cathode to the anode. The hydrogen then exits out of the cathode side while the oxygen exits out of the anode side of the stack. Since AWEs have a lower current density, they typically require a larger footprint compared to PEMs, though the technology is considered more mature for large-scale hydrogen production [2].

The AEM electrolyzer is an emerging technology that offers construction and operational advantages over both PEM and AWE types. Similar to PEM, AEM electrolyzers use an electrolyte membrane, but instead of passing a proton across the electrolyte, the OH^-anion is passed, similar to AWE. AEMs also use less expensive (nonplatinum group) electrocatalysts than PEM and can potentially operate at higher operating pressures to reduce system size [3].

SOE stacks have high conversion efficiencies compared to PEM/AWE, primarily because they operate at higher temperature (i.e., 600–850°C) where thermodynamics and reaction kinetics are favored. Additionally, SOEs can be used for the direct electrochemical conversion of steam, carbon dioxide (CO_2), or both, into hydrogen, carbon monoxide (CO), and/or synthesis gas (i.e., syngas, which is a mixture of hydrogen and CO). SOEs are further capable of converting captured $CO_2$and water into synthetic natural gas (i.e., methane or CH_4), gasoline, methanol, or ammonia. This type of electrolyzer consists of two porous electrodes surrounding a dense ceramic electrolyte capable of conducting oxide ions (O^{2-}). In addition to the efficiency benefits, SOE technology is also characterized by low-cost materials of construction compared to PEM electrolyzers, particularly in their use of non-noble metal electrocatalysts. Additionally, many SOEs in development are also thought to show greater promise than existing, commercialized electrolysis technologies for reversible operation.

Table 20.1 shows the relative performance characteristics of each of the electrolysis technologies considered.

20.2.2 Reforming

Reforming technologies use a hydrocarbon feedstock to produce a syngas, which is then converted and purified to achieve a hydrogen product. Traditionally, natural gas or coal is used as feedstock. Reforming reactions take place in reformers, usually in the presence of catalysts, at high temperatures and net endothermic conditions. However, reforming systems can have exothermic steps as part of their reaction mechanism, depending on the reaction taking place. The resulting syngas contains CO, which is then converted to hydrogen and carbon dioxide through a water gas shift (WGS) reaction. Finally, the hydrogen is purified, typically through a pressure swing adsorption (PSA) process to remove any unreacted components and impurities (primarily water, CO, CO_2, and nitrogen, depending on the process) from the hydrogen product. There are three primary reforming technologies: steam methane reforming (SMR), partial oxidation (POX), and autothermal reforming (ATR).

SMR is the most mature hydrogen production process, having been widely used for industrial-scale applications throughout the twentieth and the beginning of the twenty-first centuries. In an SMR process, natural gas is reacted with steam at temperatures of approximately 810–870°C in a catalyst-filled tube in the reformer to produce syngas. The SMR process is highly endothermic, requiring large amounts of heat

Table 20.1: Electrolysis technologies' performance comparison.

Attribute	Proton exchange membrane (PEM)	Alkaline water electrolysis (AWE)	Solid oxide electrolysis (SOE)	Anion exchange membrane (AEM)
Technology maturity	Commercially available	Commercially available	Developmental	Developmental
Technology scale	Small- to large-scale	Small- to large-scale	Small-scale	Laboratory-scale
Key features	Solid polymer electrolyte that is highly conductive for positively charged ions and resistive to negatively charged ions [6]	Positively charged anode and negatively charged cathode immersed in liquid electrolyte [6]	Electrolyzes water as steam and includes two porous electrodes on either side of a dense ceramic electrolyte capable of conducting O^{2-} ions [4]	Solid polymer electrolyte that is highly conductive for negatively charged ions and resistive to positively charged ions [3]
Key characteristics	– Smaller footprint – Modest current density – Faster dynamic response, high turndown capabilities – Higher pressure operation – Higher purity product gas (99.99%) [7]	– Large footprint – Low current density – Lower capital expenses (CAPEX) and operating expenses (OPEX) compared to PEM – Older, most established technology [6]	– High conversion efficiency and current density – Lower CAPEX compared to PEM/AWE – Abundant raw materials for cell components [4]	– Smaller footprint – Lower CAPEX compared to PEM/AWE – Higher pressure operation – Higher purity product gas (99.99%) [3]

Disadvantages	– Higher CAPEX and OPEX compared to AWE [6] – Expensive platinum electrocatalyst [4]	– Less suitable for space-constrained facilities [6] – Lower gas purity, relative to PEM (>99.5%) [7] – Requires bulk chemical storage of hazardous, corrosive electrolyte	– New emerging technology – Small stack capacities (< 10kW) [2] – More complex thermal management [2]	– New emerging technology – Lower current density compared to PEM – Decreased durability – Increased degradation rates [3]
Normalized degradation[1]	1.5% per year	0.9% per year	4.2% per year	Unknown
Life expectancy and stack replacement schedule [9]	20 + Years [5] 20,000–60,000h [7]	20 + Years [8] 60,000–90,000h [7]	Unknown Life < 10,000h [7]	Unknown Life and stack replacement
Stack energy efficiency[2] [7]	54–71% 47–61 kWh/kg	51–71% 47–66 kWh/kg	< 94% > 36 kWh/kg	58–62% 53–58 kWh/kg

Table Notes:

1) Estimated based on 95% annual capacity factor and assuming moderate operating conditions (i.e., stable temperature/loading and minimal cycling). No specific formula for degradation exists and is based on a complex set of factors that have been modeled and discussed in the literature.
2) Efficiency calculated based on lower heating value of hydrogen.

input. This heat is supplied to the SMR by fired burners surrounding the tubes, with the heat from the exiting flue gas recovered in the convection section of the SMR. The convection section preheats the natural gas and steam for the process, along with usually producing steam to recover as much heat as possible. Hot syngas from the SMR is cooled in a waste heat boiler and then passes through a series of WGS reactors to convert the CO in the syngas to hydrogen. Finally, it is purified (typically in a PSA system) to achieve the final desired hydrogen purity. Primarily due to the high temperatures required for SMR, the hydrogen production process produces large amounts of CO_2 and yields hydrogen with a high carbon intensity. Hydrogen produced through SMR is often referred to as “gray” hydrogen. However, if the SMR process is coupled with carbon capture, utilization, and sequestration (CCUS) technologies to reduce the emissions of CO_2, the product is referred to as “blue” hydrogen.

POX reforming also uses a natural gas feedstock but employs a different reaction pathway than SMR to produce syngas. Rather than using steam as a reactant, a limited amount of oxygen is supplied to partially oxidize natural gas to supply heat to the endothermic reforming reactions. Oxygen can theoretically be supplied from air but is usually supplied as a high purity oxygen stream produced from an air separation unit (ASU). The amount of oxygen supplied is limited in stoichiometric quantity to about 30–35% of that required for complete oxidation of the hydrocarbon feedstock to CO_2 and water. The CO is then converted to hydrogen in a WGS reaction and purified in a PSA. Since the POX process is exothermic, fired heaters are not required, as with SMR, thereby requiring smaller reactor sizes. However, the POX reaction pathway produces less hydrogen than SMR for the same quantity of fuel input, due to the lower amount of hydrogen produced from the steam. Additionally, using air as the oxygen source creates a product with large amounts of nitrogen, increasing the purification costs (via nitrogen rejection in a PSA) of the final hydrogen product [10].

ATR can be thought of as a hybrid process between SMR and partial oxidation, employing aspects of both reaction pathways and using both steam and oxygen as inputs to the catalyzed process with natural gas. Natural gas and steam are preheated in a fired heater before combining and entering the reformer burner, where it is mixed with an oxygen input (which can be air or a pure oxygen stream). The oxygen partially oxidizes the hydrocarbon components in the natural gas in an exothermic process, reducing the energy input to achieve the required high temperatures of the reforming process. The hot, partially oxidized stream flows downward over a catalyst bed that helps drive the reforming reactions. Exiting the reformer, the syngas is cooled in a waste heat boiler and is then upgraded and purified through a WGS reaction and PSA process. Using air as the oxygen source will result in a syngas product with a high nitrogen concentration, which must be removed during purification to achieve the desired hydrogen purity.

Table 20.2 shows the relative performance characteristics of each of the reforming processes considered.

Table 20.2: Reforming processes: performance comparison.

Attribute	Steam methane reforming (SMR)	Partial oxidation (POX)	Autothermal reforming (ATR)
Feedstocks	Natural gas, steam	Natural gas, oxygen (pure or air)	Natural gas, steam, oxygen (pure or air)
Required heat input	High	Little to none	Low to intermediate
Hydrogen production per unit of fuel	Highest	Lowest	Intermediate
Reformer size	Largest	Smallest	Intermediate
Primary impurities removed	CO_2, CO, water	CO_2, CO, water, nitrogen (if air used)	CO_2, CO, water, nitrogen (if air used)

20.2.3 Gasification

Gasification refers to the conversion of a solid biomass to syngas through a series of reactions, which can broadly be separated into four stages. The first of these stages is a drying phase (typically occurring at temperatures in the range of 100–320°C), in which the moisture in the biomass fuel evaporates. The second stage is pyrolysis, in which the volatile matter trapped in the solid matrix of the fuel is volatilized and released as a vapor. Pyrolysis occurs at temperatures ranging from just above drying to 590°C. The third stage of the gasification process is the gasification of the remaining solids, which are mostly fixed carbon by reaction with water, hydrogen, or CO_2 to produce CO, hydrogen, and CH_4. These reactions occur at temperatures above 650°C and are endothermic. The fourth stage is a combustion stage, where oxygen reacts with carbon and CO to produce CO_2. This reaction is exothermic and produces the heat required for the endothermic gasification reactions. The ordering and speed of these stages varies, based on the gasifier reactor design and operating conditions.

Following the gasification process, the cleaning and treating processes for the resulting syngas depend upon the operating characteristics of the gasification system as well as the composition of the biomass. To achieve the final hydrogen product, the gasification-produced syngas, often, must undergo an acid–gas removal process, and then proceed to a syngas upgrading and purification process, similar to that described in the reforming section.

In an entrained flow gasifier, solid feedstocks (e.g., coal or petroleum coke) are ground to a fine consistency and injected into the gasifier with water or steam and oxygen or air. There are generally no limitations on the feedstocks to an entrained flow gasifier, but high ash and high moisture feedstocks increase the oxygen consumption more than in other gasification types because of the high operating temperatures of entrained flow gasifiers. Additionally, biomass cannot be used in an entrained flow gasifier due to its inferior friability, unless when blended or pretreated.

In a fluidized-bed gasifier, the feedstock is introduced near the bottom of the gasifier and is fluidized by gas feedstocks (typically steam and oxygen or air) to the gasifier. This method may result in poor carbon conversion, as smaller particles tend to become entrained in the gas flow and will exit the gasifier before they have the chance to be gasified. Relative to entrained flow gasifiers, fluidized-bed gasifiers operate at lower pressures and temperatures, have longer fuel residence times, and have lower capacities. The temperature of the gasifier is typically kept below the ash softening point to prevent agglomeration and bed defluidization. Specific types of fluidized-bed gasifiers include circulating fluidized bed and bubbling fluidized bed.

In a moving (or fixed)-bed gasifier, the feedstock is introduced at the top of the gasifier and slowly descends by gravity toward the bottom of the gasifier. Liquids can be injected into the gasification zone of moving-bed gasifiers. The most common arrangement for gas flow through moving-bed gasifiers is countercurrent to the bed flow, where the feedstock at the top of the gasifier is dried and then pyrolyzed by heat exchange with the exiting gases. This typically generates a syngas with a high CH_4 content. Compared to entrained-flow gasifiers, moving-bed gasifiers operate at lower temperatures, have much longer fuel residence times, and have lower capacities. There are many small-scale fixed-bed gasifiers that focus on biomass or coal.

Indirectly heated gasifiers do not use partial oxidation to provide the heat required to drive the gasification reactions. Instead, heat is added to the gasifier indirectly by a heated medium by heat transfer through the vessel walls or by electric current. Indirectly heated gasifiers can also be entrained flow, fluidized bed, or moving bed; however, they have special importance in smaller biomass applications where ASUs are not economically feasible. Indirectly heated gasifiers can produce a medium heating value syngas that is mostly free of nitrogen (a portion of the fuel-bound nitrogen will end up in the syngas).

Table 20.3 shows the relative performance characteristics of the first three gasifier technologies presented.

Table 20.3: Gasifier technology performance comparison.

Attribute	Entrained flow	Fluidized bed	Moving (or fixed) bed
Solid feedstock particle size	<0.127mm	6.35–12.7mm	6.35–50.8mm
Residence time	Seconds to minutes	Minutes	Hours
Relative operating temperature	High	Low	Low
Relative operating pressure	High	Low	Low

20.2.4 Methane pyrolysis

Methane pyrolysis is a process that has been discussed since the 1970s when the temperatures needed were thought to be too high for commercial production without the use of catalysts. CH_4, usually in the form of natural gas, is heated to high temperatures (>1,000°C), in the absence of oxygen, splitting the carbon and hydrogen atoms through radical chain dehydrogenation and methylation, or "cracking" [11]. The hydrogen atoms naturally coalesce to form hydrogen gas, while the carbon forms solid carbon. Through various solid-gas separation and filtration processes, the solid carbon and any other stray gases can be removed from the hydrogen stream, producing a high-purity hydrogen gas known as "turquoise" hydrogen. Recovered carbon can be further processed and sold for a variety of end-use applications (e.g., steelmaking, battery electrodes, and composites) to decrease total cost of producing hydrogen by this method. However, if methane pyrolysis became a significant source for hydrogen production, the supply of coproduct solid carbon would quickly outpace the global carbon demand. To tackle this, additional markets would be needed [12].

The temperature required to crack the CH_4 can be reduced if a catalyst is used. Many different catalysts can be used, with transition metals such as nickel, iron, and cobalt being the most common. While these catalysts can reduce the required temperatures to as low as 500°C, the product carbon tends to coat the reactor, thereby deactivating the active sites on the catalysts and reducing the rate of reaction to zero over time. One solution to this is to use carbon-based catalysts, which are usually less active than the metal catalysts and require higher temperatures but tend to have longer catalyst lifetimes. In both cases, the catalyst must be regenerated. Currently, the only large-scale regeneration processes produce carbon emissions through the combustion or gasification of the catalyst, reducing the benefits gained from the clean production of hydrogen [13].

Even with the high temperatures required for noncatalytic thermal decomposition, methane pyrolysis is a competitive production method from an energy requirement standpoint. In SMR, the energy to convert liquid water into steam and react the steam and CH_4 is 63 $\frac{kJ}{mol\ H_2}$. According to standard reaction enthalpies, using methane pyrolysis requires 37.8 $\frac{kJ}{mol\ H_2}$. This energy could be produced by combusting natural gas or other low-carbon fuels, generating as little as 0.05 $\frac{mol\ CO_2}{mol\ H_2}$, much less than the 0.43 $\frac{mol\ CO_2}{mol\ H_2}$ generated from SMR. If part of the hydrogen produced is combusted instead, carbon emissions could be prevented altogether [14].

20.3 Hydrogen storage and transportation

Due to the fact that hydrogen is the lightest element, it can be challenging to store large quantities in terms of both mass and volume. Methane is about eight times denser than hydrogen at standard conditions on a gravimetric basis, so the pressures and temperatures required to store hydrogen in an economical manner are more extreme than that of natural gas. Two potential hydrogen storage methods are compressed gaseous hydrogen storage and cryogenic liquid hydrogen storage. Selection of storage type may depend on a variety of factors, including the amount of hydrogen stored and the relative cost of each storage option.

For over-the-road hydrogen transportation, hydrogen storage density, relative to the weight of the transporting vehicle, is especially important. Liquid hydrogen deliveries provide the densest hydrogen state, however the production of liquid hydrogen as well as the handling can make it cumbersome for short ranges with low capacities. For most hydrogen transportation applications, the distance traveled and quantity delivered between the production and the consumption of hydrogen will drive the economics of liquid hydrogen, high-pressure compressed hydrogen, or low-pressure compressed hydrogen delivery. Over long distances, transportation through pipelines is possible, and is discussed below.

20.3.1 Compression

Compressed hydrogen storage is the most common method of storage for today's industrial hydrogen consumers. Depending on the amount of hydrogen being stored, pressures can range from 140 to 690bar, with small cylinders used in the transportation sector being more suitable for the high end of the range than the large bulk tanks for industrial users. Depending on the pressure and storage volume, many smaller vessels may be more economical than one large bulk tank. Because of its very low molecular weight and small molecule size, hydrogen presents an issue with leakage, and some compressed storage applications may require special materials to line the inside of the vessel for leakage prevention.

To compress and store hydrogen for later use, a combination of compressors, heat exchangers, tanks, and balance of plant piping are required. Depending on the final storage pressure, there may be several intercooled compressor stages. The balance between capital and operations and maintenance (O&M) costs would need to be optimized to determine the best means of compression for any given situation. Lower pressure storage would require less power consumption to achieve the storage pressures, but at a cost of a greater number of tanks and a larger footprint to hold the same mass of hydrogen. Higher pressure storage would require greater power consumption to achieve the storage pressures, but offers the benefit of fewer tanks and smaller footprint.

20.3.2 Liquefaction

Hydrogen liquefaction is more energy intensive than compressed storage but may be an attractive option, depending on the amount of hydrogen storage needed. The larger the quantity of hydrogen being stored, the more economical liquefaction becomes, relative to compressed storage on a mass basis. Consider the density of liquefied hydrogen compared to compressed hydrogen: density of liquefied hydrogen is approximately 70.8 kg/m^3, while density of compressed hydrogen ranges from 2.56 to 50 kg/m^3, depending on the pressure. The storage volumes for liquefied hydrogen would be much smaller than the storage volumes for compressed hydrogen of the same mass. However, liquefied hydrogen requires far more complex auxiliary equipment.

To liquefy hydrogen, extremely cold temperatures (i.e., −250°C) need to be maintained. This requires a vapor-compression cycle, with liquid nitrogen as a refrigerant. Boil-off compressors are also required to reliquefy the hydrogen that boils off during storage. However, depending on the amounts of storage required, liquefaction can still be more economical than compressed storage. This is particularly true at large scales.

When considering liquefaction equipment, one must evaluate thermal cycling and ramp time. Cycling from ambient to extremely low temperature thermally stresses the equipment. The equipment associated with liquefaction is designed for a certain number of thermal cycles over its lifetime, and frequent cycling will significantly reduce the useable life of the equipment. Ideally, the liquefaction equipment is run continuously to minimize thermal cycles and maximize the life of the equipment. Additionally, the start-up/shutdown times associated with the liquefaction train are in the 4–8h range (from cold standby) or up to 24h (from ambient temperature). Liquefaction is best suited for continuous operation or seasonal operation at a minimum. Daily cycling of the liquefaction equipment is not considered feasible; however, designing a system for very low turndown may offer some additional operating flexibility [15].

Current methods for liquefying hydrogen use a vapor-compression cycle, which requires various equipment, including trains of intercooled compressors, brazed aluminum heat exchangers, parallel closed-loop nitrogen refrigerant cycles, and boil-off compressors. The amount of equipment required makes the process of liquefying hydrogen quite energy intensive.

20.3.3 Pipeline transportation

Pipelines are the most cost-efficient way to transport large quantities of hydrogen over long distances. Hydrogen gas pipelines are considered mature technologies and can typically cost approximately up to 10% more than a traditional natural gas transmission pipeline. For dry hydrogen service, the use of carbon steel is perfectly acceptable for the typical temperatures/pressures associated with most low-carbon hydrogen projects. In instances where corrosive contaminants or moisture condensate are present, a stainless-steel pipeline material would be selected instead, which can increase costs [16].

One attractive option is to blend hydrogen into the existing United States (US) natural gas pipeline network, which includes over 644,000 km of infrastructure. It is estimated that at typical pressures and diameters associated with natural gas pipelines, approximately 19 metric tons of hydrogen could be stored per linear mile. Hydrogen is generally thought to be limited to 5–10% blending throughout most of the United States, primarily due to safety and pipeline integrity concerns. While greater percentages may be possible if natural gas pipelines supporting infrastructure and the end users of the blended gas are evaluated and converted for use with hydrogen, these costs and the required modifications are the subject of much research and development [17].

20.3.4 Geophysical storage

Geophysical storage takes advantage of existing geological formations such as salt caverns, rock caverns, and depleted gas fields, which present an opportunity to store large volumes of hydrogen in existing features. Conceptually, hydrogen is compressed and stored in an existing geological formation and then withdrawn for later use. In practice, the application of geophysical hydrogen storage is incredibly site-specific.

Salt caverns present the most suitable geological storage feature, followed by rock caverns and then, depleted gas fields as the least suitable of the three. Since hydrogen is the lightest gas, it has the fastest molecular velocity compared to any other gas at the same conditions. Depending on the geological feature, upgrades, such as a liner, may need to be added to minimize leakage. Another consideration associated with geological storage is contamination. Depending on the geological formation, other compounds such as CH_4 or water may be present. Additional clean-up equipment upon discharge of hydrogen may be required, depending on the geographic location and the hydrogen user quality requirements.

Geophysical storage presents an attractive method to store large quantities of hydrogen for seasonal variations but is highly dependent on the location. Upgrades to the geological formation or additional clean-up equipment may drive the effective cost above that of traditional compressed storage or liquid storage [15].

20.3.5 Metal hydrides and chemical hydrides

Hydrogen can also be stored in metal hydrides or chemical hydrides that function via chemical bonding. Metal hydrides store hydrogen in a solid material, while many chemical hydrides are stored in a liquid phase. The chemical bond is much stronger than adsorption bonds and thus, often requires more energy in synthesis and dehydrogenation. However, the strong chemical bonding allows hydrogen to be stored at a higher density at ambient conditions relative to physical hydrogen storage as compressed gas or cryogenic liquid. Releasing the hydrogen from the hydride carrier also requires energy or releases energy, depending on the chemical bond. This typically warrants balance of plant thermal management equipment and can further lower the round-trip efficiency of the storage. Despite metal hydrides having a high volumetric energy density, their low gravimetric energy density typically limits their applicability in certain end uses.

Current hydride storage techniques are typically more expensive compared to gaseous and liquid storage methods. However, due to the high hydrogen density at ambient conditions, there is still significant interest in both metal and chemical hydride storage. A few specific chemicals that are produced from hydrogen and could be considered chemical hydrides are discussed in greater detail in Section 20.4.

20.4 Power to products and chemicals

Power to products and chemicals (also known as "power-to-X" or "P2X") uses electricity from renewable energy resources to produce products such as hydrogen, ammonia, methanol, and renewable natural gas (RNG). Hydrogen is used as feedstock in many industrial chemical processes; thus, when produced using electrolysis, paired with renewable electricity, the use of "green" hydrogen in these processes decarbonizes the final product. The most common chemical, ammonia, is discussed first. Then, an important companion process to deploying hydrogen derived products – CCUS – is discussed, followed by several products thus enabled.

20.4.1 Green ammonia

Ammonia is considered a leading hydrogen carrier chemical that can overcome the challenges associated with storage and transportation of hydrogen, while taking advantage of existing infrastructure. It stores hydrogen in an energy-dense form as NH_3. Although ammonia has historically been used exclusively as a precursor chemical feedstock for a variety of applications (e.g., fertilizers, pharmaceuticals, and cleaning products), its role as a hydrogen carrier is also envisioned to support a variety of new applications such as transportation fuel, energy storage, and power generation. Thus, after ammonia is synthesized, it can be transported from regions with abundant renewable energy resources to those without such resources. Once there, it can be converted back to hydrogen via cracking, for the end user to use directly.

While ammonia was first discovered in the late nineteenth century, the modern ammonia process that is attributed to commercial-scale production was developed by German chemist, Fritz Haber, in the early twentieth century [18]. This process produced ammonia from the reaction of nitrogen gas and hydrogen gas over a catalyst at high temperature and high pressure, per the following chemical equation:

Ammonia synthesis reaction:

$$3H_2 + N_2 \rightarrow 2NH_3$$

The patents to Haber's process were purchased by BASF, and the process was further commercially developed, including catalyst improvements contributed by the company chemist, Carl Bosch, from whom, the second name in the Haber-Bosch ammonia process originated. The first commercial production of ammonia using the Haber-Bosch process began at a German BASF plant in 1913. The Haber-Bosch ammonia process has been modernized to increase process efficiency and ammonia production. Ammonia production technology is well proven and commercial designs range from about 45 metric tons per day (TPD) to 3,000 TPD production rates. Ammonia is typically produced using hydrogen generated from natural gas, which is readily available in large quantities and allows for large-scale ammonia production.

20.4.1.1 Synthesis and purification

"Green" ammonia synthesis requires hydrogen from electrolysis and nitrogen. Nitrogen can either be purchased from a pipeline, if nearby, or obtained from the air by a PSA unit or a cryogenic air separation unit. To produce ammonia, hydrogen and nitrogen are compressed in a molar ratio of 3:1 to a pressure of about 150bar. The mixture then goes through a feed/effluent heat exchanger to heat the gas before it goes to the ammonia converter, which typically has three catalyst beds in the pressure vessel, with the gas flowing in an axial/radial direction. Between the beds, heat exchangers heat the incoming feed and cool the gas coming out of each catalyst bed. The gas exiting the converter is typically at a temperature of 370–475°C and is cooled when it is used to produce steam. The gas then passes to a waste heat boiler, boiler feed water heater, and then the feed/effluent exchanger. It is then cooled against cooling water, and then, a refrigeration condenser to condense more ammonia.

The refrigerant used in the process is an open-loop ammonia system driven by a refrigeration compressor. The effluent from the refrigeration condenser then goes to a vessel to separate the liquid ammonia from the unreacted gases. The unreacted gases are recycled back to the converter for near-complete conversion of the hydrogen/nitrogen fed to the system. The only unreacted gas is the relatively small amount that is dissolved in the liquid ammonia and is flashed off at lower pressure. Utilities required for ammonia production include water for steam production and cooling water. The synthesis of ammonia is exothermic, and steam is produced downstream of the converter. This steam is used to drive the compressor required to compress the hydrogen/nitrogen to the required pressure.

20.4.1.2 Storage and transportation

Pure ammonia is a gas at atmospheric pressure/temperature and can be stored as a liquid at moderate pressure or as a liquid at −33°C under atmospheric pressure. Gas storage is favorable for small-scale applications where large quantities of product are not required. Ammonia production plants typically store ammonia in the refrigerated liquid phase, as the increased density allows them to store larger volumes for offtake and distribution of their product. The ammonia liquefaction process is also less expensive and challenging when compared to hydrogen liquefaction, because only a temperature of −33°C needs to be reached for liquefied ammonia storage, whereas liquefied hydrogen requires a temperature of −253°C. Thus, the roundtrip energy efficiency of producing ammonia and converting it back to hydrogen is greater than for hydrogen liquefaction, which when coupled with the superior volumetric energy density of ammonia is a strong enabler for use in energy applications.

20.4.2 Carbon capture

CCUS is an important technology in the hydrogen economy, as it allows decarbonization of conventional, fossil-based hydrogen production methods by capturing CO_2 generated from the process for either use or sequestration. Furthermore, captured CO_2 can be used as a feedstock for the production of chemicals such as methanol in combination with low-carbon hydrogen. For purposes of this chapter, post-combustion capture and direct air capture (DAC) technologies are discussed.

20.4.2.1 Post-combustion capture

Post-combustion technologies remove CO_2 from the flue gas produced out of combustion of a fuel such as in a natural gas or coal-fired power plant. The two primary methods of post-combustion CO_2 capture include solvent absorption and solid adsorption. Other emerging post-combustion technologies include chemical looping, separation membranes, and cryogenic separation.

The solvent absorption technology, as shown in Figure 20.1, uses a solvent with a high affinity for CO_2 (typically amine-based) to selectively capture CO_2 in a multi-tray or packed absorber column. The rich amine is regenerated in a stripper tower where the CO_2 is released, and the amine is recirculated to the absorber. Amine absorption systems are widely demonstrated with well-understood costs and reliability [19]. However, because of the large capital investment, solvent absorption systems are best suited for large-scale CO_2 capture implementation.

Solid adsorption systems capture CO_2 with a selective adsorbent material that bonds to CO_2 in the flue gas. As the flue gas passes over the sorbent surface, the sor-

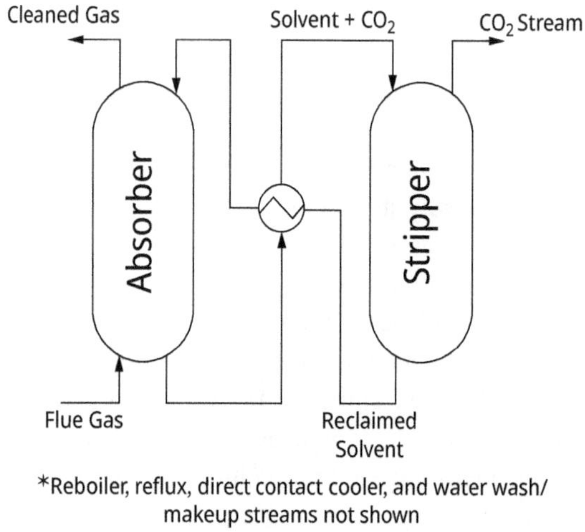

Figure 20.1: Representative solvent absorption CO_2-capture process flow diagram.

bent selectively retains CO_2 until it reaches saturation, at which point it must be regenerated to continue removal. Regeneration can occur via a temperature-swing, pressure-swing, or vacuum-swing process. To maintain a semicontinuous operation, the gas flow is typically switched to a secondary column, while the first column is regenerated. Adsorption systems are a newer technology compared to absorber systems and, generally, are more easily modularized, with the potential for reduced operating expenses compared to absorber systems. For this reason, adsorption systems are better suited for smaller-scale operations. Figure 20.2 depicts a high-level process flow arrangement for a solid adsorption system.

20.4.2.2 Direct air capture

DAC technology captures CO_2 directly from the atmosphere using solid and liquid sorbents. The solid DAC technology uses sorbent filters that chemically bind with the CO_2. The system operates by adsorbing the CO_2 and desorbing it at low pressure and temperatures between 80°C and 100°C. The liquid DAC technology uses two closed chemical loops, wherein the first loop contacts air with an aqueous solution that captures the CO_2 and the second loop releases the CO_2 from the solution and operates at high temperatures between 300°C and 900°C [20].

Captured CO_2 is sequestered in underground geological formations such as saline formations, oil and natural gas reservoirs, unmineable coal seams, organic rich shales, and basalt formations. The type of formation selected must be both safe and environmentally sustainable. To minimize the storage volume required, the CO_2 is stored underground as a supercritical fluid above temperatures of 30°C and pressures above 70bar. To prevent leakage, the formation is sealed using structural trapping, residual trapping, solubility trapping, or mineral trapping [21].

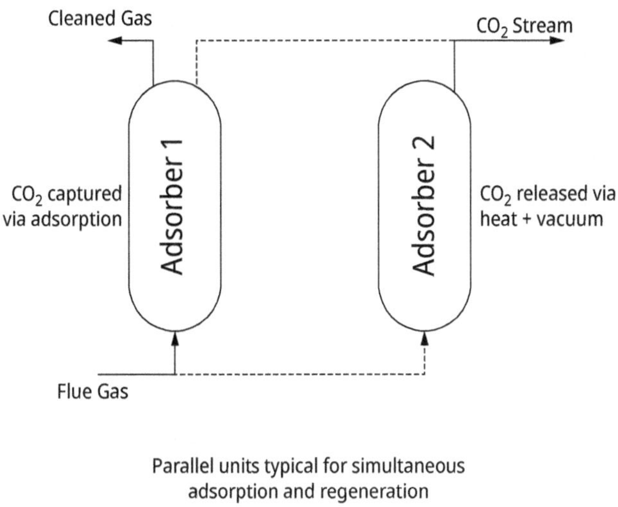

Figure 20.2: Representative physical adsorption CO_2-capture process flow diagram.

20.4.3 Green methanol

Alcohols, such as methanol, are organic compounds with at least one hydroxyl functional group that acts as an essential building block for the chemical industry. Currently, methanol is produced primarily using fossil resources; however, pathways for low-carbon methanol production using "green" or "blue" hydrogen, in combination with captured CO_2 or other "renewable" carbon sources (e.g., biomass) are of significant interest [22]. As with low-carbon ammonia, methanol is viewed as a promising hydrogen carrier that can take advantage of existing infrastructure for storage and transportation.

Methanol is produced using syngas feedstock, which is composed of CO_2, CO, and hydrogen. Traditional sources of syngas are produced using autothermal reforming, steam methane reforming, and coal gasification [23]. In a conventional gas-phase process, the syngas is fed to a methanol converter where the CO, CO_2, and hydrogen are reacted in a highly exothermic manner over a catalyst in a gas-phase fixed-bed reactor at pressures between 40 and 120 bar and at 200–320°C [24]. Once crude methanol is generated in the converter, it is distilled to remove water and byproducts of the synthesis process. Methanol is typically stored as a liquid at ambient conditions and transported to end users by ship, rail, and truck. Pipeline transportation of methanol is also possible [22].

20.4.4 Methanation and renewable natural gas (RNG)

Methanation is the hydrogenation of either CO or CO_2 to produce CH_4 and water vapor. When using hydrogen produced via electrolysis in the hydrogenation of CO or CO_2, the CH_4 produced is referred to as RNG. Both reactions are exothermic and are often used to produce steam that is needed in other unit operations in the overall RNG production facility. Methanation thermodynamic equilibrium experiments suggest that higher pressures and moderate temperatures favor the production of CH_4. Syngas methanation projects have been developed primarily since the 1970s and 1980s, as a result of interest in the production of "substitute"/"synthetic" natural gas from coal, which has given rise to several demonstration projects as well as a couple of commercialized concepts [25].

The most noteworthy commercial methanation scheme is the adiabatic fixed-bed concept. Fluidized-bed methanation (FBM) concepts have also been pursued in the past but have found limited success due to issues such as catalyst attrition, mechanical stresses during transients, and de-fluidization. Biological methanation technologies are also being developed as a means of using microorganisms to produce CH_4 at low temperatures and in stirred-tank or trickle-bed reactors [25].

In a traditional process, RNG is produced from the cleaned and shifted syngas via the hydrogenation of CO and CO_2 in a series of fixed-bed reactors according to the exothermic methanation reactions shown in the following equations:

CO methanation reaction

$$CO + 3H_2 \rightarrow CH_4 + H_2O, \quad \Delta H = -206.2 \text{ kJ/mol}$$

CO_2 methanation reaction

$$CO_2 + 4H_2 \rightarrow CH_4 + 2H_2O, \quad \Delta H = -165 \text{ kJ/mol}$$

To protect the methanation catalyst from sulfur poisoning, the feed is first passed through a sulfur guard bed to remove traces of sulfur compounds. The syngas is then mixed with recycled gas to control the maximum temperature rise of the reactor and then passed to the methanation reactor(s). The exothermic methanation reaction results in a high outlet temperature, which allows generation of superheated high-pressure steam in downstream heat exchangers. After cooling, the partially methanated syngas passes through additional methanation reactors in series for complete conversion of the CO into CH_4. The process stream leaving the final methanation reactor is cooled and sent to the final drying unit and compressed to meet pipeline specifications for natural gas [26].

In an RNG production operation that uses FBM, some noteworthy simplifications can be made. According to the Paul Scherrer Institute (PSI), FBM can enable fewer tar removal units, eliminate the need for olefin and carbonyl sulfide hydrotreating, and replace the WGS unit and series of methanation reactors, with a single FBM reactor, resulting in up to 33% reduction in capital costs [27]. As with many fluidized-bed reactors, FBM allows reactions to proceed at suboptimal conditions: lower operating temperatures, less than stoichiometric hydrogen to CO/CO_2 ratios, and less carbon deposition on catalyst surfaces [28, 29]. Prior research has also shown that the addition of three to four percent steam and maintaining a CO_2:CO ratio of less than 0.52 in the feed to an FBM process positively impacts the carbon deposition avoidance and increases the conversion of hydrogen, respectively [30]. However, some literature notes a couple of drawbacks for FBM compared with fixed beds for RNG production, such as limitations in scale-up and lower potential for the economic cogeneration of RNG and power [28, 31].

The RNG generated from methanation can be used onsite where it is generated for vehicle fuel, power generation, or thermal applications. More often, RNG is injected into a natural gas pipeline network for distribution to numerous users, assuming it meets the specifications of that particular gas utility or pipeline owner/operator [32].

20.4.5 Other fuels and chemicals

Liquid organic hydrogen carriers (LOHCs) are defined by reversible hydrogenation and dehydrogenation reactions. The benefit of LOHCs is that they are chemically stable and easy to be transported long distances because they exist in the liquid state at ambient conditions [33]. One of the more established examples of LCOHs is methyl cyclohexane (MCH) and toluene, where MCH is the hydrogenated form and toluene is the dehydrogenated form of the LCOH pair. Toluene, like most dehydrogenated LCOHs, is aromatic with a relatively low gravimetric hydrogen storage density. Once hydrogenated, MCH has a high volumetric hydrogen storage density of 47 kg/m^3. The process for dehydrogenation of MCH and other cyclic hydrocarbons is well established in industry and uses noble metal catalysts such as platinum or palladium. The process is endothermic and requires a heat source to reach the high temperature requirements, which for MCH is 270–290°C. To hydrogenate toluene into MCH, hydrogen is supplied at elevated temperatures (130–200°C) and pressures (10–50 bar) to a reactor with toluene. The hydrogenation reaction also uses a noble metal catalyst; ruthenium is most commonly used for hydrogenated toluene [15].

Hydrogenation of other compounds, including naphthalene and benzyl toluene, is possible. As a result, other potential LOHC systems exist, including decalin, perhydro-dibenzyltoluene, perhydro-benzyltoluene, dodecahydro-*N*-ethyl carbazole, 1-methylperhydroindole, 2-methylperhydroindole, 1,2-perhydrodimethylindole, perhydro-phenazine, and perhydro-2-(*N*-methyl benzylpyridine) [34]. However, these chemicals are still somewhat developmental and are in the early stages of commercialization; thus, their performance and economic competitiveness with toluene/MCH, as well as other hydrogen carriers, are not yet known.

20.4.6 End-use applications

P2X products, synthesized using hydrogen, have a wide range of end-use applications. Ammonia is primarily used as a nitrogen fertilizer but does show potential as a fuel and energy carrier, as discussed previously. Similarly, methanol is a building block for a host of different products, including acetic acid, methyl methacrylate, silicone, olefins, and formaldehyde. These chemicals are used to produce a variety of products, ranging from adhesives and paints to plastics and more. Additionally, methanol can be used as a fuel. RNG produced via methanation has similar end-use applications as fossil-based natural gas and is used as a fuel in both heat and power generation applications. LOHCs are expected to continue to be commercialized and eventually play a role in the large-scale transportation of hydrogen in a more stable form.

20.5 Project development and delivery

The development and delivery of hydrogen and P2X projects mirrors the steps that are undertaken in many different parts of large, complex infrastructure projects, such as in the power generation, chemical process, and the bio products industries. The development of low-carbon hydrogen and P2X infrastructure projects has increased dramatically, of late, as interest in hydrogen as a versatile fuel, feedstock, and energy carrier has increased manifold. The International Energy Agency (IEA), among other authorities, maintains a database of global infrastructure projects to produce hydrogen from low-carbon resources for use across the decarbonization landscape [35]. Some of the major commercial-scale projects that have reached a final investment decision (FID) since 2020 are summarized in Table 20.4.

"Green" hydrogen and P2X projects are mainly being pursued for their potential benefits in reducing key air emission criteria pollutants and greenhouse gas (GHG) emissions. These benefits are realized primarily in reducing or displacing the use of fossil-based fuels and feedstocks. However, GHG emissions reductions are highly dependent on the feedstock source, conversion process, transportation logistics, and end-use application. The life cycle GHG emission avoidance of these products are computed based on a rigorous life cycle analysis, thereby resulting in a carbon intensity score assigned to the specific fuel or feedstock. Most "green" hydrogen and P2X technologies offer a reduction in carbon intensity of greater than 50%, and their versatility makes them highly complementary with electrification and bioproducts initiatives of several industries.

Table 20.4: Major global hydrogen and P2X infrastructure projects.

Project name	Location	Capacity, investment, and end use	Status
Air Products Net-Zero Hydrogen Energy Complex [36]	Edmonton, Alberta, Canada	520 ktpa $1.3B (CAD) Merchant, Power Generation	FID
Inner Mongolia Green Hydrogen Project [37]	Ordos, Inner Mongolia, China	67 ktpa Mobility, Chemicals	FID
HYBRIT [38]	Svartöberget Luleå, Sweden	58 ktpa 200M (SEK) Iron/Steel	Construction
Advanced Clean Energy Storage [39, 40]	Delta, Utah, USA	36 ktpa $500M (USD) Energy Storage	Construction

20.5.1 Project development life cycle

Major hydrogen and P2X projects often begin with several critical strategy and planning activities, which can vary quite a bit, depending on the target markets for product offtake, the geography, and set of enabling technologies that must be employed to ensure success. After the conception of the project, a developer will often focus on establishing the technical and financial viability via a feasibility study. Such endeavors focus on the market forces driving the end-product sales and the principal technologies, systems, and processes that are needed to support a profitable asset. For example, many "green" hydrogen project developers will first explore the design of an electrolysis plant to be integrated with a renewable energy asset, what the associated capital or O&M costs are, and whether the "levelized" cost of hydrogen from the facility supports their target offtake markets and price points.

Once the feasibility of a project is established, the developer will typically turn his/her attention to advancement in the design to support financing, offtake contract negotiations, permits, and final budgetary authorization to proceed. This can involve an assortment of activities with some variation in roles, deliverables, and arrangement, but is often referred to in the chemical process industry as the front-end loading (FEL) process. FEL-1 through FEL-3 refer to a stage-gate procedure that includes a standardized set of technical deliverables (and associated improved accuracy in cost estimates at each stage) to advance the project through front-end engineering design, until it reaches the point where it can be firmly bid for full execution. Figure 20.3 illustrates the full life cycle of an infrastructure project within the context of the broader decarbonization industry, including hydrogen, P2X, and bioproducts.

20.5.2 Technology scaling and project delivery

For projects that involve an advanced technology, it is critical that appropriate measures are taken to scale that technology, so it can be seamlessly integrated into the infrastructure asset when the project is ready for execution. For example, an RNG project may involve a cutting-edge gasification or carbon capture technology, or a "green" hydrogen project may involve an advanced electrolyzer technology; in both cases, they would need to be scaled and de-risked in parallel with their implementation into full-scale facilities. This requires a combination of strategic planning and research, development, and demonstration (RD&D) activities to ensure success. A developer often will need to make critical decisions on how to manage their intellectual property and determine a strategy for the commercialization of the advanced technology (i.e., spin-off, start-up, and licensing). As the technology progresses through various technology readiness levels during RD&D, it will become critical to consider how the technology will be manufactured and integrated into pilot, demonstration, and commercial-scale facilities, including how each phase will be financed [41].

Upon successful scaling of the technology and during the late stages of an infrastructure project development, attention can be turned to the critical quality variables and associated costs for the manufactured product. These will have a significant impact on formulating a project execution strategy, including the form of contract with the engineering, procurement, and construction (EPC) contractor, the performance risks to be shared between the stakeholders of the project, and the associated impacts on cost and schedule. Figure 20.4 shows a few project delivery contracting alternatives and how risk transfer, flexibility, and cost/schedule can be balanced. However, note that this is a very simplified depiction of the various considerations, and the selection of a project execution strategy is a complex topic that should be carefully studied prior to moving forward with a specific approach [42].

Once a project is financed and permitted, and an EPC contractor is hired, it may proceed into detailed engineering, equipment/material procurement, and construction. Throughout execution, the roles of the project owner, owner representatives, the EPC contractor, and all other stakeholders will need to be carefully managed. The EPC contractor will have rigorous procedures for execution of EPC, and then, commissioning and facility turnover that will be administered through meetings and various project control tools, in which the owner and their representatives play a key monitoring and approval role. Environmental health and safety, quality management, and risk management best practices will also be employed by project management professionals in a synergistic manner throughout the execution.

Strategy (current state assessment, technology scenarios, market and regulatory view, provider landscape, economic modeling across business, project road mapping); investment planning (ranking of project investment options, co-optimize cost saving with CO_2 savings as applicable, identification and decision-making support for initial investments); preliminary feasibility and design (project feasibility, detailed site as-

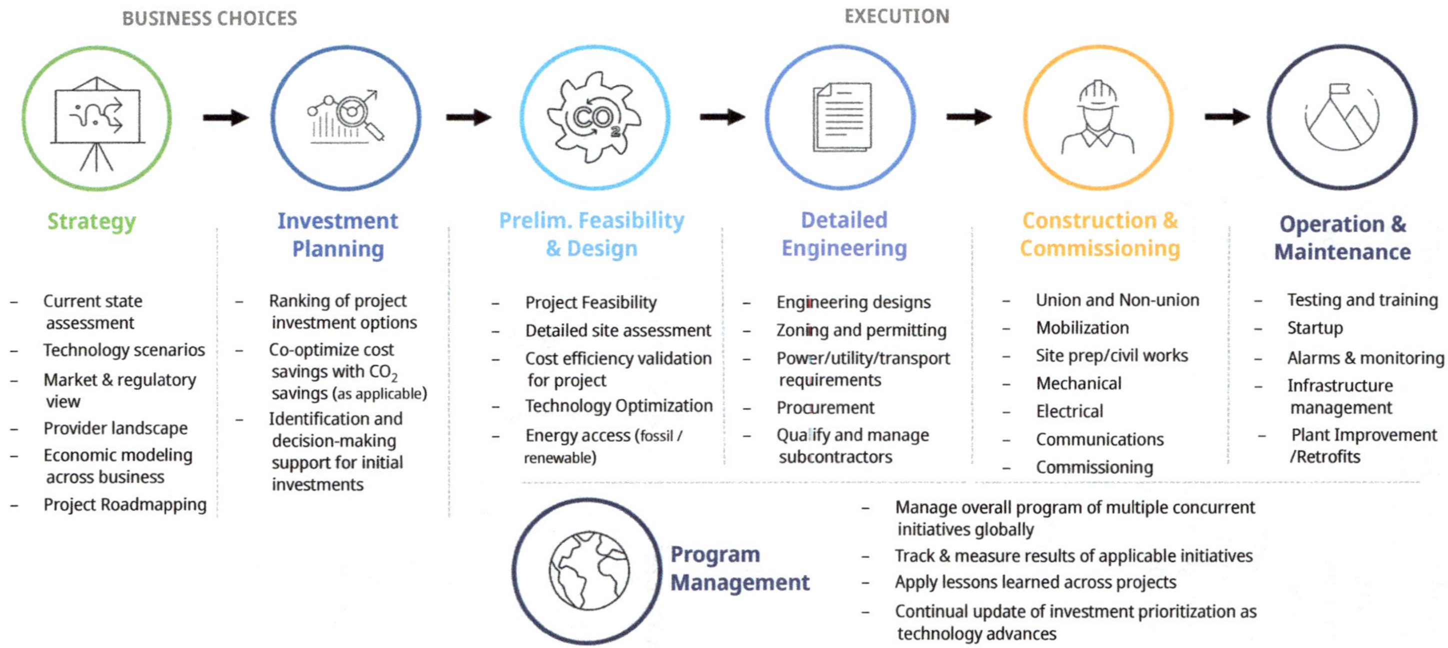

Figure 20.3: Infrastructure project development life cycle.

sessment, cost-efficiency validation for project, technology optimization, energy assessment – fossil/renewable); detailed engineering (engineering designs, zoning and permitting, power/utility/transport requirements, procurement, qualify and manage subcontractors); construction and commissioning (union and non-union, mobilization, site prep/civil works, mechanical, electrical, communications, commissioning); O&M (testing and training, start-up, alarms and monitoring, infrastructure management, plant improvement/retrofits); and program management (manage overall program of multiple concurrent initiatives globally, track and measure results of applicable initiatives, apply lessons learned across projects, continual update of investment prioritization as technology advances).

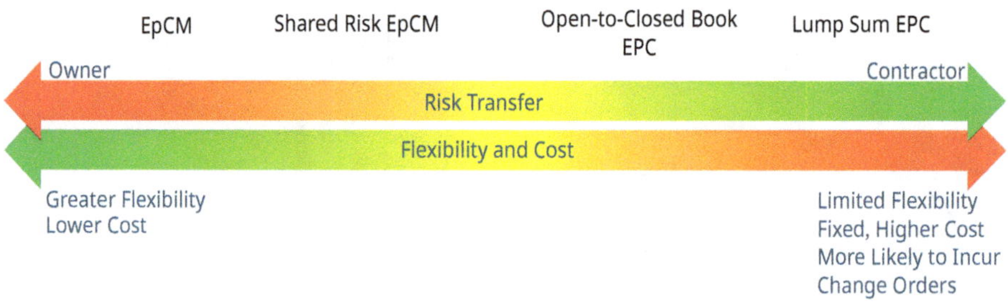

Figure 20.4: Project delivery contracting alternatives.

20.5.3 Operations and maintenance (O and M)

After the project owner has accepted the handover from the EPC contractor, the longest portion of the project life cycle begins via operations. This initially entails training of operators and maintenance staff so that a safe and profitable work environment can be maintained throughout the asset life. For complex hydrogen and P2X projects, operations can sometimes rely on numerous automated functions, but will frequently require ongoing asset monitoring and intervention, particularly as equipment begins to age and failures are experienced. A carefully executed management plan is key to successful operations and planned/unplanned maintenance outages. As new technologies are developed and commercialized, it is also advantageous to explore opportunities for plant improvements and retrofits to ensure that the asset profitability is maximized. At the end of the project life, which is typically 20–30 years for hydrogen and P2X projects, the plant will be retired and decommissioned, after which the major equipment can be recycled or repurposed and the site can be used for new means.

20.6 Conclusions

Hydrogen is one of the most abundant and diverse chemical substances in the universe, in terms of its production and consumption. These fundamental qualities have driven significant interest in hydrogen as a means for the decarbonization of numerous industries across the chemical, energy, and heavy industrial landscape. The synergies between hydrogen and the bioproducts industry are innumerable and vast, with respect to the supplantation of fossil-based resources in favor of the union of renewable carbon from biomass and/or carbon capture with renewable electric power resources. The delivery of infrastructure associated with low-carbon hydrogen and bioproducts also has many noteworthy similarities, further affirming their cross-applicability.

References

[1] Vichard L, Harel F, Ravey A, Venet P, Hissel D Degradation prediction of PEM fuel cell based on artificial intelligence. Int J Hydrogen Energy 2020;29(45):14953–14963.

[2] Brauns J, Turek T Alkaline water electrolysis powered by renewable energy: A review. Processes 2020;8:248.

[3] Pozio A, Bozza F, Nigliaccio G, Platter M, Monteleone G Development perspectives on low-temperature electrolysis. Enea 2021;1:66–72.

[4] Hauch A, Küngas R, Blennow P, Hansen AB, et al. Recent advances in solid oxide cell technology for electrolysis. Science 2020;370:6513.

[5] Kuckshinrichs W, Ketelaer T, Koj TC Economic analysis of improved alkaline water electrolysis. Front Energy Res 2017;5.

[6] Christensen A Assessment of Hydrogen Production Costs from Electrolysis: United States and Europe, Washington: International Council on Clean Transportation; 2020.

[7] Schmidt O, Gambhir A, Staffell I, Hawkes A, Nelson J, Few S Future cost and performance of water electrolysis: An expert elicitation study. Int J Hydrogen Energy 2017;52(42):30470–30492.

[8] Felgenhauer M, Thomas Hamacher T State-of-the-art of commercial electrolyzers and on-site hydrogen generation for logistic vehicles in South Carolina. Int J Hydrogen Energy 2015;5 (40):2084–2090.

[9] Marcelo C, Fritz DL, Mergel J, Stolton D A comprehensive review on PEM water electrolysis. Int J Hydrogen Energy 2013;12(38):4901–4934.

[10] Hydrogen production: Natural gas reforming. Office of Energy Efficiency & Renewable Energy. (Accessed June 2022 at https://www.energy.gov/eere/fuelcells/hydrogen-production-natural-gas-reforming).

[11] Chen CJ, Back MH, Back RA Mechanism of the thermal decomposition of methane. Am Chem Soc Symp Ser 1976;1–16.

[12] Dagle RA, Deagle V, Bearden MD, Holladay JD, Krause TR, Ahmed S An Overview of Natural Gas Conversion Technologies for Co-Production of Hydrogen and Value-Added Solid Carbon Products, USDOE Office of Energy Efficiency and Renewable Energy; 2017.

[13] Sánchez-Bastardo N, Schlögl R, Ruland H Methane pyrolysis for zero-emission hydrogen production: A potential bridge technology from fossil fuels to a renewable and sustainable hydrogen economy. Ind Eng Chem Res 2021;60:11855–11881.

[14] Ahmed S, Aitani A, Rahman F, Al-Dawood A, Al-Muhaish F Decomposition of hydrocarbons to hydrogen and carbon. Appl Catal A Gen 2009;359:1–24.

[15] Andersson J, Grönkvist S Large-scale storage of hydrogen. Int J Hydrogen Energy 2019;23(44):11901–11919.

[16] Tan-Peng C, Kelly B Hydrogen Delivery Infrastructure Options Analysis, US Department of Energy Hydrogen Program; 2006. Accessed June 2022 at. www.hydrogen.energy.gov/pdfs/progress07/iii_a_1_chen.pdf.

[17] Domptail K, Hildebrandt S, Hill G, Maunder D, Taylor F, Win V Emerging Fuels – Hydrogen State of the Art, Gap Analysis, and Future Project Roadmap, Chantilly: Pipeline Research Council International Inc; 2020.

[18] Venkat P, Richardson J Introduction to Ammonia Production, New York: American Institute of Chemical Engineers; 2016.

[19] Meeting the Dual Challenge: A Roadmap to at-Scale Deployment of Carbon Capture, use, and Storage. Chapter 5 – CO_2 Capture. National Petroleum Council; 2021.

[20] Budinis S Direct air capture: A key technology for net zero. Int Energy Agency 2022;16.

[21] Carbon Storage FAQs. National Energy Technology Laboratory 2022. (Accessed June 2022 at https://www.netl.doe.gov/carbon-management/carbon-storage/faqs/carbon-storage-faqs).

[22] Kang S, Boshell F, Goeppert A, Prakash SG, Landälv I Innovation Outlook: Renewable Methanol, Abu Dhabi: International Renewable Energy Agency; 2021.

[23] Methanol Production. Methanol Institute 2016. (Accessed June 2022 at https://www.methanol.org/wp-content/uploads/2016/06/MI-Combined-Slide-Deck-MDC-slides-Revised.pdf).

[24] Syngas Conversion to Methanol. National Energy Technology Laboratory, 2022. (Accessed June 2022 at https://netl.doe.gov/research/coal/energy-systems/gasification/gasifipedia/methanol).

[25] Rönsch S, Schneider J, Matthischke S, et al. Review on methanation – From fundamentals to current projects. Fuel 2016;166:276–296.

[26] Schmider D, Maier L, Deutschmann O Reaction kinetics of CO and CO_2 methanation over nickel. Ind Eng Chem Res 2021;60:5792–5805.

[27] Biollaz SMA, Schildhauer TJ, Jansohn P Status PSI – R&D on Fluidised Bed Methanation, Würenlingen: Paul Scherrer Institute; 2016.

[28] Rönsch S, Kaltschmitt M Bio-SNG production – Concepts and their assessment. Biomass Convers Biorefin 2012;2:285–296.

[29] Liu B, Ji S Comparative study of fluidized bed and fixed bed reactor for syngas methanation over Ni-W/TiO2-SiO2 catalyst. J Energy Chem 2013;22:740–746.

[30] Liu J, Cui D, Yu J, Su F, Xu G Performance characteristics of fluidized bed syngas methanation over Ni-Mg/Al2O3 catalyst. Chin J Chem Eng 2015;23:86–92.

[31] Heyne S Bio-SNG From Thermal Gasification – Process Synthesis, Integration, and Performance, Chalmers University of Technology Sweden; 2013.

[32] Renewable Natural Gas. Environmental Protection Agency United States, 2022. (Accessed June 2022 at https://www.epa.gov/lmop/renewable-natural-gas).

[33] Kurosaki D Introduction of Liquid Organic Hydrogen Carrier and The Global Hydrogen Supply Chain Project, Chiyoda Corporation, Advanced Hydrogen Energy Chain, Association for Technology Development; 2018.

[34] Rao PC, Yoon M Potential liquid-organic hydrogen carrier (LOHC) systems: A review on recent progress. Energies 2020;13.

[35] Hydrogen Projects Database. International Energy Agency, 2021. (Accessed June 2022 at https://www.iea.org/reports/hydrogen-projects-database).

[36] Air Products Announces Multi-Billion Dollar Net-Zero Hydrogen Energy Complex in Edmonton. Alberta, Canada: Air Products Inc; 2021. Accessed June 2022 at, https://www.airproducts.com/news-center/2021/06/0609-air-products-net-zero-hydrogen-energy-complex-in-edmonton-alberta-canada.

[37] China approves renewable mega-project for green hydrogen. The Straits Times, 2021. (Accessed June 2022 at https://www.straitstimes.com/business/economy/china-approves-renewable-mega-project-for-green-hydrogen).
[38] HYBRIT: SEK 200 million invested in pilot plant for storage of fossil-free hydrogen in Luleå. Vattenfall AB, 2019. (Accessed June 2022 at https://group.vattenfall.com/press-and-media/pressreleases/2019/hybrit-sek-200-million-invested-in-pilot-plant-for-storage-of-fossil-free-hydrogen-in-lulea).
[39] US DOE Closes $504.4 Million loan to advanced clean energy storage project for hydrogen production and storage. Mitsubishi Power Americas, 2022. (Accessed June 2022 at https://power.mhi.com/regions/amer/news/20220609).
[40] World's Largest Green Hydrogen Hub Selects Black & Veatch as EPC Provider. Black & Veatch Corporation, 2022. Access June 2022 at, https://www.bv.com/news/hydrogen-hub-selects-black-veatch.
[41] Technology Readiness Level Definitions. National Aeronautics and Space Administration United States 2017. (Accessed June 2022 at https://esto.nasa.gov/files/TRL_definitions.pdf).
[42] Shahani G Comparative analyses of common project execution alternatives. Hydrocarbon Processing, 2019. (Accessed June 2022 at https://www.hydrocarbonprocessing.com/magazine/2019/april-2019/project-management/comparative-analyses-of-common-project-execution-alternatives).

Section V: **Bioplastics and biopolymers**

Ashok Adur

Chapter 21 Renewable green biopolymers and bioplastics

21.1 Introduction

21.1.1 What are biopolymers and bioplastics?

Biopolymers are polymeric materials produced from renewable biomass, mostly plant-based sources found in nature. Natural biopolymers are produced by the cells of living organisms. Examples of these include corn starch, straw, sawdust, woodchips, recycled food waste, as well as various vegetable fats and oils called lipids from either plants or animals [1]. As covered in earlier chapters in this book, some feedstocks for biopolymers are obtained by chemical processing directly from polysaccharides (e.g., starch, cellulose, chitosan, and alginate) and proteins (e.g., soy protein, gluten, and gelatin), or biologically generated by fermentation of sugars or lipids. Others are chemically synthesized or derivatized from sources like lactic acid or various cellulose derivatives. On the other hand, common traditional plastics, such as fossil-fuel plastics (also called petro-based polymers) are derived from petroleum or natural gas or in a few cases from coal. In any case, all polymers are made of repetitive units called monomers [2]. So, how can one distinguish between biopolymers and synthetic polymers? One option is based on their molecular weight distribution in their chemical structure. Most natural biopolymers typically contain similar sequences and numbers of monomers and thus all have the same chain length and hence mass or a very narrow molecular weight distribution (M_w/M_n) of 1. This phenomenon is also called monodispersity in contrast to the polydispersity encountered in synthetic polymers. As a result, many but not all biopolymers have a dispersity of 1 [3].

Natural biopolymers can be classified into three main families according to the monomers used and the structure of the biopolymer:

- Polynucleotides [4], such as RNA and DNA, are long polymers composed of 13 or more nucleotide monomers.
- Polypeptides [5] and proteins are polymers of amino acids and some major examples include collagen, actin, and fibrin.
- Polysaccharides are linear or branched polymeric carbohydrates and examples include starch, cellulose, and alginate [6].

Ashok Adur, Everest International Consulting, LLC, Jacksonville, FL 32256, USA,
e-mail: aadur@outlook.com

 | https://doi.org/10.1515/9783112212141-021

From an organic chemistry perspective, biopolymers can be classified based on the type of repeat chemical bonds in each monomeric part or in the way each monomer unit is linked to the next one in the polymer structure:

- Polyesters: Examples are polylactic acid (PLA), polyhydroxyalkanoate (PHA), and poly-3-hydroxybutyrate (PHB).
- Polyamides: Examples are proteins and polypeptides like silk, collagen, fibrin, actin, and zein.
- Cellulosics: Examples include cellulose, starch, chitosan, and alginate.
- Unsaturated olefinics: Examples include pinenes and terpenes as well as natural rubber (polyisoprene).

Nature is far more complex to allow for simple classifications like those above. Examples of biopolymers found in nature with complex structures are suberin, cutin, and lignin, which are complex polyphenolic polymers, cutan which is a complex polymer of long-chain fatty acids, and melanin, which is a broad term for a group of natural pigments found in most organisms, produced through a multistage chemical process involving oxidation of the amino acid tyrosine followed by polymerization.

While most people think biopolymers and bioplastics are recent inventions, excavations have shown that people in Meso-American cultures (Olmec, Maya, Aztecs) used natural latex and rubber to make balls, containers, and make their clothes waterproof around 1500 BCE [7].

Many natural biopolymers are further processed and chemically modified to make them useful for various applications. Also, some industrial polymers produced today use bio-based monomers while some petro-based polymers and some bio-based polymers use bio-based plasticizers, lubricants and other additives, which are compounded in to modify the properties of both bio-based and petro-based polymers (plastics and rubber/elastomers) to make them suitable for large scale applications.

Biopolymers that can be processed like thermoplastics or thermosets are called bioplastics. A thermoplastic is a polymer material that becomes pliable or moldable at a certain elevated temperature and solidifies upon cooling, so it can be shaped using polymer processing techniques such as injection molding, blow molding, compression molding, calendaring, or extrusion processes, while a thermoset is produced by irreversible chemical bonds during the curing process and does not melt when heated, but typically decomposes and does not reform upon cooling. While all bioplastics are biopolymers, all biopolymers are not bioplastics. For example, wool and silk are biopolymers but not bioplastics. So, bioplastics are actually a subset of biopolymers but not synonymous, although many people wrongly use both terms interchangeably.

This chapter covers biopolymers that are used as plastics with emphasis on those not covered in other chapters in this book. This includes a few non-bio-based polymers but are biodegradable and/or compostable, even though they are sourced from fossil fuels like polyvinyl alcohol (PVOH), poly(butylene succinate) (PBS), and polybutylene adipate-terephthalate (PBAT). PBS and PBAT are covered in another chapter.

Some bio-feedstocks that are not covered in other chapters in this book will also be covered in this chapter, especially those that are used for bioplastics. In addition, natural biopolymers that are chemically modified like cellulose acetate, cellulose butyrate, and other cellulose derivatives as well as vulcanized natural rubber, which are of industrial importance will also be covered briefly. These sections will be followed by a discussion of limitations of biopolymers and challenges for growth in the future. Despite these limitations, biopolymers do have a bright future, especially if some of these limitations can be overcome. This chapter will end with a perspective on future trends and prospects for growth.

In most cases, polymers (plastics and rubber) whether they are bio-based or not, are rarely used without the addition of additives like plasticizers, lubricants, antioxidants and stabilizers, fillers, colorants, reinforcements, etc., in order to meet various customer requirements all the way downstream in the value chain. These additives are added during a compounding process step. Such compounds based on biopolymers as well as biocomposites, which are compounds where reinforcements are used with biopolymers will be covered in a separate chapter, whether the polymer and/or the reinforcement is bio-based in the composite.

There are numerous academic and other research institutions, primarily funded by governments and foundations, which publish valuable information on biopolymers. A vast majority of these biopolymers may not get commercialized. The reasons for this are numerous. In many cases, the researchers who have worked on conducting such research either are not interested in overcoming or are not able to overcome some of the important but necessary challenges that exist for commercialization. These include how to overcome some of these limitations of their biopolymers besides challenges of starting a new business as a start-up like scale-up processes, financing, time needed to develop new products, applications and markets, adequate capital until their sales grow to generate sufficient profits, or the contacts needed. In some cases, it is critical for a new entrant to meet an unmet need, to replace an existing product with a new entrant, or to offer better performance per unit net price to succeed. So, some of these biopolymers that are unlikely to get commercialized will be referenced but the emphasis will be commercialized or products likely to be commercialized. Many already are.

21.1.2 Global production capacities of bioplastics

With a mounting number of materials, applications, and products, along with development of technology, the number of manufacturers, converters, and end-users are also rising steadily. Significant financial investments have been made in production and marketing to guide and accompany this development, especially in Europe.

Figure 21.1 shows the estimated global production capacities of bioplastics in 2021 and 2024 by region [8]. The Asian region shows the largest production capacity, followed by Europe and then North America. On a per capita basis, due to the much higher population in Asia, the growth rates there will continue to be higher than the rest of the world. China's bio-based polymer capacity is projected to grow from 765,000 MT in 2023 to 2.53 million MT by 2026, driven by policy incentives and infrastructure investment [9]. India's bioplastics production capacity is growing steadily, though the estimated 25,000 tons still represents a small fraction of the global output, mostly PLA, PHA and starch blends, driven by bans on single-use plastics and rising awareness of sustainability [10].

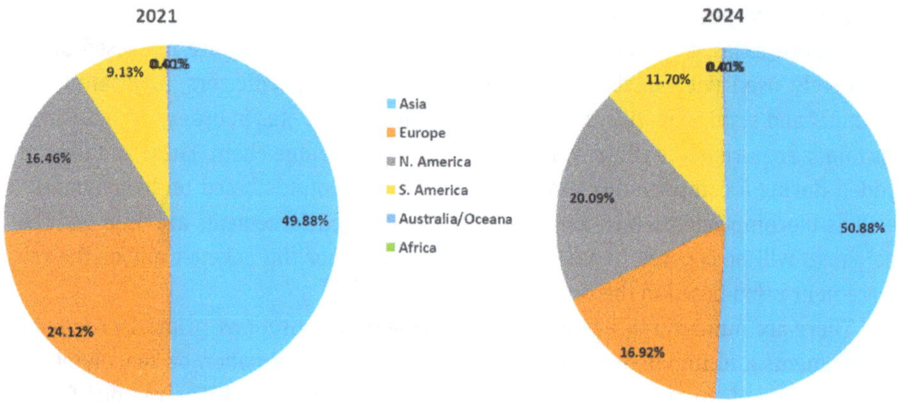

Figure 21.1: Global capacity of bioproducts by region.

The latest figures available show the global production volume of bio-based polymers is listed by the European Bioplastics e.V for 2024 is 2.469 million tons [8] but if cellulose acetate is added to this figure, the total jumps to 4.4 million MT in 2023, representing about 1% of total polymer production with a forecasted CAGR of 17% from 2023 to 2028 [9], significantly outpacing fossil-based polymers, which partly due to the latter being nearly 2 orders of magnitude higher, grew 2–3%.

Figure 21.2 shows the breakdown of biodegradable plastics by market segment based on estimates. For biodegradable plastics, packaging is the largest segment with flexible packaging being almost double that of rigid packaging, followed by agriculture and horticulture and consumer goods.

Figure 21.3 shows the global production capacities of bioplastics separated by each of the major specific biopolymers in 2021 for various market segments.

Figure 21.4 shows the production capacities by type of biopolymer comparing the quantities from 2024 and the projected values for 2029, which shows significant growth expected in those five years with PLA and PHA growing much faster than all of the other bioplastics shown. The capacity for producing PHA (polyhydroxyalkanoates) increased by 40%, despite moderate utilization rates [11]. Two others, casein polymers and PEF (polyethylene furanoate) are also gaining traction, with expected significant growth by 2029 [11].

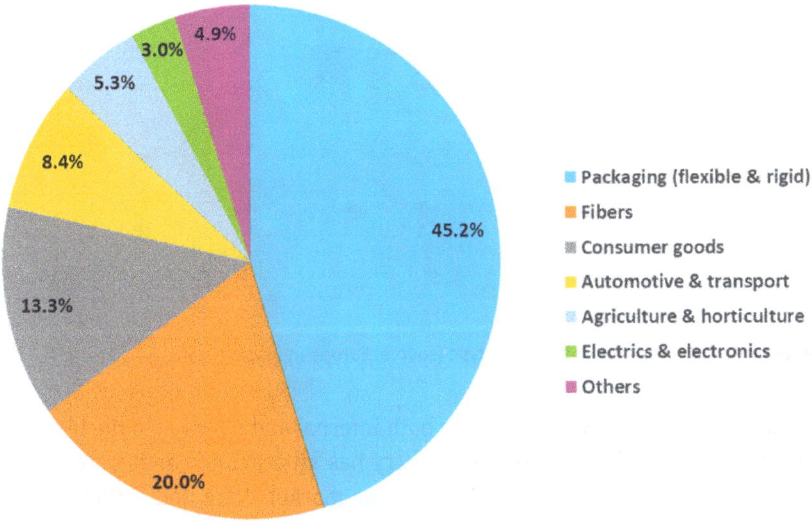

Figure 21.2: Applications of biodegradable plastics by market segment.

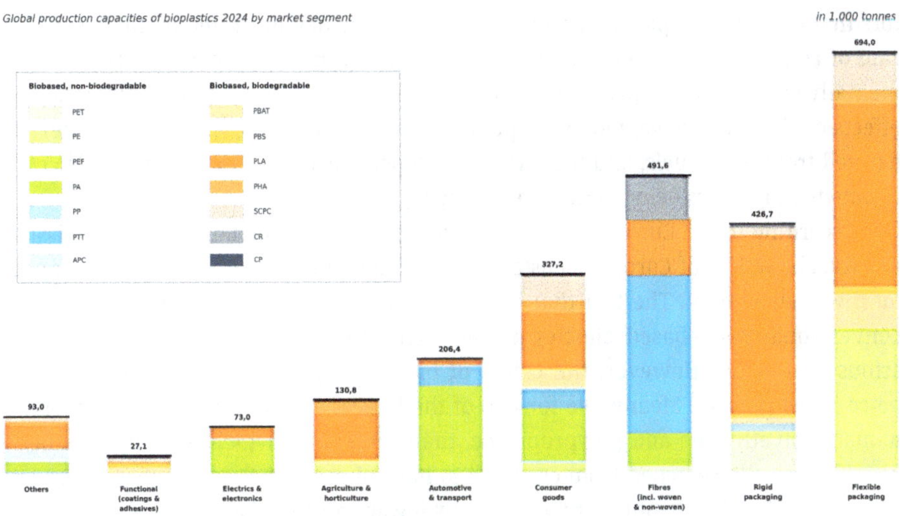

Figure 21.3: Biodegradable plastics in 2024 by market segment and biopolymer type.

This set of latest market data from all four figures are from www.bio-based.eu or based on their data. It shows and demonstrates the contributions of the industry in moving towards a sustainable future with reduced environmental impact. Their forecast also predicts that the budding bioplastics industry will continue to show enormous economic potential growth over the coming decades [8].

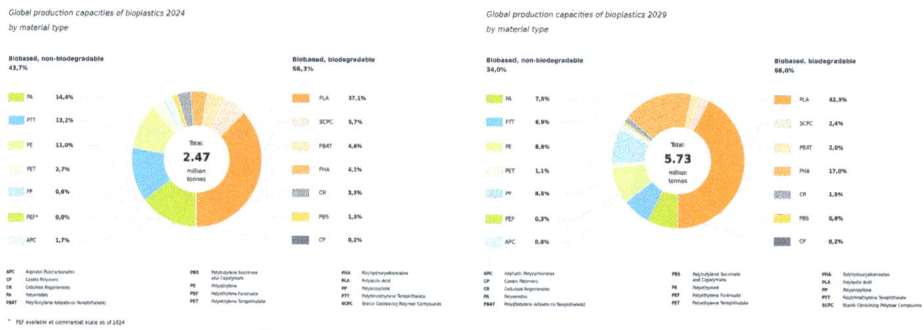

Figure 21.4: Biodegradable plastics by bioplastic type (source: European Bioplastics).

The factors driving market development are both internal and external to the biopolymer industry. Obviously, the biopolymer industry has an economic and professional bias to drive market development and increase the market share and acceptance of biopolymers and bioplastics globally. Internal factors are based on the growth of technology that is being developed along with various numbers of players at all stages of the value chain covered in the first paragraph of this section. There are external factors that drive the bioplastics growth too. This can be demonstrated by the escalating rate of consumer acceptance, as the general population has become more environmentally conscious, not just in Europe but also globally. The extensively publicized effects of climate change, mounting prices of coal, oil and natural gas, and materials derived from fossil fuels, and the increasing dependence on fossil resources also contribute to bioplastics being viewed favorably [8].

According to the Eurobarometer Survey conducted by the European Commission in 2020, about 90% of European customers want to buy products with minimal impact on the environment. The fact that bio-based plastics display clear advantages over conventional petro-based plastics makes them attractive to environmentally conscious customers. However, the prices of bio-based products are higher than the petro-based plastics. Meanwhile for two of the largest plastics that are not bio-based, namely polyethylene and polypropylene, the source from which they are produced has dramatically switched from crude oil to natural gas, which is relatively more environmentally friendly but more importantly lower cost for producing the monomers.

Bioplastics are believed to efficient and technologically mature materials by European Bioplastics e.V. because they are able to improve the balance between the environmental benefits and the environmental impact of plastics. Life cycle analyses (LCAs) demonstrate that bioplastics can significantly reduce carbon dioxide (CO_2) emissions compared to conventional plastics (depending on the material and application). Furthermore, the increasing utilization of biomass in bioplastic applications has two clear advantages: renewability and availability [8]. However, the LCA data also show that other factors such as energy consumption and water requirements of bioplastics are much higher over the full “cradle-to-grave” life cycle on a per MT basis and especially on a per cubic meter basis, because the average bioplastic has a higher specific gravity than a polyolefin, typically produced from fossil-based source.

According to a recent market study by Markets and Markets™ Inc., the global bioplastics and biopolymers market size had projected to grow at a compound annual growth rate of 23% from USD 11 billion in 2021 to USD 30 in 2026 [12].

Companies like NatureWorks (USA), Braskem (Brazil), BASF (Germany), Total's Corbion and Synbra (both in the Netherlands), Novamont (Italy), Biome Bioplastics (UK), Mitsubishi Chemical Holding Corporation (Japan), CJ Bio (Korea), Biotec and KFUR Kunstoff (both in Germany), Cardia BioPlastics and Plantic Technologies (both in Australia), Toray Industries (Japan), Futerro (Belgium), PTT Global Chemical Public Company Limited (Thailand), and Zhejainag Hilsun Biomaterials and Tianan Biologic Materials (both in China) are among those producing these bio-based polymers. NatureWorks expanded PLA and lactic acid capacity with a new plant in Thailand (175,000 MT/year), while Futerro (Belgium) is emerging as the second-largest PLA producer, with new facilities in China and France [11]. A 160,000 tons/year PLA production plant in the UAE was announced by Emirates Biotech. Several other major investments underway in Europe, China, the Middle East, and the USA, targeting both biodegradable and durable bio-based polymers. However, there is doubt about how much of the biopolymer capacity is actually being utilized. The nova-Institute's March 2024 report shows that bio-based non-biodegradable polymers show high utilization (~90%), while biodegradable polymers like PLA have lower utilization (~65%). The same report also shows the chart shown in Figure 21.5 of utilization rates of bioplastics in 2023 and 2024 – with the ratio of utilization to production at 68% in 2023 and dropping to 58% in 2024 [9].

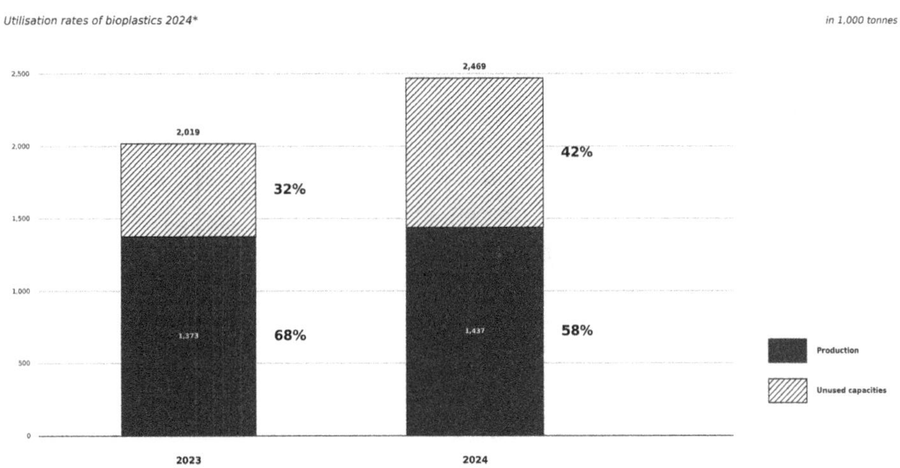

Figure 21.5: Biodegradable plastics by market segment (source: Nova-Institute for Circular Plastics NL).

21.1.3 Are biopolymers and bioplastics green, environmentally friendly, and/or sustainable?

The general belief is that all bioplastics are "green" and that anything produced from a renewable source is green. Is this a scientific "fact?" It is debatable, because the word "green" from an environmental perspective is thrown around without a strict definition. There are standardized LCA tests that have been developed and used. The results of these tests will surprise a lot of people. The common perception that because bioplastics are derived from biological sources they are therefore biodegradable and compostable, is not a fact. While some of these bioplastics are biodegradable and a few are compostable, one cannot make such a generalization for all bio-sourced materials. Those interested in understanding the differences between biodegradable, compostable, and oxo-biodegradable better can refer to this well-written article [13]. This outcome of biodegradability is achievable because the microorganisms in nature and the environment are able to break down the chemical structures making the bioplastic biodegradable under certain conditions. This is a result that is considered to be eco-friendly compared to the traditional plastics that we throw away every day. Biodegradability of a polymer depends on the chemical structure of the polymer but not on the feedstock source used for the monomers' collection.

Bio-based plastics can also be divided into four principal groups based on the source of their feedstock:

A. Bioplastics that are based on renewable resources and are also biodegradable, like starch plastic, cellulose polymers, proteins, lignin and chitosan plastics, polylactic acid (PLA), polyhydroxy alkanoates (PHAs), but also polyhydroxybutyrates (PHBs), polyhydroxyvalerate (PHV), and their copolymers in different percentages called (PHBV).
B. Biopolymers such as PVC, PE, PP, PET, nylon, and polyamides (PA), named as bioplastics because the starting monomers are obtained from biological resources.
C. Bioplastics based on petroleum resources, which are 100% biodegradable, like polycaprolactone (PCL), polybutylene succinate (PBS), polybutylene adipate (PBA), and its copolymers with synthetic polyesters like polybutylene adipate-terephthalate (PBAT) and polyvinyl alcohol (PVOH), although some of these have also been recently shown in the lab to be produced from bio-based sources.
D. Bioplastics obtained by using monomers coming from mixed biological and petroleum resources like polyesters obtained with petroleum-derived terephthalic acid and biologically derived ethanol, 1,4-butanediol and 1,3-propanediol, such as polybutylene terephthalate (PBT), polytrimethylene terephthalate (PTT), polyethylene-co-isorbite terephthalate (PEIT), polyurethane (PUR) and epoxy resins (thermoset plastic). Major biodegradable polymers that also fall within this group include PLA; PHAs, including PHB and related copolymers; cellulose diacetate; regenerated cellulose; copolyesters such as PBAT and polybutylene succinate adipate (PBSA); PBS; PCL; polyglycolic acid (PGA); and starch compounds (mixtures of starch with other biodegradable polymers such as PBAT or PLA) [14].

Most of these in Group A as well as bio-polybutylene succinate (bio-PBS), bio-polyethylene furanoate (bio-PEF), bio-polyethylene terephthalate (bio-PET), and bio-polybutylene adipate terephthalate (bio-PBA) will be covered in this book in Chapter 22.

On the other hand, there are a few polymers that are derived from petro-based sources but are biodegradable and some are even water soluble, especially when heated. Examples include PVOH, which is produced from vinyl acetate, PCL, which is produced from ε-caprolactone, a cyclic monomer, and PBS and some of its copolymers produced by polycondensation reactions of glycols, such as ethylene glycol and 1,4-butanediol, with aliphatic dicarboxylic acids, such as succinic and adipic acid [15]. Caprolactone as well as the monomers for PBS can be bio-based in smaller quantities or petro-based in larger quantities.

Biopolymers that do decompose in compost can be marked with a "compostable" symbol, under European Standard EN 13432 (2000) [16]. Packaging marked with this symbol can be put into industrial composting processes and will break down within six months or less. An example of a compostable polymer is PLA film under 20 µm thick. Films that are thicker than that do not qualify as compostable, even though they are "biodegradable" because they take too long to decompose [17]. In fact, many biodegradable and compostable bottles will not degrade much and will behave similar to petro-based nondegradable plastics, unless both sufficiently high temperature and the right range of humidity as well as the right bacteria and/or fungi are present. Too much or too little humidity or insufficient temperature can slow down the composting rate. Certain biodegradable plastics need very specific bacteria in order to decompose. In Europe, there is a home composting standard and associated logo that enables consumers to identify and dispose of packaging in their compost heap [17]. In Europe, most countries have readily accessible large scale composting facilities. The ASTM D6400 industrial composting testing and certification standard was adopted in North America in 2002, but in the USA and most other countries around the world, such composting facilities are not available everywhere and cannot be readily accessed. In such circumstances, recycling is the best option.

Biopolymers can be sustainable, carbon-neutral, and are always renewable, because they are made from plant materials that can be grown indefinitely. The growth cycle for each plant source may be different. If these plant materials come from agricultural nonfood crops, the use of biopolymers would create a sustainable industry. However, when we evaluate all materials for being green or sustainable, it is important to also determine other important factors such as water and energy usage, as well as what happens when the biopolymer degrades. In many cases, biopolymers form carbon dioxide when they degrade and use a lot more energy and water than petro-based polymers as comparable "cradle-to-grave" life cycle studies have shown.

There are some bio-sourced plastics that are produced by a fermentation process from sugar (sugarcane, beets, corn, or some other sugary bio-source, which could be from food crop), converted to ethanol, then to ethylene and polyethylene that is undistinguishable from the polyethylene produced from ethylene from natural gas or crude oil. However, if we pursued this approach, food prices to the public would increase. To avoid using bio-feedstocks that come from food crops, other bio-sources need to be used such as agricultural or food waste. A lot of work is ongoing to do just that.

Some "biopolymers" are produced from a combination of a bio-feedstock and a non-bio-feedstock, yet are claimed to be biopolymers. So, it is critical we stay away from such generalizations like all "biopolymers or natural polymers are biodegradable or compostable," or "all synthetic or oil and gas or coal-based products live forever or at least a few thousand years and not biodegradable and definitely not compostable." Each biopolymer or bioplastic material or composite or article has to be judged not on such generalizations, but on scientifically based LCA testing.

21.2 Biofeedstocks for biopolymers

21.2.1 Introduction

Many of the common biofeedstocks for producing biopolymers are covered in much more detail in Chapters 11, 22, and 24 in much more detail. Other bio-based chemicals of interest for use in biopolymers covered in this chapter are listed below.

21.2.2 Castor oil and castor oil derivatives

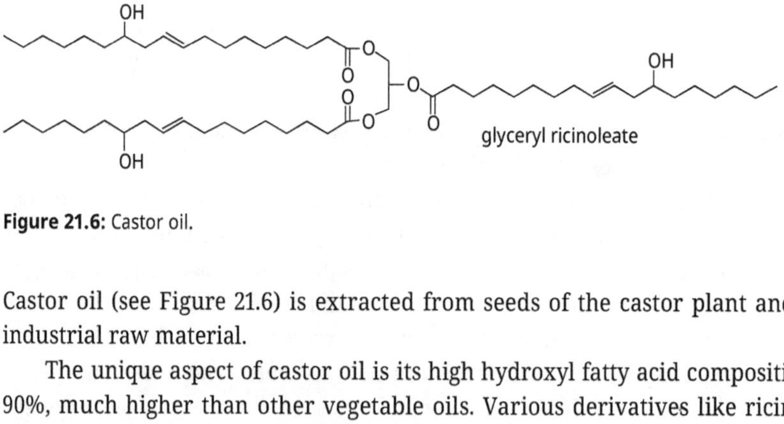

Figure 21.6: Castor oil.

Castor oil (see Figure 21.6) is extracted from seeds of the castor plant and is a vital industrial raw material.

The unique aspect of castor oil is its high hydroxyl fatty acid composition of over 90%, much higher than other vegetable oils. Various derivatives like ricinoleic acid (12-hyroxyoctadec-9-enoic acid), undecenoic acid (10-undecinoic acid), sebacic acid (octanedicarboxylic acid), azelaic acid (1.7-nonanedioic acid) and oleic acid (9-octadec -9-enoic acid) (see Figure 21.7), are produced from this glyceryl ricinoleate [18, 19]. Castor oil is used to produce polyols that are used in the production of many different specialty polyurethanes. India is the largest producer of castor seeds and accounts for over 80% of the global production, followed by China and Brazil.

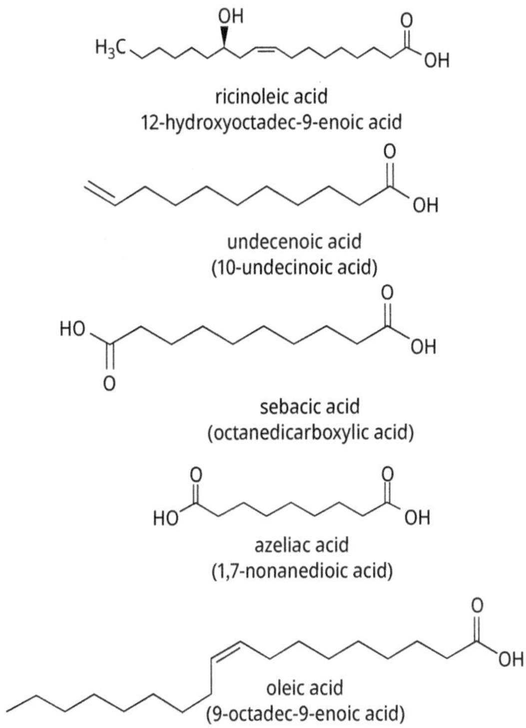

Figure 21.7: Ricinoleic acid, undecenoic acid, sebacic acid, azeliac acid, and oleic acid.

Other vegetable oils are also used to produce polyols but the polyurethanes produced from them have different properties. Other polyols can obviously be produced from petro-based sources. Producers of polyols include BASF, Cargill Inc., MCNS, Emery Oleochemicals, Croda, Alberdingk Boley, Jayant Agro-Organics Limited, Maskimi, Stahl, Polylabs, Xuchuan Chemical, Vertellus, NivaPol, MCPU Polymer, Global Bio-Chem Technology Group, and EDB Poliois Vegetais. The overall global market is estimated to be about USD 7.6 billion in 2021 with the bio-based segment to be about 11% of the total but growing at double digit rates [20].

Castor oil is converted also to ricinoloeic acid (12-hydroxyoctadec-9-enoic acid) by transesterification. From ricinoloeic acid, undecenoic acid (10-undecinoic acid) is produced by thermolysis. This 11-carbon acid can then be used to produce polyamide-11. Ricinoloeic acid is also converted to sebacic acid (octanedicarboxylic acid) and then is used to produce polyamide copolymers like 6–10, 5–10, 4–10, and 10–10.

Similarly, oleic acid (9-octadec-9-enoic acid) is produced from peanut oil. Oleic acid is converted to nonanoic acid (1-octanedioic acid), which can be polymerized to produce the same polyamide as sebacic acid. Oleic acid can also be converted to azeliac acid (1,7-nonanedioic acid) and in a few more steps, eventually polyamide-9,9 and PA-9T.

1,12-Dodecanedium acid is typically derived from an oil transformation process, but can also be obtained through palm kernel oil. This 12-carbon acid is used to produce polyamide-12 and its copolymers like PA-6,12.

21.2.3 Bio-ethylene

While ethylene itself is not found in nature, it is produced on a large scale from ethane, which is separated from natural gas and also by catalytic cracking of longer alkanes present in natural gas as well as in petroleum for producing different types of polyethylenes (HDPE, LDPE, LLDPE, and VLDPE), polypropylene, and ethylene copolymers with other olefins and comonomers like vinyl acetate and other unsaturated esters. However, due to the push for using bio-based sources, ethylene is also produced from glucose or sucrose, which is extracted from sugarcane, sugar beets, starch crops coming from corn, wheat, or other grains and lignocellulosic materials. The 12–13% solution of sucrose is anaerobically fermented in order to obtain ethanol. The ethanol is then distilled in order to remove water, giving an azeotropic solution of at 95.5 volume percentage of ethanol and water [21]. The bio-ethanol is dehydrated to obtain ethylene gas [21–24].

21.2.4 Bio-propylene

There are multiple routes to produce bio-propylene that have been documented by Venkatraman [25]. One way is by fermentation of glucose to iso-butanol, which is then converted to bio-butylene and later to bio-propylene. The first company that produced bio-PP on a pilot-plant scale was Braskem, but their process route has not been disclosed but believed to be based on metathesis of ethylene and butene produces propylene monomer. The most promising route to obtain propylene is believed to be through methanol, which is further processed to obtain propylene monomer via Lurgi's methanol-to-propylene (MTP) process or UOP's methanol-to-olefins process [25], using the same industrial plants as for petrochemical methanol. Another route to bio-propylene could be producing 1,2-propanediol or acetone through fermentation followed by conversion to 2-propanol and then dehydration to propylene [25, 26]. The thermochemical route uses a wide variety of bio-based feedstocks including corn, grass, and agricultural wastes, which are carbon-rich and can be gasified to produce syngas. Biodiesel and vegetable oils can also be used in the production of green propylene. Production cost is a very important factor in the choice of the production route. The capital costs for the fermentation route are significantly lower compared with the gasification route but the fermentation route is usually much slower [26].

Another route developed at Osaka Metropolitan University in Japan recently to produce bio-propylene uses ruthenium-iridium and ruthenium-rhodium catalysts by hydrogenolysis and electrolysis processes, respectively, to convert allyl alcohol produced from biomass-derived glycerol to bio-propylene. Glycerol is a major byproduct of biodiesel production, while hydrogen used for the reduction reaction can be green hydrogen produced from solar energy, making this whole process very sustainable [27].

21.2.5 Bio-caprolactone

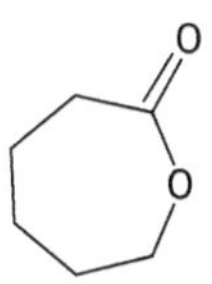

ε-caprolactone

Figure 21.8: ε-Caprolactone.

Caprolactone (see Figure 21.8) can be prepared from petro-based sources [28] or from renewable resources via chemical treatment of saccharides. The bio-caprolactone production process involves two steps: In the first step, the saccharide is converted to ethanol and acetic acid by fermentation. In the second step, ethanol is converted to cyclohexanone by using chromic acid. Then the generated cyclohexanone is oxidized by a Bayer-Villiger reaction to get ε-caprolactones covered in a Daicel patent [28], and also covered in a review by Grandhe Rani and Suraj Sharma [29].

21.2.6 Bio-glycolic acid

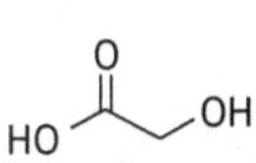

glycolic acid

Figure 21.9: Glycolic acid.

Glycolic acid (Figure 21.9) is mainly produced chemically from petrochemical resources in a process where toxic formaldehyde is needed. However, from a sustainable point of view a biotechnological production route, it is also produced commercially by three to four different fermentation methods listed [30]. Glycolic acid is used to produce PGA which is covered in the next section.

21.2.7 Bio-butadiene (bio-BD), bio-butylene glycol (bio-BG), and bio-1,4-butanediol (bio-BDO)

These chemicals that are usually produced from crude oil distillates are now being produced also from plant-based sources.

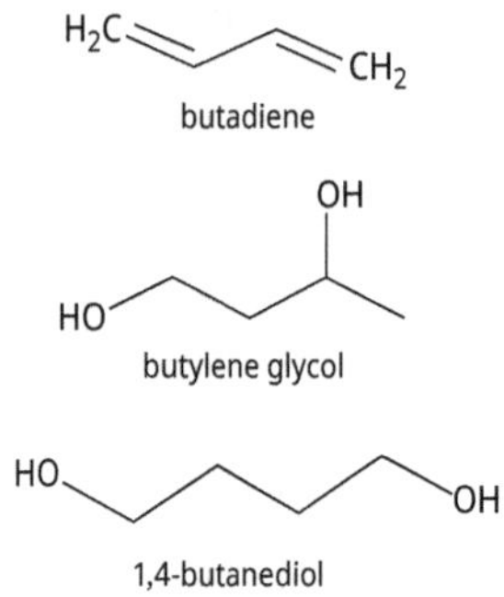

Figure 21.10: Butadiene, butylene glycol, and 1,4-butanediol.

Genomatica, a company based in San Diego, California, USA, developed manufacturing processes that enable the production of intermediate and basic chemicals from renewable feedstocks. It has made progress since its first product, bio-based 1,4-butanediol (bio-BDO). BDO is an industrial chemical that can be used as a solvent in its own right. The largest use of BDO, around 40%, is to produce tetrahydrofuran (THF), which is then used to manufacture polytetramethylene ether glycol (PTMEG), from which Spandex™ fibers are produced. BDO is also used as an intermediate to produce polyurethanes. Another use for BDO is to make single use compostable bags in Europe. As a biotechnology partner to the chemical industry, Genomatica's business model is as a technology enabler and not a producer of large-volume chemicals. Its patents cover the use of fermentation and other biotech technologies to produce bio-butadiene (bio-BD) and bio-butylene glycol (see Figure 21.10). These technologies have been licensed to several companies in the USA, Europe, and Japan, who have built plants in their home countries as well as in China and elsewhere [31].

21.2.8 Bio-hexamethylenediamine (HMDA), bio-caprolactam, and bio-adipic acid

In 2014, Genomatica announced that it had entered the bio-based nylon intermediates market focusing on hexamethylenediamine (HMDA), caprolactam, and adipic acid (see Figure 21.11) [31]. This company would develop complete process technologies for the bio-based production of these intermediates, which it will then license to major firms in the nylon value chain. These three intermediates are used primarily in the production of nylon 6 (or polyamide-6) and nylon-6,6 (PA-6,6). At that time these two

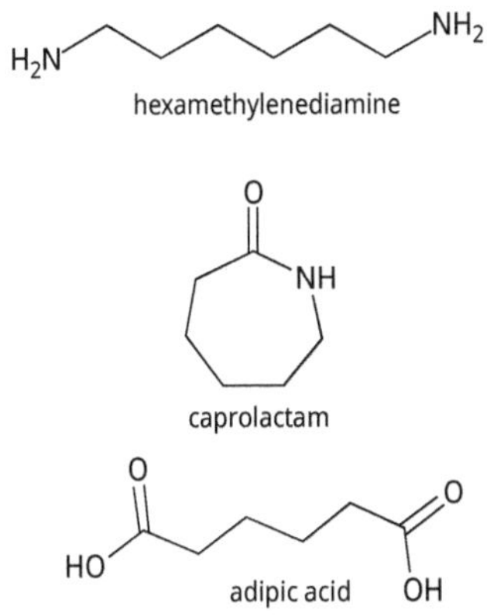

Figure 21.11: Hexamethylenediamine, caprolactam, and adipic acid.

polyamides were 100% produced from petro-based intermediates at a value of over $18 billion per year. Genomatica's announcement claimed that they had issued eight patents. The current status of this will be discussed later in the section on bio-polyamides.

21.2.9 Bio-1,5-pentamethylene diamine

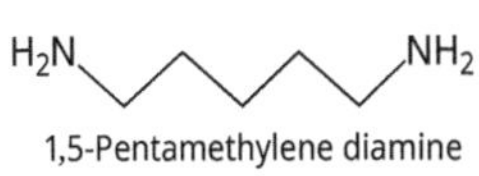

Figure 21.12: 1,5-Pentamethylene diamine.

A Japanese patent application discloses a biomass resource-derived 1,5-pentamethylene diamine (see Figure 21.12), to produce a polyamide resin composition having a good color tone. The 1,5-pentamethylene diamine in which a total of the contents of an acetamide, 2,3,4,5-tetrahydropyridine, 2-piperidone, α-amino-ε-caprolactam, lysine, and ornithine that the 1,5-pentamethylene diamine contains as impurities is less than or equal to 150 weight ppm [32]. This diamine is then used to produce several different bio-based polyamides, which have five methylene groups and are covered in the bio-polyamides section later in this chapter.

21.2.10 Natural rubber

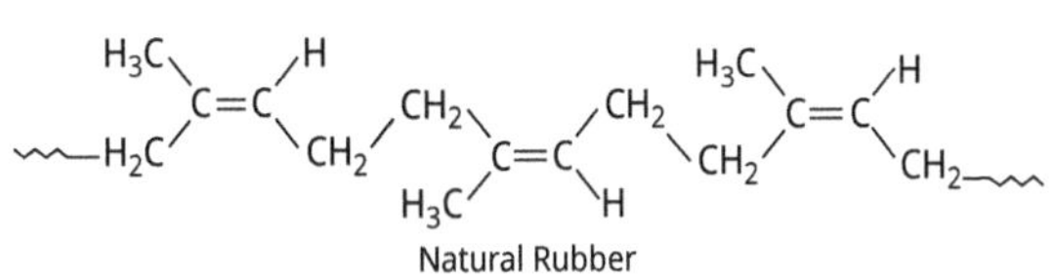

Figure 21.13: Natural rubber.

Natural rubber (see Figure 21.13) is obtained from rubber trees in a latex form. Such rubber trees are grown in the tropics in countries like Thailand, Indonesia, Malaysia, Sri Lanka, Southern India, as well as parts of equatorial countries in Africa and Latin America. The raw latex is not easy to use and does not have desirable properties to be used directly except for coatings. Chemically, it is an emulsion of polymers of isoprene. Rubber latex is a sticky, milky, and white colloidal emulsion, which is allowed to drip off by making incisions in the bark and collecting the fluid in vessels in a process called "tapping." The latex itself is difficult to use directly and has limited utility, except to be used as waterproof coating on fabric. So, the latex is allowed to coagulate in the collection cup. These coagulated lumps are collected and processed into dry forms for sale. Natural rubber is used extensively in many applications and products, either alone or in combination with other materials. In most of its useful forms, it has a large stretch ratio and high resilience and also is waterproof. Significant progress was made after the invention of vulcanization of natural rubber, which will be discussed in a section called "Vulcanized Natural Rubber" later in this chapter.

21.3 Biopolymers (not covered elsewhere)

21.3.1 Introduction

While many biopolymers are covered in Chapters 22 and 23, some are not. This section will cover those. This list includes bio-polyethylene (bio-PE), bio-polypropylene (bio-PP), polycaprolactone (PCL); PGA; bio-polytrimethylene terephthalate (bio-PTT), bio-polyamides (Bio-PA), cellulose derivatives like cellulose acetate, and starch compounds (mixtures of starch with other biodegradable polymers such as PLC or PLA and with synthetic polymers like polyolefins). Other renewable resin chemistries such as collagen and chitosan are not covered because their roles are not well-developed, at least in industrial plastics. Other bio-based polymers have also been developed like bio-poly(acrylic acid) (bio-PAA), bio-poly(methyl methacrylate) (bio-PMMA), bio-polystyrene (bio-PS), bio-poly(vinyl acetate) (bio-PVA), and bio-based poly(vinyl chloride) (bio-PVC) but with insufficient quantities being produced, if at all, are also not covered here.

21.3.2 Bio-polyethylene

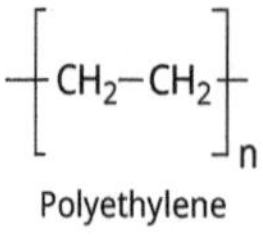

Figure 21.14: Polyethylene.

Polyethylene (PE) (Figure 21.14) is the largest volume plastics material used in the world. The starting monomer used for synthesis is ethylene, which is also used to produce other polymers like PVC and PS, and it is generally obtained from petroleum feedstock by distillation, especially in Europe. Almost everywhere else, it is separated from natural gas or by cracking of natural gas liquids. To instead obtain ethylene from biological resources a lot of research and development work has been put in. The only approach that has been successful so far has been to obtain the ethylene monomer by dehydration of bio-ethanol, obtained from glucose. Once ethylene gas is produced from a bio-source, it is indistinguishable from the ethylene produced from a petro-source and the same polyethylene grades can be produced from ethylene, irrespective of source. Consequently, the subsequent polymerization of bio-ethylene monomer is exactly the same as if the ethylene is derived from petroleum or natural gas, and the corresponding bio-polymer is identical in its chemical, physical, and mechanical properties to fossil-based PE, also with regards to the mechanical recycling processes [19, 20, 33].

By various polymerization methods that are used for polymerizing non-bio-ethylene, various types of bio-PE could be produced: bio-high-density polyethylene (bio-HDPE) with a low degree of short-chain branching (about seven branches per 1,000 carbon atoms), bio-LLDPE with a high degree of short-chain branching and bio-low density polyethylene (bio-LDPE) with a high degree of short-chain branching + long-chain branching (about 60 branches per 1,000 carbon atoms). Ethylene copolymers with other olefins like propylene, butene, hexene, octene-1, and 4-methyl pentene-1 could be produced and based on the level of the comonomer could be linear low-density polyethylene (LLDPE) grades or more elastomeric very low-density polyethylene (VLDPE), also called flexomers, or even ethylene-propylene rubber. Also, other ethylene copolymers with polar comonomers could be produced to produce ethylene-vinyl acetate (EVA), ethylene-methyl acrylate (EMA), ethylene-ethyl acrylate (EEA), ethylene-butyl acrylate (EBA), ethylene-acrylic acid (EAA), ethylene-methacrylic acid (EMAA), etc. as well as bio-polyethylene wax using bio-ethylene.

In 2010, Braskem, the company based in Brazil, became the first company to produce bio-PE on a commercial scale [19–21, 33, 34] for applications in food packaging, cosmetics, personal care, automotive, and toy applications. Even though in 2007, a joint venture between Dow and Crystalsev [35] to produce bio-based linear low-density polyethylene (bio-LLDPE) was announced, it did not go anywhere. In 2020, Dow said that their bio-PE produced from tall oil, a waste product from paper production is being used by Thong Guan to produce cling film [36]. Solvay, Nova Chemicals, and other chemical companies, which are in the bio-PE market, were supposed to contribute to the production of about 20% of the world's bio-based plastics production. At that time, Braskem was expected to cover 10% of the worldwide plastic market by the year 2020 [33], but that did not happen. Since then, Braskem has acquired many other chemical operations, especially polymer production facilities, most of which are not bio-based. Today, bio-polyethylene is a small part of Braskem's polymer production.

When production of bio-ethylene began, it was not considered to be cost-competitive when compared with petrochemical-derived ethylene [33, 34]. As bio-ethanol production has ramped up the gap has narrowed somewhat but even today the price of ethanol is about 35% higher than that of crude oil. The price of 1 kg of bio-PE is currently about 30% higher than petrochemical PE.

In May 2021, Neste, Toyota Tsusho Corp., and Mitsui Chemicals, Inc. announced a joint effort to increase the industrial-scale production of renewable chemicals and plastics, including polyethylene, from 100% bio-based hydrocarbons [37, 38]. Some reports indicate that bio-PE is also being or likely to be produced by SABIC, Solvay, Neste, and Nova Chemicals.

The main selling point and advantage is that bio-ethylene is a drop in with no changes needed downstream for producing the same articles, films, bottles or parts using the same equipment and the same processing conditions compared to those for conventional petro-based polyethylene. However, bio-PE has two major challenges to larger adoption: higher cost and insufficient availability [27].

The global bio-PE market was estimated to be around USD 1.5 billion in 2021 and expected to exceed USD 2.2 billion by 2029, generating CAGR values of 5.23% during the forecasted period in a market report by Data Bridge Market Research [39]. In reality, it is rare that either production or sales of such commodity resins grow in a straight line.

21.3.3 Bio-based polypropylene

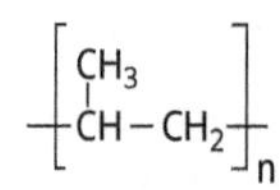

Polypropylene

Figure 21.15: Polypropylene.

Polypropylene (PP) (Figure 21.15) is the second largest plastic produced globally after polyethylene. Yet it is the most important organic building block for poly-olefin production as it has much wider set of critical applications in a multitude of end applications. Bio-PP produced from bio-propylene, using conventional polymerization methods used for non-bio-based polypropylene production. Once the bio-propylene monomer is produced, polymerization technologies to produce high molecular weight polypropylene with identical properties on a commercial scale are exactly the same as those for petro-based PP. At this stage it needs to be emphasized that green PE and PP may be bio-based, and identical to their fossil fuel counterparts (assuming the same polymerization catalysts and process are used). However, they are not biodegradable and of course, not compostable.

Braskem was the first company that produced bio-PP on a pilot-plant scale [33]. It made sense for Braskem was to leverage their R&D and capital investment in the bio-

PE plant to the production of bio-PP. Yet developing the process technology to produce bio-PP that can compete in the marketplace close to cost parity with fossil fuel-based propylene has been challenging for potential bio-PP producers. So, in spite of the know-how Braskem had, commercial production of bio-PP did not take place.

Then again, bio-PP derived from bio-based naphtha has been on the commercial market for several years in a process led by SABIC and Dow at Terneuzen, the Netherlands using Neste's ability to supply bio-based naphtha as feedstock. It is understood that they use the biomass balance approach to market their bio-derived polypropylene [39, 40].

In 2019, Neste and Lyondell-Basell announced their first commercial production of both bio-PP and bio-LDPE. This joint project uses Neste's renewable hydrocarbons derived from sustainable bio-based raw materials like waste and residue oils. The project successfully produced several thousand tons of bio-based plastics, which are marketed under Circulen™ and Circulen™ Plus, the new family of LyondellBasell circular economy product brands and are approved for the production of food packaging [40]. This achievement used the combination of Neste's unique renewable feedstock and Lyondell-Basell's technical capabilities and their cracker flexibility at its Wesseling, Germany site, which was converted directly into bio-based polyethylene and bio-based polypropylene. According to the announcement, an independent third party tested the polymer products using carbon tracers and confirmed that these products contained over 30% renewable content. Lyondell- Basell sold some of the renewable products produced in the initial trials to multiple customers like Cofresco to create sustainable food packaging materials, and with brands like Toppits and Albal, which are Europe's leading supplier of household film [40, 41].

21.3.4 Polycaprolactone (PCL)

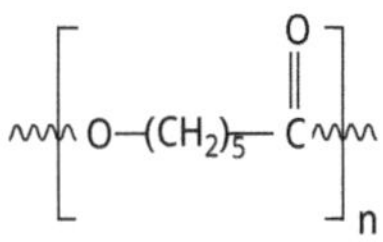

polycaprolactone

Figure 21.16: Polycaprolactone.

Polycaprolactone (PCL) (Figure 21.16) is a synthetic biodegradable polyester that is partially crystalline, but has a low melting point of about 60 °C and a glass transition temperature of around −60 °C. It is an aliphatic polyester with a high tolerance to liquids, solvents, and gasoline. This polymer is prepared by ring-opening polymerization of ε-caprolactone using a catalyst such as stannous octanoate [42]. The ε-caprolactone monomer could be produced from bio-based or from bio-petro sources.

In general, PCL is biocompatible, nontoxic, and cost-effective. Its biodegradability is due to its much lower glass-transition temperature (T_g = −60 °C) than other biodegradable polymers, despite its high degree of crystallinity, typically 50%. However, unlike other biopolymers, many different grades of PCL can be produced by adjusting the polymer's molecular weight, cross-linking, degree of crystallization, and by preparing PCL blends to vary mechanical, barrier, thermal, chemical resistance, and biodegradability rates [26, 41]. This flexibility makes polycaprolactone more adaptable for a multiple range of end applications. PCL's exceptional mechanical, chemical, and outstanding bioresorbable properties have led to its complete commercial production for biomedical applications [22, 24, 42].

The biodegradation rate of PCL is very fast when exposed to air, but is slower when not exposed to air. Biodegradation of PCL occurs by a specific thermo-tolerant microorganism called *Aspergillus* sp. at 50 °C [41, 42]. PCL is degraded by the hydrolysis reaction of its ester linkages in the human body and has hence been accepted for use as a very important implantable biomaterial, specifically for use in the preparation of long-term implantable devices.

Ingevity offers polycaprolactone derivatives for soft thermoplastic polyurethane (PU) elastomers, adhesives, and coatings under the trade name CAPA. Some of the other commercial manufacturers and trade names are BASF's Capromer™, eSun, Daicel, Perstorp's CAPA™ (previously Solvay). Dow's Tone™ has apparently been discontinued.

Some grades of PCL have elastic properties. PCL is used in various forms such as film, fibers, and microparticles. In Sweden, bags had been produced from PCL and PCL-starch bags, but they degraded very quickly even before reaching customers [43].

Overall, polycaprolactone is biocompatible, nontoxic, and cost-effective. Consequently, PCL has been used in a wide variety of applications in the medical field such as preparation and formulation for pharmaceutical purpose, nanoparticles coating, drug conjugates, as well as craniofacial bone repair, bone fixation, drug release, and even for "in vivo" applications in the biomedical industry. Another application is encapsulation of a variety of drugs within PCL beads for controlled-release and targeted drug delivery. To lower the cost, PCL is often diluted with starch to obtain a good biodegradable product produced with mixed biopolymers. Products formulated with PCL and PCL-based have exhibited promising results in controlled drug release, formation of scaffold, artificial organs, and nerve regeneration. Development of such advanced techniques encourages the progress in the field of bio-based polymer science [41, 42, 44]. Biodegradable PCL scaffolds also have been fabricated using 3D printing [46].

Despite all these advantages that PCL possess characteristics for therapeutic applications such as biodegradability, biocompatibility, and high permeability, it also has some drawbacks like burst release, low encapsulation efficiency, and low bioactivity toward the tissue because of its hydrophobic nature. To overcome this, more research is needed and is being carried out for developing modified PCL and PCL-based materials for broadening the range of effective applications even further in the biomedical field [41, 42, 44–46].

21.3.4 Bio-polytrimethylene terephthalate (bio-PTT)

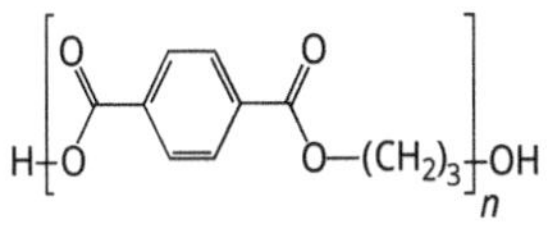

Figure 21.17: Polytrimethylene terephthalate.

Polytrimethylene terephthalate (PTT) (Figure 21.17) is a polyester that was polymerized and patented as long ago as 1941. It can be produced by condensation polymerization or by transesterification reactions from two monomers, 1,3-propanediol and either terephthalic acid or dimethyl terephthalate [47].

More economical and efficient methods to produce 1,3-propanediol in the 1980s by Degussa, via acrolein, and by Shell using the hydroformylation of ethylene oxide have greatly increased PTT's value as a commercial polymer. DuPont has since also commercialized the production of this polymer via a different bio-based method using 1,3-propanediol generated by fermentation to produce bio-PTT with the trade name Sorona™. Similar to bio-PE and bio-PP, once bio-PTT is produced, it is indistinguishable from the same polymer produced from petro-based monomer. These developments may allow PTT to effectively compete against PBT and PET in some applications and markets, two polyesters that have been far more successful than PTT to date. On March 20, 2009, the US Federal Trade Commission approved a subclass to this polyester. The US Federal Trade Commission had last approved an extension for residential carpet in 1959. The PTT fiber used in Mohawk's SmartStrand carpet and branded Sorona™ by Dupont can be labeled Triexta™. Mohawk Industries and DuPont had applied jointly for FTC approval of the Triexta polyester subclass in 2006 [48]. Other companies that also involved with PTT include Teijin Frontier, Shenghong Group, and Glory. The factors for the rising demand for the growth include wider diversity of products, increasing applications in the textile industry, as well as a larger number of applications for manufacturing monofilament, films, and engineering plastics. This demand was expected to boost the growth of the bio-PTT at 6.4% to reach an estimated volume of about 320,000tons by 2027, in spite of higher cost of raw materials and manufacturing as well as patented process acting to restrain growth [49]. Newer estimates by marketwatch.com take into account the lower volumes due to COVID but higher CAGR values after that but overall attain similar forecasts for estimated volume of production and applications in the 325,000–350,000 ton range by 2028 [50].

21.3.5 Polyglycolic acid (PGA)

PGA is also known as polyglycolide. Figure 21.18 is a thermoplastic biodegradable polymer with a linear aliphatic polyester chemical structure. It can be polymerized

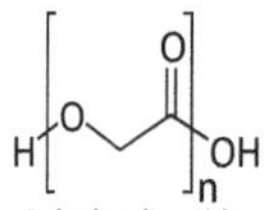

Polyglycolic acid

Figure 21.18: Polyglycolic acid.

from glycolic acid by either polycondensation or ring-opening polymerization. Other methods of production use the reaction of formaldehyde, carbon monoxide, paraformaldehyde, or trioxane or other related compounds in presence of an acidic catalyst. PGA has been known since 1954 as a tough fiber-forming polymer. However, due to its hydrolytic instability, its use was limited for several decades [51].

Such hydrolytic instability is because of the ester linkage in its main chain itself. When exposed to physiological conditions, polyglycolide is degraded by random hydrolysis, and it seems to decompose by certain enzymes, especially those with esterase activity. When this happens, the initial degradation product, glycolic acid, is nontoxic, and this acid gets metabolized to oxalic acid, which limits its use. However, in the absence of this enzyme, PGA apparently does not decompose this way.

Currently polyglycolic acid polyglycolide and its copolymers like with lactic acid called poly(lactic-co-glycolic acid), with ε-caprolactone called poly(glycolide-co-caprolactone), and with trimethylene carbonate) called poly(glycolide-co-trimethylene carbonate) are widely used as materials for the synthesis of absorbable sutures [51, 52]. This traditional role of PGA and now its copolymers as materials for biodegradable sutures has led to their evaluation in other biomedical fields. Implantable medical devices have been produced with PGA, including anastomosis rings, pins, rods, plates, and screws [52]. It has also been explored for tissue engineering or controlled drug delivery.

DuPont's PGA grade called TL F 6267 is a crystalline polyester with a molecular weight of approximately 600 and a specific gravity of 1.58 g/mL, and a melting point of 93–99 °C. It is sold as a finely ground tan powder with an average particle size of 20 μm, which is insoluble in both water and most organic solvents, but is used as a dispersion in oil and gas applications as a time-release agent for corrosion inhibitors, dispersants, decomposition inhibitors for lubricants in moving equipment in the wellbore, and other channels of the formation. This product was developed by DuPont but is now available from The Chemours Company and is claimed to be useful in such applications [53].

Kureha Chemical Industries has commercialized high molecular weight polyglycolide for food packaging applications under the tradename of Kuredux™ produced at its Belle, West Virginia, USA [54]. Its high degree of crystallization creates a tortuous path mechanism for low permeability and hence its use as a barrier material. It is expected that the high molecular weight version will have to be used as an interlayer between layers of polyethylene terephthalate to provide improved barrier protection for perishable foods, including carbonated beverages and foods that lose freshness on prolonged exposure to air. This polyglycolide interlayer technology could also enable thinner plastic bottles, which still retain desirable barrier properties.

According to a recent market report from Global Industry Analysts, Inc., the global polyglycolic acid market is expected to achieve sales of $453.4 million by 2027 while growing at a CAGR of just over 8.5% over the 2020–2027 period. While medical applications are projected to record an 8.5% CAGR and reach USD 185.2 million by 2027, growth in packaging applications was revised to 8.6% CAGR for the same period from an earlier estimate [55].

21.3.6 Bio-polyamides (bio-PA)

A polyamide is a polymer with repeating units connected by amide bonds and is produced by the polymerization of a cyclic amide or reaction of a diamine with a dicarboxylic acid or acid derivative like an anhydride or a dicarboxylic acid chloride. While there are several natural polyamides like silk, wool and proteins, nucleic acids like DNA and RNA, there are no known commercial paths from these to bio-polyamides. Synthetic polyamides especially polyamide-6 and polyamide-6,6 are the largest polyamides produced globally and used in numerous applications but they are produced from petro sources.

This chapter is intended to provide a basic background on the origin and importance of bio-based polyamides, different synthetic routes of a wide variety of starting monomeric materials (that were covered in an earlier section) obtained from bio-sources, and a brief summary of properties and applications of some common aliphatic, semi-aromatic, and fully aromatic bio-polyamides from both technical and commercial perspectives.

Just like bio-polyethylene and other biopolymers, once the polymerization is carried out with a bio-sourced monomer or a non-bio-sourced monomer, the bio-polyamide formed from one source is indistinguishable from the other. With some bio-polyamides the bio-sourced product is lower cost, while with others it is not. Purchasers, as you would expect, purchase whichever one is the lower cost, unless their customer specifies a bio-source. In some cases, only one of the comonomers is bio-sourced while the other is petro-based.

What is common among bio-polyamides is that they all have an amide group in their repeat unit, but the differences among them are in the length and type of the rest of the repeat unit. In general, the longer the aliphatic part within the repeat unit, the less crystalline the polyamide is and more hydrophobic. The lower crystallinity translates into a lower melting point and a softer polymer. If an aromatic part is in the repeat unit, the polyamide becomes more rigid and has a higher melting point.

The 11-carbon carboxylic acid discussed earlier in this chapter, undecenoic acid, is used to produce polyamide-11 (see Figure 21.19). The first application of this approach was by Organico in 1947 to produce polyamide-11 [56]. It is now sold by Arkema under the brand name Rilsan-11 [57]. Since then, other partly or fully bio-sourced polyamides have been commercialized for a very wide range of applications.

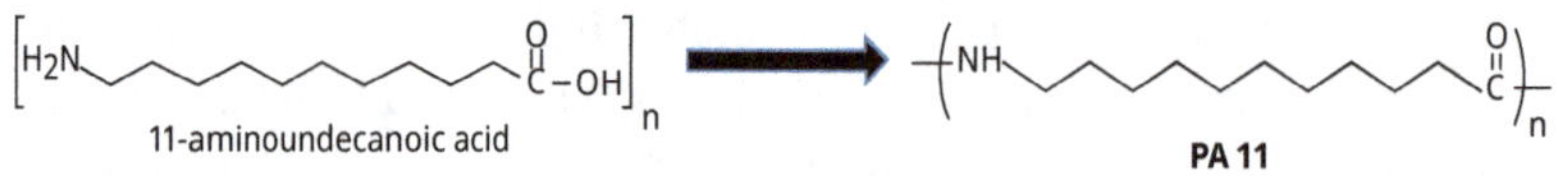

Figure 21.19: 11-Aminoundecanoic acid to polyamide-11 (PA11).

Arkema also uses bio-based feedstocks to make other polyamides, some of which are thermoplastic elastomers, all under the Pebax™ trade name [57].

Ricinoloeic acid is also converted to sebacic acid (octanedicarboxylic acid) to produce various polyamides that have 10 carbon atoms such as polyamide-10–10, and other polyamide copolymers like polyamide-6,10, polyamide-5,10, polyamide-10,10, and polyamide 4-10 [56–60].

Similarly, oleic acid (9-octadec-9-enoic acid) is produced from peanut oil, and then converted to azeliac acid (1,7-nonanedioic acid) and in a few more steps, eventually polyamide-9,9 and bio-based copolyamides PA-6,9, PA-10,9, and PA-12,9 [61–63]. Oleic acid can also be converted to nonanoic acid (1-octanedioic acid), which can be polymerized to produce polyamide-10,10 [57–65].

In 2014, Genomatica announced that it has entered the bio-based nylon intermediates market focusing on hexamethylenediamine (HMDA), caprolactam, and adipic acid. This company has licensed their technology to produce various bio-monomers for producing bio-polyamides to several companies around the world. The list of such companies includes Aquafil, BASF, DSM, Solvay, Evonik, Asahi Kasei, Covestro, Cargill, Helm, Daicel, Braskem, as well as retail brands like H&M, Lululemon, etc. [31, 66].

Poly(pentamethylene oxamide) (PA-5,2) is a unique high-molecular-weight bio-based polyamide, prepared from dibutyl oxalate and renewable monomer of 1,5-pentanediamine using a two-step polymerization method. The results of analytical chemical testing using Fourier transform infrared spectroscopy and proton nuclear magnetic resonance spectroscopy (H^+-NMR) showed that it was indeed PA-5,2. Other evaluations using methods including differential scanning calorimetry, thermogravimetric analysis, and water uptake measurements for the PA-5,2 showed that this biopolymer has excellent properties like high temperature resistance, excellent crystallization performance, and low water absorption [67, 68].

Cathay Industrial Biotech, a company listed on the Shanghai Stock Exchange, has been promoting its new renewable polyamides PA-5,6, PA-5,10, PA-5,12, and PA-5,14 (as well as their copolymers) marketed under the Terryl™ brand for textile applications and as Ecopent™ for applications in automotives, electronics and electricals, cable ties, film, and other engineering materials downstream end markets. These biopolyamides are based on Cathay Industrial Biotech's 100% bio-based 1,5 pentamethylenediamine monomer called C-BIO N5 produced under a proprietary technology process using sugarcane for feedstock [68]. The company claims the 5-carbon diamine has the potential to substitute HMDA (hexamethylenediamine), a key raw material in the production of traditional petrochemical-based nylons, but many of the properties seem to be not the same as those of the corresponding 6-carbon-containing polyamides. The company is also working on PA-5,11 (which reportedly contains 31% renewable-based monomers) and PA-5,13 (28–100% renewable). Cathay Industrial Biotech has been producing other long-chain α,Ω-dicarboxylic acids for polyamides such as series of products from sebacic acid (DC 10) to octadecanedioic acid (DC 18) and including undecanedioic acid (DC 11), dodecanedioic acid (DC 12), brassylic acid (DC 13), tetradecanedioic acid (DC 14), and hexadecanedioic acid (DC 16). These longer chain acids can be used to produce more hydrophilic, softer and lower melting bio-polyamides like PA-1012. See Figure 21.20.

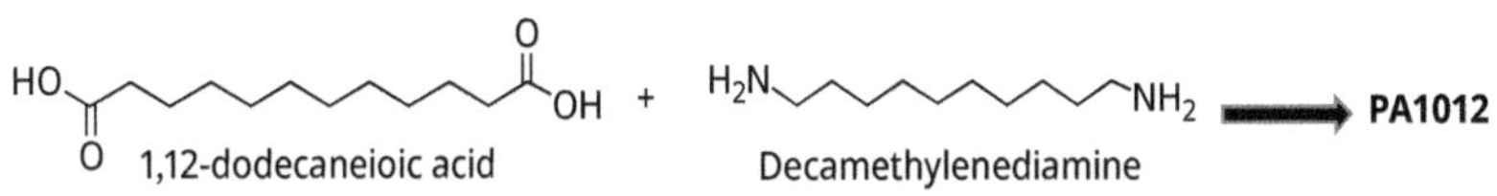

Figure 21.20: 1,12-Dodecaneioic acid with decamethylenediamine to PA1012.

The company's dibasic acids are produced from paraffin wax using fermentation methods followed by purification compared to multistep chemical processes used for petrochemical-based materials. The paraffin-based products are obviously not bio-based but the ones from sugar and other cellulosics are. Since paraffin wax comes from crude oil, which is a natural resource, the company considers it to be a natural product. But it is important to remember that just because a product is natural, it does not mean that it is eco-friendly or sustainable. It all boils down to how the product is collected, produced, and disposed of.

Polyamide-6,10 (PA- 6,10) is a semi-crystalline polyamide that is produced by the polymerization of hexamethylene diamine with a dibasic acid, i.e., sebacic acid this this case. Melting point of PA-6,10 is 223 °C. Tests have shown that nylon-6,10 has lower moisture absorption than nylon-6 or nylon-6,12 and is stronger than nylon-11, nylon-12, and nylon-6,12. Nylon-6,10 has excellent room temperature toughness even at low temperatures, much better than nylon-6 and nylon-6,6. Nylon-6,10 also has good resistance to most solvents and to dilute mineral acids. It also resists the environmental stress-cracking action of salts like zinc chloride. It is believed that a patent assigned to DuPont that expired in 2023 [69] covers their process for making partially aromatic polyamides is provided in which an aromatic dicarboxylic acid component, at least 20–100% by weight of the dicarboxylic acid in the acid component is in the form of an alkylated ester, and a diamine component, comprising a diamine having from 6 to 12 carbon atoms, are admixed in the presence of water and with heating to form polyamide having 1–100% on a molar basis of *N*-alkylated amide and amine groups.

Table 21.1 shows the renewable content of the newer bio-polyamides and their performance comparison with those of the two largest volume petrochemical-based PA-6,6 and PA-6.

Table 21.1: Properties and percent bio-content of bio-polyamides.

Polyamide	Specific gravity (g/mL)	Tensile modulus (GPa)	Tensile strength @ yield (MPa)	Moisture absorption (%)	Max. service temperature (°C)	Thermal coeff. of expansion (μm/m °C)	Coefficient of friction	% bio Content*
PA-6	1.14	2.59	71.0	4.60	123	82	0.277	100%
PA-6/30%GF	1.37	7.89	150.0	2.99	155	27	0.364	70%
PA-6,6	1.17	3.71	72.3	3.72	136	83	0.352	100%
PA-6,6/30%GF	1.38	8.46	155.0	2.02	165	32	0.310	70%
PA-6,10	1.26	6.32	114.0	1.39	132	68	0.400	100%
PA-6,10/30%GF	1.31	8.60	160.0	1.20	135	26	N/A	70%
PA 4.6	1.21	2.36	136.0	8.62	163	85	0.388	100%
PA-4,6/30%GF	1.27	9.00	120.0	4.87	163	30	0.300	70%
PA-11	1.04	1.17	35.2	1.24	102	98	0.350	100%
PA-12	1.36	4.78	42.8	1.12	120	239	0.312	100%
PA-5,6**	1.13	2.95	70.0	N/A	N/A	N/A	N/A	47%
PA-5,10	1.16	2.34	26.0	N/A	N/A	N/A	N/A	100%
PA-9 T	1.17	2.56	80.0	0.25	N/A	66	N/A	100%
PA-10,10	1.05	0.50	38.0	1.50	N/A	170	N/A	100%
PA-6,12	1.05	1.80	50.0	1.00	N/A	270	N/A	100%

*Biocontent (assumes all monomers used are bio-based); **HMDA was not bio-based.

Polyamide 6,10 is particularly useful in the manufacture of products intended for use at elevated temperatures or in applications where retention of flex modulus and tensile strength at elevated temperatures is required along with retention of good impact properties at low temperature. Such requirements are targeted especially therefore for "under the hood" parts in the engine compartment of automobiles and for various electronic appliances. These polyamides can also be formed into films and fibers for other applications.

However, nylon-6,10 and most of these other smaller volume polyamides are much more expensive than both nylon-6 and nylon-6,6. Nylon 6–10 compounds are used in a wide range of applications including zip fasteners, electrical insulators, precision parts, and filaments for brushes. DuPont™ markets PA-6,6 under the Zytel™ and longer chain PA-6,10, PA-10,10 and PA-6,12 under the Zytel™ LC brand names [70]. They also market a specialty polyamide produced from 1,4-benzenedicarboxylic acid, with 1,6-hexanediamine and 2-methyl-1,5-pentanediamine under the Zytel™ HTN [70, 71]. On February 18, 2022, DuPont announced that it has entered into a definitive agreement with Celanese Corporation to divest a majority of the Mobility & Materials segment including the Engineering Polymers business line and select product lines within the Performance Resins and Advanced Solutions business lines for $11.0 billion in cash. Zytel falls within the sale but details of whether their other products covered in this chapter are also included in the sale is not known at this time [72].

A Japanese patent [73] describes polyamide as having excellent heat resistance, water resistance, and impact resistance. With a glass transition temperature is 125 °C, it can withstand sufficiently heat even in the automobile engine compartment which is called "under the hood" in some applications such as electrical components (jacks, switches, terminals) and engine component housings and covers. This polyamide is produced from an aromatic dicarboxylic acid like a terephthalic acid and an aliphatic diamine. This patent is believed to be the patent that discloses Kuraray's new semi-aromatic Genestar™ polyamide-9T. The 9-carbon diamine could be from a bio-based source. With a glass transition temperature of 125 °C, it can withstand sufficient heat even in the automobile engine compartment, which is called "under the hood" in some applications such as electrical components (jacks, switches, terminals) and engine component housings and covers. Polyamide-9T has a material with low water absorption properties, even among other polyamides. This gives it excellent dimensional stability and excellent stable mechanical properties over a wide range of temperatures. This material exhibits low outgassing during injection molding. It provides excellent formability with very few burrs created even for their high flow grade compared to other long-chain polyamides.

The same company also was granted an earlier patent [74] that has an alicyclic dicarboxylic acid like a 1,4-hexane dicarboxylic acid reacting with an aliphatic diamine to produce alicyclic-aliphatic polyamides, which are more flexible than those with the aromatic-aliphatic polyamides from the patent discussed in the previous paragraph.

In most cases, these bio-polyamides like other polymers (plastics and rubber) are rarely used without the addition of additives and/or reinforcements in order to meet various customer requirements all the way downstream in the value chain. For meeting specific applications, it may be necessary to reinforce these bio-polyamides with glass fibers, glass beads, and carbon fibers to improve their mechanical and thermal performance. For applications like bearing materials where low friction and improved wear resistance is important, they can be compounded with additives like PTFE and molybdenum disulfide. Such compounds and reinforced composites of bioplastics will be covered in later chapters in this book.

Rilsan™ Clear G850 Rnew and Rilsan™ Clear G120 Rnew are partially bio-based materials with excellent chemical and fatigue resistance, which will provide advantages in the optics, medical, and consumer electronics markets. The company has also added PA-10,10, PA-10,12 and PA-6,10 to its portfolio. These resins offer optimized properties between that of short- and long-chain polyamides with the added benefit of having significant bio-sourced content, providing solutions for automotive, monofilament, and industrial applications. Pebax™ thermoplastic elastomers (TPEs), which have been well-known in the sports and medical industries for years, are now breaking ground in breathable, monolithic applications. New grades of Pebax TPEs are now used in house wrap and breathable textile applications by producing tough and tear-resistant, waterproof barriers while remaining breathable, plasticizer-free, and monolithic (non-porous) [57, 75, 76].

Polyamide 10/10 (PA-10,10) is the product of 1,10-decamethylene diamine and 1,10-decanedioic acid (sebacic acid). Both monomers can be extracted from castor oil beans, which can make it 100% bio-based. PA-10,10 has properties that are between those of higher performance polyamides such as PA-12 and lower cost PA-6 and PA-6,6. PA-10/10 absorbs less water than PA-6 and PA-6,6; hence, its properties are less affected by humidity. PA-10,10 has also excellent mechanical properties including high tensile strength, flexibility, toughness and good resilience, and abrasion resistance. It offers higher impact strength and improved chemical resistance to greases, oils, fuels, hydraulic fluids, and salts over PA-6 and PA-6,6 with a higher melting point, and a much higher elongation at break.

This section ends with recently published data on the growth of the global market of bio-based polyamides to emphasize the economic importance of this family of biopolymers and bioplastics [76]. In recent times, production volume and applications of these bio-based polyamides have been rapidly growing. Besides Arkema, other producers of such polyamides include, Radici, Evonik, DuPont, DSM, EMS-Grivory, DOMO Group, Cathay Industrial Biotech, BASF SE, Dupont, and Lanxess AG. Other players include AdvanSix, Honeywell International Inc., SK Group, and Asahi Kasei, who also produce some grades of these specialty polyamides. While these polyamides are produced from fully or partially bio-based sources, they are generally not biodegradable. The total global market for polyamides was around USD 30 billion growing at a CAGR of 4.5% to USD 3 billion by 2027. Within that the global bio-polyamide market was valued at about USD 130 million in 2021 and will grow with a CAGR of 10.58% from 2021 to 2027, based on research in a newly published report [77].

On the one hand, most governments globally are looking for ways to reduce their CO_2 emissions and dependence on oil-based products and bioplastics are increasingly being viewed as one option to accomplish these goals. On the other, support from companies like Coca-Cola, Ford, Mercedes, Suzuki, Pepsi, H.J. Heinz, AT&T, and Toyota, as well-known global conglomerates, is providing the bio-sourced plastics industry with market pull. Prominent big brands that have introduced bioplastic packaging include Danone™, Actimel, Activia, Volvic), Coca-Cola's PlantBottle), and Ecover™ cleaning products. Carrefour, Sainsbury, Billa, Spar, and Hofer, which are some of the European supermarket chains, offer different packaging products and/or shopping bags made of bioplastics. In the leisure/sport sector Puma, Nike, and Adidas are among companies that use bioplastics compounds in their products. Ford, Toyota, Honda, Hyundai, Kia, Suzuki, and Mercedes are among the automotive manufacturers that have introduced various bioplastic components in several car models. Fujitsu is a well-known brand in the consumer electronics market that uses bioplastics in some of its products [77, 78]. Applications in the packaging market are covered in more detail by Dr Kulkarni in Chapter 23.

21.3.7 Bio-polycarbonates (bio-PC)

Polycarbonates (PC) are a group of thermoplastic polymers containing carbonate groups in their chemical structures. Polycarbonates used in engineering applications are strong, tough materials, and many grades are optically transparent. They are easy to work with and can be molded and thermoformed. Consequently, polycarbonates find many applications. Products made from polycarbonate can contain the precursor monomer bisphenol A (BPA), which has been banned in most countries for applications where it is in contact with hot water or milk, due to endocrine disruptions and metabolic complications. Bio-polycarbonate, also called bio-PC was developed by is an eco-friendly bio-plastic that can replace conventional polycarbonate, BPA. Until now, Mitsubishi Chemical Corp. of Japan is the only one who has successfully commercialized bio-polycarbonate sold as Durabio™, which is partially bio-based and a truly durable engineering plastic. It is made from isosorbide, which is derived from sorbitol, a bio-feedstock [79].

A more recent development is from Korea Research Institute of Chemical Technology (KRICT), which has developed another semi-bio-carbonate isosorbide with nanocellulose as a bio-derived reinforcing agent with much higher tensile strength [80] using two patent applications from Samyang Corporation [81].

In addition, many other companies are evaluating the use of waste cooking oil as a source of producing feedstocks, especially for producing bio-based polycarbonates. This is a relatively new development and in the next 2–5 years, we should see some major announcements from at least four companies, because replacement of BPA has been a significant driver.

21.4 Non-bio-based polymers, but biodegradable and/or compostable

21.4.1 Introduction

There are a few polymers that are produced from fossil fuel sources but happen to dissolve in water and are biodegradable and/or compostable. Typically, these are not followed by most researchers who work with bio-based materials because their source is not bio-based. However, if the goal is finding suitable biodegradable and compostable products, they are at least worth taking a look at. A list of these materials includes polyvinyl alcohol (PVOH) and some of its copolymers, DBS, and PBAT. BDS and PBAT are covered in Chapter 22 because recently these polymers have been prepared from monomers derived from bio-sources. Hence these two polymers will not be covered in this chapter.

19.4.2 Polyvinyl alcohol (PVOH)

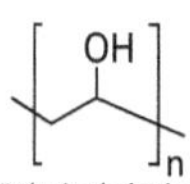

Polyvinyl alcohol

Figure 21.21: Polyvinyl alcohol.

Polyvinyl alcohol, also known as PVOH or PVAL (see Figure 21.21) is a synthetic polymer that is soluble in water. Sometimes it is mistakenly called PVA, which can be confusing since it is also used for polyvinyl acetate from which it is produced. Unlike most vinyl polymers, PVOH is not produced by polymerization of the corresponding monomer, because vinyl alcohol monomer is thermodynamically unstable and tautomerizes to acetaldehyde. Instead, PVOH is prepared by hydrolysis of polyvinyl acetate, which is produced by polymerizing vinyl acetate.

PVOH solution has excellent film-forming, emulsifying, and adhesive properties. It has no odor and is not toxic, and is resistant to grease, oils, and solvents. Its film is ductile but strong, yet flexible. The thermoplastic film typically produced using a small amount of mixture of glycerin and water as the plasticizer is transparent and has excellent barrier properties to oxygen and flavor. However, PVOH films lose part of their barrier as a function of higher humidity because water acts as a plasticizer, which limits its use in many applications in packaging that it otherwise could have been used. When opaque films are produced, rather than using minerals like in polyolefin films, starch-polyvinyl alcohol blends are used.

PVOH is mostly used to produce water-soluble film in applications like detergent pods and in the field of biomedical engineering, such as artificial cartilage, drug delivery systems, microorganism enwrapping, cell micro-capsulation, anti-thrombin materials, and biomedical sponges [82]. Over time, the PVOH dissolves as the wound heals.

In Japan and North Korea, it has been used to produce Vinylon™ fiber for low-cost clothing, which is stiff and difficult to dye but much cheaper than nylon, cotton, or polyester. This polymer is not biodegradable due to chemical modification. It is also used to produce polyvinyl butyral (PVB). PVB has amazing adhesive properties and is a water-resistant plastic film. So, one of the largest applications for this PVB film is for laminating inner layer between two glass or polycarbonate sheets used as safety windshield glass for vehicles.

Companies producing PVOH include Sekisui, Nippon-Gohsei, Kuraray, Sinopec, and Wacker Chemie with the global polyvinyl alcohol (PVA) market estimated to be around USD 1.1 billion this year growing to about USD 1.2 billion by 2025 [83].

21.5 Chemically modified natural biopolymers

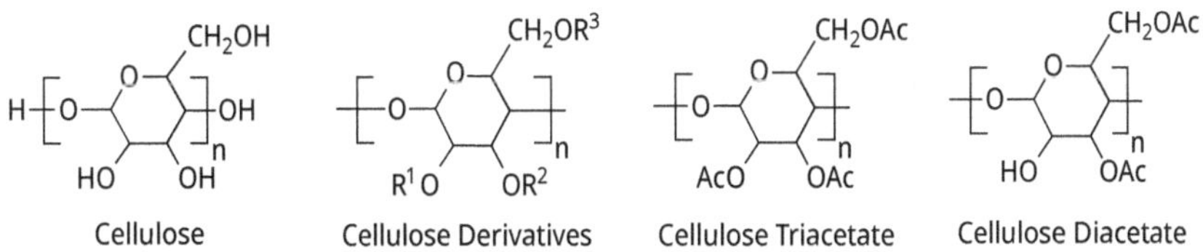

Figure 21.22: Cellulose, cellulose derivatives, cellulose triacetate, cellulose diacetate.

21.5.1 Cellulose derivatives

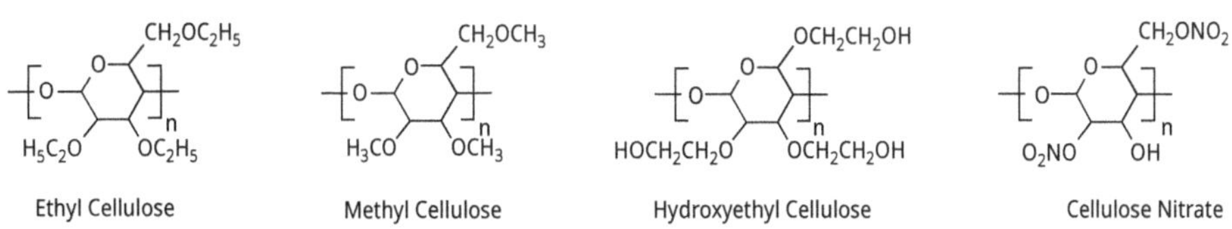

Figure 21.23: Ethyl cellulose, methyl cellulose, hydroxyethyl cellulose, and cellulose nitrate.

Cellulose is the largest organic polymer found in nature. It has been covered in the chapter by Dr Ghosalkar. Cellulose derivatives like cellulose acetate, cellulose butyrate, and ethyl cellulose have the general structure shown here. Amongst these three derivatives, cellulose diacetate has two acetate groups at the R2 and R3 position while R1 has a hydrogen, while cellulose triacetate has three acetate groups replacing all three Rs as shown here. Representations can be seen in Figures 21.22 and 21.23.

The largest application of cellulose is to produce paper and paperboard. Compared to this application, other relatively smaller quantities of cellulose include converting it to semi-synthetic cellulose derivatives, such as cellophane, rayon, and cellulose acetate and cellulose ethers. Cellulose acetate, which is widely used for industrial applications, is the largest cellulose derivative from a commercial perspective. When converted from cellulose, it can be either cellulose diacetate and cellulose triacetate or a mixture. This cellulose derivate has many important uses, which include fibers and threads for quality fabrics used in textiles, plastic films such as optical film for LCD technology and anti-fog goggles, and consumer products such as cellulose-based filters, and labels.

Of the other derivatives, methyl cellulose (MC) is the most commercially important cellulose ether. It is water-soluble and gels when exposed to heat. Unlike other water-soluble polymers, films made from methyl cellulose have a unique characteristic that they usually retain their strength and do not become tacky when exposed to humidity. Polymer films of both methyl and ethyl cellulose have excellent strength (60–70 MPa) and low elongation (5–15%) at room temperature, but their strength decreases rapidly with increasing temperature. These nonionic polymers also have excellent UV, oil, and solvent resistance. These useful properties of this bio-polymer and its ability to form gels easily have enabled numerous applications such as water-soluble films used for packaging products that dissolve in water like medical capsules, tooth pastes, detergent powders, bubble bath capsules and liquid products, rat poison and bread dough, as well as a thickener in food products for these cellulose ethers, especially for the methyl derivative [84, 85].

Cellulose nitrate or nitrocellulose (called cellophane or celluloid) has been in use for over 160 years. It was really the first transparent plastic film invented. Due to its good tensile strength and insolubility in water and many other solvents, it gained acceptance in general packaging use and for celluloid film till it was eventually replaced by cellulose acetate, polyolefin, polyester, and PVC films. Cellulose derivative manufacturers include AkzoNobel, Daicel, Eastman, Mitsubishi Rayon, Shin Etsu, Solvay, and Tembec.

21.5.2 Vulcanized natural rubber

Another natural polymer is natural rubber, which is obtained from rubber trees as a latex. The raw latex is not easy to use and does not have desirable properties. Chemically it is *cis*-1,4-polyisoprene. Natural rubber is used extensively in many applications and products, mostly after compounding with other additives and involving chemical modification, and very rarely by itself. In most of its useful forms, it has a large stretch ratio and high resilience and also is waterproof. The only reason this section exists is to illustrate the wide variety of bio-based applications.

To make it more useful, it is vulcanized or "cured" using sulfur or other cross-linkers along with various additives. Vulcanization is the process of formation of cross-links between polymer chains for hardening rubber material [86]. Covalent cross-links provide a three-dimensional elastic network with "junctions." The degree of cross-linking can be controlled by modifying the formulation and the process conditions of the vulcanization process. The higher the degree of cure measured by its cross-link density, the less elastic and more durable the chain is.

There are numerous applications where the flexibility of rubber is advantageous. These include rubber bands, tubing, hoses, tires, and rollers for a wide range of devices ranging from domestic and office supplies to large industrial machinery. Its elasticity and resistance to impact even at low temperatures makes it invaluable for various kinds of shock absorbers and for specialized machinery mountings vibration damping. Once natural rubber is fully cured, it becomes a thermoset and it cannot be melted or its shape changed. Its relatively good gas impermeability makes it useful in the manufacture of articles such as air hoses, balloons, balls, and cushions. The low water absorption and resistance of cured rubber to water and to most fluid chemicals has led to its use in rainwear, diving gear, and chemical and medical tubing and as a lining for storage tanks, processing equipment, and railroad tank cars. Rubber's electrical resistance, poor heat conduction, and low water absorption are reasons why soft rubber goods are used for thermal and electrical insulation and for protective gloves, shoes, and blankets. Hard rubber is used for articles such as telephone housings and parts for radio sets, meters, and other electrical instruments, although in a lot of these types of applications today rubber is being replaced with thermoplastic elastomers. Another unique application not well known to the general public is its use for power-transmission belting, highly flexible couplings, and for water-lubricated bearings in deep-well pumps due to the coefficient of friction of rubber, which is high on dry surfaces and low on wet surfaces [86]. Due to shortages of natural rubber, synthetic rubber was developed primarily based on styrene-butadiene copolymers. Several other elastomeric polymers, including thermoplastic elastomers, have been also developed since then. The largest application for rubber has been its use in tires. Automotive tires are mostly produced using synthetic rubber, whereas tires for trucks need to have a significant ratio of natural rubber to synthetic rubber for its far superior tear and fatigue resistance. As with any thermoset, vulcanized natural rubber is not biodegradable. It can only be degradable by controlled pyrolysis.

21.6 Limitations of biopolymers and challenges for growth

On the one hand, the production and use of bioplastics instead of nonbiodegradable and oil-based plastics reduce emissions of carbon dioxide based on using plants that absorb it during their growth stage. The technology of bioplastics also provides products from renewable and/or biodegradable sources, and can offer a promising alternative for the destination of solid biomass residues for production of these materials. Hence, these products can be considered one of the mitigating methods to minimize global warming [87].

However, bio-based products have several limitations, which provide challenges for growth and fast adoption on a really large scale, except in few cases where they have found niches, many of which are growing at high growth rates or have grown already. Most biopolymers used have practical shortcomings that include high moisture permeability and consequent poor dimensional stability, poor oxygen and flavor permeability, fragility, meagre thermal stability resistance, poor mechanical properties especially when wet, vulnerability to biological, ultraviolet and thermal degradation which are critical for durable applications, and very narrow temperature window for plastics processing. For some applications, biopolymers are inherently too absorbent.

Other commercial issues that also need to be addressed that prevent widespread adoption include:

- High costs of the biopolymer in comparison to synthetic plastics derived from fossil sources
- Large-scale production of bioplastics for different applications is limited by limited rate of production
- Cost of production of most bioplastics on average is at least 25% higher than conventional plastics that they are trying to replace
- Ability to provide consistency of quality
- Ability to supply globally prevents global consumer companies from switching to bio-based products

To address many of the functional disadvantages listed earlier in this section, it has been necessary to address each of these, to improve the range of applicability, and processability. This is carried out by either polymer modification or compounding in various additives, many of which are bio-based. The bio-based additives, whether they are bio-fibers, bio-lubricants or plasticizers, are also used in compounds where the carrier resin polymer is non-bio-based. Such compounds have found many different applications for a wide variety of end-use markets. The bio-based compounds where reinforcements are used are called bio-composites. Both bio-based plastics compounds and composites will be discussed in two separate chapters later in this book.

21.7 Will bioplastics solve the world's plastics problem?

Jim Robbins wrote a thought-provoking article entitled "Why Bioplastics Will Not Solve the World's Plastics Problem" in 2020 [88]. In his article, he has used an example of a large consumer company like Coca Cola, which has developed PlantBottle™ a new type of recyclable plastic container, produced from a blend of 30% of a biopolymer sourced from sugarcane and other plants and 70% made from traditional oil-based plastic. It is understandable that Coca Cola and its competitors like Pepsi are responding to public pressure to solve the problem of "plastic pollution" in the oceans. It is an extremely difficult challenge to come up with an organic plastic that satisfies all the cost and product requirements from production to packaging to handling to storage to the consumer, maintaining shelf life and, after its use, becomes biodegradable quickly as part of nature again. Bioplastics, which make up part of Coke's PlantBottle, have been touted as an important solution to the world's plastic pollution problem. Yet we know for a fact that the thicker the bioplastic, the slower the biodegradation rate is. Also, when it is a minor volume phase the biopolymer is not on the surface and difficult for the bio-organisms to degrade it. Yet the performance requirements for such a bottle of carbonated beverages needs to meet specific permeability and strength requirements, which dictate what the minimum thickness of such a bottle must be. So, while such efforts by companies need to be applauded as a start, they are not a solution! As my friend Dr Ramani Narayan, University Distinguished Professor in the Department of Chemical Engineering and Materials Science at the School of Packaging at Michigan State University in the USA has said and I quote, "The concept that we could use it, throw it away, and it doesn't matter where you throw it, and it's going to safely disappear, that does not exist. Nobody could engineer something like that, not even nature."

Instead, many experts believe the solution to plastic waste mainly is not in developing better bioplastics, but instead in overhauling the world's economy to recycle far greater quantities of plastic than currently are being reused. A recent two-year study called "Breaking the Plastic Wave" by Pew Charitable Trusts and SYSTEMIQ, found that despite the efforts of industry, governments, and NGOs, the plastic problem is getting much worse [89]. It is a fact that though the production of biopolymers and bioplastics is growing at a higher rate, it will take a long time for them to match the volume of production of petro-based plastics.

There are other such presentations including one by Dr Geoffrey Mitchell presented at the 8th World Congress on Biopolymers & Bioplastics in 2018 [91]. He said that an immense global demand for sustainable materials exists, especially for those which biodegrade in sea water to avoid the current problems of plastics in the oceans. He rightly observed that "before we can transform the materials supply chain it is helpful to reflect on the particular properties which have made polyolefin-based polymers so successful." We all know that just having a suitable material is not sufficient; we need manufacturing processes, which successfully produce objects as well as the appropriate designs for those materials that meet all the performance requirements of that product for each stage of the value chain. We cannot generalize that all petro-based plastics can be replaced. For certain applications high-performing polyolefins and other petro-based plastics are essential for applications like the continued safe supply of water and electrical power and parts for automobiles and other transportation end uses that need light weighting for better fuel efficiency. For other applications like some packaging, but not all, we certainly can deliver on the promise of a greener future. It is the case that we need to be more discriminating in our use of materials through appropriate design based on the application requirements for material substitution. Dr Mitchel cautioned us against seeing bioplastics as a straight swap with existing plastics. He proposed that "embracing the whole life cycle analysis approach to product design and material choice provides a sounder basis for future developments. It is timely that we embrace the concept of the circular economy whilst we deal with the plastics in the ocean disaster" [91].

The same 154-page Pew report concludes that the only solution to this proliferating problem is an enormous overhaul of the global plastic ecosystem so that these materials are reused and recycled plastic using a circular economy [89]. Bioplastics have a smaller role in this but are not the ultimate solution. If all of the report's recommendations are adopted, "plastic waste could be reduced by 80% over the next two decades" is the claim. The report proposes a combination of all of the following recommendations along with many other suggestions:

a) elimination of plastic packaging wherever possible, substituted with paper or compostable material;
b) designing products for effective recycling;
c) increasing mechanical recycling;
d) scaling up collection and recycling efforts in moderate- and low-income countries, where the vast majority of ocean plastic originates; and
e) ending exports of waste plastic, which would force countries where the waste is generated to come up with solutions to the plastics problem.

This prescription of a long-term comprehensive solution to address this issue makes a lot of sense, except for point (a) above, because numerous cradle-to-grave life cycle studies have shown, one report after another, that plastics are greener and more sustainable than paper, paperboard, metal, and glass [90]. By the way, Dr DeArmitt's unsponsored work (meaning unbiased by any outside organization for or against plastics) shows that plastics are a tiny 0.5% of global waste and despite fake photographs in social media that tug on the viewers' emotions, there is no scientific evidence to prove that it is toxic [90]. Meanwhile many "pro-environmental" organizations push policy proposals to ban plastics without studying scientifically obtained life cycle data based on either ideology or preconceived conclusions with the knowledge that most journalists, politicians, and indeed the general population will not seek out such life cycle data.

Plastics serve many useful purposes in our daily lives today. Imagine if everything globally that uses plastics were eliminated today. We would consume more energy because transportation would be heavier. Shipping costs would jump 50%. We would need food supply to grow at least 30% if proper plastics packaging that ensures sufficient shelf life for food and medicine would disappear. What about our phones, computers, appliances, electronics, medical devices, and everything else we use every day? Yet it is easy for journalists, politicians, bureaucrats, and especially “fake environmentalists” to blame someone, or in this case a set of materials, instead of putting in the work to find out what scientific data has already been generated, which are facts, determine what actions actually make sense, allocate money to take actions, and follow through on implementation to ensure this complicated issue is resolved!

As many others have pointed out, there are numerous cases of wooden and metal ships and many other materials, which have lasted over 500 or even thousands of years in the ocean. Yet, no government or organization in over 210 countries on this planet has said or done much about it. There are so many hazardous chemicals and metals like lead and mercury that are over a million times more dangerous that end up in our oceans, yet there is no action or even a serious discussion about these materials. Plastics are less than 1% of materials and the other 99% of materials are ignored. Yet LCA data shows that usually plastics are the greenest solution; greener than alternatives such as paper, glass, and metal in applications that they are used in terms of carbon dioxide generation and consumption of water and energy and usually the lowest cost option. Data shows that replacing plastics with these alternatives would generate 5–15 times the amount of CO_2. If we banned plastics, we would increase the amount of carbon dioxide into the atmosphere with other alternatives and consume more water, energy, and other resources, increase food spoilage, and it would cost us all a lot more money with every other alternative. But it is easy to blame materials that can’t talk, protest, or vote!

Meanwhile, we have a situation that is getting complicated, in Europe, which could also happen in the future elsewhere. On the one hand, European policy frameworks are increasingly favoring biodegradable materials, accelerating demand and innovation [11]. Also, the EU mandates higher and higher ratios of recycled content; there is a limit as to how high each application can handle the level of recycled content to meet the required specifications. In addition, there is a debate of recycling fossil-derived plastics versus bio-based plastics, which are not compatible with recycling methods typically used for fossil-based plastics. As we have seen earlier in this chapter, not all bio-based polymers are biodegradable or compostable. For now, the EU has taken a nuanced stance on the recycling debate between fossil-derived and bio-based plastics, especially when compatibility issues arise. The EU recognizes the environmental potential of bio-based plastics but cautions that their incompatibility with conventional recycling systems poses logistical and sustainability challenges. According to the European Commission (2022), biodegradable plastics should only be deployed in applications where reuse and recycling are impractical, and must be clearly labeled to avoid contamination of fossil-based recycling streams [92].

21.8 Future trends and prospects for growth

An article, which provided a good review on future trends for bio-based polymers, was published over a decade ago in 2013 and talked about their high growth potential [93]. A recent report from ResearchAndMarkets.com that is more relevant today contends that the global bioplastics market volume in metric tons would grow from an estimated 2.2 million metric tons in 2021 to about 3.3 million metric tons by 2026, with a compound annual growth rate (CAGR) of 8.3% for this 5-year forecasted period; from a regional perspective the bioplastics market volume is expected to grow at a CAGR of 8.4% from 0.93 million metric tons in 2021 to 1.4 million metric tons in 2026 in the same period while the Asian bioplastics market catering much larger population is estimated to grow from 0.6 million metric tons in 2021 to 0.97 million metric tons in 2026 at a CAGR of 9.6%, a higher growth rate but much lower per capita consumption. Based on these reports, the future is certainly bright for bioplastics [8, 9, 11, 93]!

Another recent review article lists 532 biopolymer start-ups in several promising areas where new innovative developments are being carried out into new business ventures [94]. It highlights the top nine biopolymer trends and 18 promising start-ups. Even if only 50% of these succeed, they will have shown the way to others and portend a lot of promise to future growth. Some of the examples from this article showing these innovations are:

- Protein-based biopolymers and some bio-polyamides have advantages of being biocompatible and biodegradable along with possessing superior gas barrier and mechanical properties compared to cellulosics. Hence, they are one of the most promising classes of biomaterials. Due to their possibility of interaction with a wide range of bioactive molecules, many protein biopolymers and polycaprolactones find applications in the medical industry for tissue engineering and tablet coating.
- Recent advances in starch-based materials have enabled films to be made antimicrobial to prevent microbial grow for packaging of food products. Many start-ups are expected to adopt this technology to meet the growing demand for starch in applications including agriculture, packaging, cosmetics, medicine and textiles.
- Pivot Materials, an American start-up, uses bamboo and rice waste fiber among other bio-materials in their products. It specializes in manufacturing natural fiber-based lightweight, sustainable, and durable composite plastics for injection and extrusion molding to improve specific material properties. The end markets for such sustainable plastics are suitable for home goods, packaging, and transportation.
- A Chinese start-up called WAVE has developed a cassava-based polymer compound in pellet form, which is then converted to bags. The natural cassava starch is water-soluble, composts naturally, and is edible by land and aquatic animals. The start-up company produces a bio-based plastic compound that safely returns

to nature in the form of water and carbon dioxide, which in turn supports sustainable agriculture. This solution is an eco-friendly alternative to regular plastics, as the starch-based bags use only sustainable natural resources. One caveat is that the bags need to stay dry, otherwise they fall apart!

- Seaweed is used as a source material by two start-up companies, UK-based SoluBlue and Norway-based B'Zeos:
  - Solublue's product is suitable for FDA-approved food packaging because it absorbs moisture and prevents mold growth on fresh food, thereby extending its shelf life. This alternative to single-use petro-based plastics also captures a vast amount of carbon and locks it in the soil when composted on land. If dropped in the ocean, their seaweed biopolymer produced film breaks down within weeks and can also be safely ingested by marine life.
  - B'Zeos's bio-based packaging materials using seaweed biopolymers is targeted to replace disposable plastic articles in the food service industry. This start-up partners only with seaweed cultivators and harvesters that work with sustainable and ethical practices. The eco-sourced seaweed uses their green processing technology to extract valuable compounds to create a packaging solution that is circular, home-compostable, and even edible.
- An Italian start-up called Earth Bi uses biomass-derived natural waste as a source to produce biodegradable and compostable polymers. They also use renewable energy for their production process and do not require pure materials. Not only is their biopolymer product suitable for traditional processing of plastic without extensive modifications through its distinct heating process, they claim it to be recoverable and recyclable.
- Starting from mushrooms, a start-up called Mushroom Material based in New Zealand has developed a bio-based foam to replace Styrofoam™, which is based on petro-based polystyrene. They take the vegetative part of the mushroom called mycelium and combine it with fibrous agricultural waste to create this unique product which is waterproof, fire-resistant, durable, and compostable, even at home for sustainable packaging applications with the same protective properties as Styrofoam.
- An American start-up called Verde Bioresins produces bio-PET under the brand name PolyEarthylene™. This plant-based sustainable and renewable biopolymer is claimed to be environmentally friendly, economically more feasible, and lighter compared to petroleum-based polymers for applications for production of durable consumer goods. While they are not competitive with PET for applications like bottles, they do compete with PLA, PHA, green PE, and other bio-based materials.
- Another US-based start-up called Biorgani is offering a bio-based resin called Faunnus™ for packaging. It is designed to be used as a blend with petro-based polyolefins by matching the pure petro-based polyolefin's weight resistance and performance for applications like vegetable wraps in the produce section of gro-

ceries and for grocery bags with a lower carbon footprint. The same company also manufactures what they call a "zero polyethylene resin" called Solum™ Compost Resin, which processes on conventional plastic processing equipment needing less temperature for processing and hence, saving energy and reducing carbon emissions during manufacturing compared to petro-based polyethylene resin.

These are just a few examples of the progress being made, a little at a time, by many companies around the world to try and mitigate the issue of the existing and growing pile of plastics in landfills, oceans, and elsewhere. While plastics ending up in waste are less than 1% of total global waste by weight (the rest is wood, metal, glass, minerals, etc.) they seem to have more visibility. Every day we are adding more plastics that are used globally. So, there is a crying need for finding sustainable alternatives as well as for solving this issue.

Bio-based and biodegradable materials have shown strong growth across industries over the past years, mainly in packaging, driven by consumer market trends toward greener packaging and waste reduction. As several examples in this chapter have shown, there are numerous applications in the medical industry for drug delivery and tissue engineering and packaging as well as in other industries. On our way to a more sustainable future, innovations in this field of bioplastics to improve biomaterials properties further, reduce their cost, and make production processes environmentally friendly are sorely needed. Although bioplastics represent just about 1% of the total amount of plastics today, as more innovative techniques are developed to overcome the various limitations listed earlier in this chapter, the field of bioplastics will grow faster, especially as the larger volumes make them more cost competitive and as fossil fuel prices increase further and more bioplastics are produced from waste materials, rather than from food crops or from sugarcane, which raises consumer prices for the growing global population.

One recent development is the use of very large amount of citrus waste generated by the citrus industry. Peels and seeds are often underutilized despite their high cellulose and pectin content. Extraction methods for recovering these products are energy-intensive and generate harmful by-products. The innovation here is that a CytroCay process has been developed using cavitation that uses sound waves to generate bubbles in liquid that converts citrus waste into a type of micronized cellulose and highly bioactive pectin, using only water and electricity, operating at room temperature, making it a model circular economy practice [95].

Finally, to put things in perspective, Figure 21.24 shows how small bio-based plastics and biofuels are in terms of utilization of global agricultural land compared to other much larger uses such as pasture and land used for producing food and fuel [96]. However, as global population and briskly increasing economic progression continues, the growing demand for materials, chemicals and energy resources will be required worldwide. A big chunk of such resources currently comes from fossil fuels. A

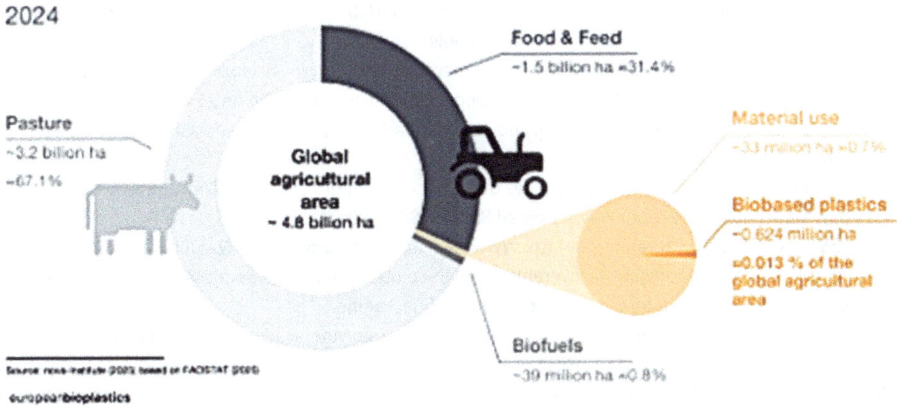

Figure 21.24: Global agriculture land distribution in 2024 and small bio-based plastics (source: European BioPlastics).

recent review estimates "8% of fossil fuels are used in polymer production and expected to surge to 20% by 2050. This mounting demand for polymers, coupled with escalating processing costs, poses a formidable long-term challenge in meeting global polymer requirements" [97]. Based on that, using biomass as well as recycling will remain the only sustainable ways to deal with this issue.

References

[1] Thielen M. "Bioplastics – Basics. Applications. Markets", Polymedia Publisher GmbH, Mönchengladbach, Revised English Edition, Germany: 2020. 3rd, ISBN 978–3-9814981-4-1.

[2] Young RJ. "Introduction to Polymers", London, United Kingdom: Chapman & Hall; 1987. ISBN 0–412-22170-5.

[3] "Natural Polymers and Biopolymers", at https://iscamapweb.chem.polimi.it/citterio/it/education/coursetopics/

[4] "Polynucleotides" at the US National Library of Medicine Medical Subject Headings (MeSH).

[5] Hamley IW. "Introduction to Peptide Science", Hoboken, NJ, USA: Wiley; 2021. ISBN 9781119698173.

[6] "Polysaccharide – Definition, Examples, Function and Structure", at https://biologydictionary.net/polysaccharide.

[7] Barrett A, "The History of Bioplastics", Bioplastics News, at https://bioplasticsnews.com/2018/07/05/history-of-bioplastics/

[8] "EUBP Facts and Figures – European Bioplastics Documents" at https://www.european-bioplastics.org/;www.bio-based.eu/markets/

[9] Bio-based Building Blocks and Polymers – Global Capacities, Production and Trends 2023–2028. nova-Institut GmbH (editor), Hürth, Germany, 2024-03; https://doi.org/10.52548/VXTH2416/

[10] https://biobiz.in/s/opp/bioplastics-biopolymers-indian-market-scenario

[11] Status and Outlook: Bio-based Polymers worldwide –, Green Chemicals Blog; https://greenchemicalsblog.com/status-and-outlook-bio-based-polymers-worldwide/.

[12] https://www.marketsandmarkets.com/PressReleases/biopolymers-bioplastics.asp/

[13] Barrett A, Bioplastics News, https://bioplasticsnews.com/2019/04/13/what-is-the-difference-between-biodegradable-compostable-and-oxo-degradable/

[14] (a) Vroman I, Tighzert L, "Biodegradable Polymers", at https://www.ncbi.nlm.nih.gov/pmc/articles/PMC5445709/; (b) European Bioplastics – Report Bioplastics. [(accessed on 25 May 2020)]; at https://docs.europeanbioplastics.org/publications/market_data/Report_Bioplastics_Market_Data_2019.pdf/

[15] Fujimaki T "Processability and properties of aliphatic polyesters", "Bionolle", synthesized by polycondensation reaction. Polym Degrad Stab 1998;59:209–214.

[16] "BS EN 13432:2000 Packaging" at https://www.en-standard.eu/bs-en-13432-2000-packaging.-requirements-for-packaging-recoverable-through-composting-and-biodegradation.-test-scheme-and-evaluation-criteria-for-the-final-acceptance-of-packaging/

[17] Ranjan S, "Biopolymers & Their Impact", 2017, at https://polymeracademy.com/biopolymers-their-impact/

[18] Ogunniyi DS, "Castor oil: A vital industrial raw material", at https://www.sciencedirect.com/science/article/abs/pii/S0960852405002026/

[19] Vasishtha AK, et al. Sebacic acid and 2-octanol from castor-oil. J Am Oil Chem Soc 1990;67:333–337. at. https://doi.org/10.1007/BF02539685.

[20] "Global Green and Bio Polyols Market Analysis 2022" by Syngene Research LLP. Bangalore, India: May, 2022, at, https://www.marketresearch.com/Syngene-Research-LLP-v4190/Global-Green-Bio-Polyols-31908029/.

[21] Siracusa V, Blanco I. Bio-Polyethylene (Bio-PE), Bio-Polypropylene (Bio-PP) and Bio-Poly(ethylene terephthalate) (Bio-PET): Recent Developments in Bio-Based Polymers Analogous to Petroleum-Derived Ones for Packaging and Engineering Applications. Polymers (Basel) Aug. 2020;12(8):1641. at. https://www.ncbi.nlm.nih.gov/pmc/articles/PMC7465145/#B9-polymers-12-01641/.

[22] Morschbacker A Bio-Ethanol based Ethylene, Polym Rev 2009;49:79–84. doi: 10.1080/15583720902834791.

[23] Report B. "Development of bio-based Olefins" (accessed on 22 August 2017); at http://www.inda.org/BIO/vision2014_659_PPT.pdf/

[24] Kahn J. Green PE: The value proposition and success in the marketplace and green PP: Update on commercialization. In Bioplastics: The Reinvention of Plastics Conference, March 6 2013. Las Vegas, NV; at. https://polymerinnovationblog.com/bio-based-polypropylene-multiple-synthetic-routes-under-investigation/.

[25] Venkataraman V. Propylene Supply, Go Green, On Purpose Technologies for the Future, at, www.slideshare.net›beroe.inc›propylene-supply-go/.

[26] Chen GQ, Patel MK. Plastics derived from biological sources: Present and future: A technical and environmental review, Chem Rev 2012;112:2082–2099. doi: 10.1021/cr200162d/.

[27] Kawai K, Takemoto S, Matsuzaka H, et al. Bimetallic Ru–Ir/Rh complexes for catalytic allyl alcohol reduction to propylene. Chem Commun 2024;60:9424–9427. https://doi.org/10.1039/D4CC01711K.

[28] Ohara E, Kawazumi K, United States Patent 6,936,724, "Process for producing epsilon-caprolactone", assigned to Daicel Chemical Co. issued August 30, 2005.

[29] Rani GU, Sharma S. Biopolymers, bioplastics and biodegradability: An introduction. In Reference Module in Materials Science and Materials Engineering, Elsevier Inc; 2021.

[30] (a) Koivistoinen OM, Kuivanen J, Barth D, Turkia H, Pitkänen JP, Penttilä M, Richard P. Glycolic acid production in the engineered yeasts Saccharomyces cerevisiae and Kluyveromyces lactis. Microb Cell Fact. 2013 Sep 23; 12:82. 10.1186/1475-2859-12–82/ (b) Soucaille, P, Canadian Patent CA-2654182-A1, "Glycolic Acid Production By Fermentation from Renewable Resources", Owner: Metabolic Explorer; PCT/EP2007/055625 & WO2007/141316.

[31] https://greenchemicalsblog.com/tag/polyamides/;www.genomatica.com/;https://greenchemicals blog.com/2014/08/08/genomatica-enters-bio-polyamides-market/;https://www.icis.com/subscriber/ news/home/viewarticles/
[32] Murata Y, Kuribayashi T, Onishi F, Hiura T. 1,5-Pentamethylene diamine and Production Method Thereof. In United States Patent Application 20160289165 Filed Nov. 18, Assigned to Toray Industries Inc. and Ajinomoto Co., Inc; 2014 published Oct. 6, 2016.
[33] Alvarenga RAF, Dewulf J. Plastic vs. fuel: "Which use of the Brazilian ethanol can bring more environmental gains?. Renew Energy 2013;59:49–52. doi.org/10.1016/j.renene.2013.03.029. [CrossRef] [Google Scholar].
[34] De Jong E, Stichnothe H, Bell G, Jørgensen H, "Bio-Based Chemicals A 2020 Update", at https://www. academia.edu/42073840/Bio_Based_Chemicals_A_2020_Update/
[35] https://www.marketwatch.com/story/dow-chemical-crystalsev-to-form-brazil-joint-venture/
[36] https://www.prnewswire.com/in/news-releases/dow-and-thong-guan-introduces-sustainable-plastic -stretch-cling-films-made-with-renewable-feedstock-in-asia-pacific-836547588.html/
[37] https://www.biofuelsdigest.com/bdigest/2021/05/23/neste-mitsui-chemicals-and-toyota-tsusho-to-start-japans-first-production-of-renewable-plastics-from-100-bio-based-hydrocarbons/
[38] McKeen LW. Plastic films in food packaging – materials, technology and applications. In Ebnesajjad S, editor. Elsevier, 2013, Hardcover ISBN: 9781455731121, eBook ISBN: 9781455731152, McKeen, LW, "Permeability Properties of Plastics and Elastomers", (Third Edition), Elsevier, 2012.
[39] https://www.databridgemarketresearch.com/reports/global-bio-based-polyethylene-pe-market.
[40] "Neste and LyondellBasell produce bio-PP and bio-LDPE at commercial scale", https://www.bioplas ticsmagazine.com/en/news/meldungen/20190618Neste-and-LyondellBasell-produce-bio-PP-and-bio-LDPE-at-commercial-scale.php/; LyondellBasell, Neste launch new bio-based plastics (chron.com)
[41] McKeen LW. "The Effect of Sterilization on Plastics and Elastomers" (Third Edition), Elsevier Inc.; 2012. doi: https://doi.org/10.1016/B978-1-4557-2598-4.00012-5.
[42] Dhanasekaran NPD, Muthuvelu KS, Arumugasamy SK. Recent advancement in biomedical applications of Polycaprolactone and Polycaprolactone-based materials. Ref Modul Mater Sci Mater Eng January 2022; doi: 10.1016/B978-0-12-820352-1.00217-0.
[43] Arif U, Haider S, Haider A, Khan N, Alghyamah AA, Jamila N, Khan MI, Almasry WA, Kang IK. Biocompatible polymers and their potential biomedical applications: A review. *Curr Pharm Des* **2019**;25(34):3608–3619.
[44] Flieger M, Kantorová M, Prell A, Rezanka T, Votruba J. Folia Microbiol (Praha). 2003, 2003. 48 1, 27–44. doi: 10.1007/BF02931273. https://www.researchgate.net/publication/358654485/.
[45] "Caprolactone", at https://www.sciencedirect.com/topics/chemical-engineering/polycaprolactone/
[46] Singh MK, Singh R, Dhami M. Biocompatible Thermoplastics as Implants/Scaffold. In Reference Module in Materials Science and Materials Engineering, Elsevier; 2020. https://www.sciencedirect. com/science/article/pii/B9780128203521000122/.
[47] https://bioplasticsnews.com/polytrimethylene-terephthalate-ptt/
[48] "FTC approves federal register notice establishing new fiber name and definition" (Press release). Federal Trade Commission. 2009-03–20.] Triexta has several advantages over polyester, including better stain resistance and softness. [Simmons, Cheryl (8 August 2019). "Triexta PTT Carpet Fiber". The Spruce.; https://www.thespruce.com/triexta-ptt-carpet-fiber-2908799/
[49] https://www.databridgemarketresearch.com/reports/global-bio-based-polytrimethylene-terephthalate-ptt-market/
[50] https://www.marketwatch.com/press-release/polytrimethylene-terephthalate-ptt-polymers-market -2022-size-2022-investigation-report-by-global-industry-share-demand-key-findings-regional-analysis-key-players-profiles-future-prospects-and-forecasts-to-2028-2022-06-16/

[51] Gilding DK, Reed AM Biodegradable polymers for use in surgery – Polyglycolic/poly (lactic acid) homo- and copolymers: 1. Polymer 1979;20(12):1459–1464. doi: 10.1016/00323861(79)90009-0.

[52] (a) Middleton J, Tipton A, "Synthetic biodegradable polymers as medical devices". Medical Plastics and Biomaterials Magazine, (March 1998);(b) Middleton, J; Tipton, A, "Synthetic biodegradable polymers as orthopedic devices", Biomaterials, Dec. 2000, 21(23):2335-46.(c) Pişkin E, "Biodegradable polymers as biomaterials", J Biomater Sci Polym Ed. 1995;6(9):775–95. 10.1163/156856295x00175/

[53] http://www2.dupont.com/Oil_and_Gas/en_CA/assets/downloads/DuPont_Polyglycolic_Acid_Sheet.pdf

[54] Kuredux® Polyglycolic Acid (PGA) Resin Archived 2020-12–09 at the Wayback Machine. www.kureha.com. accessed 4 December, 2021.

[55] Polyglycolic Acid – Global Market Trajectory & Analytics. ID: 5309880 Report April 2021, Dublin, Ireland: Research And Markets.

[56] "Polyamide 11", https://www.etigo.fr/en/blog/p-polyamide-11

[57] "Rilsan", "Pebax", Specialty Polyamides", all at https://www.arkema.com/global/en/search/?qc=search&q=polyamides

[58] Khedr MSF. Bio-based polyamide. Phys Sci Rev https://doi.org/10.1515/psr-2020-0076/.

[59] Bio-based polyamides – Current trends and overview. In paper presented by Doug Weishaar, Evonik Corporation at the 10th Annual World Congress on Industrial Biotechnology, Montreal, Canada: June 16–19 2013. https://www.bio.org/sites/default/files/legacy/bioorg/blogs/the_new_age_of_bio-based_polyamides_doug_weishaar.pdf/.

[60] Humphreys R, "Bio-Polyamides: Where Do They Come From?" https://polymerinnovationblog.com/bio-polyamides-where-do-they-come-from/

[61] Devaux J-F, Le G, Pees B, "Application of Eco-profile Methodology to Polyamide 11", https://www.docsity.com/en/polyamide-11-and-its-properties-for-hose/538543.

[62] Thielen M. Basics of bio-polyamides. Bioplastics Mag March 2010;5:50–53.

[63] "Bio-polyamides for Automotive Applications", https://www.tu-braunschweig.de/en/presse/wissenschaft-medien/homepage.

[64] Tao L. Preparation and properties of biobased polyamides based on 1,9-azelaic acid and different chain length diamines. Polym Bull 2020;77:1135–1156.

[65] Kuciel S, Kuzniar P, Liber-Knec A. Polyamides from renewable sources as matrices of short fiber reinforced biocomposites. Polimery 2012;57:627–634.

[66] www.genomatica.com/press_releases/

[67] Mutua FN, Cong Cheng C, Dong Y, Zheng C, Zhu B, He Y. Synthesis and properties of bio-based Poly (pentamethylene oxamide). Polym Eng Sci 2018;58(Issue 5): doi: 10.1002/pen.24596.

[68] https://www.cathaybiotech.com/

[69] Ng H, "Partially aromatic polyamides and a process for making them", United States Patent 6,355,769, assigned to DuPont Canada, Inc., issued March 12, 2002.

[70] "Zytel®- A Portfolio of High-Performing Nylon Resin Products", https://www.dupont.com/products/zytel.html/

[71] "Zytel® HTN -A family of cost-effective, high-performance polyamides, https://www.dupont.com/products/zytel-htn.html/

[72] https://www.dupont.com/news/dupont-announces-agreement-to-divest-majority-mobility-and-materials-segment-to-celanese.html

[73] Tamura K Polyamide, United States Patent 10,544,262, assigned to Kuraray Co. Ltd., issued Jan. 28 2020.

[74] Takeda H. Polyamide Resin, United States Patent 10,465,071, assigned to Kuraray Co. Ltd., issued Nov. 5 2019.

[75] LaPree J. High-performance polymers in the spotlight. In Chem. Engg, Access Intelligence, LLC, Rockville, MD, USA: Publisher, July 1, 2015 122 7, pp. 20–23.
[76] Market B-P. Research Report 2022, Professional Edition; https://galleonnews.com/uncategorized/283974/bio-polyamide-market-research-report-2022-professional-edition/.
[77] "Global Bioplastics Technology Markets Report 2021-2026: Industry Landscape, History of Bioplastics, Recent Developments, Pros and Cons of Bioplastics". Dublin, Ireland: Research and Markets; 2021.
[78] "Global Markets and Technologies for Bioplastics", Wellesley, Massachusetts, USA: BCC Research; 2021, November.
[79] (a) https://www.mcpp-global.com/en/mcpp-america/products/brand/durabioTM/;(b) Sasaki, H; Tanaka, T, United States Patent 8,889,790, "Polycarbonate resin composition and molded article thereof", assigned to Mitsubishi Chemical Corp., issued November 18, 2014.
[80] "South-Korea To Compete With Japan on Bio-Polycarbonate", at https://bioplasticsnews.com/2019/10/29/south-korea-japan-bio-polycarbonate/
[81] (A) Jin SC, Chang YJ, Kwon YD, "Isosorbide-aromatic polycarbonate copolymer", Korean Patent Applications 10-2015-0004084;(b) Jin, SC; Chang, YJ; Rhee, HC; Kwon, YD, "[Poly(isosorbide carbonate- aromatic carbonate)]-[polycarbonate] block copolymer", Korean Patent Application 10-2015-0004085.
[82] Guo D, Xu K, Application of Polyvinyl Alcohol in biomedical engineering (Article in Chinese) at www.pubmed.ncbi.nlm.nih.gov/16013269/
[83] "Polyvinyl Alcohol (PVA) Manufacturing Industry" report published by NIIR Project Consultancy Services (NPCS), New Delhi, India: www.entrepreneurindia.co.
[84] Heinze T, El Seoud OA, Koschella A, "Cellulose derivatives: Synthesis, structure, and properties" (Springer Series on Polymer and Composite Materials) 1st ed. 2018 Edition.; at https://polymerdatabase.com/polymer%20classes/Cellulose%20type.html
[85] "Polymers and Composites Material Consulting" at https://www.polymerexpert.biz/services/materials/
[86] "Natural Rubber", at https://en.wikipedia.org/wiki/Natural_rubber/
[87] Abe MM, Martins JR, Sanvezzo PB, et al. Advantages and disadvantages of Bioplastics production from starch and Lignocellulosic Components. Polymers (Basel) 2021;13(15):2484. doi: 10.3390/polym13152484/.
[88] Robbins J, "Why Bioplastics Will Not Solve the World's Plastics Problem", August 31, 2020, https://e360yale.edu/features/why-bioplastics-will-not-solve-the-worlds-plastics-problem/
[89] Reddy S, Lau W, "Breaking the Plastic Wave: Top Findings for Preventing Plastic Pollution", https://www.pewtrusts.org/en/research-and-analysis/articles/2020/07/23/breaking-the-plastic-wave-top-findings/
[90] De Armitt C. The plastics paradox: Facts for a brighter future. In Phantom Plastics, Terrace Park, OH, USA: LLC, ISBN 978–0-9978499-6–7.
[91] Mitchell G, Mahendra V, Pinheiro J, Sousa D, Abdulgahni S, Dias J, Faria P, Gaspar F, Mateus A. "Challenges with Biopolymers" presented at 8th World Congress on Biopolymers & Bioplastics, Berlin, Germany: June 28–29, 2018. https://www.longdom.org/proceedings/challenges-with-biopolymers-19169.html/.
[92] https://environment.ec.europa.eu/topics/plastics/biobased-biodegradable-and-compostable-plastics_en
[93] Babu RP, O'Connor K, Seeram R. Current progress on bio-based polymers and their future trends. Prog Biomater 2013 Mar 18;2(1):8. doi: 10.1186/2194-0517-2-8.
[94] "Innovation Map outlines the Top 9 Biopolymer Trends & 18 Promising Startups",at https://www.startus-insights.com/innovators-guide/biopolymer-trends-innovation/

[95] Ciriminna R, Angelllotti, Li Petri G, Meneguzza F, Riccucci C, Di Carlo G, Pagliaro M,J Bioresour Bioprod 2024;9(Issue 4):486–494. Nov. doi: 10.1016/j.jobab.2024.09.002/.

[96] Nova-Institute, (2016), "Land use for biobased materials", based on FAOSTAT data, retrieved from European Bioplastics: https://www.european-bioplastics.org.

[97] Liang X, Gao S, Kuang X, Wan X, Liu L, Zhang F, Jiang G, Cong H, He H, Tan SC. Recent advancements and perspectives on processable natural biopolymers: Cellulose, chitosan, eggshell membrane, and silk fibroin. Sci Bull 2024; doi: 10.1016/j.scib.2024.08.023.

Pramod Kumbhar*, Anand Ghosalkar, Yogesh Nevare

Chapter 22 Bio-derived polymers and plastics

22.1 Introduction

Bio-based polymers have received great attention in recent years due to their sustainability, the global movement towards circular economy, and uncertainties associated with cost and availability of fossil fuel [1]. The emerging demand for bio-based products has motivated researchers to develop novel technology solutions to produce a wide variety of biopolymers and improve their characteristics. Multiple bio-based polymers have been developed, with a wide range of properties, suitable for applications in different sectors, including food, pharmaceuticals, cosmetics, packaging, agriculture, and consumer goods [2]. Bio-based polymers have the potential to replace synthetic petroleum-derived polymers and can play a major role in the sector of polymer industries. For replacement of fossil-based molecules by renewable counterparts in an environmentally friendly manner, twelve principles of green chemistry should be followed, along with sustainable chemical alteration and polymerization techniques. Use of renewable raw material along with different biocatalytic or chemocatalytic processes has been investigated to produce carbon-neutral products [3]. Similar to the twelve principles of green chemistry by Anastase and Warner [4], Slootweg and coworkers formulated the twelve principles of circular chemistry [5].

Biopolymers are generally categorized as: (i) bio-based polymers, directly extracted from biomass, such as starch, cellulose, and guar gum [2, 6]; (ii) bio-based polymers obtained from bio-derived monomers such as poly(L-lactic acid) (PLLA), poly (butylene succinate) (PBS), Bio-poly(ethylene) (PE), bio-poly(ethylene terephthalate) (PET); and (iii) bio-based polymers produced directly by natural or genetically modified organisms such as poly(hydroxy alkanoates) (PHAs) [2]. The term "bioplastic" is sometimes used for fossil-based polymers that are biodegradable, such as poly(butylene adipate-co-terephthalate) (PBAT), poly(caprolactone) (PCL), and polyvinyl alcohol (PVA). However, use of this term is misleading because they are not produced from renewable feedstocks and do not provide the benefits of greenhouse gas (GHG) reduction. Table 22.1 shows classification of different bio-based and fossil-derived polymers and categorizes them as per their biodegradation characteristics.

*Corresponding author: **Pramod Kumbhar**, Praj Matrix-R&D Center (Division of Praj Industries Ltd), GAT#1098, 402, 403, Urawade, Pirangut, Mulshi, Pune 412115, Maharashtra, India, e-mail: pramodkumbhar@praj.net

Anand Ghosalkar, Yogesh Nevare, Praj Matrix-R&D Center (Division of Praj Industries Ltd), GAT#1098, 402, 403, Urawade, Pirangut, Mulshi, Pune 412115, India

 | https://doi.org/10.1515/9783112212141-022

Author's note: This chapter covers the following topics and complements subject matter covered in Chapters 21, 23, 24, and 25. To the extent possible, attempt is made to minimize duplication and repetition of topics covered in these chapters without sacrificing flow of the subject matter.

Table 22.1: Classification of bioplastics (see Chapters 21, 23, and 24 for acronyms used in tables and Figures).

Origin	Biodegradable	Nonbiodegradable
Bio-based	Starch and cellulose, guar gum, xanthan gum, PLA, PHA, bio-PBS	Bio-polyamide, bio-polycarbonate, bio-PET, bio-PE, bio-PP
Fossil based	PBS, PBAT, PCL, and PVA	PET, PE, PP, PS

The word "bio-derived" denotes a polymer that is made fully or partially from renewable, biological, or organic waste biomass, and the word "biodegradable" denotes a material's ability to degrade into natural by-products such as CO_2, H_2O, and biomass by the action of microorganisms. The polymer that is bio-based and chemically analogous to their petrochemical-derived counterpart, for example, Bio-PE, Bio-PP, and bio-PET, are known as "drop-in polymers" [2].

Like any chemical or biochemical, feedstocks selection plays an important role in biopolymer production on a large scale. Figure 22.1 shows the classification of feedstocks that can be used to produce different bio-based products. Extensive research has been done to expand feedstock options, including first- (sugar and starch based) and second-generation feedstocks (biomass-based sugars and lignin), to produce biopolymers. Plant-based bio-oils, including waste cooking oils (triglycerides and fatty acids) and gaseous feedstocks like biogas (CH_4 and CO_2), produced from anaerobic digestion, have also been demonstrated as sustainable sources for conversion into biopolymers.

Development and production of biopolymers is at different stages and currently, some of the biopolymers are successfully produced at commercial scale (Figure 22.2). Apart from the direct extraction of biopolymers from plant-based resources, the synthesis approaches mostly revolve around production of monomers from multiple bio-based feedstocks and subsequent polymerization of monomers via a single or multi step process. Today, almost every monomer required to produce drop-in polymers can be obtained from renewable feedstocks. Bio-based feedstocks also support direct synthesis of novel polymers like PHAs, using microbial fermentation [7, 8].

In this chapter, we provide an overview of the most commercially relevant biopolymers, which can be used as bio-derived thermoplastics. Apart from chemical composition and mechanical properties, the chapter includes major feedstocks and technological approaches reported for production of biopolymers. A brief overview of EOL scenarios have been covered, along with current global market and applications of bio-derived thermoplastics.

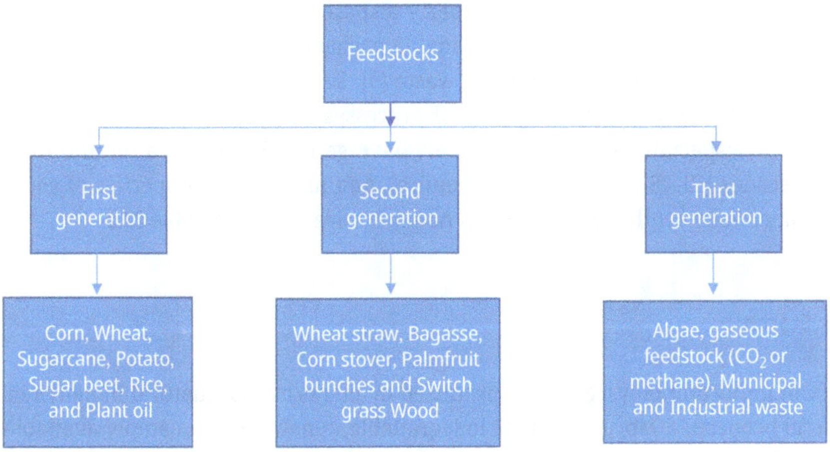

Figure 22.1: Classification of feedstocks.

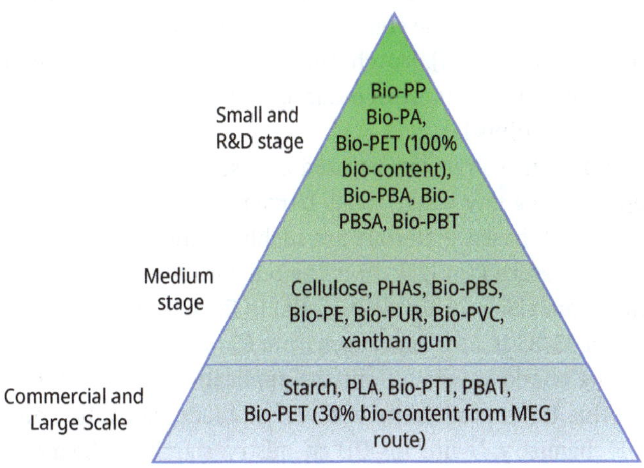

Figure 22.2: Development stages of bioplastic production.

22.1.1 Bio-derived thermoplastics: chemistry, properties, feedstock, technologies, and applications

The polymers that can be softened, reshaped, and molded into a new article by the application of heat and pressure are called thermoplastics. Due to their great versatility and distinctive properties, especially in the field of packaging (~40% thermoplastic), the thermoplastic market is growing continuously, since the 1950s. However, only a small amount of these materials is recycled, and the majority is landfilled. This has resulted in generation of huge amount of plastic waste in the form of macroplastics and microplastics, since they are not properly disposed at the end of their life. As a consequence, the development of biodegradable packaging has emerged as an attractive option to deal with the thermoplastic waste [9]. Several synthetic aspects such as atom-efficient methods, green catalysis, and processes for the synthesis of bio-based thermoplastic, especially aliphatic polyester, have gained increased attention over past few decades [10–15]. The physical, chemical, and mechanical properties of some of the commercially synthesized bio-based polymers are shown in Table 22.2.

22.1.1.1 Starch

Starch is an important polysaccharide produced by plants and is a natural biopolymer with useful properties. Moreover, it is biodegradable, renewable, and amply available at low cost [20]. Starch can be produced from several sources, including corn, wheat, potato, rice, tapioca, tam, and barley, but the major source for commercial production is maize [21]. Starch is made up of two D-glucose homopolymers: amylose (20–30%) and amylopectin (70–80%) (Figure 22.3) [9]; the composition of these components can affect the properties of starch-based films [21]. Starch constitutes a substantial portion of food waste and is used for nonsynthetic, starch-based bioplastics produced by direct processing of the starch into films [1].

From the processing perspective, starch can be used as a substitute for polystyrene (PS) and exhibits properties like low water vapor barrier, high brittleness, and poor mechanical properties. Starch-based materials are highly brittle in nature and have high glass transition temperature (Tg) [22]. Tg is the temperature at which the material shows the transition from glassy or brittle material to flexible rubbery state. Starch shows T_g in the range of 35–85 °C and Tm ranges from 65 to 115 °C (Table 22.2), indicating a narrow processing window, which limits its applications [16]. Starch is gaining commercial interest due to its availability at low cost, biodegradability, and abundant availability of –OH chemistry. Owing to –OH chemistry, starch can be modified in a number of ways by chemical or melt processing in the presence of plasticizers [21, 22]. Melt processability of starch is limited by its hydrophilicity, thermal and mechanical characteristics, quick degradability, and strong intra- and intermolecular hydrogen bonding of the polymer chains. Therefore, modifications are required to alleviate these drawbacks, while introducing new desirable qualities [23].

Thermoplastic starch (TPS) appears to be a good biopolymer because it can be processed using standard synthetic plastics manufacturing technology (extrusion and injection molding). TPS can be made using a plasticizer such as water, glycerol, or sorbitol, in combination with proper temperature and pressure treatment. The plasticizer penetrates starch granules, disrupting the initial crystallographic structure. The mechanical properties of TPS can be influenced by the temperature of starch production and water content, type and amount of plasticizers, and ancillary materials [24]. Plasticizers such as glycerol or glycol act as diluents and reduce the interaction be-

Table 22.2: Thermal and mechanical properties of commercial bio-based polymers.

Parameter	Starch	Cellulose	PHA	PLA	PBS	PEF	PET	PBAT
Glass transition temperature, T_g (°C)	35–85	493–518	2	55–60	−32	88	69–115	NR
Melting temperature, T_m (°C)	65–115	Not reported (NR)	160–175	125–178	114	210–230	245–265	115–125
Tensile strength (MPa)	2.1–46.8	15–90	15–40	10–60	34	90–100	70–177	21
Elongation at break (%)	1–1200	1–74	1–15	1.5–380	560	NR	70–180	670
Tensile/Young's modulus (MPa)	1,000–2,900	3,300–10,800	1,000–2,000	300–3,800	NR	3,600	1,700–3,500	126
References	[16]	[16]	[8]	[16]	[17]	[18]	[16]	[19]

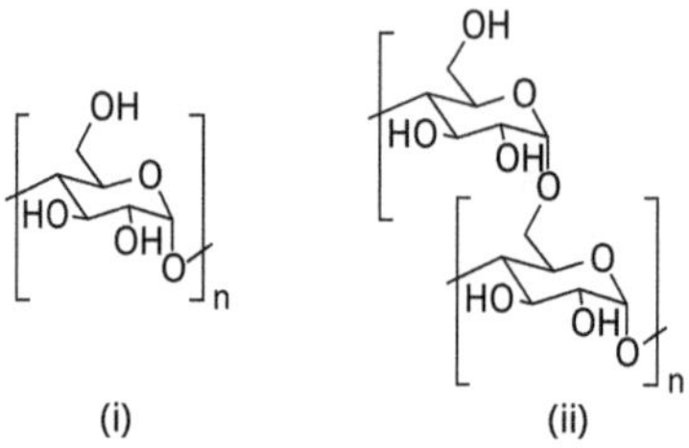

Figure 22.3: Chemical structure of starch: (i) amylose and (ii) amylopectin.

tween molecules; subsequently, they reduce tensile strength and improve the macromolecular flexibility, leading to an increase in elongation at break. The T_g of most TPS ranges from –75 to 10°C. TPS has modest tensile strength, typically below 6 MPa. Synthetic polymers, other biopolymers, nano clay, and fiber can be added to increase tensile strength and water resistance [25].

Corn, potato, wheat, and tapioca are generally used as feedstock for starch production [9, 21]. More than 85% of the corn starch is produced by wet milling and very high purity starch (99.5%) can be obtained, as compared to dry milling. Wet milling aims to fractionate different components of the corn kernel and release starch granules with minimal mechanical damage.

The industrial process involves chemical, biochemical, and mechanical operations to separate pure starch fractions from corn kernels. Various products are isolated, including starch (about 68.0%), corn bran (12.0%), corn germ (7.5%), steep liquor (6.5%), gluten (5.6%), and others (0.04%) [26]. Different studies demonstrate the applicability of starch bioplastics in shopping bags, food packaging, agriculture, and medicine [9].

22.1.1.2 Cellulose

Cellulose is the most abundant, renewable natural polymer [24] that can be extracted from plant biomass or from specific cellulose-producing bacteria, also called bacterial cellulose (BC). Cellulose is also a polysaccharide chain of D-glucopyranose units linked by β-1,4-glycoside (Figure 22.4) [27]. Cellulose can be processed into food packaging materials or used as a nano-filler additive with other bioplastics to improve barrier and tensile properties for food packaging applications. Cellulose is a highly crystalline structured material, which is useful for film application. Since cellulose is insoluble in water as well as in common organic solvents, it is essential to modify cellulose into water-soluble derivatives. Cellulose acetate, which is a derivative of cellulose, is used to make cigarette filters, membranes, and in food packaging [21]. Due to acetylation, cellulose becomes hydrophobic and degrades extremely slowly in the environment. Cellulose is an interesting material because of its outstanding mechanical properties and biodegradability. Yet, regenerated cellulose, which is popular as "viscose" or "rayon" textile fibers, constitutes 60% of seabed microplastics [1].

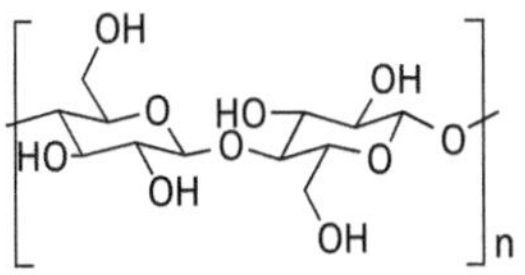

Figure 22.4: Chemical structure of cellulose.

Although around 100 years old, the viscose process for the production of cellophane (films/membranes) and rayon (fibers) is still in use [28]. The cellulose is alkalized and derivatized to cellulose xanthogenate by the addition of CS_2, resulting in a metastable intermediate. Derivatized cellulose can be dissolved in an aqueous sodium hydroxide solution to form the so-called viscose. High-purity cellulose is regenerated by removing the substituent and processed into fibers or films in a wet process. The major disadvantage of the viscose process is the requirement of sulfur and heavy metal compounds [29]. The CarbaCell process provides an alternative to viscose spinning technology and avoids use of hazardous sulfur-containing compounds and provides relatively high stability at room temperature [30]. In the Carbacell process, an aqueous or organic urea solution is used to convert the native cellulose into cellulose carbamate, which can be dissolved and processed identically to viscose. However, the CarbaCell process has not yet been industrially established due to certain requirements such as a catalyst, organic solvents, long reaction times, and high temperatures. Recently, techniques for in vitro cellulose synthesis were reviewed and critically compared, with a special focus on more recent developments [31].

The potential of cellulose can be further extended when bundled cellulose chains with highly ordered regions can be isolated as nano-particles, known as cellulose nanomaterials or nanocelluloses [32]. A number of nanocellulose forms like cellulose nanocrystals (CNCs), cellulose microfibrils (CMF), cellulose nanofibrils (CNF), amorphous nanocellulose (ANC), and cellulose nanoyarn (CNY) can be produced using different methods and from various cellulosic sources [33–35]. Size, morphology, and other characteristics of each nanocellulose can be varied depending on the cellulose origin, isolation, and processing conditions. Although several methods have been investigated to synthesize nanocellulose, acid hydrolysis using sulfuric acid is the oldest process and is most widely used. A typical approach starts with alkali and bleaching pretreatments, followed by acid hydrolysis. Cellulose nanocrystals, usually produced by acid hydrolysis, are rod-like cylindrical nanoparticles with 4–70 nm width, 100–6,000 nm length, and 54–88% crystallinity index [36]. Some of the other processes reported for CNCs include mechanical treatment, oxidation methods [37], enzymatic hydrolysis [38], ionic liquid treatments [39], and subcritical water hydrolysis [40].

Currently, nanocellulose can be employed in several fields, including continuous fibers and textiles, food coatings, barrier/separation membranes, antimicrobial films, paper products, cosmetics, biomedical products, supercapacitors, batteries, catalytic supports, electroactive polymers, and many more emerging applications [41, 42]. Despite significant industrial interest in R&D of nanocellulose, as evident from nanocellulose patents, it has been produced at limited capacities and very limited technologies have transitioned from laboratory to pilot scale. Beyond pilot plant-scale, capacities of CNCs include CelluForce (Canada, 1,000 kg/day), American Process Inc. (USA, 500 kg/day), Melodea/Holmen (Sweden, 100 kg/day), and Blue Goose Biorefineries (Canada, 10 kg/day) [43, 44].

22.1.1.3 Guar gum

Guar gum is a naturally occurring seed gum made from the endosperm of *Cyamopsis tetragonolobus* [45, 46] or *Cyamopsis psoraloides* [46], a member of the Leguminosae family. As illustrated in Figure 22.5, the primary component of guar gum is a high molecular weight polysaccharide of galactomannans that is a linear chain of (1→4)-linked β-D-mannopyranosyl units with (1→6)-linked α-D-galactopyranosyl residues as side chains [45]. Guar gum is a high molecular weight hydrophilic polysaccharide that is soluble in water but often insoluble in organic solvents like ketones, esters, lipids, and alcohols. Guar gum's physical qualities are determined by its chemical and structural composition: The majority of it is composed of a polysaccharide called galactomannan, which has a mannan backbone and side groups of galactose in a ratio of about one galactose unit for every two mannose units (Figure 22.5) [47]. The structure and behavior of guar gum have been studied using a variety of methods, including physical (optical rotation, stress-strain measurement, and x-ray analysis,), chemical (acid hydrolysis, ethylation, periodate oxidation, and creation of tolyl sulphonyl derivatives), enzymatic (by means of selective enzyme hydrolysis), and analytical (FTIR, NMR, XRD, DSC, TGA, SEM, MALDI-TOF-MS, and chromatography) [48].

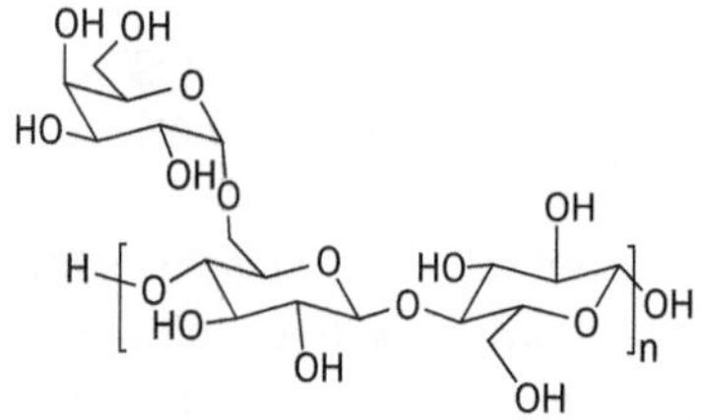

Figure 22.5: Chemical structure of guar gum.

When guar gum is dispersed, it swells and liquefies in polar liquids, creating powerful hydrogen bonds, while it creates fragile hydrogen bonds in nonpolar liquids [49]. Guar gum is used for a variety of purposes in the textile, culinary, petrochemical, mining, and paper sectors, where it serves as a suspending, emulsifying, gelling, and stabilizing agent. Because of its low degree of viscosity and hydration, guar gum has a limited range of applications in its basic forms. Its industrial applications are expanded by chemical modification, which leads to the formation of new types of superabsorbent with enhanced qualities. Guar gum-based films provide excellent barrier properties, high mechanical strength, and antibacterial- or microbial-resistant abilities [46]. Although raw gum has significant limitations and cannot always meet the needs of a particular application, incorporating newer functional groups into raw gums results in hybrid derivatives. Chemical, derivatization, and grafting copolymerization methods are used for the modification of guar gum, which helps eliminate their deficiencies that may restrain its use in a wide range of applications [48]. A major industrial application for guar is in oil and gas well fracturing.

Typically, guar gum is obtained by grinding the endosperm portion of a leguminous plant, *Cyamopsis tetragonolobus*, which has drought-resistant characteristics. The guar gum-producing plants are mostly grown in India, Pakistan, and semi-arid zones in Brazil, Southern USA, Australia, and South Africa due to favorable climatic conditions required for growth [45]. Processing involves removal of guar seeds from the pods, followed by dehusking and gum extraction. The germ portion of the guar seed is a major by-product of guar gum processing and is used as cattle feed. Due to its low cost, it is mostly popular in the food industry and also has some beneficial properties for health, like high fiber content; its consumption has been reported to reduce cholesterol levels.

22.1.1.4 Xanthan gum

Xanthan gum is also a natural polysaccharide known for inherent properties such as nontoxicity, biodegradability, and biocompatibility. Xanthan gum is mainly produced as an extracellular polymer by the *Xanthomonas campestris* bacteria [50, 51]. Xanthan gum is a long-chain pentasaccharide comprising glucose, mannose, and glucuronic acid that are repeated in the molar ratios of 2:2:1. A D-glucuronic acid unit is sandwiched between two D-mannose units that are linked to an alternate glucose unit in the main chain in the trisaccharide side chain, whereas the main chain is made up of β-D-glucose units linked at the 1 and 4 positions, resembling the structure of cellulose (Figure 22.6) [52, 53]. The D-mannose unit that is associated to the main chain is linked to an acetyl group, while the D-mannose unit at the end is linked to a pyruvic acid residue. The presence of acetic and pyruvic acid imparts an anionic character to xanthan gum. Stability of xanthan gum is affected by the structure of xanthan chain, especially the type of functional group at the outer mannose unit. Among acetyl and pyruvyl units, greater acetylation leads to more stability.

Xanthan gum is produced on a commercial scale by bacterial fermentation of sugar-based feedstocks like glucose or sucrose by *Xanthomonas campestris*. Industrial feedstocks like molasses, cheese whey, and starch derived from corn and cassava have also been used for the production of xanthan gum [54]. Xanthan gum is recovered from the fermentation broth by steps of cell separation, followed by alcohol precipitation and drying [55]. Being a fermentation-based product, selection of microbial strain, media, and fermentation conditions directly affect the composition of the func-

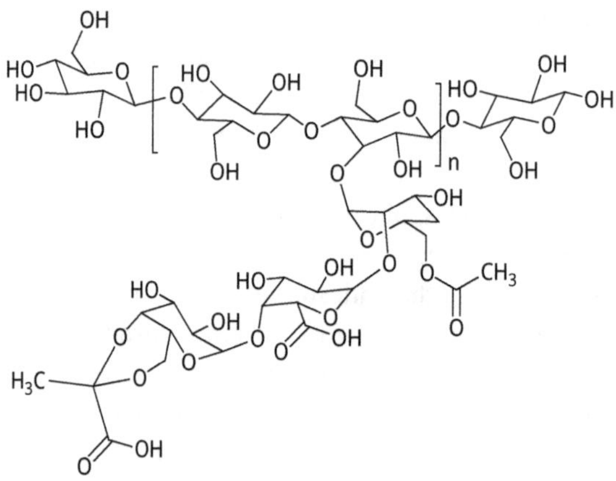

Figure 22.6: Chemical structure of xanthan gum.

tional groups in xanthan gum [39]. The concentration of pyruvic acid and acetyl depends on the type of bacteria and other aspects of the fermentation process like medium composition, variations in pH, temperature, and agitation rate [55]. The viscosity of xanthan gum solutions is affected by pyruvyl content, which has a significant impact on properties like gel formation.

Xanthan gum mainly, has applications in food, cosmetics, cleaning products, paints, ceramic glazes, inks, and formulas for water-based drilling fluids. *The* most important properties of xanthan include its high viscosity at low concentrations, high solubility in hot and cold water, viscoelastic behavior, resistance to enzymatic degradation, high stability in a wide range of pH, temperature, and salt solutions, and interaction with other polymers, as well as its straightforward processing mechanism [52]. Being a bio-based polymer, xanthan gum has also been investigated for matrices for tablets, nano- or microparticles, tissue engineering, and transdermal patches. However, the native xanthan gum has limitations with regard to its poor thermal and mechanical stability, inadequate water solubility, unusable viscosity, and susceptibility to microbial contamination. These limitations can be overcome by physical or chemical modification, or cross-linking of xanthan gum by esterification, etherification, amidation, oxidation, and acetylation of xanthan gum [56].

22.1.1.5 Poly(hydroxyalkanoates) (PHAs) (see Chapter 24 for a more complete review)

PHAs are biosynthetic, biodegradable, thermoplastic aliphatic polyesters obtained from renewable resources (Figure 22.7) [21]. PHAs can be synthesized directly by microbial fermentation of different kinds of sugars, glycerol, fatty acids, and gaseous feedstocks like biogas (methane) [57, 58]. Various bacteria, for instance, *Cupriviadus, Pseudomonas, Bacillus*, and methanotrophic strains have been widely explored for PHA synthesis.

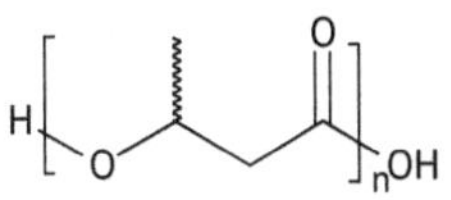

Figure 22.7: Chemical structure of poly(3-hydroxybutyrate) (PHB).

PHAs are classified as short-chain (3–5 carbon atoms), medium-chain (6–14 carbon atoms), or long-chain (>14 carbon atoms) lengths, based on the number of carbon atoms in the PHA repeating unit of monomer [59, 60]. Different material properties of PHAs can be adjusted by varying the chain length of comonomer or side chain functionality, for example, rigid and brittle PHB (Figure 22.7) and softer and flexible poly (3-hydroxybutyrate-co-3-hydroxyhexanoate) (PHBH). The flexibility to change the monomer chain length increases the design space for PHAs, as compared to other biopolymers. PHAs would be most suitable replacements for packaging materials of fossil-based plastics such as PE and PP due to their good mechanical [24] properties and barrier properties to O_2 and CO_2 [21, 61]. Most PHAs degrade rapidly as compared to PLA and, hence, are most attractive for applications in which rapid biodegradation is desired [1].

PHA feedstocks can include first-generation or second-generation cellulosic sugars, vegetable oils, organic waste, municipal solid waste, and fatty acids. In addition to glucose, other lignocellulosic monomer sugars, such as xylose, arabinose, mannose, galactose and rhamnose, and biomass-derived lignin can also be employed as PHA feedstocks. Furthermore, it is possible to produce PHAs directly from gaseous substrates like methane. An American company, Newlight Technologies' Air Carbon technology uses methane-based feedstocks for the production of PHB [62]. During fermentation, PHAs are produced by bacteria under nutrient-limited conditions. Some of the microbes can accumulate PHA intracellularly, up to 80% of the cell volume, using different fermentation strategies [63]. The intracellular PHA granules can be extracted using different extraction methods, including solvent extraction, chemical digestion assisted by surfactant and alkali, and biological extraction assisted by enzymes [64]. With increasing focus on sustainable processes, biological extraction methods (called bio-extraction methods) are gaining more attention in PHB extraction. With advances in synthetic biology and bioprocessing technologies, different types of PHAs can be produced with improved functional properties. Detailed studies regarding recent trends and future perspectives of microbial PHA production from different feedstocks, including lignocellulosic biomass, can be referred from recent reviews [60].

The primary reason for the limited production of PHAs is the higher cost of production. To reduce the costs further, it is required to use low-cost feedstocks or different agro-industrial waste materials as a substrate for PHA production [21].

22.1.1.6 Poly(L-lactic acid) (PLLA)

PLLA is a synthetic, compostable, transparent, and aliphatic thermoplastic polyester obtained from different renewable resources by a combination of fermentation and chemical route [21]. PLA (Figure 22.8) can be prepared by ring opening polymerization of cyclic diester of lactic acid, namely, L-lactide (3,6-dimethyl-1,4-dioxane-2,5-dione) or direct polycondensation of lactic acid, either in melt or solution. The former method is used successfully for the production of PLA on industrial scale. PLA can be used as potential substitute for fossil-based polyolefin films and polystyrene foams, including single-use plastics. The presence of methyl group and small repeating unit makes PLA more brittle and slow to crystallize. Physical and mechanical properties of PLA can be further improved by blending with other bio-based polymers, plasticizers, and nucleating agents before processing of the materials [1].

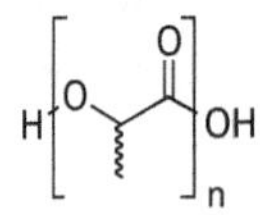

Figure 22.8: Chemical structure of poly(lactic acid).

Both sugar- and starch-based feedstocks are being used today to produce lactic acid, which is subsequently polymerized to PLA. Different carbohydrate sources can be used, including sugarcane juice or sugar industry waste like molasses, starch-rich materials like corn, cassava, or rice, and lignocellulosic materials like agricultural and forestry residues [65]. Fermentation pathway from methane to lactic acid is also being developed [66]. The fermentation process is based on a methanotrophic organism that feeds off methane and, through synthetic biology, has been modified to convert methane to lactic acid. Production of high-purity L-lactic acid is critical for the polymerization process. Optically, pure L(+)-lactic acid is produced by fermentation of sugars by lactic acid bacteria or genetically engineered yeast species [67]. The advantage of the yeast-based process lies in the direct production of lactic acid at lower pH. Bacterial production of lactic acid commonly involves the generation of by-product, gypsum (calcium sulfate), which poses disposal challenges. L-Lactide is a widely used intermediate for the industrial production of PLLA. Lactide is obtained via a two-step process consisting of firstly polymerizing lactic acid under reduced pressure to produce an oligomer or prepolymer with molecular weight range of about 500–5,000 Da, and then depolymerizing the produced prepolymer under reduced pressure, in the presence of a catalyst (Figure 22.9) [68].

Different catalytic processes for the synthesis of lactide have been extensively discussed in the literature [69, 70]. Chemical and optical purity of lactide is most desired for synthesis of high-quality and high-molar-mass PLLA [71, 72]. The conventional method of lactide production involves multistep synthesis, and purification by crystallization leads to high energy consumption, relatively lower yield, and lower optical

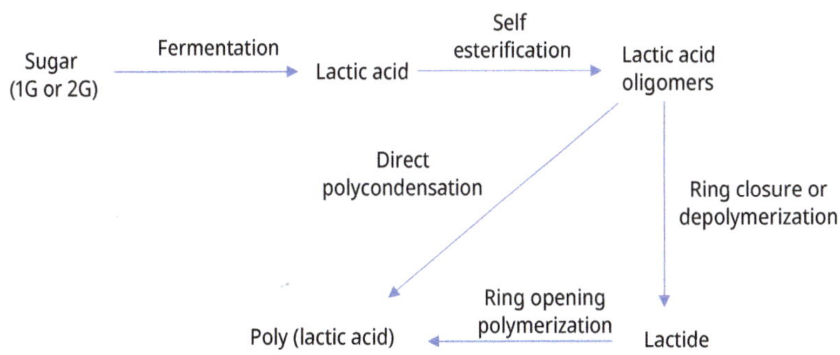

Figure 22.9: Process paths to prepare poly(lactic acid).

purity [73, 74]. Lactide to PLA polymerization proceeds through a ring-opening polymerization process, catalyzed by an organometallic catalyst at high temperatures and under vacuum [69]. This multistep process is commonly employed for PLA production, despite challenges related to higher capital and operating expenditure related to a multistep process. Research is ongoing to develop novel synthetic approaches to lactide and improve it to the desired quality [75–77].

To improve the crystallization kinetics for certain fabrication procedures and applications, D-lactic acid units are added to L-PLA. Tailoring D-isomer content in PLA has shown to significantly affect the melting temperature (T_m) of PLA. The optical comonomers alter the natural helical shape of the PLA by adding "kinks" and "defects" to the crystal arrangement, which lowers the Tm and the amount of crystallinity that can be achieved as well as slows down the crystallization process. D-Lactide and L-lactide in equimolar concentrations are known as PDLLA or poly(rac-lactide) [68, 78].

Today, PLA is used in a variety of applications like packaging and food service ware and is increasingly favored for more durable applications in automotive, electronics, and textiles. PLA products with different molecular weights are available today. They are suitable in a variety of polymer processing applications like injection molding, fiber spinning, extrusion/thermoforming, and as a nucleating agent. PLA-based products comply with industrial composting standards (ASTM D6400, EN13432) and have been approved for food contact applications in the USA, Europe, and China [79].

22.1.1.7 Bio-poly(butylene succinate) (bio-PBS)

Bio-PBS (Figure 22.10) is an aliphatic polyester that can be prepared by step-growth polymerization of succinic acid and 1,4-butanediols (BDO) in the presence of a catalyst at high temperature and vacuum. Although PBS is generally produced from fossil-based feedstock, both of its monomers, succinic acid and butanediol, can be produced by the fermentation of renewable feedstock (Figure 22.11) [21].

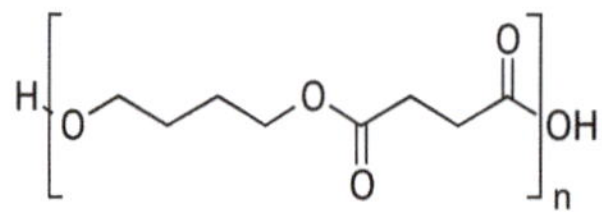

Figure 22.10: Chemical structure of PBS.

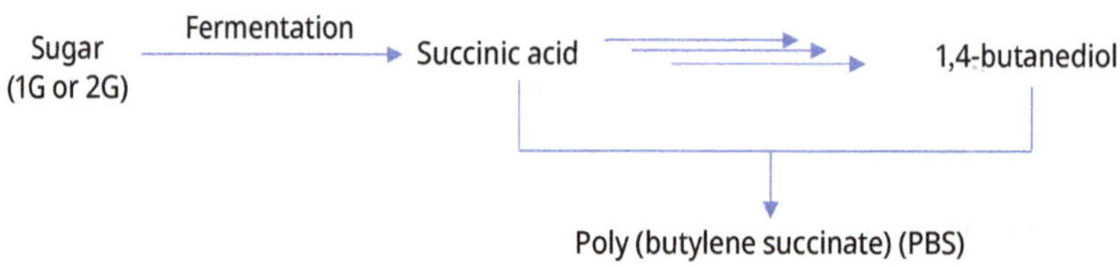

Figure 22.11: Process pathways to prepare poly(butylene succinate).

Bio-PBS has good thermal and mechanical properties that are analogous to PE and PP and is useful in numerous applications such as shopping bags, food packaging, agriculture mulch film, and hygiene products. PBS is a white, semicrystalline thermoplastic, with a T_m of ~115 °C and T_g of about −30 °C. It is characterized by good thermal stability and decent mechanical properties like tensile yield strength of 30–35 MPa (similar to that of polypropylene), Young's modulus of 300–500 MPa, and high flexibility [80]. PBS has longer chain ester linkage as compared to PLA and PHB and, thus, has more flexible molecular structure. Good oxygen barrier properties of PBS can substitute PP and its application in films, bags, or boxes, for both food and cosmetic packaging.

For the production of bio-PBS, sugar-based feedstocks are being used for the production of monomers like succinic acid as well as butanediol. PBS synthesis can be done into two steps: the first stage is the esterification of succinic acid and BDO to obtain oligomers. The second step is the polycondensation of oligomers to achieve high-molecular-weight PBS [17]. The synthesis of high-molecular-weight aliphatic polyesters by conventional polycondensation is difficult due to the simultaneous competing reactions of condensation and degradation [81]. For this reason, to obtain polyesters with useful mechanical properties, it is necessary to introduce side chains with aromatic units and chain extenders to increase the molecular weight or, more frequently, catalysts to accelerate the kinetics [82].

In 2016, Novamont (Italy) inaugurated the first bio-based plant for the industrial-scale production of 1,4-butanediol directly from sugar, using a genetically engineered bacteria [83]. Although, bio-based succinate production suffers from poor productivity and high downstream processing costs, several companies are trying to scale up bio-succinate production processes [84]. Mitsubishi Chemical (Japan) has industrialized succinic acid derived from biomass, in conjunction with Ajinomoto, to put bio-based PBS onto the market. In parallel, DSM and Roquette are jointly developing a feasible fermentation process to produce succinic acid, 1,4-butanediol, and successful achievement of bio-based PBS [85]. The high cost of bio-based PBS production demands further developments to make it more competitive with petrochemical-based PBS.

22.1.1.8 Bio-poly(ethylene furanoate) (bio-PEF)

Bio-PEF is a bio-based, highly recyclable biopolymer with superior performance properties, as compared to petroleum-based packaging materials. Bio-PEF (Figure 22.12) has a semi aromatic structure and shows similar properties to that of PET. Bio-PEF can be prepared by polycondensation of bio-based mono ethylene glycol and 2,5-furane dicarboxylic acid (FDCA) or through ring-opening polymerization of cyclic PEF oligomers, which can decrease reaction times and improve the control over molecular weight [18].

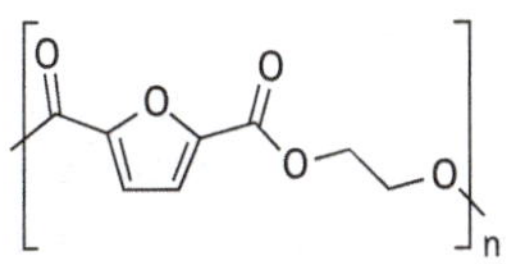

Figure 22.12: Chemical structure of PEF.

There are many similarities between bio-PEF and PET in terms of synthesis by polycondensation, properties, degradation, and EOL solutions. Bio-PEF has high T_g, higher tensile strength, and higher gas diffusion barrier properties, which may be advantageous for long-shelf-life packaging. Bio-PEF is thermally sensitive and needs more cautious processing but it also biodegrades relatively quicker (after 9 months) under suitable industrial composting conditions [1]. Molecular modification and copolymerization of PEF with different comonomers are being explored to tune T_g, mechanical strength, and degradability of PEF. FDCA, which is used a monomer for the synthesis of PEF is produced in two steps. The first step involves dehydration of fructose to 5-HMF, and in the second step, 5-HMF is oxidized to FDCA (Figure 22.13) [60].

Fructose ⟶ 5-HMF ⟶ 2,5-FDCA ⟶ PEF

Figure 22.13: Process path to prepare poly(ethylene furanoate).

Besides biomass, CO_2 is the only renewable carbon source for material synthesis. CO_2 can be converted into value-added products or monomers such as oxalic acid and glycolic acid and their polyesters. Avantium is a Dutch technology company in renewable chemistry focused on development of building blocks (monomers) from renewable sources (plant-based sugars and CO_2). Several aspects of Avantium YXY® Technology (Bio-PEF) technology, the process of bringing a new monomer and polymer to market, including process and application development, life cycle assessment (LCA), by-product valorization, EOL scenario, and recyclability and circular economy, have been reviewed recently [18]. However, the production of FDCA as a cost-effective monomer remains a challenge [86].

22.1.1.9 Bio-poly(ethylene terephthalate) (bio-PET)

Typically, PET is prepared by polycondensation of terephthalic acid and ethylene glycol in the presence of a catalyst at high temperature and under high vacuum [1]. Bio-based PET (Figure 22.14) can be produced using ethanol, which is first converted into mono ethylene glycol, followed by reaction with petroleum-derived terephthalic acid (Figure 22.15) [87]. Bio-PET is an analogous drop-in to petroleum-based PET, and hence exhibits all the equivalent properties, which makes it suitable for multiple applications, for instance, in beverages and in textile markets [1]. Polyesters like PET can be recycled with constant molecular weight of the recycled polymer after multiple mechanical recycling, without the need to go back to monomers or polymer precursors [18]. These drop-in polymers are partly bio-based and nonbiodegradable and are chemically equivalent to the petroleum-based polymer.

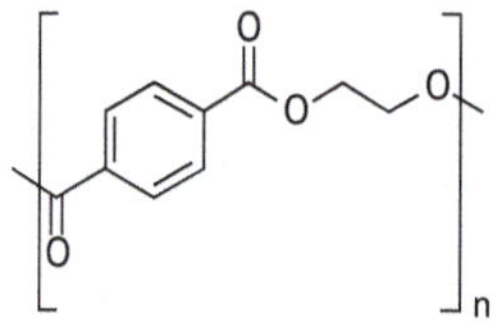

Figure 22.14: Chemical structure of PET.

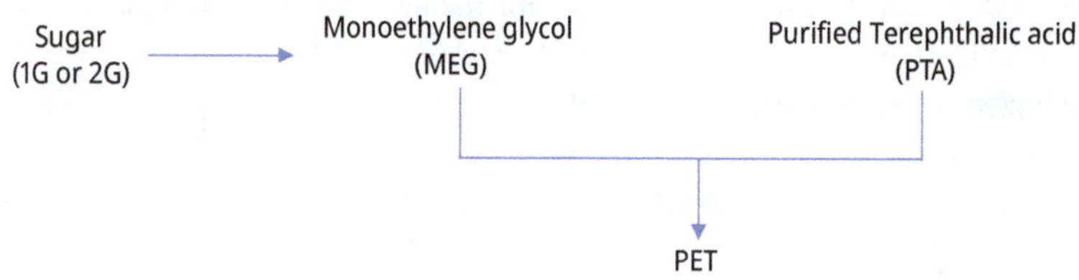

Figure 22.15: Process path to prepare poly(ethylene furanoate).

Coca-Cola Company adopted sustainable packaging in 2009, with the launch of PlantBottle™, which is made using fully recyclable PET plastic bottles, made partially from plant-based source. Sugarcane was used as feedstock for bio-PET, which was employed widely in soft drink and water bottles, beer bottles, mouthwash bottles, peanut butter and salad dressing containers, ovenable packaging films, and ovenable preprepared food trays. Similar to PET, its bio-based counterpart has the same clarity, strength/toughness, barrier to gas and moisture, and resistance to heat properties.

Both the monomers, ethylene glycol (also known as MEG in industry) and purified terephthalic acid (PTA), can be obtained from biomass or carbohydrates by catalytic reactions.

US Patent 7960594 described a process, wherein ethylene glycol is produced from cellulose. The process involves catalytic degradation and hydrogenation reactions under hydrothermal conditions. Avantium has developed a single-step hydrogenolysis process to convert glucose to MEG [88].

About 70% of terephthalic acid used in PET manufacturing is produced from the oxidation of *para*-xylene [89]. Bio-based para xylene can be produced by bio-isobutanol in a three-step reaction involving dehydration, oligomerization, and dehydrocyclization [90]. Several cutting-edge technologies for synthesis of MEG and PTA have been reviewed and can be referred for details.

22.1.1.10 Poly(butylene adipate-co-terephthalate) (PBAT)

Although this chapter is focused on bio-derived thermoplastics or polymers derived from renewable resources, the term bioplastics has also been used for fossil-based plastics that are biodegradable. PBAT was synthesized by conventional step-growth polymerization (polycondensation) of 1,4-butanediol, terephthalic acid, and adipic acid monomers in the presence of zinc, tin, or titanium catalysts. The reported studies suggest that the condensation reaction is usually performed at high temperature >190 °C, under high vacuum, and it requires long reaction time. The structure of PBAT includes terephthalic acid, adipic acid, and 1,4-butanediol as comonomers. Bio-based BDO, produced by industrial fermentation, can directly replace petrochemical-derived BDO in PBAT. Adipic acid, produced from castor oil, has been used as a monomer to make copolyesters [91]. PBAT is a degradable aliphatic-aromatic copolyester, sold as Ecoflex by BASF (Germany) and under other trade names by several suppliers in Asia. BASF, Germany produced Ecoflex, a PBAT-based product, with the highest production capacity of 60,000 tons/year. Kingfa, China, Novamont, Itali DuPont, and Jinhui, China supply PBAT under the trade names Ecopond, Origo-Bi, Biomax, and Ecowords, respectively, with a production capacity of 20,000–50,000 tons/year [19]. PBAT (Figure 22.16) is widely used in agricultural mulch films, which can degrade in soil over a period of >9 months [1].

PBAT is known for being flexible and tough, which makes it ideal for combinations with other bio-based polymers that have high modulus and strength, but are very brittle. PBAT is widely used in applications like garbage bags, wrapping films, and disposable plastic products (lunch boxes, dishes, cups, etc.) [19].

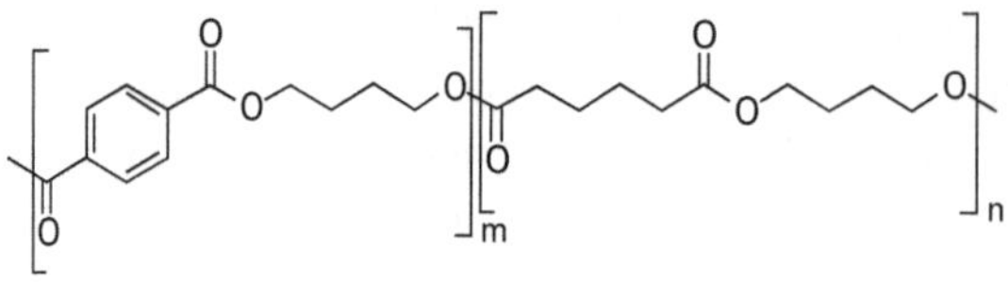

Figure 22.16: Chemical structure of PBAT.

22.2 End-of-life scenario of bio-derived thermoplastics

Author's note: see also Chapters 23, 24, and 25 for additional details.

The longevity of polymeric materials and the lack of proper collection and recycling technologies at the end of product life have led to continuous accumulation of plastic-related waste and incorrect product disposal practices. Extensive usage of single-use plastics and disposable PPEs during the COVID-19 outbreak further worsened the plastic waste problem. Most fossil-based plastic materials do not have pathways or mechanisms for natural degradation, and the stability provided by the polymeric structure further leads to their accumulation in the environment. Oceans, freshwater lakes, and soil tend to accumulate plastics, which are constantly damaging the environment and ecosystem [1]. There have been several reports on accumulation of microplastics in the environment, since the word was first coined in 2004. Microplastics have been found everywhere, including in deep oceans; in Arctic snow and Antarctic ice; in shellfish, table salt, drinking water, and beer; and drifting in the air or falling with rain over mountains and cities [92]. According to the European Chemical Agency, any type of plastic particle less than 5 mm is considered as microplastic.

Plastic trash can be managed in one of three ways: it can be recycled for reuse, managed through landfilling and incineration, or directly disposed into the environment. To avoid leakage of contaminants in the environment or the formation of microplastics, complete digestion of polymeric materials is necessary. Biodegradation is an end-of-life (EOL) option that can be achieved for easily hydrolysable polymers, such as aliphatic esters like PLA and PHA. The terms "biodegradation" and "composting" refer to the metabolic breakdown of polymeric material into CO_2, H_2O, and other inorganic compounds by a variety of microorganisms under favorable conditions. Disintegration by different physical techniques like fragmentation and particle size reduction can augment biodegradation process and enzymatic breakdown [1]. Degradation of polymeric materials by oxidative processes, commonly known as oxo-degradation, has been banned in the European Union and many other countries due to lack of evidence of the complete degradation of oxo-degradation particles and products by biological systems [1].

Another most sought-after EOL option for polymeric compounds is recycling for new products. The easiest, most affordable, and most popular type of recycling is mechanical recycling, which often involves separating plastic waste by polymer type, removing labels, washing, mechanically shredding, melting, and remolding into new shapes. Re-extrusion of different plastic has been attempted; however, mechanical recycling is currently not feasible for bioplastics at commercial scale. Deterioration in quality has been observed when PLA and PHA are mechanically recycled, with loss of tensile strength and molecular weight. To create polymers of the desired quality, plastic products must first be depolymerized into their monomeric subunits. Recycling by chemical processes like solvolysis and thermolysis are other EOL alternatives for polymeric materials. The only extensively used technology for treating plastic solid waste on a significant scale is mechanical recycling. The primary procedures involve washing away organic waste, followed by shredding, melting, and remolding the polymer, which is frequently merged with virgin plastic of the same type to create a substance with advantageous qualities for manufacture. As each type of plastic reacts to the process differently based on its chemical composition, mechanical behavior, and thermal qualities, mechanical recycling technologies have limitations [93, 94]. Enhancing chemical recycling efficacy [95, 96] and selectivity through catalyst development, reducing the requirement for sorting through compatibilizer design, and expanding recycling beyond thermoplastics are three current research approaches that show significant promise in augmenting plastic recycling methods. The reason that chemical recycling with thermolysis offers a recycling approach through decomposition of a polymer to lower-molecular-weight streams is that most polymers are immiscible with one another, yielding phase-separated mixes with diminished characteristics. The properties of recycled material may change ever so slightly when one type of plastic is contaminated with another, which could make it harder to use. By creating compatibilizers that regulate the phase behavior of polymer mixtures, this issue can be solved [93]. For plastics with petrochemical origin, recycling is well-established, but, for bioplastics, an efficient and effective recycling technology has not yet been developed. For majority of bioplastics that are in use at reasonable commercial scale today (starch, PLA, PBS, PHA, etc.) segregation may be required, based on their chemical make up for recycling into new products. Landfilling is another commonly used option for most of the single-use plastic products. Most of the fossil-derived plastics do not degrade in soil and, therefore, the sizes of landfills increase every year. It is imperative now to develop sustainable materials like bio-based plastics that can fulfil the growing demand, but, at the same time, are benign to the environment and human health. Figure 22.17 shows the ideal scenario to achieve circularity of plastic-based products, with different feedstocks and production processes. Among all EOL options, biodegradation is the most efficient method for treating bioplastics with bio-based origin. Bioplastics are expected to have little impact on the environment, because their biodegradation does not induce the release of any toxic compounds. However, widespread use of existing and newly developed bioplastics need regulations that verify and certify important properties like origin and bio-based content, biodegradability, and compostablity to ensure their safety in use and the lack of any harm to the environment [21]. Experts from various fields and countries have been collaborating with global organizations to develop standards that may be used and implemented to examine important characteristics of various bioplastics, under repeatable conditions. The International Organization for Standardization (ISO), the European Committee for Standardization (CEN), and the American Society for Testing and Materials (ASTM) are the principal organizations involved in developing these standards. The most popular and often-cited plastics standards from various organizations are included in Table 22.3 [21]. Critical evaluation of sustainability of different bioplastics is another important aspect and multiple LCA studies have been published with different boundary conditions. There are international standards in place, such as ISO 14040, to guide the LCA structure, conduct, and assumptions made with limitations. EN 16760 specifically provides guidelines for LCA requirements and guidance for bio-based products, excluding food, feed, and energy. However, heterogeneity in LCA approaches and assumptions persist and require a systematic framework for bioplastic LCAs that can improve meaningfulness for these studies and inform decision makers better [97].

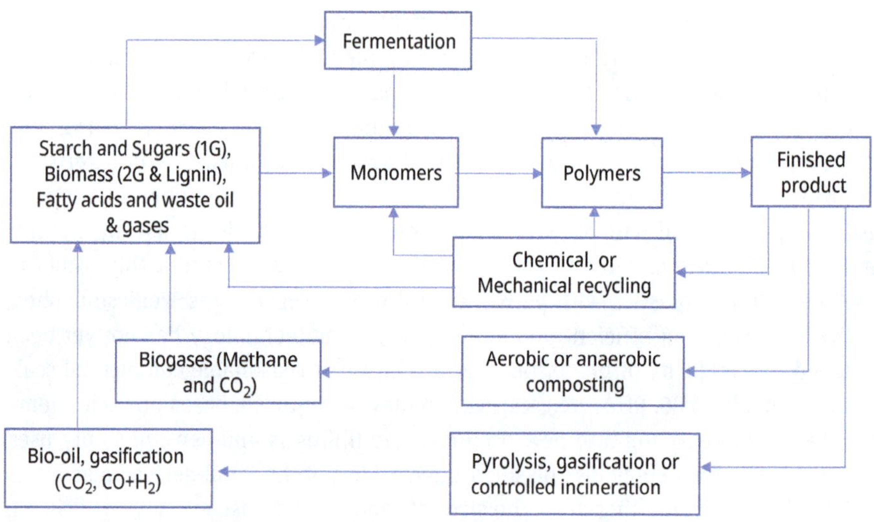

Figure 22.17: Circular economy of bio-plastics.

Table 22.3: Standard test methods for aerobic and anaerobic biodegradation of plastics and bio-based content.

Parameter	Test method	Disposal	References
Aerobic biodegradation	ASTM D6400-04, EN13432, ASTM D5338-15, ASTM D5338-98, ASTM D5511-18 and ISO 15985 ISO 14855, ASTM D6400-19, ASTM D7081 and ASTM D6868, ISO17556, ASTM D5338-98, ASTM D6002-96 and AS 4736–2006, Soil (ISO 17556, ASTM D5988, ISO 11266), and water (ISO 14851, ISO 14852)	Compostable plastics or OK biodegradable or OK compost	[96, 98–101]
	ASTM D5929	Natural or home composting	[99]
	EN 13432	Industrial composting	[98]
	ASTM D6691-17, ASTM D7473-12, ASTM D7991-15, ASTM D7081-05, ISO 14851, ISO 14852 and ASTM D6692	Marine biodegradation	[99, 101]
	ASTM D6400 and ASTM D6868	US biodegradable product institute (BPI)	[96]
	EN17033	Biodegradation of mulch film (>90% degradation within 2 years)	[98]
	ASTM D5526-18 and ASTM D7475-20	Landfill biodegradation	[99]
Anaerobic biodegradation	ASTM D5210-92 and ISO 14853, ISO 13975	Sewage sludge biodegradation	
	ASTM D5511 and ISO 14853, ISO 15985	Anaerobic digestion biodegradation	[102]
	ASTM D5526-18, ASTM D7475-20, and ASTM D5526-18	Accelerated landfill biodegradation	[99]
OK bio-based polymer	CEN/TS 16,137 and ASTM 6866		[96]
	ASTM 6866	USDA-certified bio-based product	[102]
	ASTM 6866 and CSN/TS 16137	Din Gepruft bio-based polymer (DIN CERTCO)	

22.3 Global market and application of bio-derived thermoplastics

Author's note: Some of the market and applications are also covered in Chapters 21, 23, 24, and 25

The global production of 100% bio-based polymers is currently more than 2.4 million tons with two-thirds of it contributed by biodegradable polymers [103, 104]. In comparison to current fossil-based plastic production of 380 million tons, the fraction of biopolymers is very little and has significant potential to increase over the years. Sustainability is the main driver for increase in demand for biopolymers. In the last couple of decades, it has become clear that biopolymers cannot be sold at much higher prices as compared to petrochemical counterparts; however, there is continued interest from large corporations, as biomaterials can help in achieving emission reduction targets and help in transitioning to net zero commitments by 2050. A recent survey by McKinsey in ten countries has found that consumers see sustainability as increasingly important, and the vast majority of consumers showed willingness to pay more for sustainable packaging. Further, policy makers and regulatory bodies across the globe are pushing for sustainability through initiatives such as the "European Green Deal" and "Single-use plastic directives" of the European Union. Two key regulatory objectives are reducing CO_2 emissions and environmental leakage of plastics that do not biodegrade. Both the objectives can be addressed through biomaterials, and increased demand and recycling-oriented regulations might create further incentives. Food packaging and fast-moving consumer goods are the largest markets for short-lived to medium-lived plastics and, therefore, also for bioplastics. Global bioplastics production capacity is set to increase significantly from around 2.41 million tons in 2021 to approximately 7.59 million tons in 2026 [105, 106]. Out of the current total capacity in 2021, up to 40% biopolymers belong to both bio-based and biodegradable (PLA, PHA, starch blends, and cellulose films). Among different biopolymers, PLA is the most established and accepted material, with total global production capacity beyond 300 kilotons/annum (KTA), and another 100 KTA capacity expansion is under progress. The demand for PLA has steadily increased beyond 190 KTA, with greater than 60% demand coming from the packaging sector alone. It is estimated that by 2030, the potential demand for PLA may cross 600 KTA. From approximately 380 million tons of total plastics consumption, replacement of even 1% by bioplastics would require about 1.4 million tons of additional capacity. Improvements in chemical recycling and availability of biodegradation facilities can further pave the way for bioplastics, such as PHAs, which are more hydrolysable and, therefore, more compatible with existing composters [99]. Table 22.4 shows the current production capacities and applications of main biopolymers in use today. According to report by World Economic Forum, investor interest in environmental, social, and governance assets is growing, with

86% believing that these will be better long-term investments. Up to $30 trillion are now spent on sustainable assets globally (one-third of total investments), with plastic and climate change topping the list of sustainable investor interests [107]. The Alliance to End Plastic Waste, comprising major chemical companies and brand owners, has promised to spend a total of $1.5 billion for sustainable plastic-related projects. Companies from multiple sectors like Nestle from food and nutrition, and Toyota and Peugeot Citroën SA from the automotive sector have pledged to introduce renewable polymers in their products [108].

Table 22.4: Overview of bio-based polymers.

Bio-polymers	Producer	Cost US$/kg	Capacity (tons/year)	Applications	References
Starch	Plantic® Australia, Mater-Bi® and Novamont Italy, BIOPAR® Portugal, Biofase® Mexico and Sonaly® Canada, Ecofoam National Starch	0.5–2	20,000	Filler, thickening, stiffening, gluing, and binding agent, precursor to dextrins, glucose, ethanol, and sugars for processed food, in the production of biofuel and as adhesive in the papermaking process. Agriculture, medicine, engineering, food, and packaging, food packaging, shopping bags, agriculture, medicine, disposable domestic utensils (cup and plates), even automotive industry, controlled release of fertilizer, efficient bone cement, and drug release film in medical industry.	[109–111]
Cellulose	Eastman Chemical Company produced cellulose acetate	1.8–4		Drug delivery, cell and tissue growth, scaffold, implants, clothing, decorative, and automobile industries.	[109, 112]

Table 22.4 (continued)

Bio-polymers	Producer	Cost US$/kg	Capacity (tons/year)	Applications	References
PHA	Shenzhen Ecoman Biotechnology, China Danimer Scientific USA Kaneka, Japan Bio-on, Italy Tianan Biological material polyone, China Biomer, Germany Mitsubishi Gas, Japan PHB Industrial, Brazil Metabolix, USA P&G Chemicals, USA/Japan Telles LLC, USA Jiangsu Nantian Graup, China Goodfellow Cambridge Ltd, UK Tepha Inc., USA Cofco, China Mango Materials, USA Newlight Technologies, USA PhaBuilder, China RWDC Industries Ltd. USA Bluepha Co. Ltd, China	2.4–5.5	100–10,000	Packaging film and food services (shopping bags, paper coatings, cups diapers, carpets, PHA coated boxes, paper, sheets, boards), agricultural (biodegradable plastic film, seeds encapsulation, encapsulation of fertilizers, and controlled release of pesticides and insecticides), Medical application, drug delivery (controlled delivery of antibiotics), tissue engineering (scaffolds, pins, films, sutures), and wound dressing (swabs, fleece, lint). Biofuels or fuel additives, protein purification specific drug delivery, healthy food additives, fine chemical industries, printing, and photographic industries.	[113–115]
PLA	Bionolle Starcla™ - Showa Denko Japan, BIOFRONT-Teijin Japan, Cargill-Dow LLC, IngeoTM-Nature Works USA, WeforYou Austria, Total-Corbion Netherlands/Thailand	4–6	140,000	Food packaging, Cups, fiber, tissue engineering scaffolds, bio-absorbable medical implants, dermatology, cosmetics, targeted delivery of bioactive compounds, fixation rods, plates, pins, screws, and sutures, suitable for bone, cartilage, ligament, skin, blood vessels, nerves, construction, electronics, textile, and agriculture applications.	[79, 109, 116]

Table 22.4 (continued)

Bio-polymers	Producer	Cost US$/kg	Capacity (tons/year)	Applications	References
PBS	DSM and Roquette, Mitsubishi and Ajinomoto, Japan, BASF, Germany Hexing Chemicals, China IPC-CAS, China IRE Chemical, Korea Mitsubishi Gas Chemical, Japan Showa, Japan	4–10	3,000	Food packages, bottles, supermarket bags, flushable hygiene products, mulch film and compost bags, agricultural mulch film, dishware, fishery, forestry, construction, electronics, drug encapsulation systems, orthopedic, and coffee capsules.	[17, 109, 117]
PEF	Avantium, BASF,	0.15–1.75	1,000	Film, bottles, and fibers for textiles.	[18, 118]
PET	Coca-Cola-Gevo Venture, PepsiCo-Virunt Venture Toyota Tsusho Corporation, Japan			Fibers, packaging, bottles, containers, and films.	[119]
PBAT	BASF, Germany Kingfa, China Novamont, Italy Tunhe, China Xinfu, China Jinhui, China	3.8–5.8	20,000–60,000	Rigid and flexible packaging, shopping and garbage bags, sacks, cutlery, mulch film, fibers, paper coating, and foodservice.	[19]

22.4 Summary and future perspectives

Significant technological advancements have been made in the last few decades in developing bio-based alternatives of fossil-derived polymer. With ever-increasing focus on circular economy and sustainable products, biopolymers can provide multiple options with similar or better thermal and mechanical characteristics. However, in comparison to fossil-based polymers, bio-derived polymers are mostly produced in small quantities at relatively high cost, and in some cases, do not always have substantial environmental benefits. To increase production of biopolymers to a commercial scale, a few limitations need to be addressed, including the cost of production, availability of raw material, EOL status, biodegradability, and compatibility of the end products. Substantial growth in biopolymer sales would also depend on their engineering application and expansion in consumer goods. Low-cost feedstock is the key to further expand production of bio-derived polymers and besides first-generation feedstocks, lignocellulosic materials, including agricultural and other bio-wastes, including gaseous streams and present abundant and viable feedstock from a future sustenance perspective. It is expected that bio-refinery processes will improve in efficiency, and with the application of green chemistry principles (such as using nontoxic chemicals and reducing the energy demand), supply of monomers or polymer building blocks will be increasingly available at cost-competitive and sustainable manner. Advances in synthetic biology are expected to improve monomer yields, biopolymer diversity, and synthesis rates (e.g., PHAs and xanthan gum). New approaches of enzyme discoveries like metagenomic approaches will improve biomass deconstruction rates and will also help in biological depolymerization for recycling. In the long run, just the biodegradability of biopolymers will not be sufficient, and recycling technologies, including chemical and biological, will be required to handle mixed waste streams of poor quality as well as recycled material from multiple sectors. Apart from recycling of polymeric materials, consumer preferences toward renewable materials and strict usage of renewable energy for polymer and plastic production remain essential strategies to mitigate plastic waste and carbon emissions. Harmonization of bioplastic waste and disposal regulations and homogenization of methodology standards to make LCAs more consistent and comparable will help in identifying sustainability limitations of different biopolymers.

References

[1] Rosenboom J-G, Langer R, Traverso G. Bioplastics for a circular economy. Nat Rev Mater 2022;7:117–137. https://doi.org/10.1038/s41578-021-00407-8.

[2] Moshood TD, Nawanir G, Mahmud F, Mohamad F, Ahmad MH, AbdulGhani A. Sustainability of biodegradable plastics: New problem or solution to solve the global plastic pollution?. Curr Res Green Sustain Chem 2022;5:100273. https://doi.org/10.1016/j.crgsc.2022.100273.

[3] Llevot A, Dannecker P-K, Von Czapiewski M, Over LC, Soeyler Z, Meier MAR, et al. Renewability is not enough: Recent advances in the sustainable synthesis of biomass-derived monomers and polymers. Chem Eur J 2016;22:11510–11521. https://doi.org/10.1002/chem.201602068.

[4] Anastas PT, Warner JC. Green Chemistry: Theory and Practice, Oxford University Press; 1998.

[5] Keijer T, Bakker V, Slootweg JC. Circular chemistry to enable a circular economy. Nat Chem 2019;11:190–195. https://doi.org/10.1038/s41557-019-0226-9.

[6] Coppola G, Gaudio MT, Lopresto CG, Calabro V, Curcio S, Chakraborty S. Bioplastic from renewable biomass: A facile solution for a greener environment. Earth Syst Environ 2021;5:231–251. https://doi.org/10.1007/s41748-021-00208-7.

[7] De Donno Novelli L, Moreno Sayavedra S, Rene ER. Polyhydroxyalkanoate (PHA) production via resource recovery from industrial waste streams: A review of techniques and perspectives. Bioresour Technol 2021;331:124985. https://doi.org/10.1016/j.biortech.2021.124985.

[8] Bugnicourt E, Cinelli P, Lazzeri A, Alvarez V. Polyhydroxyalkanoate (PHA): Review of synthesis, characteristics, processing and potential applications in packaging. Express Polym Lett 2014;8:791–808. https://doi.org/10.3144/expresspolymlett.2014.82.
[9] Morinval A, Averous L. Systems based on biobased thermoplastics: From bioresources to biodegradable packaging applications. Polym Rev 2021;1–69. https://doi.org/10.1080/15583724.2021.2012802.
[10] Albertsson A-C, Varma IK. Aliphatic Polyesters: Synthesis, Properties and Applications. Degrad. Aliphatic Polyesters, Berlin, Heidelberg: Springer Berlin Heidelberg; 2002. Vol. 157, pp. 1–40. https://doi.org/10.1007/3-540-45734-8_1.
[11] Pang K, Kotek R, Tonelli A. Review of conventional and novel polymerization processes for polyesters. Prog Polym Sci 2006;31:1009–1037. https://doi.org/10.1016/j.progpolymsci.2006.08.008.
[12] Maisonneuve L, Lebarbé T, Grau E, Cramail H. Structure-properties relationship of fatty acid-based thermoplastics as synthetic polymer mimics. Polym Chem 2013;4:5472–5517. https://doi.org/10.1039/c3py00791j.
[13] Stempfle F, Ortmann P, Mecking S. Long-chain aliphatic polymers to bridge the gap between semicrystalline polyolefins and traditional polycondensates. Chem Rev 2016;116:4597–4641. https://doi.org/10.1021/acs.chemrev.5b00705.
[14] Debuissy T, Pollet E, Avérous L. Biotic and abiotic synthesis of renewable aliphatic polyesters from short building blocks obtained from biotechnology. Chem Sus Chem 2018;11:3836–3870. https://doi.org/10.1002/cssc.201801700.
[15] Nomura K, Binti Awang NW. Synthesis of bio-based aliphatic polyesters from plant oils by efficient molecular catalysis: A selected survey from recent reports. ACS Sustain Chem Eng 2021;9:5486–5505. https://doi.org/10.1021/acssuschemeng.1c00493.
[16] Filiciotto L, Rothenberg G. Biodegradable plastics: Standards, policies, and impacts. Chem Sus Chem 2021;14:56–72. https://doi.org/10.1002/CSSC.202002044.
[17] Xu J, Guo BH. Poly(butylene succinate) and its copolymers: Research, development and industrialization. Biotechnol J 2010;5:1149–1163. https://doi.org/10.1002/biot.201000136.
[18] De Jong E, Visser HA, Dias AS, Harvey C, Gruter G-JM. The road to bring FDCA and PEF to the market. Polymers 2022;14:943. https://doi.org/10.3390/polym14050943.
[19] Jian J, Xiangbin Z, Xianbo H. An overview on synthesis, properties and applications of poly(butylene-adipate-co-terephthalate)–PBAT. Adv Ind Eng Polym Res 2020;3:19–26. https://doi.org/10.1016/J.AIEPR.2020.01.001.
[20] Apriyanto A, Compart J, Fettke J. A review of starch, a unique biopolymer – Structure, metabolism and in planta modifications. Plant Sci 2022;318:111223. https://doi.org/10.1016/j.plantsci.2022.111223.
[21] Melchor-Martínez EM, Macías-Garbett R, Alvarado-Ramírez L, Araújo RG, Sosa-Hernández JE, Ramírez-Gamboa D, et al. Towards a circular economy of plastics: An evaluation of the systematic transition to a new generation of bioplastics. Polymers 2022;14:1203. https://doi.org/10.3390/polym14061203.
[22] Janssen L, Moscicki L. Thermoplastics starch. Biodegrad Polym Their Pract Util 2009;1–29.
[23] Ojogbo E, Ogunsona EO, Mekonnen TH. Chemical and physical modifications of starch for renewable polymeric materials. Mater Today Sustain 2020;7–8:100028. https://doi.org/10.1016/j.mtsust.2019.100028.
[24] Peelman N, Ragaert P, Ragaert K, De Meulenaer B, Devlieghere F, Cardon L. Heat resistance of new biobased polymeric materials, focusing on starch, cellulose, PLA, and PHA. J Appl Polym Sci 2015;132:42305. https://doi.org/10.1002/app.42305.
[25] Zhang Y, Rempel C, Liu Q. Thermoplastic starch processing and characteristics – A review. Crit Rev Food Sci Nutr 2014;54:1353–1370. https://doi.org/10.1080/10408398.2011.636156.

[26] Zhang R, Ma S, Li L, Zhang M, Tian S, Wang D, et al. Comprehensive utilization of corn starch processing by-products: A review. Grain Oil Sci Technol 2021;4:89–107. https://doi.org/10.1016/J.GAOST.2021.08.003.

[27] Babu RP, O'Connor K, Seeram R. Current progress on bio-based polymers and their future trends. Prog Biomater 2013;2:8. https://doi.org/10.1186/2194-0517-2-8.

[28] Klemm D, Heublein B, Fink HP, Bohn A. Cellulose: Fascinating biopolymer and sustainable raw material. Angew Chem Int Ed 2005;44:3358–3393. https://doi.org/10.1002/ANIE.200460587.

[29] Rose M, Palkovits R. Cellulose-based sustainable polymers: State of the art and future trends. Macromol Rapid Commun 2011;32:1299–1311. https://doi.org/10.1002/MARC.201100230.

[30] Turunen O, Mandell L, Eklund V, Ekman K, Huttunen JI. Procedure for producing soluble cellulose carbamate. FI61033 1982.

[31] Lehrhofer AF, Goto T, Kawada T, Rosenau T, Hettegger H. The in vitro synthesis of cellulose – A mini-review. Carbohydr Polym 2022;285:119222. https://doi.org/10.1016/J.CARBPOL.2022.119222.

[32] Foster EJ, Moon RJ, Agarwal UP, Bortner MJ, Bras J, Camarero-Espinosa S, et al. Current characterization methods for cellulose nanomaterials. Chem Soc Rev 2018;47:2609–2679. https://doi.org/10.1039/C6CS00895J.

[33] Phanthong P, Reubroycharoen P, Hao X, Xu G, Abudula A, Guan G. Nanocellulose: Extraction and application. Carbon Resour Convers 2018;1:32–43. https://doi.org/10.1016/J.CRCON.2018.05.004.

[34] Pires JRA, Souza VGL, Fernando AL. Valorization of energy crops as a source for nanocellulose production – Current knowledge and future prospects. Ind Crops Prod 2019;140:111642. https://doi.org/10.1016/J.INDCROP.2019.111642.

[35] Salimi S, Sotudeh-Gharebagh R, Zarghami R, Chan SY, Yuen KH. Production of nanocellulose and its applications in drug delivery: A critical review. ACS Sustain Chem Eng 2019;7:15800–15827. https://doi.org/10.1021/ACSSUSCHEMENG.9B02744/ASSET/IMAGES/MEDIUM/SC9B02744_0021.GIF.

[36] Naz S, Ali JS, Zia M. Nanocellulose isolation characterization and applications: A journey from non-remedial to biomedical claims. Bio-Design Manuf 2019 23 2019;2:187–212. https://doi.org/10.1007/S42242-019-00049-4.

[37] Sun B, Hou Q, Liu Z, Ni Y. Sodium periodate oxidation of cellulose nanocrystal and its application as a paper wet strength additive. Cellul2015 222 2015;22:1135–1146. https://doi.org/10.1007/S10570-015-0575-5.

[38] Enzymatic preparation of nanocrystalline and microcrystalline cellulose, TAPPI JOURNAL May 2014 n.d. https://imisrise.tappi.org/TAPPI/Products/14/MAY/14MAY35.aspx (accessed August 11, 2022).

[39] Lazko J, Sénéchal T, Bouchut A, Paint Y, Dangreau L, Fradet A, et al. Acid-free extraction of cellulose type i nanocrystals using brønsted acid-type ionic liquids. Nanocomposites 2016;2:65–75. https://doi.org/10.1080/20550324.2016.1199410.

[40] Novo LP, Bras J, García A, Belgacem N, Aas CD. A study of the production of cellulose nanocrystals through subcritical water hydrolysis. Ind Crops Prod 2016;93:88–95. https://doi.org/10.1016/J.INDCROP.2016.01.012.

[41] Moon RJ, Schueneman GT, Simonsen J. Overview of cellulose nanomaterials, their capabilities and applications. Jom 2016 689 2016;68:2383–2394. https://doi.org/10.1007/S11837-016-2018-7.

[42] Thomas B, Raj MC, Athira BK, Rubiyah HM, Joy J, Moores A, et al. Nanocellulose, a versatile green platform: From biosources to materials and their applications. Chem Rev 2018;118:11575–11625. https://doi.org/10.1021/ACS.CHEMREV.7B00627/ASSET/IMAGES/MEDIUM/CR-2017-00627Z_0029.GIF.

[43] Xie H, Du H, Yang X, Si C. Recent strategies in preparation of cellulose nanocrystals and cellulose nanofibrils derived from raw cellulose materials. Int J Polym Sci 2018 2018; https://doi.org/10.1155/2018/7923068.

[44] Trache D, Tarchoun AF, Derradji M, Hamidon TS, Masruchin N, Brosse N, et al. Nanocellulose: From fundamentals to advanced applications. Front Chem 2020;8:392. https://doi.org/10.3389/fchem.2020.00392.

[45] Mudgil D, Barak S, Khatkar BS. Guar gum: Processing, properties and food applications – A review. J Food Sci Technol 2014;51:409–418. https://doi.org/10.1007/S13197-011-0522-X/TABLES/3.
[46] Sharma G, Sharma S, Kumar A, Al-Muhtaseb AH, Naushad M, Ghfar AA, et al. Guar gum and its composites as potential materials for diverse applications: A review. Carbohydr Polym 2018;199:534–545. https://doi.org/10.1016/j.carbpol.2018.07.053.
[47] Theocharidou A, Mourtzinos I, Ritzoulis C. The role of guar gum on sensory perception, on food function, and on the development of dysphagia supplements – A review. Food Hydrocoll Heal 2022;2:100053. https://doi.org/10.1016/J.FHFH.2022.100053.
[48] Thombare N, Jha U, Mishra S, Siddiqui MZ. Guar gum as a promising starting material for diverse applications: A review. Int J Biol Macromol 2016;88:361–372. https://doi.org/10.1016/j.ijbiomac.2016.04.001.
[49] Adimule V, Kerur SS, Chinnam S, Yallur BC, Nandi SS. Guar gum and its nanocomposites as prospective materials for miscellaneous applications: A short review. Top Catal 2022;1:1–14. https://doi.org/10.1007/S11244-022-01587-5/TABLES/2.
[50] García-Ochoa F, Santos V, Casas J, Gómez E. Xanthan gum: Production, recovery, and properties. Biotechnol Adv 2000;18:549–579. https://doi.org/10.1016/S0734-9750(00)00050-1.
[51] Becker A, Katzen F, Pühler A, Ielpi L. Xanthan gum biosynthesis and application: A biochemical / genetic perspective. Appl Microbiol Biotechnol 1998;50:145–152. https://doi.org/10.1007/S002530051269.
[52] Bhat IM, Wani SM, Mir SA, Masoodi FA. Advances in xanthan gum production, modifications and its applications. Biocatal Agric Biotechnol 2022;42:102328. https://doi.org/10.1016/j.bcab.2022.102328.
[53] Abu Elella MH. Synthesis and potential applications of modified xanthan gum. J Chem Eng Res Updat 2021;8:73–97. https://doi.org/10.15377/2409-983X.2021.08.6.
[54] Habibi H, Khosravi-Darani K. Effective variables on production and structure of xanthan gum and its food applications: A review. Biocatal Agric Biotechnol 2017;10:130–140. https://doi.org/10.1016/J.BCAB.2017.02.013.
[55] Palaniraj A, Jayaraman V. Production, recovery and applications of xanthan gum by xanthomonas campestris. J Food Eng 2011;106:1–12. https://doi.org/10.1016/j.jfoodeng.2011.03.035.
[56] Patel J, Maji B, Moorthy NSHN, Maiti S. Xanthan gum derivatives: Review of synthesis, properties and diverse applications. RSC Adv 2020;10:27103. https://doi.org/10.1039/D0RA04366D.
[57] Winnacker M. Polyhydroxyalkanoates (phas): Recent advances in their synthesis and applications. Eur J Lipid Sci Technol 2019;1900101. https://doi.org/10.1002/ejlt.201900101.
[58] Pérez V, Mota CR, Muñoz R, Lebrero R. Polyhydroxyalkanoates (PHA) production from biogas in waste treatment facilities: Assessing the potential impacts on economy, environment and society. Chemosphere 2020;255:126929. https://doi.org/10.1016/J.CHEMOSPHERE.2020.126929.
[59] Kopf S, Åkesson D, Skrifvars M. Textile fiber production of biopolymers – A review of spinning techniques for polyhydroxyalkanoates in biomedical applications. Polym Rev 2022:1–46. https://doi.org/10.1080/15583724.2022.2076693.
[60] Isikgor FH, Becer CR. Lignocellulosic biomass: A sustainable platform for the production of bio-based chemicals and polymers. Polym Chem 2015;6:4497–4559. https://doi.org/10.1039/C5PY00263J.
[61] Intasian P, Prakinee K, Phintha A, Trisrivirat D, Weeranoppanant N, Wongnate T, et al. Enzymes, in vivo biocatalysis, and metabolic engineering for enabling a circular economy and sustainability. Chem Rev 2021;121:10367–10451. https://doi.org/10.1021/acs.chemrev.1c00121.
[62] Newlight technologies, Inc. | AirCarbon, n.d. https://www.newlight.com/. accessed July 3, 2022.
[63] Blunt W, Levin D, Cicek N. Bioreactor operating strategies for improved polyhydroxyalkanoate (PHA) productivity. Polymers 2018;10:1197. https://doi.org/10.3390/polym10111197.

[64] Haddadi MH, Asadolahi R, Negahdari B. The bio extraction of bioplastics with focus on polyhydroxybutyrate: A review. Int J Environ Sci Technol 2019;16:3935–3948. https://doi.org/10.1007/S13762-019-02352-0.

[65] Abedi E, Hashemi SMB. Lactic acid production – Producing microorganisms and substrates sources-state of art. Heliyon 2020;6e04974. https://doi.org/10.1016/j.heliyon.2020. e04974.

[66] Henard CA, Smith H, Dowe N, Kalyuzhnaya MG, Pienkos PT, Guarnieri MT. Bioconversion of methane to lactate by an obligate methanotrophic bacterium. Sci Rep 61 2016;6:1–9. https://doi.org/10.1038/srep21585.

[67] Miller M, Suominen P, Aristidou A, Hause BM, Van Hoek P, Dundon CA. Lactic acid-producing yeast cells having nonfunctional l-or d-lactate: Ferricytochrome c oxidoreductase gene. US8137953B2 2006.

[68] Drumright RE, Gruber PR, Henton DE. Polylactic acid technology. Adv Mater 2000;12:1841–1846. https://doi.org/10.1002/1521-4095(200012)12:23<1841::AID-ADMA1841>3.0.CO;2-E.

[69] Cunha BLC, Bahú JO, Xavier LF, Crivellin S, De Souza SDA, Lodi L, et al. Lactide: Production routes, properties, and applications. Bioengineering 2022;9:164. https://doi.org/10.3390/bioengineering9040164.

[70] Meng X, Yu L, Cao Y, Zhang X, Zhang Y. Progresses in synthetic technology development for the production of l-lactide. Org Biomol Chem 2021;19:10288–10295. https://doi.org/10.1039/D1OB01918J.

[71] Lorenzo D, Synthesis MLAR. Structure and Properties of Poly(lactic acid). Cham: Springer International Publishing; 2018. 279. https://doi.org/10.1007/978-3-319-64230-7.

[72] Masutani K, Kimura Y. Chapter 1. PLA synthesis. From the monomer to the polymer. In Jiménez A, Peltzer M, Ruseckaite R, editors. RSC Polym. Chem. Ser, Cambridge: Royal Society of Chemistry; 2014. Vol. 12, pp. 1–36. https://doi.org/10.1039/9781782624806-00001.

[73] Yoo DK, Kim D, Lee DS. Synthesis of lactide from oligomeric PLA: Effects of temperature, pressure, and catalyst. Macromol Res 2006;14:510–516. https://doi.org/10.1007/BF03218717.

[74] Yarkova AV, Novikov VT, Glotova VN, Shkarin AA, Borovikova YS. Vacuum effect on the lactide yield. Procedia Chem 2015;15:301–307. https://doi.org/10.1016/j.proche.2015.10.048.

[75] Ghadamyari M, Chaemchuen S, Zhou K, Dusselier M, Sels BF, Mousavi B, et al. One-step synthesis of stereo-pure l,l lactide from l-lactic acid. Catal Commun 2018;114:33–36. https://doi.org/10.1016/j.catcom.2018.06.003.

[76] Xu Y, Fang Y, Cao J, Sun P, Min C, Qi Y, et al. Controlled synthesis of l-lactide using Sn-beta zeolite catalysts in a one-step route. Ind Eng Chem Res 2021;60:13534–13541. https://doi.org/10.1021/acs.iecr.1c02448.

[77] Dusselier M, Van Wouwe P, Dewaele A, Jacobs PA, Sels BF. Shape-selective zeolite catalysis for bioplastics production. Science 2015;349:78–80. https://doi.org/10.1126/science.aaa7169.

[78] Baratian S, Hall ES, Lin JS, Xu R, Runt J. Crystallization and solid-state structure of random polylactide copolymers: Poly(l-lactide-*co*-d-lactide)s. Macromolecules 2001;34:4857–4864. https://doi.org/10.1021/ma001125r.

[79] Balla E, Daniilidis V, Karlioti G, Kalamas T, Stefanidou M, Bikiaris ND, et al. Poly(lactic acid): A versatile biobased polymer for the future with multifunctional properties – From monomer synthesis, polymerization techniques and molecular weight increase to PLA applications. Polymers 2021;13:1822. https://doi.org/10.3390/polym13111822.

[80] Celli A, Colonna M, Gandini A, Gioia C, Lacerda TM, Vannini M. Polymers from Monomers Derived from Biomass. Chem. Fuels from Bio-Based Build, Blocks, Weinheim, Germany: Wiley-VCH Verlag GmbH & Co. KGaA, 2016. pp. 315–350. https://doi.org/10.1002/9783527698202.ch13.

[81] Sugihara S, Toshima K, Matsumura S. New strategy for enzymatic synthesis of high-molecular-weight poly(butylene succinate) via cyclic oligomers. Macromol Rapid Commun 2006;27:203–207. https://doi.org/10.1002/MARC.200500723.

[82] Takiyama E, Fujimaki T, Seki S, Hokari T, Hatano Y. Method for manufacturing biodegradable high molecular aliphatic polyester. US 5310782 1994.

[83] Sisti L, Totaro G, Marchese P. PBS makes its entrance into the family of biobased plastics. Biodegrad Biobased Polym Environ Biomed Appl 2016;225–285. https://doi.org/10.1002/9781119117360.CH7.

[84] Aeschelmann F, Carus M. Biobased building blocks and polymers in the world: Capacities, production, and applications–status quo and trends towards 2020. Ind Biotechnol 2015;11:154–159. https://doi.org/10.1089/ind.2015.28999.fae.

[85] Sanford K, Chotani G, Danielson N, Zahn JA. Scaling up of renewable chemicals. Curr Opin Biotechnol 2016;38:112–122. https://doi.org/10.1016/J.COPBIO.2016.01.008.

[86] Gallezot P. Conversion of biomass to selected chemical products. Chem Soc Rev 2012;41:1538–1558. https://doi.org/10.1039/C1CS15147A.

[87] Thuo M, Gregory P, Banerjee S, Du C. 1 Introduction: Biopolymers and Biocomposites, Compos B. De Gruyter; 2021. 1–26. https://doi.org/10.1515/9781501521942-001.

[88] Van Der Waal JC, Gruter GJM. Process for preparing ethylene glycol from a carbohydrate. WO 2016; 2016/114658 A1.

[89] Li M, Niu F, Zuo X, Metelski PD, Busch DH, Subramaniam B. A spray reactor concept for catalytic oxidation of p-xylene to produce high-purity terephthalic acid. Chem Eng Sci 2013;104:93–102. https://doi.org/10.1016/J.CES.2013.09.004.

[90] Peters MW, Taylor JD, Jenni M, Manzer LE, David H. Integrated process to selectively convert renewable isobutanol to p-xylene. 2012;US20110087000A1.

[91] Kanwal A, Zhang M, Sharaf F, Li C. Polymer pollution and its solutions with special emphasis on poly (butylene adipate terephthalate (PBAT)). Polym Bull 2022:1–28. https://doi.org/10.1007/s00289-021-04065-2.

[92] Lim XZ. Microplastics are everywhere – But are they harmful?. Nature 2021;593:22–25. https://doi.org/10.1038/D41586-021-01143-3.

[93] Garcia JM, Robertson ML. The future of plastics recycling. Science 2017;358:870–872. https://doi.org/10.1126/science.aaq0324.

[94] Lamberti FM, Román-Ramírez LA, Wood J. Recycling of bioplastics: Routes and benefits. J Polym Environ 2020;28:2551–2571. https://doi.org/10.1007/S10924-020-01795-8.

[95] Häußler M, Eck M, Rothauer D, Mecking S. Closed-loop recycling of polyethylene-like materials. Nature 2021;590:423–427. https://doi.org/10.1038/s41586-020-03149-9.

[96] Niaounakis M. Recycling of biopolymers – The patent perspective. Eur Polym J 2019;114:464–475. https://doi.org/10.1016/j.eurpolymj.2019.02.027.

[97] Spierling S, Knüpffer E, Behnsen H, Mudersbach M, Krieg H, Springer S, et al. Bio-based plastics – A review of environmental, social and economic impact assessments. J Clean Prod 2018;185:476–491. https://doi.org/10.1016/J.JCLEPRO.2018.03.014.

[98] Hottle TA, Bilec MM, Landis AE. Sustainability assessments of bio-based polymers. Polym Degrad Stab 2013;98:1898–1907. https://doi.org/10.1016/j.polymdegradstab.2013.06.016.

[99] Meereboer KW, Misra M, Mohanty AK. Review of recent advances in the biodegradability of polyhydroxyalkanoate (PHA) bioplastics and their composites. Green Chem 2020;22:5519–5558. https://doi.org/10.1039/d0gc01647k.

[100] Fakirov S.. Biodegradable Polyesters, Weinheim, Germany: Wiley-VCH Verlag GmbH & Co. KGaA; 2015. Vol. 5. https://doi.org/10.1002/9783527656950.

[101] Helanto K, Matikainen L, Talja R, Rojas OJ. Bio-based polymers for sustainable packaging and biobarriers: A critical review. BioResources 2019;14:4902–4951. https://doi.org/10.15376/biores.14.2.Helanto.

[102] Greene JP. Sustainable Plastics, Hoboken, NJ, USA: John Wiley & Sons, Inc.; 2014. https://doi.org/10.1002/9781118899595.

[103] Global production capacities of bioplastics 2021-2026 Global production capacities of bioplastics 2021 (by material type) Global production capacities of bioplastics 2026 (by material type) n.d. (accessed August 11, 2022)

[104] PlascticsEurope. Plastics-the Facts 2021 An analysis of European plastics production, demand and waste data. n.d.; accessed August 11, 2022.

[105] Plastics – The Facts 2021 • Plastics Europe. n.d. https://plasticseurope.org/knowledge-hub/plastics-the-facts-2021/ (accessed August 11, 2022).

[106] Global bioplastics production will more than triple within the next five years – World Bioeconomy Forum. n.d. https://wcbef.com/news-bulletin/global-bioplastics-production-will-more-than-triple-within-the-next-five-years/ (accessed August 11, 2022).

[107] The business case for investing in sustainable plastics | World Economic Forum. n.d. https://www.weforum.org/agenda/2020/01/the-business-case-for-investing-in-sustainable-plastics/ (accessed July 29, 2022).

[108] Size BM. Share & Growth Analysis Report, n.d. https://www.bccresearch.com/market-research/plastics/global-markets-and-technologies-for-bioplastics.html. accessed July 29, 2022.

[109] Gross RA. Biodegradable polymers for the environment. Science 2002;297:803–807. https://doi.org/10.1126/science.297.5582.803.

[110] Singh AA, Genovese ME. Green and sustainable packaging materials using thermoplastic starch. In Sustain, Food Packag. Technol., Wiley: 2021. pp. 133–160. https://doi.org/10.1002/9783527820078.ch5.

[111] Shafqat A, Tahir A, Mahmood A, Tabinda AB, Yasar A, Pugazhendhi A. A review on environmental significance carbon foot prints of starch based bio-plastic: A substitute of conventional plastics. Biocatal Agric Biotechnol 2020;27:101540. https://doi.org/10.1016/j.bcab.2020.101540.

[112] Mishra S, Singh PK, Pattnaik R, Kumar S, Ojha SK, Srichandan H, et al. Biochemistry, synthesis, and applications of bacterial cellulose: A review. Front Bioeng Biotechnol 2022;10:https://doi.org/10.3389/FBIOE.2022.780409.

[113] Koller M, Mukherjee A. A new wave of industrialization of PHA biopolyesters. Bioengineering 2022;9:74. https://doi.org/10.3390/bioengineering9020074.

[114] Sharma V, Sehgal R, Gupta R. Polyhydroxyalkanoate (PHA): Properties and modifications. Polymer 2021;212:123161. https://doi.org/10.1016/j.polymer.2020.123161.

[115] Chen G-Q. A microbial polyhydroxyalkanoates (PHA) based bio- and materials industry. Chem Soc Rev 2009;38:2434–2446. https://doi.org/10.1039/b812677c.

[116] Xiao L, Wang B, Yang G, Gauthier M. Poly(lactic acid)-based biomaterials: Synthesis, modification and applications. Biomed Sci Eng Technol 2012;902. https://doi.org/10.5772/23927.

[117] Aliotta L, Seggiani M, Lazzeri A, Gigante V, Cinelli P. A brief review of poly (butylene succinate) (PBS) and its main copolymers: Synthesis, blends, composites, biodegradability, and applications. Polymers 2022;14:844. https://doi.org/10.3390/polym14040844.

[118] Loos K, Zhang R, Pereira I, Agostinho B, Hu H, Maniar D, et al. A perspective on PEF synthesis, propserties, and end-life. Front Chem 2020;8:585. https://doi.org/10.3389/fchem.2020.00585.

[119] Siracusa V, Blanco I. Bio-polyethylene (bio-PE), bio-polypropylene (bio-PP) and bio-poly(ethylene terephthalate) (bio-PET): Recent developments in bio-based polymers analogous to petroleum-derived ones for packaging and engineering applications. Polymers 2020;12:1641. https://doi.org/10.3390/polym12081641.

Sandeep Kulkarni

Chapter 23 Green bioplastics in sustainable packaging for the emerging bioeconomy

23.1 Introduction

Bioplastics present a significant opportunity within the general umbrella of "bioproducts," particularly when specific conditions, including sources, end-of-life options, and global standards are met. This chapter enhances the coverage of the bioplastics topics also covered in Chapters 21, 26, and 27. Here, we explore definitions, sources, end-of-life options, and global standards. We also examine drivers for bioplastic use in packaging, and common types of bioplastics for packaging applications. Finally, this chapter also provides additional insight into current challenges and opportunities.

23.2 Definition of bioplastics

Bioplastics, according to the most widely accepted definition, are plastics that are partially or fully bio-based and/or are biodegradable. There are two key terms in this definition:

a. A bio-based bioplastic has some or all of its carbon derived from a renewable source.
b. Biodegradable bioplastics are those that degrade into carbon dioxide (CO_2), methane (CH_4), water (H_2O), and biomass through biological action in a defined environment and in a defined timescale. The defined environments typically include composting and anaerobic digestion. (See Section 23.5, as well as Chapters 21 and 26 for further details.)

It should be noted that the term "biodegradable" is an open-ended term and does not mean much unless it is accompanied by a description of the conditions of degradation (temperature, humidity, and nature of microbes) as well as the time frame of degradation. Explored later in Section 23.4, the unqualified use of "biodegradable" as a marketing claim for packaging is misleading, and is either restricted or banned by many global trade authorities and governments.

Sandeep Kulkarni, KoolEarth Solutions, Inc., Alpharetta, GA, USA,
e-mail: skulkarni@koolearthsolutions.com

 | https://doi.org/10.1515/9783112212141-023

The two important considerations related to bioplastics are where they come from/originate ("beginning of life" or source) and what happens to them at the end of their useful life ("end of life"). Table 23.1 provides examples of some common commercial bioplastics, categorized by their beginning and end of life. Acronyms: PLA (polylactic acid), PHA (polyhydroxyalkanoate), PBS (polybutylene succinate), PET (polyethylene terephthalate), PEF (polyethylene furanoate), and PBAT (polybutylene adipate terephthalate).

Table 23.1: Beginning and end of life of some commercial bioplastics.

	End: **biodegradable**	***End:*** **nonbiodegradable**
Beginning: **Bio-based (corn, sugarcane, stover, wheat straw, etc.)**	PLA, PHA, PBS	PET, PEF
Beginning: **Non-bio-based (petroleumbased)**	PBAT	Not classified as bioplastic

As can be seen from this table, different bioplastics can have different combinations of beginning-of-life and end-of-life features. A common misconception is that all bioplastics are also biodegradable. However, this is not true. As indicated in the Table 23.1, bioplastics such as PET and PEF can be derived from biological/renewable sources but are not biodegradable (with the caveat of biodegradeable, at least in any reasonable time frame).

23.3 Sources of bioplastics (or *where do bioplastics come from?*)

Bioplastics can come from variety of plant-based and other renewable sources, called "feedstocks." These feedstocks are generally classified into first, second, and third generations.

First-generation (or Gen1) feedstocks include crops (such as corn and sugarcane), traditionally grown for human consumption. The use of such feedstocks for producing bioplastics has given rise to concerns that this will result in reduced availability as food sources and also that it will result in land being used exclusively for growing crops for bioplastic production. These concerns can be largely mitigated by sustainable sourcing of feedstocks and sustainable farming practices. It should also be noted that as per information summarized by European Bioplastics [1], the feedstock currently used to produce bioplastics relies on less than 0.02% of the global agricultural area, compared to 91% of the area used for pasture and the production of food and feed. Despite the predicted continued growth in the bioplastics market at the current stage of technological development, the share of global agricultural area used to grow feedstock for the production of bioplastics will only slightly increase – to still below 0.06% in 2026. This clearly demonstrates that there is no competition between food/feed and industrial production.

Second-generation feedstocks (Gen2) are ones that are traditionally not used for human consumption. These are the byproducts remaining after harvesting of the primary food crop, and include materials such as corn stover, wheat straw, rice husks, and oat hulls.

Third-generation (Gen3) feedstocks are non-plant-based sources such as algae and waste gases, such as CO_2 and methane from steel mills or landfills.

Due to the concerns and perception issues with Gen1 feedstocks, there is a great deal of focus and interest in using Gen2 and ultimately Gen3 feedstocks for bioplastics.

23.4 End-of-life options for bioplastics (*where do they end up and what happens to them?*)

The options for processing bioplastic-based packaging, after the end of its useful life, depends (as shown in Table 23.1) on whether the bioplastic is biodegradable or nonbiodegradable. For nonbiodegradable bioplastics (such as bio-PE, bio-PET, and PEF), the only end-of-life option available is recycling. These materials should be recycled through the appropriate recycling stream available for their petrochemical counterparts. For instance, bio-PET can be recycled along with conventional PET (petroleum-based) recycling processes.

For biodegradable bioplastics, there are few different end-of-life options available.

23.5 Composting

Composting is defined as the biological degradation process of heterogeneous solid organic materials under controlled moist, self-heating, and aerobic conditions to obtain a stable material that can be used as organic fertilizer [2].

As is clear from the above definition, composting is a specific type of biodegradation that takes place under controlled conditions and in a fixed time frame (usually 90–120 days). Compost is rich in nutrients. It is used, for example, in gardens, landscaping, horticulture, urban agriculture, and organic farming. The compost itself is beneficial for the land in many ways, including as a soil conditioner, a fertilizer, as a source of vital humus or humic acids, and as a natural pesticide for soil. Composting can be carried out on an industrial scale as a manufacturing process or on a smaller scale (such as home/backyard composting).

23.5.1 Industrial composting

These processes use well-defined temperature, humidity, and processing conditions. The following describes the properties of an ideal compost pile:
- Carbon-to-nitrogen ratio is roughly 30:1.
- Presence of oxygen in the decomposition process is necessary to oxidize the carbon.
- Moisture levels of 40–50% are necessary to promote growth of natural microorganisms that break down the waste.
- Temperature in the range of 120–170 °F during the active composting period.

There are three major industrial composting processes used by compost manufacturers in North America, including windrow (Section 23.5.1.1), aerated static pile (Section 23.5.1.2), and in-vessel composting (Section 23.5.1.3).

23.5.1.1 Windrow composting

In this process, the organic waste is laid in long piles called windrows. The piles are aerated periodically by either manually or mechanically turning them. The height of the pile is large enough to generate the required temperatures, yet is small enough to allow air flow. This process produces compost within 90–120 days.

23.5.1.2 Aerated static pile (ASP) composting

In this process, organic waste is mixed in a large pile. To aerate the pile, layers of loosely piled bulking agents (e.g., wood chips and shredded newspaper) are added so that air can pass from the bottom to the top of the pile. The piles can also be placed over a network of pipes that deliver air into or draw air out of the pile. Air blowers might be activated by a timer or temperature sensors. The ASP process produces compost in 3–6 months.

23.5.1.3 In-vessel composting

In-vessel composting can process large amounts of waste without using as much space as the windrow method and it can accommodate virtually any type of organic waste. This method involves feeding organic materials into a drum, silo, concrete-lined trench, or similar encasing. This allows control of the environmental conditions such as temperature, moisture, and airflow. The compost is mechanically turned or mixed to make sure the material is aerated. This method produces compost in just a few weeks. It takes a few more weeks or months until it is ready to be used because the microbial activity needs to stabilize, and the pile needs to cool.

As discussed above, depending on the type of the industrial composting process, the timing for completion of the composting process can be as short as 90 days and as long as 180 days. Most compost manufacturers attempt to produce the compost in as short time as possible, since this allows them to increase their throughput, and hence revenue. This can pose a challenge for packaging targeted to be compostable in an industrial composting process, as the packaging may not completely degrade within the short time span and may get rejected or screened out of the final compost product.

Another consideration is that compostable packaging by itself provides no value (adds no nutrients) to the final compost product. The value of compostable food packaging comes from the fact that the packaging can divert more food waste/food scraps to composting facilities, since food waste is desirable for composters. Also, it keeps food waste and food scrap from ending up in landfills where it can generate methane, a highly potent greenhouse gas (GHG). Further information on locations can be found in a study published by the Sustainable Packaging Coalition (SPC) in 2017 [3], where they examined a variety of scenarios, such as a farmer's market, a concert event, an all-day festival, a quick-service restaurant, and a national grocery chain.

23.5.2 Home/backyard composting

Home or backyard composting can refer to a variety of processes used by households to compost their food waste, yard trimmings, and other organics. These systems can range from simple bins or drums to more advanced systems where the organic waste is turned regularly and where high temperatures are employed. Unlike in the case of industrial composting, there are no standard processes or practices employed, and one home composting system can be completely different from another. Further, there is typically no control of temperature and humidity in the waste pile in home composting systems. Finally, most home composting systems are unable to achieve the high temperatures employed in industrial composting piles, making the process of composting slower and longer.

This lack of control and lack of standardization in home composting systems makes it very difficult for compostable packaging to degrade in home composting environments within a reasonable time frame. For this reason, packaging that is designed to be compostable in an industrial composting environment cannot typically be composted in home composting systems. Furthermore, as will be discussed in the subsequent section, there are almost no standards and certifications available for home compostable packaging, making it very difficult to design packaging to be home compostable.

23.5.3 Marine and soil/land degradation

Over the past decade, there has been a significantly heightened awareness and concern about litter and pollution due to plastic packaging (particularly single-use packaging), both on land and in the oceans. One of the approaches in addressing this issue has been the development of packaging that could potentially biodegrade in marine as well as in land environments. The thought process behind this approach is that if any packaging is inadvertently and improperly disposed on land or is washed into the ocean, it will degrade within a relatively short time frame (and therefore, with the goal of not harming marine life or life on land).

The operative word in the above discussion is "**inadvertently**" since the ability of a package to degrade, either on land or in the ocean, should not be taken as a license to litter.

Like the situation with home composting, there are almost no global standards and certifications for marine and soil/land degradable packaging. The primary reason for this is that soil and marine conditions vary tremendously across the globe. Marine temperatures, pH, types of microbes, and so on can be very different for different oceans and marine bodies, making it extremely difficult to define a standard set of marine conditions for testing degradation of packaging.

23.6 Global standards and certifications for biodegradation

23.6.1 Need for standards

As mentioned earlier, the term "biodegradable" is an open-ended term that does not convey much information and can, in fact, be misleading to a consumer. Many suppliers of packaging or other plastic items have used a "biodegradable" claim/label, without any qualification, in an attempt to portray that the packaging is environmentally friendly. This approach is in fact considered "greenwashing" (disinformation dissemination by an organization, so as to present an environmentally responsible public image). As a result, many countries and regions across the world have either severely restricted or banned outright, the unqualified use of the term "biodegradable" in marketing claims. For instance, the US Federal Trade Commission states the following in its Green Guides:

> **Marketers may make an unqualified degradable claim only if they can prove that the "entire product or package will completely break down and return to nature within a reasonably short period of time after customary disposal." The "reasonably short period of time" for complete decomposition of solid waste products is one year. Items destined for landfills, incinerators, or recycling facilities will not degrade within a year; so unqualified biodegradable claims for them shouldn't be made.**

A company can be subject to significant financial penalties as well as loss of credibility and market share for violating these guidelines.

To address and eliminate greenwashing and to further provide an avenue for legitimate biodegradation claims, a few different global standards, certifications, and testing protocols have been developed. The following sections outline these standards and certifications.

23.6.2 Standards and certifications

ASTM D6400 is the primary standard used in the United States. This standard covers plastics and products made from plastics that are designed to be composted under aerobic conditions in municipal and industrial aerobic composting facilities where thermophilic conditions (i.e., microbes that live at high temperatures between 60 and 108 °C) are achieved.

To be considered compostable by ASTM D6400, the product (packaging, plastic item, etc.) must be tested and they should demonstrate the following:

1. Disintegration: The product must disintegrate into a defined particle/fragment size over a 12-week period.
2. Biodegradation: 90% of the organic carbon must be converted to carbon dioxide by the end of the test period, when compared to the positive control (cellulose).
3. No ecotoxicity: No adverse effect on the quality of the compost.

The **Biodegradable Products Institute (BPI)**, based in the USA, provides industrial compostability certification for packaging that meets the ASTM D6400 standard, based on lab testing for disintegration, biodegradation, and eco toxicity. The following is an example of the type of label and communication that can be used on packaging that has been certified by BPI, as explained in the BPI CERTIFICATION MARK USAGE REQUIREMENTS Version 2 – September 2022, shown in Figure 23.1 [4]:

Figure 23.1: BPI certification label for compostable in industrial facilities.

Starting December 1, 2025, BPI is also launching a Commercial & Home Compostable Certification program. This certification offers a scientifically rigorous framework to clearly identify products that safely break down in home compost piles. This will allow packaging that meets the new BPI commercial and home composting standards, to be labeled as "Home Compostable," using the following label shown in Figure 23.2 [5].

Figure 23.2: BPI label indicating compostable for commercial and home.

The **EN-13432** is the primary standard used in Europe. This standard specifies that packaging may be deemed to be compostable only if all the constituents and components of the packaging are compostable. During the certification process, an assessment is made not only of the basic materials but also of the various additives and product properties.

The key requirements for the EN-13432 certification are:
- Chemical composition
- Biodegradation
- Disintegration
- Quality of final compost and ecotoxicity

TUV Austria, a leading Testing, Inspection, Certification & Training company based in Austria, provides industrial compostability certification for packaging that meets the EN-13432 standard, based on lab testing (either by TUV itself or by another independent lab). The following is an example of the type of label and communication that can be used on packaging that has been certified by TUV Austria, as outlined on the TUV Green Marks website [6] and shown in Figure 23.3.

Figure 23.3: TUV Austria's industrial OK compost label.

In addition to industrial compostability certification, TUV Austria also provides certification for home compostability (called OK Compost Home). The OK compost HOME certification program does not explicitly refer to a specific standard but details all the technical requirements that a product must meet in order to obtain the certification. In fact, the requirements of the OK compost HOME program, defined in 2003, have served as the basis for the drafting of several standards such as:

- Australia: AS 5810 (2010) – Biodegradable plastics – Biodegradable plastics suitable for home composting
- France: NF T 51800 (2015) – Plastics – Specifications for plastics suitable for home composting
- Europe: prEN 17427 (2020) – Packaging – Requirements and test scheme for carrier bags suitable for treatment in well-managed home composting installations

The label shown in Figure 23.4 communicates that packaging has met the TUV Austria requirements for home compostability:

Figure 23.4: TUV Austria home compost label.

In addition to these composting standards, TUV Austria also provides certifications for biodegradation in soil, marine, and freshwater environments. These labels are shown in Figure 23.5.

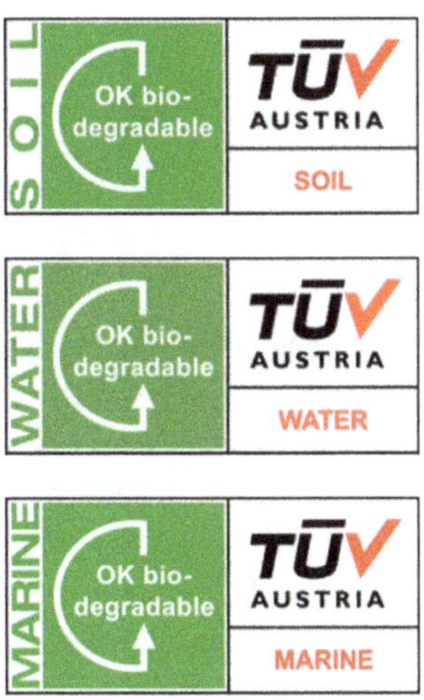

Figure 23.5: TUV Austria soil, water, and marine, OK biodegradable labels.

These last three certifications are not common, given the fact that there currently are no globally accepted standards for biodegradation in soil, marine, or natural freshwater environments. TUV provides these certifications based on testing and standards developed by them internally; however, these methods and standards are not yet widely accepted. Also, there is a risk that such labels can encourage consumers to dispose packaging improperly on land or in oceans, leading to increased littering. Nonetheless, as mentioned earlier, the ability of a plastic or packaging to biodegrade on land or in marine environments in a reasonably short time frame can be useful in avoiding/managing any plastics that might **inadvertently** end up on land or in oceans as litter.

23.7 Examples of bioplastics used in packaging

The global production capacity for bioplastics in 2021 was roughly 2.5 million tons, as discussed in Chapter 21 by Adur. Further, this capacity is generally expected to grow to 7.5 million tons by 2026. This chapter will discuss the packaging applications of some of these bioplastics (note that the chemistry, production, and properties of these bioplastics are discussed in Chapter 21 by Adur and will not be repeated here).

23.7.1 Polylactic acid (PLA)

Polylactic acid (PLA), shown in Figure 23.6, is a compostable polyester made from renewable feedstock. With lactic acid as the raw material, PLA is produced by fermentation of glucose or sucrose, and is refined to a high purity.

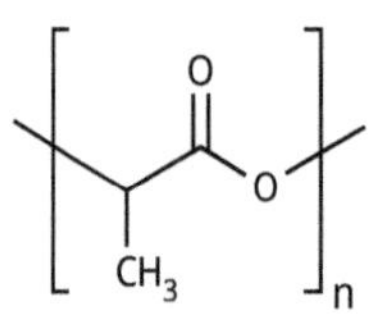

Figure 23.6: Structure of PLA.

Currently, the two major global producers of PLA are NatureWorks LLC and TotalEnergies-Corbion. NatureWorks is based in the USA and uses corn as a feedstock for their Ingeo^R resins. TotalEnergies Corbion is based in Thailand and produces their Luminy^R PLA resins, primarily from sugarcane sugars. Recently, Emirates Biotech, a UAE-based start-up, has announced plans to establish the word's largest PLA production facility in Dubai, UAE. Emirates Biotech plans to initially source lactic acid on the open market for their PLA production, and will later switch to other feedstocks such as corn. Additional details on these enterprises can be found on their respective websites and through relevant business development media coverage.

Over the past decade or so, PLA has found several applications in food and food service packaging industry, such as:

- Cups and lids for hot and cold beverages
- Carry-out food containers
- Single-use plates and cutlery (forks, spoons, knives, etc.)

In addition to these rigid packaging applications, PLA has also been used in flexible packaging applications such as packaging for snack foods (chips, confectionery, and so on).

For both rigid and flexible packaging applications, it is necessary to compound PLA with appropriate additives to allow it to be processed using existing processing equipment (since current equipment is optimized for "conventional" plastics such as PE and PP).

As discussed in other chapters, PLA inherently exhibits a low barrier to penetration of gases such as oxygen and water vapor. For this reason, many of the flexible packaging applications of PLA require that it be used a part of a multilayered structure where the non-PLA layers can impart moisture vapor and oxygen barrier to the packaging structure. By appropriately selecting the non-PLA barrier layers, it is possible to maintain the compostability of the package while providing appropriate protection to the food being packaged.

It should be noted that current commercially available PLA resins are only compostable in an industrial composting environment, and not in home composting or other uncontrolled environments (such as soil, ocean). The reason for this is that the PLA requires elevated temperature as well as specific moisture and microbial conditions for biodegradation, and these conditions are only found in an industrial composting situation.

One of the areas where PLA-based packaging has found good success is in situations where PLA-based packaging can drive increased diversion of food waste into composting (and hence away from landfills). For instance, in the case of some Quick Service Restaurants (QSRs) and sports arenas, it has been possible to collect food-service ware (plates, bowls, and clamshells), along with leftover food scraps, and send the combination to a local composting facility. However, it is important to ensure that the compostable packaging used in such cases is certified compostable by a recognized body (such as BPI in the USA) and demonstrated to degrade completely within the 60–90-day time frame used by most composters. This avoids the contamination of the compost by any packaging that has false/misleading compostability claims.

PLA has also found application as a liner for fiber-based coffee cups, as an alternative to polyethylene, which is currently used in most coffee cups. Many of these PLA-lined cups are certified as compostable in industrial composting. However, currently there is a major lack of collection infrastructure as well as lack of widespread availability of industrial composting facilities (particularly those that accept compostable packaging). This has led to a large number of compostable cups ending up in landfills.

An example of the use of PLA in the flexible packaging arena is the compostable SunChips bag introduced by PepsiCo in early 2010. This bag was composed of a 3-layer all-PLA structure, with metallization on the innermost layer for moisture vapor and oxygen barrier. This bag was certified compostable by BPI.

While the bag was initially favorably received in the market, consumers discovered that the bag was extremely "noisy." This was due to the higher stiffness of the PLA structure (versus the traditional BOPP film construction), which resulted in a loud crackling sound when the bag was handled or pressed. Due to the noise issue as well as the higher cost of PLA, PepsiCo/Frito-Lay pulled the compostable bags out of the market toward end of 2010.

In late 2021, PepsiCo/Frito-Lay introduced an industrially compostable bag for its Off the Eaten Path brand. These bags are available at Whole Foods and select retailers. In order to promote composting of these bags, Frito-Lay has teamed up with Terracycle. In this process, consumers can either mail in the bags or locate a local composting facility where they can drop off the bags. The bags are presumably made of PLA although Frito-Lay has not provided any details of their construction.

An example of the use of PLA in rigid packaging applications is yogurt cup by Danone for its Activia product line. This PLA-based yogurt cup was introduced in Germany in 2011 and Danone used Ingeo PLA from NatureWorks for this cup. The new yogurt cup was the result of a close cooperation between Danone, WWF (World Wildlife Fund)-Germany, and NatureWorks. Covered in multiple packaging news sites, Danone claimed (based on a LCA study) that this move resulted in 25% reduction of carbon footprint and 43% less fossil resources use, versus their conventional polystyrene cups.

One application area where PLA is notably absent is bottles for beverages such as water. The reason for this is that PLA bottles can be easily confused with PET bottles by a consumer (since both are clear). The recycling of PLA bottles, along with PET, causes significant disruption to the PET recycling process, and for this reason, there was significant pushback from PET recyclers to the initial introduction of PLA water bottles in the market. Many of the large PLA producers, such as NatureWorks, have therefore not marketed their resins for use in beverage bottles.

23.7.2 Polybutylene succinate (PBS)

Polybutylene succinate (PBS) is polyester produced by copolymerization of butane diol and succinic acid, shown in Figure 23.7.

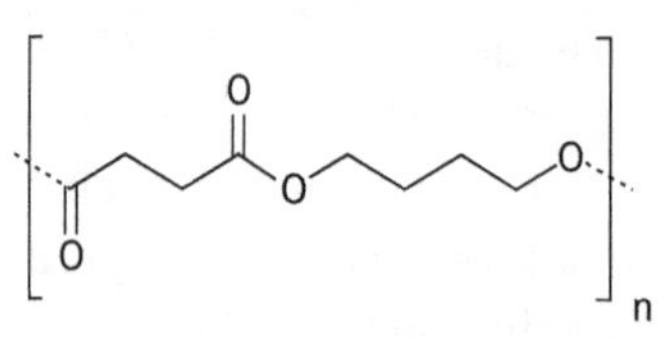

Figure 23.7: Structure of PBS.

While succinic acid is typically derived from renewable sources (usually corn), currently, most butane diol is based on petrochemical sources. For this reason, currently available grades of PBS are only partially bio-based. There are, however, some companies (such as Genomatica) that have developed butane diol from biological sources and are in the process of scaling up production. However, given the price premium for bio-sourced butane diol versus its petroleum-based counterpart, it is unlikely that 100% bio-based PBS will be commercially available unless there is strong pull from major consumer brands.

Similar to PLA, PBS is compostable in an industrial composting environment. In addition, some grades of PBS, specifically those based on polybutylene succinate adipate, are also home compostable.

Some key advantages of PBS in flexible packaging applications are:

- Excellent heat sealability with low heat seal temperature
- Better oxygen barrier than LDPE
- Good to retain aroma, such as limonene
- Suitable for food contact, such as weak acidic, fatty, and oily food
- As sealant, this can be used by coextrusion or dry lamination

PBS has found commercial applications in compostable coffee pods/capsules and coffee pouches, and as liner for coffee cups. While PBS exhibits a fairly good barrier to oxygen ingress into a package, it has a very low barrier against water vapor. Most food packaging applications typically require a high degree of moisture barrier in order to prevent the food from getting soggy. For this reason, PBS film is typically laminated to a film with high moisture barrier (such as NatureFlex regenerated cellulose film from Futamura), along with a layer of vapor-deposited aluminum oxide (in a process known as metallization)

PBS was one of the technologies selected as the winner of the NextGen Cup Challenge [4] (https://www.closedlooppartners.com/nextgen/impact/packaging-innovations/cup-challenge/), launched in 2018 by Starbucks and McDonald's, together with other leading global food service and coffee brands. The goal of the challenge was to identify and develop recyclable and compostable liner technologies to replace the polyethylene coating in current coffee and cold-drink cups. As a result of the development work, Starbucks conducted a market trial with cups lined with the bioPBS coating (from PTTMCC Biochem) in early 2020 [7].

Eastman Chemical Company recently introduced Solus™ 7,000 series high-performance additive and masterbatch portfolio, specifically developed for use with biodegradable and compostable polymers, such as PBS, for extrusion coatings on paper. Further details from the company can be found on the company website (eastman.com). The main applications of this product is in food service (for example, cups, to-go boxes, and wraps), as well as other food packaging uses. Eastman claims that while many biopolymers have challenges in processing, modification of lines, and end-use, the Solus™ additives represent a new and superior product through polymer-to-paper adhesion, improving coating speeds, decreasing polymer coat weight, and decreasing scrap rates. Further, they claim both beginning- and end-of-life profiles with being biobased, with optional recycled content, and recyclability and compostability. While these are large claims, this provides a compelling example of product development in this area, as well as the priorities of the industry and potential future directions (for example, perhaps home composting).

23.7.3 Polybutylene adipate terephthalate (PBAT)

PBAT, shown in Figure 23.8, is a copolymer of adipic acid, butane diol, and terephthalic acid (TA). It should be noted that the structure of PBAT is a random copolymer of the blocks shown in the figure above. PBAT is currently derived completely from petroleum (nonrenewable) sources and yet is industrially compostable (some grades are also home-compostable).

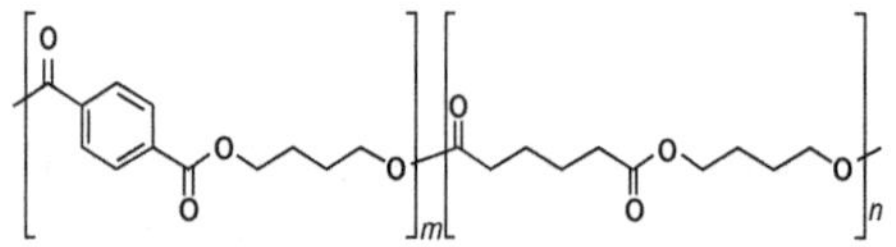

Figure 23.8: Structure of PBAT.

One of the most well-known commercial PBAT resins is Ecoflex, marketed by BASF, for use in flexible packaging as a replacement for polyolefin films (PE, PP). However, typically, PBAT is not used by itself but is blended with PLA to provide properties such as flexibility and toughness. Also, the incorporation of PLA makes the blend partially bio-based/renewable. Ecovio from BASF is an example of a PBAT/PLA blend available commercially. PBAT can also be blended with other materials such as thermoplastic starch (such as the MaterBi resin from Novamont).

As discussed in the case of PBS, PBAT/PLA films are typically combined with other barrier films as well as a layer of metallization in order to achieve high-moisture-barrier packaging. An example of such a structure is an Ecovio film, laminated with NatureFlex film. With appropriate selection of the film layers and the laminating adhesive, it is possible to achieve a high-barrier structure for demanding applications and yet have the package be fully compostable (industrially and potentially home-compostable).

In addition to flexible packaging applications, PBAT/PLA blends have also been evaluated as paper coatings for applications such as liners for paper cups. However, there are currently no commercial cups that use the PBAT/PLA coatings.

23.7.4 Polyhydroxy alkanoate (PHA)

PHAs, represented in Figure 23.9, are a unique family of bio polyesters produced by a bacterial synthesis process, starting from nonedible oils and other nonfood feedstocks.

PHA is the only bioplastic that is biodegradable under a wide range of conditions – industrial composting, home composting, and soil and marine degradation. As the chemistry and properties of PHAs are described in detail elsewhere in this edition, these will not be repeated here.

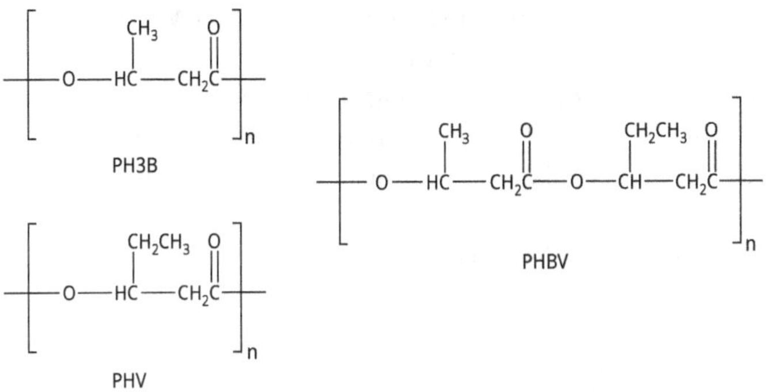

Figure 23.9: Structure of PHA.

One of the leading companies developing PHAs is Danimer Scientific. Danimer and PepsiCo announced a joint venture a few years back to develop flexible packaging based on Danimer's Nodax PHA. Similarly, Mars Wrigley announced a 2-year partnership with Danimer Scientific to develop home-compostable snack and candy packaging for Mars.

As covered by multiple news and media outlets, Danimer Scientific filed for Chapter 11 bankruptcy in March 2025, citing a critical liquidity shortage and an inability to secure additional financing. Subsequently, the manufacturing assets of Danimer were acquired by Teknor Apex in May 2025. Teknor intends to restart production of Danimer's PHA resins at their Kentucky pilot plant.

An example of a flexible packaging structure containing PHA as shown below in Figure 23.10.

In addition to the earlier work of Danimer, several other companies such as Full Cycle Bioplastics, RWDC Industries, Mango Materials, and so on are also working on developing PHAs from a variety of sources such as used municipal solid waste, used cooking oil, and methane from landfills.

Danimer and RWDC had also initiated development of water-based PHA dispersion coatings for use in paper-based packaging applications, providing packaging producers and addition option to apply PHA on to paper (apart from the traditional ex-

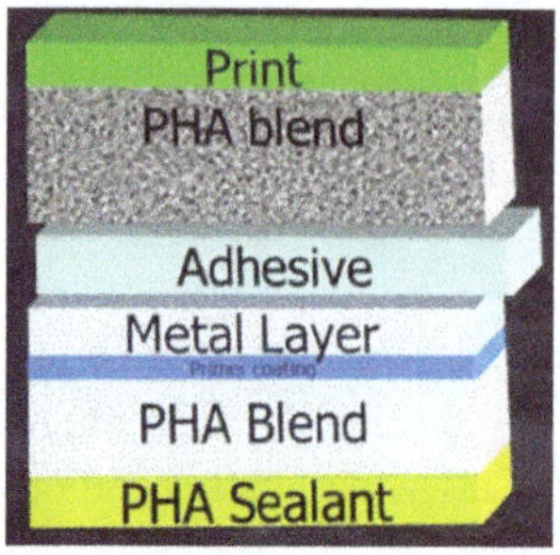

Figure 23.10: PHA flexible film packaging structure.

trusion-coating approach). Danimer had partnered with Kemira and RWDC with Trinseo for the development of these dispersion coatings.

5Gyres, a well-known nonprofit organization that focuses on reducing plastics pollution (by conducting primary research on plastics in the ocean), has recently completed an extensive study on the degradation of various PHA articles (such as films, forks, and bowls) in actual ocean environment. These studies help establish the time frame for degradation of PHA-based food service items in the ocean.

23.7.5 Bio-polyethylene terephthalate (bio-PET)

Bio-PET, represented in Figure 23.11, like conventional PET polymer, is produced by the copolymerization of ethylene glycol (EG) with TA. In the case of bio-PET, however, either one or both monomers (EG and TA) can be derived from renewable sources.

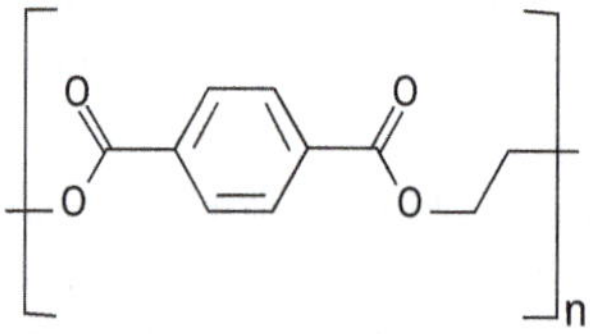

Figure 23.11: Structure of bio-PET.

Currently, the very limited commercially available bio-PET resin is partially bio-based where the EG monomer has been derived from plant sources. One of the most prominent examples of a package, based on this partially bio-based PET, is the Coca Cola Plant Bottle that was first introduced into the market in 2009.

As discussed earlier, this bottle was partially bio-based (up to 30% bio-content), where the EG monomer was derived from ethanol produced from sugarcane grown in Brazil. Coca Cola announced in 2015 that they had developed a fully bio-based PET bottle but this bottle was not commercially launched.

In addition to Coca Cola, several other global beverage brands (PepsiCo, Suntory, and Nestle Waters) have ongoing initiatives as well as alliances to develop 100% bio-based PET bottles.

Bio-PET (whether partially or fully bio-based) is identical in every way to petroleum-based PET and can be recycled using standard PET recycling processes.

23.7.6 Bio-polyethylene (bio-PE)

Bio-PE is identical to petroleum-based PE, represented in Figure 23.12, except that it is derived from renewable sources. Currently, the only commercial producer of bio-PE is Braskem, based in Brazil. Braskem manufactures its bio-PE by polymerizing ethylene derived from sugarcane ethanol.

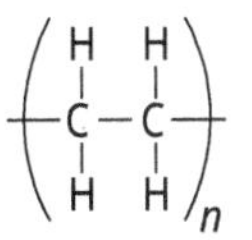

Figure 23.12: Structure of bio-PE.

Bio-PE can be used as a "drop-in" replacement for conventional PE in all applications. The use of bio-PE, instead of petroleum-based PE, allows a brand owner to reduce the GHG impact of their packaging and also to claim renewable content in their packaging.

Given the fact that bio-PE is indistinguishable from petroleum-based PE, it can be recycled through established PE recycling processes.

23.7.7 Polyethylene furanoate (PEF)

PEF, represented in Figure 23.13, is a polyester synthesized by the copolymerization of EG and furan dicarboxylic acid (FDCA). While the EG monomer can be derived from either petroleum or renewable sources, FDCA monomer can be obtained only from natural sources such as certain carbohydrates. In fact, FDCA was identified by the US Department of Energy as one of 12 priority chemicals for establishing the "green" chemistry industry of the future (see Chapter 2 for additional discussion).

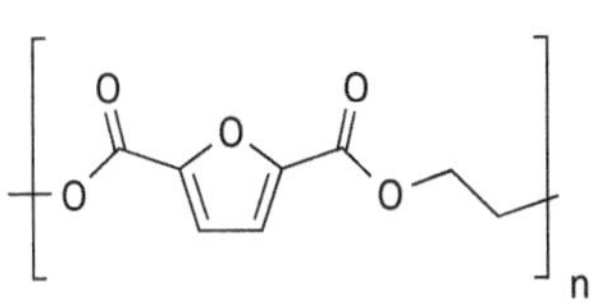

Figure 23.13: Structure of PEF.

FDCA exhibits significantly higher barrier to oxygen and carbon dioxide gases when compared to PET (as much as 6–8 times better gas barrier). This results in significantly longer shelf life for juices and carbonated drinks packaged in PEF bottles, as against those packaged in PET containers, for the same weight of the bottle. Alternatively, significantly lighter weight PEF containers can provide similar level of shelf life as their heavier PET counterparts. PEF also has superior mechanical properties (tensile strength, hoop strength, and so on) when compared to PET, which in turn allows lighter weight bottles to be produced while maintaining good mechanical properties and stability.

Given the benefits of PEF over PET, there has been significant interest in developing this resin as well as beverage bottles from this material. Currently, two leading companies, Avantium and Corbion, are working on developing and scaling up the production of PEF as well as its applications. However, the higher cost of PEF, when compared to PET, has somewhat hampered the speed of development of PEF applications in beverage packaging. One of the approaches currently being explored is to create a multilayered bottle, with PEF as a thin layer "sandwiched" between thicker PET layers. This is expected to keep the cost increase reasonable, as against a bottle made fully of PEF, while still obtaining some level of shelf life (barrier) improvement.

One of the concerns with replacing PET with PEF in beverage bottles is the impact of PEF in the PET recycling process. PEF bottles will need to be collected and recycled through the existing PET recycling stream since a dedicated recycling process for PEF will not be economically (and logistically) viable, at least initially when the volumes of these bottles will be relatively small. Initial studies conducted by PEF producers as well as by some brand owners have indicated that small percentage of PEF bottles in the PET stream (likely 10% or less) will likely have little or no impact on the quality of the recycled PET.

23.7.8 Bioplastics derived from algae/seaweed

Over the past decade, there has been a growing interest in utilizing algae/seaweed as a source of bioplastics. This is an interesting potential – and growing – area of bioproduct research. Bioplastics can be made from algae by using polysaccharides (such as alginate and agar) or proteins and other polymers found within algae cells, or by blending algae biomass with other polymers. Algae is a promising feedstock because it can be a carbon-neutral resource, with rapid growth, using nonarable land, and avoiding competition with food crops. Further, algae can be a useful way to, at least temporarily, sequester some carbon dioxide from the atmosphere. In the longer term, this may be less of a benefit, as the carbon may be released during product breakdown and not recaptured.

Several companies, detailed below, are developing or producing bioplastics from seaweed, focusing on sustainable packaging and low-carbon materials. These companies are able to leverage the natural abundance of seaweed, the potential biodegradability, and carbon-sequestering properties to create alternatives to petroleum-based plastics. Innovations in this research and development space range from edible and compostable packaging, to thermoplastic pellets, and water-soluble films. These are designed to reduce plastic pollution and GHG emissions, while also having specific waste benefits that are intended to mitigate harm in marine ecosystems and coastal economies.

Some of the companies working on developing seaweed-based bioplastics are:

- **PlantSea Ltd** – A UK-based company that is developing low-carbon, seaweed-derived bioplastics and packaging materials, focusing on being recyclable and compostable, while also supporting seaweed aquaculture and circular economy initiatives.
- **Marea** – An Icelandic start-up creating THARAPLAST, a product that is a seaweed-based bioplastic film and thermoplastic pellets, with ongoing efforts around certification for biodegradability, as well as partnerships in macroalgae research.
- **Sway** – A biomaterial design studio producing regenerative seaweed-based packaging, designed to replace single-use plastics, while also emphasizing ecological benefits and scalability.
- **Biopac (Seaweedtama Biopac Indonesia)** – A pioneering organization in edible-grade, seaweed-based bio-packaging that is home-compostable, vegan-friendly, and safe for marine life, with applications across multiple consumer sectors.
- **Notpla** – Develops seaweed-based packaging, for example: Ooho (edible liquid pods), Notpla Coated boxes (compostable takeaway packaging), Notpla Film (biodegradable flexible film), and Notpla Rigid (seaweed-based cutlery)

23.8 Sustainability benefits of bioplastics

Bioplastics can provide several sustainability benefits for a consumer packaged goods (CPG) company. Some of these benefits/attributes have been discussed in other chapters; so this section will briefly touch on four key drivers from a packaging standpoint.

23.8.1 Greenhouse gas (GHG) reduction

Most major CPG companies have announced targets for reducing their GHG emissions across the board. Packaging is typically a fairly large contributor to the carbon footprint of a company. The use of bio-based bioplastics can help significantly reduce the GHG impact from packaging. The reason for this is that renewable sources for bioplastics, such as Gen1 or Gen2 feedstocks, are in fact “carbon sinks,” i.e., they absorb and sequester CO_2 during their growing process. The benefit of this carbon sequestration gets "passed on" to the bioplastics produced from these renewable sources. In fact, the use of bioplastics is by far the largest "lever" a company can have for achieving 20% or higher GHG reduction from its packaging.

23.8.2 Decoupling from fossil sources

There are growing concerns with the use of nonrenewable fossil resources (such as petroleum) for the production of plastics, given the finite quantities of these resources as well as the pollution and emissions involved in the production of conventional plastics. The use of bio-based bioplastics allows decoupling of plastic production from fossil sources. As discussed earlier, various types of renewable feedstocks (Gen1, Gen2, or Gen3) can be used for producing bioplastics, and this allows appropriate selection of feedstocks, based on region/geography, to avoid large transportation distances (for instance sugarcane and bagasse in Brazil, corn and corn stover in midwestern USA, wheat straw in western Europe).

23.8.3 End-of-life for nonrecyclable (or hard to recycle) plastic packaging

Compostable bioplastics are a viable solution for plastic packaging that is not currently recyclable or is difficult to recycle for either or both of the following reasons:

a. Packaging that is composed of multiple layers of plastic (such as snack food pouches, sachets, etc.) that cannot be recycled through existing recycling streams
b. Packaging that does not have a viable collection and recycling infrastructure (such as single-use cups, plastic cutlery, straws, etc.)

As discussed earlier in this chapter, there have been some notable initiatives and successes in the use of biodegradable bioplastics for packaging applications, such as the fully compostable SunChips bag from PepsiCo. As reported by multiple outlets across sustainability and sporting media, the Seattle Mariners stadium (T-Mobile Park) in Seattle, Washington, provides a compelling example of a location driving sustainability initiatives. They were able to achieve over 80% waste diversion through converting all food service ware (forks, spoons, cups, etc.) to PLA-based compostable plastics, collecting these items, and sending them to a composting facility.

23.8.4 Performance benefits

Some bioplastics can have significant performance advantages over comparable petroleum-based plastics. As discussed, PEF, which can be completely derived from plant sources, has eight times better oxygen barrier as well as six times better CO_2 barrier, when compared to PET. This can significantly increase the shelf life of juices and carbonated beverages, thereby reducing spoilage and waste.

23.9 Myths/misconceptions regarding bioplastics

While packaging based on bioplastics has some key benefits, as outlined in the previous section, there are some widespread misunderstandings with regard to how bioplastics are perceived by consumers. Many of these misconceptions have in fact been perpetrated by misguided (or in some cases, downright unethical) suppliers of bioplastic-based packaging in an attempt to gain market share by portraying their packaging as "environmentally friendly." Here, we dive into the myths and realities:

Myth: All bioplastics will break down and "disappear" if discarded on the ground, for instance on roadsides or in open fields, thereby solving land litter problem.

Truth: As discussed in a previous section, compostable bioplastics will typically degrade ONLY in specific composting environments (elevated temperatures). Nonbiodegradable plastics (PET, PEF) will not degrade under ANY conditions.

Myth: All biodegradable bioplastics will break down and "disappear" in marine environments (thereby solving marine litter issue).

Truth: Currently, ONLY one class of biodegradable bioplastics (PHA) appears to have fairly rapid marine degradability (6 months), based on lab studies. There are currently studies underway to determine the rate of biodegradation of various PHA articles, such as bags, cutlery, and so on, in actual marine environments. Other biodegradable plastics, such as PLA, can take significantly longer to degrade in a marine environment.

Myth: Packaging based on bioplastics is always the most sustainable option.

Truth: While bioplastics do offer several sustainability benefits, they may not be the most sustainable choice for all packaging applications. For instance, as discussed earlier, replacing PET bottles with PLA bottles can be detrimental to the PET recycling stream.

Biodegradable bioplastics cannot be and should not be considered a panacea or a silver bullet to solve the plastic litter issues (either on land or in the oceans). Plastic litter can only be addressed through consumer education, proper waste management, and through diversion of plastic “waste,” either into recycling or composting processes.

23.10 Key consideration in developing packaging based on bioplastics

Development of packaging based on bioplastics (either for a completely new packaging application or as a replacement of existing petroleum-based plastic packaging) requires significant upfront thought and analysis. Some of the considerations a brand must keep in mind are:

Goal/objective: It is important to clearly define what the sustainability goal, or combination of goals, is (GHG reduction, end-of-life, and renewable content) and how bioplastics can help achieve these.

Packaging format and performance: It is critical to define the desired packaging format (rigid and flexible) and product protection needed (moisture barrier, oxygen barrier, and CO_2 barrier). As discussed previously, a single bioplastic may not be able to meet the performance requirements, and in such a case, a combination of bioplastics (along with a non-bioplastic such as aluminum) may be needed.

End-of-life: End-of-life scenarios for the package (composting/recycling) should be clearly defined, along with strategies regarding how the packaging can be appropriately collected and disposed off.

Value chain partnerships: In order to successfully commercialize packaging based on bioplastics, a brand needs to work closely with bioplastic resin suppliers, compounders, converters, and so on. This collaboration is critical since many bioplastics can perform very differently, compared to conventional plastics, in their processing characteristics, package formation, filling, and so on.

Supply chain development: Given the fact that most bioplastics are produced from renewable/biological sources, new supply chains may need to be developed to produce the intermediates needed for production of these bioplastics. For instance, for the production of the Plant Bottle, Coca Cola had to work with partners to develop a completely new supply chain for the production of bio-based EG.

23.11 Regulations and plastic bans

Growing public awareness and concern regarding plastic pollution, in particular, marine litter, has prompted several countries worldwide to implement sweeping legislations and bans focused on plastics, specifically single-use plastics (such as straws, cutlery, cups, lids, and so on). As widely covered in global conversation and media, for instance, India has announced bans of various single-use plastics. Similarly, the state of California in the USA has passed a law that will ensure a 25% drop in single-use plastic by 2032. The law also requires that at least 30% of plastic items sold or bought in California are recyclable by 2028, and establish a plastic pollution mitigation fund. Other states in the USA, as well as some countries in Europe, have announced bans against PE shopping bags.

Many of these regulations/bans specifically exempt compostable bioplastics such as PLA. This has prompted brand companies to switch to PLA-based straws, cups, lids, and other single-use items. This has led to significantly higher demand for PLA resin over the past few years. It is worth noting that many of the countries banning conventional single-use plastic packaging have little or no infrastructure for collecting and composting PLA-based (or other compostable plastic) items. As a result, it is very unlikely that the litter and pollution issues will be solved simply by switching from conventional plastics to compostable bioplastics for single-use applications.

23.12 Summary

Bioplastics are an important and growing segment of the flexible and rigid packaging industry worldwide to address the need to mitigate GHG emission, municipal waste, and the attendant marine pollution. Progress is evident on several fronts, including the availability of more number of bioplastic candidates, government and legislative actions, biodegradability standardization, increased awareness and acceptance by consumers, and major brand owners investing to develop and commercialize eco-friendly green packaging solutions.

References

[1] https://www.european-bioplastics.org/faq-items/how-much-agricultural-area-is-used-for-bioplastics/.
[2] Lobo MG, Dorta E. Postharvest Technology of Perishable Horticultural Commodities, 2019. 639–666.
[3] https://greenblueorg.s3.amazonaws.com/smm/wp-content/uploads/2018/05/Value-of-Compostable-Packaging.pdf.

[4] NextGen Cup Challenge https://www.closedlooppartners.com/nextgen/impact/packaging-innovations/cup-challenge/
[5] https://bpiworld.org/resources/Documents/BPI%20Documents%202020/BPI%20Certification%20Mark%20Usage%20Requirements.pdf.
[6] https://www.tuv-at.be/green-marks/.
[7] https://stories.starbucks.com/stories/2020/starbucks-trials-a-nextgen-cup-solution/.

Ray Bergstra

Chapter 24 PHAs: the all-purpose green bioplastic for the emerging bioeconomy

24.1 Introduction

Polyhydroxyalkanoates, or PHAs, as they are most widely known, are one of the most promising biopolymer technologies that are coming into commercial prominence as the globe struggles with greenhouse gas emissions and plastic waste. It is a bioplastic. But what does this mean and why are PHAs considered to be so important?

Fundamentally, PHAs can address problems with plastic waste, which is being generated at a rate of more than 350 million tons per year [1]. As a result, government and industry are led by strong consumer and public pressures to develop alternatives to petroleum-based plastic products and the product sector, referred to as "bioplastics" or biomaterials, is expanding.

Any discussion of bioplastics often includes the term "single-use plastics" or SUPs, which generally refer to "disposable" plastics, largely used for things like takeaway/ fast food and consumer goods packaging. This desire to replace existing plastic products has been taken on by packaging technology companies, which are responding in areas such as coated paper and paperboard, renewable plastics containers, biodegradable bottles, and a range of other approaches. Plastic straws, beverage containers, and cutlery are other items targeted for replacement with new biomaterials deemed to be more sustainable.

The sustainability movement has also significantly influenced corporate governance and the financial industry, as a 2023 survey of a full range of industries and companies by McKinsey found over 90% of respondents reported that ESG subjects were on their organization's agenda [2]. And, it is well known that the world's major brand owners, whose packaging materials are the leading contributor to plastic waste, are all active in trying to implement sustainable technologies and practices in order to reduce waste and improve the environmental sustainability of their operations. These companies include an ever-increasing number of major brands across a range of industries such as Coke, Pepsi, Unilever, Nestle, Mondelez, and Mars.

In this chapter, we will first provide an overview of bioplastics markets and the technologies, and then move on to technical details of PHAs, their role in the emerging bioeconomy, advantages and disadvantages of PHAs, commercial developments, and what can be expected of PHAs as a viable alternative to petroleum-based plastics. Ad-

Ray Bergstra, MTN Consulting Associates, Edmonton, Canada, e-mail: rjbergstra@mtnconsulting.ca

 | https://doi.org/10.1515/9783112212141-024

ditional details and further explorations of bioplastics are available also in other chapters of this book, including particularly Chapter 21.

24.2 What are bioplastics?

There are numerous types and approaches to bioplastics that are best described by European Bioplastics (EUBP), an organization that represents the bioplastic industry in Europe. Publications by EUBP [3] highlight that bioplastics are not a single material but rather a family of plastics materials that are bio-based, biodegradable, or both.

Bio-based polymer plastics are those produced from biologically sourced carbon, such as starch, sugar, cellulose or vegetable oils, where the carbon in these natural substances can be converted into plastics. Within this category are two production approaches. Some bio-based plastics are produced via chemical synthesis using bio-based monomers, while others (including PHAs) are produced directly by microbes through their natural metabolic processes.

Biodegradable plastics degrade naturally in the environment, without leaving behind residues. Typically, this implies microbial degradation, where microbes consume the plastic material and convert it into carbon dioxide and other natural carbonaceous materials that do not contribute to pollution.

It should be noted that throughout the literature, the terms biopolymer and bioplastic are often used interchangeably, and this is mainly because in the context of functional materials, they quite often mean the same thing.

Each bioplastic can be categorized, as shown in Figure 24.1, as published by European Bioplastics. Please see Chapters 21 and 25 for further information and acronyms used in Figure 24.1 and Table 24.1. There are four quadrants in the chart that outline the relationship between a material's bio-based and biodegradability characteristics:

- Petroleum- or fossil-based and nonbiodegradable, which includes common conventional plastics such as polyolefins and PET
- Petroleum-based and biodegradable, such as PBAT and PCL
- Bio-based and nonbiodegradable, where conventional type polymers are produced using bio-based monomers such as ethylene, derived from corn ethanol
- Bio-based and biodegradable – most environmentally sustainable bioplastic category that includes PHAs, thermoplastic starch (TPS), some PBS polymers, and PLA.

Further to the categorization of biopolymers and bioplastics, a comparison of the different types is important for determining how they fit into the marketplace. A brief summary is provided in Table 24.1.

Each type of bioplastic has a place in the market. In many instances, the sustainability goals of a company can by met by using bio-based plastics that are not rapidly biodegradable, in order to maintain specific performance specifications existing in

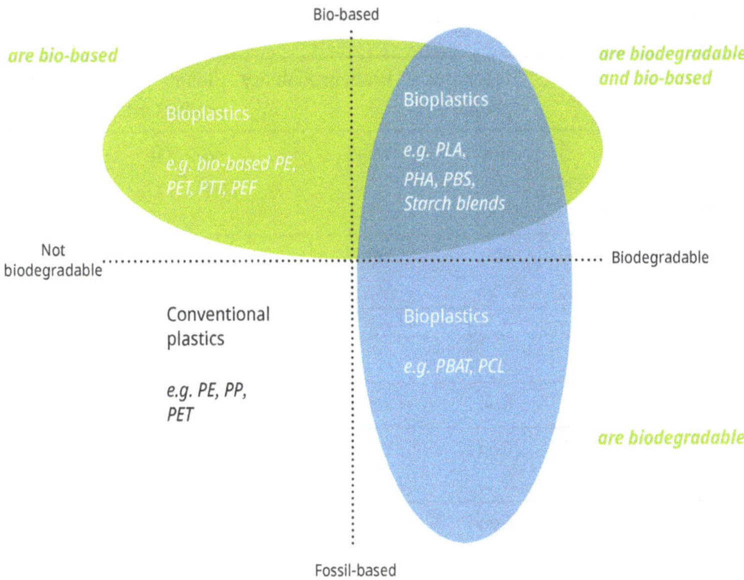

Figure 24.1: Categories of bioplastics.
Source: European Bioplastics, used with permission.

the market. This refers to critical properties such as gas barrier properties for food packaging or mechanical strength and durability for packaging or functional components needed in manufactured goods. For these reasons, there is a strong emphasis on developing bioplastics with "good" mechanical properties, which includes the ability to design a material with the required tensile strength and elasticity.

In other applications, fast biodegradability may be the top priority, such as in horticulture, where seedling containers can be planted and left to biodegrade in the ground. In these instances, biopolymers, like starch, with poorer mechanical properties can be used. And of course, some markets are more price sensitive than others and are less prepared to take on the higher cost of a premium bioplastic.

Table 24.1 also indicates the typical feedstocks used to produce each type of plastics. Bio-based polymers produced using fermentation, including PHAs, can be produced from glucose (starch or cellulose) and/or fatty acid sources. Bio-based, but non-biodegradable plastics, such as bio-based PE are generally derived from sugar-based ethanol, which is then chemically converted to ethylene. For this point, any ethylene-derived chemical can be manufactured using this bio-based carbon. Bio-based polyamide (PA) is another unique example, as it is produced chemically from hydroxy fatty acid present in castor oil.

PHAs occupy a leading position in terms of overall feasibility and viability. They can be produced from for a variety of organic feedstocks; a range of properties can be built into the polymer design; they are bio-based, biodegradable, biocompatible, and

Table 24.1: Comparison of bioplastics.

Technology (pure polymer)	Carbon source	Mechanical properties	Biodegradability	Global production, T	Relative cost
PHA	Cereal sugars, vegetable oils, and organic waste	Good	Yes	<200,000	High
Thermoplastic starch	Cereal starches	Poor	Yes	<200,000	Low
Bio-based PE	Sugarcane and cereal sugars	Good	No	300,000	Moderate
PBAT	Petrochemical	Good	Yes	<200,000	High
PBS	Cereal sugars and petrochemical	Good	Yes	<200,000	High
Polyethylene furandioate	Cereal sugars and petrochemical	Good	No	Limited	High
PLA	Cereal sugars	Good	Low	1,000,000	Moderate
Biobased PA	Castor oil	Good	No	400,000	Moderate
Conventional polyolefins	Petrochemical	Good	No	Very high	Low

compostable; they can address a very wide range of product applications; and their production costs are expected to be moderate as production volumes grow.

As the bioplastic sector works to establish itself as a significant segment in the overall plastics industry, each type of biopolymer has challenges to overcome. First and foremost, in all cases, is the inertia of the existing industry where supply systems, processing technology, and product specifications are firmly ensconced at low cost and high production efficiencies. The massive $700 billion plastics market and the associated infrastructure represent formidable barriers to any proposed changes to the industry.

New bio-based and/or biodegradable polymers face the upfront financial challenges. Investments in the billion-dollar range are required over what is usually more than a decade to develop a single polymer production technology and adapt that biopolymer into final plastic products that meet the detailed and often complex requirements of different applications in existing markets. Securing these investments is further complicated by the fact that biopolymer production processes are inherently more costly than petrochemical processes, simply based on feedstock costs, limited scale, and availability. Nonetheless, biopolymer market feasibility is gaining increased support from the overwhelming global market demand for plastics and other materials, as they can be sustainably produced, minimize GHG generation, reduce impacts on climate change, and do not persist in the environment.

Governments and industry are responding to strong consumer and public pressures to develop alternatives to petroleum-based plastic products. Support is justified by more comprehensive and inclusive estimates of product costs, where direct feedstock and production costs of biopolymers are balanced by the social and direct expenses related to waste management and environmental consequences of petrochemicals.

EUBP provides projections of global biopolymer production capacities based on the impacts of known investments in each area. This is discussed further in Chapters 21 and 22. The data indicates that in the near term, the largest growth will come from new bio-based and biodegradable polymers such as PHAs, but capacity in all bio-based, durable, nonbiodegradable polymers will continue to build industry share.

Another critical observation is that while significant jurisdictions in North America (Canada, California, and other states) Europe and China have made commitments to eliminate the use of conventional single use plastics, the production of 7.5 million tons of all bioplastics by 2026 would represent a market share of less than 2%. If one were to take a position that a significant impact from any individual industry practice isn't realized until a new technology has a 10% share, the bioplastics industry will require an exponential growth rate in financial infusion, to see environmental impacts within the next 10–15 years.

24.3 PHA technology: market drivers and opportunities

In terms of the potential for bioplastics, high level assessments of some key applications are provided in Table 24.2 (note Chapter 21 has further details on markets). These estimates are general, and are made using an overall aggregation of reports, press releases, and government policy announcements.

Multipurpose consumer goods packaging is a dominant use of plastics. It remains to be seen how all manufacturers of consumer goods adjust their practices and demand for plastics to meet sustainability goals – goals that may be diverse and unique to each company, industry wide, or within geopolitical jurisdictions. Nonetheless, there are many significant market sectors being targeted by biopolymers, including PHAs.

Agricultural plastic, adhesives, sealants, personal care, and general packaging are all important markets. Plastics for agricultural and food applications are a natural fit for biodegradable polymers, based on their short-term use, probability of mixing with food waste, and the existing trend toward composting or biodigestion of waste organics. Similarly, used adhesives, sealants, and personal care products are very difficult, if not impossible, to aggregate and recycle and can be considered as total loss applications. This means that they can contribute to the build-up of microplastics in the environment and therefore are prime targets for use of PHAs. General packaging is also important because of the volume potential where alternative waste management practices would reduce the volume of materials going to landfills.

Table 24.2: Plastics market size estimates.

Market	Total addressable	Serviceable by PHAs
Worldwide $800 B		,000 tons
Agricultural and food [4]	$70 b/20 m ton	20,000
Adhesives/sealants [5]	$75 b/12 m ton	6,000
Personal care [6]	$8 b/1 m ton	250
Packaging [7]	$400 b/150 m ton	120,000

Various market reports continue to paint a positive picture of the real growth in the bioplastics market, both due to announced investments and production plant constructions, as well as growth trends in relevant markets. As already noted, bio-based and/or biodegradable plastic production capacity is projected to grow from 2.1 million tons in 2020 to 7.5 million tons by 2026. Other reports indicate that the agricultural mulch film market is projected to grow by 6.5% annually, which represents $200 million in new annual sales. Another interesting example is the 3D printing plastics market, which has been projected to grow by around 20% annually from $600 million in 2021 to nearly $6 billion by 2030. And finally, the global flexible packaging market is estimated at $300 billion. It is a sector where major brand owners are increasingly making commitments and have the resources to drive growth in sustainable plastic packaging technology and reduce the negative environmental impact of their product packaging.

Where does PHA technology fit into this picture? At a high level, PHAs are of interest for two primary reasons:

1. They can be naturally produced using renewable carbon sources, which of course means that PHAs do not contribute to introducing new petroleum-based carbon into the Earth's environment.
2. PHAs will naturally degrade under a range of conditions with no residues, which means that even if a PHA material ends up in the environment, the long-term impact can be managed and is minimal.

Many review papers have been written over the years that discuss the history and fundamentals of PHAs and several recent ones are referred to here that provide excellent assessments of their market opportunities. Koller and Mukherjee's 2021 published paper, "A New Wave of Industrialization of PHA Biopolyesters" [8] includes the history of the development of PHAs as well as a summary of the commercial activities of a range of companies producing PHAs and related products.

Dublin-based O'Conner and Padamati published an extensive PHA technology review in 2022, entitled "A Review on Biological Synthesis of the Biodegradable Polymers Polyhydroxyalkanoates and the Development of Multiple Applications" [9]. The paper provides a thorough description, with more than 300 literature references of the biological processes, performance properties, PHA-based biomaterials, processing for applications, and end-of-life management.

Wageningen Food & Biobased Research and Invest-NL published a report entitled "Paving the way for bio-based materials – a roadmap for the market introduction of PHAs," which discusses all aspects of what is required to expand the PHA industry. The report focuses on a number of commercial or near-commercial grades of PHA, and includes the potential carbon sources, microorganism for each type of PHA, production processes and unit operations, typical properties, market shares, production capacities, as well as an assessment of where each type of PHA technology can have a best fit as replacement of petrochemical plastics.

The BioSinn report entitled "Products for which biodegradation makes sense" [10], published in 2021 by the Nova Institute (funded by Germany's Ministry of Food and Agriculture), includes fact sheets for 25 applications where biodegradation is a viable end-of-life option. The applications listed focus on specific products where the probability of loss to the environment is very high and even 100% by design. This includes products such as organic waste bags, geotextiles, tea bags, fertilizer coatings, brush bristles, and marine ropes. The report estimated the market potential to be in the 1 million tons per year range in the EU alone.

From this perspective, PHA technology can readily address many market needs by providing bio-based plastic materials and packaging products designed to function in industrial settings, and then either biodegrade in-place or be easily composted or biodigested, as required by the user. These four recent reports highlight how PHAs are ideally placed to provide sustainable plastic materials.

When discussing the management of waste plastics, recycling continues to be promoted by the existing industry. Unfortunately, recycling has been shown to be largely ineffective, as more than 85% of the world's plastics end up in landfills and the value of the collected plastics is very low due to contamination and poor performance. The objective of plastics recycling is to blend the collected materials with virgin plastics, but the recycled content reduces the final product quality. This limit on recycled content therefore perpetuates the use of petroleum plastics, and we have ended up with a build-up of the collected and unused dirty plastics.

Nonetheless, recycling efforts will continue despite the shortcomings of the current systems, due to the effort already invested, along with ongoing support from existing producers with very large resources, like ExxonMobil, Sinopec, and Dow.

PHA and some other bioplastic technologies have significant potential to help manage plastic waste over the long term in terms of reducing greenhouse gas emissions (GHGs), enabling alternatives to landfills, and minimizing the impact of waste plastic contamination of the environment (land and water):

- Companies and jurisdictions have ongoing efforts in place to improve the sustainability of their operations and waste management systems and to stabilize the global carbon cycle.
- GHG reductions targets are global, and bio-based plastic use can contribute by growing the circular carbon economy.
- There are a range of plastic applications where biodegradation in place provides operational benefits for industries – agriculture and forestry in particular – where used plastics would not need to be recovered and disposed.
- Plastic products, especially consumer packaging (including adhesives and sealants), often become part of mixed waste streams, and the contaminated plastics cannot be efficiently recycled. On the other hand, biodegradable plastics can be included in mixed organic waste streams and sent for composting instead of filling landfills.

The level of interest in PHAs from the market has never been higher and the primary reason is their robust sustainability profile. Being bio-based is favorable due to the increase in regulatory pressure from various government policies related to reducing the generation of waste plastics and GHGs. Compared to combustion of hydrocarbons as fuel, plastics are not a major contributor of GHG emissions, but a 300-million-ton global plastics market represents a major portion of the carbon economy. PHA's strong environmental story includes waste management practices of bioplastics, which can contribute to reduced GHG emissions. As landfills reach capacity (and new landfills are increasingly difficult to commission) and recycling continues to be problematic, more and more waste plastics are likely to be incinerated for energy recovery, but this practice would generate CO_2 emissions in the 1 trillion ton range every year. In the case of PHAs, incineration results in the return of CO_2 to the environment, which is the essence of the circular carbon economy [11].

Globally, the current PHA manufacturing capacity globally is approaching 100,000 tons per year, whereas just 2% of the plastic market is in range of 6 million tons. It will take decades to build PHA production capacity to a point where its replacement of petroleum plastics begins to have an environmental impact based on renewability, biodegradability, and compostability. Nonetheless, during this time, PHAs and other bioplastics will begin to demonstrate that they are viable and sustainable options to the petroleum plastic manufacturing and waste management systems that exist today.

24.3.1 What are PHAs?

PHAs are polyhydroxyalkanoates, which are polymerized linear hydroxy acids. There are many versions of PHAs under development. As the industry seeks out PHAs with optimal properties, there are technically an infinite variety of hydroxy acids that can be polymerized. Even the well-known polylactic acid (PLA) and polycaprolactone (PCL) belong to the PHA group of materials.

Author's note: Defining polymers in chemical terms is not straightforward. It is not simple to name a polymer and automatically communicate the function and usefulness. For example, the simplest (chemically speaking) polymer is polyethylene, and it comes in three main forms. The three forms – high density, low density, and linear low density – have different functions based on their elasticity, strength, and other properties. Similarly, polyesters generally mean polyethylene terephthalate (PET), but there are other polyesters. PBT (polybutylene terephthalate) is used for entirely different applications due to its toughness. The same goes for nylons or polyamides – there are numerous types, and each has its preferred uses that range from clothing to industrial materials, to military-grade protective equipment.

So, what are we really talking about when we discuss PHAs? In the industry today, PHAs refer to linear polyesters that are:

1. derived from hydroxy fatty acids as monomers, and
2. naturally produced using microorganisms that consume natural carbon sources (sugars, CO_2, methanol, or methane) and/or nutrients (vegetable oils, starches, waste plant material, etc.) and biologically convert the carbon sources into polymers of hydroxy acids.

The most common variations of PHA are shown in the Figure 24.2, which shows the length of the polymer side chains that can be varied through the biological design of the production process. If the side chain is one or two carbons in length, the PHA is referred to as a short-chain length PHA, and if between three and eight carbons, it is known as medium-chain length PHA.

It is important to note the origin of this conventional nomenclature is the fatty acid industry. The chain length indicated for PHAs is not the length of the polymer chain. Rather, it is the length of the acid monomer. Historically, natural acids with 6–12 carbons are known as the medium-chain length fatty acids. Less than 6 are the short-chain length acids, and more than 12 are the long-chain length fatty acids.

The most common PHA is polyhydroxybutyrate, which is an scl-PHA, normally known as PHB (Figure 24.3). A quick comparison of the mechanical properties of PHB and mcl-PHAs highlights the range in properties that are achievable from PHAs overall. PHB is rigid and strong, while mcl-PHAs are softer and flexible. This range of properties forms the basis of the bioplastic product development work discussed later in this chapter, where commercial ventures have worked to take advantage of the structural options available in order to produce high-quality and functional PHA plastics.

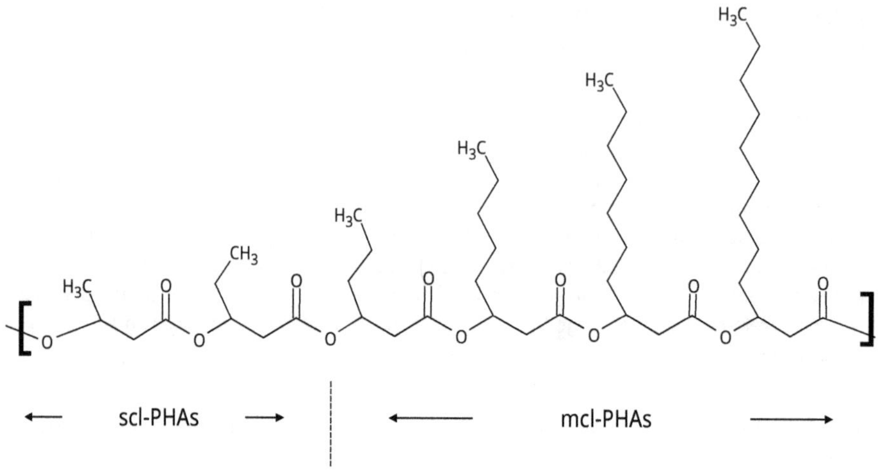

Figure 24.2: PHA polymer structure.

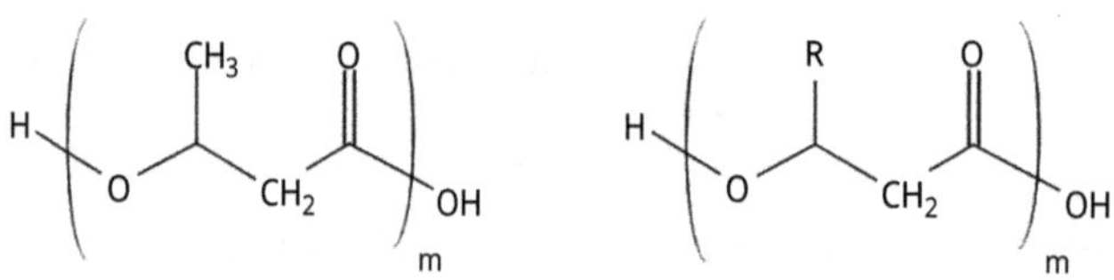

R = C_3H_7, C_5H_{11}, C_7H_{15}, C_9H_{19}

PHB

Poly**h**ydroxy**b**utyrate
- Rigid, crystalline, high strength, and toughness
- Bio-based and readily biodegradable, and compostable
- Often compounded with other biopolymers for conventional plastics applications

mcl-PHA

Medium-**c**hain **l**ength **p**oly**h**ydroxy**a**lkanoates
- Flexible, reduced crystallinity, softer
- Bio-based, readily biodegradable, and compostable
- Reduced melting temperature
- Targeted for coatings, adhesives, and sealants

Figure 24.3: Structure of PHB compared to mcl-PHAs.

24.3.2 Origins of PHA technology

Maurice Lemoigne was a French researcher, who is largely credited with the first discovery of biodegradable plastic, based on his work with *Bacillus megaterium*. He was able to isolate poly-3-hydroxybuturate (PHB), but this work was relatively unknown until the 1950s when researchers in the United States and Great Britain once again found PHB, and there was interest in its biodegradability. The oil crisis in the 1970s generated a high level of attention on bio-based technologies as alternatives to petroleum-based products. However, when crude oil prices dropped back to more acceptable levels, PHA interest faded. It was in the 1990s when a slowly growing but sustained interest in environmentally friendly products caught hold – first within the research community, but gradually moving into industrial operations. Polylactic acid (PLA) was the first bio-based polyester to achieve industrial-scale commercial status, and now in 2022, we are indeed seeing more and more ventures moving forward with major investments and commercial plays in PHA.

Dr Guo Qiang (George) Chen [12] is a well-known leader in developing PHA technology since the early 1990s. His research has focused on microbial materials, PHA metabolic engineering, and PHA biomaterials application. Dr Chen's affiliation with research conducted in Graz, Austria, during the early 1990s forms the foundation for PHB production using c. necator. After joining Tsinghua University in 1994, he has been actively promoting microbe-derived material industries related to the production of PHB, PHBH, mcl-PHAs, and P3HB-4HB in China. Dr Chen has produced 200 international publications and his technologies and patents have been utilized by numerous companies that have succeeded in scaling up production of PHAs.

Another leading commercial development of PHA production technology is a convoluted tale. ICI, once a globally leading and broad-based chemical company, was an original developer of microbial technology for the production of PHBV. A key aspect of this material was the advantage of PHBV being of mixed monomers (hydroxybutyric acid and hydroxyvaleric acid). During ICI's demerging activities in the early 1990s, this PHA technology and production of BIOPOL® was sold to Zeneca, and then a few PHB polyhydroxybutyrate, years later to Monsanto. In 1999, this business was sold to Metabolix, a company that was a global leader in PHA innovation during the 2000s. Full-scale production of PHAs from corn sugars went ahead via a joint venture with ADM, but this effort was abandoned and in 2016, the PHA production and related technology was sold to CJ Bio. In 2022, CJ Bio announced its plan to go ahead with commercial production of P3HB4HB.

Another major PHA technology that is being commercialized is known as Nodax®, an scl/mcl PHA, originally developed by Dr Isao Noda and patented by Proctor & Gamble. This PHA combines some of the properties of PHB with the elastomeric properties of mcl-PHAs by using vegetable oils as feedstock, resulting in a range of lengths of the side chains. The Nodax technology and tradename were owned and marketed by Danimer Scientific, and now by Teknor Apex. A related PHA, known as PHBH (copolymer of hydroxybutyric acid and hydroxy hexanoic acid) is produced by Kaneka.

A leading PHA production process developer specializing in mcl-PHA is Dr Bruce Ramsay, founder of PolyFerm Canada and formerly of Queen's University in Canada. Dr Ramsay has more than 80 publications related to microbiological systems, the majority of which have made a significant contribution to mcl-PHA technology in the areas of metabolic pathways, feeding strategies, and PHA performance.

Koller and Mukherjee provide a list of other biopolymer companies that have been focused on process development for the production of basic PHA polymers from organic waste. While these processes are attractive from a waste utilization standpoint, feedstock aggregation is often a challenge and the overall yields tend to be low, making it difficult to meet the demands of many industry sectors where biopolymers are required in high volume and at low cost in order to make an impact.

24.3.3 Common types of PHAs

Developing a single new polymer material nearly always requires additional design and engineering to produce a new plastic product with the desired functional properties that the market can use. This is a major challenge for new bio-based PHA plastics, as this emerging industry is competing against 100 years of petroleum plastics technology development, evolution, and advancement.

24.3.3.1 PHB

Poly-3-hyroxybutyrate is the most widely available and most well-known of the PHAs. There are numerous companies that are manufacturing small commercial quantities, including TianAn, Tianjin, Biomer, and others. For high volume and efficient production, the preferred bacterium is *Cupriavidus necator*, with glucose as the feedstock.

There are numerous other routes to PHB production, including using methanotrophs, such as Newlight Technologies using methane or TerraVerdae Bioworks, which demonstrated the potential scaleup of Canada's National Research Council technology for PHB production using *Methylorubrum extorquens* and renewable methanol as feedstock. As noted previously, there are numerous other early-stage PHB production ventures that are producing PHB from organic waste streams for pulp mills, waste cooking oils, and food waste, although these processes often require recombinant microbes that are undesirable systems for PHA applications in personal care markets.

Pure PHB is highly crystalline, which on the positive side, results in high strength and impact resistance, and good barrier properties, but the material is generally not sufficiently elastic to be used on its own in high-volume flexible packaging applications. Therefore, it is normal to compound PHB with other biopolymers, biomaterials, or additives to enhance the processing and mechanical property requirements, while maintaining the biodegradability and compostability of the resulting compounded resin.

24.3.3.2 PHBV

PHBV is the term used for polyhydroxybutyrate-valerate copolymer, although it is important to note that this is a random copolymer, according to the microbe's metabolic pathways. PHBV is commonly accepted to have a desirable blend of properties, where the highly crystalline properties of pure PHB are appropriately compromised by the valerate content, resulting in a lower melting temperature and higher elasticity. This makes the material more amenable for flexible packaging applications, as long as the hydroxyvalerate content is sufficient (more than 15%). Currently, PHBV with elevated valerate content, is not widely available in the industry.

24.3.3.3 mcl-PHA

Pure mcl-PHAs have seen limited industrial activity to date even though their preparation has been known for decades. Using natural organic acids as feedstock and *Pseudomonas putida* as the microbe, Polyferm Canada has developed the production process and licensed the technology to TerraVerdae Bioworks, which is developing applications such as coatings and adhesives, where lower melting points, low crystallinity, and high tackiness provide a range of performance advantages.

24.3.3.4 PHBH/PHBHHx

Similar to PHBV, random, naturally produced copolymers of hydroxybutyrate and hydroxyhexanoate or other mcl hydroxyalkanoates have a long history, which includes the Nodax technology. The microbe is recombinant *C. necatur*, and the feedstock is mainly vegetable oil. These copolymers have a range of desirable properties in areas of elasticity, processability, and melting temperature. Because of this, PHBHs have received a high level of commercial interest and indeed, Danimer, RWD, and Kaneka are three companies that have moved forward to achieve industrial-scale production capacity.

24.3.3.5 P3HB-4HB

P3HB-4HB is another copolymer that exhibits high elasticity. This PHA was developed by Metabolix and is now being commercialized by CJ Bio. It is produced by recombinant *E. coli* using sugar feedstock, along with butanediol, which introduces the 4-HB component (see Figure 24.4). The random copolymer produced has sufficient disruption of the inherent crystalline nature of P3HB, such that P3HB-4HB is amorphous, which is an important advantage for adhesive and coating applications.

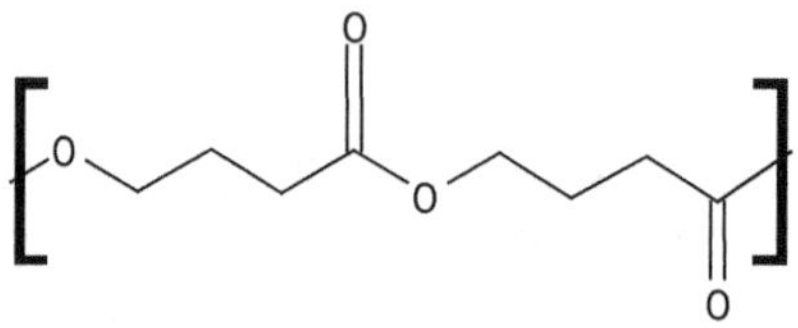

Figure 24.4: Poly-4-hydroxybutyrate (P4HB).

24.3.3.6 Compounded PHA resins

Developing a single new polymer material nearly always requires additional design and engineering to produce a new plastic products with the desired functional properties that the market can use. This is a major challenge for new bio-based PHA plastics, as this emerging industry is competing against 100 years of petroleum plastics technology development, evolution, and advancement. Therefore, in addition to PHA polymers, current offerings to industry often involve PHA compounded resins, particularly with rigid PHAs like PHB and PHBV. Compounding PHB with other bioplastics such as PLA, PBAT, PBS, and thermoplastic starches provide opportunity to balance the properties of the final compounded resins, with PHB/PHBV providing improved biodegradability, bio-based content, and barrier properties, depending on the targets and expertise of the compounder.

24.4 Commercialization

In terms of manufacturing feasibility, there are two main points that support the use of PHAs. First, PHAs are produced using industrial fermentation technology – similar general processes and equipment that produce foods, ingredients, biobased fuels and chemicals. When establishing market costs for PHA, high production process cost is a key aspect facing the industry. Governments and industry, nonetheless, have generally come to accept that optimized fermentation and related bioprocessing using natural plant-based materials as the carbon source represent the future for non-petroleum-based fuels, chemicals, and polymers.

The commercial prognosis for PHAs continues to improve, with favorable tail winds from myriad fronts, such as improved technologies, large scale fermentation processes, and investments in major companies, as shown in Table 24.3.

PHB production capacity is the most advanced within the field of PHAs, led by plants in China, such as BluPHA, TianAn, and Tianjin. One of the largest producers of other PHA resins has been Danimer Scientific, with a reported 65 million pounds (30,000 tons) annual capacity from two US plants. In 2025, Danimer filed for bankruptcy protection and was subsequently acquired by Teknor Apex, a global leader in plastic material science. Other PHA ventures are moving forward quickly.

Table 24.3: Snapshot of global bioplastic producers.

Company	Primary technology	Commercialization stage
Natureworks	PLA	Full commercial
BASF	PBAT	Full commercial
Kaneka	PHA	Small commercial
Teknor Apex	PHA	Small commercial
Mitsubishi	PBS	Small commercial
Novamont	TPS	Small commercial
Avantium	PEF	Pilot
Total Corbion	PLA	Full commercial
TianAn Biopolymer	PHB	Small commercial
RWDC Industries	PHA	Pilot
Tianjin Guoyun Biomaterials	PHB	Small commercial
CJ Bio	P3HB4HB	Small commercial
Newlight Technologies	PHB	Small commercial
TerraVerdae Bioworks	PHB, mcl-PHA	Pilot

RWDC Industries announced a funding round of $95 million in 2021. According to the press release, RWDC intends to expand its production of PHA in Athens Georgia to 50,000 tons per year, as well as establish a production facility in Singapore. In 2023, RWDC announced a partnership with Lummus Technology, a process technology company for the expansion of RWDC's Solon® PHA manufacturing, using waste cooking oils. Groundbreaking for the new production facility is expected in 2025.

CJ Bio commissioned its 5,000 ton per year production plant in Indonesia during 2022 and had planned an expansion to 65,000 tons per year in 2025. CJ Bio had also anticipated sufficient demand to 300,000 tons per year by 2030 of both its amorphous P3HBP4HB and scl-PHA.

Kaneka completed construction of its 5,000 ton per year PHBH plant in Japan in 2020, and in 2024, the capacity was expanded to 20,000 tons. Kaneka had projected demand to support up to 300,000 tons of production per year by 2030.

There are many other PHA production technologies that are in the process of scaling up production as they gain support for commercialization. Many are working from the standpoint of carbon waste utilization using carbon sources such as methane, food waste, and pulp mill streams but conventional fermentation systems are also expanding. Examples include Nafigate, Mango Materials, Genecis Bioindustries, Bosk Bioproducts, TerraVerdae Bioworks, and Full Cycle Bioplastics.

The commercial activities of PHA received important support from GO!PHA [13], a PHA industry organization cofounded by Anindya Mukherjee, Jan Ravenstijn, and Rick Passenier – it has been working to educate the market on how PHAs can help solve waste plastic problems. This effort is supported by both PHA technology developers and final product manufacturers. GO!PHA published an industry review of the PHA world in 2022, "Mimicking Nature The PHA Industry Landscape Latest Trends and 28 producer profiles," which detailed the state of the industry in areas of applications and general growth of the industry [14]. Among other activities, GO!PHA has published numerous articles that discuss how PHAs fit into the demand for bio-based and biodegradable plastics. The main objectives are to align the PHA industry with government priorities for clean technology policy, reduce the perceived risks with the financial sector, and to demonstrate to major global corporations and brand owners that PHAs will help them meet ESG goals (environmental social governance).

The work is based on a recognition that while PHAs are polymers and plastics, companies that have been processing polyolefins for decades need to recognize and understand that adjustments will be required in terms of logistics, operations, and material cost as they look to implement PHA-based products. During transitions to new product technologies, some level of short-term impact on operational efficiencies and revenues must be expected; GO!PHA's message is that it can be managed successfully.

Overall, PHAs and PHA-based resins are designed such that they can be processed using equipment that is currently in use by polymer and plastic processors. Operating conditions may require adjustment and optimization, but the plastic product supply industry is not expected to require major investments into new equipment in order to transition toward PHAs. This is a critical message from GO!PHA supporters to the incumbent plastics industry. As a result, there are currently numerous longer-term strategic moves being played out in the industry, as exemplified by large global brand owners, particularly in the food and beverage sector.

24.5 Summary and conclusions

The future for PHAs can be summarized by a SWOT (strength-weakness-opportunity-threats) analysis.

Strengths

- Bio-based and inherently sustainable: What does this mean in practical terms? If we consider that 50% of the worlds demand for plastics were to be met by PHAs, this will require nearly 200 million tons of polymers, which at a high level, would likely require around 400 million tons of starch, cellulose, and natural oils to produce. If purpose-grown crops are used, and estimating an average production in the 5 ton per hectare range, these natural organic feedstocks might require in the neighborhood of 80 million hectares, which represents around 1.6% of the existing arable land globally. This level of arable land use for nonfood applications can be managed without a significant impact on food supply and should be achievable with overall beneficial results, based on the potential reduction of GHG emissions in the 100 million ton CO_2 eq. range per year.

- Non polluting at end of life: There is a plethora of data supporting the inherent microbial degradation of PHAs, which is a critical piece to reducing the impact of plastic waste in the environment. While the timelines for biodegradation will depend on the specific PHA products, PHAs are biodegradable within months and/or years time frames, whereas petroleum-based plastics degrade only over hundreds of years. The long-term impacts fully support a cleaner environment, particularly the oceans, through use of biodegradable PHAs.
- Offers a range of sustainable solutions: PHAs can be biodigested, composted, or left to degrade in the environment by design (i.e., mulch films, erosion control products, etc.), depending on the application and regional waste management systems. Even if PHAs end up in an incineration stream, the escaped CO_2 simply returns into the natural carbon cycle and does not represent the additional petroleum-derived CO_2 into the Earth's environment, since the PHA is produced from CO_2 taken from the environment via photosynthesis.
- Production feasibility has been demonstrated: The recent advancements of production capacity are just the first steps to demonstrate feasibility. As this activity continues to gain momentum, efficiencies and streamlining will only further support the feasibility of PHAs.
- Brand owners with major existing market presence support product development: The combined financial impact of major global corporations supporting sustainable plastics like PHA is more than sufficient to facilitate market demand and product roll-out from perspective of sustainable commercial operations.
- Solid financial backing: As noted above, with major global corporations and governments in support, the investments to grow the capacity of a new PHA plastics industry are more than sufficient. Even if a few of the global chemical companies (think of Dow, ExxonMobil, BASF, and Sabic) were to take on the challenge, by each investing $10 billion over 10 years, this would be sufficient to build production capacity to the 10 million ton per year range.
- First-generation PHA polymer and resins are currently being introduced to the market: Refer to previous section regarding CJ Bio, TianAn, RWDC, BluPHA, Kaneka, and others.
- Able to address diverse product applications: The existing variations of PHA polymer chemistry is just the first generation of the potential sophistication of PHAs. As the industry grows, new "by-design" PHAs will be developed that provide tailored performance. While there are likely to be certain specific applications where highly engineered chemical plastics are required, PHAs' potential has already been demonstrated to include a broad spectrum of applications, particularly from high-volume market sectors such as flexible and rigid packaging, coatings, films, and adhesives.

Weaknesses

- Constrained by limited chemical manipulations: There is the potential that chemical modifications to achieve performance requirements in areas of processability, barrier properties, durability, etc. could compromise biodegradability, which is one of the inherent strengths of PHAs.
- Native properties can be challenging to improve: Biological systems (fermentations) are inherently more difficult to control and manipulate than chemical processes since the microorganism needs to survive and thrive during production – altering temperature, pH, and chemical reagents are usually not viable options.
- Higher costs, compared to current market pricing for most plastics: This is strictly from a product cost perspective, as the world currently operates based on the low-cost structure of the petrochemical industry.
- Challenging to achieve the high-performance capabilities of existing fossil-based polymers, such as polyurethanes, polyamides, and engineered polyesters.

Opportunities

- Provide solution to organizations and jurisdictions to meet sustainability goals.
- Enable a wide range of industries to implement sustainable (low carbon footprint) products, including very high-volume (and waste-generating) applications, such as single-use plastic packaging.
- Supply technology with direct benefits to GHG reductions, healthier soil, and cleaner water
- Reduce municipality waste management costs related to transportation to landfills.

Threats

- Idealistic policies that do not favor any manufactured polymers, even if they are bio-based. Some markets and policy makers view all manufactured plastics as potentially harmful to the environment and are pushing agendas supporting natural materials only, such as those based on cellulose and starch.
- Poorer economies not willing to make long-term investments into new plastics technology. Existing plastics industry sways government policy away from bioplastics, based on a strong lobbying position to maintain existing industries.
- Plastics industry does not have experience with PHAs and is resistant to change.

Overall, positive attributes and impact of PHAs are very encouraging and bode well for their commercialization. This positive picture is further boosted by increased awareness by end-use customers for bioplastics, especially when mixed waste streams are inevitable, as with food packaging. A strong push to address global warming, greenhouse gas emissions, microplastic accumulation, and the circular economy are very favorable for the strong growth of PHAs. However, PHA industry growth will not take place without business potential and the related financial support. Recent manufacturing capacity increases and investment announcements provide the real evidence that PHAs can achieve a significant position within the emerging bioeconomy.

References

[1] https://iucn.org/resources/issues-brief/plastic-pollution#:~:text=Over%20460%20million%20metric%20tons,and%20contributes%20to%20climate%20change.
[2] McKinsey & Company. ESG Momentum: Seven reported traits that set organizations apart, 2023.
[3] https://www.european-bioplastics.org/news/publications/
[4] https://www.towardspackaging.com/insights/plastic-food-packaging-market-sizing;https://www.marketresearch.com/DataM-Intelligence-4Market-Research-LLP-v4207/Global-Agricultural-Plastics-34371077/.
[5] https://www.specialchem.com/adhesives/news/adhesives-sealants-market–reach-usd-79-bn-000228415
[6] https://dataintelo.com/report/global-polymer-ingredients-for-personal-care-market
[7] Estimated from online market summaries from Fortune Business Insights, Grand View Research, Coherent Market Insights and others
[8] Koller M, Mukherjee A. A new wave of industrialization of PHA biopolyesters. BioEng 2022;9:74.
[9] Dalton B, Bhagabati P, De Micco J, Padamati RB, O'Connor K. A review on biological synthesis of the biodegradable polymers polyhydroxyalkanoates and the development of multiple applications. Catalysts 2022;12:319.
[10] https://renewable-carbon.eu/publications/product/biosinn-products-for-which-biodegradationmakes-sense-pdf/.
[11] https://www.european-bioplastics.org/bioplastics/environment/
[12] Currently Professor, School of Life Sciences, Tsinghua University. Profile: https://life.tsinghua.edu.cn/lifeen/info/1034/1088.htm
[13] www.gopha.org
[14] Ravenstijn J, GO!PHA, https://renewable-carbon.eu/publications/product/mimicking-nature-the-pha-industry-landscape-latest-trends-and-28-producer-profiles/

Ashok Adur

Chapter 25
Bio-based plastic compounds for the emerging bioeconomy

25.1 Introduction

Concerns about global warming and other environmental challenges among the global population have increased both interest and demand for a clean and pollution-free environment with reduction of carbon dioxide resulting from fossil fuel usage. Based on this, a lot of attention has been focused on research and development to replace petro-based (which includes oil and gas) commodity plastics by biodegradable materials sourced from biological and renewable resources. Biopolymers, which are polymers derived from natural sources either chemically from a biological material or biosynthesized by living organisms, are claimed to be suitable alternatives for addressing these issues [1]. Polymers (plastics and rubber, thermoplastic or thermosets) whether they are bio-based or not, are rarely used without the addition of additives like plasticizers, lubricants, antioxidants and stabilizers, fillers, colorants, and reinforcements, in order to meet various customer requirements all the way downstream in the value chain. Some bio-based plasticizers are covered in Chapter 18. Biocompounds are compounds that contain one or more biopolymers. These additives are added during a compounding process step to overcome these limitations and to increase the range of performance characteristics required in the target end use applications. This chapter covers several emerging and commercial bio-compounds, while the addition of reinforcements to form plastics bio-composites is covered in the next chapter.

25.2 Processing to produce bio-based compounds

Similar to other polymers with high viscosity, compounding of biopolymers needs sufficiently high shear and temperature to bring down viscosity to a range where adequate mixing is easier, whether the ingredients to be mixed in are in liquid, powder, or pellet form. The process to add additives to withstand thermoplastic processing, to make the polymers processable downstream and to modify their properties into useful articles is called compounding. When one polymer is blended with another poly-

Ashok Adur, Everest International Consulting, LLC., Jacksonville, FL 32256, USA,
e-mail: aadur@outlook.com

 | https://doi.org/10.1515/9783112212141-025

mer to obtain a combination of properties, it is called a polymer blend, especially if it forms multiple phases. This is typical either if the viscosities of the two polymers at the mixing temperature are vastly different and mixing is inadequate or if the two polymers are chemically incompatible. Incompatible polymers can be made compatible by adding small amounts of compatibilizer as additives during compounding [2]. If the powders are not fed at a constant rate that can keep up with the production rate, special accessory equipment may be needed to ensure that it is uniformly introduced into the compounding equipment.

The major difference with bio-based ingredients in compounding is the need to ensure adequate moisture control because most bio-based materials are very hygroscopic and absorb water. This can cause foaming, voids, and other issues. In a few end products, this sort of foaming and voids in the end product could be useful. Nevertheless, additional equipment to control moisture is needed in most applications. A few petro-based polymers like nylons and some polyesters also have this issue, but their moisture absorption levels are much less compared to many biopolymers.

Examples of equipment to process compounds are no different than what is used for non-bio-based compounds, and includes Banbury mixers, Farrell continuous mixers (FCMs), various kinds of extruders that are single, twin, multiple screw, or oscillating types, heated roll mills, etc. Each type of equipment provides different shear profiles during mixing as well as differences in temperature buildup and residence times. It is important to choose the right equipment and processing conditions for each set of compounding, whether the product has biopolymers or biofibers, some of which are more sensitive to hot spots during processing, so screw optimization may be critical. In some cases, antioxidant additives are needed just to prevent degradation during compounding, especially true for polymers that are susceptible to oxidation. Typical compounding equipment converts the mixture of the polymer with the additives and reinforcement into either pellets (also called granules) or directly into sheet or profiles. The pellets are then molded by an injection-molding or a blow-molding process into films, sheets, parts, bottles, or other articles. The shape and size of the extruded sheets can be controlled to the required thickness and width and cut in the same process equipment into the desired lengths for the intended application or further thermoformed or machined into other shapes in a secondary step. During extrusion, most fibers orient in one direction, typically the direction of the extrudate flow. This leads to different properties in the extrusion direction versus the perpendicular or transverse direction. This can happen a little during the compounding stage and a lot more at the subsequent extrusion stage or the injection-molding stage. In injection molding, the design of the gate determines the direction of the fiber orientation. In order to obtain the same properties in both directions, various downstream methods can be used or in some cases, rather than feeding the fibers into the compounding equipment, a different set of processes are employed. Sometimes, depending upon the end application, the fiber is first converted into a cloth or a nonwoven fabric, which is then processed using a compression molding or similar process to convert the cloth or fabric into a two-dimensional sheet of the fiber-reinforced composite. Compression- or lamination molding and a variety of cast processes are more commonly used with thermoset type of plastics. Figure 25.1 shows processing steps for bio-fiber plastics compounds and composites.

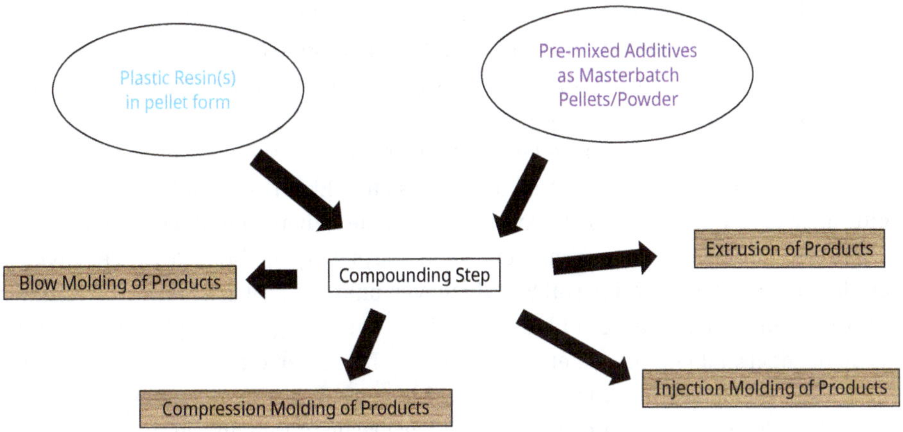

Figure 25.1: Processing steps for bio-plastics compounds.

25.3 Unique challenges

In Chapter 21, and in other chapters in this book, it has been made clear that biopolymers have advantages of outstanding biodegradability and if some of them are not degradable or compostable, at least they are renewable, and in many cases can be produced into useful form starting from plant sources or even from waste products like industrial or consumer waste. However, they have characteristics and properties, as listed below, which pose some challenges compared to conventional petro counterparts [3]:

- Some of them are not degradable or compostable, but renewable.
- Poor mechanical properties
- Higher sensitivity to moisture and water absorption
- Poor resistance to fatigue life
- Little chemical resistance
- Low long-term durability
- Issues related to production in large scale and supply chain
- Limited processing capability

Most of these disadvantages of biopolymers can be overcome at least partially by compounding them with other polymers to form compounds and/or with reinforcing fillers or in some cases nanofillers. Nanofillers are fillers with at least one of its dimensions as low as 100 µm to qualify to be in the nanometer range [3]. Bio-composites are discussed in the next chapter.

Some bio-based polymers have thermal degradation points lower than those of petro-based polymers. In a few cases, their degradation points are lower than or very close to their melting point. So, one needs to increase the processing window of the biopolymer by plasticizing it with a plasticizer. An extreme example of this is: starch with water or other polar additives such as glycerin in it can be extruded while dry starch just degrades and turns brown.

Another challenge that has to be overcome is that, as we have seen in Chapter 21, just because a bioplastic is sourced from a bio-source like trees, plants, algae, or agricultural or food waste, it is not always biodegradable. When a nonbiodegradable plastic is added to biopolymer to form blends and alloys, only the biodegradable polymer can degrade or can be compostable. Whichever phase is larger in terms of volume, not weight, only one phase can biodegrade if the nonbiodegradable plastic is encased inside the degradable biopolymer matrix and for biodegradability to occur. So, at the end of the bio-blend's life and it is time to biodegrade it; the outer biopolymer degrades first, leaving bits and pieces of the other plastic to remain. On the other hand, if the nonbiodegradable polymer is the matrix as the larger volume phase, biodegradation slows down and what results over a longer period is a film or bottle with lots of holes in it.

In order to obtain better properties, it may be necessary in biopolymer blends to either chemically bond or "couple" the other polymer to the bioplastic matrix. This is carried out by modifying the chemistry of the biopolymer or the other polymer by some sort of radiation treatment (ozone or corona or even high-energy radiation like beta or gamma radiation) or chemically modified. Such modification is called "functionalization." The choice of such functionalization is based on the specific chemistry of each of the major components of the blend. Such functionalization is used with conventional petro-based material multiphase systems too. Once compatibilized, the morphology of these blends changes to the extent that on a microscopic scale the two phases intermingle enough that the domains of the minor phase get dispersed better and get smaller and, in a few cases, can be seen in photomicrographs to diffuse into each other's phases [2].

When such chemical bonding of one phase to the other occurs, it results in significant improvement to performance opening up a wider range of applications, but makes biodegradability even more challenging. Yet, from a net carbon dioxide balance perspective such bio-compounds would be beneficial.

So, rather than concentrating on biodegradability or compostability of bioplastics, this chapter will cover compounds made from biopolymers to turn them into usable bioplastics and bio-based compounds to include combinations of polymers where at least one of the combining materials is bio-based.

25.4 Bio-based compounds

25.4.1 Bio-based compounds: introduction

In this section, compounds of biopolymer are discussed to cover biopolymers that have been compounded to modify their properties using various bio-based and non-bio-based additives like impact modifiers, nucleating agents, and other polymers. There are again numerous academic publications and several patents granted. Some applications have commercialized, and the use of such blends is growing fast globally.

However, it is important to remember that the biodegradability of articles or parts produced from such blends of starch with nonbiodegradable plastics like polyethylene and polypropylene depends on several factors such as the part's thickness, morphology of the blend in the part and of course on the specific microorganism that will consume the part. The morphology of the blend in the part depends on factors such as the type of starch and particle sizes of both the starch and the other plastic (which is the dispersion of one phase in the other), the ratio of the starch to the non-biodegradable plastic, how well the plastic is dispersed in the starch phase using compatibilizers (if any), and the equipment as well as the process conditions of the compounding process. The mechanical properties also depend on the morphology along with how well the starch is bonded to the other plastic. If the volume ratio of the plastic is higher than that of the starch, after biodegradation, you will be left with the film or other article with holes where the starch was. If this ratio is not low enough you could be left with tiny pieces of plastic that are turned into microplastics. While there is a big hue and cry from some in the environmental community about microplastics, there is no data yet to show any harm from microplastics residue in humans or animals from such microplastics. However, appropriate formulation and compounding can surmount most issues. The optical properties of these blends differ from transparent individual plastic material. While starch blends are processed similar to mineral-filled plastics except for lower temperature and shear, almost all starch blends are opaque and not transparent, because it is difficult to match the refractive index of starch with other polymers.

25.4.2 Starch compounds

Author's note: see Chapter 22 from Praj Industries for additional information.

Biodegradable compounds are disclosed in a patent where compositions comprising a biodegradable resin and starch, wherein fat or oil treated starch or its gelatinized product is utilized as the starch and thereby its production cost and biodegradability is improved without losing its mechanical properties. This patent also covers biodegradable compositions comprising a gelatinized product of fat or oil-treated starch and a biodegradable resin and the process for the production of the biodegradable composition. The process involves heating and kneading fat or oil-treated starch and a biodegradable resin in the presence of water or water and a plasticizer [4].

Biome Bioplastics Limited in the U.K. has developed a thermoplastic, biodegradable polymer blend having a high surface energy and good adhesion without the need for pretreatment, with starch in the range 10–60% by weight of the polymer blend, at least one further carrier polymers in the range 40–90% by weight of the polymer blend, and one or more process aids in the range 0–0.5% by weight of the polymer blend [5]. The same company also has developed copolymers of pyridinedicarboxylic acid (PDCA) and of furandicarboxylic acid (FDCA), both monomers derived from lignin-sourced furanic sugars. These copolymers are also biodegradable and useful in treating microbial infection.

Since starch-based plastic blends can be incorporated with various biopolymers and/or petroleum-based polymers to create unique bio-compounds, they can be used in a whole host of applications. These bio-compounds can then be extruded, blow-molded, or injection-molded using traditional processing equipment. Because starch is relatively cheap, starch-based plastics compounds are more cost competitive than alternative bioplastics. Based on formulating them appropriately, such compounds can be compounded to meet a wide range of physical properties that other bioplastics lack, such as tensile strength, low specific gravity, and heat resistance. To reduce cost further and increase sustainability, these starch composites can also be formulated by using recycled plastics. When starch is incorporated to replace traditional petro-based polymer resins, such starch-based compounds can reduce the carbon footprint of where they can replace petroleum-based polymers with natural ones. If the other polymer in such compounds is biodegradable or compostable as well, the compound also becomes highly degradable, that is, such compounds will also be very degradable and can be used as a compostable plastics material.

An analysis by Green Dot Plastics of a market research report stated that "starch blends are expected to account for the largest share in the market" from 2015 to 2020. In 2017, out of the 2.05 million tons of bioplastics produced worldwide, 18.8% were starch-based blended compounds. In fact, European Bioplastics stated that "Bioplastics are used in an increasing number of markets, from products, consumer electronics, automotive, agriculture/horticulture and toys to textiles and a number of other segments." Additional opportunities will come from a continued capacity for growth and consumer demand for environmentally friendly products over and above the efforts to reduce pollution and to diminish our reliance on petroleum-based products in the future [6].

Rising demand results in growing supply and the Chinese government's increasing support for bioplastics as a way to improve the standard of living in the country's rural regions is just one example of this. Major indigenous producers of starch plastics and other types of bioplastics are rapidly changing marketplace dynamics. Municipal governments in central, eastern, and southern China are planning for strong growth in all sectors of this industry and developing countries such as Thailand and Brazil are also emerging as bioplastic-producing powerhouses [7].

25.4.3 Polycaprolactone (PCL) compounds

Polycaprolactone (PCL) has a low degradation temperature. Hence for it to be used, its processing temperature profile must be lowered to allow it to be manufactured into films, sheets, or articles molded with injection- and blow-molding. PCL is also much more expensive than starch.

Hence compounding PCL by adding lower cost biopolymers like starch as well as compounding in mineral fibers, animal fibers, and natural fibers as reinforcements and talc or calcium carbonate as low-cost fillers make PCL compounds more cost-effective. However, such mineral fillers are left behind after the biodegradable plastic parts of such compounds get degraded into carbon dioxide and water. Here are a few selected examples from literature:

- PCL is often used in biodegradable polymer blends such as PCL with wheat-based thermoplastic starch [8] and PCL/PLA blends [9]. Since PCL and PLA are both synthesized by ring-opening polymerization, PCL/PLA multiblock copolymers can readily be produced, to give biodegradable thermoplastic elastomers [10].
- There are many examples in academic and patent literature to produce PCL-based biodegradable blended plastics compounds such as starch/PCL, soy/PCL, chitin/PCL, lignin/PCL, cellulose acetate/PCL, PCL with cellulose derivatives like hydroxyethylcellulose acetate, cellulose acetate and ethylcellulose, etc. It has been shown that the mechanical and thermal properties of blends of maleated PCL and starch polymer alloys, summarized as "PCL-g-MAH/starch" were studied to show significant improvement over similar blends without the maleation. What caused this improvement? The greater compatibility of PCL-g-MAH with starch resulted in much better dispersion and homogeneity of starch in the PCL-g-MAH compound (as determined by microscopic examination of the compound's morphology due to the formation of an ester carbonyl group) [11]. While Wu's title of his paper calls it composite, from a polymer chemistry perspective, this is a polymer alloy compound and not really a composite.
- In Sweden, bags had been produced from PCL and PCL-starch bags, but they degraded very fast even before reaching the customers [12]. However, a proprietary blend of PCL and starch has been successfully used by Yukong Company for making trash bags in Korea but they use some polyethylene with a coupling agent in their patented formulation to obtain good enough mechanical properties along with biodegradability [13].
- Two synthesis routes to produce composites of graphene and PCL were investigated and the properties of the resulting composites compared [14]. In these composites, the graphene was also covalently bonded to the carrier biopolymer. Since

in vitro cytotoxicity testing of the compounds showed good biocompatibility along with good conductivity, products molded from these promising materials for applications as conducting substrates for the electrically stimulated growth of cells hence claimed to be suitable for use in tissue engineering. This type of compounds has been used for several different applications in the biomedicine and medical device markets [15]. *(Editor's note: See Chapter 28 for greater discussion of bioproducts in the medical market.)*

25.4.4 Polylactic acid (PLA) and polyhydroxy alkanoate (PHA) compounds

Both biopolymers, PLA and PHA, have been covered in Chapters 22, 23, and 24. PLA has several limitations when used by itself such as slow biodegradation rate, hydrophobicity, and low impact toughness. Also, PLA by itself or as a blend with starch has a major limitation of relatively low glass transition temperature (typically between 111 and 145 °F depending upon the ratio of its dextro isomer to its levo isomer). This makes it fairly unsuitable for temperature applications. For example, when you put a PLA straw in a cup with boiling water, the straw will instantly deform, making it unusable. Therefore, PLA straws are only suitable for cold drinks, while hot drinks require the use of crystallized PLA as one of the options to extrude high-temperature straws. Parts produced from plastics with low heat distortion temperature can soften and deform in conditions like a hot car in the summer. In addition, PLA has higher permeability than other plastics, making both moisture and oxygen pass more easily through it than other plastics. This will lead to faster food spoilage, so PLA is not recommended for long-term food storage applications [16].

PLA and PHA are now well-known biodegradable biopolymers. Films produced from PLA offer sufficient gloss and transparency, comparable to polystyrene (PS) films. They also have a decent level of heat resistance and mechanical strength and also enable the creation of biodegradable plastics with superior antibacterial and antifungal properties. Once such biopolymers are compounded, many of their limitations can be overcome. Blending such biopolymers with additives and other polymers are some of the common successful ways to improve properties to overcome such limitations and produce novel biopolymer blend compounds for target applications. For example, a Japanese patent disclosed compositions comprising polylactic acid (PLA) and 0.05 to 5 parts by weight of silicone oil per 100 parts by weight PLA, with significantly improved mold-release properties, elongation at break, impact strength, and tensile strength [16].

Another patent [17] discloses degradable materials, which include a nontoxic hydrolytically degradable polymer like PLA and a nontoxic modifier, wherein the modifier is compatible with the polymer and the modifier is nonvolatile and nonfugitive. The modifier is selected from the group consisting of lactic acid, lactide, oligomers of lactic acid, oligomers of lactide, and mixtures thereof, wherein said oligomers of lactic acid and oligomers of lactide include more specifically plasticizers and modifiers like acetyl tributyl citrate, lactide, glycolide, lactic acid esters, dimethyl adipate, diethyl adipate, caprolactone, acetyl triethyl citrate, bis-2-ethyl hexyl sebacate, and bis-2-ethyl hexyl adipate, dibutyl sebacate, and triethyl citrate. Also disclosed are processes for forming the various degradable materials, which include films, molded products, laminates, foams, powders, nonwovens, adhesives, and coatings. The disclosed materials and processes are particularly useful for the production of commercial and consumer products in high volumes, which are suitable for recycling after use or which are discarded into the environment in large volumes.

Another option to overcome the significant disadvantage of insufficient heat resistance for PLA for several applications is by compounding in inorganic fillers or polymers with higher HDT. The addition of talc at about 1% improves heat resistance as measured by HDT by increasing the nucleation rate of its crystallization and its crystallinity [18]. However, it makes products molded or extruded from such PLA compounds too stiff and brittle. This lack of toughness of talc-nucleated PLA can be solved by adding compatible elastomers and/or plasticizers. If the elastomer is not compatible, functionalized products or coupling agents can be used to obtain compatibility. For example, the brittleness and low HDT can be overcome by the use of styrene-maleic anhydride (SMA) copolymer in combination with an epoxy functional styrene-acrylate oligomeric chain extender. Such a composition also often exceeds a threshold of 65 °C in heat deflection temperature as shown in this patent by Avakian et al. [19].

Another significant disadvantage of the use of PLA is the lack of flame retardance, which has been overcome in two different inventions. The first patent discloses the use of polyphosphonate-co-carbonate in combination with an impact modifier, a drip suppressant, and optionally, an epoxy functional styrene-acrylate oligomeric chain extender. The compound achieves a UL 94 rating of V-0 or V-1 and a Notched Izod value of more than about 876 Newtons/m. The compound also exceeds a threshold of 100 °C in HDT in one patent [20]. In the other patent the same researchers at PolyOne, now called “Avient” overcame the lack of flame retardance of PLA by using specific combinations of either polycarbonate or polyphosphonate-co-carbonate in combination with non-halogenated flame retardants of polyphosphazene or phosphate ester such as resorcinol bis(diphenyl phosphate) or metal hypophosphite, a drip suppressant, and optionally an inorganic synergist of either zinc borate or talc or both and optionally other ingredients. The compound achieves a UL 94 rating of V-0 or V-1 at 1.6 mm [21].

At the SPE ANTEC held in Charlotte, N.C., USA on May 14–16, 2022, Prof. Raymond Pearson presented a paper entitled “Enhanced Compost Rate and Simultaneous Toughening by Multifunctional Additives in Polylactic Acid.” In this work, a specially designed microencapsulated particle additive was added to simultaneously improve ductility and biodegradation behavior of PLA. This unique approach involves the use of encapsulation technology to create degradation-promoting additives while limiting any breakdown of the matrix during melt extrusion and service life. These microencapsulated particles were dispersed within the PLA matrix by an extrusion process to 3D printer filament. The results showed that elongation at break was improved compared to PLA with only a slight loss of yield strength. The degradation rate in the composting step of these compounds accelerated and decoupled from environmental conditions by embedding the degradant material into the PLA matrix itself. This rate was helped by encapsulation technology that isolated and protected the degradant. This technology of dual use microencapsulated particles has the potential to broaden the uses of PLA by reducing total composting time by over 65% [22].

At the same conference, another paper was presented, which showed that adding small quantities of bio-based and fully biodegradable amorphous polyhydroxyalkanoate (a-PHA) copolymer increased the impact strength of PLA without compromising the compostability and bio-based carbon content of the final product. These blends exhibited good toughness and clarity when they were injection-molded and extruded into sheet and into blown film. What is significant is that this technology by fine-tuning the blend composition, to achieve the targeted levels of toughness increases the flexural modulus reduction needed for the application. Such impact-modified PLA were highlighted for applications such as film, thermoforming, and injection-molding using these new amorphous PHA commercial grades [23].

A South Korean patent disclosed PLA resin compositions and molded product comprising the same with excellent environmental friendliness and biodegradability, has excellent heat resistance characteristics due to high heat deformation temperature (HDT), and may be injection-molded within a commercially reasonable cycle time due to high crystallinity and crystallization rate, is suitable for the preparation of a molded product. They accomplished this by compounding a PLA resin, which includes a hard segment containing a PLA repeating unit and a soft segment containing a polyurethane polyol repeating unit in which polyether-based polyol repeating units are linearly linked via a urethane bond, a poly-D-lactic acid (PDLA) resin as a nucleating agent, and talc [24].

A series of highly flexible biodegradable poly(ethylene oxide)/poly(l-lactic acid) (PEO/PLA) (PELA) multiblock poly(ether-ester-urethane) polymers, which perform as thermoplastic elastomers were synthesized and characterized [25].

PHA is a class of biopolymers, which chemically are classified as polyesters and can be rigid thermoplastic or elastomeric with melting points ranging from 40 to 180 °C. Polyhydroxy butyrate (PHB) and other PHAs with shorter aliphatic chains are much more rigid and not ductile while the ones with longer chains are more elastomeric. Ideally, these two types can be blended to get to the desired set of mechanical properties. Alternatively, the rigid PHA can be made ductile as disclosed in a patent, by compounding it with macrocyclic poly(alkylene dicarboxylate) oligomer to produce a ductile biopolymer compound compared to the original PHA without the macrocyclic poly(alkylene dicarboxylate) oligomer. Optionally, the compound also can include additional polymers and functional additives to modify physical properties of the compound to suit the required properties of the molded or extruded plastic articles, which can be produced from the compound [26].

A report (specifically, see table 3 in [27]) has summarized the progress of the large number of applications of PLA, PHA, PHB, and their blends with other biopolymers in various fields like medical devices and other biomedical as well as textiles, automotive spare and parts, packaging films, electrical appliances, commodity containers, floor mats, and mobile phone housings from 1999 to 2013 [27].

An American patent assigned to Arkema Inc. disclosed the use of compatibilizers like their Lotader™ and Lotryl™ copolymers, which are functionalized olefin-methacrylic and olefin-acrylic copolymers, to compatibilize blends of polyolefins and a biodegradable polymer such as PLA and PHB. Such alloys had improved processibility, melt strength, and melt elasticity [28].

A large number of compounding companies like Sukano, Americhem, Tosaf, AF-Color, Avient, and others have developed grades of masterbatches (also called concentrates) using PLA, PHA, PHB, and other biopolymers as carrier resins for use downstream in extrusion and molding applications for a wide range of applications. Types of such masterbatches include color pigments, flame retardants, antioxidant packages, and other additives.

More than 100 eco-friendly companies around the world, large and small, are manufacturing various cutlery and other articles using PLA compounds, some of which also contain minerals like talc and calcium carbonate, which are added to provide additional stiffness, whiter color, and lower cost. A few of biodegradable plastic applications compiled in an article in 2018 [29] are listed below:

- Papelyco sells Plantable™ paper plates produced from bio-based plastic compounds that are not just biodegradable but also incorporate important minerals in the plate to provide their customers' new sapling with the nutrients it needs to grow. So, customers can use their plate, place it in the ground, and a plant will literally grow out of it. This is a cool green invention and sustainable in several ways.
- A company called Ecoware in New Zealand has developed their biodegradable dishware and bioplastic packaging that has an 80% lower carbon footprint than similar products produced from petro-based plastics. They claim to be a carbon neutral company, because they offset all of their energy output by investing in clean technology.
- The US-made line of compostable cutlery and packaging from a company called WorldCentric includes a very broad range of biodegradable bags, dishware, and cutlery for household and corporate use. Their products produced from bio-based plastics compounds are sourced from perennial plant fibers with long roots that help store carbon underground and are certified-compostable disposables. In addition, 25% of their profits are donated to worthy causes.

- Compostable packaging is being brought to consumers by Be Green Packaging by partnering with companies such as Google, Whole Foods, Gilette™ from Procter and Gamble, Samsung, Virgin America, and others.
- A line of biodegradable flexible plastics like bags with zippers and garbage bags is produced by a company called Tipa. Such bioplastics are heavily preferred and used by consumers and has the potential to grow significantly to a more sustainable future.

25.4.5 Other compounds

Compounds comprising by weight, 70–96% of at least homopolyamides or copolyamides with 11 and/or 12 carbon atoms, 4–10% of a plasticizer, and 0–25% of an NBR or H-NBR elastomer, such that the sum of the amount of plasticizer and the elastomer is between 4% and 30% were patented [30]. Such compounds are used in pipes in the operation of offshore oil and gas fields due to their aging resistance as well as in pipes in motor vehicles due to their temperature resistance, especially in the presence of automotive fluids for applications in the engine compartment. In some cases, such compounds are used as a layer in multilayer extruded pipes.

A patent disclosing bio-resin compositions containing cellulose acetate and a biodegradable secondary polymer in a ratio of at least 10:1 by weight of the total weight of the composition was issued in 2015 [31]. The secondary polymer is an aliphatic polyester such as polybutylene succinate or an aliphatic co-polyester such as polybutylene succinate adipate, or an aliphatic/aromatic polyester such as polybutylene adipate terephthalate. The secondary polymer may be present in an amount up to 10% by weight, more preferably up to 6% by weight, and still more preferably in the range 2% to 3% by weight of the total weight of the composition, where the combined amount of secondary polymer and cellulose acetate may be up to 85% of the total weight of the composition.

Blends of thermoplastic starch with other polymers were reviewed by authors from Malaysia for applications mostly in cutlery and food packaging [32].

Diglycidyl ethers of isosorbide (DGEI) derived from starch via glucose converted into sorbitol and then isosorbide have been used to produce epoxy resins, which have a V-shaped diol with rigid fused tetrahydrofuran rings. These compounded resins have high thermal stability up to 270 °C. The advantage of DGEI is that it is renewable and a nontoxic alternative to bisphenol A. Applications include electronics and protective coatings and sustainable composites and adhesives [33, 34].

Citrate esters like acetyl tributyl citrate (ATBC), triethyl citrate (TEC), etc. as well as succinate esters of succinic acid with various alcohols work well as plasticizers because they have excellent miscibility with PVC and PLA. These plasticizers are nontoxic, biodegradable, and REACH-compliant. In such plasticized compounds with these plastics, they offer low migration and leaching, making them suitable for sensitive applications, such as medical devices (e.g., blood bags, tubing), food packaging, and children's toys. Another application is in plastisol formulations [35–38].

Castor oil is a naturally occurring polyol, which is covered in the previous chapter. Glycerol from biodiesel by product streams can be converted into a polyol. Both these bio-derived polyols have been used to produce flexible, flame-retardant, and impact-resistant polyurethane foams for insulation, automotive, and furniture applications [39, 40].

25.4.6 Specialty compatibilizer additives

A recent development since the first edition of this book is that a Mexican company called Addico, based in Toluca has developed and is marketing specialty compatibilizer additives, which can be used to compatibilize recycled, compostable polyester streams. Using AddiCo 9,000, 9,050, 3,910, or 9,080, they claim you can compatibilize and improve the viscosity of recycled polyester streams such as blends of PLA/PBAT; PLA/PHA, etc. AddiCo additives will chemically react with these polyesters and bring together chains of these polyesters together, creating a homogeneous material with improved viscosity [41]. These additives compatibilize the ester groups present in these polymers. However, this compatibilization technology cannot be used to blend even a small amount of post-consumer recycled stream of bio-based plastics like PLA, PHA, PBAT, etc. into a bottle-grade PET recycle stream because these polymers degrade at typical processing temperatures that such PET streams are recycled.

25.5 Limitations of bio-based plastic compounds and challenges for growth

Research into bio-based polymers has been active as evidenced by the large number of patents, technical publications, and new products. Globally environmental and sustainability consciousness have risen to a stage where consumers desire products. This in turn has encouraged efforts for developing materials based on bio-based compounds for different end-use applications.

Conversely, these bio-based plastics compounds are not direct substitutes for petro-based plastics compounds without any negative issues or problems. They have several drawbacks and limitations such as poor moisture resistance (hydrophilicity), supply logistic issues, low thermal stability, flammability, poor electrical properties, and highly anisotropic properties. They also have challenges related to various parts of the value chain including extraction, processing, surface modification, machining, manufacturing, and characterization [42]. Consequently, due to challenges related to the properties of these materials compared to synthetic petro-based polymers, such as their resistance to chemicals or weather, applicability and usage have not dramatically increased yet in many end-use markets compared to petro-based polymers as was originally forecasted. For demanding applications as well as for those where durability is essential, such drawbacks will preclude biopolymers and bioplastics as well as products produced from their compounds that cannot displace incumbent non-bio-based products.

As stated earlier, another major issue is that since these biopolymers are all found in nature, there is a lot of inconsistency from one batch grown at one location to another grown elsewhere. This occurs due to differences in their growth due to differences in soil, temperature, and rainfall, or water available at each location. Thus, consistency of mechanical performance is more difficult to maintain compared to non-bio-based polymers. Some of these issues can be mitigated by innovative compounding techniques we have discussed in this chapter as well as by improving quality control of incoming raw materials and outgoing quality assurance of products.

Besides the limitations that are discussed earlier in this chapter, there are challenges in bio-based plastic compounds and products produced from them in trying to make them compatible with existing industry practice, equipment, capital investment, etc. In some cases, some modifications in process conditions or reduction of production rate may be needed. Lowering of processing temperature, on one hand, can reduce energy consumption, while the reduction of production rate can lead to higher costs.

However, such bio-compounds may be 100% bio-sourced but suffer from the drawback that life cycle analysis results show they consume a lot more of two precious resources, namely energy and water, than petro-based composites and sometimes cost more, based on data from life cycle analysis [43, 44]! One prominent scientist-researcher who has reviewed over 400 scientific articles on the subject of plastics, bioplastics, and their degradation chemistry states in his book that "Every LCA study I have found clearly states that these biodegradable polymers are worse for the environment than standard plastics like PE and PP. So, they are not green at all, despite the marketing claims of the manufacturers" [45].

One more factor that needs to be considered for the long-term viability of bio-based polymers and compounds, as Michael et al. have pointed out in 2011 very eloquently, is the expectation that there will be feedstock competition as global demand for food and energy increases over time. Currently, a large number of renewable feedstocks used for manufacturing bio-based monomers and polymers often compete with requirements for food-based products [46]. With the current invasion of Ukraine by Russia and the resultant looming food shortage in many parts of the world, this factor is even more relevant. The expansion of first-generation bio-based fuel production especially could place unsustainable demands on biomass resources and is as much a threat to the sustainability of biochemical and biopolymer production as it is to food production [47].

Ultimately customers want value and to get there for bioplastics in general and bio-based plastic compounds in particular, these products need to demonstrate value to break into the marketplace. Rather than trying to replace existing products that are not bio-based, the use of bio-based products needs to be more discriminatory rather than seeing bioplastics as a straight swap with existing plastics by embracing circular economy and whole life cycle analysis in making decisions on materials, according to Prof. Mitchell [47].

25.6 Conclusions

While biopolymers and bioplastics compounds have several limitations, it is clear from this chapter that interest in them is very high, not just related to research and creative ways to overcome some of these limitations, but also to move forward with scaling up these innovations to commercial products. Issues related to moisture absorption and dimensional stability are handled by encasing the bio-layer in a plastic material that provides a moisture barrier to protect the layer. Another limitation of poor interfacial adhesion is overcome with the use of coupling agents and compatibilizers, using techniques covered in a couple of online courses and a blog [2, 48].

As more and more bio-based polymers and compounds are being adopted, applications will grow and the total volume of such bio-materials should see an attractive annual increase, estimated by several commercially available market reports to grow at CAGRs in the range of 7–12%. The growth in bio-based compounds is driven by their adoption in growing end markets like pharma, textiles, automotive, and packaging.

References

[1] Fakirov S, Bhattacharyya D. Handbook of engineering biopolymers – homopolymers, blends and composites. Hanser Publ 2007;1–1. ISBN: 1-68015-262-9, 1-56990-405-7.

[2] Adur A. Compatibilization of Polymer Alloys Blends and Composites, online course taught on specialchem.com, 2002–2004, available from SpecialChem, Paris, France.

[3] Díez-Pascual AM, Cinelli P. Synthesis and Applications of Biopolymer Composites, Basel, Switzerland: MDPI; 2019, ISBN 3039211323 & 9783039211326.

[4] Shitaohzono T, Muramatsu A, Hino J, Biodegradable Compositions, United States Patent 5,691,403, issued November 25, 1997 assigned to Nihon Shokuhin Kako Co., Ltd. (Japan).

[5] (a) Law, PW; Longdon, T; Perez, AD; Gomis, B; Gomis, M, United States Patent 9,205,963, "Biodegradable Material" issued December 8, 2015, assigned to Biome Bioplastics Ltd. (GB). (b) https://biomebioplastics.com/applications/

[6] https://www.greendotbioplastics.com/starch-based-plastics/]

[7] Global Bioplastics Industry Report 2021–2026 with Focus on Plastics Made from Renewable Resources Such as Biomass or Food Crops, Research and Markets, https://www.prnewswire.com/news-releases/global-bioplastics-industry-report-2021-2026-with-focus-on-plastics-made-from-renewable-resources-such-as-biomass-or-food-crops-301458684.html/

[8] Averous L, Moro L, Dole P, Fringant C. Properties of thermoplastic blends: Starch–polycaprolactone. Polymer May 2000;41(11):4157–4167. https://www.sciencedirect.com/science/article/abs/pii/S0032386199006369/.

[9] Przybysz-Romatowska M, Haponiuk J, Formela K. Poly(ε-Caprolactone)/Poly(Lactic Acid) Blends compatibilized by peroxide initiators: Comparison of two strategies. Polymers (Basel) 2020 Jan 16;12 (1):228. https://doi.org/10.3390/polym12010228.

[10] Cohn D, Salomon AH. Designing biodegradable multiblock PCL/PLA thermoplastic elastomers. Biomaterials May 2005;26(15):2297–2305.

[11] Wu C-S. Physical properties and biodegradability of Maleated-Polycaprolactone/Starch composite. Polym Degrad Stab April 2003;80(1):127–134.

[12] Flieger M, Kantorová M, Prell A, Rezanka T, Votruba J. Biodegradable plastics from renewable sources. Folia Microbiol (Praha) 2003;48(1):27–44. https://doi.org/10.1007/BF02931273/.

[13] Lee BH, Jung KS, Kim YW, Bang SG, Cho WY, Jo B, Cheon U, Ki N, United States Patent 5,861,461, Biodegradable plastic composition, method for preparing thereof and product prepared therefrom, assigned to Yukong Ltd., issued January 19, 1999.

[14] Sayyar SS, Murray E, Thompson BC, Gambhir S, Officer SL, Wallace GG. Covalently linked biocompatible graphene/polycaprolactone composites for tissue engineering, Carbon: February 2013. Vol. 52, pp. 296–304.

[15] Dhanasekaran NPD, Muthuvelu KS, Arumugasamy SK. Recent advancement in biomedical applications of Polycaprolactone and Polycaprolactone-based materials. Ref Modul Mater Sci Mater Eng January 2022;https://doi.org/10.1016/B978-0-12-820352-1.00217-0.

[16] Barrett A, Advantages and disadvantages of PLA, June 9, 2020, https://bioplasticsnews.com/2020/06/09/polylactic-acid-pla-dis-advantages/

[17] Tanifuji Y, Tokushige Y, United States Patent 5,691,398,, "Polylactic acid composition", issued November 25, 1997, assigned to Shin-Etsu Chemical Co., Ltd.

[18] Sinclair RG, Lipinsky ES, United States Patent 5,502,158, "Degradable polymer composition", issued March 26, 1996 assigned to Ecopol, LLC, Golden, CO, USA.

[19] Zhu S, Avakian RW, United States Patent 8,765,865, "Heat resistant polylactic acid compositions", issued July 1, 2014, assigned to PolyOne Corp.

[20] Zhu S, Avakian RW, United States Patent 9,062,201, "Flame retardant polylactic acid compounds", issued June 23, 2015, assigned to PolyOne Corp.

[21] Zhu S, Avakian RW, United States Patent 9,534,116, "Flame retardant polylactic acid compounds", issued January 3, 2017, assigned to PolyOne Corp.

[22] Pearson R, Enhanced compost rate and simultaneous toughening by multifunctional additives in Polylactic acid, paper presented at SPE's ANTEC 2022, Charlotte, NC, USA.

[23] Krishnaswamy R, Impact modification of PLA (Poly Lactic Acid) by blending small amounts of Amorphous PHA (Polyhydroxyalkanoate) copolymers, paper presented at SPE's ANTEC 2022, Charlotte, NC, USA.

[24] Kim M-Y, Kim T-Y, Hwang -J-J, Chung J-I. United States Patent 11,118,051, Poly(lactic acid) resin composition and molded product comprising same, assignee: SK Chemicals Co., Ltd; issued September 14 2021.

[25] Cohn D, Hotovely-Salomo A. Biodegradable multiblock PEO/PLA thermoplastic elastomers: Molecular design and properties. Polymer 10 March 2005;46(7):2068–2075.

[26] Avakian RW, United States Patent 7,968,657, Thermoplastic polyhydroxyalkanoate compounds, Assigned to PolyOne Corp. issued June 28, 2011.

[27] Babu RP, O'Connor K, Seeram R. Current progress on bio-based polymers and their future trends. Prog Biomater 2013 Mar 18;2(1):8. https://doi.org/10.1186/2194-0517-2-8/.

[28] Donnelly Z, United States Patent Application, 2012, 0035,323 1, Polyolefin/polylactic acid blends, assigned to Arkema Inc.

[29] 5 Bio-Degradable plastic companies for a greener future, Jan 19, 2018, https://www.goodnet.org/articles/5-biodegradable-plastic-companies-for-greener-future/

[30] Jacques B, Pees B, Werth M, Polyamide-based composition for flexible pipes containing oil or gas, United States Patent 6,913,043, July 5, 2005, assigned to Arkema.
[31] Longdon T, Law PW, Sillence K, Wetters DJ, Perez DA, United States Patent 9,062,186, Bioresins, assigned to Biome Bioplastics Limited (UK), issued June 23, 2015.
[32] Diyana Z, Jumaidin R, Selamat M, Ghazali I, Julmohammad N, Huda N, Ilyas R. Physical properties of thermoplastic starch derived from natural resources and its blends: A review. Polymers 2021;13 (1393):https://doi.org/10.3390/polym13091393.
[33] Belgacem MN, Gandini A, editors. Monomers, polymers and composites from renewable resources, Elsevier; 2008.
[34] Liu Y, Zhang Y, Wang R, Li J. Synthesis and characterization of isosorbide-based epoxy resins for sustainable coatings. ACS Sustain Chem Eng 8(4):1723–1731. doi: https://doi.org/10.1021/acssuschemeng.9b06543/.
[35] https://kanademy.com/citrate-plasticizers-as-a-replacement-of-dop/
[36] (a) Miaza, M; Benaniba, M.T.; Hamam, A; Messaoudi, M; Bouchoul, B; Plasticizing of Polylactic acid (PLA) biopolymer with citrate esters, paper presented at First Arab Conference on Mechanical Engineering (ARCME'17); Biskra, Algeria, December 10–11, 2017 (b) Miaza, M; Benaniba, M.T.; Hamam, A; Massardier-Nageotte, V, J. Polymer Engg., 36, no. 4, 2016, pp. 371–380. https://doi.org/10.1515/polyeng-2015-0140/
[37] Rahman M, Brazel CS. The plasticizer market: An assessment of traditional plasticizers and research trends to meet new challenges. Polym Rev 2004;44(4):427–467. https://doi.org/10.1081/MC-200027955.
[38] European Chemicals Agency (ECHA). (n.d.). Substance information: Citric acid esters and succinate derivatives. Retrieved from https://echa.europa.eu
[39] Ionescu M. Chemistry and technology of polyols for polyurethanes, Rapra Technology Ltd; 2005.
[40] Patel M, Crank M, Dornburg V, Hermann BG, Roes L, Hüsing B, van Overbeek L, et al. Techno-economic feasibility of large-scale production of bio-based polymers in Europe. Renew Sustain Energy Rev 2016;16(2):745–759. https://doi.org/10.1016/j.rser.2011.09.020.
[41] https://www.addico.com.mx/polyesters/
[42] Andrew JJ, Dhakal HN. Sustainable biobased composites for advanced applications: Recent trends and future opportunities – A critical review. Composites 2022;Part C 7:100220. doi: https://doi.org/10.1016/j.jcomc.2021.100220.
[43] Hottle TA, Bilec MM, Landis AE. Biopolymer production and end of life comparisons using life cycle assessment. Resour Conserv Recycl 2017;122:295–306. doi: https://doi.org/10.1016/j.jclepro.2020.121158/.
[44] Walker S, Rothman R. Life cycle assessment of bio-based and fossil-based plastic: A review. J Clean Prod 2020;261(07–10):121158. doi: 10.1016/j.jclepro.2020.121158/.
[45] De Armitt C. The plastics paradox: Facts for a brighter future. In Phantom Plastics LLC, Terrace Park, OH, USA: ISBN 978-0-9978499-6-7.
[46] Michael C, Dirk C, Harald K, Jan R, Joachim V. Policy paper on bio-based economy in the EU: Level playing field for bio-based chemistry and materials, 2011. https://renewable-carbon.eu/publications/product/nova-paper-1-on-bio-based-economy-level-playing-field-for-bio-based-chemistry-and-materials-%E2%88%92-full-version/.
[47] Mitchell G, Mahendra V, Pinheiro J, Sousa D, Abdulgahni S, Dias J, Faria P, Gaspar F, Mateus A; Challenges with Biopolymers presented at 8th World Congress on Biopolymers & Bioplastics June 28–29, 2018, Berlin, Germany, https://www.longdom.org/proceedings/challenges-with-biopolymers-19169.html/
[48] Bicerano J, A practical guide to polymeric compatibilizers for polymer blends, composites and laminates, SpecialChem, December 2005; http://polymerexpert.biz/blog/135-compatibilization-of-immiscible-polymers/

Ashok Adur

Chapter 26
Bio-based fiber plastic composites for the emerging bioeconomy

26.1 Introduction

Biocomposites are compounds that contain a biopolymer and/or a reinforcing biofiber using some sort of compounding equipment. Typically, reinforcements are added to plastics to improve their mechanical properties and thermal resistance. As is covered in the previous chapter on biocompounds, biocomposites would also need the use of various additives to ensure that the biopolymer does not degrade during the compounding step and can meet various end-use application and handling requirements. They may contain one or more bio-based major ingredients, meaning the matrix polymer, the reinforcement, or both could be bio-based.

These additives and the reinforcements are added during a compounding process step to overcome inherent limitations of biopolymers and biofibers in composites and to increase the range of performance characteristics required by customers in various applications. This is what drives the need and hence the practice of using additives and reinforcements using the right compounding equipment and process to achieve target specification requirements for each application.

This chapter covers various commonly used biofibers and several emerging and commercial biocomposites. So, this chapter will dive into what is common and what is different for biofibers versus non-bio-sourced fibers like glass or carbon fibers, as well as for biocomposites versus conventional petro-based plastic composites. Challenges in bio-based plastic composite products in trying to make them compatible with industry practice, equipment, capital investment, etc., and how specific biocomposites have been successful and what factors led to their success will also be discussed. Many chapters of a 2007 book [1] covered many combinations of such composites but there has been a lot of progress and developments in bio-based plastic composites since then.

Ashok Adur, Everest International Consulting, LLC., Jacksonville, FL 32256, USA, e-mail: aadur@outlook.com

 | https://doi.org/10.1515/9783112212141-026

26.2 Processing to produce bio-based plastic composites

Similar to other polymers with high viscosity, compounding of biocomposites needs sufficiently high shear and temperature to bring down the viscosity of the polymer matrix to a range where adequate mixing is easier, and additives and reinforcements to be mixed into the molten polymer should be in liquid, powder, pellet or fiber form. The process to add additives to make the polymers processible downstream and modify their properties into useful articles is called compounding. This includes adding reinforcements to polymers to produce plastic composites. When one or more reinforcements are incorporated in a polymer, it always forms two or more phases. It is more challenging then when two polymers are blended, because the reinforcement does not melt but similar to other bio-sourced polymers, can degrade during processing.

When reinforcements are added to polymers using compounding equipment, care must be taken to ensure that the reinforcing fiber is uniformly distributed into the polymer. This starts with ensuring adequate and uniform feeding of the fibers as well as other additives in liquid, pellet, or powder form into the compounding equipment. If the fibers are not fed at a constant rate that can keep up with the production rate, special accessory equipment may be needed to ensure that it is, whether the fibers are bio-based or not. The larger the aspect ratio of the fibers, the more difficult it is to feed them at a consistent rate. Also, if the fibers stick to one another, the difficulty of feeding increases.

Similar to biopolymers, almost all biofiber reinforcements are very hygroscopic and absorb water, and hence need adequate drying to minimize the moisture content, prior to the compounding process. The moisture absorption of biofibers has been discussed for each biofiber in the section on biofibers. Too much moisture can cause foaming, voids, and other issues. In a few end products, this sort of foaming and voids could be useful to reduce the specific gravity of the composite part, but this is more of an exception. Some of the moisture can be driven out through the hopper by a proper selection of the extruder screw configuration. Nevertheless, additional equipment is needed to control moisture to the right level needed for the application. Non-biofibers like glass fiber and carbon fiber do not have this issue at all, since they are not hygroscopic and do not absorb large quantities of water. Hence, the small amount of water in these other fibers does not need a pre-drying step like biofibers do, and gets converted to steam some through back-venting through the feed hopper and/or the die.

The equipment used to produce biocomposites is similar to that of non-bio-based composites and is not repeated here. However, the process parameters are usually a little different. Besides the additional equipment to ensure moisture control, temperature and shear control need to be moderated to ensure the fiber does not degrade, both from mechanical chopping, which can reduce the fiber aspect ratio too low to affect mechanical properties, and from thermal or oxidative degradation. Figure 26.1 shows the processing steps for bio-based plastic composites.

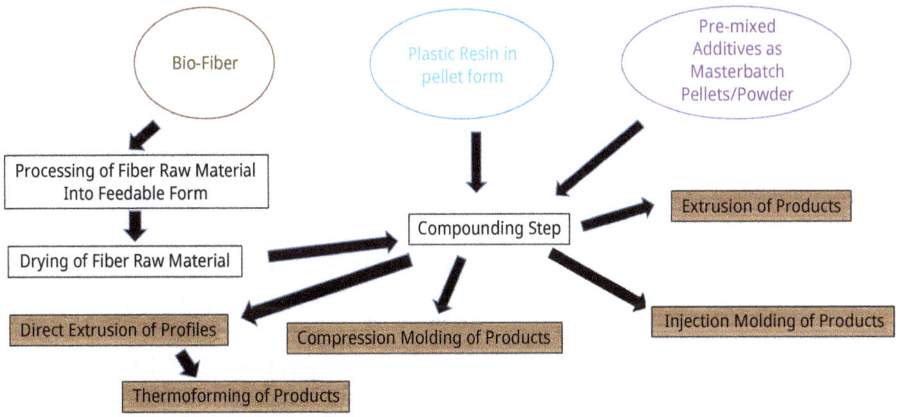

Figure 26.1: Processing steps for bio-based plastic composites.

In addition, the aspect ratio of the reinforcement needs to be considered. Usually, the reinforcing fiber has a higher length than its width or depth, like a rod, i.e., a higher aspect ratio compared to mineral reinforcements, which could be spherical, round, or flat in shape. A fiber is long only in one direction but in some cases, it is woven into a fabric or converted into nonwoven from the natural fiber itself, so it is larger in two dimensions than the other, and then used as reinforcement. During extrusion, most fibers orient in one direction, typically in the direction of the extrudate flow. This leads to different properties in the extrusion direction versus the perpendicular or transverse direction. This can happen a little during the compounding stage and a lot more at the subsequent extrusion stage or the injection molding stage. In injection molding, the design of the gate determines the direction of the fiber orientation. In order to obtain the same properties in both directions, various downstream methods can be used, or in some cases, rather than feeding the fibers into the compounding equipment, a different set of processes are employed. Sometimes, depending upon the end application, the fiber is first converted into a cloth or a nonwoven fabric, which is then processed using compression molding or similar process to convert the cloth or fabric into a two-dimensional sheet of the fiber-reinforced composite. Compression or lamination molding and a variety of cast processes are more commonly used with thermoset type of plastics. Two-dimensional reinforcements are also used for end-use applications with thermoplastic composites where mechanical strength in more than one direction is insufficient. In some cases, the biofiber is powdered into flour, especially for wood plastic composites (WPCs), which will be covered in detail later in this chapter.

In most cases, the reinforcement is chemically incompatible. Incompatible polymers can be made compatible by adding small amounts of compatibilizer as additives during compounding.

26.3 Unique challenges

Similar to biopolymers, biofibers have advantages due to their outstanding biodegradability and if some of them are not degradable or compostable, at least they are renewable, and in many cases, can be produced into useful form, starting from plant sources or even from waste products like industrial or consumer waste. However, they generally possess poor mechanical properties, higher sensitivity to moisture and water absorption, short fatigue life, low chemical resistance, poor long-term durability, and limited processing capability compared to petro-based plastics [2]. Most of these challenges and disadvantages of synthetic polymers and biopolymers can be overcome at least partially by compounding them with other polymers to form compounds and/or with reinforcing fillers or in some cases, nanofillers, to form composites [2]. A nanofiller is small enough that at least one of its dimensions is in the nanometer range or below 100 µm.

Similar to the inherent attractiveness of biopolymers, bio-fiber materials also have a positive environmental impact because they are low-cost, renewable, reusable, and biodegradable, and, hence, environmentally friendly and sustainable. A common disadvantage almost all of these fibers share with biopolymers, covered in the previous chapter, is that the fibers are also hydrophilic polar polymers themselves and absorb water and lose some of their mechanical strength, measured as tensile strength, bending strength, and especially rigidity, which is usually measured as flexural modulus, when wet.

Being biopolymers themselves, biofibers also have a fairly low degradation temperature, especially when compared to non-biofibers like glass fiber or carbon fiber. However, in general, biofibers are slower to degrade than biopolymers. Some of this is due to the presence of lignin in many biofibers as well as the inability to melt in most cases, and in some cases, their crystallinity.

Since these biofibers are all found in nature, there is a lot of inconsistency from one batch grown at one location to another grown elsewhere. This occurs due to differences in their growth due to differences in soil, temperature, and rainfall or water available at each location. Even biofibers from the same plant or tree and at the same location, exhibit variability. So, consistency of mechanical performance is more difficult to maintain, as compared to synthetic fibers, which are more precisely controlled through process control. On average, the standard deviation of tensile strength of dry bio-fibers is in the range of 5–10 times higher than that of glass fibers. When biofibers are wet, they lose dimensional stability and this range can widen up to ten times more. Most large-scale producers of biocomposites have learned to handle this variability by ensuring that before the fibers go into their compounding equipment, their moisture content is below a threshold that their equipment can handle, to still meet customer requirements, and also widen the specifications compared to those without bio-based content. Some manufacturers also carry out post-blending of pellets after the compounding stage to minimize variability.

Another unique issue about biofibers is that due to the higher lignin levels in many biofibers, the color of the composite gets darker and, more specifically, browner as a function of heat applied to it during its processing, both during the production of the biofiber and the processing steps of compounding it into biocomposites and further converting into the final product. The color darkens as a function of its heat history beyond a threshold temperature, which varies based on each type of biofiber. To prevent excessive darkening, monitoring and controlling the maximum temperature it sees during processing is even more critical. This is not a major issue with reinforcing fibers from non-bio-based sources.

Due to higher stiffness and rigidity that reinforcing fibers provide to biocomposites, these composites typically are more brittle and less ductile than the polymer matrix on its own. This is true for non-bio-based fiber-reinforced composites as well. In many cases, higher stiffness in composites, compared to unreinforced plastics, means lower impact properties as well. However, with adequate increase in interfacial adhesion between the reinforcing fiber and polymer matrix, this see-saw effect of rigidity versus impact can be overcome and a synergistic combination of properties can be achieved.

In order to obtain better properties and minimize the see-saw effect, it may be necessary to either chemically bond or "couple" the biofiber to the bioplastic matrix [3], or the surface of the reinforcing fiber is ozone or corona treated or chemically modified, which is called "functionalization" [4]. This type of surface modification and use of coupling agents are also commonly used with non-biofibers such as glass fiber for petro-based thermoplastics. The choice of such functionalization is based on the specific chemistry of the matrix and the reinforcement, similar to conventional petro-based materials.

For example, polypropylene does not form good bonds with glass fiber. So, a polypropylene grafted with maleic anhydride is used as a coupling agent during the compounding process to bond to the amine groups on the aminosilane on the glass fiber surface and, on the other side, the polypropylene chain on the coupling agent is able to co-crystallize with the polypropylene matrix to provide that critical interfacial adhesion between the glass fiber and the polypropylene, thus giving the composite significantly higher tensile strength, flexural modulus, impact strength, long-term fatigue performance, and improved heat distortion temperature [5].

Another example is where during the production process of producing these specialized glass fibers, the hot glass fibers are dipped into a sizing solution containing alternating copolymer of ethylene and maleic anhydride. This incorporates anhydride groups on the glass surface, which during the subsequent compounding process downstream binds to the amine end groups of nylon. This reaction results in covalent bonding between the glass fiber and the nylon matrix to improve interfacial adhesion between the reinforcing fiber and the polymer matrix and improve all mechanical properties, increased heat distortion temperature and resistance to harsh chemicals, like radiator fluid used in automotive parts in the engine compartment of automobiles [6–8]. Before this this type of technology – surface treatment of the glass fiber to produce glass fiber-reinforced nylon-6,6 composites – was developed, radiators used to leak frequently. In the last five years, such occurrences are very rare.

Similar technology can and has been used with biofibers to improve interfacial adhesion between the biofiber and the polymer matrix. When such chemical bonding of the biofiber to the biopolymer occurs, it results in significant improvement in performance, opening up a wider range of applications, but makes biodegradability even more challenging. Yet, from a net carbon dioxide balance perspective, such biocomposites would be beneficial. However, such biocomposites may be 100% bio-sourced but suffer from the drawback that life cycle analysis results show they consume a lot more of two precious resources, namely energy and water, than petro-based composites, and sometimes cost more!

So, rather than concentrating on biodegradability or compostability of bioplastics and their composites, this chapter will cover compounds made from biopolymers to turn them into useable bioplastics and bio-based composites, to include combinations of polymers reinforced with fibers, where at least one of the combining materials is bio-based.

26.4 Biofibers and other bio-reinforcements

26.4.1 Introduction to biofibers and other bio-reinforcements

A biofiber is a fiber sourced from biological matter and has either plant or animal origin. Sometimes, it is used naturally without further processing or more often processed further in some sort of a subsequent process. The main difference between a biofiber and a non-biofiber is the biological source for the former. Modern technology has blurred this distinction somewhat, because the production of fibers with a chemical structure that is identical to those found in nature through chemical and non-chemical modification is now possible [9]. Unlike chemically modified fibers, biofibers are abundant in nature in fibrous materials, especially those in the cellulosic family. Hence, some prefer to call biofibers “natural fibers.”

The promise of biofibers is the preservation of natural resources, with a focus on the utilization of renewable raw materials that, in many cases, are biodegradable at end of life. Besides the fact that they contribute to the consumption of carbon dioxide gas during their growth stage, this advantage of natural fibers is neutralized if at the end of their lives, they are burned, generating the same amount of the CO_2 emission. The low abrasive nature of biofibers reduces wear of processing equipment and hence they are greener, compared to more abrasive carbon or glass fibers [10]. In spite of these benefits, finding applications that consume sufficiently large quantities of these materials to enable price reduction and allow biofibers to compete economically in the market remains a continuing challenge [9]. The major exception to this is the area of WPC, where wood scrap converted to wood flour from the wood industry is used as a relatively cheap fiber-reinforcing material, but mostly with synthetic petro-based polymers, which again are relatively cheaper than biopolymers.

From an environmental and sustainability perspective, biofibers typically are believed to have a preferred life cycle that is typically measured from "cradle to grave" from an environmental impact perspective, which includes sourcing, manufacturing, use, and end of life, when compared to non-bio-based polymer fibers like glass or carbon fibers. However, there are many key performance indicators in each aspect of the life cycle. Two of the most often considered are greenhouse gas emissions in production and end of life. However, this type of analysis is somewhat flawed because it ignores other factors like resource usage, which is another important factor to consider, particularly when evaluating biofiber production. Biofibers typically require a huge amount of water for both irrigation and processing. In terms of environmental biodegradability, biofibers typically break down more quickly than petro-based polymers, which is a positive. But from a cost perspective, except for a few exceptions, biofibers have a disadvantage over petro-based polymer or glass fibers due to the issue of scale, covered earlier, as well as the cost of the water and energy needed for irrigation, cleaning, processing, and drying [11].

There are many naturally occurring biofibers that are used in fiber–polymer composites. Among them are cellulose fiber, hemp fiber, kenaf, flax, bamboo, jute, coir, rice husk, corn husk, wood fiber, leaf fibers (like those from sisal, abaca and banana leaves), pineapple fiber, crop residues, etc. While a lot of these fibers are cellulosic in nature, there are other chemical compounds present, depending on the source of these fibers. The chemical and mechanical structure of each type of fiber gives each of these, unique set of performance characteristics when they are used in composites.

Most of these biofibers can be used after cutting them into the desired fiber length, keeping in mind that inherently during compounding and further process stages, the fiber length and hence the aspect ratio gets trimmed, a feature that is common to fiber compounding, bio-based or not.

In order to use natural fiber reinforcements and achieve isotropy in two directions, the fibers need to be converted into a textile preform with many different structures to suit each end application, similar to manmade reinforcements. The most widely used structure is the nonwoven preform, which is usually comingled (blended) with 50 wt.% polypropylene so that it can be directly compression molded for making automotive parts. The polypropylene intertwined between the biofiber allows for good moldability in this process. Due to the random fiber orientation obtained in this process, high fiber packing and consequently high fiber volume fraction are achieved. Nonwoven reinforcements can be cost effective and at the same time, utilize hard fibers that otherwise would be difficult to spin. Also, they provide a porous structure with up to 30% void content, which is critical in achieving the levels of thermal insulation and vibration damping required by the transportation industry and specifically, by the automotive industry. A few of the important suppliers to tier 1 automotive in the United States of these type of composites were FlexForm and Ecotechnilin in Europe a couple of years ago [11]. Today, with the rapid adoption of bio-based polymers and biofibers of more diverse types of biofiber-reinforced composites by many different industries, there are several more in many countries around the globe.

These natural fibers also include fibers from various parts of plants such as leaf, seeds, and grass as well as branches, leaves, and trunks of trees as well as from products like wood scrap, waste paper and other bio-derived sources. Such biofibers are being used as alternative materials as fillers in the conventional petro-based polymer matrices and in bioplastics. Usually, fibers that are easily available locally, such as bamboo, rice straw, sugar palm fiber, kenaf, flax, hemp, roselle, and pineapple. Other natural fibers that are derived from waste products, forest clippings, and remnants after crop harvest is completed can also be used as a biosources for use in production of biocomposites as well. Maintaining consistency in the quality for such fibers is inherently difficult for routine production of composites because there is no guarantee that the fiber properties will be similar from one day to the next or even within a batch of fiber. This makes producing composites with the same properties daily or even hourly very difficult [12]. There are several good reviews of these cellulose-based natural fibers available in literature [10–17], which cover many different perspectives, from sourcing through converting to various polymer composites to a myriad of end applications.

Figure 26.2 shows a few photographs of the various biofibers that are used to produce plastic composites in both biopolymers and petro-based polymers.

Figure 26.2: Various biofibers (from [11]).

Table 26.1 shows the chemical composition of several common biofibers [11–15]. Cellulose is typically the largest component in biofibers, with varying amounts of hemicellulose, lignin, and other minor components like waxes, pectin as well as inorganic and water-soluble compounds. While cotton fiber is listed in these tables, it is rarely used because of its demand for clothing, hence higher cost, and in spite of a lot of research data available in literature.

Table 26.1: Composition of natural biofibers.

Fibers	Cellulose (wt.%)	Hemicellulose (wt.%)	Lignin (wt.%)	Waxes (wt.%)
Abaca	56–63	20–25	12–131	–
Bagasse	55.2	16.8	25.3	–
Bamboo	26–43	30	21–31	–
Banana	63–64	–	5–11	–
Coir	32–43	0.15–0.25	40–45	–
Corn husk	31–39	34–41	14-Feb	–
Cotton	85–90	5.7	–	0.6
Flax	71	18.6–20.6	2.2–20.6	1.5–1.7
Hemp	68–74	15–22.4	3.5–10	0.8
Jute	61–71.5	13.6–20.4	12–13	0.5
Kenaf	45–72	20.3–21.5	8–13	–
Pineapple	80.5	17.5	8.3	–
Ramie	68.6–76.2	13–16	0.6–0.7	0.3
Rice husk	36–40	0–4	28–32	–
Sisal	65–78	10–14	9.9–14	2

Table 26.2 shows the mechanical and physical properties of these natural fibers, respectively. The values of these properties can be used to calculate the estimated values of composites produced from them, with relevant assumptions or, even better, microscopic measurements of fiber lengths in the final part.

Table 26.2: Mechanical properties of natural biofibers.

Fibers	Tensile (MPa)	Young's modulus (GPa)	Elongation at break (%)	Specific gravity (g/mL)
Abaca	400	12	3.0–10.0	1.50
Bagasse	350	22	5.8	0.89
Bamboo	290	17	1.5–3.0	1.25
Banana	529–914	27–32	5.9	1.35
Coir	220	6	15–25	1.25
Corn husk	145–165	4–5	18–24	N/A
Cotton	400	11.8	3.0–10.0	1.51
Flax	800–1,500	60–80	1.2–1.6	1.40
Hemp	550–900	70	1.6	1.48
Jute	410–780	26.5	1.9	1.48

Table 26.2 (continued)

Fibers	Tensile (MPa)	Young's modulus (GPa)	Elongation at break (%)	Specific gravity (g/mL)
Kenaf	930	53	1.6	1.20–1.45
Pineapple	413–1627	60–82	14.5	1.44
Ramie	500	44	2.0	1.50
Rice husk	25–75	2.5–4	0.5–1.0	0.95
Sisal	610–720	10–40	2.0–3.0	1.3–1.5

Dimensions of fiber length and diameter for some of the biofibers are shown in Table 26.3.

Table 26.3: Fiber dimensions, growth cycle, and performance benefits of biofibers.

Fiber type	Fiber length (mm)	Fiber diameter (μm)	Growth cycle (month)	Performance benefits
Bagasse	0.65–2.2	20–30	10–15	Good durability, low sp. gr.
Bamboo fiber	60–200	6–10	3–6	Antimicrobial, UV-resistant, and high tensile strength
Banana fiber	60–80	15–25	10–15	Lightweight
Brown coir	200–350	12–25	3–6	Resistance to microbes and salt water
Flax	90–125	18–22	2–4	Low growth cycle
Hemp	120–200	16–50	10–20	Low sp. gr., good thermal and acoustic insulation
Jute	1,000–4,000	15–20	4–5	Strong fiber, antistatic
Kenaf	2.5–4	15–25	3–7	Stiff fiber
Rice husk	5–7		4–6	High silica content
Sisal	1.0–2.0	20–25	24–36	Good strength and stiffness
Wood fiber	60–200	6–14	12–24	Good flexibility

Both fiber length and diameter contribute to the mechanical properties of the fiber and of the composite. In the discussion of composites, the significance of chemical composition, fiber length and diameter, cell structure, original polymer matrix attributes, interaction of the fiber with the matrix, and other characteristics and their effect on composites will be better understood. The growth cycle is related to the economics of renewal and hence the raw cost of the biofiber, and ultimately of the bio-based plastic composite. The longer the growth cycle, the more expensive the fiber obviously becomes, assuming that all the other factors remain the same.

Another factor that needs some discussion here is the challenge of providing sufficient volumes of both biopolymers and biofibers. Production of both materials, based on natural sources, is limited by factors like weather issues and the current bottlenecks in supply chain locally and globally. With respect to issues like scale-up and availability of supply, it varies significantly with each individual biofiber. Some are already produced in large quantities for other lower value applications. Also, the growth cycle of many common biofibers, shown in Table 26.3, is relatively shorter than the time needed to set up a manufacturing plant for biopolymers. Consequently, for scale-up of composites that are produced, starting with biopolymer reinforced with biofibers, the time to develop large-scale manufacturing is determined by the speed at which the production of the biopolymer can be scaled up.

Table 26.4: Annual production and origin of biofibers.

Fiber	Annual production (tons)	Origin
Abaca	70,000	Leaf
Bamboo	10,000,000	Stem
Banana	200,000	Leaf/stem
Coir	100,000	Fruit
Cotton lint	18,500,000	Stem
Flax	810,000	Stem
Jute	2,500,000	Stem
Kenaf	770,000	Stem
Ramie	100,000	Stem
Rice and corn husk	Abundant	Fruit/grain
Sisal	4,500,000	Leaf
Wood	1,750,000,000	Stem

Table 26.4 shows the annual global production of different biofibers in terms of usage in composites, by source of fiber. The global production of natural fibers was estimated to be around 25 million tons per year in the mid-2000s [15]. During the past decade, global annual natural fiber production has ranged between 28 million tons and 35 million tons, with most of the variation from 1 year to the next caused by changes in yields linked to weather. This data was collected by DNFI, in cooperation with the Food and Agriculture Organization of the United Nations (FAO), on 14 categories of natural fibers, ranging from abaca to wool and other animal fibers is available for public viewing [18]. This shows that the production of such fibers has not kept up with growth in global GDP. This is likely because the large majority of these fibers are used for clothing and relatively a small fraction for use as reinforcement in biocomposites.

26.4.2 Cellulose fiber

Cellulose fibers, used for reinforcement of composites, are fibers made with some type of processing, whether it is derivatives of cellulose ethers or esters of cellulose, which can be obtained from the bark, wood, or leaves of plants, or from other plant-based material. Such derivatives have been covered in other chapters in this book. Besides cellulose, the fibers typically also contain hemicellulose and lignin, with different percentages of these components. This variability alters the mechanical properties of the fibers, depending upon many factors such as the species of the plant or tree it came from, how much processing was carried out to separate the cellulose from lignin, hemi-cellulose, and other extractables. While there are patents that cover cellulose fiber composites, in order to cover as many of the biofibers, rarely do they use pure cellulose, when used with plastics for such composites. So, this chapter will cover each of the major types of biofibers that have cellulose individually, as much as possible.

26.4.3 Jute fiber

Jute is an herbaceous plant that is cultivated for jute fiber with edible young shoots and is primarily grown in India and Bangladesh. It has since been cultivated in many tropical countries. The name originates from the Bengali *jhuto*, meaning "matted hair". Jute is a long, soft, shiny bast fiber, composed primarily of the plant materials cellulose and lignin. Jute is renewable, versatile, nonabrasive, porous, hydroscopic, viscoelastic, biodegradable, combustible, compostable, and reactive. The fiber has a high aspect ratio, has high strength-to-weight ratio, is low in energy conversion, and has good insulation properties. Previously, while jute was used for low-value products, such as gunny bags, twine, and carpet backing, with many considering jute as a low-quality resource and considered its biodegradability and combustibility as disadvantages. Biodegradability provides a means of predictable and programmable disposal, not easily achieved with other resources [19].

When wet, jute, like other cellulosic fibers, absorbs water and swells but retains enough tensile strength to be used when twisted into rope. However, for most industrial applications, minimizing water is critical to ensure dimensional stability as a function of humidity. Using bio-based polymers that are hygroscopic to produce such composites is not practical for most industrial applications. For example, biodegradable jute fiber-biopolymer composites, such as those covered by S. R. T. Reddy, et al. [20], are hygroscopic and articles produced from such composites are not very useful for many applications due to high moisture absorption, poor dimensional stability, and susceptibility to insect and microbial attack.

Fibers like jute can also be used after processing them into a cloth. The processes to use such cloth reinforcements of jute or other reinforcements include thermal lamination, compression molding, and a variety of cast processes.

26.4.4 Coir fiber

Coir is the fibrous material found between the hard, internal shell and the outer coat of a coconut and is sometimes called coconut fiber. Commercial coir is a natural fiber extracted from the outer husk of coconut. There are two types of coir fiber. The more commonly used brown coir is from mature coconuts while white coir is from unripe tender coconuts with finer fiber and low lignin content. Among vegetable fibers, brown coir fiber has one of the highest concentrations of lignin, making it stronger but less flexible than cotton and is unsuitable for dyeing. The tensile strength of coir is lower than that of abaca or bamboo, but it has good resistance to decomposition from microbial action and salt water damage and needs no chemical treatment. This key advantage of coir is used for manufacturing durable products. Coir fibers measure up to 35 cm in length, with diameters of 12–25 µm. Coconut trees are grown on over 10 million hectares of land throughout the tropics. A coconut harvest occurs once in about 45 days. From 1,000 coconuts, it would be possible to extract 10 kg of coir [21]. Coir is a fibrous material that is widely used to overcome the problem of erosion. This works based on weaving them into geotextiles and placed on areas in need of erosion control, where it promotes new vegetation by absorbing water and preventing the top soil from drying out. Such geotextiles retain the natural ability of coir to retain moisture and protect from the Sun's radiation, just like natural soil, and unlike geo-synthetic materials, it provides good soil support for up to three years, allowing natural vegetation to grow.

Coconuts are typically grown by small-scale farmers, who use local mills for fiber extraction. Globally, around 650 000 tons of coir are produced annually, mainly in India and Sri Lanka, where the coir industry is fully developed. They, are also the main exporters, followed by Thailand, Indonesia, the Philippines, Malaysia, Vietnam, and Brazil. There are many other countries that grow coconuts but they have not moved to commercialize coir fiber. About 80% of the coir produced is exported in the form of raw fiber, while smaller quantities are exported as yarn, mats, matting, and rugs. Research and development efforts are continuing to focus on the use of coir in geotextiles and other new applications like coir plastic composites as the market shows promising prospects. Coconuts are grown in more than 93 countries globally and hence there is considerable scope to develop coir industry in more countries [21].

26.4.5 Hemp fiber

Hemp fiber is derived from the *cannabis sativa* plant. This fiber is located in the stalks of hemp plants. The stalks are surrounded by thick skins, with bundles of fibers present directly under this outermost layer. Hemp is also one of the bast fibers known from ancient times. Recent excavations in Egypt have shown hemp to be one of the oldest cultivated fiber plant but is currently finding renewed interest holistically, including all parts of the plant's fibrous material. Thus, hemp is both historical and holistic! It has been widely cultivated as a source of bast fibers in many countries that have decent rainfall, a mild climate with humid atmosphere, and a rainfall of at least 65 cm per year. Hemp is an annual plant that can grow on a wide variety of soils, but well drained, nitrogen-rich, and nonacidic soil is essential for ample hemp cultivation.

The chemical composition of hemp stems shows a significant difference between the outer bast and the woody core. The chemical composition of the outer bast is 60–70% cellulose, 15–20% hemi-cellulose, below 5% of lignin, and much smaller amounts of pectin, fats, and waxes. So, only the bast is used for fiber production rather than the hemp woody core, which is very similar to that of hardwood, with roughly 40% of both cellulose and hemi-cellulose and 20% lignin, and hence more difficult to process. The fibers are removed from the hemp plant, similar to the process of flax fiber production [22]. Chapter 27 has more details on hemp.

Hemp has been grown for many end uses – from textiles, matting, ropes and sail cloth – for its fiber properties. Hemp has gotten a bad rap, especially in the United States and was banned for most of the twentieth century because of its similarity to the marijuana plant. This negative reputation is undeserved. In the United States, we have recently started exploring the potential of the hemp plant once again. The CBD industry, for instance, would not exist without hemp. Industrial hemp generally has less than 1% of THC, which is a psychoactive chemical. However, the two varieties appear very identical, which often causes confusion. Today, it is one of the fastest growing plants, as CBD usage has grown [23].

26.4.6 Bagasse fiber

Bagasse is the dry pulpy fibrous material that remains after crushing sugarcane or sorghum stalks to extract their juice. Chemical analysis of washed and dried bagasse shows it has 45–55% cellulose, 20–25% hemicellulose, 18–24% lignin, and smaller quantities of ash and waxes [24]. Bagasse fiber length is 0.65–2.17 mm and its width is 21–28 μm [25].

For every 100 tons of sugarcane crushed, a sugar factory produces nearly thirty tons of wet bagasse. It is challenging to use this byproduct directly as a fuel because of its high moisture content, typically 40–50%. Instead, bagasse is typically stored prior to further processing. Products from bagasse biodegrade quickly and can be composted without releasing any harmful chemicals. It is considered a waste product, meaning it requires no additional resources to produce, and the largest current application for bagasse is as fuel to generate electricity, but only after drying [26]. One company has used it to produce what is claimed to be the world's best disposable plates [27].

26.4.7 Bamboo fiber

Bamboo, (subfamily *Bambusoideae*), is part of tall tree-like grasses that belong to the Poaceae family, comprising more than 115 genera and 1,400 species. Bamboos grow in tropical and subtropical to mild temperate regions, with the heaviest concentration and largest number of species in East and Southeast Asia and on islands of the Indian and Pacific Oceans. A few species are native to the southern United States along riverbanks and in marshy areas. Bamboos are typically fast-growing perennials. Depending on the species, some grow as much as 30 cm/day. The smallest species of bamboos grow to just 10–15 cm (about 4–6 inches) in height while others grow to more than 40 m [28].

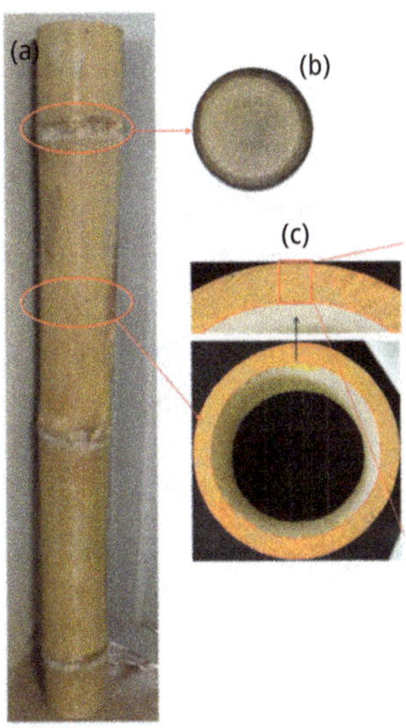

Figure 26.3: Bamboo and its nodes.

The woody ringed stems, known as culms, are typically hollow between the rings (nodes) and grow in branching clusters from a thick rhizome (underground stem). Figure 26.3(a) is the bamboo stem, with the first red circle being the node, (b) shows the cross section of the node, and (c) shows the cross section of the hollow bamboo where there is no node [29].

Bamboo is usually extracted into fibers and this fiber is then used as a reinforcement material in the polymer matrix. Various properties of bamboo-fiber-based composites have been investigated extensively and will be covered in this chapter in the next section. The reason bamboo fiber (BF) is selected as reinforcement is because of its good mechanical and thermal properties, facile extraction and fiber treatment, low cost, environment friendly nature, and ability to be used as a product in the industry. BF is identified to have strength and stiffness, and it contains microfibrillar angles that give it very high aspect ratio naturally, and its thick cell walls make BF an ideal reinforcing fiber. However, these fibers themselves have poor compressive strength. Most researchers have separated the BFs from the nodes before using them. But the nodes were left in place by one researcher that will be discussed later in the section on BF plastic composites, where most of the limitations of bamboo were mitigated and where bamboos were used in an application where the high tensile strength was useful. Water absorption reduced significantly and its poor compression set was compensated by using it in as reinforcement in concrete, where the concrete's high compression strength was used for synergistic effect.

26.4.8 Kenaf fiber

Kenaf is a plant that is native to Southern Asia, though its exact origin is unknown. The name also applies to the fiber obtained from this plant. These fibers are in many ways similar to fibers of jute and show similar characteristics. It is an annual or biennial herbaceous plant (rarely, a short-lived perennial) and has to be replanted after 1 or 2 years. It grows to 1.5–3.5 m tall with a woody base. The stems are 1–2 cm in diameter, and are often, but not always, branched. The leaves are only 10–15 cm in length and vary in shape, with leaves near the base of the stems being deeply lobed, while leaves near the top of the stem are shallowly lobed or unlobed lanceolate [30].

The bast (bark) is 40% of the plant and the core (wood) is where the fibers in kenaf are present. "Crude fiber," separated from the bast is multicellular, consisting of several individual cells stuck together [30]. More details about cell size of the fibers are available elsewhere [31].

The stems produce two types of fiber: a coarser fiber in the outer layer (bast fiber), and a finer fiber in the core. Kenaf matures in 100 to 200 days. First grown in Egypt over 3,000 years ago, the leaves of the kenaf plant were a component of both human and animal diets, while the bast fiber was used for bags, cordage, and the sails for Egyptian boats. Even today, bast fibers are used to make ropes. This crop was not introduced into southern Europe until the early 1900s. Today, while the principal farming areas are China and India, kenaf is also cultivated for its fiber in Bangladesh, southern part of the United States, Indonesia, Malaysia, South Africa, Vietnam, Thailand, parts of Africa, and to a small extent, in southeast Europe [30, 31].

26.4.9 Rice and corn husk fiber

The rice husk, also called rice hull, is the dry outer covering of a grain of rice. Lab tests have shown that the husk is about 40% cellulose-based fiber, 30% lignin group, and 20% silica. The lignin and silica are hard materials, which have the natural function to protect the seed during the growing season. Rice husk is the exterior shell of rice grain, which is always removed during the milling of rice. It is of no direct nutritional value as food for mankind, and in most mills, it is often discarded or allowed to rot away [32, 33]. Rice husk can absorb water, ranging from 5% to 16% by weight, and the unit weight of rice husk is 83–125 kg/m^3 [34]. The rice husk is removed in the milling process, which accounts for 20% of the world's total rice production. With about 760 million tons of world paddy production in 2017, according to the FAO data [35] and data from knoema.com [36], this can be calculated to about 152 million tons of rice husk being available globally.

Similarly, corn husk is the outer covering of an ear of corn, also called maize. Commercially bought corn husks are commonly the whole husk and are dried in sunshine, or by air, or in an oven. Corn husk fiber has low crystallinity and rough surface, with hollow cross-sectional area. Cornhusk fiber shows less tensile strength and more elongation compared to other popular lignocellulosic fibers like jute and cellulosic cotton, due to lower lignin and higher cellulose and hemi-cellulose content. Despite wide availability in many countries around the world, the only applications appear to be in various culinary recipes and in tribal handicrafts; not much as a biofiber [37].

26.4.10 Wood fiber

Wood is a porous and fibrous structural tissue found in the stems and roots of trees and other woody plants. It is a natural composite of cellulose fibers that is strong in tension and embedded in a matrix of mostly lignin and hemicellulose, which together resists compression.

Wood fibers are usually cellulosic elements that are extracted from trees and used to make materials, including paper and paperboard. This is one of the largest three traditional applications, the others being furniture and building construction. The end paper product dictates the species, or blends of two or more species used, that is best suited to provide the desirable sheet characteristics of the paper or paperboard, and also dictates the required fiber processing. In North America, virgin wood fiber is primarily extracted from both hardwood and softwood trees for producing paper and paperboard. The wood fiber is either extracted as a primary product, or collected during the milling of lumber [38, 39]. Wood fibers can also be recycled from used paper materials. The furniture and building construction industries also use the right species of wood that is cut at lumber mills. Most of the wood fibers used for

WPC are from lumber mill waste that comes mainly from these three industries. Depending upon the type and species of wood and what part of the tree it is derived from, the fiber can be 1–3 mm in length and 20–30 μm [39].

26.4.11 Leaf fiber, especially from sisal, abaca, and banana leaf

Various leaves from trees and plants can also be used, but those from sisal, abaca, and banana leaves are the most important. Sisal fiber has one of the largest cross-sectional diameters of natural fibers, which gives it higher tensile strength and stiffness, which makes it useful for its current applications for cords, power transmission, marine ropes, baler, binder twine, sacks, paper filter, and other industrial uses. While some research work has been carried out to use sisal fiber as a reinforcement for plastic composites, which is covered later in this chapter, it has not been commercialized for such applications, at least in a large scale, similar to fiber from abaca and banana leaves.

Abaca fibers, also called manila hemp, are extracted from the leaf sheath around the trunk of the abaca plant (*Musa textilis*). The plant is native to the Philippines and widely found in the humid tropics close to the Equator. The process of harvesting abaca is labor-intensive because each stalk must be cut into strips that are scraped to remove the pulp, and fibers are washed and dried. The leaf fiber is composed of long slim cells that form part of the leaf's supporting structure. The high lignin content of 15% is the reason that abaca fiber is prized for its mechanical strength, resistance to saltwater damage, and long fiber length, up to 3 m. The best grades of abaca are fine, lustrous, light beige in color, and very strong. It is pulped to make sturdy manila envelopes. Today, it is still used to make twines, ropes, fishing lines and nets, as well as coarse cloth for producing sacks. There is also a flourishing niche market for abaca clothing, curtains, screens, and furnishings, but paper-making is currently the main use of the fiber, especially in Indonesia [40].

Banana fibers, obviously from the banana trees, are concentrated near the outer surface. The fibers are extracted by hand scraping, chemically, by retting, or using raspadors. They can also be extracted by boiling leaf sheaths in sodium hydroxide solution [41]. The use of these fibers for textiles has been proposed, based on research in using them in composites, but no large commercial applications are known yet [42].

26.4.12 Fibers from crop residues and other biomass

There have been a lot of statements from many companies around the world about using crop residues and biomass; only a few large-scale productions have started using what is called a biorefinery, except for generating biofuels. This consists of various conversion processes and equipment that convert biomass into fuels, chemicals, and power [43]. Deep insight of chemistry, production, and conversion technology is required to transform biomass to useful products. If the process temperature is controlled sufficiently, the residue left over contains a lot of useful fiber that could be used for producing composites. This is especially true if agricultural crop residues or waste products come from paper-making or other industries, which essentially use natural biosources that have strong fibers.

Crop residues are the materials left over after a crop harvest, in the form of leaves, stalks, stems, seeds, etc. The amount of crop residue depends on the type of the crop and growth of the crop, as well as methods of tillage and harvesting. Cereal crops produce large quantities of residues, which are useful in adding soil organic matter. They are also used as feed, fodder, fiber and construction material, and as fuel for energy [44]. Use of such biomass as feedstock for biopolymers has been covered elsewhere in this book, including in Chapters 2 and 27. Post-harvest residues, which are used for soil amendments, like husk and bagasse, will not be covered here because they have already covered earlier. In Europe, brewer's spent grain, olive pomace, and residual pulp from fruit juice production and other agri-food wastes, which are produced annually in huge quantities, pose a serious problem, economically and environmentally, because landfilling biowaste is expensive and unsustainable in the long run. All such lignocellulosic matter can be converted into fibers, along with forest residues, yard clippings, wood chips, and even municipal solid waste, as per the objectives of the circular economy [45]. Industrial biomass can also be grown from numerous types of plants, including miscanthus, switchgrass, wheat, hemp, corn, poplar, willow, sorghum, sugarcane, bamboo, and a variety of tree species, ranging from eucalyptus to palm oil, after valuable oils, other bio-chemicals, and in many cases, biofuels have been already extracted. The only issue with biomass sometimes is that besides carbon, hydrogen, oxygen and nitrogen, other elements like alkali, alkali earth, and heavy metals are also present in such biofibers [45], which might restrict their use in some downstream applications.

26.5 Bio-based fiber plastic composites

26.5.1 Introduction to bio-based fiber plastic composites

Fiber composites are compounds where a polymer is reinforced with a reinforcing agent like a fiber, and the fiber provides reinforcement and rigidity in the composite. As such, composite structures are generally a combination of two or more materials at the macroscopic level and typically have an interface layer between the two main phases [3, 46, 47]. This section covers composites, whether the polymer and/or the reinforcement is bio-based in the composite. In some cases, the carrier polymer used is bio-based; in some composites, biofibers are used to reinforce non-biopolymers; and in others, both the reinforcing fiber and the polymer matrix are bio-based. Natural fiber is used in these composites as a reinforcement material embedded in a polymer matrix, where the polymer is either a thermoplastic or thermoset. Composites containing natural biofiber and polymers have been called by many different names like natural fiber composites (NFC) [48, 49], and natural-fiber-reinforced polymer composites (NFPs) [50].

A bio-based composite's properties are influenced by a number of major variables, including the fiber type, environmental conditions (where the plant fibers are sourced), processing methods, and any modification of the fiber. Ultimately, the properties of the composite are determined by the individual properties of the reinforcing fiber and those of the polymer matrix, as well as the interfacial interaction between the two phases. The properties of the composite are dependent on the following factors:

- aspect ratio, measured by length of the fiber to its diameter,
- length of the fiber,
- mechanical strength of the fiber,
- volume ratio of the fiber to the matrix,
- distribution and dispersion of the fiber in the matrix,
- orientation of the fiber within the matrix,
- interfacial adhesion of the fiber to the matrix,
- the mechanical strength of the matrix, which itself is dependent on its crystallinity, its orientation, and its molecular weight, among other factors.

Most natural bio-based polymers and bio-based fibers are polar, hygroscopic, and have high-enough water absorption. This factor restricts their use without chemical modification of those polar groups and/or sufficient coverage by nonpolar polymers, almost all of which are petro-based like polyethylene and polypropylene.

One of the greatest challenges in working with biofiber-reinforced plastic composites either experimentally in trials or in large-scale production is the large variation in properties and characteristics of biofibers [51]. Natural fibers vary a lot, compared to synthetic fibers that consist of metal, glass, or synthetic polymer, as measured by standard deviation.

Over the past few decades, there has been a significant increase in the production of a wide range of products that use natural-fiber-reinforced polymer composites. These composites are one of the best alternatives to produce environmentally friendly materials by combining bio-based and non-bio-based polymers and natural fibers in various products for many different applications [48–52]. Using these natural fibers has a high impact on the manufacturing industry, as these materials are readily available, are of low cost, and are easy to design, and increase productivity [50]. Natural fiber sources are increasingly gaining attention and acceptance for use in fiber-reinforced polymer composites, such as with biopolymer matrixes like:

- polyhydroxyalkanoates (PHA) [53],
- polylactic acid (PLA) [54],
- chitosan [55],
- polycaprolactone [56], and
- thermoplastic starch [57].

Kuciel et al. published their results on short-fiber-reinforced biocomposites produced using polyamides from castor oil produced from renewable sources in 2012 [58]. Jumaidin et al. reported on their work on processing and characterization of banana leaf fiber-reinforced thermoplastic cassava starch composites [59]. Another detailed review was published in a book entitled “Natural polymers, biopolymers, biomaterials, and their composites, blends, and IPNs” in 2021 [60]. While a few applications of banana fiber-reinforced epoxy composites in automotive and aircraft parts have been pursued, there is no confirmation available in literature.

Numerous applications exist currently for bio-based fiber polymer composites in packaging, automotive parts, furniture, building and construction, medical devices, and other end-use markets. For example, Mercedes Benz has used a composite produced from abaca yarn in polypropylene rein matrix for molding automobile body parts. Replacing glass fibers by natural fibers can reduce the weight of automotive parts and facilitates more environmentally friendly production and recycling of the parts, typically using maleic anhydride-grafted polypropylene as the coupling agent. Due to the high mechanical strength of the fiber as well as its length, application of abaca even in highly stressed components offers great potential for this and other industrial applications [61].

26.5.2 Cellulose fiber composites and cellulose nanocomposites

Cellulose fiber has been used as reinforcement for polymer composites with many different plastics matrices, for example, with dimer fatty acid-based polyamides [62]. At the Society of Plastics Engineers (SPE) meeting, hosted by the Automotive Composites Conference (ACCE) in Novi, Michigan, in September 2015, Dr. Alper Kiziltas, lead research scientist at the Research & Innovation Center of Ford Motor Co. discussed blends of bio-based polyamide (PA-6,10 and 10,10) reinforced with cellulose fibers, sourced from sustainable forestry [63].

What is not well known is that the strongest part of a tree lies not in its trunk or its sprawling roots, but in the walls of its microscopic cells. A single wood cell wall is constructed from fibers of cellulose (nature’s most abundant polymer), and the main structural component of all plants and algae. Within each fiber, there are reinforcing cellulose nanocrystals (CNCs), which are chains of cellulosic polymers arranged in nearly perfect crystal arrays or patterns. At the nanoscale, CNCs are stronger and stiffer than Kevlar™. The idea was that if the crystals could be worked into materials in significant fractions, CNCs would provide a route to stronger, more sustainable, naturally derived plastic composites. Recently, a team at the Massachusetts Institute of Technology in the USA has developed a composite made mostly from CNCs mixed with a bit of synthetic polymer. The bio crystals make up about 60 to 90% of the material – the highest fraction of CNCs achieved in a composite to date. The researchers found that the cellulose-based composite is stronger and tougher than some types of bone, and harder than typical aluminum alloys. They found that this engineered material has a brick-and-mortar microstructure that resembles nacre, the hard inner shell lining of some mollusks [64].

26.5.3 Kenaf fiber polymer composites

Kenaf fibers also have been used as reinforcements in various plastic composites. Examples are Anand Sanadi's work with polypropylene, as far back as 1995 [65], and a review covering kenaf fiber-reinforced plastic composites with many polymers by Akil et al. in 2011 [66]. A Japanese patent developed kenaf fiber-reinforced biodegradable resin composition for molded articles such as parts for electrical and electronic equipment [67]. However, it is not known if any commercial breakthroughs have happened with kenaf fiber-based polymer composites.

26.5.4 Bagasse fiber polymer composites

Bagasse fibers are low-cost waste products from sugar production, after sugar cane juice is extracted from it. These fibers have been evaluated as a reinforcing filler to enhance mechanical properties, reduce product weight, and enhance environmentally friendly characteristics of a wide range of thermoplastics and thermosetting matrices, such as polypropylene, polyesters, polyvinyl chloride, polyurethane, epoxy, and high-density polyethylene (HDPE), to form composites.

Bagasse has also been used to produce polymer composites with significant performance characteristics using gamma radiation to produce in situ thermoset composites using unsaturated polyester and styrene monomers as well as styrene–acrylonitrile monomers [68].

Bagasse fiber has been used along with another waste product from coal-powered power plants called fly ash, to produce bagasse-flyash polymer composites, using the same radiation processing method, and a patent was granted [69]. The combination of bagasse, a low-cost raw material, and flyash, which power plants pay companies to haul away and actually has a negative cost, enable products made from this combination to be used for composites, which is an amazing synergy of performance and cist for producing products with excellent performance-to-cost ratio.

More recently, a review was published on bagasse-fiber-composites with various polymers by Verma et al. [70].

26.5.5 Jute fiber polymer composites

Using jute fiber for composites has many advantages. Jute is renewable, versatile, nonabrasive, porous, hydroscopic, viscoelastic, biodegradable, combustible, compostable, and reactive. The fiber has a high aspect ratio, high strength-to-weight ratio, is low in energy conversion, and has good insulation properties.

A comparative analysis of jute fiber polymer composites with similar composites using jute and bamboo was published in 2017 [71].

Production of surface-modified biocomposite jute fiber hybrid and differing polylactide and PCL fractions has been recorded with improved rigidity and durability. This type of biocomposites was fabricated by hot pressing of solvent-impregnated prepregs. The addition of PCL led to an improvement in the biodegradation rate of biocomposites, thus making them more environmentally friendly [72]. The effect of compatibilizer to improve properties of short jute fiber-reinforced polypropylene composites has also been studied [73]. However, such composites are no longer biodegradable.

26.5.6 Bamboo polymer composites

Fundamentals and processes of bamboo-based thermoset composites have been reviewed recently [74]. Most of these have been with BFs and some in combination with wood fibers. Many of the initial commercial applications used phenol-formaldehyde, polyurethane, polyvinyl acetate, and epoxy resins, mostly to produce oriented strand board for housing and construction applications.

Both thermoset and thermoplastic bamboo fiber-reinforced thermoset and thermoplastic composites have been reviewed recently. These references cover properties, fabrication, preparation, and chemical modification, including methods for improving bonding of the bamboo, impregnation methods used for thermoset polymers, process details, and applications [75–79].

BFs were compounded with polypropylene in a twin-screw extruder and injection molded into a test specimen for determining mechanical and other properties, by Lee et al. [80]. They found that properties like tensile modulus, flexural modulus, and water absorption were enhanced by increasing the bamboo fiber loading. They also determined that the addition of tetramethoxy orthosilicate (TMOS) and aminopropyltrimethoxysilane (AS), after alkali pretreatment of the fibers, increased tensile, flexural, impact strength, and water desorption and other properties of the resultant composites, due to improved interfacial adhesion between the biofibers and the PP matrix. The melting temperature, melting enthalpy, crystallization enthalpy, and crystallinity were decreased by the increase of fiber loading and the "AS + TMOS" treatments. On the other hand, the crystalline temperature was increased by the addition of BF, AS, and TMOS. This data proved that AS and TMOS are effective coupling agents for the bamboo-polypropylene thermoplastic composite systems.

A 1944 patent describes that the first impregnation of bamboo fibers was carried out with synthetic group of resins like phenolic aldehyde resins and urea aldehyde resins, to form thermoset composites [81]. The first use of radiation processing to prepare bamboo-polymer composites was reported by Wang [82, 83]. Some preliminary work on preparation of bamboo-polymer composites using radiation processing was also published in a paper by Mani et al. [84].

Several years ago, a unique method was developed using sections of bamboo instead of first separating fibers and throwing away the rigid nodes. In this method, the long bamboo stems were instead cut along the length of the bamboo, into sections, with the nodes left in place. The sections were placed in a specially designed chamber to remove moisture and air, to a vacuum of 4 mm of mercury for 4 h. Liquid monomers were added into the impregnation container, so the bamboo could soak the monomer mixture overnight. Two monomer mixtures were used: styrene-acrylonitrile-benzene hexachloride (St-AN) and unsaturated polyester-styrene-chlorinated paraffin (St-PE) in the ratios of 60:40:3 and 20:80:30, respectively, in separate set of experiments. Chlorinated paraffin and benzene hexachloride were used as free radical chain propagators to accelerate the polymerization and minimize the radiation dosage needed. The chlorinated paraffin also helped to reduce the viscosity of the unsaturated polyester, to improve the impregnation into the bamboo. The container was then slowly restored to ambient pressure. This technique to get fibrous material to incorporate monomers or other liquids is called vacuum impregnation [85–87]. The soaked samples were then packed in polyethylene and irradiated with cobalt-60 gamma radiation at a dose rate of 0.11 MRad per hour. From prior work, the exact radiation dosage was determined to achieve full polymerization. The total dosage used for the St-AN mixture was 2.12 MRads and for the St-PE mixture, it was 2.49 MRads, based on preliminary work to determine dosage needed to get to the required gel ratio. The microstructure of the composites showed that the polymer covered the cellulosic fiber of the bamboo using phase-contrast photo-microscopy and electron microscopy [85, 88].

Test specimens were all prepared using machine-cut strip pieces of bamboo that were 50 cm. in length and a cross-sectional area of 1 sq. cm. The test results showed improvements in tensile strength, rigidity, modulus of elasticity, and reduction of water absorption, even after 24 h and 3 months [85, 89]. Huge improvements were observed in resistance to attack by fungus, bacteria, and insects without using biocides over a 6-month period by which time, untreated bamboo fell apart in indoor open air [85].

At a polymer loading of 30% on three different commercially available bamboo species, water absorption was reduced to the 1–2% range and dimensional stability in water was adequate for the intended application for these bamboo-polymer composites was for replacing steel rebar with these composites in concrete for building structures [85, 86, 90]. The properties of these composites were such that they had a higher strength-weight ratio compared to steel rebar, which is used for reinforcement of concrete. Testing of these composites was carried out by a large housing government corporation and found to be viable, based on higher bonding strength to concrete and other properties, at a time when there was a huge shortage of steel in India [90]. However, the economics fell apart, with the tripling of monomer prices in 1972–1974. The significance of this research is that, with these bamboo plastic composites, several of the limitations of bamboo, like high water absorption, poor dimensional stability, poor resistance to rotting, etc. were mitigated, and bamboos were used in an application where the high tensile strength of the bamboo was synergized with concrete's high compression strength, and the poor compression set of bamboo was not a factor in it by using it as reinforcement in concrete. Moreover, water absorption reduced significantly, helping to ensure bamboo's other disadvantages were overcome in these unique composites [90].

In the United States, Cali-Bamboo has been selling its BamDeck™ composite, made from 60% recycled bamboo fibers and 40% recycled plastics, for decking, fencing, poles, edging, flooring, etc. [91, 92].

26.5.7 Wood polymer composites (WPCs)

Wood fiber can either be extracted as a primary product or collected during the milling of lumber. Wood fibers can also be recycled from used paper materials. Most of the wood fiber used for WPCs are from lumber mill waste. Of all of the biofiber-based composites, this is the one with the largest sales volume and revenue globally and in most countries around the world. A total of 10,866 patents have been granted, just in the USA, related to wood polymer composites, also called wood plastic composites, which just shows how much interest there is in this area. There are several good reviews available on this topic [93, 94].

Wood plastic composites can be divided into three types:

(1) Traditional WPC, such as particle board, fiber-based board, plywood, oriented strand board (USB), etc., in which the wood particles, fibers, and veneers are held together with a thermoset adhesive, used at 4–15% [60, 94].
(2) Nonadhesive-based WPC covers both wood that is impregnated with monomer systems and polymerized in situ to form thermoset WPC [95, 107].
(3) Nonadhesive-based WPC, which are produced from wood fiber, powder/flour, wood micropowder, or wood-based source that is recovered from waste materials [96].

Both Type (1) and Type (2) are reviewed in this eBook on "Wood Polymer Composites" [93]. Type (2) are materials in which wood is impregnated with monomers that are later polymerized in the wood to tailor the material for special applications. This polymerization technique is called "in situ polymerization." The resulting performance properties of these materials, like low specific gravity, boosted mechanical properties and superior sustainability, has created a rising number of applications in building and construction as well as automotive and other transportation industries. This eBook review covers a wide variety of topics such as the manufacture of wood-polymer composites, how their properties can be evaluated and enhanced and their range of uses, as well as key aspects of manufacture, including raw materials, manufacturing technologies, and interactions between wood and synthetic polymers. The remaining chapters discuss mechanical and other properties such as durability, creep behavior, and processing performance [93].

Prof. Li's book reviewed well over 33 different publications covering various monomers polymerized in situ with peroxides and other initiators from 1968 to 2011 [94].

Many of the initial WPC products were produced without using too many additives or optimized for long-term applications. They had problems like resistance to moisture and exhibited swelling after contact with water. As a result, loss of properties was observed when the composite parts got wet and later resulting in rotting due to insect and microorganism attack, among other issues. Over time, many of these were mitigated or eliminated by using pretreatment of the wood or adding additives [95].

Some of the initial work on high-energy radiation processing of WPC with cobalt-60 gamma radiation was carried out with methyl methacrylate, styrene, and other monomers after impregnating these monomers into the wood with vacuum impregnation, in Taiwan and the USA [96, 97]. Later on, several papers were presented and published based on work carried out on WPC using vacuum impregnation, followed by radiation processing, at the Bhabha Atomic Research Center in India [98–100].

Most current commercial production of WPC is using thermoplastic plastics like high-density polyethylene (HDPE), low-density polyethylene (LDPE), linear low-density polyethylene (LLDPE), a mixture of these polyethylenes, or polyvinyl chloride (PVC), by several companies around the world. Recycled HDPE from blow-molded milk bottles is used, sometimes mixed with virgin HDPE and some LLDPE, to improve impact performance, when it is needed for some specific end applications. For some end indoor applications, recycled LDPE, usually from post-consumer film waste, works if the application is such that it is not likely to see temperatures above 25 °C. In most cases, the process used is shown in Figure 26.4, which shows a typical WPC compounding process to produce WPC pellets.

The main carrier resin (or mixture) is added in the main hopper in pellet form, while the additive masterbatch (concentrate containing a coupling agent, antioxidants, flow aids, lubricants, biocides, etc.) is added midstream, as shown in Figure 26.4. Biocides are used to ensure resistance to rotting, especially for decking, fencing, boat docks and

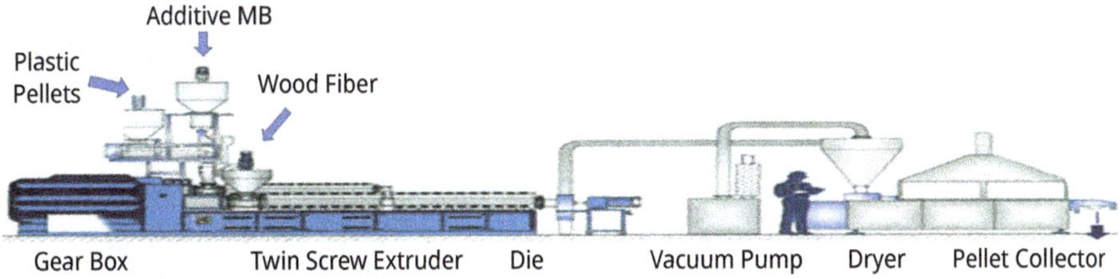

Figure 26.4: Typical WPC compounding process to produce WPC pellets.

other products which can get wet. The wood fiber, ideally in pellet form, is added also into the main hopper, but if feeding it is slower than the required production rate, sometimes it is added through two separate feeds or with a mechanical stuffer, through a secondary feed. The wood fiber in powder/flour form is added downstream to ensure that it is added, once the plastic pellets melt. It is typical that the ratio of plastic to wood is 50:50 by weight, which ensures that the volume ratio of the plastic is higher than that of the wood due to the former's lower specific gravity. The wood fiber needs to be sufficiently dry to ensure that voids are not created, which can reduce the tensile, impact, and bending strength, and other types of mechanical properties of the WPC. The remaining moisture in the wood is removed using the dryer and vacuum pump system, as shown in the figure.

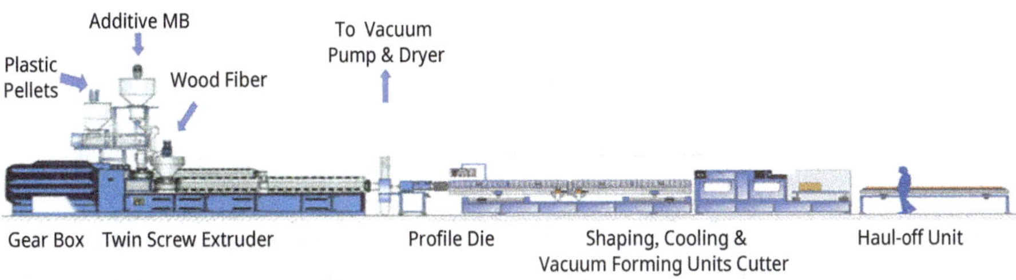

Figure 26.5: Typical WPC compounding process to produce WPC profiles.

In cases where WPC pellets are produced, they are converted by the customer into whatever products the customer needs in a secondary process, which was discussed in Section 26.2 of this chapter. In most cases, large companies that produce profiles in large volume, for decking, boat docks, or along water fronts, extrude the profiles directly in a single step. Instead of a pelletizing die, as shown in Figure 26.4, they have a profile die-and-cooling system that can also indent specific designs or print colors or designs, as shown in Figure 26.5.

Many WPC companies use a combination of virgin and recycled plastics to produce their products with many different proprietary designs, especially when viewed from a cross-sectional perspective, as well as various colors and surface designs and groves.

Based on a patent granted to Trex Co., a leader in the WPC industry, it uses a different method to produce their polyethylene-based WPC. Rather than using polyethylene pellets, they use specialized equipment to first compress post-consumer polyethylenes in films, bottles, and other forms, and stuff into a thermoplastic component, with a particle size of less than about 3/4 inch (19 mm). The process involves also grinding the wood to a particle size of less than about 600 microns, proportioning the said wood component and the said first thermoplastic component in a weight ratio of wood component to thermoplastic component in the range of about 65/35 to about 40/60, to form a wood-thermoplastic mixture, pelletizing the said wood-thermoplastic mixture into a first stable non-separating high bulk density feedstock, wherein the said first feedstock is capable of being readily air transported. These steps are repeated to form a second stable non-separating high bulk density feedstock, after which the second high bulk density feedstock, comprised of a second thermoplastic component having different physical characteristics from the said first thermoplastic component, is then mixed with the first feedstock and the second feedstock, to form a blended high bulk density feedstock [101]. This ensures use of recycled materials with wood-based powder to produce profiles that have among the highest recycled and bio-sourced product in the market.

Other WPC companies like Timbertech™ (in 2023, it was part of Azek, which recently merged with James Hardie) and Fiberon™ (now owned by Fortune Brands) use a more traditional method of production, similar to what is shown in Figure 26.4, except that rather than producing pellets, the molten WPC is directly converted using a profile die into decking boards, fence posts, or other profiles. These products are sold with a 25- to 50-year warranty. In some products, HDPE or PVC is used. Anderson Windows & Doors sells Fibrex™ windows produced from PVC and reclaimed wood.

The combination of using recycled wood from construction and demolished structures with recycled or even virgin plastics forms wood-polymer composites (WPC) have a very wide scope of usage. Such recycled composites have advantages such as very low environmental impact in terms of abiotic potential, global warming potential, and greenhouse potential. Another significant benefit is flexibility, because WPCs can be compounded and processed, thus making them easy to be modified to meet customers' predetermined strength values needed for each end application. If the WPC is thermoplastics, conventional polymer composite manufacturing techniques such as injection molding and extrusion can be used, but for thermoset WPCs, such processes are not feasible, without precisely controlling the process parameters. Many rheological characterization techniques need to be determined to ensure the quality of final WPCs, based on the effect of formulation and process parameters [102].

Even though wood-plastic composites can reduce plastic waste mass in landfills and can improve the physical and mechanical properties of wood, it also creates new problems in its life cycle when WPC products are disposed of. It increases the volume of plastic waste, when including the volume of wood in it. It is difficult to recycle, especially WPC is produced using thermoset polymers [103]. At least, scrap thermoplastic-based WPC can be recycled at percentages below 25% but such products acquire a darker color than the original WPC.

However, modern WPC products can last 30–50 years, depending on the brand and many large producers provide multidecade warranties, compared to treated wood, which is only good for 6–10 years. WPC profiles from reputable companies are also maintenance-free, whereas treated wood for decks needs to be re-stained every year after the first 2 years. This is the value that WPC products provide and this value is why this type of bio-based fiber plastic composites use has grown so large all over the world. The amount of WPC produced keeps increasing. The wood plastic composite market is projected by researchandmarkets.com to grow at a CAGR of 5.75%, to reach a market size of $7.533 billion by 2026 from $5.092 billion in 2019 [104].

The growing demand is primarily in the building and construction end market, with applications like decking, molding and siding, and fencing. In the furniture market, WPCs outperform wood, which has been the main traditional furniture-making material for centuries. These composites are resistant to rotting, moisture, and infestation and can be extruded or molded into a variety of forms, as required. Over time, the volume of recycled materials (combination of biofiber and plastics) has gradually increased. In April 2019, a WPC was produced with up to 95% recycled material by weight, including building and demolition waste by Conenor, a Finnish business specializing in composite extrusion techniques, based on technology developed by Technalia, a prominent R&D center in Spain [104]. Why is this sort of progress important? The ratio of renewable/recycled to nonrenewable resources has a direct influence on the environmental impact of WPCs and the added benefit of reducing the quantity of materials going into landfills. This shows that the use of recycled materials is growing with WPC, both by using recycled plastics and recycled wood fibers. Fossil fuel-based polymers, which were widely utilized in many such applications a couple of decades ago, based on nonrenewable raw materials and typical plastics used for these WPC applications, are nonbiodegradable and considered undesirable, whereas wood used before that time would need maintenance each year and replacement every five to ten years.

However, cost is a challenge, where the consumer needs to be educated because surface-treated wood and recycled plastic profiles are of much lower cost, while WPC has a significantly high initial cost [104].

Another factor that needs to be considered is that during installation, the material undergoes thermal expansion and shrinkage. Traditional wood planks used in these applications expand and contract due to their ability to absorb water. Since composites are a blend of wood fibers and plastics and are not completely natural, this should be considered. Many projects cause the thermal expansion of the floor to curve, since there is no gap between the boards. Moreover, since a few plastics have greater chemical heat content and can melt, some WPC compositions have more fire hazard qualities than wood alone, unless fire retardants are used in the compounding of the WPC.

26.5.8 Other Biofiber composites

Besides the biofiber composites discussed in Sections 26.6.2–26.6.7, other biofibers have also been used to prepare and, in a few cases, produce biofiber plastic composites. For example biodegradable polymer nanocomposites have been prepared and found to be biodegradable [105]. Another example is that of a German patent [106], which disclosed biologically reinforced degradable polymer compositions, based on thermoplastic starch or a polymer mixture containing thermoplastic starch and at least one hydrophobic biologically degradable polymer, reinforced by natural fibers that are incorporated in the polymer, and the polymer mixture furthermore contains a phase mediator for the molecular coupling of the starch phase with the hydrophobic polymer phase. The reinforcing natural fiber is at least one that is selected from the group consisting of ramie, cotton, jute, hemp, sisal, flax, linen, silk and abaca. The data shows that their substantially biologically degradable polymer composite is prevented from at least partially losing its biological degradability when reinforced with the natural fibers such as, in particular, sisal or ramie fibers.

Recycled newspaper fibers were used to produce fiber polypropylene thermoplastic composites [107]. Thermoset composites based on polyester reinforced with corn husk fibers have been studied by q group of researchers from Indonesia, Malaysia, and India [108].

Hemp fiber composites are starting to get commercialized in Europe and the United States. A U.S.-based startup, The Hemp Plastic Company, produces a 100% plant-based PLA composite reinforced with 25% hemp content. The material substitutes conventional plastics without significant capital or operating expense, making the transition easier in injection molding applications [109]. A Canadian company called INCA Renewtech™ has developed some proprietary technology and made inroads into using their thermoset hemp-based composites for applications in wind energy and in the automobile industries [110].

Several researchers have also studied hybrid biofiber plastic composites, in which natural fiber in combination with artificial fibers show interesting and sufficiently high enough properties, over only using biofibers, to make them suitable for certain engineering applications [111].

There are several publications that summarize biofiber-reinforced plastic composites for different end-use market applications:

- Ballistic applications are reviewed by Odesanya and his co-writers [112].
- A comprehensive review on recent developments in biopolymer-based composites and their use in drug delivery and biomedical applications was published in 2017 [113].
- Medical applications of chitosan-based biofiber composites are among the several topics covered in this review by researchers at M.G. University in Kottayam, Kerala, India [114].

- Biopolymers in electronics are reviewed in a 2017 book, which addressed the challenges posed by the increased use of biopolymeric materials in electronic applications for applications such as sensors, actuators, optics, fuel cells, photovoltaics, dielectrics, electromagnetic shielding, piezoelectrics, flexible displays, and microwave absorbers [115].
- At the Texas Tech University, a novel composite material using a new composite material formulation of a compound, combining two biopolymers polycaprolactone (PCL) and polyglycolide (PGA), was used to produce miscible 50/50 PCL-PGA blended electro-spun fibers. These biofibers were compounded into a PCL matrix to form these unique bio-based plastic composites. This bio-based composite was developed for fabricating temporary implanted orthopedic devices, which can be produced from these biopolymer composites that biodegrade and can be replaced by natural tissues. In order to further increase strength, single-walled carbon nanotubes (SWNTs) were purified and wrapped with double-stranded deoxyribonucleic acid (dsDNA) and introduced in the fibers. This design utilizes the long degradation rate of PCL while acquiring the strength of PGA, along with increasing the interfacial bonding between the fibers and matrix, while the dsDNA is used to improve the dispersion of the SWNTs in the fibers, which enables the mechanical properties of the composite to be enhanced. These devices can be used for permanent applications, such as total knee replacements, where the device is essential for the lifetime of the patient, or for temporary applications, such as bone fractures, where the device is no longer needed once the patient has healed, with modification of the formulation and process [116]. There are numerous and growing examples of such bio-based fiber-reinforced composites in the biomedical industry and are discussed in a book published in 2018 [117].
- Many of the medical industry applications of a new product called Fibertuff™ are covered in the chapter by Robert Joyce in this book (see Chapter 28).

26.6 Limitations of bio-based plastic composites and challenges for growth

Research into bio-based polymers has been active, as evidenced by the large number of patents, technical publications, and new products. The rising environmental and sustainability consciousness has motivated efforts for developing many different bio-based composite materials for different end-use applications and as a novel and more sustainable alternative to conventional nonrenewable synthetic fibers such as glass and carbon-reinforced composites.

On the other hand, biocomposite materials are not a problem-free direct substitute substitutes, and they have some drawbacks such as poor moisture resistance (hydrophilicity), fiber/matrix incompatibility, low thermal stability, flammability, poor electrical properties, extraction, processing, surface modification, machining, manufacturing, supply logistic issues, and characterization-associated challenges, along with highly anisotropic properties [118]. However, due to challenges related to the properties of these materials, compared to synthetic polymers, such as their resistance to chemicals or weather, applicability and usage have not dramatically increased yet in many end-use markets compared to petro-based polymers, as was originally forecasted. For demanding applications as well as for those where durability is essential, such drawbacks will preclude biopolymers, bioplastics, and biofibers from displacing incumbent non-bio-based products.

As stated earlier, a common disadvantage almost all of these fibers have is that they are hydrophilic polar polymers and absorb water. When the fibers get wet, they lose some of their mechanical strength, especially tensile strength and rigidity, measured as flexural modulus. Another issue is that each of these has a fairly low degradation temperature, especially when compared to non-biofibers like glass fiber or carbon fiber. A third issue is that since these are all found in nature, there is a lot of inconsistency from one batch grown at one location to another grown elsewhere. This occurs due to differences in their growth due to differences in soil, temperature, and rainfall or water available at each location. So, consistency of mechanical performance is more difficult to maintain compared to synthetic fibers. Some of these issues can be mitigated by innovative compounding techniques that have been discussed in this chapter as well as by improving quality control of incoming raw materials and outgoing quality assurance of products.

An unappreciated complexity that Prof. Midani points out is that there are few in the composites industry with enough experience and knowledge of natural fiber reinforcements, to work with them confidently. He says that there is a large gap between different stages of the value chain. For example, processors of fibers often operate with comparatively low-tech equipment, while composites manufacturers are relatively high-tech. Given this difference, there is a compelling need to bridge this gap between the different value chain actors: at the one end, to educate the farmers, fiber processors, and fabric makers on this new end-use of natural fibers, and at the other end, to educate composites manufacturers of composites on this new type of reinforcement and how it differs from synthetic fibers, whether they are petro-based or glass [119]. We can assume that over time this issue will resolve itself.

In the fiber-reinforced composite industry, in most cases, glass and carbon fibers have poor adhesion and chemical compatibility to the plastics polymer matrix and hence chemical modification of these reinforcing fibers either through surface modification or with use of coupling agents during production processes are used to provide the adhesion and compatibility between the two phases. For using biofibers, several projects targeting the improvement of compatibility of natural fiber reinforcements to various matrixes (bio-based and non-bio-based) have resulted in achieving the goal of providing off-the-shelf reinforcements, similar to glass and carbon fiber. Examples of such projects included using biofiber-reinforced plastic composites using low-twist rovings, unidirectional tapes, woven fabrics, spread-tow fabrics, and prepregs, and most of these have involved flax fibers provided by Bcomp Ltd. (based in Fribourg, Switzerland) and Composites Evolution (based in Chesterfield, UK). Another approach to improve compatibility of the biofiber with the resin matrix is the water-based acrylic resin Acrodur, developed by BASF in the United States. Midani points out that there is also a project called QualiFlax, which is responsible for specifically developing to control and ensure consistent technical performance, and reliable supply and pricing of natural fiber reinforcements, targeted primarily at flax. In addition, other projects like Fibragen in Spain aim to develop genetically selected flax with improved and consistent properties [119].

Table 26.5: Fiber production by country.

Country	Share (%)
China	23%
India	21%
USA	16%
Brazil	10%
Bangladesh	6%
Sri Lanka	5%
Turkey	3%
Uzbekistan	2%
Australia	2%
Mexico	2%
Pakistan	1%
Turkmenistan	1%
Others	8%

Many of the supply chain issues on biofibers have been covered except for one that has not been addressed so far. The latest data available for global fiber production by country is from the 2018 data from Discover Natural Fibres Initiative [120], along with data from other sources, which is compiled and shown in Table 26.5. As Midani has pointed out, natural fiber reinforcements are mostly derived from plants grown mostly in China, India, Brazil, Bangladesh, and Sri Lanka, while till a few years ago, most of the larger manufacturers of composites were mostly located in Western Europe, the United States, and Japan. Such a geographic barrier made it difficult for composites manufacturers to easily access natural fiber reinforcements for large-scale production and even for research. Moreover, the wide ban on growing hemp in certain countries significantly limits its availability, both for trials and production. Hence, composite manufacturers had to develop composites based on biofibers available locally, such as flax, which is mostly grown in France and Belgium, close to regions where many manufacturers of composites are located. In fact, the easy and quick access to flax in Western Europe has significantly increased its development as a viable biofiber for trials, and now flax has become the most widely used natural fiber reinforcement [119, 120]. Recently, the use of the natural fiber composites based on locally available has been growing in India, China, Brazil, and SE Asia. The world production of natural fibers is estimated at about 34 million tons in 2022 by Discover Natural Fibres Initiative [119]. If we look at the global bio-based composites market, BCC Research has estimated that the global bio-based composites market grew from $5.9B in 2023 to the projected $9.0B by 2029, a CAGR of 7.4%, driven by demand in automotive, construction, and consumer goods sectors [121].

26.7 Conclusions

While biopolymers and biofibers have several limitations, it is clear from this chapter that interest in them is very high, not just related to research and finding creative ways to overcome some of these limitations, but also to move forward with scaling up these innovations to commercial products. Issues related to moisture absorption and dimensional stability are handled by encasing the biofiber in a plastic material that provides a moisture barrier to protect the fiber. Another limitation of poor interfacial adhesion is overcome with the use of coupling agents and compatibilizers, using techniques covered in a couple of online courses and a blog [3, 122].

Biofiber reinforcements have many advantages over glass fibers. For example, bio-fibers have low density, high specific properties, are biodegradable, are derived from renewable resources, have a small carbon footprint, and provide good thermal and acoustical insulation. These relative advantages must be emphasized when developing new products from bio-based fiber plastic composites. This is the main reason why the automotive industry is at the forefront of adoption of such composites these days. OEMs of transportation vehicles are taking advantage of such benefits, as more and more specification requirements for modern-day vehicles are demanding such adoptions because of the growth of electric vehicles and the ever-stricter environmental regulations [11].

Furthermore, there are many enthusiastic researchers, engineers, architects, and designers who are exploring new horizons for b-based plastic composites. For example, in the Netherlands, an entire façade of a building was constructed using a bio-composite, using bio-based epoxy resin as the matrix, with hemp as the reinforcing fiber. Also, in the summer of 2021, two composite pavilions were built from bio-based bioplastic composites in Germany. The first one, produced using flax-reinforced bio-based epoxy using a mold-less filament winding process, was called Livmat™. The second was the Biomat™ composite pavilion, which was constructed with flax and hemp pultruded rods. Other interesting applications of biofiber-reinforced composites include:

- Hardcase travel bags by a Danish company called Projektkin, available from their online store for US$450.
- Applications for printed circuit boards are being pursued by another UK-based company called Jiva, using fully sustainable substrates.
- Greenboats, a German-based company, which is famous for its iconic flax 27 Daysailer™, 2021 installed the first NFC nacelle in the Netherlands in summer in a test program to study the performance of such new materials in the wind industry [9].
- A specific application on hemp-fiber based thermoset plastic composites is also covered in the chapter by Saltman and Saltman, with successful commercialization for wind blade and automotive applications [109]. See Chapter 27.

As more and more bio-based polymers and biofibers as well as compounds and composites are being adopted, applications will grow and the total volume of such biomaterials should see an attractive annual increase, estimated by several market reports to grow at CAGRs in the range of 7 to 12%.

References

[1] Stoyko F, Debes B. Handbook of Engineering Biopolymers – Homopolymers, Blends and Composites, Cincinnati, OH, USA: Hanser Gardner Publications, Inc.; ISBN: 1-68015-262–9, 1-56990-405–7 2007.

[2] Díez-Pascual AM, Cinelli P. "Synthesis and Applications of Biopolymer Composites", Basel, Switzerland: MDPI; ISBN: 3039211323 & 9783039211326 2019.

[3] Adur A. "Compatibilization of Polymer Alloys Blends and Composites", online course taught on specialchem.com, 2002-2004, available from SpecialChem, Paris, France.

[4] Marechal E. "Surface Modification of Natural Fibers: Chemical Aspects", Chapter 2 in book, "Handbook of Engineering Biopolymers – Homopolymers, Blends and Composites", Cincinnati, OH, USA: Hanser Gardner Publications, Inc; pp. 49–77. ISBN: 1-68015-262–9, 1-56990-405–7.

[5] Constable RC, Adur AM. Chemical Coupling of Glass-Fiber Reinforced Polypropylene Using Acid or Anhydride-Modified Polypropylenes. New York City, NY: SPE 49th ANTEC"91 at, April 2–5 1991; Preprints 1892–1896.

[6] Adur AM, "Surface modification of glass fibers & other reinforcements using Ethylene-Maleic Anhydride alternating copolymers", InnoPlast's Polymer Compounding for Innovations in Plastics Industry, Hilton Airport Hotel, Atlanta, GA, USA Sep. 23–25, 2014; https://www.prlog.org/12348262

[7] Adur AM. Using ZeMac® copolymers to modify surface chemistry of glass fibers to improve performance of glass-reinforced polyamides. In 2017 International Congress on Glass Annual Meeting, Istanbul, Turkey: Oct. 22 to 25 2017.

[8] Adur AM. Modifying surface chemistry to improve performance of polyamides reinforced with glass fibers and other functional additives for tailoring polyamide properties to meet automotive industry requirements. In AMI's 2nd Performance Polyamides USA Conference, Pittsburgh, PA, USA: 2018, Nov. 6–7, 2018.

[9] Plotz C. What is a Biofiber?. Int Fiber J 2021;1:40–42. https://fiberjournal.com/what-is-a-biofiber/.

[10] Bledzki AK, Gassan J. Composites reinforced with cellulose based fibres. Prog Polym Sci 1999;24:221–274. https://doi.org/10.1016/S0079-6700(98)00018-5/.

[11] Midani M, Elseify LA. Natural fiber composites – A practical guide for industrial utilization. Int Fiber J 2022;1: https://fiberjournal.com/natural-fiber-composites-a-practical-guide-for-industrial-utilization/.

[12] Ngo T-D. "Natural Fibers for Sustainable Bio-Composites", in book "Natural and Artificial Fiber-Reinforced Composites as Renewable Sources", Günay E, editor. IntechOpen, London, UK: editor 2018, ISBN:978–1-78923-061–1/.

[13] Ilyas RA, Sapuan SM, Ibrahim R, Abral H, Ishak M, Zainudin E, Asrofi M, Atikah MSN, Huzaifah MRM, Radzi AM, Azammi AMN, Shaharuzaman MA, Nurazz NM, Syafri E, Sari NH, Norrrahim MNF, Jumaidin R. Sugar palm cellulosic fiber hierarchy: A comprehensive approach from macro to nano scale. J Mater Res Technol 2019;8:2753–2766. https://doi.org/10.1016/j.jmrt.2019.04.011/.

[14] Luhar S, Suntharalingam T, Navaratnam S, Luhar I, Thamboo J, Poologanathan K, Gatheeshgar P. Sustainable and renewable bio-based natural fibers and its application for 3d printed concrete: A review. Sustainability 2020;12(10485): https://doi.org/10.3390/su122410485/.

[15] Latif R, Wakeel S, Khan NZ, Siddiquee AN, Verma SL, Khan ZA. Chemical composition of natural fibers" in "Surface treatments of plant fibers and their effects on mechanical properties of fiber-reinforced composites: A review. J Reinf Plast Compos 2019;38:15–30. https://doi.org/10.1177/0731684418802022.

[16] Gholampour A, Ozbakkaloglu T. A review of natural fiber composites: Properties, modification and processing techniques, characterization, applications. J Mater Sci 2020;55:829–892. https://doi.org/10.1007/s10853-019-03990-y/.

[17] Taj S, Munawar MA, Khan S. Natural fiber-reinforced polymer composites. Carbon 2007;44:129–144. Google Scholar.

[18] Townsend T, "Natural Fibers and the World Economy, July 2019, https://dnfi.org/coir/natural-fibers-and-the-world-economy-july-2019_18043/

[19] Rowell RM. Potentials for jute-based composites. In paper presented at Jute India '97 Conference & Expo, New Delhi, India: October 20–22 1997, https://www.fpl.fs.fed.us/products/publications/specific_pub.php?posting_id=13212/.

[20] Reddy SRT, Prasad AVR, Ramanaiah K. Tensile and flexural properties of biodegradable jute fiber reinforced poly lactic acid composites. In Materials Today: Proceedings, 2020. 44, Part 1, December, pp. 917–921.

[21] "Future fibres", https://www.fao.org/economic/futurefibres/fibres/coir/en/

[22] "Hemp Fiber: Properties, processing and uses", https://www.textileblog.com/hemp-fiber-properties-processing-and-uses/

[23] "What is hemp fiber?", https://secretnaturecbd.com/blogs/cbd/what-is-hemp-fiber/

[24] https://en.wikipedia.org/wiki/Bagasse

[25] http://www.paperpulping.com/application/bagasse-pulp-making.html\

[26] "Biofuel and electricity cogeneration from sugarcane", https://tractorexport.com/biofuel-and-electricity-cogeneration-from-sugar-cane/

[27] Rathore R, "Bagasse FAQ", June 30, 2020 https://www.biogreenchoice.com/blogs/blog/bagasse-faq/

[28] https://www.britannica.com/plant/bamboo/

[29] Kadivar M, Gauss C, Ghavami K, Savastano H Jr. Densification of Bamboo: State of the Art" September 2020, Materials. Densification Bamboo State Art 13(19):4346.

[30] https://en.wikipedia.org/wiki/Kenaf/

[31] Nanko H, Button A, Hillman D. "The World of Market Pulp", WOMP, LLC, Appleton, WI, USA: 2005. pp. 258. ISBN 0–615-13013-5.

[32] "Rice Husk", https://www.dairyknowledge.in/article/rice-husk/

[33] Phonphuak, N N, Chindaprasirt K. Chapter 6 – Types of waste, properties, and durability of pore-forming waste-based fired masonry bricks", in "Eco-Efficient Masonry Bricks and Blocks, Pacheco-Torgal F, Lourenço PB, Labrincha JA, Kumar S, Chindaprasirt P, editors. Sawston, Cambridge, UK: Woodhead Publishing; 2015. pp. 103–127, ISBN 9781782423058.
[34] Mansaray KG, Ghaly AE. Thermo gravimetric analysis of rice husks in an air atmosphere. Energy Source 1998;20:653–663.
[35] FAO, XXI 1, April 2018, https://www.fao.org/markets-and-trade/commodities/rice/en/
[36] https://knoema.com/atlas/World/topics/Agriculture/Crops-Production-Quantity-tonnes/Rice-paddy-production
[37] "Corn Husk", https://www.specialtyproduce.com/produce/Corn_Husk_473.php#
[38] Philip Joseph Burton PJ. "Towards Sustainable Management of the Boreal Forest", New York, USA: NRC Research Press, Great River; 2003. pp. 759. ISBN 978-0-660-18762-4.
[39] Rowell RM, editor. "Handbook of wood chemistry and wood composites", Boca Raton, FL, USA: CRC Press; 2005.
[40] "Abaca", https://www.fao.org/economic/futurefibres/fibres/abaca0/en/
[41] Chand N, Fahim M. Tribology of natural fiber polymer composites. Second Ed 2020 Elsevier Ltd ISBN 978–0-12-818983-2.
[42] Debnath S. Chapter 3 – "Sustainable production of bast fibres", in "Sustainable Fibres and Textiles. In Muthu SS, editor. The Textile Institute Book Series, Sawston, Cambridge, UK: Woodhead Publishing; 2017. pp. 69–85. ISBN 9780081020418.
[43] Zafar S, "Biorefinery as a Source of Advanced Biofuels", August 29, 2020, www.cleantechloops.com/biorefinery/
[44] https://www.farmpractices.com/crop-residues-types-management-uses/
[45] "Biomass- Types, Uses, Merits and Demerits of Biomass", byjus.com/biology/biomass/
[46] Huang Y, Petermann J. Interface layers of fiber reinforced composites with transcrystalline morphology. Polym Bull 1996;36:517–524.
[47] Go KL, Aswathi MK, De Silva RT, Sabu T, editors. "Interfaces in Particle and Fibre Reinforced Composites", Amsterdam, The Netherlands: Elsevier Ltd; 2020.
[48] Lau AK-T, Cheung KHY. "1 – Natural fiber-reinforced polymer-based composites" in book "Natural Fiber-Reinforced Biodegradable and Bioresorbable Polymer Composites", Lau AK-T, Hung AP-Y, editors. Sawston, Cambridge, UK: Woodhead Publishing; 2017. pp. 1–18, ISBN 9780081006566.
[49] Faruk O, Bledzki AK, Fink H-P, Saini M. Biocomposites reinforced with natural fibers, 2000-2010. Prog Polym Sci 2012;37(11):1552–1596. https://doi.org/10.1016/j.progpolymsci.2012.04.003/.
[50] Ilyas RA, Sapuan SM. Biopolymers and biocomposites: Chemistry and technology. Curr Anal Chem 2020;16:500–503. do.org/10.2174/157341101605200603095311.
[51] Rajak DK, Pagar DD, Menezes PL, Linul E. Fiber-reinforced polymer composites: Manufacturing, properties, and applications. Polymers 2019;Oct 12 11(10):1667. https://doi.org/10.3390/polym11101667.
[52] Saheb DN, Jog JP. Natural fiber polymer composites: A review. Adv Polym Technol 1999;18:(4):351–363.
[53] Gómez-Gast N, Cuellar MDRL, Vergara-Porras B, Vieyra H. Biopackaging potential alternatives: Bioplastic composites of Polyhydroxyalkanoates and vegetal fibers. Polymers 2022;Mar 10 14(6):1114. https://doi.org/10.3390/polym14061114.
[54] Ilyas RA, Zuhri MYM, Aisyah HA, Asyraf MRM, Hassan SA, Zainudin ES, Sapuan SM, Sharma S, Bangar SP, Jumaidin R, Nawab Y, Faudzi AAM, Abral H, Asrofi M, Syafri E, Sari NH. Natural fiber-reinforced polylactic acid, polylactic acid blends and their composites for advanced applications. Polymers 2022;14(202): https://doi.org/10.3390/polym14010202/.

[55] Ilyas RA, Aisyah HA, Nordin AH, Ngadi N, Zuhri MYM, Asyraf MRM, Sapuan SM, Zainudin ES, Sharma S, Abral H, Asrofi M, Syafri E, Sari NH, Rafidah M, Zakaria SZS, Razman MR, Majid NA, Ramli Z, Azmi A, Bangar SP, Ibrahim S. Natural-Fiber-Reinforced Chitosan, Chitosan Blends and their nanocomposites for various advanced applications. Polymers 2022;14(874): https://doi.org/10.3390/polym14050874.

[56] Ilyas RA, Zuhri MYM, Norrrahim MNF, Misenan MSM, Jenol MA, Samsudin SA, Nurazzi NM, Asyraf MRM, Supian ABM, Bangar SP, Nadlene R, Sharma S, Omran AAB. Natural fiber-reinforced Polycaprolactone green and hybrid biocomposites for various advanced applications. Polymers 2022;14(182): https://doi.org/10.3390/polym14010182/.

[57] Mohammed AABA, Omran AAB, Hasan Z, Ilyas RA, Sapuan SM. Wheat biocomposite extraction, structure, properties and characterization: A review. Polymers 2021;3(3624): https://doi.org/10.3390/polym13213624.

[58] Kuciel S, Romanska P, Liber-Knec A. Polyamides from renewable sources as matrices of short fiber reinforced biocomposites. Polimery 2012;57:627–634. https://www.researchgate.net/publication/274882725/.

[59] Jumaidin R, Diah N, Ilyas R, Alamjuri R, Yusof F. Processing and characterization of banana leaf fiber reinforced Thermoplastic Cassava Starch composites. Polymers 2021;13:1420. https://doi.org/10.3390/polym13091420.

[60] Natural polymers, biopolymers, biomaterials, and their composites, blends, and IPNs. In Thomas S, Ninan N, Mohan S, Francis E. editors, Print Book, Toronto, Canada: Apple Academic Press, 2021.

[61] "Abaca", https://www.fao.org/economic/futurefibres/fibres/abaca0/en/

[62] Hablot E, Matadi R, Ahzi S, Averous L. Renewable bio-composites of dimer fatty acid-based polyamides with cellulose fibers: Thermal, physical and mechanical properties. Compos Sci Technol 2010;70:504–509. https://doi.org/10.1016/j.compscitech.2009.12.001.

[63] Kiziltas A. Bio-based polyamides reinforced with cellulose nanofibers-processing and characterization. In paper presented at SPE's Automotive Composites Conference & Exhibition (ACCE), Novi, Michigan, USA: Sept 2015, https://www.researchgate.net/publication/283505156/.

[64] (a) "MIT team engineers plant-derived composite with potential for stronger, tougher applications", March 4, 2022, https://www.compositesworld.com/news/mit-team-engineers-plant-derived-composite-with-potential-for-stronger-tougher-applications/ (b) Rao, A; Divoux, T; Owens, CE; Hart, AJ, "Printable, castable, nanocrystalline cellulose-epoxy composites exhibiting hierarchical nacre-like toughening", Cellulose 29, 2387–2398 (2022). https://doi.org/10.1007/s10570-021-04384-7/ (c) https://www.goodnewsnetwork.org/new-plant-derived-sustainable-plastic-is-tough-as-bone-mit/

[65] Sanadi AR, Caulfield DF, Jacobson RE, Rowell RM. Renewable agricultural fibers as reinforcing fillers in plastics: Mechanical properties of Kenaf Fiber-Polypropylene composites. Ind Eng Chem Res 1995;34(5):1889–1896. https://doi.org/10.1021/ie00044a041/.

[66] Akil HM, Omar MF, Mazuki AAM, Safiee S, Ishak ZAM, Abu Bakar A. Kenaf fiber-reinforced composites: A review. Mater Des 2011;32:4107–4121. https://doi.org/10.1016/j.matdes.2011.04.008/.

[67] United States Patent 7,445,835, Serizawa S, Inoue K, Iji M. "Kenaf-fiber-reinforced resin composition", Tokyo, Japan: NEC Corporation; issued November 4, 2008, assigned to.

[68] Adur AM, Majali AB, Iya VK. Development of radiation processed wood-polymer and bagasse-polymer composites. In paper presented at the High Polymer Symposium, IIT, Kanpur, U.P., India: January 21–23 1972. Preprints, pp. 135.

[69] Indian Patent 139,260. Iya VK, Adur AM, Majali AB. Bagasse-Flyash polymer composites. In Iya VK, Majali AB, Adur AM, Chem. Abstr, 1980. 92, 77703.

[70] Verma D, Gope PC, Maheshwari, Sharma RK. Bagasse fiber composites-A review. J Mater Environ Sci 2012;3(6):1079–1092. https://www.researchgate.net/publication/284625466/.

[71] Bansal S, Ramachandran M, Raichurkar P. Comparative analysis of bamboo using jute and coir fiber reinforced polymeric composites. Mater Today Proc 2017;4(2): Part A 3182–3187. https://www.sciencedirect.com/science/article/pii/S22147853173041210/.

[72] Goriparthi BK, Suman KNS, Mohan Rao N. Effect of fiber surface treatments on mechanical and abrasive wear performance of polylactide/jute composites. Compos Part A Appl Sci Manuf 2012;43 (10): October 1800–1808. https://www.sciencedirect.com/science/article/abs/pii/S1359835X12001649/.

[73] Rana AK, Mandal A, Mitra BC, Jacobson R. Short jute fiber-reinforced polypropylene composites: Effect of compatibilizer. J Appl Polym Sci 1998.

[74] Nkeuwa WN, Zhang J, Semple KE, Chen M, Xia Y, Dai C. Bamboo-based composites: A review on fundamentals and processes of bamboo bonding. Compos Part B Eng 2022;235:109776. https://www.researchgate.net%2fpublication%2f359018108/.

[75] Tahir PM, Ahmed AB, SaifulAzry S, Ahmed A. Retting process of some bast plant fibres and its effect on fiber quality: A review. BioResources 2011;6(4):5260–5281.

[76] Popat TV, Patil AY. A Review on bamboo fiber composites. Iconic Res Engg J Aug 2017;1 (2):54–72. IRE Journals, 1700041.

[77] Kaur N, Saxena S, Gaur H, Goyal P. A review on bamboo fiber composites and its applications. In paper presented at 2017 International Conference on Infocom Technologies and Unmanned Systems (Trends and Future Directions), ICTUS; https://doi.org/10.1109/ICTUS.2017.8286123/.

[78] Muhammad A, Rahman M, Hamdan S, Sanaullah. Recent developments in bamboo fiber-based composites: A review. Polym Bull 2019;76:2655–2682. https://doi.org/10.1007/s00289-018-2493-9/.

[79] Radzi AM, Sheikh Ahmad Zaki SA, Hassan MZ, Ilyas RA, Jamaludin KR, Daud MYM, Aziz SA. Bamboo fiber-reinforced thermoset and thermoplastic polymer composites: A Review of properties, fabrication, and potential applications. Polymers 2022;Apr 14(7):1387. https://doi.org/10.3390/polym14071387;https://www.ncbi.nlm.nih.gov/pmc/articles/PMC9003382/.

[80] Lee S-Y, Chun S-J, Doh G-H, Kang I-A, Lee S, Paik K-H. Influence of chemical modification and filler loading on fundamental properties of bamboo fibers reinforced polypropylene composites. J Compos Mater 2009;43:1639–1657. https://doi.org/10.1177/0021998309339352/.

[81] Shannon. Method of Impregnating Bamboo with Synthetic Resin", United States Patent 2352740, Issued May 14, 1944, assigned to Bakelite Corp., New York City, NY, USA.

[82] Wang UP. papers presented at Study Group meeting on "Impregnated Fibrous Materials", held at Bangkok, Thailand, Proceedings, Vienna, Austria: I.A.E.A.; 1968. pp. 35. 293.

[83] Hao, PLC; Wang, UP, Report on I. A E. A. contract # 565/RI/RB, Vienna, Austria: I.A.E.A; 1970.

[84] Mani RS, Jayaraman MT. Preparation of bamboo-plastics composites. Indian J Technol 1969;7 (1):16–22.

[85] Adur AM. "Studies on the Radiation Processing of Bamboo- Polymer Composites, Part IV of M.Sc, Thesis, Univ. of Bombay: 1974.

[86] Adur AM, Nigam SK. Studies on Cobalt-60 Gamma radiation processing of bamboo polymer composites. Isotpenpraxis 1975;11(1):21–24. https://www.osti.gov/biblio/4219041-cobalt-gamma-radiation-processing-bamboo-polymer-composites/.

[87] Adur AM. Gamma radiation processed bamboo-polymer composites I. Preparation, process parameters and product microstructure. J Radiat Curing 1977;4(4):2–17.

[88] Adur AM. Gamma radiation processed bamboo-polymer composites II. Mechanical properties, moisture content and water absorption characteristics. J Radiat Curing 1978;5(2):4–12. https://www.osti.gov/biblio/5242471.

[89] Adur AM. Gamma radiation processed bamboo-polymer composites III. Possible applications for tensile reinforcement of concrete. J Radiat Curing 1978;5(4):9–16. https://www.osti.gov/biblio/5378997.

[90] "CaliBamboo Introduces First Composite Deck Made from Bamboo", https://www.multihousingnews.com/cali-bamboo-introduces-first-composite-deck-made-from-bamboo/

[91] https://www.calibamboo.com/composite-decking/bamdeck-3g.html/

[92] Oksman K, Bengtsson M. "Wood Fiber Thermoplastic Composites: Processing, Properties, and Future Developments" in "Handbook of Engineering Biopolymers – Homopolymers, Blends and Composites", Stoyko F, Bhattacharyya D, editors. Munich, Germany: Hanser Publishers; 2007. pp.655–671.

[93] Niska KO, Sain M, editors. Wood-Polymer Composites, 1st, Sawston, Cambridge, U.K: eBook, Woodhead Publishing; 2008. ISBN: 9781845694579.

[94] Li Y, "Wood-Polymer Composites", September 2011; .org/10.5772/17579/ in book "Advances in Composite Materials – Analysis of Natural and Man-Made Materials

[95] Ergun Baysa E, Kemal MY, Altinok M, Sonmez A, Peker H, Colak M. Some physical, biological, mechanical, and fire properties of wood polymer composites (WPC) pretreated with boric acid and borax mixture. Constr Build Mater 2007;21:1879–1885. https://www.academia.edu/20153251/.

[96] D; Levins IF, Perry PL, Jr FG, Lindstrom RS. "Technical and economic consideration for an irradiated wood-plastic material", United States Atomic Energy Commission. TID-21434.

[97] Narayanamurti D, Jayaraman MT, Thacker R, Anantanarayanan S. "Note on radiation processed wood-plastic materials. Wood Sci Technol 1970;4:226–236. https://doi.org/10.1007/BF00571857/.

[98] Iya VK, Majali AB, Adur AM, Nigam SK. Radiation processed wood polymer composites. In paper presented at Symposium on Wood Panel Products at, Bangalore, India: 1974.

[99] Adur AM, Majali AB, Thacker RC. "Radiation Processing of WPC and Drierite Plastic Combinations", paper presented at Chemistry symposium; Madras, India (25 Nov 1970) of Proceedings of the Chemistry Symposium, 1970. II. Dept. of Atomic Energy (1970) pp 31–41.https://www.osti.gov/biblio/4655925

[100] (a) Adur, AM; Nigam, SK; Patil, ND, "On the Electrical Properties of Radiation Processed Wood Polymer Composites", paper presented at DAE Chemistry Symposium, Aligarh, U.P., India, December 21–23, 1972, Preprints, I, pp. 333–339.(b) Adur, AM; Nigam, SK, "Electrical Conductivity of Radiation Processed Wood Polymer Composites", J. Radiat. Curing, 1979, 5, (1) pp. 18-24, https://www.osti.gov/biblio/5343202.

[101] Gustafsson K-A, Muller JJ, Wittenberg RA, United States Patent 5,746,958, "Method of producing a wood-thermoplastic composite material", issued May 5, 1998 assigned to Trex Co.

[102] Ramesh M, Rajeshkumar D, Sasikala G, Balaji D, Saravanakumar A, Bhuvaneswari V, Bhoopathi R. A critical review on wood-based polymer composites: Processing, properties, and prospects. Polymers 2022;14(3):589.

[103] Taifor A. A review of wood plastic composites effect on the environment. J Univ Babylon Pure Appl Sci Aug 2016;25(2):360–367. https://www.researchgate.net/publication/317344750.

[104] "Wood Plastic Composite Market – Forecasts from 2021 to 2026", market report by ResearchAndMarkets.com https://www.globenewswire.com/news-release/2022/03/04/2397099/28124/en/Global-Wood-Plastic-Composite-Market-2021-to-2026-Rising-Demand-from-the-Construction-Industry-is-Driving-Growth.htm/

[105] Mclauchlin AR, Thomas NL. "Biodegradable polymer nanocomposites". In "Advances in Polymer Nanocomposites", Sawston, Cambridge, UK: Woodhead Publishing; 2012. pp. 398–430. https://scholar.google.com/citations?view_op=view_citation&hl=en&user=wMwpY_8AAAAJ&pagesize=80&sortby=pubdate&citation_for_view=wMwpY_8AAAAJ:35N4QoGY0k4C/.

[106] Tomka, I, United States Patent 5,663,216, "Reinforced biodegradable polymer ", assigned to Bio-Tec Biologische Naturverpackungen GmbH (Germany) issued September 2, 1997.

[107] Sanadi AR, Young RA, Clemons C, Rowell RM. Recycled newspaper fibers as reinforcing fillers in thermoplastics: Part I-Analysis of tensile and impact properties in polypropylene. J Reinf Plast Compos 1994;13(1):54–67. https://www.researchgate.net/publication/237535647/.

[108] H SN, Pruncu CI, Sapuan SM, Ilyas RA, Catur AD, Suteja S, Sutaryono YA, Pullen G. The effect of water immersion and fiber content on properties of corn husk fibers reinforced thermoset polyester composite. Polym Test 2020;91:106751. https://doi.org/10.1016/j.polymertesting.2020.106751/.
[109] "Innovative biocomposite and bioplastic plastic alternatives that can reduce your carbon footprint and plastic use", https://hempplastic.com/products/
[110] Kistaiah N, Chavan UK, Reddy GR, Rao, Rao MS. Mechanical characterization of hybrid composites: A review. J Reinf Plast Compos July 2014;33(14):1364–1372. https://doi.org/10.1177/0731684413513050/.
[111] K.O O, Ahmad R, Jawaid M, Bingol S, Adebayo GO, Wong YH. Natural Fiber-reinforced composite for Ballistic applications: A review. J Polym Environ 2021;29:3795–3812. https://doi.org/10.1007/s10924-021-02169-4/.
[112] Jana S, Maiti S, Jana S. "Biopolymer-based composites: Drug delivery and biomedical applications". eBook, 2017, Duxford, UK: Woodhead Publishing, an imprint of Elsevier.
[113] Thomas S, Ninan N, Mohan S, Frances E, editors. Natural polymers, biopolymers, biomaterials, and their composites, blends, and IPNs. In Book, Toronto, Canada: Apple Academic Press; 2021.
[114] Sadasivuni KK, Ponnamma D, Kim J-W, AlMaadeed MA, Cabibihan -J-J, editors. Biopolymer composites in electronics. In Print Book, New York, USA: Elsevier; 2017.
[115] Swain Spearman S. Fabrication of biodegradable biopolymer composites for orthopedic applications", Ph.D. Thesis Dissertation, Texas Tech University; https://ttu-ir.tdl.org/2016-06-20T20:01:08Z/.
[116] "Biodegradable and biocompatible polymer composites: Processing, properties and applications". In Shimpi NG, editor. Woodhead Publishing, an imprint of Elsevier, Print Book, United Kingdom: Duxford; 2018.
[117] Andrew JJ, Dhakal HN. Sustainable bio-based composites for advanced applications: Recent trends and future – A critical review. Compos Part C Open Access 2022;7:100220. https://doi.org/10.1016/j.jcomc.2021.100220/.
[118] Midani M, "Natural fiber composites: What's holding them back?", https://www.compositesworld.com/articles/natural-fiber-composites-whats-holding-them-back/
[119] (a) Data from various publications from Discover Natural Fibres Initiative (DNFI), https://dnfi.org/ (b) "World Natural Fibre Update", August 2022, https://dnfi.org/statistics/dnfi-world-natural-fibre-update_34610/
[120] Townsend T. 1B – World natural fibre production and employment" in book, 'Handbook of Natural Fibres. In Woodhead Publishing Series in Textiles, 2nd Edition, 2020. pp. 15–36. https://doi.org/10.1016/B978-0-12-818398-4.00002-5/.
[121] BCC Research. Global bio-based composite material market analysis report: Trends, forecasts, and segmentation by fiber type, resin type, and end-use industry, BCC Research; 2024. https://www.bccresearch.com/market-research/advanced-materials/bio-composites-market.html.
[122] Bicerano J. A Practical Guide to Polymeric Compatibilizers for Polymer Blends, Composites and Laminates. SpecialChem: December 2005, http://polymerexpert.biz/blog/135-compatibilization-of-immiscible-polymers/.

Camille Sobrian Saltman and David Saltman

Chapter 27 Turning Biomass into Business™

27.1 Introduction

As political landscapes shift, regulatory frameworks evolve, and economic conditions fluctuate, ESG (environmental, social, and governance) has continued to be a major force in global finance, attracting trillions of dollars in assets under management.

Bloomberg projects global ESG assets are expected to surpass USD $40 trillion by 2030 [1]. This presents an exponential growth opportunity internationally for bio-based products, even as the US market faces headwinds due to political divisions and legal challenges.

While Europe continues to lead the charge, Asia-Pacific is emerging as a powerhouse, with countries such as Taiwan and China driving strong ESG investment flows. Bloomberg expects Canada as well as expanding Asia-Pacific markets like Japan and Australia to play a leading role in supporting these gains.

Several major trends are driving dollars into ESG investing, which is, in turn, driving the market growth of bioproducts and biocomposites. These include:

- Government regulations in the EU, Asia-Pacific countries, and Canada to reduce plastics pollution, sequester carbon, and achieve higher life cycle performance
- Investor appetite for companies that have a competitive advantage due to being at less risk from climate change, better positioned for the long term and better prepared for uncertainty
- Industry mandates to lower production costs, reduce waste, and reduce carbon emissions thereby increasing the life cycle performance of their products.
- Reducing risk from supply chain disruptions
- Consumer demand for less toxic, more environmentally responsible products

According to NielsenIQ, 72% of global consumers are willing to pay more for sustainable products.[2]

These powerful market trends have led to significant innovations in bio-based polymer chemistry and natural fiber-reinforced materials; however, there continue to be two significant barriers to wide-scale industrial adoption: price and performance. Many biopolymers, including polyhydroxyalkanoates (PHA) and polylactic acid (PLA), are more expensive than petroleum-derived incumbents such as polypropylene (PP) and polyethylene (PE). This is mainly due to scale. New materials have to compete

Camille Sobrian Saltman, INCA Renewtech, 250-300-5254, e-mail: csaltman@incarenewtech.com
David Saltman, INCA Renewtech, 250-300-5253, e-mail: dsaltman@incarenewtech.com

 | https://doi.org/10.1515/9783112212141-027

with the enormous, vertically integrated petrochemical companies that have invested billions of dollars in manufacturing infrastructure. Higher prices impede the formation of the major industrial product development partnerships necessary to develop high-performance alternatives. Until bio-based materials can compete on price, they will remain niche products.

The automotive sector is the one major exception and has been incorporating bio-based materials for the last 30 years. Components that incorporate bio-based materials have been used by many automobile manufacturers around the world, including Audi, BMW, Chrysler, Fiat, Ford, General Motors, Honda, Mazda, Mercedes Benz, Peugeot, Renault, Toyota, Volkswagen, and Volvo. The most common components that have been made with bio-based materials include door panels, seat backs, package trays, dashboards, headliners, trunk liners, and other interior trim parts.

Stricter fuel economy and emissions standards compelled automakers in the early 2000s to develop and adopt lighter weight alternatives to glass-reinforced plastics. This led to exponential growth of the composites industry. Natural fibers further reduced composite weight, which directly improving fuel efficiency, and thereby reduced emissions. See Chapter 26 for further discussion of biofibers.

The global adoption of electric vehicles (EVs) has further driven adoption as manufacturers look to tackle the weight of battery packs, which can add between 450 and 1,000 pounds (about 200–450 kg) to a vehicle's total weight. According to Mordor Intelligence, in 2025, the automotive composites market size is estimated at $10.6 billion with an extremely high projected growth rate of 12% between now and 2030. The Asia-Pacific region currently holds 40% of the marketshare. The American Composites Manufacturing Association (ACMA) reported that in 2024 the global demand for lightweight vehicle composite materials hit a record high of 4.9 billion pounds (2.22 billion kg).

INCA Renewable Technologies is a Canadian-based corporation with the mission to replace glass-reinforced plastics and balsa wood with higher-performance, lower-cost natural fiber composites for our industrial customers. The company's novel products are not only less expensive, lighter, and stronger, they are also far more easily recyclable representing a step-change innovation in the marketplace.

This disruptive advancement was made possible by an internationally recognized team of experts in plant agronomics, fiber processing, biocomposites innovation and manufacturing. The team has a considerable track record, proven over decades, to transform low-cost renewable resources into high-value products destined for the automotive, recreational vehicle (RV), consumer products, wind energy, and marine industries.

At the heart of INCA's business strategy is vertical integration. By utilizing sustainably grown agricultural resources rather than petroleum, fiberglass or wood alternatives, INCA can reduce plastics pollution, sequester carbon, preserve rainforests, and help the world transition from a hydrocarbon to a carbohydrate-based economy. Just as the petrochemical industry leverages its expertise in exploration, extraction, and refining to maximize the return on each barrel of oil, INCA will leverage its expertise in fiber processing and composites manufacturing to maximize the return on each bale of hemp.

27.1.1 Vertical integration

To accomplish this vertical integration, INCA is building a state-of-the-art fiber processing and composites manufacturing facility in Vegreville, Alberta, Canada. This facility will be the world's first vertically integrated composites company. The decision to build in one location reduces capital and operating costs, dramatically increases logistical efficiencies, and delivers significantly expanded economic, social, and environmental benefits to the province.

INCA will purchase the biomass of hemp currently being grown in the Prairie for plant-based protein. Hemp has been legally cultivated in the Canadian Prairies for over 20 years, primarily for that purpose. The remaining stalks contain some of the strongest natural fiber on earth. If properly refined and manufactured, hemp fiber can function as a direct replacement for glass fiber in composites. Because the hemp bales that enter INCA's factory are 45% carbon by weight, over the first 5 years, this operation is projected to sequester over half a gigaton of carbon.

The company will process the biomass in-house and then directly transform this low-cost, renewable resource into a set of proprietary high-value products for major industrial customers. There will be a seamless integration of INCA's operation – from farm gate to factory floor. INCA will acquire this "waste" biomass in the form of baled straw and decorticate it, a process that mechanically separates the long outer bast fiber from the short inner core (called hurd). The bast fiber will be further refined to create the ultraclean material required for its products.

The hurd will be used to manufacture INCA BioBalsa™, a direct replacement for the balsa wood used in boats and wind turbine blades and BioPlastics™ for the consumer products industry. The refined long bast fiber will be manufactured into two additional products: INCA BioPanels™ for the RV industry and INCA PrePregs™ for the automotive industry.

Domestically sourcing and processing hemp lowers raw material and logistics costs and reduces supply chain challenges versus procuring and importing expensive jute or kenaf from Southeast Asia, as other biocomposite manufacturers do. Locating the fiber processing operation close to the hemp fields enables quality control of raw material.

27.1.1.1 A brief history of hemp

Hemp (*Cannabis sativa* L.) was one of the first crops cultivated by humans. Recorded use dates back over 8,000 years. Hemp cultivation began in what is now China and spread quickly across Asia, Europe, Africa, and later South America. It was such a strategic crop that in 1533, King Henry VIII fined English farmers if they did not raise hemp. Literally, Britannia could not have ruled the waves without hemp rope and canvas. Until the late 1800s, most of the paper in the world was produced from hemp pulp.

Hemp was originally introduced to North America in 1606 where it was grown for papermaking, cordage, lamp fuel, and clothing. However, at the turn of the twentieth century, powerful forces set out to delegitimize the cultivar. It was seen as a direct competitor by the pulp and paper industry, and by DuPont, which invented nylon as a petroleum-based substitute for hemp textiles. Backed by powerful industry lobbyists, hemp was included in the 1938 Opium and Narcotic Drug Act in Canada, and in the 1937 Marijuana Tax Act in the US. This legislation essentially killed the hemp industry.

The good news is that hemp is now experiencing a major rebirth in North America. In 1998, it became legal to grow it for commercial production. Canada's Industrial Hemp Regulations, part of the Cannabis Act 2018, allows for whole plant utilization of the crop. Health Canada controls the production, processing, transportation, and sale of industrial hemp. In 2020, there were approximately 53,000 acres of hemp being cultivated, and this is expected to expand dramatically as the market for plant-based protein and fiber grow.

In 2009, the Government of Alberta made a strategic investment in the nascent hemp industry by building a large fiber processing plant at the InnoTech Alberta research facility in Vegreville. INCA is partnering with InnoTech to further advance hemp genetics and agronomics, focusing on cultivators that can provide both seed and fiber yield.

The Alberta government is investing in the hemp industry for the following reasons:

- Hemp matures in as little as 90 days and can be grown with low water and chemical inputs.
- As a rotational crop, it breaks disease cycles and has a long tap root that aerates the soil.
- It yields 4× the biomass of a forest in one season, compared to 25 years of tree growth.
- Every portion of the plant can be commercialized, creating multiple revenue streams.

Thanks to this early government support as well as the long hours of sunshine, which the Canadian Prairie affords, Alberta currently grows 40% of Canada's hemp. INCA has partnered with companies that are breeding high yield hemp seed. By providing seed and signing long-term offtake agreements with the farmers for the residual biomass after protein is harvested, the dual income will make hemp one of the most profitable rotational crops in the region.

27.2 Advanced materials from hemp carbohydrates

Today, many of the same products that are synthesized from hydrocarbon molecules derived from oil and natural gas can be produced from carbohydrates derived from agriculturally grown plants. This includes plastics, textiles, fertilizers, building materials, furniture, papermaking, packaging, industrial fillers in paints, plastics and adhesives, pharmaceuticals, nutraceuticals, and cosmetics.

The potential benefits of bio-based materials are very promising. Petroleum-based products are energy-intensive to refine, generate harmful byproducts in the process, and are often difficult to recover and recycle at the end of their useful life cycles. Plant-based products sequester carbon, generate no toxic byproducts, and are much more easily recovered, recycled, or even composted. INCA has chosen a product development strategy focused on producing materials that can meet its customer's price, performance, and environmental objectives. This work has led to the development of four proprietary products:

1. INCA PrePregs™: A direct replacement for the balsa wood and plastics used in boats and wind turbine blades
2. INCA BioPlastics™: A replacement for glass-reinforced plastics in the consumer products industry
3. INCA BioCore™: A replacement for natural and glass-fiber-reinforced plastics (GRPs) for auto interiors
4. INCA BioPanels™: A direct replacement for hardwood plywood used in the RV industry

All four of INCA's products solve major industry challenges with products that are higher performing, less expensive, far more easily recycled, and with significantly lower carbon, energy and resource impacts. In addition, the company is engineering recycling pathways into INCA's production lines to handle trim waste as well as post-consumer waste once the final products manufactured by our customers reach the end of their useful life. While we are initially focused on four vertical markets, there is no question that INCA's products have applications in a wide range of industrial applications.

27.2.1 INCA product lines

27.2.1.1 INCA PrePregs™

The automotive industry utilizes resin-infused panels (prepregs) to mold three - dimensional interior trim components such as door panels, seat backs, headliners, and package trays. Prepregs can incorporate glass, polymer, or carbon fiber combined with thermal melt resin systems. Prepregs can come in the form of flat panels or non-woven fabrics such as those produced by FlexForm Technologies and Carver Non-Woven Technologies. (Both these companies were formerly built and led by INCA's senior management.) Prepregs will represent a $12.6 billion global business by 2026 according to MarketsandMarkets [3]. The automobile industry was an early adopter of these materials due to their need for lower cost and high strength-to-weight ratio, which resulted in better fuel mileage and customer safety.

Manufacturing non-wovens involves blending natural fiber and polymer fiber such as PP and then depositing this mat on a moving belt. The material is consolidated, carded, cross-lapped, and sent through a needle-punch system to interlace the final product. Automotive manufacturers heat these prepregs in hot presses, surface them with cover stock, and compression mold them into final 3D components. While non-wovens have proven to be of great value to the automotive industry, they depend upon expensive natural fiber, such as jute or kenaf imported from Southeast Asia. They also use a high percentage of polymer fiber, adding considerable cost.

INCA has developed an entirely new methodology to deposit multi-layered mats of hemp and polymer fiber, and consolidate the material on our twin-belt press to produce thermoformable prepregs for Toyota North America. These prepregs are stronger, less expensive, improve side-impact resistance and reduce vehicle weight, improving mileage. Any trim waste generated can be recovered and recycled back into final parts. INCA PrePregs™ use far less petroleum-based polymer, further improving environmental performance. Independent life cycle analysis performed by the research firm GreenStep Solutions demonstrates that INCA PrePregs™ will reduce carbon emissions by 76%, waste production by 89%, and water consumption by 82% compared to conventional biocomposite automotive prepregs (see Figure 27.1).

These factors are of great importance to Toyota, which has engaged INCA in helping formulate their Vision 2030 program, a mission that broadly seeks "to make earth a better place to live, enrich lifestyles, and promote a compassionate society." More specifically, the INCA team is working with Toyota's engineers to develop proprietary prepregs for interior trim components on their 2027 model year EVs. Our products will capture carbon, reduce vehicle weight, improve side-impact resistance, and be formulated to be recovered and reprocessed back into future car parts.

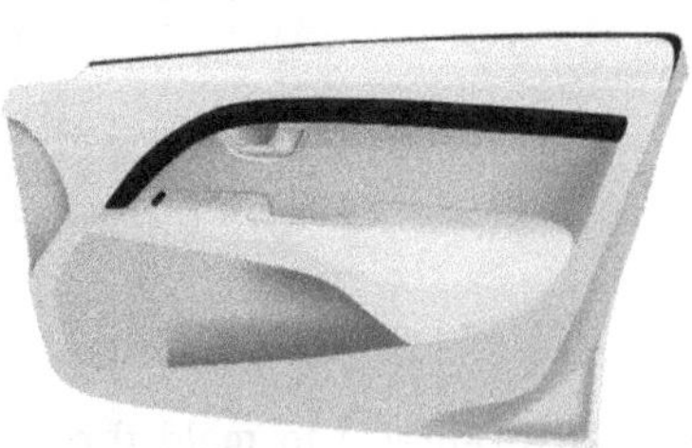

Figure 27.1: INCA PrePregs™ will be molded by Toyota into 3D interior trim components (© 2026 INCA Renewtech).

27.2.1.2 INCA BioPlastics™

Plastics are a $580 billion global industry led by some of the largest petrochemical companies in the world. The top polymer producers include Dow Chemical, Lyondell Basell, Exxon Mobil, SABIC, INEOS, and BASF.

Thermoplastics such as PP, high-density polyethylene (HDPE), and polyvinyl chloride (PVC) are used in packaging, furniture, flooring, roofing membranes, automobile interiors, medical devices, and the countless consumer products that surround us. They are used in virtually every industrial segment of the economy because at relatively low cost they offer high moisture resistance, light weight, corrosion resistance and ability to be heated and shaped using injection, compression, or extrusion molding machines.

Most such polymers are made from oil and natural gas. Petrochemical extraction and refinement is carbon-intensive and generates air, water, and land pollution. Only 9% of plastic products manufactured from oil and gas are recycled. The remainder covers the earth and fills the seas with pollution. Petroleum-based plastics do not break down, they just get smaller. In addition, polymers often contain toxic chemicals that bio-accumulate in humans and animals. Microplastics, with literally thousands of different potentially dangerous chemicals, are now ubiquitous in humans and in all living creatures.

For strength, thermoplastics are reinforced with glass fiber and mineral fillers, which have major environmental impacts of their own. Compounding with glass fibers reinforces the polymer matrix and enhances structural properties, however, there are trade-offs including the energy intensity to manufacture, increased cost and weight, and high tensile strength but low impact strength. Glass is also abrasive on processing equipment and difficult to recycle once products have reached the end of their useful life cycles.

Bio-based plastics currently account for a small fraction of this market but they are growing at a compound annual growth rate (CAGR) of over 20%, driven by legislation and industrial and consumer demand for more sustainable, less toxic products. MarketsandMarkets projects the global bioplastics market was already a $29.7 billion industry in 2025.

Europe has been the major hub for bio-based innovation, and currently produces 55% of all biopolymers, but capacity now is expanding rapidly in Asia and North America. Many of the companies leading the charge are agricultural bioscience companies, or petrochemical companies making the transition to plant-based chemistry.

Leading manufacturers of bio-based PLA, a biodegradable and compostable bioplastic include Nature Works (US), Total Energies Corbion (Netherlands), and Futerro (Belgium). Other notable companies involved in PLA production or related technologies include BASF SE (Germany), COFCO (China), Sulzer Ltd. (Switzerland), and Zhejiang Hisun Biomaterials Co. Ltd. (China).

While these polymers show great promise, they tend to be significantly more expensive than incumbent petroleum-based polymers and lower in structural performance. INCA has developed a novel manufacturing methodology to replace glass fiber with natural hemp fiber as a reinforcement in polymeric composites. In INCA's process, hemp fiber is refined to precise geometries and then blended with various polymers and additives under relatively low heat and pressure. This preserves the structural properties of the natural fiber as well as that of the polymer.

INCA formulates materials to meet customer specifications. For example, one company may want to utilize recovered PP to lower product costs and increase recycled content, while a second customer may specify a bio-based polymer such as PLA or PHA to create 100% bio-based products. These formulations are brought together in the form of compounded "master batch" pellets that can be shipped directly to manufacturers for use in their injection or compression or extrusion molding operations with little or no modifications to their existing lines. According to a life cycle analysis provided by the research firm GreenStep Solutions INCA's BioPlastics will reduce carbon emissions by 23%, waste by 68%, and water consumption by 18% versus glass-reinforced plastics.

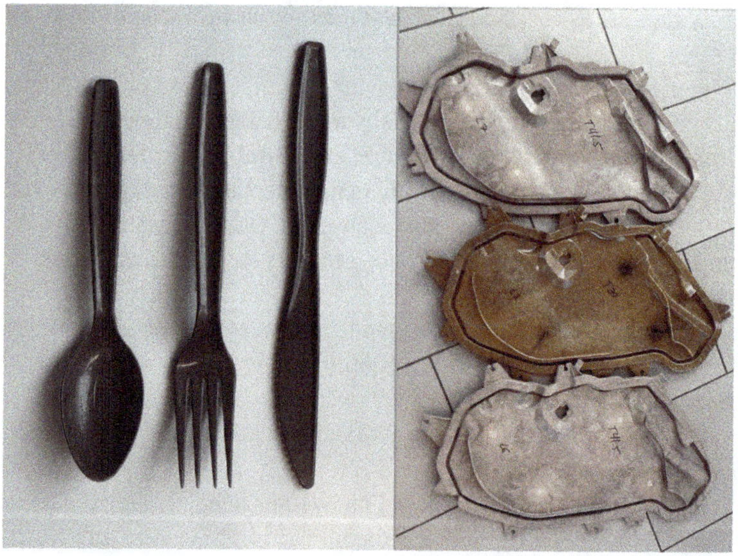

Figure 27.2: INCA BioPlastics™ can be injection molded into (on the left) strong 100% bio-based cutlery or combined with polyprop to create interior automotive trim components such as door panels (on the right), seat backs, headliners, and package trays (© 2026 INCA Renewtech).

Test results demonstrate that product benefits of INCA BioPlastics™ compounded polymers with refined natural fiber can double structural properties. The product is not only stronger, it is less expensive and offers improved stiffness and impact strength, and lower wear on processing equipment. Unlike glass fiber, refined hemp fiber is more amenable to multiple recycling episodes without significant loss in structural properties (see Figures 27.2 and 27.3).

27.2.1.3 INCA BioCore™

The balsa tree (Ochroma pyramidale) is a large, fast-growing species, native to the rainforest regions – from southern Mexico to southern Brazil. Although classified as a hardwood, balsa is exceptionally soft and lightweight. This is because the trunk of the tree has large cells that store water. When cut and dried, these empty cells retain their structure, giving the resulting lumber exceptionally low density and high compressive strength. As a result of these properties balsa is widely used as a core material in automotive, aerospace, and marine applications and even surfboards.

Recently, balsa has become one of the most sought-after materials in the construction of wind turbine blades where it is used to reduce blade weight and increase blade stiffness, particularly in the root section where the mechanical forces are most demanding. This is accomplished by placing shaped balsa in the core of the blade and encapsulating it with styrene-resonated glass fiber to form stressed skin panels. Blade skins are put into tension on one side and compression on the opposite side, and held in opposition by the balsa core to resist collapse.

The global balsa wood market size was approximately USD $169.18 million in 2024 and is projected to reach around USD $308.73 million by 2034, with a CAGR of approximately 6.2% between 2025 and 2034 [4]. The US research lab, formerly known as the National Renewable Energy Laboratory, found that a single 300-foot turbine blade requires 5,300 cubic feet of balsa [5].

Historically, most balsa wood came from the rainforests of Ecuador, but the species has been largely clear-cut, with devastating impacts on ecosystems and indigenous communities. Sixty percent of balsa wood is now plantation-grown. However, without the many years required to grow mature trees large enough to produce solid lumber, the industry has adopted end-grain balsa – essentially slicing logs across the grain, milling rectangular pieces, and gluing them back together to form board stock. This has added considerable cost to the product, reduced consistency, and lowered technical properties. As a result, many blade manufacturers are turning to high-density polyethylene terephthalate (PET) type of closed cell thermoplastic structural foam – for the tips and mid-sections of the blades. However, PET foam lacks the compressive and sheer strength to replace wood in the root sections. It is also more expensive and petroleum-based.

INCA engineers have developed a novel methodology to transform refined hemp hurd into a balsa-like material with high, uniform compressive strength, and a density of 10 pounds per cubic foot – the ideal weight for turbine blades. Unlike balsa wood, it can be formulated to increase moisture resistance, and “tuned” to be compatible

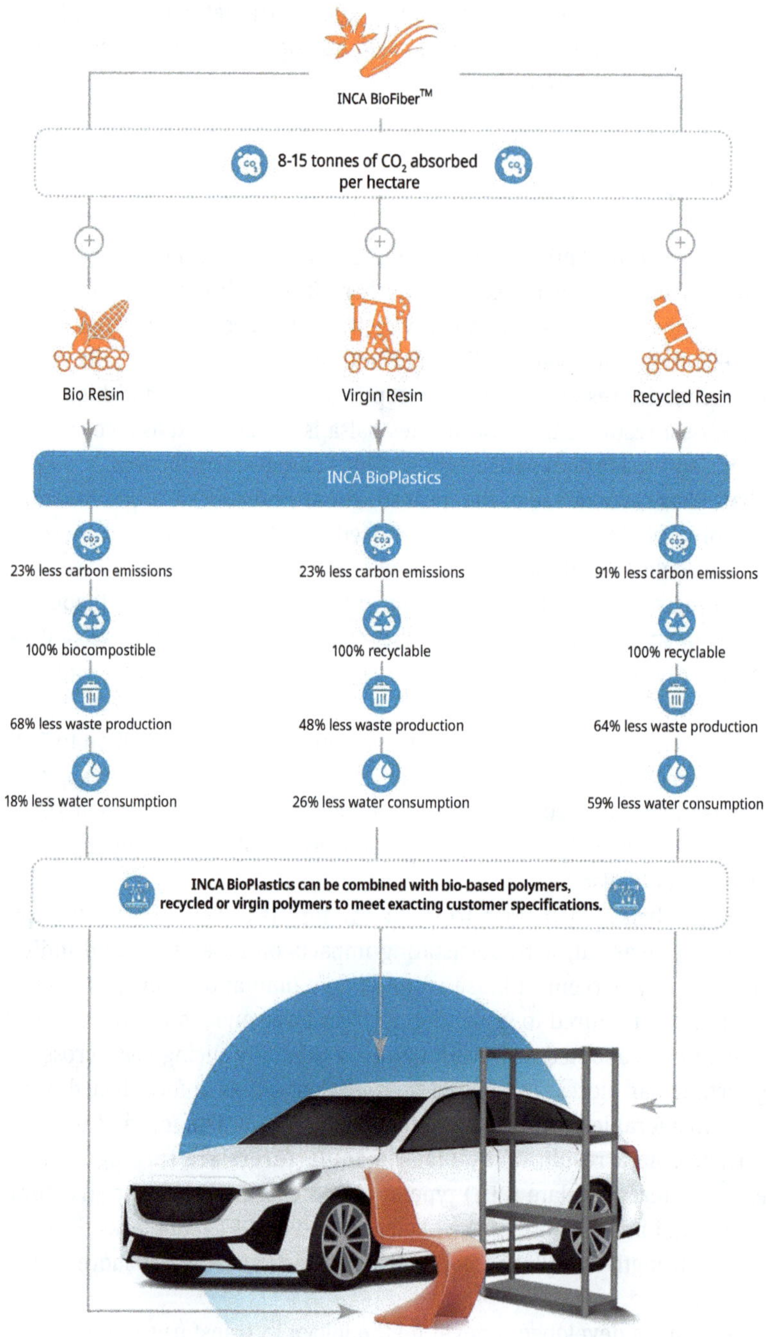

Figure 27.3: INCA BioPlastics™ combined with bio-polymers, recycled polymers and virgin polymers versus conventional glass reinforced petroleum based plastics (© 2026 INCA Renewtech).

with specific resin systems. Unlike PET and PVC foams, it has the strength to be used in all sections of the turbine blade. These characteristics provide significant advantages to engineers who are no longer forced to compensate for variations in the material properties of plantation-grown trees.

Constructed from hemp hurd cellulose, INCA BioCore™ presents a product of comparable density but with higher compressive strength, thanks to the randomness of fiber placement – which delivers uniform properties in all directions. Unlike balsa, it can be formulated to reject moisture and attract specific binding resins. It has the shear strength required for installation in all sections of the wind turbine blades, including the root section.

INCA has signed a strategic development agreement with Gurit to develop and manufacture BioBalsa™. Gurit develops and manufactures advanced composite solutions for the wind energy, aerospace, and marine industries, and their portfolio of structural core materials, prepregs, adhesives, and engineering services is recognized world-wide. The industries Gurit serves are actively seeking an alternative core material that can deliver the compressive and shear strength of balsa as well as meet their aggressive sustainability goals.

GreenStep Solutions compared INCA BioCore™ to balsa wood and high-density PET foam. The study considered the environmental impacts of raw material extraction, processing, manufacturing, transportation, and recovery. BioCore™ has 164% less carbon impact than PET and 107% less than balsa wood. The manufacturing of BioCore™ produces 93% less waste than cutting and milling wood, and 99.6% less than PET. It also consumes 93% less water during manufacturing. Each cubic meter absorbs 262 kg of CO_2e (see Figure 27.4).

27.2.1.4 INCA BioPanels™

Since the 1970s, the RV industry has been dependent upon rainforest hardwood plywood from Southeast Asia to construct side walls and roofing systems. Hardwood plywood's strength, light weight, and smooth surface make it a product in high demand not only for RVs but also for transport trailers, furniture, decorative paneling, and even Hollywood movie sets.

In 2021, the RV industry alone used 620 million board feet of lauan plywood [4]. Unfortunately, the popularity of this material has led to devastation of primordial rain forests throughout Indonesia, Malaysia, Borneo, and the Philippines. Entire ecosystems are being lost, species endangered, and indigenous communities displaced. Plywood prices have skyrocketed, quality has fallen, and manufacturers are actively seeking alternatives.

To construct an RV, sheets of 4′×8′ and 5′×10′ plywood are seamed together like stick frame housing. Exterior wall panels are laminated with aluminum or fiberglass, and inte-

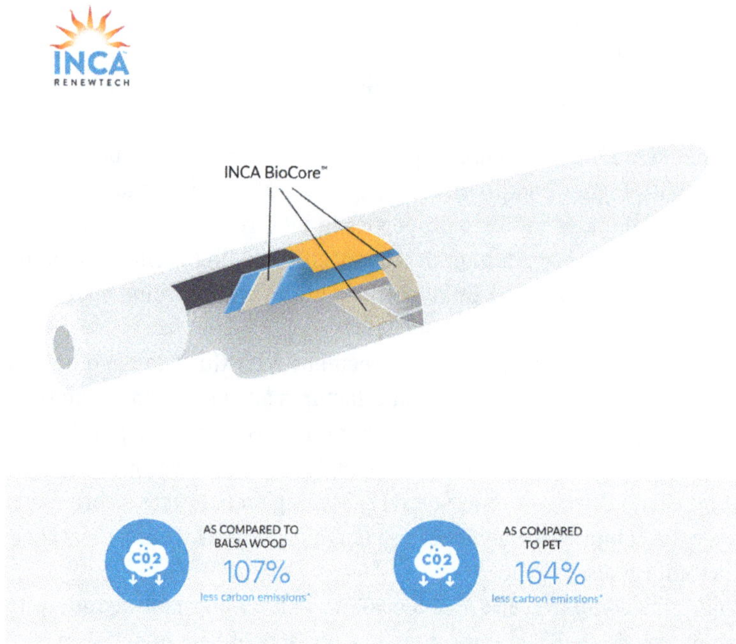

Figure 27.4: INCA BioCore™ versus conventional alternatives (© 2026 INCA Renewtech).

rior walls with paper or vinyl. Urethane foam insulation is sandwiched between the inner and outer walls.

However, plywood presents significant performance problems for RV owners. It off-gasses toxic formaldehyde, absorbs moisture, and burns easily. After just a few years, seams telegraph through the exterior walls, and the strength and visual appearance of the vehicles are compromised.

Oriented strand board is being used as a plywood substitute, but it is much heavier, and it too absorbs water and off-gasses formaldehyde. Fiberglass-reinforced plastic (FRP) panels of various types are being used in higher-end RVs. Manufacturers such as Crane Composites and Vixen Composites create multi-layered wall systems with expanded polystyrene cores, FRP skins, and aluminum framing. In addition to being much heavier, more expensive and labor-intensive, FRP panels offer their own performance challenges, including delamination, warping, and poor insulation properties.

INCA has developed a revolutionary panel product made from a blend of hemp fiber and a proprietary thermoset resin system. Panels will be produced on a twin-belt press capable of making boards up to 10 feet wide and 45 feet long. Rather than having to seam together sheets of plywood, BioPanels™ will save significant labor costs and create lighter, stronger, and more durable unibody vehicles. INCA BioPanels™ will be fire rated, moisture resistant, lighter in weight, and available with class A automotive finishes, saving manufacturers the significant cost of painting (see Figure 27.5).

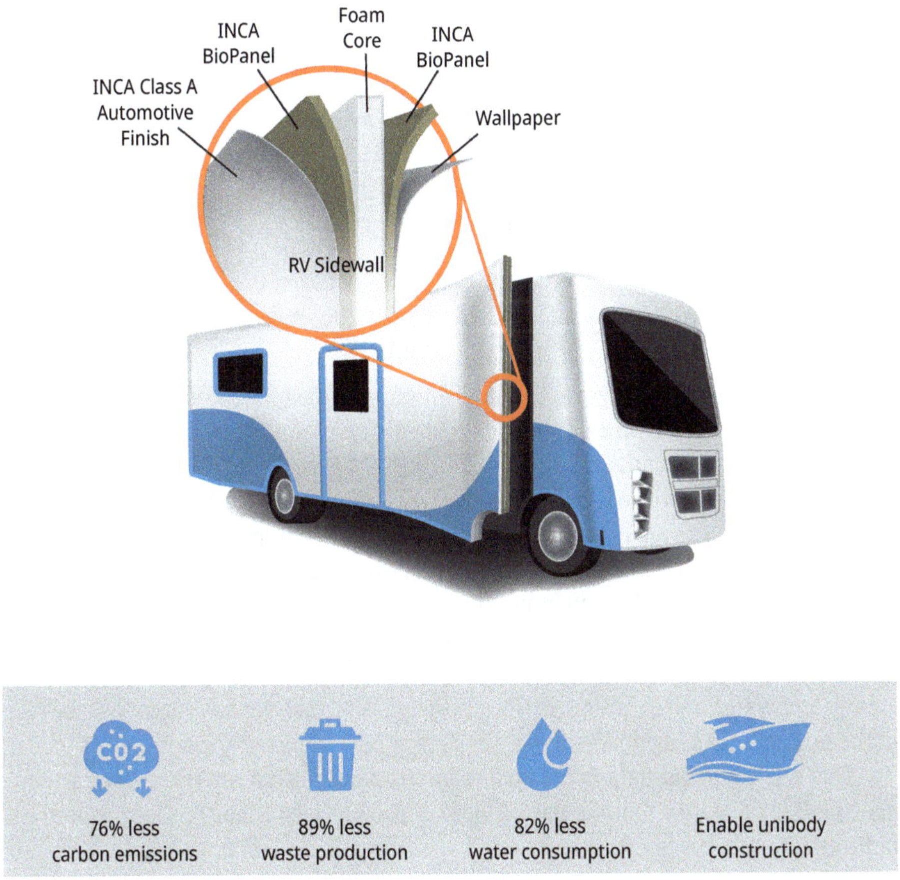

Figure 27.5: INCA BioPanels™ versus rainforest plywood (© 2026 INCA Renewtech).

INCA has signed an exclusive sales agreement with Elkhart-based Genesis Products, Inc., which designs, engineers, and manufactures a range of products for the RV, construction, transportation, and furniture markets. Their customers include Thor Industries, Winnebago, and Forest River. Consolidation is the name of the game in the RV industry. Forest River purchased Coachman, which was in turn acquired by Berkshire Hathaway. Thor Industries now owns Airstream, Heartland, Jayco, and Keystone. Winnebago owns Newmar. These public companies have implemented aggressive ESG targets that include sourcing alternatives for rainforest plywood.

Genesis supplies one-third of the plywood used in the North American RV industry. The company currently operates multiple hot and cold lamination lines in nine factories, with over 1 million square feet of production, assembly, and warehousing space. As the exclusive distributor of INCA BioPanels™, Genesis has facilitated a product development agreement with Winnebago for the Company's entire panel production capacity. Winnebago is partnering with INCA to help them achieve their stated environmental sustainability goals. These include:

- Achieving net-zero greenhouse gas emissions by 2050
- Diverting 90% of all manufacturing waste by 2030
- Reducing fresh water use by 30% in 2050
- Including eco-friendly options on all products by 2025

27.2.2 Life cycle analysis

INCA commissioned GreenStep Solutions, an independent research and certification organization, to undertake a product life cycle analysis of our entire product line versus conventional alternatives.

GreenStep is a Certified B Corporation and has helped over 3,000 private, public, non-profit, and academic institutions to understand the full impact of their operations along the entire life cycle. They utilize the internationally recognized Greenhouse Gas Protocol Product Life Cycle Accounting and Reporting Standard to conduct their work.

Their in-depth research analysis includes consideration of the impact of industrial hemp farming, product formulations, raw material extraction and processing, energy and water inputs, product manufacturing, and fuel consumption at each stage of the process including downstream transportation and distribution.

This science-based data is invaluable in positioning INCA within the larger market trends that are driving the use of bio-based materials. Highlights of the analysis indicate:

INCA PrePregs™ will reduce carbon emissions by 27%, waste production by 30%, and water consumption by 68% compared to competitive natural fiber non-wovens. One square meter will sequester 4.27 kg of CO_2e.

INCA BioPlastics™ when compounded with recycled PP will reduce carbon emissions by 91%, waste production by 64% and water consumption by 59% compared to GRPs. One cubic meter will absorb 517.96 kg of CO_2e.

INCA BioCore™ has 164% less carbon impact than PET and 107% less than balsa wood. Production results in 93% less waste than balsa and 99.56% less than PET. It also means 93% less water consumption during manufacturing. One cubic meter of BioCore absorbs 262.31 kg of CO_2e .

INCA BioPanels™ will reduce carbon emissions by 76%, waste production by 89%, and water consumption by 82% compared to lauan rainforest plywood. One square meter will sequester 4.27 kg of CO_2e.

27.2.3 Sustainable Development Goals

The Sustainable Development Goals (SDGs) are a collection of 17 interlinked global goals designed to be a "shared blueprint for peace and prosperity for people and the planet, now and into the future." The SDGs were developed in 2015 by the United Nations General Assembly and are intended to be achieved by 2030. As a corporation, INCA Renewable Technologies supports all 17 of these goals. Directly the company advances nine of them (see Figure 27.6).

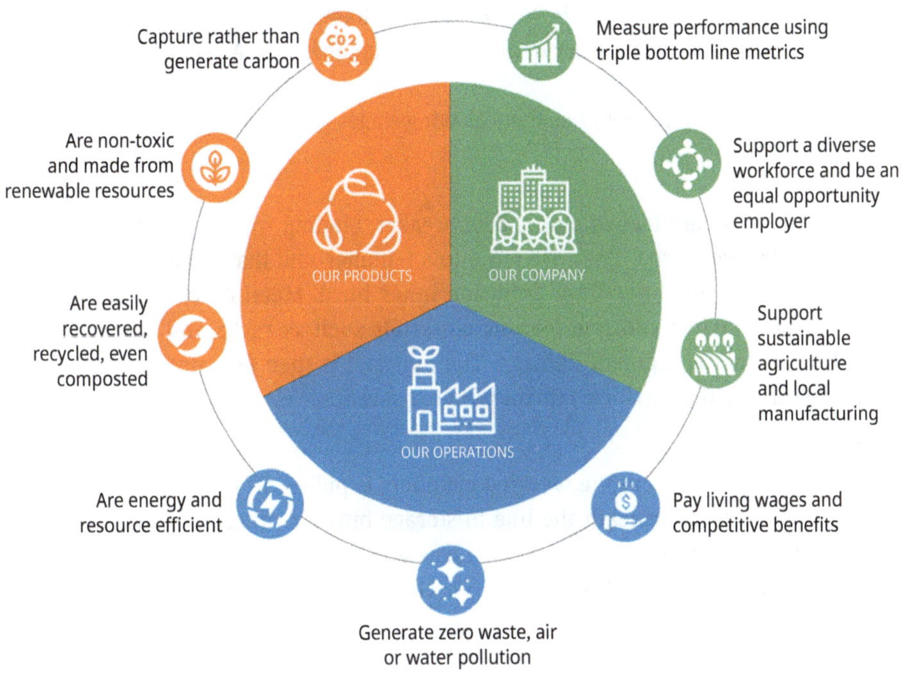

Figure 27.6: INCA Renewtech ESG Commitment (© 2026 INCA Renewtech).

27.3 Manufacturing operations

27.3.1 Fiber processing

INCA will build a 127,000 square foot building on a 23-acre industrial development owned by the Town of Vegreville, in Alberta, Canada. This location is within 150 kilometers of large industrial farmers growing hemp as a rotational crop. It is adjacent to the cross-continental highway and CN Rail line, enabling efficient logistics of incoming raw material via truck, and outgoing product shipped via rail.

Here INCA is installing a state-of-the-art decortication and fiber processing line supplied by DILO Group with the capacity to process 8,960 tons per year of hemp biomass. This will yield approximately 2,240 tons of clean bast fiber, 4,928 tons of hurd, with the remaining mass consisting of excess moisture that will be used to humidify the building. Any field dirt and unusable organic material will be composted on site.

Our engineering team has worked with DILO to design a uniquely configured line capable of producing the ultra-clean, textile grade long fiber for INCA PrePregs™ and INCA BioPanels™, the short fiber (5–9 mm) for INCA BioPlastics™, and clean inner core, called hurd, that will be pulped, dried, and pressed to create INCA BioCore™. No other fiber processing system in North America has the level of sophistication required to produce the exact fiber qualities and geometries INCA requires to make advanced biocomposites.

The decortication process has the following stages:

Stage 1
Whole hemp bales are loaded onto a bale breaker, binding straps removed, and bundled stalks decompressed. They are fed into a hammer mill that crushes the stalks to begin separating the outer fiber from the inner hurd. Metal detectors and weight scales ensure ferrous and non-ferrous materials such as rocks and stones are removed to prevent equipment damage. The material is then fed continuously into a scutching turbine that performs further fiber separation.

Stage 2
The low-density hurd from the working cylinders is pulled off by vacuum and transported by a conveyor beneath the line to storage bins for processing. The bast fiber goes through a series of aggressive step openers and wire drums, opening the vascular fiber bundles to achieve fiber that is 99% clean, a specification essential to successful composites manufacturing.

Stage 3
The refined long fiber is conveyed via overhead ducts to a dedicated storage silo for processing into INCA PrePregs™ for the automotive industry.

27.3.2 INCA PrePreg production

INCA PrePregs™ consist of a non-woven mat of natural and polymer fibers designed to be compression molded into automotive trim components such as door panels, seat backs, headliners and package trays. INCA has developed and patented a methodology to produce these products using a multi-layered scattering system that dramatically increases the strength and decreases the weight of the final parts.

Independent tests at Toyota confirmed that our product outperformed all other non-wovens including those from our former companies, increasing the strength while lowering the weight of each part by an average of 23%.

The manufacturing process has the following stages:

Stage 1

Natural and synthetic fibers are decompressed in a de-baler and passed through an in-line weight scale. The specified fiber formulation is blended using a willow card system that moves the material through rotational rollers. A sophisticated dust collection system recovers any fine fiber from this process and sends it to a hopper to be used in INCA BioPlastics™. The fiber formulation is blown up through ducts to six laying towers. As the continuous conveyor belt travels beneath the towers precise amounts of fiber are sequentially deposited to form a multi-layered mat of material.

This deposition system, designed by INCA and built by TechnoPlants in Italy is unique, patented, and provides an XYZ fiber orientation which dramatically increases the structural properties or the final product. It also enables the recycling of all trim waste back into the production line and even the utilization of post-consumer materials to be remade into new prepregs.

Stage 2

This lofted matrix travels through a set of rollers to reduce the overall volume and prepare the web for mechanical bonding using a needle punch hyper loom to achieve the desired final thickness and density. The consolidated product passes through a final density scanner supplied by Mahlo Canada to document the quality required by our automotive customers per their QA/QC certification programs. A final visual inspection is performed at a light panel to ensure consistent density a final pass through a metal detector verifies that no foreign objects are left in the web.

Stage 3

Material is edge-trimmed, slit to the final width, and either cut into blanks or rolled onto cylinders for packaging and warehousing. Edge-trim material is reintroduced into the blending process, helping us achieve our zero-waste goals. Bar codes are applied to all packaged goods. Rolls that require special liquid resin impregnation are sent to the resination line.

Stage 4

Some automotive programs require that non-woven mats be infused with higher performing resins. Essentially, liquid formulations are converted into foam and infused under pressure into the dry non-woven matrix with extreme accuracy. The material exiting this system is dried using a five-zone radio frequency oven that acts upon only upon the water molecules thus not affecting the resin chemistry. The resulting products are packaged, bar coded, and prepared for shipping to customers.

27.3.3 INCA BioPlastics™ manufacturing

INCA BioPlastics™ are essentially compounded pellets designed for use in injection molding operations. One of the challenges of natural fiber composites versus glass-reinforced plastics is that the natural fiber loses structural integrity if subject to temperatures above 190°F. INCA has developed and patented a methodology to produce these pellets at low temperature, preserving the strength of both the natural and polymer fibers.

A second feature of the product is that it can be manufactured to match customer specifications. Automotive manufacturers may specify that we create a formulation that combines our fiber with PP to meet their side-impact requirements. Another customer may require that their final products be biocompostable to meet legislative mandates. In this latter case, INCA can compound INCA BioPlastic™ with PLA, a starch-based resin, to create pellets that are 100% bio-based.

The manufacturing process has the following stages:

Stage 1
The short bast fiber generated by the decortication process is further refined to geometries suitable for reinforcing polymers destined for injection molding operations. This fiber is 5–9 mm in length, with aspect ratios (length to width) averaging 15:1.

Stage 2
Fiber is transferred from a storage hopper via loss-in-weight feeder and gravity fed into a specially configured twin counter rotating Helicoid blender. Simultaneously, the selected polymer pellets are transferred from a second storage hopper and measured via a weight loss feeder, processed through a triple roll crusher to increase surface area, and deposited into the helicoid blender. While ratios of fiber to polymer can vary per customer specification, the base INCA BioPlastic™ product is 50% fiber and 50% resin. Coupling agents and other performance additives can be applied at this stage.

Stage 3
The compounding mill contains a 32″ die with specified openings of diameter and path length to achieve a targeted pellet diameter and compression ratio. Rollers within the mill compress the fiber and polymer through the die opening. Compression creates enough friction to raise the polymer temperature to its softening point, but not to its melting point. Pellets exiting the die are then cut by knives to specified lengths and conveyed to a screen deck to remove undersized pellets, which are recycled back to the mill.

Properly sized pellets pass through an in-line cooling chamber to remove excess heat prior to drying in a vacuum assist radio frequency oven. Once pellets reach the targeted moisture content, they are conveyed to a receiving storage buffer tank, packaged in drums or gaylords, labeled, and shipped to customers. Unlike glass-reinforced plastics, INCA BioPlastics™ can be recovered and manufactured into new products sequestering carbon for generations.

27.3.4 INCA BioCore™ manufacturing

INCA BioCore™ is essentially an ultra-light fiberboard and a direct replacement for balsa wood and high-density foams. This product is entirely novel and patented. The primary equipment suppliers of this line are SiempelKamp for pulping, TechnoPlant for the transfer systems, and IPCO for the press line.

Manufacturing is accomplished in the following steps:

Stage 1
Refined hemp hurd from the decortication line is stored in several 7,000-gallon buffer tanks which allow for continuous operations of this manufacturing line. It passes through a rotary valve, which is speed-controlled by an in-line weight scale to provide a precise feed rate to the pulping line. Fibers are conveyed to the top of an 85-foot pulping tower, enabling gravity fed flow from one process to the next. At the top level, hurd enters a pre-cooker that uses recovered steam and hot water to soften the cellulose and lignin. Exposure time is short and it is measured in seconds.

Stage 2
Cooked pulp enters the mill which consists of counter-rotating plates that de-fibrillate the fiber to much smaller geometries to maximize fiber surface area. Fiber enters the center of these stainless-steel plates and is transported outward. As the hurd moves over the plate surfaces, it is constantly rolled, compressed, and vigorously manipulated to break down any bundles of fiber to their individual component. This process maximizes surface area, key to achieving the strength properties of the final product.

Stage 3
The resulting slurry now enters the blow-line. Fiber is dried by dramatically accelerating travel speed in a long steel contoured pipe. Dry or liquid resin binders are infused in this moving stream of pulp. Travel distance and turbulation ensures uniform distribution of additives. At the end of the blowline, the fiber mat is captured on a mesh belt and enters a vacuum-assisted radio-frequency drier to remove any excess moisture.

Stage 4
Resonated pulp is deposited on a continuous Teflon belt press where thickness of the mat is reduced to the specified density of the final panel product. The continuous board is edge-trimmed, crosscut, and loaded onto a rotating cooling system. Once cooled, INCA BioCore™ panels are palletized, shrink wrapped, bar coded, and shipped to customer locations.

27.3.5 INCA BioPanels™ manufacturing

INCA will install a 60″-wide twin-steel-belt press built by IPCO in Germany. IPCO is the world's leading manufacturer of double-belt presses and precision scattering systems. The press line will have the technical capabilities to produce thermoset panels for the RV industry and thermomelt prepregs for the auto industry.

The production process will have the following stages:

Stage 1
Natural and synthetic fiber bales are placed on a bale opener system to open the fibers and prepare them for blending. They are distributed onto the blend line where they are weighed and deposited on a weigh belt to create the required formulation. They are mixed using a willow card system, passed through a metal detector to remove impurities, and sent to a feed tower for further processing.

Stage 2
The blended fiber is distributed from the feed tower into a parallel laid carding system and deposited onto an apron belt where it is weighed for correct distribution. The apron belt speeds up or slows down to control the amount of material deposited on the belt. The fiber mat then enters the cross-lapping system where multiple layers of web are laid cross-directionally to vary the orientation of the fibers. The web then passes through a density scanner to automatically correct any profile issues detected on the web.

Stage 3
This lofted matrix travels through a set of rollers to reduce the overall volume and prepare the web for mechanical bonding. Mechanical bonding is achieved in a needle punch hyper loom to achieve the desired final density. The consolidated product passes through a final density scanner to document the quality of the web for quality assurance the customer certifications required by the automotive industry. A visual inspection is performed at a light panel to ensure consistent density of the web and a final pass through a metal detector verifies that no foreign objects are left in the web.

Stage 4
Material is edge-trimmed, slit to final width, and either cut into panels or rolled into cylinders for packaging and warehousing. Edge-trim material is reintroduced into the blending process to help achieve the company's zero-waste goals.

Stage 5
To create interior sidewall panels for the RV industry, rolls of this specially formulated non-woven material are transferred to the twin-belt press line. A special scrim is applied to one side to prevent glue dive during RV manufacturing. Under heat and pressure the polymer melts to encapsulate the fiber matrix. A cooling section hardens the panels and enables release from the press. Panels are edge-trimmed, cut to length, and palletized for shipment.

27.3.6 Environmental production

INCA's manufacturing facilities will be optimized in terms of location, cleanliness, safety, labor, and efficiency. The factory will recycle all waste back into the production system, and the company anticipates yield losses to be very low. INCA will use geothermal heating and cooling to save energy.

Safety procedures will be put in place, and staff trained to maintain a safe and healthy work environment. Compliance with ISO 9000 (quality) and ISO 14,000 (sustainability) standards will be verified. The ability to displace significant volumes of petrochemical plastics, reduce deforestation and greenhouse gas emissions, and contribute to a more sustainable future is at the heart of the company's mission and vision.

27.4 Commercialization strategy

Product innovation is a core competency of INCA Renewtech. The market traction that INCA has already secured from Toyota, Winnebago, Genesis, and Gurit is a direct result of a multi-decade track record developing and marketing successful biocomposite programs for major corporations.

Chairman and Chief Executive Officer David Saltman has built multiple successful cleantech companies. He and Chief Technology Officer Garry Balthes co-founded Flexform Technologies, North America's first natural fiber composites. They developed programs that include door panels, package trays, truck liners, and load floors for Honda, Toyota, GM, Ford, Chrysler, Mercedes, and Tesla. Saltman is the recipient of the Award for Excellence in Innovation (ACE) from the ACMA. In 2010, Balthes received the Toyota Global Award for Sustainable Composite Technology. He went on to design and build Carver Non-Woven Technologies, the most advanced composites manufacturing facility.

Chief Marketing and Sustainability Officer Camille Sobrian Saltman is the former President and Chief Marketing Officer of Malama Composites, manufacturers of the first rigid urethane foams made from bio-based polymers. INCA President, Paul Wybo, brings 30 years of manufacturing management expertise, most recently as the former Chief Operating Officer of ABC Industries.

Rather than invent products and then search for markets, the INCA team forms customer relationships, develops a deep understanding of the customer's needs and then works in partnership with them. Extensive line trials with equipment vendors and investment in an ASTM-compliant laboratory enable the product development team to work with customers to create customized materials capable of meeting their rigorous price and performance criteria. These collaborations lead to product innovation, intellectual property generation, and long-term relationships.

Further research and development will focus online extensions of INCA's first four products, improving their performance and modifying them to address additional industrial market segments.

INCA's commitment to research and product development will create significant barriers to entry for competitors, open new global market opportunities, and generate long-term value for shareholders.

27.5 External market drivers

Today, many of the same products that are synthesized from hydrocarbon molecules are being produced from carbohydrates derived from agriculturally grown plants. These products include plastics, textiles, fertilizers, building materials, furniture, papermaking, packaging, industrial fillers in paints, plastics and adhesives, pharmaceuticals, nutraceuticals, and cosmetics. Powerful market forces are driving this transition to bio-based materials:

27.5.1 Green purchasing guidelines

One of the most important market trends is the green purchasing guidelines being adopted by nations and corporations.

Canada's green procurement program requires that federal and provincial governments incorporate environmentally responsible guidelines in their purchasing decisions while maintaining compliance with all legislative, regulatory, and policy obligations. Canada defines environmentally preferable products as those that reduce greenhouse gases and other toxic emissions, improve resource conservation, enable reuse and recycling, and protect biodiversity.

The UK has a Green Public Procurement (GPP) program that encourages public sector buyers to integrate environmental considerations into their purchasing decisions. This program is aligned with the EU's GPP initiative and aims to direct public spending towards environmentally friendly products and services. The UK's GPP strategy includes setting targets for various sectors, such as construction, food, and information and communications technologies, to support climate action and the transition to a circular economy.

There is little doubt that INCA's products and the company's customers will directly benefit from these and other governments purchasing guidelines.

27.5.2 LEED certification

The Leadership in Energy and Environmental Design (LEED) program was originally created by the Natural Resource Defense Council and the US Green Building Council to provide third-party verification that homes and buildings were designed and built to achieve higher energy and resource efficiencies and better indoor air quality. LEED has become the de facto standard for the built environment, influencing not only design and construction methodologies, but selection of interior paneling, flooring and furniture. INCA's building products will offer significant advantages for builders and manufacturers to achieve LEED certification.

27.5.3 ESG investment

While government policies are crucial to supporting more sustainable products and services, the single most powerful force influencing industry trends is the capital markets, and the single most important shift among money managers and investors is their move toward ESG criteria in making investment decisions.

Investors recognize that ESG metrics are financially relevant to managing risk and understanding corporate purpose, strategy, and management objectives. The origins of this trend can be traced back to work done by the United Nations in conjunction with the International Finance Corporation and the Swiss government, which produced a report entitled "Who Cares Wins." This report argued that incorporating ESG considerations in evaluating investment decisions makes sound business sense. The report formed the basis for the launch of Principles for Responsible Investment at the New York Stock Exchange.

27.5.4 Consumer demand

While government regulations and ESG investment are critical to bringing about a more sustainable society, the real market makers are consumers. Sustainability is increasingly influencing consumer purchasing behavior:

- Seventy percent of global consumers are willing to pay more for sustainable products [1].
- Thirty-four percent of global consumers are more likely to buy products with sustainable credentials [2].
- Seventy-eight percent of US consumers say they feel better when they buy products that are sustainably produced [6].

- Sixty-four percent of global consumers expressed concern about climate change in 2023 [7].
- Eighty-five percent of consumers report experiencing firsthand the disruptive effects of climate change in their daily lives [8].
- Forty percent of global consumers report that waste reduction and recycling initiatives would make them more likely to purchase from a brand [8].
- Consumers are willing to pay an average of 9.7% more for sustainably produced or sourced goods, even with cost-of-living concerns [8].
- Products with ESG-related claims accounted for 56% of all growth over the last 5 years, about 18% more than expected based on their initial market share [9].
- Of 10 consumers, 8 said that they would pay up to 5% more for sustainably produced goods [10].

27.6 Conclusion

Economic factors are now playing a powerful role in the adoption of bio-based materials along with global sustainability mandates. While cost and performance continue to be an impediment to wide-spread adoption across all industries, the automotive industry remains the exception. Emissions reduction legislation and EV adoption are driving industry sector growth, particularly where cost reductions can be achieved while at the same time delivering superior performance.

INCA Renewtech has recognized that to achieve market penetration and compete with incumbents, the company requires a business model that can deliver price and performance along with circularity and carbon neutrality.

References

[1] https://www.bloomberg.com/company/press/global-esg-assets-predicted-to-hit-40-trillion-by-2030-despite-challenging-environment-forecasts-bloomberg-intelligence/
[2] https://nielseniq.com/wp-content/uploads/sites/4/2022/08/Are-companies-transparent-enough-in-their-sustainability-initiatives-to-meet-shopper-needs-1.pdf
[3] https://www.prnewswire.com/news-releases/prepreg-market-worth-12-6-billion-by-2026--exclusive-report-by-marketsandmarkets-301513599.html
[4] https://www.zionmarketresearch.com/news/global-balsa-wood-market
[5] https://www.economist.com/the-americas/2021/01/30/the-wind-power-boom-set-off-a-scramble-for-balsa-wood-in-ecuador
[6] https://www.unilever.com/files/3a8d41a0-ffba-4899-8a0b-4986be51e0d0/making-purpose-pay-inspiring-sustainable-living.pdf
[7] https://www.euromonitor.com/article/megatrends-understanding-sustainable-consumers-2023-key-insights

[8] https://www.pwc.com/gx/en/news-room/press-releases/2024/pwc-2024-voice-of-consumer-survey.html
[9] https://www.mckinsey.com/industries/consumer-packaged-goods/our-insights/consumers-care-about-sustainability-and-back-it-up-with-their-wallets
[10] https://www.pwc.com/gx/en/industries/consumer-markets/consumer-insights-survey.html

Robert Joyce

Chapter 28 The evolution of bioproducts and biomaterials that drive sustainability in a medical market

Editor's note: This chapter provides a first-person perspective on innovation and development. As such, this is presented in a narrative style while the author takes a third-person approach.

28.1 Background

During the 1980s, the oil crisis accelerated the development of sustainable materials to replace fossil fuels. These sustainable materials, based on cellulosic sources, included wood fibers, hemp, flax, and sisal, which were mixed or compounded with various thermoset or thermoplastic polymers. The innovative blends or compounds developed in the following decades helped define biocomposite materials used in a variety of applications, including medical fields.

Over the last decade, medical professionals have replaced biomaterial technologies, namely metals and synthetic polymers, with biocomposites. These biocomposite materials are identified as bioresorbable, made from polyester, or even plant-based corn. The bioresorbable products have unlimited storage capacity of fluids, degrading over time, and will not always be predictable as an implant. Further, bioresorbable implants are highly dependent on patient health, including age, diabetes, and even alcohol consumption.

There is now considerable interest in creating biomaterials that have synthetic polymers with minerals, additives often referred to as thermoplastic composites. These thermoplastic composites have some bone-like features. These bone-like features include biomimicking qualities and performance in the body and outside of the body. Further, these new bio innovations are a process that can take many years of data gathering and observation. Some of the most obvious improvements in biomaterials and processes are often overlooked in development and commercialization efforts.

An investigation of a nylon composite, reinforced with cellulose fiber branded as FibreTuff PAPC – polyamide and polyolefin with cellulose fiber – has produced a most promising biomaterial. The FibreTuff PAPC research and development started over 25 years ago with the intention of utilizing bioproducts, natural fibers with traditional molding methods, and foaming agents. Unexpectedly, the FibreTuff PAPC incor-

Robert Joyce, FibreTuff, e-mail: robert@fibretuff.us

 | https://doi.org/10.1515/9783112212141-028

porated cellulose fiber and was found to be biocompatible. The FibreTuff technology was a replacement for traditional biomaterials such as polyether ether ketone (PEEK) and bioresorbables. Further research and development of the FibreTuff PAPC composition demonstrated a 3D printable, nonresorbable biomaterial with a cellular structure for patient-specific medical applications, including but not limited to anatomical bone models, phantoms, presurgical tools and guides, and implantable bone grafts. The printable FibreTuff Technology has evolved and is formulated to be designed and manufactured to resemble natural bone. For instance, the cellulose fiber mimics collagen fiber, while the nylons have the backbone chemistry of proteins. The polyolefin helps with desired ductility and hydrophobicity. Furthermore, a combination of inorganic and organic ingredients in printed FibreTuff PAPC has demonstrated to promote bioactivity, even osteoblast adhesion.

Interestingly, Chinese scientists have been investigating Nylon66 composites to replace traditional biomaterials such as PEEK since late 2019. Robert Joyce, president and founder of Innovative Plastics and Molding, Inc., and developer of printable FibreTuff PAPC, has demonstrated a high heat polymer, Nylon66. The printable FibreTuff PAPC + Nylon66 has been identified as an amphiphilic composition to show bone likeness with a high strength-to-weight ratio, low moisture absorption with limited degradation, yet absorbs fluids to promote flexibility and impact. The development of the printable FibreTuff PAPC + Nylon66 continues as part of ongoing research aimed at regenerating cells and tissues and drug delivery. The printed FibreTuff PAPC + Nylon66 is striving to advance medical technology, specifically, improving patient care by accelerating the healing process and reducing infections.

28.1.1 Bioproducts and biomaterials: a preface

The investigation of compounding wood fiber with thermoplastics led to many bioproduct innovations by Robert Joyce of Innovative Plastics and Molding, Inc. Innovative Plastics and Molding, Inc. purchased a used Leistritz twin-screw extruder in 2003. This Leistritz twin-screw extruder included a 54″ sheet die, along with a downstream calendar wood sheet setup for thermoforming automotive parts – specifically, package trays. Robert Joyce successfully produced a wood fiber calendar sheet in a 14,000 sq. ft. facility he helped build in Dundee, Michigan. To support production efforts after product development of the wood sheet, a series of communications with an automotive giant revealed that wood sheet production was a commodity business with slim profit margins – typically 15% or less. Furthermore, Innovative Plastics and Molding, Inc. was looked upon as a vehicle to lower prices for wood fiber composite sheets in the market. It also became apparent that increasing profit margins would only be possible through innovation. Thus began the pursuit of new methods for compounding and molding bioproducts using natural fibers and thermoplastics, targeting markets such as automotive, construction, building, and furniture.

There were numerous challenges with bioproducts, including high product costs, recyclability, mechanical performance, unpleasant odors, natural fiber bleaching, and susceptibility to fungal or microbial contamination. These technological hurdles were gradually overcome through a series of experiments and research aimed at building a commercialization platform. The company began exploring the injection molding of biocomposites, including gas-assist techniques. Joyce would be awarded his first patent in 2007, "Molded Article," a natural fiber gas assist claim for a molded article [1].

Innovative Plastics and Molding, Inc. would then partner with a compounding company for the commercialization of the molded article. Joyce would generate several patent applications and patent awards for the polyamide with polyolefin and cellulosic fiber composition of matter and process. The high level of sophisticated thermoplastic research with natural fibers included natural polymers, i.e., cellulose fibers melt blended with thermoplastics to help establish freedom to practice [2–4].

The value of Joyce's work would be demonstrated as valuable by a breach of the contract documents, personal attacks, unsubstantiated claims, and dismissal of a patent lawyer. The legal challenges Joyce faced would decimate his support team. Joyce's marriage wouldn't survive, and his lawyer would demonstrate a self-serving benefit. Joyce would weather the storm of the $600k legal battle and continue the operation of the business. At one point, Joyce was offered a substantial amount of money for his technology by a group of investors and a plastic processor. The investors and partner failed to get Joyce the $800k investment. Joyce would continue to knock down the barriers of injection molding biopolymers with gas to alleviate odor, degradation, and processing difficulties.

In 2015, the company pivoted toward the medical field after the FibreTuff biocomposite compound – PAPC – passed a basic cell study conducted by professors at the University of Toledo, located in Northwest Ohio, USA. Further developments would refine the FibreTuff compound into a printable alloyed composition, combining natural polymers like cellulose fiber with both polar and nonpolar thermoplastics, the nylon and polyolefin. FibreTuff development in extrusion would lead to the production of innovative FibreTuff filaments and powders for 3D printing. The FibreTuff cellulose fiber thermoplastic compounds were tested by third parties to meet stringent mechanical and biological performance requirements for medical devices and implants. These FibreTuff biomaterial products – identified as novel and nonobvious were awarded patents for 3D printing – were capable of replacing inconsistent, less efficient, and more expensive alternatives. However, commercializing these medical-grade filaments and powders proved extremely challenging, especially in securing funding to support innovation. The company would seek out Federal, State, and Local community innovation funds in the State of Ohio.

The FibreTuff Technology started to gain recognition in 2018, and licensing efforts took hold when collaboration efforts were rewarded with several collaborators from Ohio. The collaboration was awarded an Ohio Third Frontier grant project for $500k. This project helped FibreTuff development achieve 3D printing sales revenue. In the next 12 months, FibreTuff development stalled; a mere $2K in grant support was awarded from $10 million in local funding from two universities, two hospitals, and the State of Ohio. The local state representative and the local government officials would offer no practical solutions to find funding.

Shortly thereafter, Innovative Plastics and Molding, Inc. would face several more challenges to get the FibreTuff PAPC biomaterials to market. The company's efforts would get support from a 3D printing company, Ultimaker, located in the Netherlands. The Ultimaker relationship with Innovative Plastics and Molding, Inc. would be recognized by the Society of Manufacturing Engineers. The Society of Manufacturing Engineers would give Innovative Plastics and Molding, Inc. a Medical Emerging and Enabling Technology award for 3D printing FibreTuff PAPC. The 3D printable FibreTuff PAPC would continue to evolve in terms of composition, processing, and application, demonstrating both relevance and profitability in the medical market.

A point-of-care (POC) Roadmap for 3D-printed medical applications would be presented by Robert Joyce in 2019 with the FibreTuff brand. This POC roadmap with FibreTuff was part of a soft launch in the commercialization process. The roadmap helped healthcare service providers understand the benefits of 3D printing FibreTuff. An obvious benefit is efficiency; one printer and one material approach would lead to less fragmentation at the hospitals for improving patient outcomes. The bone models and tools for preclinical assessment, bone grafts, and possibly permanent implants, could be printed in hours, not days.

This roadmap with FibreTuff was more than a business strategy, but a tool for educating customers to generate revenue with 3D printing and improve patient outcomes. There was industry validation through collaboration with the US military and civilian healthcare facilities (HCFs). The FibreTuff Technology was a novel and not obvious development generating 3D-printed patents. However, there were challenges to commercialize the many 3D-printed patents. These commercialization challenges included an unprepared licensee and unethical business practices; samples were stolen, and some were used to produce copied versions. These imitations falsely claimed to offer the same features and benefits as printed FibreTuff PAPC, often with unsubstantiated data, compromising market integrity.

As FibreTuff filament and powder sales grew globally, larger customers – including universities, hospitals, medical centers, and physicians – began using it to print bone-like parts such as skulls, mandibles, foot and ankle bones, metatarsals, hands, and fingers. This soft launch of commercialization allowed continued refinement of the composition, enabling the production of filaments and powders that met the increasing demands of sophisticated customers. In late 2019, the 3D-printed FibreTuff brand had gained global recognition. The company would get a 3D printing patent with claims associated with processing FibreTuff PAPC filaments and powders [5]. However, the FibreTuff brand needed to expand for commercialization, as well as meet the technical challenges associated with bone likeness and replace the existing biomaterials. These additional challenges were identified as smaller pore sizes for ingrowth, higher strengths, better screw retention, and a higher level of biocompatibility.

In 2023, Innovative Plastics and Molding, Inc. would get a biocompatible 3D printing materials award from Start Us Insight. The US Military – specifically Walter Reed Hospital had 3D-printed FibreTuff for applications in an austere environment, and the research was published by *Military Medicine*. This piloted research study evaluated the use of FibreTuff PAPC technology for POC applications such as preclinical assessment, bandages, and instruments. This US Military research, along with an additional patent awarded to Innovative Plastics and Molding, Inc., motivated further product development. Interestingly, a research publication by Chinese scientists identified hydroxyapatite HaNylon66 as a formidable permanent implant to replace PEEK. Their work specifically identified a highly biocompatible HaNylon66 biomaterial for the spine [6].

Consequently, new FibreTuff PAPC product developments focused on the selective laser sintering (SLS) printing process and incorporating a high-heat polymer – Nylon 66. The resulting PAPC + Nylon66 composite, with improved bone-like properties, includes organic and inorganic chemistry to promote cell and tissue regeneration. This FibreTuff technology has opened many possibilities for applications in both military and civilian HCFs, including leapfrogging the HaNylon66 technology.

28.1.1.1 Entrepreneurship and small business innovation with bioproducts

An entrepreneur – as well as a business – can be short-lived, depending on the individual's perseverance, persistence, passion, drive, risk tolerance, and ability to deliver results. The author believes that a successful entrepreneur must have a clear vision for both the short and long term in order to build a sustainable business. In addition, the entrepreneur must embrace the responsibility of managing daily operations and making decisions – positive or negative – often with minimal data or input.

In many cases, the entrepreneur must generate their own data or rely on independent third parties to develop a sustainable proof of concept. This data, combined with entrepreneurial vision, informs the decisions that shape the overall strategy for achieving positive outcomes. Moreover, such outcomes are essential components of the entrepreneurial vision – a critical skill needed to attract funding or generate profitability to move the venture forward.

One cannot fully prepare for the entrepreneurial journey simply by reading textbooks. While theoretical knowledge is valuable, entrepreneurs must also develop practical skills to identify the right business partners, hire productive employees, manage personnel, and even select the right ingredients for novel compounds used in bioproducts. Furthermore, bioproducts make biomaterials, where a nascent field has limited historical data. There are significant opportunities for innovation and out-of-the-box thinking.

However, thinking outside the box can be a daunting task for innovation. Entrepreneurs must be curious and capable of clearly communicating their vision to others, which includes academia. Passion for the vision must be evident, and the time invested should feel worthwhile – ideally leading to tangible results. While the goal is often to generate revenue, entrepreneurs must also recognize the high degree of volatility in the path ahead.

Finally, the pursuit of entrepreneurial goals often requires personal sacrifices, including challenges that may impact family and personal relationships. The entrepreneurial journey of Robert Joyce, founder of Innovative Plastics and Molding, Inc., began through a combination of education and life experience. He acquired both practical and theoretical knowledge from books, college, family, and professional roles. Joyce's entrepreneurial spirit emerged in his early 30s while working in sales and plastic applications for a large industrial controls company.

He recalls a defining moment when his entrepreneurial drive surged – fueled by curiosity, impatience, and a realization that he no longer fit the role of a traditional employee working for a large company. Joyce believed his idea to reuse and recycle industrial controls could be profitable, fulfilling, and fun – though undeniably risky. He left his job and launched his first business centered on this innovative reuse-and-recycle concept.

Within 6 months, Joyce faced financial challenges after leaving his former employer. To survive, he shifted focus to repairing and selling higher-value plastic machinery and products. His entrepreneurial journey would later evolve to include operating plastic machinery, extruding, and melt-blending wood fiber with thermoplastics to produce bioproducts.

Joyce would eventually need to replace the wood fiber and chose a cellulose fiber to melt blend with polyamide and polyolefin. The cellulose fiber thermoplastic alloy would lead to medical and nonmedical applications. Joyce would continue his journey, filled with hardship, some success, and passion, which he would pursue for more than three decades.

28.2 Early bioproducts manufacturing

In about 2003, Innovative Plastics and Molding, Inc. purchased from Ford Motor Company a twin-screw compounding extruder with a 54" sheet die and downstream calendar for thermoforming automotive parts – package trays. The company proceeded to install the twin-screw extruder calendar sheet line in a 14,000 sq. ft. facility built in Dundee, MI. Robert Joyce would operate the compounding sheet extrusion line but faced a few processing challenges related to dispersion, odor, and brittleness. He found a way to connect with the former extrusion operator and chemist who retired from Ford Motor Company. After several unsuccessful compounding and sheet production trials with the chemist, Joyce made several adjustments to the formulation and operation to produce a 54" wood composite sheet for thermoforming. Joyce proceeded to produce the package trays to show customers for possible partnerships and collaboration. He would identify a few possible automotive customers, including an OEM. After a few discussions and quotations, it was determined that a 15% margin to supply package trays for the automotive interior application wasn't a possibility. The automotive company chose to take its business overseas, while Joyce would innovate to create a better margin and value for his company.

Figure 28.1: Nitrogen-injected wood fiber with nylon and polyolefin.

Innovative Plastics and Molding, Inc. began to explore innovation for reducing the bioproduct costs and increasing mechanical performance. Joyce's work led to an invention – a molded article that utilized his biocomposite and gas-assist technology (see Figure 28.1). The gas-assist process would use nitrogen, a gas with a low water content. It was introduced through gas pins in an injection-molded tool. Further, the biocomposite and nitrogen gas would be mixed to produce one or more cavities and porosity in thick-walled parts, creating an impervious skin. Depending on the processing technique, heat, and pressure, you could get different results. While Joyce was producing the biocomposite, he decided to mix two types of synthetic polymers, a polyamide and a polyolefin, with wood fibers to make a biocomposite compound for injection molding. This biocomposite compound could then be sold to produce a porous, thick-walled bio product for the automotive, furniture, and building industries.

In 2007, Innovative Plastics and Molding, Inc. would shortly thereafter partner with a compounding entity to help commercialize the wood fiber gas-assist technology. Joyce would first submit a provisional patent application, sign a confidential agreement and contract, and share his knowledge and ideas for biocomposite compounding and a commercialization strategy. The partnership worked for approximately 12 months, and the commercialization of biocomposite compounds showed profitability in several industries. Joyce's entrepreneurial spirit and innovation created value, recognized by the business partner. Unfortunately, the contractual agreement would be breached, and the commercialization strategy would be derailed. Joyce prevailed in the litigation, but a most devastating event would include a divorce during the tail end of the litigation. His support wall began to disappear, including his lawyer. The litigation cost was $600k to acquire the biocomposite technology invented by Joyce.

The 2 years of litigation would slow Joyce's bioproducts research and development. He did manage to continue testing and validating the novelty of his bioproducts and biomaterial compound utilizing natural fibers. Based on customer requirements and necessity, Joyce began to investigate replacing natural cellulosic fibers with cellulose fiber to blend with thermoplastics. This innovative cellulose fiber thermoplastic composition, branded as FibreTuff, developed by Joyce, focused on reducing impurities, such as lignin and hemicellulose. He would achieve a superior product by eliminating offensive odor, no visible degradation, and improving process ability. Joyce would further refine his compound to be identified as biocompatible, not harmful to cells and tissues. His journey would start to evolve in manufacturing 3D-printed cellular structures for devices and implants in the medical market.

28.3 Bioproducts manufacturing evolving to biomaterials and new innovations

Innovative Plastics and Molding continued its efforts to demonstrate sustainability and value with biocomposite and bio-based products made from cellulosic fibers. The development of FibreTuff worked toward identifying opportunities to injection mold cellulosic fiber thermoplastic bioproducts for automotive companies and their suppliers. The FibreTuff biocomposite compound was successfully injection molded for car interior applications.

Two significant challenges emerged: colorability and part odor. In response, FibreTuff was reformulated, and the biocomposite compound needed to replace the original cellulosic fibers with cellulose fibers melt blended with thermoplastics. Additional modifications to the formula included new types of additives to meet the material requirements for interior automotive parts. This innovation led to the patented technology for claims associated with a low-moisture composite pellet suitable for molding.

Innovative Plastics and Molding, Inc. attempted to collaborate and license technology with a compounding company in Sterling Heights, Michigan. Both companies would sign a mutual nondisclosure agreement. Robert Joyce of Innovative Plastics and Molding, Inc. would put forth considerable time and effort in discussing ingredients, compounding to commercialization. Joyce's proprietary work with cellulose fiber, nylon, and polyolefin was used not only by the compounder but also included an automotive manufacturer. The automotive company would successfully injection mold an instrument panel using Joyce's biocomposite with his active participation and not with the compounding partner. The resulting part exhibited no cellulose fiber agglomeration, no offensive odor, and a 20% shorter production cycle time compared to existing parts.

The automotive company would continue its investigation of product testing and qualification of the FibreTuff composition. Over the next 16 months, numerous meetings were held between Joyce and the automotive company to discuss product qualification. In one meeting, the company's engineers acknowledged that FibreTuff technology was significantly more innovative than "Thrive," a biocomposite produced by a major wood products company. Innovative Plastics and Molding, Inc. was considered too small to be a qualified supplier. Subsequently, Joyce discovered that the automotive engineers he had been working with to commercialize FibreTuff and compounding partner were concurrently developing and patenting a similar biocomposite using flax and polyolefin. He later filed a lawsuit against the compounder for breaching the mutual nondisclosure agreement. Joyce would be successful in getting reimbursed for his work.

Several years later, the Society of Plastic Engineers (SPE) published a bioproduct research with polyamide, polyolefin, and cellulose fiber performed by advanced researchers employed by the automotive company. This automotive company had been helped by Joyce to injection mold an instrument panel with FibreTuff composition of a polyamide, polyolefin, and cellulose fiber. The advanced manufacturing researchers from the automotive company would win a SPE award in 2015, not referencing Joyce. In parallel with the automotive investigation, Joyce began exploring collaborations with universities and nonprofit organizations to secure nondilutive funding and grants. A university in Northwest Ohio tested the FibreTuff biocomposite for potential medical applications. After conducting a basic cell study, the university returned favorable results. In fact, they tested the compound and parts twice to verify the unexpectedly positive outcomes. These results opened new possibilities, prompting Innovative Plastics and Molding, Inc. to pursue the ambitious goal of replacing PEEK in implant applications.

The basic cell study results would refocus FibreTuff development. The company would pivot from automotive manufacturing and focus on higher-value medical applications. The FibreTuff biocomposite would be further studied at the University to include both biocompatible and mechanical investigations. The company worked closely with several bioengineering professors and collaborated with senior university students, who conducted tests comparing FibreTuff to other biomaterials. Joyce would procure a profile sample with aligned fibers to benchmark against PEEK's mechanical properties. The data collected indicated that FibreTuff demonstrated exceptional mechanical performance suitable for spinal implant applications. Unfortunately, after this extensive work, Joyce was informed that his extruded samples had disappeared. This incident underscored the value of his innovation and motivated him to further investigate medical applications using FibreTuff – a cellulose fiber thermoplastic composition. This led him to focus on developing 3D printable filament made from the material.

In a short period of time, FibreTuff was successfully 3D printed into complex geometries. Several FibreTuff formulations were extruded and 3D printed. Eventually,

FibreTuff exhibited bone-like mechanical performance and passed stringent biocompatibility tests, meeting expectations for strength and flexibility. Joyce would discover additional bone likeness with the printed FibreTuff material exhibited radiopacity, or radio density – a property not commonly found in most polymers (which are typically radiolucent) or metals (which are radiopaque). This radiopacity, measurable in Hounsfield Units via CT scans, became a key differentiator. It allowed clinicians to determine the location and integration of implants, as actual bone appears brighter than printed FibreTuff in imaging. This could help assess healing progress. With repeatable and accurate data, the use of artificial intelligence in analyzing FibreTuff biomaterials became a real possibility.

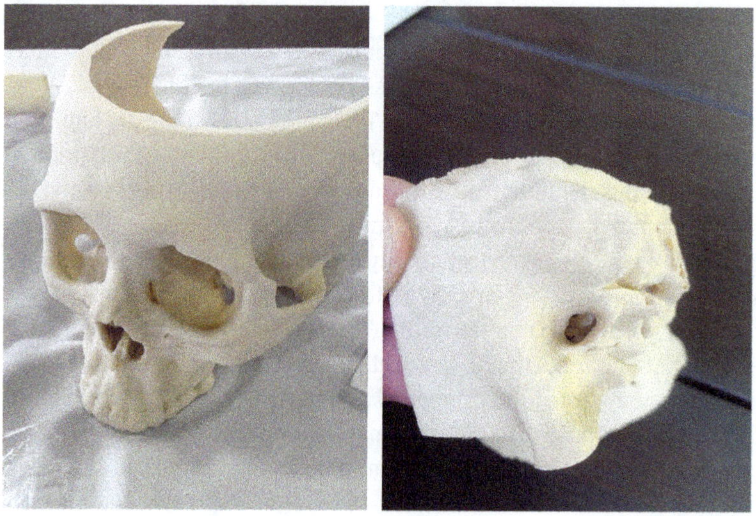

Figure 28.2: 3D-printed FibreTuff PAPC skull and temporal bone.

As FibreTuff filament and powder sales grew globally, larger customers – including universities, hospitals, medical centers, and physicians – began using it to print bone-like parts such as skulls, mandibles, foot and ankle bones, metatarsals, hands, and fingers. This soft launch of commercialization allowed continued refinement of the composition, enabling the production of filaments and powders that met the increasing demands of sophisticated customers. The sophisticated users who were early adopters include Dr Edward Strong, Professor and Vice Chairman, Department of Otolaryngology at the University of California, Davis; Dr Michael Harmon, a cranio maxillofacial surgeon, founder and owner of Digital Provisionalization Technologies in Mechanicsburg, PA; and Dr Ross Salary of Marshall University, Department of Mechanical Engineering. For instance, Marshall University professor Dr Ross Salary would write about 3D fabrication of bone structures using FibreTuff. A most important commercial challenge for FibreTuff filaments occurred in the late summer of 2019, when eight printed skulls were requested by Dr Edward Strong for a cranio maxillofacial event held in Tampa, FL, in November (see Figure 28.2). This event would have 50 surgeons attend an educational and training seminar where the printed FibreTuff skulls would demonstrate an orbital placement. Fortunately, the exercise would prove to be most beneficial in helping promote FibreTuff creditably. The second example of an early adopter includes Dr Michael Hartman, who began printing prototype surgical models with FibreTuff filaments. These prototype surgical guides were for mandible reconstruction. This application would help FibreTuff filaments define suggestive uses for POC. Another early adopter of FibreTuff filaments included Dr Ross Salary at Marshall University. Dr Salary would investigate printed FibreTuff, producing posters and publications: 3D Fabrication of Bone Structures, Based on FibreTuff Research Group: Dr Ross Salary, Division of Engineering, Marshall University, Huntington, WV [7, 8].

In late 2021, further mechanical validation and testing on FibreTuff would be performed at Drexel University's Implant Research Center by Abigail Tetteh, Dr Daniel MacDonald, and Dr Steven Kurtz. They had written an abstract [7], Anisotropy in Fused Filament Fabrication Printed FibreTuff for Synthetic bone models. The testing confirmed that the alignment of cellulose fibers would increase the compressive strength in printing bone-like models. The abstract validated that the compressive strength was higher in the z-direction, which could be because of directionally arranged cellulose fibers in the material to increase its strength. This analysis found the compressive modulus of the fused filament fabrication-printed FibreTuff, tested in both directions, to be in the range of the cancellous bone, while the compressive strength was significantly higher. It should be noted that compression testing was performed with a mechanical testing machine, which was not equipped with an extensometer, and the strain values of the experiment were calculated from recorded time values. Therefore, the reported values are the apparent compressive modulus and compressive strength. Further, 3D-printed samples with 100% infill had a significant amount of porosity, which may have affected the performance during mechanical testing. Nevertheless, given the results of this study and the printability of this material, FibreTuff remains an attractive material for patient-specific bone modeling. After introducing the bone-like printed FibreTuff models, further development would lead to new types of customized phantoms for radiographic imagery. Phantoms, by definition, are specially designed objects that are scanned or imaged in the field of medical imaging to evaluate, analyze, and tune the performance of various imaging devices. Through ingenuity and a little investigation with glue, one could mimic real bone models, having produced a bone ring with about 3,000 HU Hounsfield units (see Figure 28.3).

These attractive mechanical qualities and performance attributes helped initiate biocompatibility testing. The biocompatibility testing performed on the printed FibreTuff included NAMSA, a regulatory body located in Northwood, Ohio, USA. NAMSA performed an animal study that showed the formulation passed certification for USP Class VI. The USP Class VI is a certification for implantable devices of less than

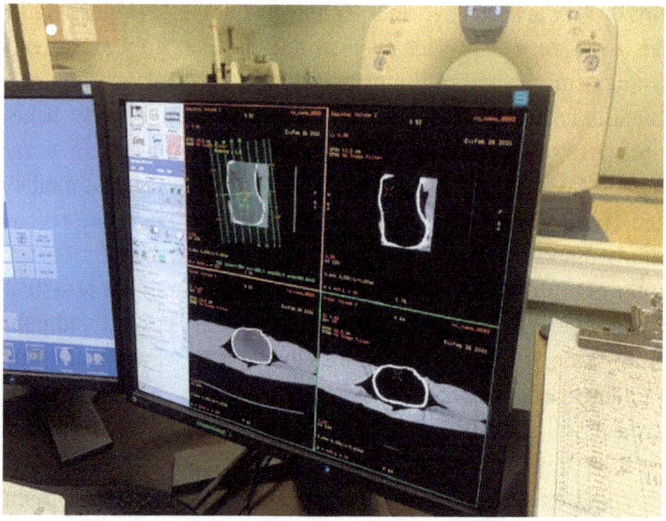

Figure 28.3: 3D-printed FibreTuff PAPC phantom – calcaneus bone.

30 days in the body. Additional biological investigations on printed FibreTuff were performed by the Northeast Ohio Medical University (NEOMed), located in Rootstown, OH. Dr Fayez Safadi, PhD, conducted two biological tests on several 3D-printed disks made with the FibreTuff composition. The investigation for 22 days produced data that showed that osteoblasts (bone-forming cells) did attach to the FibreTuff. It was also identified that FibreTuff might induce bone matrix mineralization rather than maturation, i.e., it is great to have a matrix that will enhance the mineralization of osteoblast matrix. (see Figure 28.4).

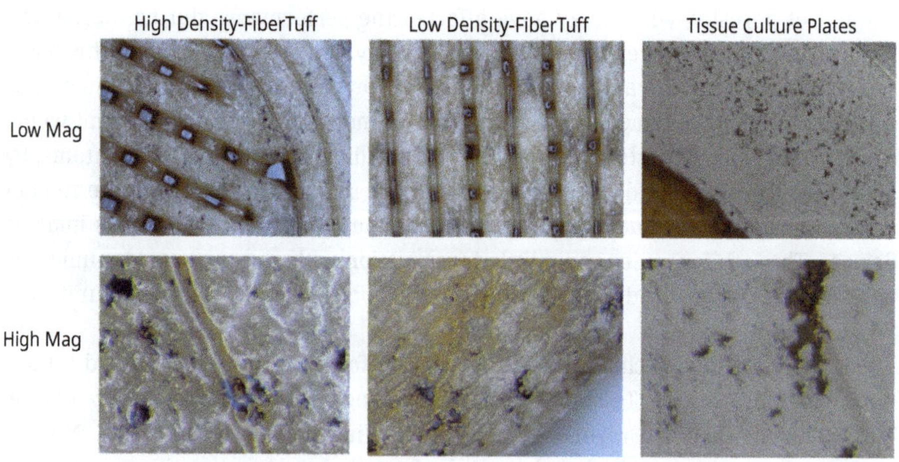

Figure 28.4: Osteoblast adhesion with 3D-printed FibreTuff.

In November 2021, further biological work was performed on a fused deposition modeling (FDM)-printed FibreTuff micropore specimen. This biological work was performed by Dr Prabaha Sikder, PhD, Assistant Professor, Mechanical Engineering Department, Washkewicz College of Engineering, Cleveland State University; and Bharath Tej Challa, Graduate Student, Mechanical Engineering Department, Cleveland State University. They titled the paper, "Results of a bioactive study using a 3D-printed cellulose fiber advanced composite." Dr Sikder sterilized and immersed the printed FibreTuff micropore cellular structure in simulated body fluids (SBF) (see Figure 28.5). The study concluded there was evidence to show cell and tissue adhesion for bone regeneration without closure of the printed micropores for cell proliferation (see Figure 28.6).

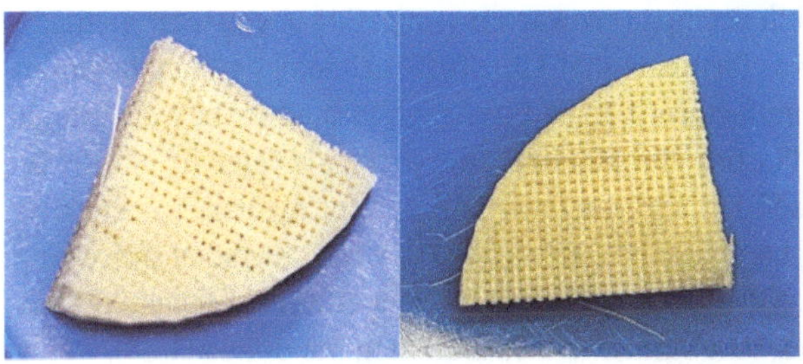

Figure 28.5: 3D-printed FibreTuff PAPC sample before immersion in SBF and 7 days after submersion.

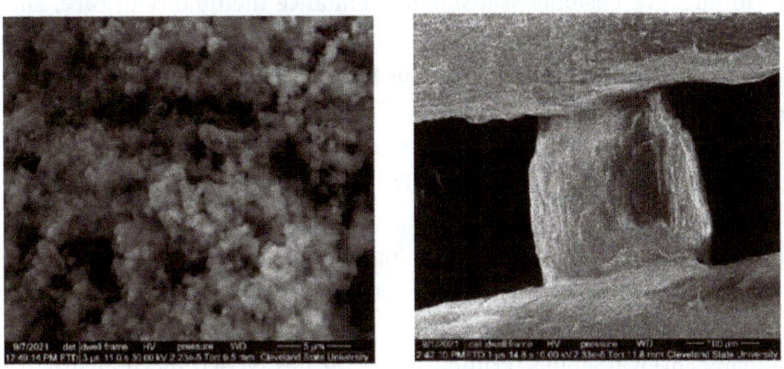

SEM pictures of SBF+DMEM immersed sample

SEM pictures of SBF+DMEM immersed sample

Figure 28.6: SEM imaging.

The high magnification SEM micrograph shows the formation of calcium-deficient apatite structures on the FiberTuff specimens after they were immersed in DMEM + SBF for 7 days. The calcium-deficient apatite is characterized by the formation of nanosized, well-defined globules, covering the strands of the porous FiberTuff specimens. The formation of such calcium-deficient apatite is an indication that FiberTuff is inherently bioactive in nature.

28.4 The "soft launch" commercialization process of biomaterials for the medical market

The commercialization journey for FibreTuff Technology involved an attempt at a company buyout and licensing agreement. Both commercial vehicles for FibreTuff development in 3D printing failed due to funding challenges. A business roadmap and strategy developed by Joyce in 2019 led the company toward investigating one printer method – FDM with FibreTuff PAPC. It was fortunate that FibreTuff Technology gained favorability by a popular 3D printer manufacturer named Ultimaker, located in the Netherlands. Ultimaker helped Innovative Plastics and Molding, Inc. capture a "global" market for FibreTuff sales, essentially a "soft launch" commercialization for medical bone models. There were challenges along the way for FibreTuff PAPC, which included processing, printer software, and hardware malfunctions. These challenges were resolved by Innovative Plastics and Molding, Inc., and Ultimaker distribution in the United States.

The 3D printing of FibreTuff filaments helped define a roadmap to deliver customer value and ensure success. This roadmap, aimed at POC applications, was designed for HCFs to improve customer satisfaction, enhance the quality of care, and reduce costs (see Figure 28.7).

This roadmap is illustrated through four silos in the commercial strategy for FibreTuff-printed products:

- **Silo 1**: Printing FibreTuff to deliver *functional* bone-like qualities, including good screw retention, the ability to cut and saw like real bone, strong adhesion/gluing properties, and fiber alignment.
- **Silo 2**: Printing customized radiographic phantoms. FibreTuff's radiopacity makes it suitable for radiographic imaging, including use in hospital-based customized phantoms.
- **Silo 3**: Producing surgical guides. Printed FibreTuff has passed biocompatibility testing, making it suitable for surgical applications. This prompted exploration into bone-like, implantable graft prototypes.
- **Silo 4**: Printing bone grafts – considered the most lucrative stage of the POC roadmap. For printed FibreTuff to enter the bone graft market, commercialization must comply with ISO 10993 regulations. The FDA classifies bone grafts as Class II or Class III medical devices. FibreTuff bone grafts will require a 510(k) submission, in accordance with FDA regulations. Once in vivo studies meet ISO 10993-1 protocols, a first-in-human study will begin to clear the path to full commercialization. This process is expected to take approximately 14–16 months.

Roadmap For Point of Care

Pre surgical planning to implantable grafts

Anatomical Bone Models	Phantoms	Surgical Guides	Permanent Implants
Versatility to print bone with tissue	X-Ray and CT Scan	Machinability with push pin	Non re sorbable having Micropore structure
Cortical bone with Cancellus bone	Superior bone like images - bone ring and tissues	Machine ability having small threads	More consistent and effective to replace costlier grafts
Excellent screw retention with gluing capability	Eco friendly coating	High speed drills	Customization will help reduce time in OR
Radiopacity w Laser cutting, sawing like bone	Improved visualization	Small hole size 120 micron	Bioactive properties
Standard sterilization methods - Autoclave	Quick turnaround	Laser cutting	Eliminate variability identified in Allograft and Bio re sorbables
Look and appearance of bone	Customization	No special drills / tooling required	Adheres to collagen

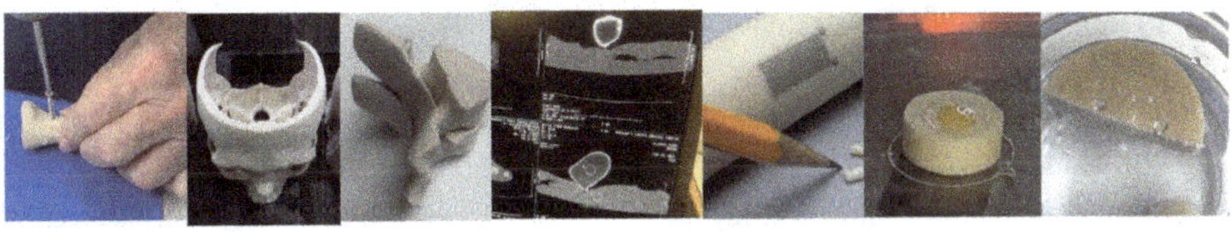

Figure 28.7: Printed FibreTuff filament having four silos for point of care (POC).

Innovative Plastics and Molding, Inc. was positioned globally to offer a four-silo commercial 3D printing solution, as defined by the POC roadmap. The estimated cost for a HCF to implement the FibreTuff POC roadmap ranged from $90,000 to $120,000 USD. However, this investment can be recouped within 6–12 months, making it financially viable.

Of these four silos, FibreTuff has generated revenue from bone models and phantoms with success. The third and fourth silos are expected to become revenue generators upon the successful completion of ISO 10993 studies. The 3D-printed FibreTuff PAPC had passed the in vivo studies performed by NAMSA, a regulatory agency in Northwood, OH.

The 3D-printed biomaterial FibreTuff PAPC had traction to replace existing toy-like plastic for preclinical assessment "bone" models. His product was functional with bone-like qualities and properties. Essentially, he was successfully replacing traditional processing methods with foaming agents for producing cellular structures with 3D printing. His research and development with 3D printing anatomical models with "bone-like" properties needed more data. He generated more data through customer feedback – mostly positive, occasionally discouraging, but often insightful.

In 2021, there was a pandemic and significant changes in the business environment. Innovative Plastics and Molding, Inc. had struggled with supplier quality and commitment. For instance, a mutual nondisclosure agreement was executed with a filament producer in Grand Rapids, MI. This company would produce FibreTuff filament for its inventory. They would inquire into branding a filament with their name. Innovative Plastics and Molding, Inc. didn't oblige, later to find the vendor had other ideas, including PAPC pellet contamination, not paying a $14,000 invoice, and reverse-engineered product to compete against FibreTuff. The vendor began using some of Innovative Plastics and Molding, Inc.'s marketing information, such as no offensive odor, radiopaque, and sold his material at two times less than FibreTuff's advertised price. The vendor entered the medical market without any prior experience other than knowledge. Additionally, other competitors entered the bone model market offering low-functionality, plastic toy-like products, which, despite their lack of realism, were attractive to bioengineering students.

Despite this promising framework, the company's commercialization strategy fell short of expectations. The fourth silo – focused on 3D-printed implantable bone grafts and permanent implants – proved to be the most challenging and costly to bring to market. Moreover, the company identified several limitations associated with the FDM printing method. These included poor layer adhesion, slow production speeds when processing thermoplastics with fillers, concerns over long-term functionality, and difficulties in achieving high-resolution prints with high-heat polymers such as Nylon 66 blended with PAPC.

Innovative Plastics and Molding, Inc. would move its business to establish a 3D printing service bureau in Grosse Ile, Michigan, USA. The objective was to create an improved business environment with a greater value chain with the patented FibreTuff technology, utilizing the SLS printing method. The company was chartered to get more FibreTuff exposure in other industries supporting medical applications, as well as meet new collaborators and find funding for further commercial advancement. Robert Joyce of Innovative Plastics and Molding, Inc., would present at several foam-related events in four months, targeting the SPE Foams and Performance Polyamides conferences. His presentations reflected the 3D printing FibreTuff PAPC + Nylon66 having a cellular structure with bone-like performance, utilizing the SLS method.

28.5 The challenges with the roadmap for POC, biomaterials, and implants

The pandemic brought new types of challenges to the further development of 3D printable FibreTuff PAPC. Innovative Plastics and Molding, Inc. had seen commercial relations – collaborations with global research partners come and go. While some universities wanted to validate 3D printing FibreTuff, others would write abstracts and even research papers. More specifically, after working 3 years with West Virginia University, Innovative Plastics and Molding, Inc. became distraught after helping professors and bioengineering students write research papers, which included his techniques and vision to print "bone-like" properties with 3D-printed FibreTuff PAPC. At one point, the company had seen an opportunity to grow, even relocate its business with university support. The first step in business relocation failed when company representatives pursued a small business grant program in West Virginia for applying to the National Science Research NSF SBIR program. Unfortunately, Innovative Plastics and Molding, Inc., the university, and small business advocates failed to work collectively as a team.

In late 2022, Innovative Plastics and Molding, Inc. officially moved back to Michigan, USA, relocating to a small island southwest of Detroit, located in Wayne County – Grosse Ile. The company owner, Robert Joyce, would continue developing the 3D printable FibreTuff for medical applications and investigate nonmedical applications where light weighting is important, i.e., mobility, drones. Joyce has approached the Grosse Ile community to establish a production facility and rejuvenate a Naval Air base owned by the Department of Defense (DOD) back in the 1960s. He has the vision to 3D print protective gear that floats, such as flak jackets, even floating drones. Joyce began investing more time into the production of 3D printing FibreTuff PAPC as an implantable. He would be awarded a 7th patent: molded parts with thermoplastic biopolymer compositions having oriented fibers for medical devices and implants, US Patent 11497837 [9]. The patent claims were printable nonresorbable articles with a cellular structure produced with polyamide, polyolefin, and cellulose fiber.

More than several DOD SBIR grant applications with academia would be written at the Grosse Ile, Michigan, location. Several universities in Michigan helped prepare proposals for investigating 3D-printed FibreTuff for permanent implantables. The DOD SBIR implant grant proposals had focused on improving patient outcomes related to orthopedic injuries and better trauma care. The needs were to eliminate the long healing times and reduce infection rates. Further, the proposal's aims were to improve upon the suboptimal results due to the poor hydrophilicity and structural limitations of traditional biomaterials, including inadequate mechanical performance and chemistry. We believe the disruptive technology was clearly presented in the grant applications, however, the results were not favorable.

In the past decade, advancements have been made in synthetic implantable biomaterials for 3D printing for trauma-related injuries. In particular, synthetic polymers have been combined with minerals such as hydroxyapatite to create implants with absorbent, "bone-friendly" osteoconductive properties. While this mineral incorporation improves biocompatibility, it does not significantly enhance cell proliferation or bone regeneration and often introduces brittleness, reducing mechanical strength [10]. Additionally, the presence of mineral fillers decreases melt flow, complicating the extrusion and processing of monofilament strands and powders used for 3D printing patient-specific implants. Moreover, when such implants are 3D printed, the lack of compressive forces during the printing process further compromises strength and elongation. These implants are typically only partially resorbed by the body, and degradation over time remains a concern. Bioresorbable implants have also demonstrated osteoconductive behavior. However, their structural strength is limited – typically below 28 MPa – and their resorption rates vary depending on the patient's health status. Clinical data indicate a typical healing window of 12 to 24 weeks. Yet, in patients with diabetes, advanced age, smoking habits, or high alcohol consumption, healing times may extend to a year or more. This unpredictability makes bioresorbables a less risk-averse option for many patients and clinicians.

One emerging solution is HaPEEK, a composite biomaterial consisting of synthetic hydroxyapatite and PEEK. HaPEEK becomes bioactive only when compounded or blended with 30–40% hydroxyapatite. While considered resorbable – with a resorption rate of 5–15% per year – this composition also results in a significant loss of mechanical strength, typically a 20–25% reduction, yielding final strengths below 50 MPa, which is near the lower limit of cortical bone performance.

HaPEEK is primarily printed using SLS, which produces structures that are stiff and brittle. While porosity can be introduced to support cell and tissue growth for scaffold applications, the mechanical performance is still limited.

In contrast, the printed FibreTuff PAPC composition has bone-likeness, where the mechanical strength can achieve a tensile strength beyond 45 MPa. The printed FibreTuff PAPC has shown greater mass and aligned fibers to exceed the scaffold strength of HaPEEK. Further, printed FibreTuff PAPC had evolved to include both organic and inorganic chemistry. The 3D printable FibreTuff PAPC amphiphilic composition is bioactive, showing apatite formation and demonstrating osteoblast adhesion. The inorganic phase plays a key role in defining cell identity and guiding tissue-specific functions. The organic phase provides a protein-rich matrix that supports cell and tissue proliferation. This biomimetic approach sets FibreTuff apart from other printed bioactive materials to date.

28.5.1 The role of cellulose fiber in FibreTuff PAPC implantable

The cellulose fiber dispersement and the melt blending of nylon and polyolefin took many years to research and develop. The compounding of cellulose fiber with the proper barrel and screw design was critical for fiber dispersion for thermoplastic alloying. Further, the processing expertise to not degrade cellulose fiber with Nylon66 and polyolefin would be disruptive. The 3D-printed FibreTuff PAPC + Nylon66 random cellular structure with a porous and nonporous surface requires a type of cellulose fiber, not nano cellulose, which can greatly affect the outcome of the printed part surface energy and interface to coatings.

A most important and underappreciated effect in biomedical polymer composites: the role of cellulose fiber polarity in enhancing interfacial compatibility between hydrophilic polyamides (like Nylon66) and hydrophobic polyolefins. The FibreTuff's PAPC composite could benefit from this synergistic "alloying" behavior, especially for permanent implant applications. Some of these components and functions are illustrated further in Table 28.1.

Table 28.1: Polar/nonpolar and key functions for cellulose fiber, polyamide, and polyolefin (e.g., PP and PE).

Component	Polar/nonpolar	Key function in composite
Cellulose fiber	Highly polar (–OH groups)	Acts as a **molecular bridge** between incompatible phases
Polyamide	Polar	Naturally compatible with cellulose via hydrogen bonding
Polyolefin (e.g., PP and PE)	Nonpolar	Normally *incompatible* with nylon, but cellulose mandates

28.5.1.1 How cellulose enables "alloying" (synergistic effects)

- Interfacial compatibility
  In polymer blends, phase separation causes mechanical and thermal weaknesses.
- But cellulose's hydroxyl groups form hydrogen bonds with nylon, including Nylon66, and its fibrous, rough surface can entangle or disperse within polyolefins – acting like a surfactant or compatibilizer. This improves stress transfer, dispersion, and melt blending between polymers.
  Tunable surface energy
- The high polarity of cellulose increases wet ability and protein adhesion – good for bone bonding. But combined with polyolefin regions, you get amphiphilic surface domains – which can be engineered to have different biological responses.
  Mechanical reinforcement
- Cellulose fibers act as microreinforcement, increasing flexural modulus, impact resistance, and creep resistance (important in long-term implants)

There are similarities between cellulose and collagen, which include the following:

Both cellulose and collagen are fibrous materials, composed of long, thread-like molecular chains. This structural similarity allows cellulose to function in biomedical applications where collagen's fibrous characteristics are beneficial, particularly in tissue engineering.

Key parallels include:

- **Fibrous structure**: Both materials can form scaffolds with a high surface area, promoting cellular attachment and tissue integration.
- **Biocompatibility**: Neither material elicits significant toxic or adverse reactions, making them suitable for implantation or regenerative therapies.
- **Cell adhesion support**: Cellulose, like collagen, can facilitate cell adhesion – a critical factor in tissue repair and regeneration.

28.5.1.2 Bone tissue engineering

Bacterial cellulose, in particular, can serve as a scaffold that mimics the collagen matrix in bone, supporting cell attachment, growth, and mineralization for bone regeneration.

- Wound healing – Cellulose-based wound dressings offer a fibrous, absorbent, and biocompatible matrix that supports tissue repair, similar to collagen's role in healing.
- Scaffolds for tissue regeneration – Cellulose materials can be engineered to replicate the extracellular matrix of tissues such as bone, cartilage, and skin. These scaffolds offer structural support for cell proliferation and tissue growth.

28.5.1.3 Key considerations for cellulose with PAPC in regenerative medicine

- Tunable properties: Cellulose can be chemically modified to enhance its performance. For example, phosphorylation can promote hydroxyapatite growth, crucial for bone tissue engineering.
- Mechanical reinforcement: While cellulose can structurally mimic collagen, its mechanical properties may not be equivalent. Synergistic blending with thermoplastics, such as Nylon, can provide the necessary reinforcement to achieve desirable mechanical performance, similar to biomaterial alloying.
- Ongoing research: While early studies show significant promise for cellulose as a collagen substitute, further research is needed to optimize its properties and better understand its behavior in various biomedical applications [11].

28.5.2 HaNylon66 is disrupting the implant market with a lower cost and high-performing solution

There had been limited published research on alternatives to nonresorbable titanium or PEEK for spinal implants. That changed when a group of Chinese scientists identified a nano-mineralized Hydroxyapatite Nylon66 (HaNylon66) as a potential replacement [12, 13]. This biomaterial demonstrated improved bone integration and a reduced inflammatory response. However, the addition of hydroxyapatite still imposed limitations, including reduced strength, and challenges associated with 3D printing patient-specific implants. These challenges include degradation, brittleness, resorbability, and processing.

In contrast, an early-stage development of a 3D-printable polyamide, polyolefin, and cellulose (PAPC) biomaterial – an advanced Nylon composite designed for additive manufacturing has shown robust implant characteristics. The 3D-printed Fibre-Tuff PAPC is biocompatible, amphiphilic, or bone-like, with a nonresorbable cellular structure that retains absorbency while exhibiting minimal, if any, strength degradation. This novel PAPC biomaterial presents a disruptive opportunity to surpass conventional materials such as Nylon 6/66, HMWPE, PEKK, and osteoconductive PEEK in the $26 billion global biomaterials market.

In February 2023, Military Medicine published piloted research titled "Additive Manufacturing for Fabrication of Point-of-Care Therapies in Austere Environments," confirming the novelty of FibreTuff PAPC [14]. The study demonstrated two key applications:

1. Human anatomic models – The 3D printing of anatomical features such as the ear and vertebral body using PAPC.
2. Bioactive bandage – A base structure printed using fused filament deposition of PAPC was overlaid with live MSC cells in a cell-laden cellulose/alginate bio-ink, neomycin salve, and adhesive in designated regions of the bandage.

The 3D printable FibreTuff PAPC shows promise in life-saving applications due to its ability to withstand standard sterilization methods (e.g., autoclaving), support antibiotic infusion, and reduce bacterial contamination. According to Army COL Dr Jack Ryan, orthopedic surgeon: "PAPC may be used far forward in combat support hospitals to span fractures with critical bone defects to maintain limb length, or as substitute structures for skull, facial bones, eye support, ears, or pelvic reconstruction."

There was preliminary research data on 3D-printed FibreTuff PAPC + Nylon66. The research was published by the American Society of Mechanical Engineers. Authors included Robert Joyce, Dr Roozbeh Salary, and Brandon Coburn from Marshall University, Department of Mechanical Engineering, with the title, "Investigation of the influence of Nylon-6 vs. Nylon-66 on the mechanical performance of composite bone tissue scaffolds." The paper was presented at the International Mechanical Engineering Congress and Exposition IMECE2023, October 29–November 2, 2023, New Orleans, LA, USA [15].

28.5.3 3D printable FibreTuff PAPC + Nylon66: the next-generation amphiphilic composite for medical implant

To meet many of the challenges of producing medical devices, including medical implants, the FibreTuff PAPC biomaterial has included a high-heat polymer Nylon66, resulting in a next-generation 3D printable composite. The FibreTuff PAPC + Nylon66 is like an "alloy" (though it is a composite). In metals, an alloy combines dissimilar atoms to create a unified structure with enhanced mechanical or chemical properties.

In FibreTuff's PAPC + Nylon66:

- Cellulose enables molecular-level integration of hydrophobic and hydrophilic polymers
- The result is a cocontinuous phase structure rather than phase-separated domains
- That is why it behaves like a polymer alloy rather than a weak composite

You are essentially getting the toughness and chemical resistance of polyolefin, the bio-affinity of Nylon66, and the bonding reinforcement of cellulose – all in one material.

Additional findings of the PAPC + Nylon66 indicate:
- Low moisture content
- Reduced degradation in vivo
- Superior biocompatibility
- Antibacterial properties
- Improved flexibility and hardness

The FibreTuff PAPC composition is an amphiphilic (hydrophilic + hydrophobic) structure, having bone likeness, and *might* perform better as a medical implant.
- Optimized biological interface
  Hydrophilic surfaces (like HA/Nylon66) support faster cell adhesion, protein adsorption, and bone bonding.
  Hydrophobic regions can reduce fluid absorption, minimize long-term swelling, and control degradation or overintegration (e.g., too much bone remodeling).
- A smart amphiphilic material can steer biological response, allowing:
- Bone ingrowth in defined zones (hydrophilic pores/surfaces)
  - Mechanical sealing or barrier effect in others (hydrophobic shell or walls)
  - Mechanical toughness + damping
    Polymers with mixed-phase domains can dissipate energy better than monolithic materials. Think of it like a composite shock absorber: the hydrophobic domain may resist fatigue better, while hydrophilic domains facilitate healing.
- Controlled porosity and bone integration
  FibreTuff's PAPC bi-layer porosity – microporous + dense – plus amphiphilicity might allow:
  - Early vascularization and integration
  - Long-term structural support without complete bone overgrowth (which can be problematic)
- Reduced foreign body reaction
  Purely hydrophobic materials like PEEK sometimes trigger fibrous encapsulation (i.e., the body walls it off). Adding hydrophilic regions can mitigate immune isolation, improving long-term integration.

These two identified improvements enable the production of a wide range of applications at Roadmap for POC, including but not limited to protective gear, prosthetic sockets, preclinical assessment tools, orthopedic implants, including meshes, vascular filters, and superior bone void filler implants – a segment valued at $3.2 billion in 2023 [16].

The regulatory pathway for PAPC + Nylon66 includes a Master File with the FDA. The company plans to develop multiple 510(k) submissions for novel medical devices and implants. Notably, data shows that a bone adhesive can bond to printed PAPC + Nylon66 and natural bone, presenting an opportunity to replace traditional pins, clips, and plates – reducing both implant costs and hospital expenses. In addition, the inorganic and organic chemistry of the printed FibreTuff PAPC + Nylon66 will promote cell and tissue regeneration with the patient-specific device and implants. The development of the printable FibreTuff PAPC + Nylon66 continues as part of ongoing research aimed at advancing medical technology to enable quicker healing and reduce infections.

The impact of PAPC + Nylon66 could be transformative for military and emergency trauma care. A wounded soldier, for example, could receive a 3D-printed, patient-specific facial implant or eye socket reconstruction within an hour, rather than waiting days. Prior to surgery, the printed implant could be sterilized and immersed in an anti-inflammatory compound (e.g., naproxen, celecoxib, meloxicam) for sustained, localized drug release – enhancing the healing response.

For instance, a cranial flap was 3D-printed from PAPC + Nylon66 in under 90 min and sterilized via autoclave with no compromise in performance. This biomaterial is hydrophilic, has both porous and nonporous surfaces suitable for coatings, and supports drug delivery mechanisms – features not commonly found in other implantable materials. The cost of a cranial flap made from PAPC + Nylon66 is approximately $2,000, compared to a $4,000 flap made from PEEK.

In 2024, further development of the 3D-printed FibreTuff PAPC + Nylon66. Innovative Plastics and Molding, included applications for DOD SBIR grants, state, and local grants. A most promising opportunity for FibreTuff would be a DOD and state funding solicitation. The solicitation was an advanced materials request for additive manufacturing from LIFT in Detroit, MI. Innovative Plastics and Molding, Inc. would propose a sound advancement to the piloted research performed by the US Military, Walter Reed Hospital, to help the wounded warfighter in the golden hour.

The proposal included a collaboration with Michigan Technological University and Mayo Clinic, the VA Recovery, and others. The proposal identified five areas where the collaboration will greatly improve upon the piloted US Military research performed in 2023: (1) A high heat Nylon 66 polymer with PAPC will produce porous parts that are stronger, have bonelike properties and qualities. (2) An SLS method is the preferred printer method for processing the PAPC + Nylon66 composite. The medical applications for SLS printed PAPC + Nylon66 will expand at the POC for the production of products not limited to novel protective gear, medical filtration, improved prosthetics, preclinical tools such as bone models, surgical guides and instruments, bandages, and implants such as bone void fillers. (3) The SLS printed PAPC + Nylon66 will provide a dense cellular structure conducive to improving upon the functionality of current protective gear, prosthetic sockets, and even implants, such as bone grafts. There will be testing of the final prototypes. (4) A bone filler investigation will lead to new implantables with printed PAPC + Nylon66, including a novel bioresorbable bone adhesive. The PAPC + Nylon66 with bone adhesive will replace implants requiring pins, clips, plates, and sutures that can slow the healing process and cause infection. (5) The printing of PAPC is a dual-use technology that will provide radiopacity inputs for machine learning, increasing efficiency at the US Military and Civilian HCF.

A letter of support for the proposal would come from Michigan Senator Peters, Congresswoman Dingell, and collaborators such as Spartan Medical, located in MD, to help service and support product distribution of the printed FibreTuff PAPC parts. FibreTuff development efforts would fail to get the grant funding from LIFT and DOD in support of innovative medical technology for the wounded warfighter at the golden hour.

28.5.4 3D printing FibreTuff PAPC + Nylon66 with selective laser sintering (SLS) method

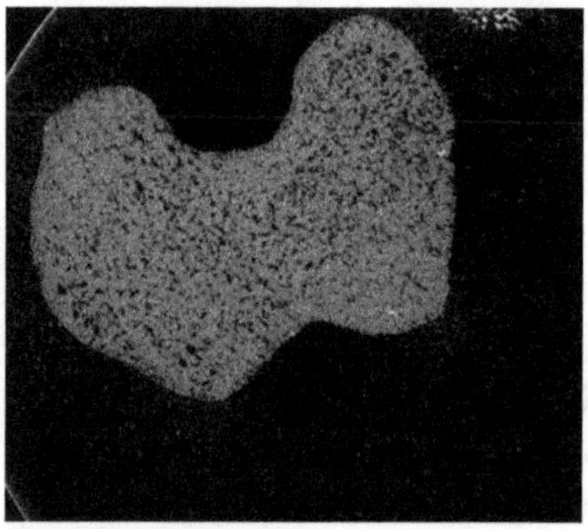

Figure 28.8: A micro-CT scan of a 3D-printed SLS and FibreTuff PAPC.

The traditional molding methods using foaming agents cannot replicate the unique structural outcomes achieved through 3D printing, particularly when using SLS with FibreTuff PAPC + Nylon66. First, as seen in micro-CT scans, the SLS-printed FibreTuff demonstrates both porous and nonporous interconnecting structures. The internal porosity features radiopacity, difficult to reproduce using conventional techniques and minerals (see Figure 28.8). Second, the SLS method, a nonpressurized process, has no shear with very little heat exposure to minimize destruction of the cellulose fibers. Third, the combination of nylon and polyolefin in FibreTuff is engineered with a compatibilizer that enhances interaction with the high polarity of cellulose fibers. The superior dispersion of the cellulose fibers, helped by the high heat blending and gentle mixing between 212 and 218 °C, further enabled the thermoplastic "alloying" process. In addition, the process may promote the addition of hydroxyapatite into the process and will not compromise the mineral composition. Lastly, the cellulose fibers in the printed FibreTuff PAPC created a surface roughness conducive to bone interface, even coatings. A nanocellulose would show a polymeric surface, challenges associated with process ability, and having a lower surface energy for a nonproductive interface.

Further, the SLS printed FibreTuff PAPC + Nylon66 is a multiuse technology. The ability to produce a lightweight bone-like cellular structure without destructive testing is applicable to many markets besides medical, such as automotive, and even aerospace. Compared to other bioactive biomaterials, FibreTuff PAPC + Nylon66 offers unique advantages:

Porous and dense cellular structure

- FibreTuff can be printed with pore sizes as small as 60–80 microns, conducive to bone ingrowth. In contrast, PEEK is typically printed with macro pores larger than 200 microns.

Improved thermal properties

- Containing cellulose, FibreTuff does not conduct heat like a neat resin, enabling faster cooling during the printing process with smaller pore sizes.

Radiopacity and monitoring

- FibreTuff has radio density, allowing clear visibility of implant placement and bone bridging under imaging. (Note: Real bone appears brighter than FibreTuff, while materials like HaPEEK remain radiolucent.) FibreTuff also supports data generation for artificial intelligence-based monitoring of healing progression.

Hydrophilic and drug absorbent

- FibreTuff incorporates hydrophilic ingredients, improving flexibility and impact resistance while reducing brittleness. It also supports antibiotic absorption and drug delivery mechanisms, which can significantly reduce infection risks.

Antibacterial

FibreTuff PAPC, with the addition of Nylon66, has provided antibacterial properties to the printed part, expanding the POC applications.

28.6 Summary

The early 2000s marked the beginning of an investigation into bioproducts that eventually evolved into the research and development of a novel, medical-grade biomaterial, now branded as FibreTuff. The developer of FibreTuff, Robert Joyce, has been on a journey filled with challenges – some unethical and even disturbing – for innovators. He believes that 3D printing will ultimately offer a lower-cost, high-performance solution for producing medical devices and implants, compared to processes like injection molding using supercritical fluids or chemical additives.

Joyce has identified clear differentiators in 3D-printed FibreTuff PAPC, which exhibits bone-like characteristics, including a novel and nonobvious hard surface that enhances bone integration to help heal patients quicker. Additionally, the radio density of 3D-printed FibreTuff PAPC improves efficiency not only in the machine learning process but also helps manufacturers qualify critical part dimensions. Furthermore, hospitals are expected to benefit from FibreTuff PAPC's radio density, as it aids in identifying implant orientation in the body and tracking bone bridging – contributing to improved patient outcomes.

The success of these 3D-printed FibreTuff PAPC innovations, pioneered by Robert Joyce, will be validated through ongoing product improvements, clinical applications, and collaborations. Joyce remains committed to finding partners and collaborators who share his vision of quicker healing, less infection, and improving patient outcomes through the creation of innovative medical devices and implants. The company and its founder extend sincere gratitude to all who have contributed to this effort and helped make this publication a meaningful part of the FibreTuff journey.

References

[1] Joyce R Molded Article, U.S. patent 7,214,240.
[2] Joyce R Wood Composite Alloy Composition having a compatibilizer that involves the ability to process and compress cellulosic fiber, U.S Patent 7,994,241.
[3] Joyce R Composition of Matter for thermoplastic composition biopolymer"FiberTuff", U.S. patent 8,546,470.
[4] Joyce R Composition of Matter for thermoplastic biopolymer: FiberTuff", U.S Patent 9,109,118.
[5] Joyce R Cellulose Fiber thermoplastic composition having a cosmetic appearance and molding thereof, U.S. patent 10,233,309.
[6] Zhipeng D, Bowen H, Yang X, Lei Wang, and Yueming Song The improved bioactive n- HA/PA66 cage versus the PEEK cage in anterior cervical fusion: results from a 6-year follow up and a case-matched study. December 2022.
[7] Mandaava KR, Day JB, Joyce R, Salary R Investigation of the mechanical properties and bioactivity of additively manufactured bone tissue scaffolds, composed of polyamide, polyolefin, and cellulose fiber, MSEC and ASME, June 27, 2022.
[8] Chaffina A, Yu M, Claudio PP, Day JB, Salary R Investigation of the functional properties of additively fabricated Triply Periodic Minimal surface-based bone scaffolds for the treatment of Osseous fractures, Aug 4, 2021– American Society of Mechanical Engineers ASME.
[9] Joyce R Molded parts with thermoplastic biopolymer compositions having oriented fibers for medical devices and implants U.S. patent 11,497,837
[10] Ma R, Guo D Evaluating the bioactivity of a hydroxyapatite-incorporated polyetheretherketone biocomposite. J Orthop Surg Res 2019;14(1):1–13. doi: 10.1186/s13018-019-1069-1.
[11] Ryan JH, Andrew P. Cellulose Biomaterials for Tissue Engineering, Frontiers in Bioengineering and Biotechnology, vol. 7, p. 1, Mar. 21, 2019.

[12] Jianxio L, et al. Insight into Osseo integration of Nano hydroxyapatite/Polyamide66 based on Radiolucent gap: comparison with Polyether-Ether-Ketone frontier materials. June 2021.

[13] Li JC, et al. Nano-hydroxyapatite/Polyamide 66 cage in anterior cervical corpectomy and fusion: A comprehensive systematic review and meta-analysis. Eur Rev Med Pharmacol Sci 2023:Oct;27 (20):9648–9659. doi: 10.26355/eurrev_202310_34136.

[14] Barnhill J, Gaston JD, Deffenbaugh PI, Wagner L, Liacouras PC, Ho VB Additive manufacturing for fabrication of point-of-care therapies in austere environments. Mil Med 2023;00:1–7. doi: 10.1093/milmed/usad007.

[15] Brandon C, Robert J, Roozbeth S. Investigation of the influence of Nylon6 versus Nylon66 on the mechanical performance of composite bone tissue scaffolds. Am Soc Mech Eng ASME Oct–November 2023. Paper No: IMECE2023-110405, V003T03A087; p.1–8.

[16] Fortune business insights / medical device / bone void fillers market. Published June 2023.

Rachel Yoho*

Chapter 29 Artificial intelligence and the next frontier of bioproducts research and innovation

29.1 Introduction

Undoubtedly, by the time this chapter moves from writing to publication, the content may be out of date – or more positively framed as – representative of a snapshot of research and innovation from late 2025. The rate of change of artificial intelligence (AI)-based technologies is immense. While all research and publications should serve as the best relevant representation of the state of the literature at the time, with the AI-based technologies, the innovation versus the lag in publications, especially those that are peer reviewed, creates a challenge for building a foundational literature base that is *truly* current.

With these complexities in mind, this chapter provides an introduction to AI applications in bioproducts research and innovation, including the need, opportunities, and consequences of AI use. First, though, we establish some definitions and shared understandings leading to the remainder of the discussion. Since AI text generators burst onto the scene in 2022, impacting work and research across industries, multiple models are currently in use. AI is an umbrella term for a number of other terms, including machine learning and natural language processing [1]. AI models can be sorted taxonomically, based on their input and output formats, for example, with inputs or outputs as text, image, audio, or other formats [2]. Examples of text-in and text-out models include the widely known ChatGPT (from OpenAI), as well as others like Copilot (from Microsoft) and Claude (from Anthropic). In the simplest sense, AI models "are programs that detect specific patterns using a collection of datasets" [1]. In general, AI models work by creating a model – some sort of algorithm – that processes and assesses datasets, then is trained by running data through this model. The training is either supervised (known classified data) or unsupervised (model assesses data). The result of this is what we observe through multiple types of AI models – the application in real-life scenarios creating outputs [1]. Essentially, for large language models (LLMs), we might consider this very generally as what is the most likely next word generated sequentially in a string of words.

*Corresponding author: Rachel A. Yoho, (formerly) George Mason University, Fairfax, Virginia, USA, (currently) College of Southern Nevada, 6375 W. Charleston Blvd., Las Vegas, Nevada 89146, USA, e-mail: rachel.yoho@csn.edu

 | https://doi.org/10.1515/9783112212141-029

29.2 Trends over time

The potential applications are enormous across research and industries. Mainstream use of AI chatbots, particularly starting with ChatGPT in 2022, has upended work and education. In this, we want to focus on the aspects of AI that may be most relevant to bioproducts research and areas. Stepping back into a larger scope, interest in "artificial intelligence + bioproducts" in Google Trends indicates a rapidly increasing interest in recent years, again since about 2022. Google Trends data for this search are included in Figure 29.1. As a note, similar terms (e.g., bioproducts, bio-based products) provided similarly shaped trendlines. Similarly, a Boolean search in Google Scholar of "artificial intelligence AND bioproducts," limiting the range to 2022–present, provides over 16,000 results. Naturally, these will be of varying relevance and quality, as with any wide search.

Figure 29.1: Google Trends data for the search "artificial intelligence + bioproducts" on October 21, 2025. Upper details data since 2004, and lower is the previous 5 years.

29.3 Applying AI to bioproducts research

29.3.1 Overview of AI in bioproducts

If we ask an AI text generator – say, Copilot – about bioproducts research and innovation, we can compare it to another, for example, Google's AI mode. This result, in Table 29.1, helps us set a foundation for a conversation by putting LLMs into practice.

With the same prompt in Table 29.1, we see that Copilot (Microsoft) and Google AI mode (Google) produce largely similar content, with some differences in the detail of information and the examples. As a fun aside, asking Copilot, "What book would I read to learn about bioproducts?" does produce the first result of the first edition of

Table 29.1: Prompt and side-by-side comparison of responses for Copilot and Google AI assistant for "Which types of AI models are useful in bioproducts research?" Resulting outputs edited for formatting, and no references/links from outputs included.

Copilot (Microsoft)	Google AI mode
AI models are increasingly valuable in **bioproducts research**, which spans areas like biofuels, bioplastics, pharmaceuticals, and agricultural innovations. Here are the key types of AI models and how they are used: 1. **Machine learning (ML) models** These are foundational for pattern recognition and prediction. – **Supervised learning**: Used for predicting yields, optimizing fermentation processes, or classifying microbial strains. – Examples: Random forest, support vector machines (SVM), gradient boosting. – **Unsupervised learning**: Useful for clustering microbial communities or discovering hidden patterns in omics data. – Examples: *K*-means, hierarchical clustering, PCA. 2. **Deep learning models** These are powerful for handling complex, high-dimensional biological data. – **Convolutional neural networks (CNNs)**: Applied in image analysis (e.g., microscopy, plant phenotyping). – **Recurrent neural networks (RNNs)/ LSTMs**: Used for time-series data like bioreactor monitoring or gene expression over time. – **Autoencoders**: For dimensionality reduction and feature extraction in genomics or metabolomics. 3. **Natural language processing (NLP) models** Useful for mining scientific literature, patents, and lab notes. – **Transformer-based models** (e.g., BERT and BioBERT): Extract insights from text, assist in hypothesis generation, and automate literature reviews.	Different AI models are used in bioproducts research, with their usefulness depending on the specific application. Key models include generative AI for designing novel molecules and deep learning models for optimizing complex biological processes. **Generative AI** Generative AI excels at creating new data, such as novel molecule sequences or structures, rather than just classifying existing data. – **Generative adversarial networks (GANs):** Used to accelerate the discovery of chemical compounds with desired properties by efficiently exploring vast chemical spaces. GANs train two networks, a generator and a discriminator, to create realistic new data points. – **Variational autoencoders (VAEs):** Generate new enzyme variants by mapping protein sequences to a lower-dimensional space and then decoding new sequences from that space. – **Language models (e.g., GPT-like models):** Treat protein sequences like a language. By training on large datasets of existing protein sequences, these models can generate novel, functional sequences for new enzymes. **Deep learning** Deep learning models use multilayered neural networks to learn hierarchical patterns from complex data, which is essential for processing biological big data like genomics and imaging. – **Convolutional neural networks (CNNs):** Excel at image analysis. In bioproducts research, CNNs can analyze medical imaging for biomaterial design, classify microbes from microscopy images, or count microbial colonies.

Table 29.1 (continued)

<table>
<tr><th>Copilot (Microsoft)</th><th>Google AI mode</th></tr>
<tr><td>4. Reinforcement learning
Applied in process optimization, such as controlling bioreactor conditions or adaptive experimental design.
5. Graph neural networks (GNNs)
Ideal for modeling biological networks like metabolic pathways, protein-protein interactions, or gene regulatory networks.
6. Generative models
Used for designing new molecules, enzymes, or metabolic pathways.
– Generative adversarial networks (GANs): For synthetic biology applications.
– Variational autoencoders (VAEs): For generating novel compounds or optimizing molecular structures.
7. Hybrid models
Combining physics-based models with AI to simulate bioprocesses more accurately (e.g., hybrid modeling of fermentation kinetics).
Would you like examples of how these models are applied in a specific bioproduct domain, like biofuels or pharmaceuticals?</td><td>– Recurrent neural networks (RNNs): Specialized for sequential data. They can analyze DNA, RNA, and protein sequences to help identify genetic variations and their impact on protein function, which is useful in strain and enzyme engineering.
– Graph neural networks (GNNs): Process data with a graph structure, which is common in biology for modeling molecular interactions or microbial co-occurrence networks. GNNs have been used to discover antimicrobial drugs and identify microbial biomarkers for disease diagnosis.
Machine learning algorithms
These models are more traditional but remain highly effective for analyzing biological data and optimizing processes.
– Support vector machines (SVMs): Used for classification and regression tasks, such as predicting protein structures or identifying biomarkers from complex omics data.
– Random forests: Classify or predict outcomes by combining the results of multiple decision trees. They are used for tasks like predicting host phenotypes from microbiome data.
– Genetic algorithms (GAs): Based on the process of natural selection, GAs are optimization techniques used to find optimal parameters for complex bioprocesses, such as nutrient formulation.
– Regression models (e.g., linear amd logistic): Predict a continuous output value (regression) or a category (logistic regression) based on input variables. They are useful for forecasting yields or estimating the probability of an event.
Hybrid and specialized models
Bioproducts research is increasingly adopting more sophisticated or combined AI approaches.
– Hybrid models: Combine data-driven AI techniques with traditional scientific principles to create more robust and adaptable models for bioprocess control.</td></tr>
</table>

Table 29.1 (continued)

Copilot (Microsoft)	Google AI mode
	– **Digital twins:** Virtual representations of physical systems, such as bioreactors, that use AI to replicate and simulate bioprocesses in real-time. This allows for virtual testing and optimization of process parameters. – **Reinforcement learning (RL):** Focuses on decision-making through trial and error. RL can be used to regulate bioprocess variables like temperature and oxygen to maximize production yield. – **Multimodal AI:** Integrates and analyzes data from diverse sources, such as medical imaging, genomics, and clinical records. This approach is used to design personalized biomaterials that are compatible with a patient's genetic profile.

this book, published by DeGruyter, along with a brief overview and a "best for" audience description. Overall, Table 29.1 provides a fairly reasonable overview of and introduction to the application of different types of AI-based models to bioproducts research. Several of the topics listed by the outputs in Table 29.1 from Copilot and Google's AI mode are explored in the specific bioproducts sections below, including machine learning techniques, generative AI, and LLMs, and other subtopics. Using this compelling and comparative foundation, in the next section, we provide specific examples of AI applications and current relevant to the content of this book. Further, Chapter 30 jumps into more specific bioproducts applications and provides examples for research strategies.

29.3.2 Applications of AI in bioproducts research

The emergence of widespread access to and use of AI-based technologies is currently shaping research across different bioproducts-related areas, however some of these strategies and applications are already in a comparatively mature research state [3], with notable applications related to the 2024 Nobel Prizes in Physics and Chemistry, including machine learning technologies with artificial neural networks (ANNs), even as early as the 1980s [4, 5]. Applications of generative AI provide opportunities for the enhancement of ongoing research opportunities, further study of complex system dynamics, and creative ways to develop new research trajectories. In particular, we note several opportunities in both process scale-up and life cycle assessment. However, many potential environmental consequences and drawbacks, detailed later in this chapter, remain as potential detractors.

An example innovation in AI application in the general bioproducts-related area is the concept of a "self-driving laboratory" where the role of the human evolves to shaping questions, defining parameters, delineating reaction protocols, and general oversight [4, 6]. Applications of AI and machine learning are particularly useful for overcoming barriers or research challenges in biomaterials design and biomass breakdown, especially by shaping experimental planning [7]. This can aid in both process optimization and *research* process optimization, by potentially saving time and resources in the lab stages. Further, machine learning applications have been seen as useful in process scale-up for bioproducts broadly, particularly again for nonlinear optimization [8]. Bioprocess engineering not only advances the industry and applications but also explores the biological systems themselves, with more recent use of machine learning and AI [9].

Another key area with significant opportunities for applying AI-based technologies in the research process is life cycle analysis (LCA). While explored in detail in other chapters and as related to policy, particularly see Chapters 5 and 6, we mention this here due to the additional AI opportunities. AI incorporation may be able to improve the accuracy of modeling and predictions from LCAs, as well as save time in the research and data collection processes [10]. The applications of LCA to inform policymaking and customer interests in environmental impact accountability create an opportunity for LLMs and generative algorithms that balance human efforts with AI integration [11]. In particular, machine learning has been applied across many fields, including environmental applications, but a significant opportunity in automated or semiautomated LCA for products and processes remains [11]. An example environmental LCA application applied ANNs and adaptive neuro fuzzy inference system for sugarcane production [12]. Sugarcane has many applications in bioproducts innovation-related areas [13]. In brief, ANNs are algorithms that function based on neurons and processing in the brain and can be trained or "learn" [14]. Adaptive neuro-fuzzy inference system, in a high-level overview, combines two machine learning techniques (neural networks and fuzzy logic) and uses weighting to map inputs to outputs [15, 16]. However, as with all AI applications, risks in the LCA application remain, including data standardization and risks of biases, in particular [10]. Again, like other research applications, input data, AI methods, and tool use, interdisciplinary knowledge and applications, and collaboration become methods for responsible AI integration into LCA [11].

29.3.2.1 Biomass

While biomass-related bioproducts topics are covered in detail in many chapters (notably, see Chapter 2 about feedstocks, Chapter 26 on fibers, Chapter 27 on business aspects, among others), we highlight several aspects of AI integration into the research space here. Key areas for innovation and application include process optimization, catalyst design, and resource management [3]. AI applications in the biomass space may allow for improved real-time data analysis, particularly in the operational space to improve performance and increase efficiency [17]. Challenges in the biomass-to-renewable energy space include feedstock recalcitrance, high expense, time-consuming processes, and similar issues [18]. AI-related opportunities, perhaps to address some of these challenges, include specific areas in resource management and sourcing, parameter optimization, modeling and managing nonlinear situations and relationships, reducing or mitigating risk, and researching or optimizing complex metabolic reactions [17]. These research developments, applying AI and machine learning, may help to improve efficiency, improve accuracy, and reduce cost in outputs [18]. Specific applications include neural networks, support vector machines, and hybrid models that can be applied to processes, resources and sourcing, and outputs [18]. In brief, a support vector machine is a complex machine learning technique that is also computationally demanding, which addresses big data classification issues [19]. Much like the other AI research applications described in other sections, as well as in other chapters, challenges for AI applications in biomass processing include datasets, model transparency, computational requirements (and consequences further explored in Section 29.4), and complexity, among others [18].

As a key example, we anticipate research advancements in this space may look like the advancements recently described for biofuel production from algae [20], in which AI tools were applied for optimization purposes around microbes, enzymes, and for scalability, alongside LCA. These AI and machine learning techniques applied included predictive modeling, ANNs, fuzzy logic, and particle swarm optimization (all described elsewhere in this chapter), and evolutionary (genetic) algorithms in order to optimize (lower) cost, improve scalability, for enzyme design, and controlling the fermentation process [20]. Genetic algorithms are based on the "survival of the fittest" concept from biology, applied in search and optimization applications [21].

Much of this conversation on biomass can be extended into specific interest areas such as bioplastics and biopolymers. Many chapters in this book include detailed information on bioplastics and biopolymers (notably, see Chapters 21 and 22). The revolutionary advancements from AI in this space are rapidly emerging. Particular applications of AI technologies focus on overcoming the barriers of conventional research in enhancing efficiencies and productivity, including nonlinear relationships and predictions of mechanical strength, glass-transition temperature, and solubility from existing datasets [4].

For further exploration, we refer readers to other published works, including the special edition of *Bioresource Technology* edited by Khanal et al. [17] for an in-depth analysis of AI and machine learning applications in bioprocesses including: anaerobic digestion, thermochemical conversion, composting, wastewater treatment, bioprocess optimization/monitoring/control, techno-economic analysis and life cycle assessment, and other topics that may be of interest to readers of this volume.

29.3.2.2 Microbial systems

While covered in more detail in Chapter 13, Section 13.10, mathematical modeling is a key area of established practice, and AI applications are of emerging use and importance. This application may serve to help with the research process due to the many unknowns in microbial systems and optimization [22].

As a particular example and related to the microbial bioproducts interest area, optimization (see also Section 29.3.2) may include both the process and microbial strain selection, monitoring, and control, as well as the scale-up already discussed [23]. Examples of specific tools applied in this area, for microbial systems and production of metabolites and enzymes, include ANNs, adaptive-network-based fuzzy inference systems, and artificial bee colony [23]. ANNs are described earlier. Adaptive-network-based fuzzy interference systems apply hybrid learning, using human knowledge with fuzzy if-thens, and defined input-output pairs. These are particularly useful in nonlinear functions, among other applications [15]. Finally, the artificial bee colony algorithm is a swarm-based algorithm, for example, applied to optimize large sets of numerical functions [24].

In the next chapter, we explore additional applications and tangible examples of AI technologies applied to selected areas of bioproducts innovation.

29.4 Challenges and consequences of AI use

Research applications of AI are significant; however, many current research challenges exist. For example, in the polymers space, this may include limited datasets, broad distributions of molecular weight, and irregular configurations of polymers [25]. In shifting the conversation from research challenges like these with polymers, we can extend the conversation out to broader challenges and consequences. Among many considerations, data privacy and data ownership are a broad challenge of AI use across fields, disciplines, and applications [3, 26].

While the applications are enormous, the potential consequences of AI applications are also immense – and pose an interesting challenge, particularly for bioproducts research and innovation. The same technologies that create opportunities for rapid analysis, innovation, and identification of opportunities are currently causing, and are projected to cause, significant environmental damage in the future. This presents a bit of a vicious cycle, in that increasing the research and innovation space and applications for bioproducts, many of which have positive environmental gains or impacts, also has negative impacts in use of AI. While this could benefit from its own version of a life cycle assessment or LCA, we recommend considering the current discussion of LCAs in bioproducts found in Chapter 5 in this book as a foundational starting point for impacts and assessments, before including AI in these considerations. Broadly, this has been called “AI for Green” [27].

As a tool, the sweeping umbrella of AI technologies requires significant computing resources, drawing extensive power demands, often from fossil fuel sources – as well as support from data centers and cloud computing facilities, which have significant carbon footprints [28]. Recent estimates of the environmental impacts and emissions from data centers indicate that these negative impacts could be comparable to those of other, existing, global industries [29]. Specifically, a 2024 assessment of 2,132 data centers in the US estimated that these demanded 4% of the US’s energy, with over half (56%) from fossil fuel sources [30]. While the national or global impacts of data centers are important, the impacts on local communities, including the local social and natural environments [31], provide additional nuance. Future projections of the annual water footprint range from 731 to 1,125 million m^3 and an additional 24 to 44 Mt CO_2-equivalent in the 2024 through 2030 period [32].

Further, there may be efforts to shape perceptions of data centers in local communities [31]. As an example, data centers played an important role in a recent local election of a Virginia state house of representatives campaign (November 2025), with Democratic candidate John McAuliff, featuring a pushback against data center development in television ads (author’s personal observation) and, prominently, on his campaign website with a tab at the same organizational level as “my story,” “priorities,” “endorsements,” and similar. Figure 29.2 shows a screenshot from this candidate’s campaign website [33] with information and emphasis on the environmental

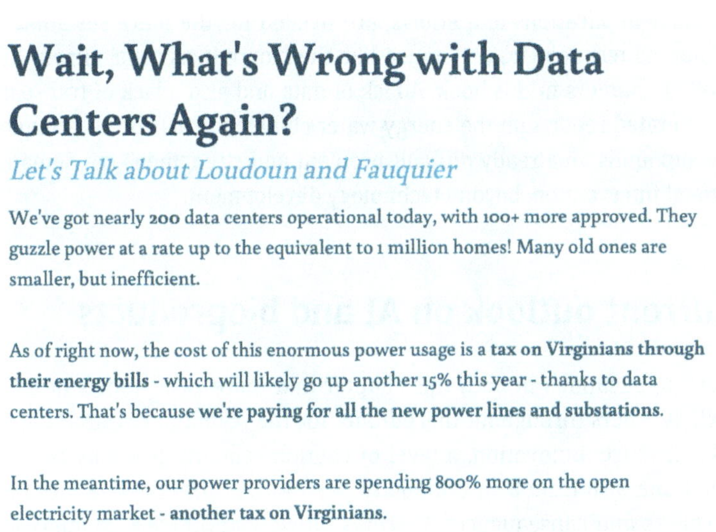

Figure 29.2: A screenshot from a portion of a campaign website discussing data centers in November 2025 for a local candidate running for state-level House of Representatives in the Commonwealth of Virginia, USA (source [33]).

impacts ("guzzle power at a rate up to the equivalent of 1 million homes"), costs ("we're paying for all the new power lines and substations"), and impacts on energy bills. In this election in November 2025, John McAuliff ousted the incumbent Republican running for re-election.

As a particularly prominent area for data centers – and continued data center construction – in the United States, Northern Virginia may be a unique case where the local and environmental impacts of AI take center stage in the local discourse and may enter the conversation for local elections. While further investigation in the social science areas of technology impact may fall a bit outside of the scope of this book, we note that important interdisciplinary research areas may continue to investigate these rapidly changing dynamics.

The application of AI-based technologies, particularly with the training of more detailed models with additional parameters, has increased greenhouse gas emissions and greater potential for negative environmental impact [27]. However, there are several spaces for innovation in the design of data centers, energy sources, cooling technologies, and other areas that *can* help decrease the energy consumption of applying AI tools [29]. Applying best practices and innovations has the potential to decrease emissions and water footprints by approximately 70–80% each, however, applications are limited by infrastructure challenges [32].

To this point, "sustainable AI" [28] remains a concept and, perhaps, a goal or ideal for humanity's next frontier in technological development. A possibility exists for a more sustainable digital future [29] when considering the compound impacts in energy-water-climate [32]. Holistic approaches combining research into technology development, as well as policy applications and efforts, are needed for the more sustainable development of AI-based technologies [29], much like the bioproducts development described in many other chapters in this book. A lack of data and also a lack of transparency in industry-generated reports on the energy-water-climate impacts of AI technologies [32] further compounds an already difficult problem and strengthens the demand for policy and societal intervention, beyond technology development.

29.5 Our current outlook on AI and bioproducts

While we see many applications in both this chapter and more case studies next in Chapter 30, as well as others throughout this edition, for the potential research applications of AI to bioproducts innovation, a level of caution remains. Not only from a research standpoint, but also from a sustainability perspective, does the field need to be aware of drawbacks and consequences. It seems possibly antithetical to apply an energy-intensive and likely negatively impactful technology to innovate in a "green" space for sustainability purposes. As such, caution and thoughtful applications are even more imperative for the field moving forward.

References

[1] Hewlett Packard Enterprise. AI Models.HPE Glossary.October 21, 2025.Available from https://www.hpe.com/us/en/what-is/ai-models.html#:~:text=AI%20models%20or%20artificial%20intelligence,and%20robotics%20and%20control%20systems.

[2] Gozalo-Brizuela R, Garrido-Merchan EC ChatGPT is not all you need. A state of the art review of large generative AI models. arXiv preprint arXiv:2301.04655 2023 Jan 11.

[3] Orejuela-Escobar, L., Venegas-Vásconez, D., & Méndez, M. Á. (2024). Opportunities of artificial intelligence in valorisation of biodiversity, biomass and bioresidues–Towards advanced bio-economy, circular engineering, and sustainability. Int. J. Sustain. Energy Environ. Res, 13, 105–113.

[4] Chen J, Christakopoulos P, Ivanov IN, Vasudevan RK, Kumar R, Sumpter BG, Advincula RC Artificial intelligence/machine learning of polymer materials and discovery chemistry. MRS Bull 2025;50:1–12.

[5] Li B, Gilbert S Artificial Intelligence awarded two Nobel Prizes for innovations that will shape the future of medicine. NPJ Digit Med 2024;7(1):336.

[6] Tom G, Schmid SP, Baird SG, Cao Y, Darvish K, Hao H, Lo S, Pablo-García S, Rajaonson EM, Skreta M, Yoshikawa N Self-driving laboratories for chemistry and materials science. Chem Rev 2024;124(16):9633–9732.

[7] Smith JC, Smith MD, Liu SH, Rukmani SJ, Mohan M, Yu Y, Goswami M Molecular simulation and artificial intelligence for the circular economy of bioenergy and bioproducts. Biophys J 2025;124:3827–3852.

[8] Sharma D, Singh K AI-enhanced bioprocess technologies: Machine learning implementations from upstream to downstream operations. World J Microbiol Biotechnol 2025;41(8):278.

[9] Singh DP, Singh S, Krishania M Introduction to industrial-scale bioprocess. In Bioprocess Engineering and Technology, CRC Press; 2025. pp. 1–22.

[10] Gachkar D, Gachkar S, García Martínez A, Angulo C, Aghlmand S, Ahmadi J. Artificial intelligence in building life cycle assessment. Archit Sci Rev 2024;67(6):484–502.

[11] Nwagwu CC, Ogorodnyk O, Sølvsberg E, Eleftheriadis RJ, Meskers C Integrating artificial intelligence into life cycle assessment: A framework for balancing automation and human expertise. J Sustain Metall 2025;11(4):3590–3605.

[12] Kaab A, Sharifi M, Mobli H, Nabavi-Pelesaraei A, Chau KW Combined life cycle assessment and artificial intelligence for prediction of output energy and environmental impacts of sugarcane production. Sci Total Environ 2019;664:1005–1019.

[13] O'Hara I, Mundree S, eds. Sugarcane-based biofuels and bioproducts, John Wiley & Sons; 2016.

[14] Krogh A What are artificial neural networks?. Nat Biotechnol 2008;26(2):195–197.

[15] Jang JS ANFIS: Adaptive-network-based fuzzy inference system. IEEE Trans Syst Man Cybern 1993;23(3):665–685.

[16] Al-Hmouz A, Shen J, Al-Hmouz R, Yan J Modeling and simulation of an adaptive neuro-fuzzy inference system (ANFIS) for mobile learning. IEEE Trans Learn Technol 2011;5(3):226–237.

[17] Khanal SK, Tarafdar A, You S Artificial intelligence and machine learning for smart bioprocesses. Bioresour Technol 2023;375:128826.

[18] Duda LJ, Woiciechowski AL, Scapini T, Letti LA, Soccol CR AI and machine learning tools for optimization of biomass thermochemical conversion and bioprocesses. In Biomass Valorization Technologies, CRC Press; 2025. pp. 340–376.

[19] Suthaharan S Support vector machine. In Machine learning models and algorithms for big data classification: Thinking with examples for effective learning, Boston, MA: Springer US; 2016, pp. 207–235.

[20] Pusuluru S, Velmurugan K, Punna SK, Ravikumar M, Kumar M, Almukhlifi HA, Khan FR, Hazazi A, Menaa F Advancing biofuel production through algal biorefineries: Optimization of polysaccharide

utilization, AI-driven innovations, and sustainability assessments. Biomass Bioenergy 2026;204:108366.

[21] Sivanandam SN, Deepa SN Genetic Algorithms. In Introduction to Genetic Algorithms, Berlin, Heidelberg: Springer Berlin Heidelberg; 2008, pp. 15–37.

[22] Gadkari S, Gu S, Sadhukhan J Towards automated design of bioelectrochemical systems: A comprehensive review of mathematicalmodels. Chemical Engineering Journal 2018;343:303–316.

[23] Dixit M, Gupta GK, Pandey D, Kumari R, Chhabra D, Kango N, Shukla P Microbial bioprocess efficiency improvement through Artificial Intelligence and Machine Learning (AI-ML) tools. In New Horizons in Halophilic Microbes and Their Enzymes, CRC Press; 2025. pp. 246–257.

[24] Karaboga D, Akay B A comparative study of artificial bee colony algorithm. Appl Math Comput 2009;214(1):108–132.

[25] Ge W, De Silva R, Fan Y, Sisson SA, Stenzel MH Machine learning in polymer research. Adv Mater 2025;37(11):2413695.

[26] Yoho R No, let's not go back to handwritten activities: Inclusive teaching strategies in the context of ChatGPT. In National Teaching & Learning Forum, 2023, October. 32, 6.

[27] Ligozat AL. Lefèvre J, Bugeau A, Combaz J. Unraveling the hidden environmental impacts of AI solutions for environment life cycle assessment of AI solutions. Sustainability 2022 Apr 25;14(9):5172.

[28] Chauhan D, Bahad P, Jain JK Sustainable AI: Environmental implications, challenges, and opportunities. Explainable AI (XAI) for sustainable development 2024;1–15.

[29] Ewim DRE, Ninduwezuor-Ehiobu N, Orikpete OF, Egbokhaebho BA, Fawole AA, Onunka C Impact of data centers on climate change: A review of energy efficient strategies. J Eng Exact Sci 2023;9 (6):16397–01e.

[30] Guidi G, Dominici F, Gilmour J, Butler K, Bell E, Delaney S, Bargagli-Stoffi FJ, 2024. Environmental burden of United States data centers in the artificial intelligence era. arXiv preprint arXiv:2411.09786.

[31] Ipsen HA, 2018. Catching the Cloud and Pinning It Down: the Social and Environmental Impacts of Data Centers(Master's thesis, Syracuse University).

[32] Xiao T, Nerini FF, Matthews HD, et al. Environmental impact and net-zero pathways for sustainable artificial intelligence servers in the USA. Nat Sustain 2025; https://doi.org/10.1038/s41893-025-01681-y.

[33] https://www.john4va.com/data-centers

Rachel Yoho, Parthiban Pannerselvam, Ram Lalgudi*

Chapter 30 AI-enabled bioproducts: sustainable feedstocks, formulation intelligence, and green by design

30.1 Introduction

Bio-based products derived from renewable feedstocks promise performance-advantaged alternatives to petrochemical incumbents while reducing greenhouse gas emissions and toxic exposures. Yet their inherent chemical diversity and feedstock variability complicate design, scale-up, and verification. Artificial intelligence (AI) accelerates discovery, improves process stability, and enhances sustainability by using predictive life-cycle analysis – techno-economic analysis LCA/TEA gates to screen candidates before pilot-scale investment [1–3]. AI is transforming bioproducts research and development by providing advanced tools and methodologies that enhance efficiency, accuracy, and innovation. AI technologies, such as machine learning (ML), deep learning, and data analytics, are being integrated across various stages of bioproduct development, from discovery through commercialization.

One of the key frontiers in bioproducts research is the application of AI to the identification of sustainable feedstocks and the development of their formulations. AI algorithms can analyze vast datasets to identify patterns and predict outcomes, enabling researchers to design and optimize formulations with unprecedented precision. AI's potential extends to identifying bioproduct replacements for harmful petroleum-derived chemicals, especially in specialized applications such as coatings, paints, and microencapsulation. By leveraging AI, researchers can systematically explore vast chemical and biological databases to pinpoint sustainable alternatives that deliver essential performance characteristics while reducing environmental impact. In coatings and paints, for instance, AI-driven discovery platforms are being used to screen and design bio-based polymers and resins that match or exceed the durability, gloss, and adhesion properties of traditional petrochemical binders. These bioproduct coatings are being tested and deployed for applications ranging from architectural paints to corrosion-resistant industrial surfaces [4].

***Corresponding author: Ram Lalgudi**, Aries Science and Technology, LLC, 1381 Kinnear Rd., Columbus, OH 43212, USA, e-mail: lalgudir@ariesst.com

Rachel A. Yoho, (formerly) George Mason University, Fairfax, Virginia, USA, (currently) College of Southern Nevada, 6375 W. Charleston Blvd., Las Vegas, Nevada 89146, USA

Parthiban Pannerselvam, University of the Cumberlands, 6178 College Station Drive, Williamsburg, KY 40769, e-mail: ppannerselvam39788@ucumberlands.edu

 | https://doi.org/10.1515/9783112212141-030

This chapter provides a practitioner-focused treatment of three frontier topics in bioproducts innovation sustainable feedstocks, formulation intelligence, and green-by-design. This supplements Chapter 29, which provided an introduction and overview, as well as key considerations to AI integration in bioproducts research and innovation. In this chapter, we synthesize current evidence, tools, and case studies to show how AI accelerates route scouting, stabilizes processes under feedstock variability, links molecular descriptors to end-use performance, and embeds safety, degradability, and circularity objectives from day 1. We conclude with implementation blueprints and curated references to support new product development, regulatory dossiers, and pilot planning.

30.2 Adapting processes to sustainable feedstocks

30.2.1 Agricultural residues and lignocellulosics

Heterogeneous composition and recalcitrance in lignocellulosic residues demand data-driven pretreatment and hydrolysis. Ensemble ML models can predict lignin destabilization and guide severity selection across residue types, with temperature often the dominant factor [5]. Bayesian multi-objective optimization in green biorefineries identifies pretreatment conditions that jointly optimize lignin yield and β-O-4 linkages, accelerating route discovery with minimal experiments [6]. Reviews emphasize ML across pretreatment, hydrolysis, mixed-sugar fermentation, and anaerobic digestion, recommending transfer learning and hybrid ML for data-sparse regimes [7]. Engineering strategies highlight integrating synthetic biology, AI/ML Ops, TEA, and LCA for yield, economic viability, and sustainability [8]. Wang et al. [9] reported that a combination of deep eutectic solvent (DES) and microwave pretreatment resulted in approximately 96% lignin removal from wheat straw in 150 s at 130 °C. A life cycle assessment (LCA) revealed a 63% reduction in environmental impact relative to traditional heating methods and a 44% reduction in life-cycle costs.

30.2.2 Vegetable oils (triglycerides): polyols, plasticizers, and bio-resins

Triglyceride oils (e.g., soy, canola, palm, and castor) are abundant and chemically versatile feedstocks. Through functionalization (epoxidation, hydroformylation, ozonolysis, and etherification), they yield natural-oil polyols for polyurethanes (PUs) and bio-resins. See Chapters 8 and 16 for additional conversation on these topics. Numerous AI and ML tools and workflows enable the identification, screening, and formulation of bio-based PU and other resin systems. The following sections highlight open-source selectors, high-throughput inverse design platforms, and data-driven formulation strategies applicable to coatings, adhesives, and structural polymers:

1) **PolymatGIQ "bio-based" monomer selection tool** [10]
   An open-source Python tool that predicts critical polymerization and material properties, propagation rate constant (K_p), reactivity ratios (R_r), glass transition temperature (T_g), and water solubility to recommend bio-based monomer pairs that can replace conventional petrochemical systems in emulsion polymerization. It is useful for pre-screening bio-based candidates prior to laboratory synthesis.
2) **PolyID (polymer inverse design): NREL** [11]
   A graph neural network (GNN) platform that screens millions of bio-based polymer designs and predicts properties such as T_g, elasticity, and thermal tolerance. PolyID demonstrated the discovery of performance-advantaged, bio-derived analogs (e.g., polyethylene terephthalate-like materials) and provides domain-of-validity analysis to support more reliable predictions, relevant to formulating bio-based resins and binders for coatings and composites.
3) **Bayesian optimization and active learning for bio-based epoxy formulations** [12]
   ML-guided formulation frameworks (Bayesian optimization and active learning) have achieved rapid improvements in multicomponent amino-acid-based epoxy systems, predicting T_g with a mean absolute error of ~4 °C after only a handful of experiments. These approaches generalize to PU and other resin families for property-targeted formulation with fewer trials.
4) **AI workflows in coatings and resin product development** [13]
   Industry reports and technical reviews describe ML models that predict resin or formulation performance prior to synthesis, use clustering/classification to mine historical data, and integrate digital twins and generative design to accelerate sustainable resin development while meeting regulatory and performance requirements.
5) **Materiom AI for bio-based ingredient discovery** [14]
   A generative AI assistant (beta) that surfaces biopolymer ingredients, additives, feedstock sourcing, and processing methods from extensive open-access literature useful for early-stage bio-based resin ideation and pairing of components for targeted performance.

30.2.2.1 Practical use cases and implementation

Recent advances in AI and ML have enabled transformative approaches to the design and formulation of bio-based PUs and resin systems. Rapid pre-screening of bio-based polyols and chain extenders using ML predictors targeting key properties such as glass transition temperature (T_g), solubility, and reactivity greatly accelerates the identification of promising candidates for PU synthesis, reducing experimental burden and focusing resources on viable alternatives to petrochemical components. In the field of binder development for coatings, inverse design frameworks leverage property predictors to generate and evaluate backbone structures that meet critical performance criteria including durability, gloss, adhesion, and volatile organic compound (VOC) constraints while prioritizing the incorporation of bio-derived feedstocks. Furthermore, active-learning-driven formulation loops enable dynamic optimization of hard and soft segment ratios in PU systems, directly targeting elasticity, Shore hardness, and thermal stability, and iteratively refining models based on experimental feedback. Ingredient discovery for bio-based resins is also being revolutionized by literature-mining AI assistants, which map compatibility among plasticizers, crosslinkers, and catalysts, thereby guiding component selection for targeted performance and sustainability. Effective implementation of these AI-enabled workflows begins with defining property targets and constraints (e.g., T_g range, modulus, bio-content percentage, VOC limits, and cost parameters), followed by using ML tools to down-select candidates and validate their performance using small-scale design of experiment (DoE) matrices. To ensure reliable predictions and avoid extrapolation outside the domain of trained chemical space, domain-of-validity or applicability checks are essential, supporting robust and generalizable outcomes in the development of next-generation sustainable PU and resin formulations.

30.3 Formulation intelligence: linking molecular descriptors to end-use performance

30.3.1 Surfactant design and detergency

AI has enabled the development of integrated generative, predictive, and reinforcement-learning workflows for surfactant design. By leveraging molecular encoding frameworks such as SELFIES (SELF-referencing embedded strings) and GNN property predictors, these workflows generate and screen novel surfactant candidates that conform to specified property constraints. Molecular dynamics (MD) simulations further validate the stability of these candidates, ensuring their suitability for practical applications [14]. Artificial neural network-quantitative structure-activity relationship (ANN-QSAR) models enable rapid prediction of aggregation numbers for cationic surfactants using compact molecular descriptor sets, thereby streamlining the design of micellar systems [15]. Advanced pipeline approaches, such as QSAR-ICE-SSD [16] (quantitative structure-activity relationships – interspecies correlation estimation – species sensitivity distributions), facilitate the estimation of freshwater hazard concentrations across surfactant categories, revealing critical structure-toxicity relationships and sensitivities among aquatic taxa an essential consideration for the dual optimization of product mildness and environmental safety. Moreover, industry-focused

AI playbooks now guide surfactant selection processes that anticipate not only mildness and preservative compatibility but also sustainability requirements, reflecting a holistic approach to modern formulation intelligence [17].

30.3.2 Rheology and micellar/colloidal behavior

The viscosity behavior of micellar systems is profoundly influenced by both ionic strength (typically modulated through salt addition) and the molecular architecture of the surfactant components. Specifically, variables such as the degree of ethoxylation, hydrophobic tail length, and the extent of polydispersity serve as critical, interpretable features for ML models tasked with predicting rheological outcomes. The ethoxylation degree affects the hydrophilic-lipophilic balance (HLB) and micelle formation, while tail length modulates packing efficiency and aggregate size; polydispersity introduces complexity in the distribution of molecular weights, thereby impacting flow properties and viscoelastic response [18].

ANNs have demonstrated efficacy in predicting thixotropy time-dependent viscosity changes under shear and other key rheological parameters in complex emulgel formulations. This capability provides a robust computational framework for optimizing formulations in home and personal care applications, where texture, stability, and sensory attributes are tightly linked to rheological performance. By leveraging experimental and simulated data, ANNs can discern nonlinear relationships among compositional variables and dynamic viscosity profiles, thereby facilitating the rational design of products with desired flow and application properties [19].

The FormulationAI platform demonstrates that integrating multi-system datasets with advanced property predictors enables cross-domain formulation design and rapid screening of candidate systems. By unifying diverse data sources spanning surfactant types, additives, and environmental conditions, FormulationAI enables the identification of transferable design rules and accelerates innovation across different product categories. Such platforms are instrumental for academic research, as they support hypothesis generation, model validation, and the systematic exploration of formulation spaces, ultimately advancing the science of complex fluids and soft materials [20].

30.3.3 Safety, tolerability, and regulatory alignment

Computational toxicology surveys provide a comprehensive synthesis of ML methodologies applied to the prediction of chemical safety and tolerability, with a particular emphasis on regulatory compliance and alignment. These surveys systematically review various supervised and unsupervised ML algorithms, including support vector machines, random forests, deep neural networks, and ensemble approaches that are employed to model toxicological endpoints such as cytotoxicity, genotoxicity, and organ-specific adverse effects. Such approaches are increasingly recognized by regulatory agencies for their potential to supplement or partially replace traditional animal-based toxicological testing, thereby accelerating the risk assessment process and supporting the development of safer chemical products [21].

A notable achievement within this domain is the deployment of highly accurate ML classifiers for the prediction of lung surfactant inhibition, a critical endpoint in evaluating the safety of aerosolized products and occupational exposures. These classifiers, often trained on curated datasets comprising molecular descriptors, physicochemical properties, and historical bioassay results, have demonstrated predictive accuracies approaching 96%. The ability to reliably forecast surfactant inhibition is invaluable for pre-market screening of inhalable formulations, reducing reliance on costly and time-consuming experimental assays. Furthermore, such predictive tools facilitate the rapid identification of chemical structures with a high probability of respiratory toxicity, enabling early-stage risk mitigation and more efficient regulatory submissions. Continued advancements in computational toxicology driven by the integration of large-scale data sources, advanced feature engineering, and explainable AI are poised to further enhance the robustness and generalizability of safety assessments, fostering innovation while upholding stringent regulatory standards [22].

30.4 Green by design: embedding degradability, toxicity, and circularity in the objective function

30.4.1 Metrics and frameworks

In the context of advancing sustainability assessment for bio-based products, hybridized first-line indicators represent an important methodological innovation. These indicators integrate the foundational principles of green chemistry such as waste minimization, use of renewable feedstocks, and design for degradation with key concepts from circular economy frameworks, including resource efficiency, material recirculation, and product life extension. By merging these domains, hybridized indicators provide a more holistic and actionable early-stage evaluation tool that complements traditional LCA approaches. This integration is particularly valuable in the screening and development phases of bio-based product design, where rapid, yet meaningful, sustainability insights are needed before detailed LCA data are available [23].

The development of a composite safe and sustainable by design – circularity indicator (CI-SSbDC) further operationalizes the European Commission's Safe and Sustainable by Design guidance [24]. This composite metric enables stakeholders to systematically triage and prioritize product portfolios based on safety, sustainability, and circularity attributes, even in the absence of complete LCA datasets. The CI-SSbDC thus serves as a pragmatic decision-support tool, facilitating the early identification of promising candidates for scale-up and commercialization while ensuring alignment with emerging regulatory and market expectations for green products.

For robust and credible environmental claims, harmonized reporting standards such as the environmental footprint (EF), product environmental footprint (PEF), and the ISO 14040/44 series provide essential methodological backbones [25]. These frameworks ensure consistency, transparency, and comparability in the communication of environmental performance, supporting both regulatory compliance and stakeholder trust. By embedding hybridized indicators and composite metrics within these harmonized reporting structures, academic and industrial practitioners can advance a more integrated and forward-looking approach to sustainability assessment, ultimately accelerating the transition toward circular and low-impact bio-based product systems.

30.4.2 ML-enabled, early-stage LCA/TEA

Process specific predictive LCA, exemplified by the APPROPRIATE methodology [1] (A PROcess-Specific Pre-assessment of Impacts for Appropriate Technologies), leverages advanced ML techniques such as Gaussian processes and the extraction of learned molecular or process descriptors to rigorously distinguish between process alternatives and quantify uncertainty especially at early technology readiness levels (TRL 2). These Gaussian process models are well-suited for small-data environments typical of emerging technologies, providing probabilistic predictions and confidence intervals that inform decision-making under uncertainty. The approach enables a nuanced comparison of candidate processes by not only estimating environmental impacts but also explicitly modeling the uncertainty associated with these predictions, thereby supporting more robust go/no-go decisions in technology development pipelines. Recent systematic reviews [2] have cataloged the diverse applications of ML across the LCA workflow, encompassing inventory data generation, impact estimation, calculation of characterization factors, and result interpretation. These reviews emphasize the importance of robust statistical modeling and the explicit treatment of uncertainty, as ML-driven models can propagate and even amplify input uncertainties throughout the LCA process. Attention to uncertainty quantification is critical for ensuring the credibility and reliability of LCA results, particularly when informing policy, regulatory submissions, or investment decisions for novel process technologies.

Prospective LCA guidance, as outlined in contemporary literature [3], highlights the pivotal role of ML in addressing future indeterminacies such as evolving electricity grid mixes, dynamic supply chain configurations, and the emergence of novel feedstocks or process routes. ML models, when integrated with scenario analysis and predictive analytics, allow for the exploration of multiple plausible futures, thereby enhancing the anticipatory power of LCA and supporting resilient technology design and deployment strategies. Case studies at the intersection of TEA and LCA further demonstrate the utility of Bayesian co-optimization frameworks. In these applications, Bayesian optimization is employed to concurrently minimize production costs and greenhouse gas emissions (global warming potential, GWP) in biomass conversion routes. By iteratively updating process parameters and model predictions based on experimental or simulated data, these frameworks identify optimal operating conditions that satisfy both economic and environmental objectives. The integration of Bayesian inference not only accelerates process development but also ensures that trade-offs between cost and sustainability are transparently evaluated and managed, advancing the adoption of green technologies in industrial practice.

30.4.3 AI-assisted sustainable polymers

Polymer informatics protocols represent a sophisticated framework for the rational design and screening of polymeric materials, rigorously aligning material discovery with sustainability imperatives and application-specific requirements. These protocols commence by defining explicit sustainability criteria tailored to the intended application domain. Such criteria may encompass environmental metrics (e.g., life cycle greenhouse gas emissions, and biodegradability), functional performance thresholds (e.g., mechanical strength and barrier properties), and regulatory considerations (e.g., absence of hazardous monomers and compliance with circularity principles). The formalization of these criteria ensures that the subsequent informatics-driven workflow remains anchored in both technical and sustainability targets.

The development of property predictors [26, 27] constitutes a core element of polymer informatics. Advanced ML models, ranging from quantitative structure-property relationship frameworks to deep neural networks are trained on curated datasets to accurately predict critical polymer attributes such as glass transition temperature, thermal stability, mechanical moduli, and permeability. These predictors employ diverse molecular representations, including descriptors derived from polymer repeat units, SMILES (Simplified Molecular-Input Line-Entry System) strings, and higher-order structural features. Once robust predictive models are established, the informatics pipeline proceeds to enumerate a large virtual library of candidate polymers, systematically generating chemical structures that span the design space relevant to the target application. To ensure candidates meet stringent safety and sustainability standards advance, filtering protocols are implemented in accordance with safe and sustainable by design (SSbD) rules. SSbD integrates hazard assessment, sustainability benchmarking, and circularity evaluation into the material selection process. Candidates are computationally screened for potential toxicity, persistence, and alignment with circular-economy strategies, thereby prioritizing polymers most likely to satisfy both regulatory and market expectations for green materials.

Recent advances in language-model-based approaches, exemplified by PolyNC [28], have further transformed polymer informatics. These models unify natural-language text prompts (e.g., desired property specifications or application contexts) with chemical string representations such as SMILES, thereby facilitating property prediction across extensive and heterogeneous polymer datasets. By encoding both chemical and contextual information, language models can generalize across diverse chemistries and usage scenarios, enabling the rapid identification of polymers that best match user-defined requirements. Moreover, the integration of multimodal pretraining where models are jointly trained on one-dimensional (1D) polymer sequences and three-dimensional (3D) molecular conformations has been shown to substantially enhance prediction accuracy and generalization capability. This multimodal strategy enables models to capture complex structure-property relationships arising from the interplay between primary chemical structure and higher-order spatial arrangements. By leveraging both sequence and conformation data, these pretrained models deliver more reliable predictions for novel or out-of-distribution polymers, thereby supporting accelerated and informed material discovery in sustainable polymer research.

30.5 Implementation blueprint

Define multi-objective scorecards early: In the context of sustainable material and surfactant development, it is essential to establish comprehensive multi-objective scorecards at the initial stages of research and process design. These scorecards should holistically capture performance metrics such as detergency efficacy, barrier properties (e.g., oxygen transmission rate and water vapor transmission rate), and defined viscosity windows alongside safety criteria rooted in advanced QSAR modeling for sensitization and inhalation toxicity. Furthermore, sustainability dimensions must be quantified using circularity and carbon metrics, including the material circularity indicator, LCA surrogates, CI-SSbDC, and greenhouse gas intensity (expressed as kg CO_2/kg product). Such early-stage, multi-dimensional scorecards ensure that candidate materials and processes are systematically evaluated against technical, safety, and environmental benchmarks, facilitating the alignment of research outputs with regulatory, market, and societal expectations.

Curate a feature registry: The development of robust predictive models and digital workflows relies on the careful curation of a feature registry. This registry should encompass a wide range of molecular descriptors including HLB, critical micelle concentration, and quantitative metrics characterizing the head and tail groups of amphiphilic molecules as well as micellar properties, spectroscopic features, and relevant process conditions. Importantly, the registry should be expanded to include ecotoxicological endpoints and biodegradation descriptors, thereby integrating environmental fate and hazard considerations directly into the informatics pipeline. Such a comprehensive feature set enables the construction of highly predictive, interpretable models that can inform both material selection and process optimization, while ensuring compliance with eco-design and regulatory mandates.

Modeling loop integration: Implementing a closed-loop modeling framework is critical for efficient experimental design and process control. Bayesian optimization and active learning strategies can be deployed to guide experimental campaigns particularly under conditions of feedstock variability by prioritizing experiments that are most informative for model refinement. The integration of soft sensors and model predictive control enables real-time process monitoring and adjustment, thereby maintaining product quality and consistency. Predictive TEA and LCA surrogate models are then used to dynamically update cost-carbon trade-off frontiers, providing real-time feedback on process economics and environmental impacts, and supporting decision-making across the development pipeline.

Scale-up guardrails: To ensure the reliability and scalability of novel processes, it is imperative to implement guardrails using hybrid digital twins, virtual representations that integrate first-principles physical models with ML components for critical unit operations such as pretreatment, fermentation, and downstream processing. These hybrid models provide transparent, explainable criteria for product release, aligned with harmonized environmental reporting standards such as the EF and PEF frameworks. By embedding these digital twins within the scale-up workflow, developers can anticipate and mitigate risks associated with process scale transitions, ensuring that both performance and sustainability objectives are met.

Bayesian optimization and digital twins in foam development: In the formulation and scale-up of polymeric foams, Bayesian optimization is utilized to systematically balance multiple, often competing, material properties such as compressive strength, thermal conductivity, and density. This approach enables the targeted exploration of formulation and processing conditions to identify optimal trade-offs that meet application-specific requirements. Concurrently, digital twins' virtual replicas of the manufacturing process are employed to stabilize curing profiles and reduce batch-to-batch variability. These digital twins integrate real-time process data and predictive models, enabling adaptive control and rapid troubleshooting.

Advances in polymer informatics and co-optimization frameworks: State-of-the-art polymer informatics leverages advanced ML techniques including language models and multimodal pretraining to accelerate and enhance the prediction of polymer properties. Language-based models can interpret both chemical representations (e.g., SMILES strings) and natural-language prompts, enabling generalized, context-aware property predictions across diverse polymer datasets. Multimodal pretraining, which integrates both 1D sequence and 3D conformational data, further boosts predictive performance and generalization to novel chemistries. In parallel, Bayesian co-optimization frameworks are increasingly adopted to simultaneously minimize production costs (via TEA) and environmental impacts (via LCA), with GWP serving as a key metric. Predictive digital twins ensure process consistency in the face of feedstock variability, maintaining both economic and sustainability targets throughout the development pipeline.

QSAR/ANN models and inverse design in bio-based polymers: For bio-based and protein-derived polymer systems, QSAR and ANN models are employed to map relationships among protein architecture, cross-linker ratios, and resultant properties, such as bond durability and moisture resistance. These models facilitate the rational design of polymer blends and cross-linking strategies via inverse design algorithms that suggest optimal formulations to achieve targeted performance attributes. Predictive TEA and LCA surrogate models are integrated early in the technology readiness pipeline (TRL 2–3) to quantify trade-offs between cost, performance, and environmental impact, thereby enabling informed go/no-go decisions and resource allocation in research and development.

References

[1] Kleinekorte J, Kleppich J, Fleitmann L, Beckert V, Blodau L, Bardow A. APPROPRIATE Life Cycle Assessment: A PROcess-Specific, PRedictive Impact AssessmenT Method for Emerging Chemical Processes. ACS Sustain Chem Eng 2023;11:9303–9319. https://doi.org/10.1021/acssuschemeng.2c07682.

[2] Romeiko XX, Zhang X, Pang Y, Gao F, Xu M, Lin S, et al. A review of machine learning applications in life cycle assessment studies. Sci Total Environ 2024;912:168969. doi: https://doi.org/10.1016/j.scitotenv.2023.168969.

[3] Blanco CF, Pauliks N, Donati F, Engberg N, Weber J. Machine learning to support prospective life cycle assessment of emerging chemical technologies. Curr Opin Green Sustain Chem 2024;50:100979. https://doi.org/10.1016/j.cogsc.2024.100979.

[4] Kim MO. AI-Driven Polymeric Coatings: Strategies for Material Selection and Performance Evaluation in Structural Applications. Polymers 2026;18:5. https://doi.org/10.3390/polym18010005.

[5] Kovačić Đ, Radočaj D, Samac D, Jurišić M. Influence of Thermal Pretreatment on Lignin Destabilization in Harvest Residues: An Ensemble Machine Learning Approach. AgriEngineering 2024;6:171–184. https://doi.org/10.3390/agriengineering6010011.

[6] Machine Learning Optimization of Lignin Properties in Green Biorefineries | ACS Sustainable Chemistry & Engineering n.d. https://pubs.acs.org/doi/10.1021/acssuschemeng.2c01895 (accessed January 13, 2026).

[7] Huang X-Y, Zhang X, Xing L, Huang S-X, Zhang C, Hu X-C, et al. Promoting lignocellulosic biorefinery by machine learning: progress, perspectives and challenges. Bioresour Technol 2025;428:132434. doi: https://doi.org/10.1016/j.biortech.2025.132434.

[8] Engineering strategies to optimize lignocellulosic biorefineries | Nature Reviews Bioengineering n.d. https://www.nature.com/articles/s44222-024-00247-5 (accessed January 13, 2026).

[9] Wang H, Chen J, Pei Z, Huang J, Wang J, Yang S, et al. Enhancing lignocellulosic biorefinery sustainability: mechanisms and optimization of microwave-responsive deep eutectic solvents for rapid delignification. Biofuel Res J 2025;12:2306–2318. https://doi.org/10.18331/BRJ2025.12.1.3.
[10] GitHub – PolymatGIQ/biobased: ML models and selector tool for bio-based monomer replacement n.d. https://github.com/PolymatGIQ/biobased (accessed January 13, 2026).
[11] Wilson AN, St John PC, Marin DH, Hoyt CB, Rognerud EG, Nimlos MR, et al. PolyID: Artificial Intelligence for Discovering Performance-Advantaged and Sustainable Polymers. Macromolecules 2023;56:8547–8557. https://doi.org/10.1021/acs.macromol.3c00994.
[12] Designing formulations of bio-based, multicomponent epoxy resin systems via machine learning | MRS Bulletin n.d. https://link.springer.com/article/10.1557/s43577-023-00504-9 (accessed January 13, 2026).
[13] Cadden M Facilitating Coatings Product Development with Artificial Intelligence 2021. https://www.paint.org/coatingstech-magazine/articles/facilitating-coatings-product-development-with-artificial-intelligence/ (accessed January 13, 2026).
[14] Surfactant-Specific AI-Driven Molecular Design: Integrating Generative Models, Predictive Modeling, and Reinforcement Learning for Tailored Surfactant Synthesis | Industrial & Engineering Chemistry Research n.d. https://pubs.acs.org/doi/10.1021/acs.iecr.4c00401 (accessed January 13, 2026).
[15] Abdous B, Sajjadi SM, Bagheri A. Predicting the aggregation number of cationic surfactants based on ANN-QSAR modeling approaches: understanding the impact of molecular descriptors on aggregation numbers. RSC Adv n.d.;12:33666–33678. https://doi.org/10.1039/d2ra06064g.
[16] Liang W, Zhao X, Wang X, Tang Z, Zhang X, Wang X. Prediction of freshwater ecotoxicological hazardous concentrations of major surfactants using the QSAR–ICE–SSD method. Environ Int 2024;185:108472. https://doi.org/10.1016/j.envint.2024.108472.
[17] Giovanni. AI-Driven Surfactant Selection and Predictive Formulation. Grand Ingredients 2025. https://grandingredients.com/ai-driven-surfactant-selection/ (accessed January 13, 2026).
[18] Flores SL, Cabry CP, Barlow H, Peterson J, Cook JL, Mihailova O, et al. Molecular architecture modulates self-assembly and micellar rheology of model ionic surfactant systems. Soft Matter 2025;21:5990–5997. https://doi.org/10.1039/d5sm00252d.
[19] Artificial neural network as a strategy to predict rheological properties in emulgel formulations | Scientific Reports n.d. https://www.nature.com/articles/s41598-026-35795-w (accessed January 13, 2026).
[20] Dong J, Wu Z, Xu H, Ouyang D. FormulationAI: a novel web-based platform for drug formulation design driven by artificial intelligence. Brief Bioinform 2023;25:bbad419. https://doi.org/10.1093/bib/bbad419.
[21] Di Guardo A, Trovato F, Cantisani C, Dattola A, Nisticò SP, Pellacani G, et al. Artificial Intelligence in Cosmetic Formulation: Predictive Modeling for Safety, Tolerability, and Regulatory Perspectives. Cosmetics 2025;12:157. https://doi.org/10.3390/cosmetics12040157.
[22] Liu JY, Peeples J, Sayes CM. Evaluation of machine learning based QSAR models for the classification of Lung surfactant inhibitors. Environ Health 2024;2:912–917. https://doi.org/10.1021/envhealth.4c00118.
[23] Lokesh K, Matharu AS, Kookos IK, Ladakis D, Koutinas A, Morone P, et al. Hybridised sustainability metrics for use in life cycle assessment of bio-based products: resource efficiency and circularity. Green Chem 2020;22:803–813.
[24] Arias A, Cinelli M, Moreira MT, Cucurachi S. A composite indicator for evaluating safety and sustainability by design and circularity in emerging technologies. Sustain Prod Consum 2024;51:385–403. https://doi.org/10.1016/j.spc.2024.09.027.
[25] Environmental Footprint Methods – Green Forum – European Commission 2024. https://green-forum.ec.europa.eu/green-business/environmental-footprint-methods_en (accessed January 13, 2026).

[26] Design of functional and sustainable polymers assisted by artificial intelligence | Nature Reviews Materials n.d. https://www.nature.com/articles/s41578-024-00708-8 (accessed January 13, 2026).
[27] Recent Progress of Artificial Intelligence Application in Polymer Materials n.d. https://www.mdpi.com/2073-4360/17/12/1667 (accessed January 13, 2026).
[28] PolyNC: a natural and chemical language model for the prediction of unified polymer properties – Chemical Science (RSC Publishing) n.d. https://pubs.rsc.org/en/content/articlelanding/2023/sc/d3sc05079c (accessed January 13, 2026).

Rachel Yoho*, Ram Lalgudi*

Conclusions

The path forward for BioProducts

With nearly 30 chapters and 700 pages, it is challenging to sum up the depth and breadth of work spanning the bioproduct-related areas.

Particularly with the second edition, we highlight the ways in which global technology innovation (e.g., artificial intelligence) and geopolitical trends and interactions have shaped the research and development space in the last couple of years. "Green" trends continue in the consumer space, as does "greenwashing" non- or less-sustainable products and practices. While these are unlikely to change, the large-scale adoption of bio-based products and processes continues in a similarly disjointed manner as previously.

A large backswing in the past couple of years, driven by changing political climates, has impacted the scientific research and development space, as well as adoption of clean energy technologies and bioproducts, and overall interest in energy sources beyond fossil fuels. With much of the returned interest in fossil fuels and nuclear energy from the United States, we see increasing geopolitical implications surrounding climate change impacts. Many nations and scientists continue to work to address climate change, the local and global impacts, and continue – as noted in the first edition – this existential issue. Many companies worldwide are navigating shifting consumer bases, US tariffs, changing emissions and regulatory policies, and consumer sentiment. While we do not wish to center the United States in this discussion, we acknowledge the impact US policies have on global emissions and other nations. Just in the past few months, largely unprecedented efforts have been made by nations to come together around economic policies. These global contexts, and many other related topics, continue to shape the global outlook on climate change impacts, fossil fuel consumption and emissions, adoption of electric vehicles for personal or fleet use, more renewable power generation, and sustainable BioProducts. As Dr Vijayendran noted in the first edition, capitalism and consumerism – as key economic drivers – continue to shape lives worldwide.

Further shifts in research and development areas include not only the advancement and development of technologies, but the context in which research happens. In the United States, the drastic reduction in federal funding for science and research is showing immediate impacts in the stoppage of grants, as well as the impacts on universities, scientists' careers (particularly early career scientists), and the environment

***Corresponding author: Rachel A. Yoho**, (formerly) George Mason University, Fairfax, Virginia, USA, (currently) College of Southern Nevada, 6375 W. Charleston Blvd., Las Vegas, Nevada 89146, USA, e-mail: rachel.yoho@csn.edu
Ram Lalgudi, Aries Science and Technology, LLC, 1381 Kinnear Rd., Columbus, OH 43212, USA, e-mail: lalgudir@ariesst.com

 | https://doi.org/10.1515/9783112212141-031

in which research occurs. The attacks on higher education, funding sources, and career opportunities (graduate funding and grant funding, in particular) are shifting the global research landscape. Already, students and trainees are reconsidering US universities, federal grant funding cutoffs are causing mass layoffs from US universities, and global institutions are seizing the opportunity to recruit talent. All of these will undoubtedly impact the global research and development landscape, and particularly in the fields related to bioproducts.

In some ways, the future may be slightly less bright than 2024 for the publication of the first edition. However, we take heart in the global community of researchers working across bioproducts areas, as well as the citizens of many nations working towards a better future. We see and acknowledge these efforts. While the path is not easy, many – across research, community, and policy – are working for the future and the collective, global good.

Finally, we honor the memory and career impacts of Dr Bhima Vijayendran with the second edition of this book.

In research, application, and solidarity,
Dr Rachel Yoho
and
Dr Ram Lalgudi

Meet the Coeditors

Dr. Rachel Yoho is a faculty member and the Inaugural Director of the Faculty Center for Teaching and Learning at the College of Southern Nevada. She joins CSN from her former position as a faculty member at the Stearns Center for Teaching and Learning at George Mason University, serving as the assistant director. She has an interdisciplinary background with expertise in biotechnology and bioproducts, as well as a particular emphasis on impacts of global climate change and social/environmental justice. She also publishes frequently in the areas of teaching and learning.

She held faculty affiliations in the Department of Bioengineering in the College of Engineering and Computing and also the Department of Environmental Science and Policy in the College of Science, both at George Mason University. She also held a research affiliation with the Center for the Advancement of Well-Being at George Mason University. She is an award-winning educator with extensive experience as a faculty member. She is a Fulbright awardee and completed a Fulbright Specialist award visiting the University of Galway (Ireland). In 2023, she received the Trailblazer Award as a university partner of the GMU College of Engineering and Computing. At a previous institution, she received the university-wide educator Rising Star Award two years in a row and an Exemplary Course Award for "Climate Change, the Environment, and the Future of Public Health," a course that she developed and taught. She also received the Diversity, Equity, and Inclusion Strategies Award for this course at the same institution. She received two faculty commendations earlier at another institution.

She received her PhD in biological design (engineering) and a graduate certificate in scientific teaching in higher education (life sciences) from Arizona State University, as well as a BA in biology from Capital University. She held a postdoctoral position at Michigan State University and also worked many years ago as an intern at Battelle with Dr Ram Lalgudi.

Ram Lalgudi, PhD, is an innovative materials scientist, whose pioneering contributions in polymer chemistry, sustainable materials, and electrochemical systems have transformed multiple technological domains. His work spans biobased polymers, high-temperature fuel-cell membranes, lithium-ion battery binders, smart self-healing coatings, and advanced encapsulation platforms. Over a distinguished career exceeding two decades, he has combined deep scientific insight with strong commercial acumen, translating complex molecular design principles into scalable, high-performance solutions that influence energy storage, environmental remediation, medical devices, and next-generation manufacturing. A prolific inventor with 46 granted US patents, four R&D 100 Awards, and a portfolio of technologies that have generated millions in funded research and industrial adoption, he exemplifies a relentless pursuit of materials that perform beyond convention conducting more efficiently, protecting more intelligently, healing autonomously, and advancing sustainability. As founder and CEO of Aries Science & Technology, he continues to lead the development of advanced functional materials, guided by the belief that thoughtfully engineered chemistry can address humanity's most pressing challenges.

 | https://doi.org/10.1515/9783112212141-032

Index

 | https://doi.org/10.1515/9783112212141-033

www.ingramcontent.com/pod-product-compliance
Lightning Source LLC
Chambersburg PA
CBHW061149240926
51846CB00006B/960